Site-Specific Management for Agricultural Systems

Related Society Publications

Soil Specific Crop Management—A Workshop in Research and Development Issues

For information on these titles, please contact the ASA, CSSA, SSSA Headquarters Office; Attn: Marketing; 677 South Segoe Road; Madison, WI 53711-1086. Phone: (608) 273-8080. Fax: (608) 273-2021.

Proceedings

of

Site-Specific Management for Agricultural Systems

Second International Conference

Editors
P. C. Robert, R. H. Rust, and W. E. Larson

March 27-30, 1994
Thunderbird Hotel
2201 East 78th St.
Minneapolis, MN

Conducted by the
Department of Soil Science and Minnesota Extension Service
University of Minnesota

Published by:
American Society of Agronomy, Inc.
Crop Science Society of America, Inc.
Soil Science Society of America, Inc.

Cover art provided by authors

American Society of Agronomy, Inc.
Crop Science Society of America, Inc.
Soil Science Society of America, Inc.
677 South Segoe Road, Madison, WI 53711 USA

Library of Congress Cataloging-in-Publication Data

Cataloging-in-publication data pending

Printed in the United States of America

TABLE OF CONTENTS

SESSION I - SOIL RESOURCE VARIABILITY

SESSION III - ENGINEERING TECHNOLOGY

Papers

SESSION IV - PROFITABILITY

Papers

SESSION VI - TECHNOLOGY TRANSFER

PREFACE

Site-specific crop management (SSCM) refers to a developing agricultural management system that promotes variable management practices within a field according to site or soil conditions. While this technology is only a few years old, various names have been used to describe the concept: farming by soil; farming soil, not fields; farming by the foot; spatially prescriptive farming; computer aided farming; farming by computer; farming by satellite; high-tech sustainable agriculture; soil-specific crop management; site-specific farming; and precision farming. Interest in this emerging concept has been featured in a variety of news media.

There is no broadly accepted definition of SSCM. We proposed the following:

> site-specific crop management is an information and technology based agricultural management system to identify, analyze, and manage site-soil spatial and temporal variability within fields for optimum profitability, sustainability, and protection of the environment.

SSCM employs a system engineering approach to crop production where inputs are made on an "as needed basis", and was made possible by recent innovation in information and technology such as microcomputers, geographic information systems, positioning technologies (Global Positioning System), and automatic control of farm machinery. It is a holistic approach to micro manage spatial and temporal variability in agricultural landscapes based on integrated soil, plant, information, and engineering management technologies as well as economics.

It is still in infancy, and will require major efforts in research, development, and education from private and public sectors but promises to deliver a combination of benefits for the 21st century: <u>a more profitable and sustainable agriculture while preserving the delicate balance between food production and the quality of land and water resources</u>.

This book contains the proceedings of the Second International Conference on Site-Specific Management for Agricultural Systems held in Minneapolis (Bloomington) March 28-30, 1994. A first conference was held in Minneapolis in April 1992. The proceedings provide an overview of recent and current research and applications related to various aspects of SSCM namely, soil resources, managing variability, technology, profitability, environment, and technology transfer. Six position papers located at the beginning of each section introduce the six aspects of SSCM just mentioned followed by volunteer oral and poster papers. Papers presented during a pre-conference workshop on Yield Monitoring and Mapping are also included. Presenters of the second pre-conference workshop on Site-Specific Crop Management Systems are listed in the Appendix I. Abstracts of poster presentations without manuscripts are at the end of each section.

On behalf of all participants, we wish to express our gratitude to sponsoring organizations for their support and to ASA-CSSA-SSSA for publishing this document. We also wish to express our appreciation to all speakers and poster participants for their presentations and manuscripts, and to all participants who made the conference a success. We look forward for a third conference in 1996.

P. C. Robert,
R. H. Rust,
W. E. Larson, co-editors

ACKNOWLEDGMENTS

Organizing Committee

Jonathan Chaplin
H. H. Cheng
Mark DeBower
Dean Fairchild
Earl Fuller
Al Giencke
William Larson
Bruce Montgomery
Pierre Robert, Chair
Richard Rust
Dave Wall

Editorial Committee

Ray Allmaras
William Larson
Robert Munter
George Rehm
Pierre Robert, Chair
Richard Rust

Sponsors

University of Minnesota
Department of Soil Science
Department of Agricultural Engineering
Minnesota Extension Service

American Society of Agronomy, Crop Science Society of America and Soil Science Society of America

American Society of Agricultural Engineers

CONTRIBUTORS

Cargill, Incorporated, Minneapolis, Minnesota

Deere and Company, Moline, Illinois

Farm Foundation, Chicago, Illinois

Minnesota Department of Agriculture -
Agronomic Services

Minnesota Pollution Control Agency

U. S. Department of Agriculture
Agricultural Research Service
Cooperative State Research Education
and Extension Service
Natural Resources Conservation Service

American Society of Agronomy, Crop Science Society of America and Soil Science Society of America
Outreach Committee

WORKSHOP I

YIELD MONITORING AND MAPPING

1 Yield Monitoring for Mapping

T. S. Colvin
D. L. Karlen
J. R. Ambuel
F. Perez-Munoz

USDA-ARS
National Tilth Lab
Ames, Iowa

Soil and crop management practices have often been applied on a field basis rather than taking into account the soil variability within the field. Decisions based on the area of greatest need may cause excess materials to be applied to other parts of the field. This can result in surface or ground water pollution. In today's society when the bottom line is so important, it makes economic sense not to have unnecessary inputs. If areas with equal yield potential could be mapped accurately and located efficiently, soil amendments and other inputs could be applied in a more optimum manner. This paper explores these considerations with information gained by producing yield maps for several locations near Ames, IA during the period 1989-1993.

PRIOR WORK

Fields are usually divided by arbitrary boundaries with little regard for variation in soil series, phase, or potential productivity (Karlen et al., 1990). Knowledge of the spatial variability of yield, yield determining factors, or some combination of these would allow farmers to better manage their operations. Grain yield maps might be generated from grain flow rate data acquired during combining to characterize within-field variability (Borgelt, 1993). Topography and, as Schweitzer (1980) suggested, fertility are important factors in determining yield variability. Planting and soil engaging operations could also be optimized with respect to energy inputs based on field variability of physical properties (Schueller, 1988).

Mapping crop yield variability is an important closing link in the site-specific decision-making control loop, either by itself or in combination with other spatial variability data. To prove this, Schueller and Bae (1987) used a portable computer, a mobile microwave distance measuring unit, and two stationary remote transponders for spatially-precise yield data acquisition in

Texas. They obtained yield estimates, based on field location, that could be used to program fertilizer application according to individual soil capability and residual fertility.

Several studies have compared crop yields between soil types within a field. Larson (1986) reported from work in Montana that spatially-variable management practices normally increased returns, although results varied. For 9 of 15 soil series comparisons, there were significant differences in yield, with 6 of 9 sites showing significantly different yields between soil series within the same field. Larson suggested that yield variability for a particular soil series may be affected by variations in slope, aspect, and shape of plot area (concave or convex). Other work in Montana found that grain yield, test weight, and returns over variable costs varied greatly between soil units in each field. The actual yields of some soils were higher or lower than the estimated goal (Carr et al., 1991).

A detailed soil map was used in a 4-yr study in South Carolina to map yield variability between soil types (Karlen et al., 1990). Yield variation within an individual map unit, however, was almost as large as variation between map units.

METHODS

We conducted a study on four adjacent 16 ha fields in Boone County, IA (Fig. 1–1). A conventional farming method consisting of a corn-soybean (*Zea mays* L.-*Glycine max.* [L.] Merr) rotation, conventional tillage, and application of commercial fertilizer and pesticides was used on the western fields. An alternative system consisting of a 5-yr corn, soybean, corn, small grain-meadow, meadow rotation with manure–sewage sludge for nutrients and ridge tillage for row crops was used on the eastern fields. Eight soil map units are represented in the study area.

The scale drawing in Fig. 1–2 shows how corn and soybean yield plots were arranged on one of the fields. The important points to observe are the small size of the plots and the relatively large distance between plots. The total number of plots on any 16–ha field ranged between 200–250 depending on the length (typically 20–30 m) used.

Corn yields were determined on eight east-west transects or lines in the two south fields using a modified Gleaner K[1] combine (Colvin, 1990) in 1989. Weight and moisture content were measured every 13 m. Data was collected on all four fields from 1990 to 1993. Soybean yields were determined with the same combine or a modified John Deere 4420. Oat and hay yields were determined on 1 m long sections of windrow. The per acre yield for each crop was matched with soil map unit.

---

[1]Trade names are provided for the benefit of the reader and do not imply endorsement or preferential treatment of the products by USDA or Iowa State University.

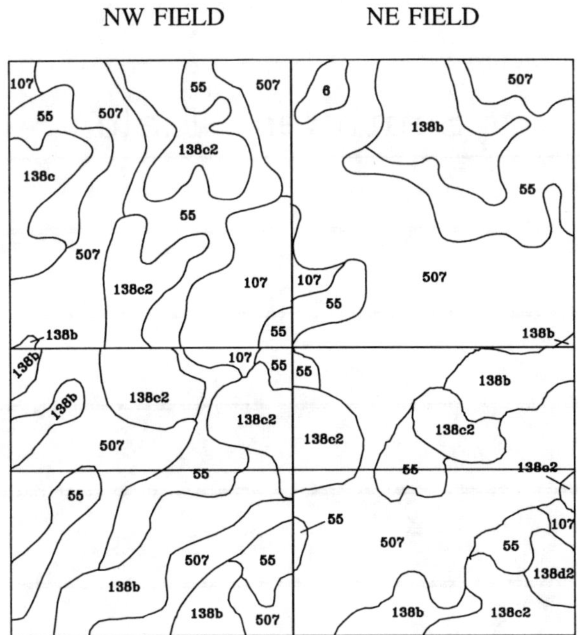

#	SOIL TYPE	% SLOPE	DESCRIPTION
6	**Okoboji silty clay loam**	**0-1**	**Level, very poorly drained**
55	**Nicollet loam**	**1-3**	**Somewhat poorly drained**
107	**Webster silty clay loam**	**0-2**	**Nearly level, poorly drained**
138b	**Clarion loam**	**2-5**	**Gently sloping, well drained**
138c	**Clarion loam**	**5-9**	**Mod. sloping, well drained**
138c2	**Clarion loam**	**5-9**	**Same as above-mod. eroded**
138d2	**Clarion loam**	**9-14**	**Strong slope-mod. eroded**
507	**Canisteo silty clay loam**	**0-2**	**Nearly level, poorly drained**

Fig. 1–1. Soil map, soil descriptions, and location of yield data locations for Fig. 1–2 and 1–3 (After USDA, 1981).

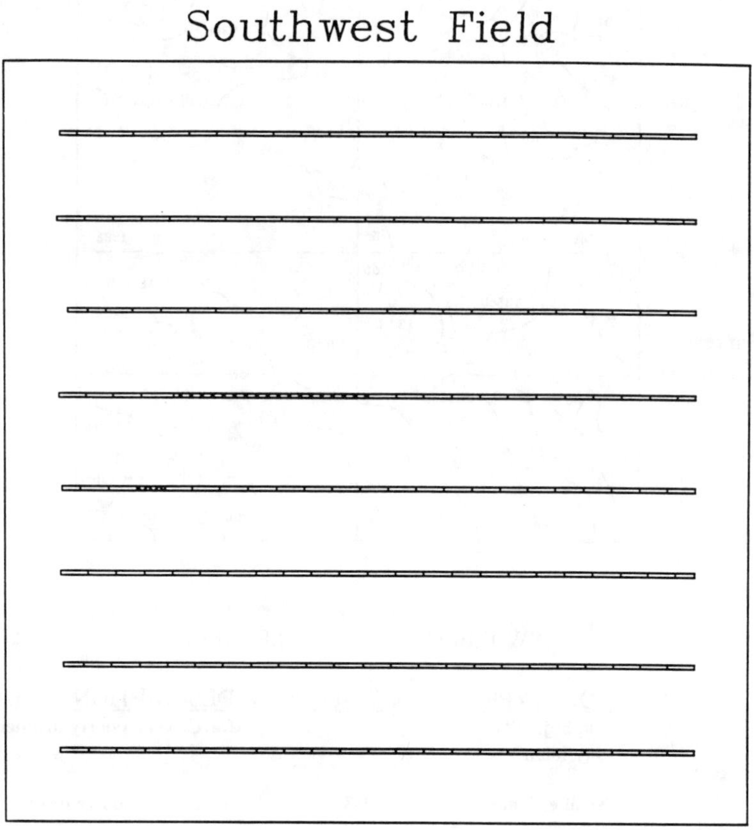

Southwest Field

Fig. 1–2. A map of the harvest plots for the southwest field showing the eight transects. The field is ≈400 m on each side. The distance between transect lines is about 45 m, and the length of each plot is about 13 m.

Position was determined by dead reckoning. This was done because GPS signals were not routinely available in 1989, and we did not have the equipment necessary to receive the signals until 1990. After we purchased hardware and software, the signals still were available only at irregular times of the day because enough satellites (four) were not often in view to allow 3-D location.

We were not able to use GPS as a normal part of our yield data acquisition with combines until 1992. This was the case because we had to coordinate our work in the cooperators fields with their harvest operations and we needed fairly long periods of time to work in the 16 ha fields. We have been using our GPS system regularly since 1990, but had to use it in experiments where we could do the entire project such as tillage of a series of plots in one to two hours to fit into one of the periods with at least four satellites available.

Though accurate, the dead reckoning method is limited to straight lines of travel. The Global Positioning System (GPS) that was purchased in 1990 was capable of providing $1/100^{th}$ of a second location precision without dependence on clear lines of vision (Ambuel et al, 1991). The system consisted of a Trimble 10X Navigation Receiver, a Zenith 184 lap top portable computer, and a GeoLink Software package (GeoResearch Co.). The GPS allows us to improve sampling efficiency by tagging the location from which a sample was taken and provides flexibility in spatial sampling. In other words, it is not as important to start on a transect and finish it to completion when conditions are impassable as they were in some fields in 1993 due to wet areas. We could start on one line; move to another to get around muddy conditions; and then return to the first transect later with the particular locations made part of the data record as combining was done.

RESULTS AND DISCUSSION

Average yields of soil map units were not clearly different from each other, based solely on the overlap of one-standard-deviation error bars. A preliminary report by Pierce and Warncke (1994) said "There was little correlation between soil test levels and yields...". Neither soil maps nor soil tests by themselves are sufficient to explain the variability. Relating yield to soil type might be done by increasing sample numbers, or by using a more detailed soil map but Karlen et al. (1990) do not support that idea.

Probably, a better yield variability model that incorporates such factors as fertility and topography with soil type is needed to refine yield projections for individual locations. Once mapped, a machine equipped with GPS or some other locator technology could apply inputs specified for locations. Decisions about rates for specific locations would be made with the help of yield maps and maps of other pertinent information (Schnug et al, 1993).

Figure 1–3 shows yields for four yr on the SW field (Fig. 1–1) on harvest line 5.

It is clear that there is a problem in the area between positions 135 and 127. In two of the four yr there was almost no yield, and in the other two yr there was at least a small amount of yield depression. That area is a low area in the field with soils that do not drain well (#'s 55 and 507 in Fig. 1–1). Generally, the area does not pond but it appears that the soil was probably too wet for good plant growth at least in some part of each season whether the general conditions were wet or dry. Just to the north of that area in the same field is a fairly steep area (5–9% slope-soil 138c2 in Fig. 1–1) that has shown the same pattern over the years of the study. This is contrary to expectations that in generally dry years the low areas would do well and in wet years the sloping or sidehill areas might excel.

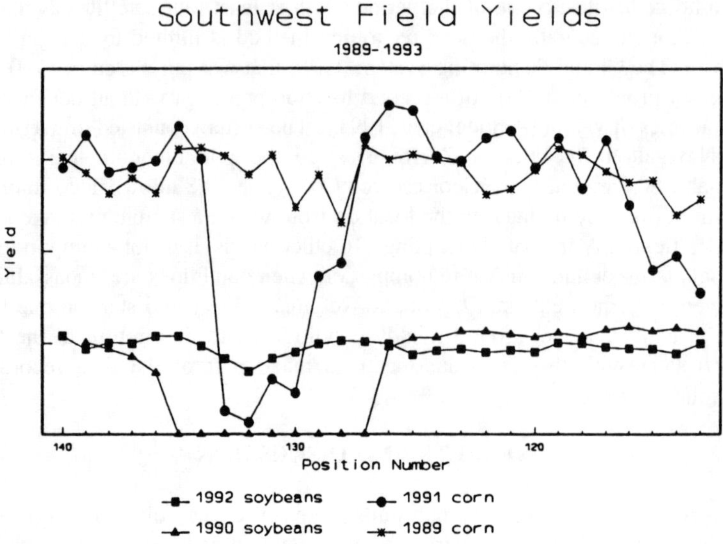

Fig. 1–3. Line five indicated on soil map (Fig. 1–1).

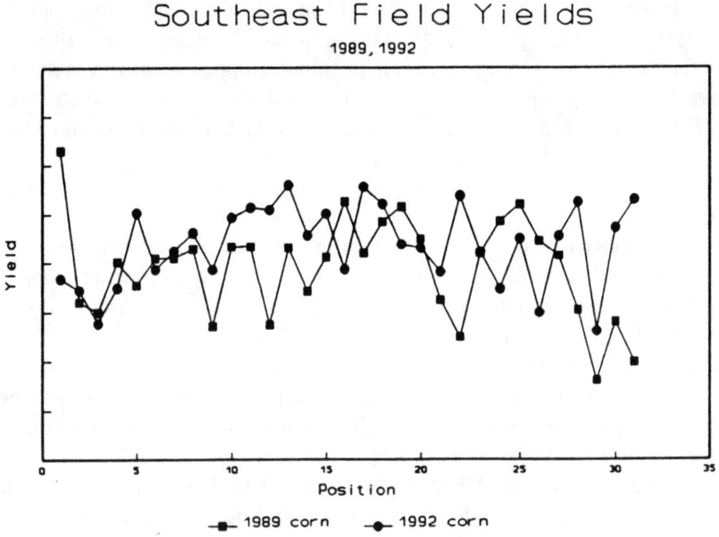

Fig. 1–4. Line five on soil map (Fig. 1–1). Relative yields.

Figure 1–4, which shows yields for the SE field for the same transect, does not show the same pattern. There are only two yr of data presented for the SE field in Fig. 1–4. There is a 5-yr rotation in that field which had only two yr of corn in the period of the study, and soybean occurred only in 1993 which was an unusual yr, and does not help understand the response of the field. For this field in general and this line in particular, the stable pattern of at least some areas being repeatably better or worse does not appear. If one looks at the soils in the area of line 5 on the SE field, one sees fewer soil map units marked. The data for the entire field do not exhibit the obvious repeating patterns that we observed for the SW field.

There are areas in the SE field that do seem to be yielding somewhat similarly in the two yr, such as positions 3, 9, and 39. At least, they are moving in the same direction each yr with respect to their neighbors. Locations 5 and 22 seem to be yielding differently in the two yr.

The yr 1993 was extremely wet in the Ames area with areas of these fields under water for significant periods of time and water logged much of the time during the growing season. There were no new insights from that yr. It has not been shown because it makes the graphs harder to read for the other years.

In these four fields, we have one which has some areas with stable yield patterns from yr to yr (SW), and the other three which seem to vary more randomly. It is also clear from Fig. 1–2 that there can be large changes in yield within short distances (20 to 40 m). This would seem to argue that long distances between samples with statistical methods for describing the unsampled areas may be unreliable if true site-specific farming is the goal. Jaynes et al. (1993) used the type of data shown in this study (transects spaced at least 50 m apart) and were able to develop good relations on some fields between kriged yield and kriged electrical conductivity measurements. In this case, the transects for both measurements were the same. It probably can not be argued for that study that statistical techniques were significant in making the patterns clear. The data was not taken at the same spacing between data points down the individual transects so the kriging was helpful (to that extent).

One question that could be asked is what happens if hybrids are changed on a field? Would this automatically make information collected with the prior hybrid unusable? There is not a definitive answer to this question from this research partly because it was not designed to answer that question, but it is possible to make some comparisons within this work. The hybrids were different between the SW and SE fields in 1989 (SW was Pioneer 3615, SE was Pioneer 3379). In 1991, the SW field had Pioneer 3417. The SE field was Asgrow RX746 in 1992. There appears to be no obvious difference in the general way the different hybrids reacted in the different years. This was particularly true in general on the SW field even though this is contrary to what can be seen in Fig. 1–2. Line 5 in Fig. 1–2 did not show as much decline in yield in 1991 as in 1989 for positions 125-133 but the soybean crop in 1990 reacted like the corn crop in 1989. This suggests that the weather in particular years may be more important than particular hybrid choices.

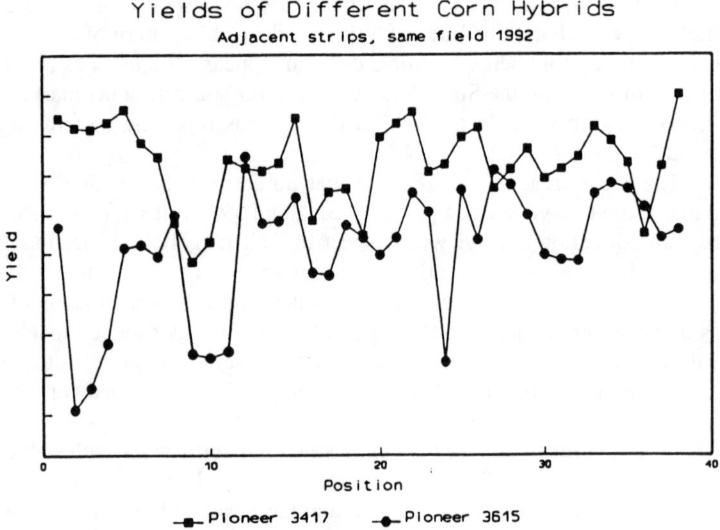

Fig. 1–5. Comparison of corn hybrid yields in the same field. Pioneer 3417, generally higher yielding, than Pioneer 3615.

Figure 1–5 is a plot of two hybrids along 800 m transects, in one field, that were about 20 m apart. This field was not at the same location as the four fields previously discussed but is on the same general soils and is in the same county. The general trends are similar but, of course, with the distance between the lines an exact match was not expected.

Another way to compare the four fields in Fig. 1–1 would be to look at the range of yield levels for the four fields. The SW and NE fields are somewhat similar in having highly variable yields within years. The SE and NW fields tend to have less variable yields on the individual fields in specific years. These similarities are across management units. The two east fields are operated under one management system and the two west fields under another. This suggests that underlying soil properties may have a larger effect than the management systems that are imposed on top of them.

Observations on yield monitoring

The large yield variability seen in the fields in this study suggest that continuous yield monitoring should be used so that all parts of the field can be sampled. Work by the authors to date has not shown a satisfactory method of taking the signal from at least one commercial yield monitor and turning it into a set of data points that correlate well with the samples taken with the stop-and-go method used on the four fields in Fig. 1–1. It required a little more than 4 hr to combine the eight transects on one of the 16 ha fields. This would be about the time required to completely harvest the same field with a 6–row combine, assuming that the combine would work steadily at ≈8 km/hr. If the data collection system slows down the harvest for any reason, including lack of a method to determine position, then that would be an added cost for data collection. It appears that GPS has sufficient availability, as of early 1994, for 24-hr per d use. The yield monitors that are being sold and further developed do not appear to hamper normal combining operations. There will undoubtedly be some data transfer required while the combine is not harvesting. That activity will have to be absorbed along with routine maintenance on the combine and data system unless it can be done unattended without interfering with combine use.

Considerable work on continuous flow monitoring has been done by Vansichen and De Baerdemaeker (1992) along with others. Problems to be resolved include noisy signals, time delay caused by the combine processes from the header to the grain tank or measuring location, and concerns about the width of cut when row crops are not being harvested. Figure 1–6 shows two transects harvested 3 m apart in 1992 in one field with a common hybrid. The boxes give the average yield for the indicated area determined with the stop-and-go method that has been used in the fields in Fig. 1–1. The line gives one–s output measurements from a continuous flow yield monitor. The continuous flow data was manipulated with running averages and other common statistical techniques but it was not possible to get a good correlation between the results and the yields determined with the stop-and-go method.

Work is being done to relate yields to specific soil properties and combinations of properties. Samples necessary to determine the Tilth Index (Singh et al., 1992) were taken in the SW field (Fig. 1–1) in 1989. The results were essentially random when the samples were taken at regular intervals along the eight transects. We plan to take similar samples from the areas that have been identified on that field as being consistently high and low yielding to see if a pattern exists or if the Tilth Index will correlate with yield as it has done on tillage experiments.

The SW and SE fields (Fig. 1–1) have been shown to have different organic matter (OM) levels (Karlen and Colvin, 1992). Although the differences are fairly small, they may be large enough to have an effect which shows up as the consistent low yields in specific locations seen on the SW field. A confounding factor in that analysis is that the SW and NW fields have been farmed as a unit for many years. They probably do not have the same differences in OM that exist between the SW and SE fields which have been farmed with different rotations. In addition, manure has been added on the SE field but not for many years on the SW field.

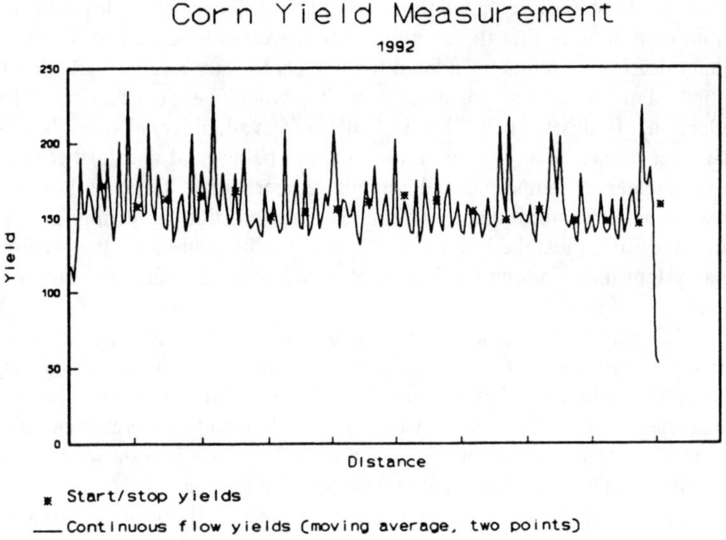

Fig. 1–6. A comparison of the stop–and–go method of yield determination with the use of a continuous flow yield monitor.

FUTURE WORK NEEDED

The positioning with GPS seems to be working well but the level of accuracy needed for site-specific activities is not known at this time. It will probably be necessary to have location accuracies of less than 3 m and probably in the sub-meter or centimeter range to be effective.

Continuous yield monitoring with a combine is a commercial reality. The question of data processing both for actual yield level and for tying the proper position to a yield remains unanswered.

At the moment, we have an art as far as site-specific farming is concerned. That means that probably some people can do it well. They might take some sort of grid samples, maybe even yield maps, or even visual observations and do an excellent job of managing a field (providing inputs by location) with good economic and environmental results. Unfortunately, the technique that they might use is not easily transferred to the general population. This will become a science when the principles of cause and effect are understood so that the transfer can be made. In general, we know principles of crop nutrition, water requirements, etc., but there is much to learn about these topics when we move from the field average to location specific.

The question is, if we want to manage an area differentially within a field, how should inputs be varied compared to other parts of the field? Until we can determine the cause of a low yield, for example, we cannot make that determination. There is also the possibility that an area might be best left alone. We don't run factories at 100% capacity to get the highest net return from the factory. We might be able to generate a higher net return from a field by leaving part of the field out of production. Locations that produce low returns repeatably are candidates for abandonment if the problem that causes the low return cannot be fixed economically. The use of the location system would be necessary to coordinate all applications of material to the crop so that no inputs would be used on sites that are effectively abandoned if that is the best choice.

ACKNOWLEDGEMENTS

The authors wish to thank Jeff Cook, Darrin Hansen, Rich Hartwig, and other members of the technical staff at the National Soil Tilth Lab for their assistance with the work used as the basis for this report. The work could not have been done without the cooperation and assistance of the landowners and operators. Timely reviews by E. J. Sadler and others were important in the preparation of this report.

REFERENCES

Ambuel, J.R., T.S. Colvin, and K. Jeyapalan. 1991. Satellite based positioning system for farm equipment. SAE Technical Pap. 911827.

Borgelt, S.C. 1993. Sensing and measurement technologies for site specific management. p. 141–157. *In* P.C. Robert et al. (ed.) Proc. of First Workshop on Soil Specific Crop Management. ASA, Madison, WI.

Carr, P.M., G.R. Carlson, J.S. Jacobsen, G.A. Neilsen, and E.O. Skogley. 1991. Farming soils, not fields: A strategy for increasing fertilizer profitability. J. Prod. Agric. 4:57–61.

Colvin, T.S. 1990. Automated weighing and moisture sampling for a field plot combine. Ap. Eng. in Agric. 6(6):713–714.

Jaynes, D.B., T.S. Colvin, and J.R. Ambuel. 1993. Soil type and crop yield determination from ground conductivity surveys. ASAE Pap. 933552.

Karlen, D.L., and T.S. Colvin. 1992. Alternative system effects on profile nitrogen concentrations on two Iowa farms. Soil Sci. Soc. Am. J. 56:1249–1256.

Karlen, D.L., E.J. Sadler, and W.J. Busscher. 1990. Crop yield variation associated with coastal plain soil map units. Soil Sci. Soc. Am. J. 54:859–865.

Larson, M.H. 1986. The influence of soil series on cereal grain yield. M.S. thesis. Dep. of Plant and Soil Sci., Montana State Univ., Bozeman.

Pierce, F, and D. Warncke. 1994. Site-specific results from Michigan. ag/Innovator Magazine, Linn Grove, IA. 1(6):8.

Schnug, E., D. Murphy, E. Evans, S. Haneklaus, and J. Lamp. 1993. Yield mapping and application of yield maps to computer-aided local resource management. p. 87–93. *In* P.C. Robert et al. (ed.) Proc. First Workshop on Soil Specific Crop Management. ASA, Madison, WI.

Schueller, J.K., and Y.H. Bae. 1987. Spatially-attributed automatic combine data acquisition. Comp. and Elec. in Agric. 2:119–127.

Schueller, J.K. 1988. Machinery and systems for spatially-variable crop production. ASAE Paper 88–1608. 1988 Winter Meeting. Chicago, IL. Dec. 13–16.

Schweitzer, B.D. 1980. Spring wheat yields on two contrasting Aridic Argiborolls in Northcentral Montana. M.S. thesis. Montana State Univ., Bozeman.

Singh, K.K., T.S. Colvin, D.C. Erbach, and A.Q. Mughal. 1992. Tilth Index: an approach to quantifying soil tilth. Trans. ASAE 35(6):1777–1785.

U.S. Department of Agriculture. 1981. Soil survey of Boone County, Iowa. USDA–SCS.

Vansichen, R.J., and J. De Baerdemaeker. 1992. Signal processing and system dynamics for continuous yield measurement on a combine. p. 340–341. *In* B. Sundell, and O. Noren (ed.) AgEng '92: Proc. of an International Conference on Agricultural Engineering. Available from the editors in Uppsala, Sweden.

2 Crop Yield Mapping: Comparison of Yield Monitors and Mapping Techniques

Stuart J. Birrell
Steven C. Borgelt

Dep. of Agricultural Engineering
University of Missouri
Columbia, Missouri

Kenneth A. Sudduth

Cropping Systems and
Water Quality Research Unit
USDA—Agricultural Research Service
Columbia, Missouri

Although the variability in soil nutrients and crop yields has been well documented since the turn of the century (Robinson & Lloyd, 1915; Fairfield Smith, 1938), the mechanization of agriculture and the trend to larger implements has led to larger areas being treated as a single unit. Recent advances, however, in machine technology and improvements in data management have made it possible to reverse this trend by implementing site specific crop management (Goering, 1993).

The implementation of spatially selective field operations is dependent on field mapping of the variations in soil and crop parameters (Stafford et al., 1991). The mapping of crop yield is especially important since the recommendation rates for many inputs are determined by the yield goal, and yield maps are necessary to assess the effects of site specific crop management. Since yield maps may be used both for the determination of management inputs and to evaluate the results of these strategies, it is important that the accuracy of the yield map be considered.

Recently, continuous grain flow monitors in conjunction with speed sensors and location systems have been used to develop crop yield maps (Searcy et al., 1989; Wagner & Schrock, 1989; Colvin, 1991, Stafford et al., 1991, Vansichen & De Baerdemaeker, 1991; Schnug et al., 1992; Auernhammer et al., 1993; Birrell et al., 1993; Pringle et al., 1993; Stott et al., 1993). When compared with the mass of grain accumulated over a certain time, the accuracies of the grain flow monitors are generally reported to be within 5 percent. All of the monitors reported on to date, however, measure the flow of grain into the grain bin. This flow must then be related to the flow of grain into the combine head, which requires modelling of the grain flow dynamics through the combine. Searcy et al., (1989) and Vansichen and De Baerdemaeker (1991) both suggested modelling the grain flow through the combine as a first order system with a time lag. Although the actual system dynamics of the combine may be of a higher order, the accuracy of the grain flow sensors probably does not warrant the use of a higher order model. Higher order models would also make the system more susceptible to noise or errors in the input data.

The development of grain yield maps requires that instantaneous grain yields be averaged or interpolated to obtain an estimate of the average yield within a certain area. The unit cell size and the mathematical methods used to generate the maps can have a significant effect on the results. Furthermore, sudden changes in the harvesting speed can cause large errors in the instantaneous yield. These errors are particularly large when the combine stops suddenly, which causes the calculated instantaneous yield to approach an infinite spike, since a finite volume of grain is divided by an area approaching zero (velocity times swath width). These spikes can have a significant effect on the calculation of yield in the local region around the point.

Objectives

1. Investigate the use of different combine grain flow models to determine instantaneous grain yield.
2. Analyze the effect of different Kriging parameters and combine models on the mapping of grain yield.
3. Investigate the effect of using evenly spaced transects instead of yield data collected continuously across the field to develop yield maps.

Equipment and Instrumentation

Two instrumented combines, a 3-row John Deere 3300 combine and a 6-row Gleaner R62 combine, were used to collect yield data within the same field. The combine yield mapping systems consisted of four components: (1) Global Positioning System (GPS) receivers to determine combine location; (2) a combine mounted grain flow sensor; (3) a sensor to measure ground speed; and (4) a data acquisition system. In addition, the John Deere combine had a weigh bin mounted in the combine grain tank that was used to measure accumulated grain from the grain sensor for calibration of the system.

Two Ashtech M-XII GPS receivers were used for position location (Harrison et al., 1992) when harvesting with either combine. The receivers were used in the post-process differential mode with the clear access (C/A) code. This mode of operation provided robust data acquisition with improved accuracy as compared to a stand-alone receiver. The receivers were set to record a position every 5 s. The fixed receiver remained over a known geo-referenced point and the rover receiver was placed in the combine, with the GPS antenna fixed on top of the combine cab. During harvest the necessary files required for post the GPS satellite signal were internally stored in the respective receivers. The receivers were able to store up to eight hours of data. The GPS files in the receivers were then downloaded into a computer for processing at the end of each day's harvest.

The John Deere 3300 3-row combine was equipped with a volumetric Claydon Yieldometer (Bae et al., 1987). This yield sensor measured the volume flow of grain from the clean grain elevator. A capacitive level sensor controlled the rotation of a six-flight paddle wheel, to maintain the level of grain above the paddle within certain thresholds. The standard yieldometer was modified by replacing the standard 2 counts/revolution sensor connected to the shaft of the paddle wheel with an angular position encoder (1024 counts/revolution). A DICKEY-john radar gun was mounted on the combine to measure ground speed and travel distance. A DMC moisture sensor was installed below a hole cut into the grain bin fill auger, with a solenoid used to control grain flow to the moisture sensor.

The Gleaner R62 combine was instrumented with an impact-based, AgLeader™ Yield Monitor 2000™. This sensor measured the force of the grain impacting against a plate situated at the top of the clean grain elevator. The force and other parameters such as elevator speed were then used to determine mass grain flow rate. A DMC moisture sensor was also installed directly into the grain bin auger, where a small section of auger flighting was removed. A magnetic pick-up on the drive train was used to measure ground speed. The AgLeader system also included a monitor which displayed instantaneous values and cumulative totals for parameters such as yield, grain moisture, grain flow, speed and distance. Although the monitor only stored the cumulative totals for each load, it could also output the following parameters to a RS-232 serial port on 1 s intervals: ground speed pulses; ground speed (mph); area count flag; elevator speed (rpm); grain flow sensor force (lb); grain flow rate (lb/sec); instantaneous yield (bu/ac); and grain moisture (% wet basis).

Both combine data acquisition systems relied on a portable computer running essentially the same software. The computer received the uncorrected GPS position and GPS time over one serial port, and the yield monitor data, either via the parallel port or a s serial port, on 1 s intervals. In the John Deere combine, the analog and digital signals from moisture sensor, volumetric yield monitor and speed sensor were input into an IOtech Daqbook™ data acquisition box and transferred to the computer through the parallel port. The Gleaner system with the impact-based yield monitor logged its output into the computer over a RS-232 serial port. Selected information such as position, total harvest area, and instantaneous yield were displayed on the computer screen. One window of the screen was dedicated to display the trajectory of the combine superimposed over the field boundaries. The program had the capability to show any selected information as text, a bar graph or a strip chart.

Procedures

The test field was located at the Missouri Management Systems Evaluation Area (MSEA) near Centralia, MO. The volumetric monitor was used to harvest a pair of transects every 100 m. The remainder of the corn (Zea mays L.) in the field was harvested using the impact-based monitor.

The radar gun and volumetric yield sensor on the John Deere combine were calibrated by flagging the beginning and end of a transect of known distance, while collecting the grain in the weigh bin situated in the combine clean grain tank. This was repeated several times and a linear least squares regression was used to calibrate the recorded counts to distance and mass respectively (Table 2-1). The magnetic pickup on the Gleaner combine was calibrated prior to harvest by travelling along a known distance at normal harvest speeds. The grain flow rate measurement was calibrated by comparing the nominal accumulated grain mass (kg/load) recorded by the impact-based monitor to the total mass of grain unloaded into the grain truck.

The GPS files stored in the receiver were post-processed to obtain a differentially corrected position coordinate for the combine every 5 s. The GPS time associated with the corrected positions was then matched to the UTC time that was logged to the computer over the RS-232 serial port during harvest. These times were used to replace the uncorrected positions with the post-processed differential positions. The positions between each differential correction were calculated using a linear interpolation based on time.

Table 2-1. Least squares linear regression for yield and distance calibration for the John Deere 3300 combine and Gleaner R62 (with impact-based monitor) combines.

	DISTANCE (meters)		YIELD (kg)	
	JD 3300	Gleaner	JD 3300	Gleaner
R Squared	0.998	-	0.997	0.993
Intercept	0	-	0	0
Std. Error of Y estimate	3.3108	-	15.00	1991.936
Regression slope coefficient	0.01000	0.02217	0.005653	2.6872
Std. Error of slope coefficient	0.00002	-	0.000023	0.035476
Degrees of Freedom	141	-	77	22

Data Analysis

For yield mapping, the measured grain flow f(t) must be related to the rate of grain flow r(t) into the combine head. The rate grain flow into the combine head can then be related to the yield since the velocity v(t), swath width (w) and position (x,y) of the combine are known. If the combine system dynamics are modelled as a simple time delay (p), then the yield y(t), can be calculated directly by dividing the measured grain flow at any instant by the area covered p seconds previously. For discrete sampling the yield can expressed as

$$y(i-p/T-s/2T) = \frac{\sum_{j=i-s/T}^{i} f(j)}{\sum_{k=i-p/T-s/T}^{i-p/T} d(k)} \cdot \frac{1000}{w} \tag{1}$$

where y(i) is the calculated yield (kg/hectare), f(i) is the measured grain flow (kg) and d(i) is the distance (m) travelled during the sampling period T. The delay time is p and s is the interval over which a running average yield is calculated.

In the continuous time domain, a first order system is expressed as

$$f(t) = r(t)(1-\exp((t-p-t_o)/\tau)) \tag{2}$$

where f(t) is the measured grain flow, r(t) is the grain flow into the combine head, t_o is the time the combine started harvesting and τ is the time constant. The first order system can be written in the Laplace domain as

$$G(s) = \frac{F(s)}{R(s)} = \frac{e^{-pTs}}{(1 + \tau s)} = \frac{1/\tau\ e^{-pTs}}{(1/\tau + s)} \tag{3}$$

where G(s) represents the transfer function from r(t) to f(t). The Laplace function must then be written in the z domain since the data collection process is discrete. There are different methods to convert to the z transform but the step-invariant transform is probably the most appropriate, since entering the crop is approximately a step input. The step-invariant transform is calculated by adding a zero order hold transform to the Laplace transfer function and then converting to the z transform.

$$G(z) = Z\ \frac{(1-e^{-Ts})}{s} \cdot G(s) \tag{4}$$

The z transform will then be

$$G(z) = \frac{(1-e^{-T/\tau})z^{-(1+p/T)}}{1-e^{-T/\tau} z^{-1}} \tag{5}$$

The discrete difference form of the z transform is

$$r(k-p/T-1) = \frac{1}{(1-e^{-T/\tau})} \cdot (f(k) - e^{-T/\tau} f(k-1)) \qquad (6)$$

where r(k) is the grain flow into the combine header (k=t/T), f(k) is the grain flow monitored, p is the time delay and τ is the time constant.

Due to the on–off method of operation of the volumetric yield monitor, the raw yield counts did not increase with every reading taken on 1 s intervals, even when there was a continuous grain flow into the monitor. The yield monitor itself acted as a sample and hold system, which exhibited a varying time constant since the interval between successive rotations was determined by the flowrate. In general, successive rotations occurred within 3 s, therefore, the original raw data was modified to obtain a single reading every 3 s by accumulating the yield counts over 3 s intervals.

Instantaneous yields were calculated from the raw impact-based monitor data and the modified volumetric monitor data using both a simple time delay model and a first order system model. These models were implemented with varying time delays and time constants for the first order system and with varying degrees of smoothing of the grain flow rate and velocity data inputs. Grain flow rate and velocity inputs were smoothed using a running average with a range of averaging times. The correlations of the calculated yield using different models were determined for each transect by comparing yields on a point-by-point basis along each transect. The John Deere combine with the volumetric yield monitor harvested six pairs of transects spaced evenly across the field. The adjacent transects harvested by the Gleaner combine with the impact-based monitor were identified.

The calculated instantaneous yields were then exported to a geostatistical package for analysis and the development of yield maps by Kriging. The accuracy of yield maps is dependent on the precision of the grain yield sensor and the process used to generate the maps. The results from the different models were used to develop maps over the field with identical cell sizes (10m). When the input data sets included all of the yield transects over the field, the Kriging parameters were set to restrict the number of nearest neighbors to a maximum of 8 or 24 neighbors. When the 6 pairs of transects were used to generate the map, a maximum of 32 neighbors was used.

Results and Discussion

While the average yield for the two monitors was similar, the calculated instantaneous yields from the impact-based monitor showed considerably less noise that those from the volumetric monitor. The difference is primarily due to the discrete operation of the volumetric yield monitor, whereas the impact-based monitor more closely approximated a continuous sampling system.

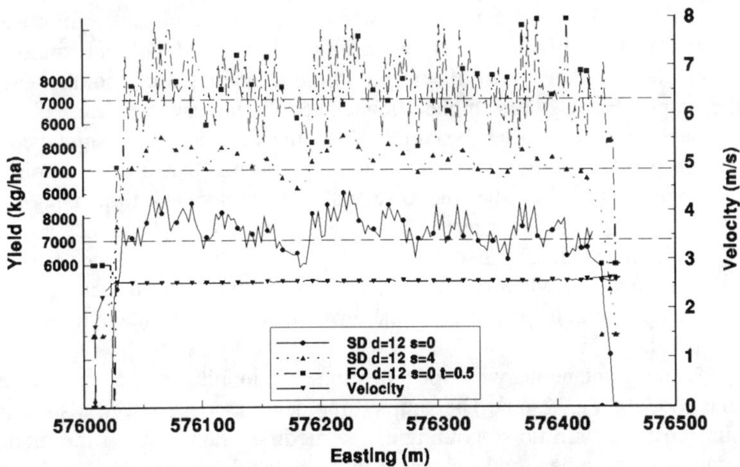

Fig. 2–1. Yield calculated from impact-based yield monitor data, using a simple time delay (SD) and a first order (FO) model with several smoothing times. (Time delay d=12s; Smoothing time s=0-4s, Time constant t=0.5s).

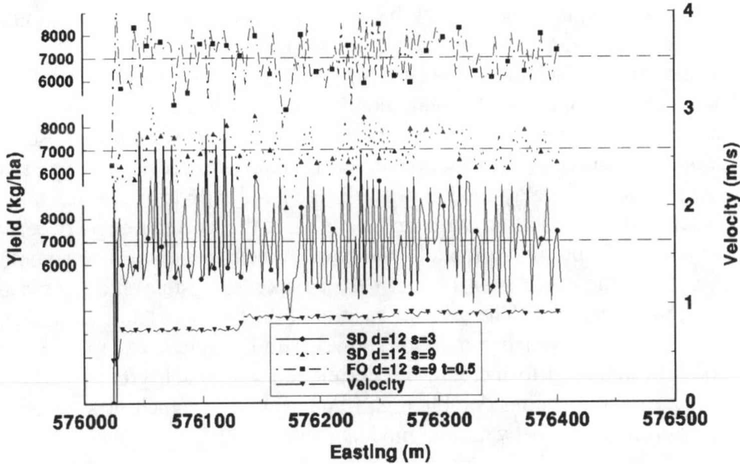

Fig. 2–2. Yield calculated from volumetric yield monitor data, using a simple time delay (SD) and a first order (FO) model with several smoothing times. (Time delay d=12s; Smoothing time s=3-9s, Time constant t=0.5s).

The instantaneous yield for the impact-based monitor was calculated using different models (Fig. 2–1). When a simple time delay model was used the calculated instantaneous yield showed little noise even with no smoothing of the raw data, and smoothing of the raw data did not significantly improve the calculated yield. When a first order model was used without smoothing, however, the calculated yield displayed a high frequency noise component due to the amplification of the higher frequencies in the raw data caused by the inversion of the first order system. When the raw data was smoothed, the calculated yield from the first order system approached that from a simple time delay model (Fig. 2–1). The smoothed first order and simple time delay model yields were very similar except when entering the crop, where the first order delay model more closely modelled the step change in yield. Since the yields calculated, however, when a large change in velocity occurred were unreliable and probably should be disregarded, there would be little advantage to modelling this step change in yield.

The instantaneous yield for the volumetric monitor was calculated using various models (Fig. 2–2). The simple time delay model showed a substantial amount of noise with no smoothing of the modified raw data. If the modified raw data was smoothed the local trends in yield could be seen (Fig. 2–2). When a first order system was used, the high frequency noise began to dominate the signal, even with smoothing of the model input data. This was caused by the discrete operation of the monitor. Theoretically, the monitor could be modelled as an additional first order system. The time constant of this model would vary with yield, however, adding another unknown parameter to the complete system model. Figures 2-3 and 2-4 show the calculated instantaneous yield for a pair of east-west transects for the impact-based monitor and volumetric monitor, respectively. As expected, the transects show similar local trends, although this was less apparent with the volumetric monitor.

The correlation coefficients between the instantaneous yields obtained using the different models were calculated for each transect. Table 2-2 shows the correlation coefficients between a single "reference" model and the other models for each monitor. The same number was used to identify a single pair of transects harvested adjacent to each other and the letters were used to separate the individual transects. The impact-based monitor showed a high correlation between simple time delay models with different smoothing intervals. When a first order system was used with no smoothing, the correlation between this model and the simple delay models was low. When the amount of smoothing was increased, the correlation between the first order system and simple time delay systems increased. The reverse was true if the time constant was increased (Table 2-2). The volumetric monitor showed similar trends, except the degree of smoothing required to increase the correlation was much greater. The first order systems showed considerable high frequency noise which was reflected in the lower correlations between the models (Table 2-2).

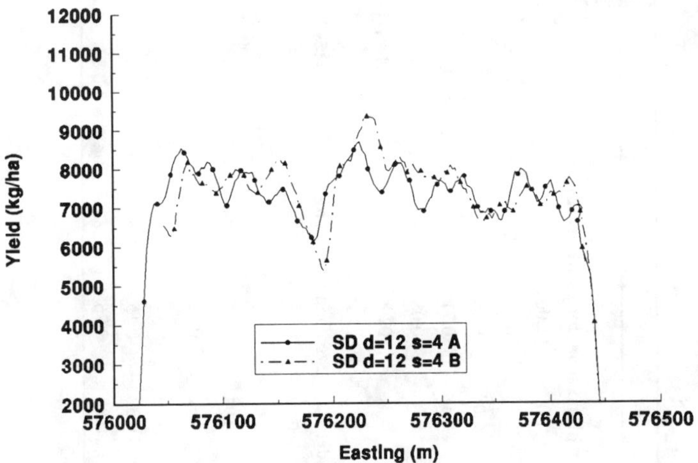

Fig. 2–3. Yield calculated from impact-based yield monitor data for a pair of side by side transects, using a simple time delay (SD) model. (Time delay, d=12s; Smoothing time s=0s)

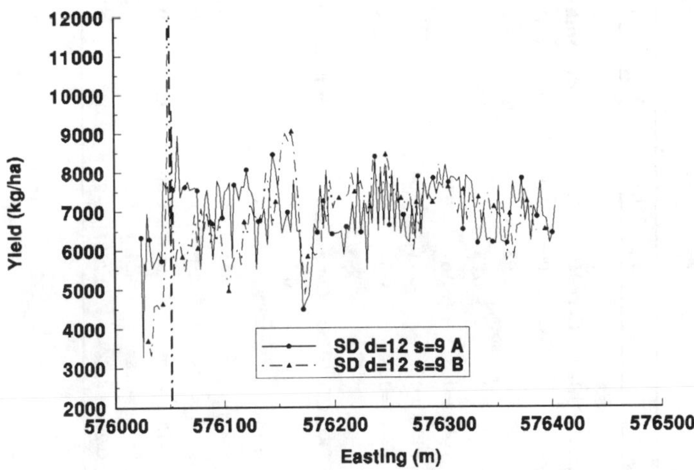

Fig. 2–4. Yield calculated from volumetric yield monitor data for a pair of side by side transects, using a simple time delay (SD) model. (Time delay, d=12s; Smoothing time s=9s)

Table 2-2. Correlations between a selected reference model and the Simple Time Delay (SD) and First Order (FO) combine models with different degrees of smoothing (Time delay(d) =12s; Time constant(t) =0.5 or 0.3s; Smoothing time(s)=0-21s).

TRANSECT	1A	1B	2A	2B	3A	3B	4A	4B	5A	5B	6A	6B	1A-6B
GLEANER COMBINE: Correlation of different models with simple time delay model with no smoothing (SD d=12s, s=0s)													
SD d=12s, s=0s	1.00	1.00	1.00	1.00	1.00	1.00	1.00	1.00	1.00	1.00	1.00	1.00	1.00
SD d=12s, s=4s	0.92	0.94	0.74	0.94	0.97	0.93	0.89	0.94	0.94	0.95	0.96	0.91	0.92
FO d=12s, s=0s, t=0.5s	0.08	0.16	-0.31	0.09	0.56	0.22	0.09	0.26	0.23	0.32	0.34	0.22	0.19
FO d=12s, s=4s, t=0.5s	0.67	0.72	0.45	0.74	0.86	0.70	0.48	0.71	0.74	0.73	0.68	0.63	0.68
FO d=12s, s=0s, t=0.3s	-0.11	-0.04	-0.41	-0.11	0.39	0.03	-0.06	0.08	0.03	0.11	0.11	0.03	0.00
FO d=12s, s=4s, t=0.3s	0.37	0.45	0.30	0.48	0.79	0.53	0.32	0.58	0.54	0.56	0.44	0.47	0.49
JOHN DEERE COMBINE: Correlation of different models with simple time delay model with 15 second smoothing (SD d=12s, s=15s)													
SD d=12s, s=3s	0.41	0.51	0.49	0.25	0.16	0.44	0.53	0.35	0.46	0.56	0.46	0.44	0.42
SD d=12s, s=15s	1.00	1.00	1.00	1.00	1.00	1.00	1.00	1.00	1.00	1.00	1.00	1.00	1.00
SD d=12s, s=21s	0.83	0.94	0.93	0.66	0.64	0.92	0.93	0.47	0.89	0.95	0.94	0.90	0.83
FO d=12s, s=3s, t=0.5s	0.30	0.44	0.44	0.05	0.15	0.28	0.41	-0.12	0.38	0.38	0.48	0.35	0.30
FO d=12s, s=15s, t=0.5s	0.65	0.86	0.82	0.39	0.37	0.84	0.85	0.15	0.82	0.86	0.86	0.72	0.68
FO d=12s, s=3s, t=0.3s	0.24	0.35	0.36	0.02	0.12	0.20	0.32	-0.15	0.30	0.29	0.42	0.28	0.23
FO d=12s, s=21s, t=0.3s	0.80	0.93	0.88	0.78	0.72	0.89	0.88	0.76	0.86	0.91	0.87	0.84	0.84

When yield semi-variograms were calculated over 300m, the best fit was obtained with a linear variogram that displayed a high nugget variance for all models. If the variograms were calculated over 100m, however, the simple time delay models and highly smoothed first order systems exhibited either an exponential or spherical semi-variogram, with low nugget variance and a definite spatial range (generally 25–30m). The first order systems with little smoothing displayed no spatial relationship. Theoretically, if a large active range is used the data should be de-trended before the calculation of the semi-variance. Since the neighborhood used during Kriging was restricted, however, and only the low lags of the semi-variogram were used, the Kriged output map essentially showed the local trend.

The complete set of raw impact-based monitor harvest data was used to generate yield maps, using different combine models and Kriging parameters (Fig. 2–5 through 2–8). A simple time delay model with no smoothing (Figs. 2–5 and 2–6) and a first order model with no smoothing were used (Fig. 2–7 and 2–8). The active range for the calculation of the semi-variance was 100m for Fig. 6 and 8 and 300m for Fig. 2–5 and 2–7. During Kriging the number of known points used for the calculation of the grid cell was restricted to 8 or 24 neighbors, for Fig. 2–6 and 2–8, and Fig. 2–5 and 2–7, respectively. The general trends were the same for all of the maps. When the number of known samples used in the Kriging process was reduced the trends did not change but the local variability increased as shown in Fig. 2–6 and 2–8. When a first order system as used (Fig. 2–7 and 2–8) the maps show a much higher yield along the edge of the field than when a simple time delay model was used (Fig. 2–5 and 2–6), due to the step response of the first order system. The calculated yields are probably higher than the actual yields, however, and the error associated with the calculation at these transition points was high.

Six pairs of transects 100m apart harvested by the Gleaner combine with the impact-based monitor were used to generate a map, using a simple time delay model with 4 s (Fig. 2–9). While much of the fine detail was lost, the general trends were very similar to the previously shown general field trends' (Fig. 2–5 through 2–8). Adjacent pairs of transects harvested by the Deere combine with the volumetric monitor, were used to generate a map, using a simple time delay model and 15 s of smoothing (Fig. 2–10). Although further detail was missing the basic trends were the same. Although there were some differences between the maps developed from transects as compared to those developed from a complete set of data, the transect maps do show the general trends and were a reasonable representation of yield trends. However the accuracy of these transect-based maps would also depend on the location of the transects relative to important changes in yield.

The correlations between the Kriged maps were compared on a cell by cell basis (Table 2–3). All of the whole-field simple time delay models exhibited a high correlation with each other. The unsmoothed first order system exhibited a low correlation when compared to the simple time delay maps, due to the high frequency component introduced. The transect maps developed from the Gleaner combine were reasonably correlated to the maps from the whole field data, but the John Deere combine transects showed a lower correlation.

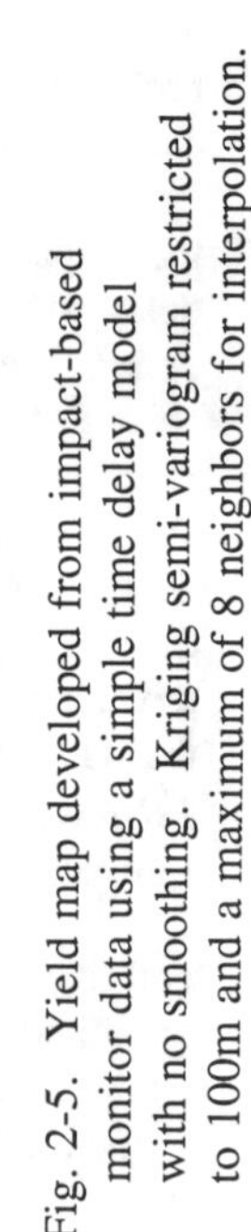
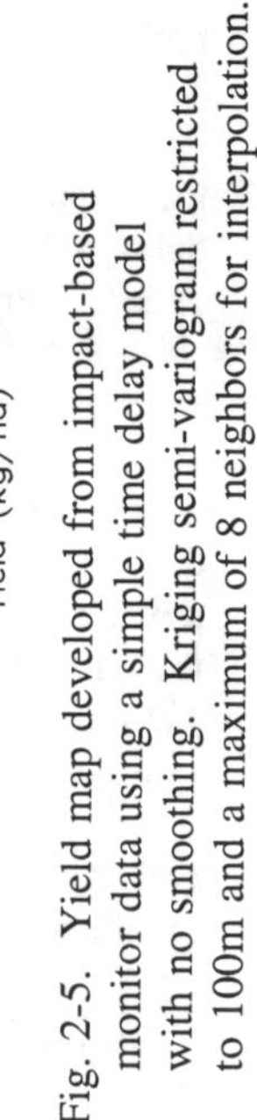

Fig. 2-5. Yield map developed from impact-based monitor data using a simple time delay model with no smoothing. Kriging semi-variogram restricted to 100m and a maximum of 8 neighbors for interpolation.

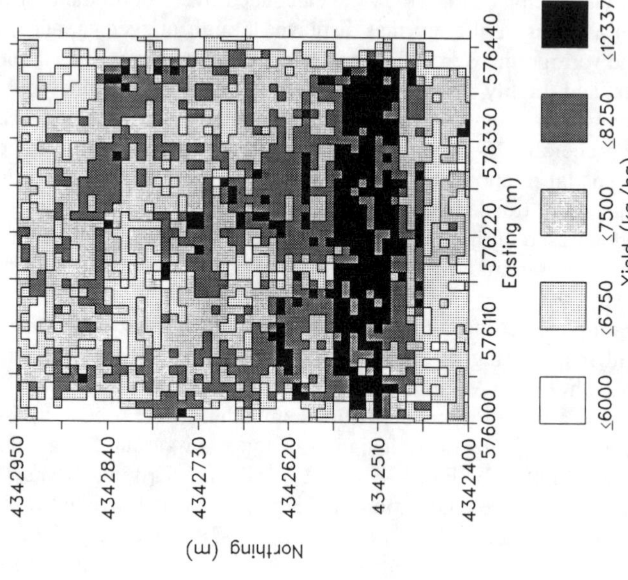

Fig. 2-6. Yield map developed from impact-based monitor data using a simple time delay model with no smoothing. Kriging semi-variogram restricted to 300m and maximum of 24 neighbors for interpolation.

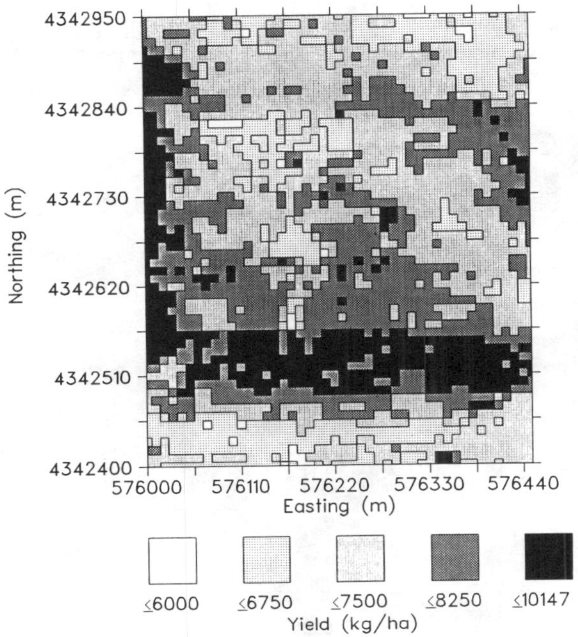

Fig. 2-7. Yield map developed from impact-based monitor data using a first order model with no smoothing. Kriging semi-variogram restricted to 300m and a maximum of 24 neighbors for interpolation.

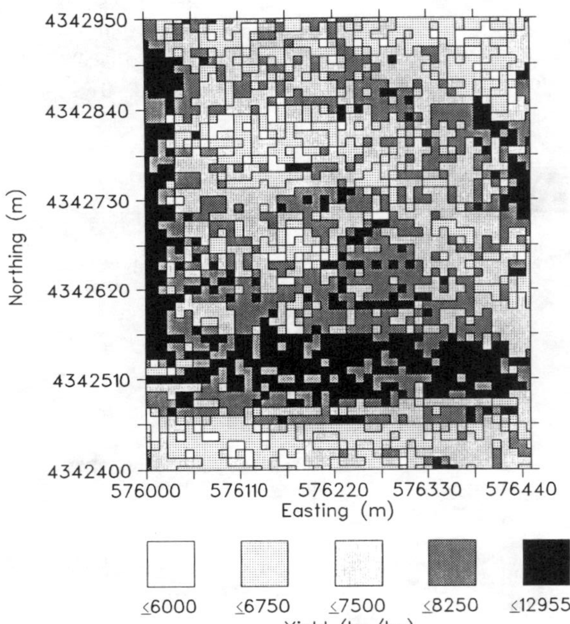

Fig. 2-8. Yield map developed from impact-based monitor data using a first order model with no smoothing. Kriging semi-variogram restricted to 100m and a maximum of 8 neighbors for interpolation.

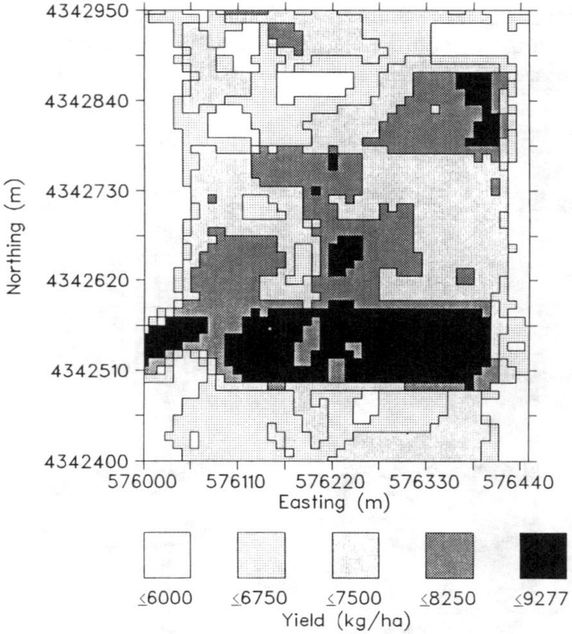

Fig. 2-9. Yield map developed using six pairs of transects of impact-based monitor data using a simple time delay model with 4 s smoothing. Kriging semi-variogram restricted to 300m and a maximum of 32 neighbors for interpolation.

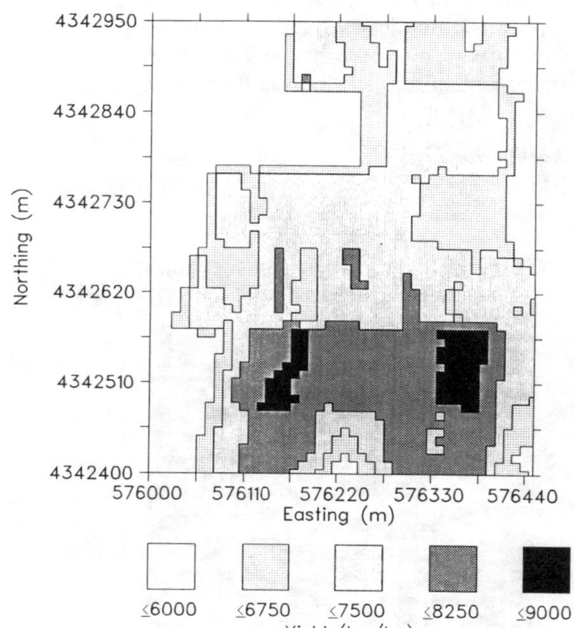

Fig. 2-10. Yield map developed using six pairs of transects of volumetric monitor data using a simple time delay model with 15 s smoothing. Kriging semi-variogram restricted to 300m and a maximum of 32 neighbors for interpolation.

Table 2–3. Correlations calculated for Kriged maps developed with different combine models and Kriging parameters. (Compared to map developed from Gleaner combine data with a simple delay model, no smoothing and 300m active Kriging range.)

Data Source	Model Type	Smoothing (s)	Time Constant (s)	Kriging Parameters		Correlation Coefficient
				Semi-variogram Active Range (m)	Number of Neighbors	
Gleaner (all data)	Simple Time Delay	0		300	24	1
Gleaner (all data)	Simple Time Delay	0		100	8	0.91
Gleaner (all data)	Simple Time Delay	4		300	24	0.9
Gleaner (all data)	Simple Time Delay	4		100	8	0.81
Gleaner (all data)	First Order System	0	0.5	300	24	0.58
Gleaner (all data)	First Order System	0	0.5	100	8	0.49
Gleaner (all data)	First Order System	4	0.3	300	24	0.68
Gleaner (all data)	First Order System	4	0.3	100	8	0.61
Gleaner (transects)	Simple Time Delay	4		300	32	0.64
JD 3300 (transects)	Simple Time Delay	15		300	32	0.41

The lower correlation for the John Deere combine transects was due to the increase in the amount of smoothing required, which removed the yield variation over short distances. In general, it appears that a simple time delay model with minimal smoothing provided the best yield maps. Maps generated from evenly spaced transects, however, showed the general yield trends and would provide useful information.

Summary

Modelling of the instantaneous yield response for two different yield monitors was investigated. Both simple time delay and first order models appeared to be reasonable models of the combine flow dynamics. The simple time delay model was less susceptible to noise but did not accurately model the step change in yield seen when entering or exiting a crop. The first order system modelled the step yield input but was highly susceptible to noise and required smoothing of the raw data. The Kriging of instantaneous yield to develop maps was fairly robust if a complete data set was used. The general trends in the data were evident even when cell values were calculated from a very localized region. Evenly spaced transects could be used to develop maps showing general yield trends although some detail is lost.

ACKNOWLEDGEMENT

We would like to thank Dr. S. W. Searcy and the Department of Agricultural Engineering, Texas A&M University, for the loan of the Claydon Yieldometer.

Mention of trade names or specific products is made only to provide information to the reader, and does not constitute an endorsement by the University of Missouri or the USDA Agricultural Research Service.

REFERENCES

Auernhammer, H., M. Demmel, K. Muhr, J. Rottmeier, and K. Wild. 1993. Yield measurements on combine harvesters. ASAE 93–1506. ASAE, St. Joseph, MI.

Bae, Y.H., S.C. Borgelt, S.W. Searcy, J.K. Schueller, and B.A. Stout. 1987. Determination of spatially variable yield maps. ASAE 87–1533. ASAE, St. Joseph, MI.

Birrell, S.J., K.A. Sudduth and S.C. Borgelt. 1993. Crop yield mapping using GPS. ASAE Paper MC93–104. ASAE, St. Joseph, MI.

Colvin, T.S., D.L. Karlen, and N. Tischer. 1991. Yield variability within fields in central Iowa. *In* Automated agriculture for the 21st Century, Proc. of the 1991 Symp. ASAE Publ. 11–91, ASAE, St. Joseph, MI.

Fairfield Smith, H. 1938. An empirical law describing the heterogeneity in the yield of agricultural crops. J. Agric. Sc. XXVIII(Jan):1–23.

Goering, C.E. 1993. Recycling a concept. Agric. Eng. 74(6):25.

Harrison, J.D., S.J. Birrell, K.A. Sudduth, and S.C. Borgelt. 1992. Global positioning system applications for site-specific farming research. ASAE 92–3615. ASAE, St. Joseph, MI.

Pringle, J.L., M.D. Schrock, R.T. Hinnen, K.D. Howard, and D.L. Oard. 1993. Spatial yield variation in grain crops. ASAE Paper 93–1505. ASAE, St. Joseph, MI.

Robinson, G.W., and W.E. Lloyd. 1915. On the probable error of sampling in soil surveys. J. Agric. Sci. 7:144–153.

Schnug, E., D. Murphy, E. Evans, S. Haneklaus, and J. Lamp. 1993. Yield mapping and application of yield maps to computer-aided local resource management. *In* Proc. of the Workshop on Research and Development Issues on Soil Specific Crop Management. 14-16 April 1992. Minneapolis, MN. ASA, Madison, WI.

Searcy, S.W., J.K. Schueller, Y.H. Bae, S.C. Borgelt, and B.A. Stout. 1989. Mapping of spatially variable yield during grain combining. Trans. ASAE 32(3):826–829.

Stafford, J.V., B. Ambler, and M.P. Smith. 1991. Sensing and mapping grain yield variations. *In* Automated Agriculture for the 21st Century, Proceedings of the 1991 Symposium. ASAE Publ. 11–91, ASAE, St. Joseph, MI.

Stott, B.L., S.C. Borgelt, and K.A. Sudduth. 1993. Yield determination using a instrumented Claas combine. ASAE 93-1507. ASAE, St. Joseph, MI.

Vansichen, R., and J. De Baerdemaeker. 1991. Continuous wheat yield measurement on a combine. *In* Automated Agriculture for the 21st Century, Proceedings of the 1991 Symposium. ASAE Publ. 11-91, ASAE, St. Joseph, MI.

Wagner, L.E., and M.D. Schrock. 1989. Yield determination using a pivoted auger flow sensor. Trans. ASAE 32(2):409-413.

3 Yield Mapping - A Guide to Improved Techniques and Strategies

Donal P. Murphy
Ewald Schnug
Silvia Haneklaus

Federal Agric. Research Center
Institute of Plant Nutrition and Soil Science
Braunschweig-Voelkenrode
Germany

The determination of the spatial variability in crop yield at the point of harvest provides a biological indicator of variation in factors affecting productivity and a valuable diagnostic tool (O'Callaghan, 1988). The Local Resource Management concept (Schnug et al., 1993) proposes that the combination of such spatially distributed yield data with data on soil and plant parameters provides the foundation of a Local Resource Information System (LORIS) that enables spatial variation in the application of fertilizers to be calculated. Yield meters introduced in the early 1980s for installation into combine harvesters were designed to give an accurate estimation of the total yield of a crop harvested. More recently, access to the Global Positioning System (GPS) enables the spot yield data from these yield meters to be combined with position providing the spatially distributed data required for yield mapping. The effectiveness of these systems for yield mapping relies on the data obtained accurately reflecting true variation in the yield of the crop grown. Changes in the rate of grain flow at the point of sensing sometimes do not directly reflect changes in crop intake. Thus, yield mapping is easy in principle but presents considerable technical difficulties if accurate yield maps are to be obtained. This contribution reviews the technology used for this task and proposes techniques that minimize errors. It is based on both observations of the operation of these systems under a wide range of conditions, and analysis of the resultant data. If our proposals are adopted, it would greatly improve the performance of combine harvesters as providers of data for yield mapping and reduce the constraints that operators must presently observe for valid data collection.

GRAIN FLOW METERS FOR YIELD MAPPING

Borgelt (1993) gives a brief summary of the sensor development. Yield mapping equipment is now commercially available in Europe and used on farms pioneering the Local Resource Management concept. Three sensors are used: the Claydon Yieldometer, 'Ceres' and 'Flowcontrol'.

Claydon Yieldometer

The Claydon Yieldometer was invented by Mr. J. Claydon (Norfolk, England) to enable the monitoring of grain yields on his farm using a Claas combine harvester. This instrument (Fig. 3–1) counts the number of times a paddle wheel, located under the outlet of the clean grain elevator, rotates to transfer the grain to the clean grain auger. A head of grain is maintained above the paddle wheel and no grain passes without its rotation. The paddle wheel, which rotates in response to grain reaching the level sensor, is driven directly by the harvester's power system, and its rotation is relatively constant at 100 rpm at normal operating speed. A harvester designed to handle 20 tons per h is fitted with a paddle wheel that transfers wheat grain (bulk density of 75 kg Hl^{-1}) at 70 tons per h. Thus, when the machine is operating at capacity, the paddle wheel is operating for < 30% of the time with no possibility of being overloaded. The volume of grain harvested per unit time is calculated by integrating the volume of grain transferred over a 4 s period. This is converted to a 'spot' yield per unit area by calculating the area cut from the harvester's forward speed. Another sensor cuts out area measurement when the table is raised. The volume of grain measured is converted to mass at a constant moisture content using an estimate of the average grain bulk density (hectoliter weight) and moisture content. The software with the associated instrumentation provides a 'spot' yield output from the rolling average of the last four calculations, i. e., 4 x 4 s intervals. Thus, short distance variability in grain yield is buffered by both the harvesting mechanism and data handling. The data shown in Fig. 3–1 reflect this: the average yield was 9.24 ton ha^{-1} with a standard deviation of 1.1. This buffering minimizes the effect of fluctuations in grain flow that are unrelated to spatially dependent changes in crop yield.

Our experience confirms that, once the mean grain bulk density and moisture content is determined correctly, the system is capable of measuring the total yield of a given area very accurately (+/- 1%), according to the manufacturer). The recording of the spot yield has provided the data for yield mapping (Searcy, 1989; Schnug et al., 1992).

'Ceres'

'Ceres', marketed by RDS Technology, is based on a patent (DE3045728) granted to Claas mbH, Germany in 1982 (Diekhans, 1985). It is based on the non-intrusive installation of standard electronic components onto the clean grain elevator (Fig. 3–2). A light source and photosensor detect the degree to which the elevator flights are loaded with grain. Thus, the sensor operates on the principle that the volume of grain flowing is related to the cross section area of

(a)

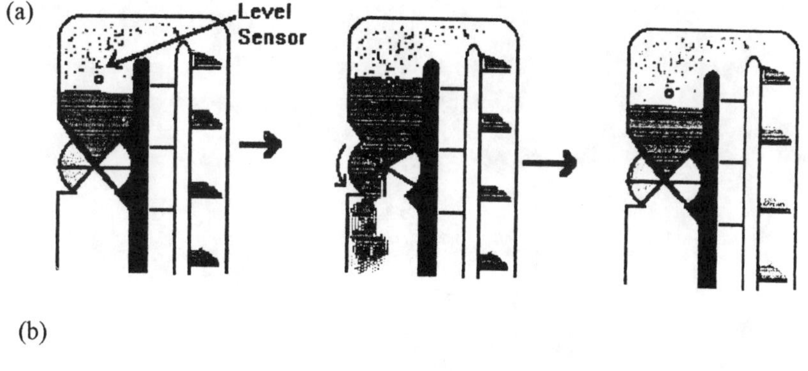

(b)

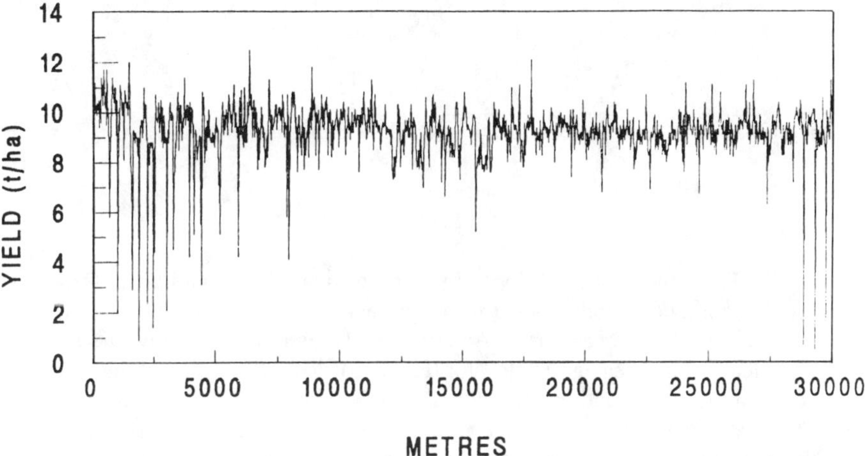

Fig. 3–1. A schematic view of the Claydon yieldometer in operation (a), and spot readings recorded every 20 m over 30,000 m (b).

the grain carried by the elevator flights, and thus, the amount of transmitted light detected at the sensor. The quantity of light detected, however, is not linearly related to grain flow, and this non-linear relationship varies between harvesters, crop species, and according to grain moisture content and the inclination of the harvester to the horizontal. The conversion of grain volume to mass requires that the bulk density of the grain is measured and the value inputted into the instrumentation. 'Ceres' is the latest of the three sensors to come on the market, and the manufacturer has not published detailed data concerning its performance. Anecdotal evidence indicates that operators may achieve errors of < 2% in the measurement of total grain yield where the instrument is carefully calibrated and grain bulk density is regularly measured.

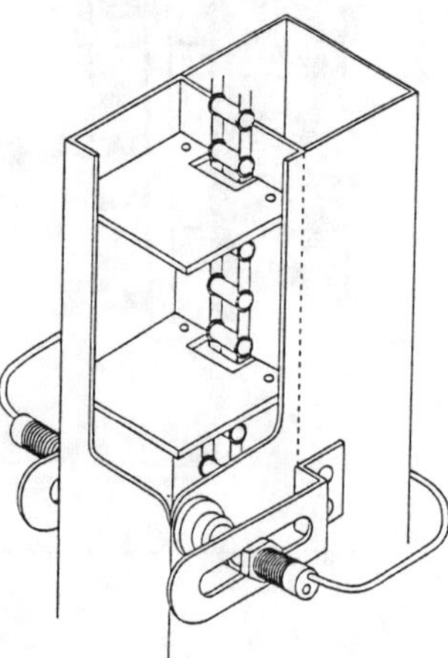

Fig. 3–2. A 'Ceres' sensor installed on the clean grain elevator.

'Flowcontrol'

This sensor was developed by Thoustrup and Overgaard a/s in Denmark for Dronningborg Industries (manufacturers of Massey Ferguson combine harvesters). A gamma source (Americium 241) is mounted below the flow of grain leaving the clean grain elevator (Fig. 3–3). The attenuation of the radiation between the source and the detector is a stable function of the mass of the grain. This relationship is not affected by grain moisture content or crop species. Thus, the radiation attenuated per unit time is related to the mass of the grain flow which is converted to mass per unit area using the harvester forward speed and cutting width. With 'Flowcontrol', the attenuated radiation is integrated over a 2 s period to give a spot yield reading.

The use of the sensor for yield mapping is greatly aided by its basic simplicity, its non-intrusion into the flow of grain, and its direct measurement of grain mass rather than volume. Its adoption for routine yield mapping has also been greatly facilitated by the 'Danavision' electronic control system that has been installed on the larger Massey Ferguson harvesters sold in Europe since about 1980. With the yield mapping prototypes tested in 1991 (Schnug et al., 1992), the data logging equipment and software were very easily integrated into the existing control mechanism that provided the data on forward speed and cutter bar position. Thus, the commercial systems that have been subsequently produced have been integrated into the harvester's mechanical and electronic structure without difficulty.

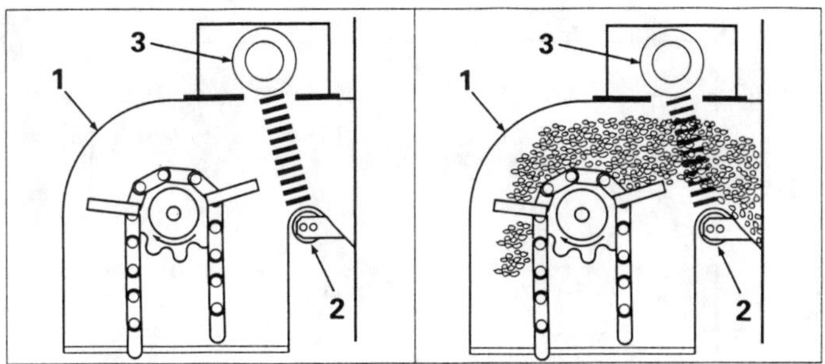

Fig. 3–3. A 'Flowcontrol' sensor in relation to the clean grain elevator (i), source (ii), and detector (iii).

At present, the 'Flowcontrol' based system is the only one available with a data logging and transfer system dedicated to producing yield maps under normal commercial farming conditions. The system is accompanied by yield mapping software based on 'SURFER' that has been specially adapted for farm use. The software controlling logging allocates spot yield to position every 2 s according to a lag of 16 s between the cutting and the yield reading. This 16 s lag is also used to block data logging following the lowering of the cutter bar.

POSITIONING

Accurate positioning is a pre-requisite for yield mapping. All Global Positioning System (GPS) receivers with differential assistance (DGPS) enable position on the earth's surface to be identified within 10 m, and their use is now standard in Europe. The purchase of a differential base station is a large capital cost associated with yield mapping equipment. The base station, however, may be used for all DGPS applications within its range, and it is possible that operators could cooperate locally to share a base station facility. A survey of prices (Anon. 1993) shows that to obtain an accuracy of 10 m with GPS (standalone) requires the purchase of equipment costing \$60,000 - \$100,000 per unit (Fig. 3–4). This contrasts with DGPS units, many of which cost < \$5,000. The purchase of the base station is additional to this. It is concluded that the use of DGPS is necessary for the economic real time positioning of farm machinery.

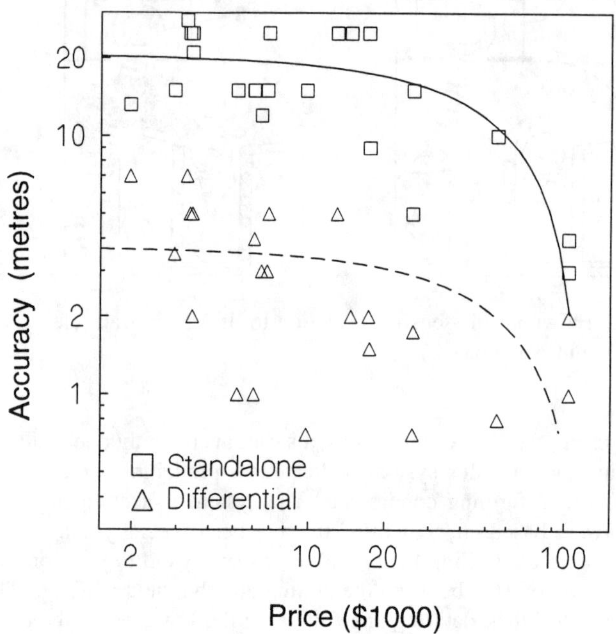

Fig. 3–4. Price and accuracy of GPS and DGPS receivers (mobile units).

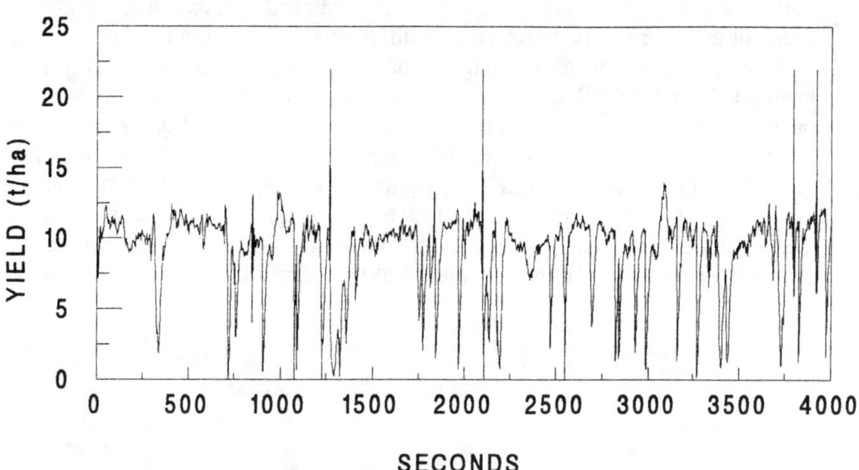

Fig. 3–5. 'Flowcontrol' yield output over 4000 s.

DATA LOGGING

Data from the first prototype commercial yield mapping system (using 'Flowcontrol') tested in Cleveland, England (1991) demonstrate the difficulties that may occur in using raw sensor data to produce yield maps (Fig. 3–5). In this sample of 2000 unprocessed yield readings, grain yield averaged 9.18 t ha^{-1}, ranged from 0–22 t ha^{-1} with a standard deviation of 2.78. Since data were logged only when the machine was actually cutting, and yields of winter barley (Hordeum Vulgare L.) over 10 to ha^{-1} are exceptional, it is clear that the raw yield data are subject to error. These errors apply to any grain flow meter, and their sources may be summarized as follows:

1. A stream of zero, near zero and low recordings following the start of each harvest run.

2. Periods during which the harvester is operating but not taking in crop material, is reversing, or is recovering from unpredictable interruptions in grain flow.

3. Sudden fluctuations in forward speed. Grain flow is combined with forward speed and cutting width to calculate spot yield per unit area. Very high readings are attributable to situations where rapid deceleration takes place while grain flow from the crop that is already cut is being processed. Thus, grain flow associated with the harvester progressing at a higher speed is linked with a reduced forward speed.

4. Grain flow and recorded grain yield is reduced if the cutting width set on the instrument is not used.

The data given in Fig. 3–6 show the pattern of raw yield data coming from the 'Flowcontrol' sensor recorded over the first 60 s of four harvest runs. These data relate to parts of the field where high and relatively stable yields are expected. A clear pattern is evident in the data. Very low readings in the first 6 s are followed by a 10–20 s period when grain flow increases rapidly to be followed by a third phase in which grain flow continues to increase slowly and systematically. The average yield readings are also plotted and a 4th order polynomial equation very accurately fits these data (r^2 = 0.9952). With this example, the most rapid increase in yield (point of inflection) occurs 16 s after the start of the harvest run, and yield stops rising systematically at about 40 s. All data associated with this start-up period are potentially invalid. Furthermore, the resultant concentration of erroneously low values at the beginning of harvest runs is such that the interpolation procedures involved in gridding the data in the mapping process are biased in favor of erroneously low grid values where harvest runs commence.

The occurrence of erroneous readings due to sudden fluctuations in harvester forward speed is less problematic since they are more likely to be randomly distributed throughout the field. Furthermore, high values associated

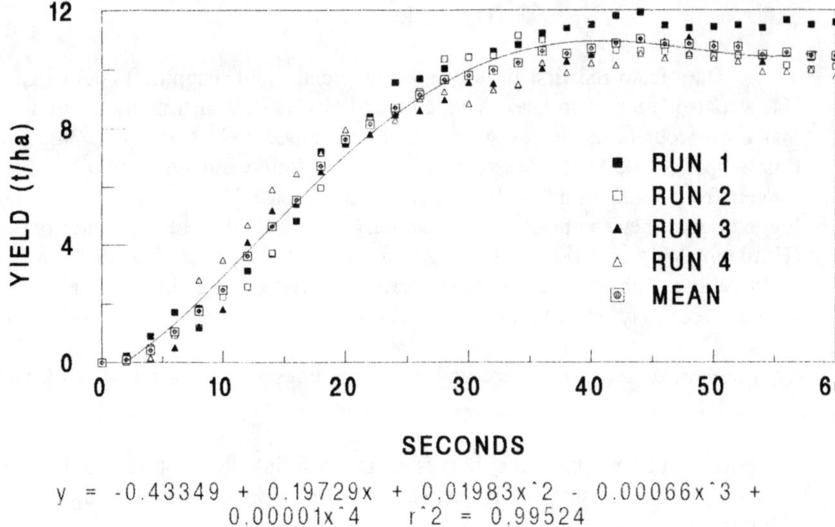

Fig. 3–6. 'Flowcontrol' yield readings over the first 60 s of four harvest runs.

with deceleration are likely to be followed by low values associated with the subsequent stabilization, or acceleration, in forward speed. The effects of these randomly distributed yields are masked by the presence of a much larger number of adjacent valid points.

MAPPING PROCEDURES

Regardless of the yield recording system used, the resultant file consists of a stream of data relating crop yield to distinct points. In order to produce a yield map, a regular spaced grid of yields interpolated from the irregular spaced harvester data must be made. It is at this point that some geo-statistical theory is employed. Theoretically, harvest data should be dense enough to give adequate data within the distance that data are spatially dependent for the calculation of grid point values. Smoothness of the resultant contours is dependent on the input data, grid density and the gridding algorithms used. Since harvest data may be subject to erroneous fluctuations (e.g., Fig. 3–6), it is clear that smoothing is an important function of the procedures used which should also ensure that valid small scale spatially related variation is not masked.

Data intensity and configuration

Our experience shows that, in western Europe, typical yield data from combine harvesters are spatially related over a range of 50-200 m. Even greater ranges are expected in geological older soils, or where crop yield is strongly influenced by soil water availability. Harvesters equipped with grain flow sensors and GPS receivers supply very dense data in relation to these ranges.

Thus, data density is not a factor limiting the selection of mapping techniques. Ideally, data should be spaced with similar lag distances in all directions. This configuration provides data that can be subjected to a wide range of gridding procedures. The very dense data sets provide ample opportunity for removing all data that are suspected to be artifacts of the harvesting procedure.

Grid Density

For yield mapping, the gridding procedure has two purposes: the production of cartographically acceptable yield maps, and the preparation of data that can be spatially synchronized with soil or other plant data coming from other, and most likely, less dense sampling procedures so that grids of the spatial variation in fertilizers or pesticides can be calculated.

Gridding Methods

A wide range of methods are available to calculate grid point data from sample data, however, in practice the inverse distance method or its variations, and kriging, are suited to the majority of harvest data sets. With both methods, a search radius is defined within which data points are used to calculate grid values.

Inverse Distance Method

Data are weighted so that their influence on the resultant grid value declines with distance. A weighting power is used to adjust the relationship between distance of a sample from the grid point and its influence on the computed grid point value. The greater the weighting power, the greater the influence of the most proximate points within the search area. Figure 3-7 shows the effect of weighting power and grid size on the yield map resulting from the harvest of winter wheat using a Dronningborg (Massey Ferguson) harvester fitted with 'Flowcontrol' at Vindum in Denmark.

The search method used to identify the data points used may be defined in response to some characteristics of the data. A normal search method specifies that the nearest points to the grid node are used. The Quadrant and Octant search methods divides the search area into four and eight sectors, respectively, and identifies a specified number of the nearest data points in each sector. The Quadrant and Octant methods are suited to some harvest data sets where data are recorded at very closely spaced intervals along wider spaced harvest runs. In this way, the influence of the direction of the harvester on the resultant grid value is controlled.

The number of nearest data points used affects the resultant grid values. A small number of nearest points will result in large variation between adjacent grid values. Similarly, as with reducing the weighting power, specifying a large number of nearest points smooths the grid data.

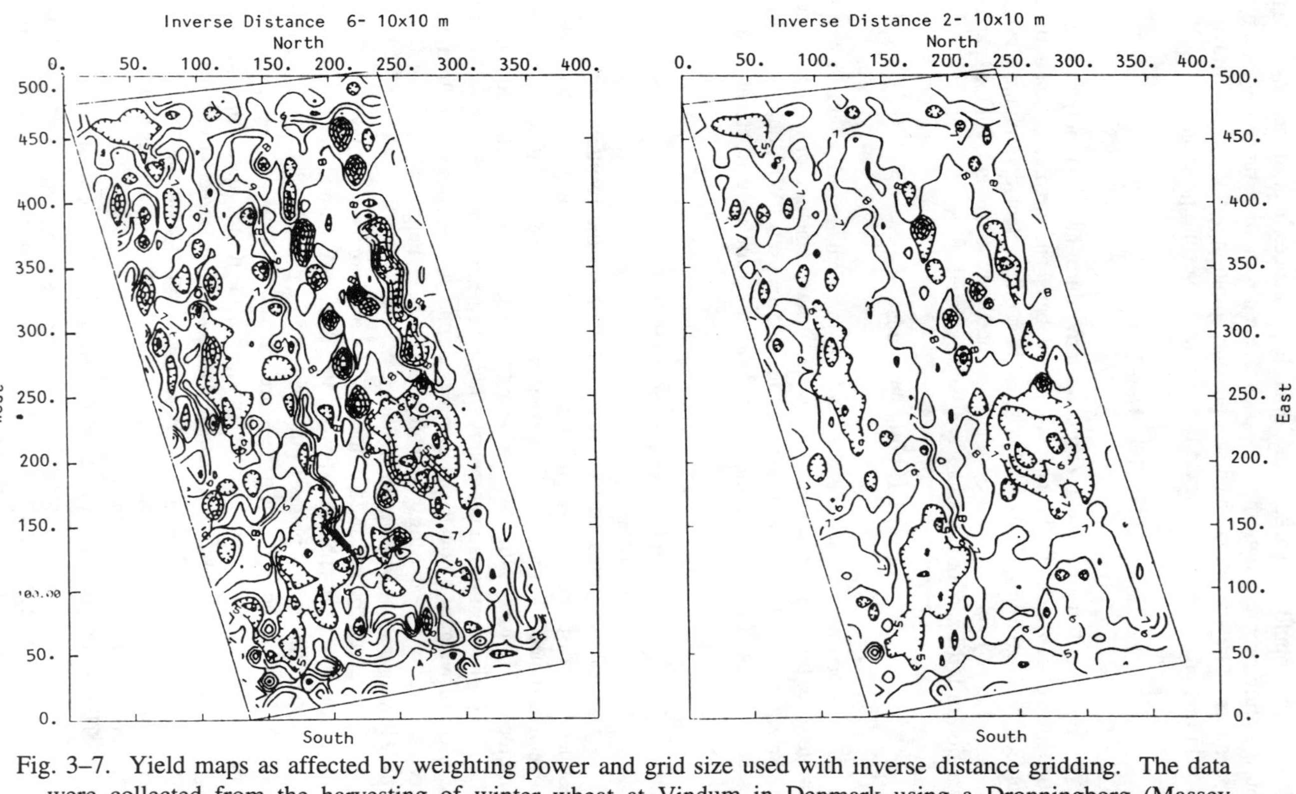

Fig. 3–7. Yield maps as affected by weighting power and grid size used with inverse distance gridding. The data were collected from the harvesting of winter wheat at Vindum in Denmark using a Dronningborg (Massey Ferguson) harvester fitted with 'Flowcontrol'.

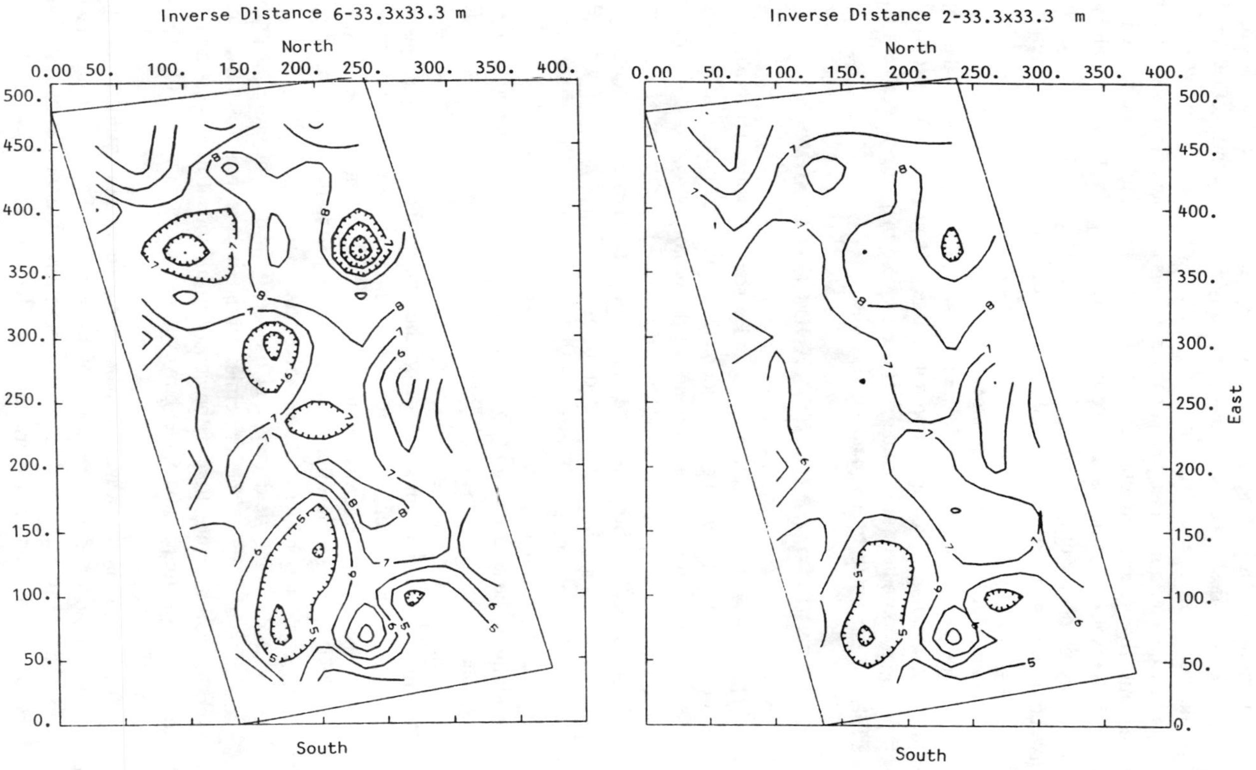

Fig. 3–7. Continued

Kriging

Kriging is based on the theory of regionalized variables (Matheron, 1971). It is fully described by Webster and Oliver (1990). In practice, kriging is similar to the inverse distance method; it differs in that weighting given to data points is determined in a statistically rigorous manner. Thus, while other mathematically based techniques are intuitively suitable for mapping, kriging employs statistical techniques to determine how sample values are weighted. The kriging procedure employs the relationship in the data between the distance between points and their variance (the semi-variogram) to calculate weightings. Thus, the criteria used in the interpolation procedure are directly related to the spatial dependence in the data set concerned.

While kriging is statistically the most acceptable way to grid data, it is the most demanding in terms of computer time and memory. With currently used personal computers, this hinders its use in yield mapping where very large data sets are generated by frequent yield readings.

RECOMMENDATIONS AND CONCLUSIONS

The yield mapping technologies and associated software and mapping procedures have been developed and adapted for yield mapping on the basis of empirical observation of the performance of the prototype systems. In the case of 'Flowcontrol', a very effective system for the rapid production of yield maps has been developed and brought into commercial use. The development of yield mapping systems has largely resulted from exploiting the spot yield data produced by the yield meters. These spot yield data are strongly buffered by the harvesting and threshing process. For yield mapping, we require the minimum of smoothing of the sensor output so that crop yield related fluctuations in grain flow are preserved. For this, the 'Flowcontrol' system is effective: grain flow is not interrupted, and the spot yield reading is related to the grain flow over 2 s. With the Claydon yieldometer, the paddle wheel, and the integration of its rotation over 4 s introduces a buffering effect which, as the instrument presently operates, is increased further by the use of the 16 s moving average. With the Claydon system, we propose that this buffering effect is minimized by logging direct sensor readings rather than averaged data.

For commercial yield mapping, the Ceres instrument is not at present fully tested technology. The principle behind its use requires that special attention be paid to calibration of the sensor in individual harvesters and crop species. It has the major advantage, however, that it can be very easily fitted to all makes of harvester.

It has been proposed that the harvester should be the primary screening tool with erroneous data removed before logging (Schnug et al., 1992). Erroneous data may be divided into two groups: errors associated with the start of harvest runs that are concentrated at headland areas, and errors randomly distributed throughout the field associated with interruptions in grain flow and sudden fluctuations in forward speed. The data presented in Fig. 3–6 show the pattern of yield readings associated with the start-up of a particular Massey

Ferguson harvester. We propose that the time between cutting and the arrival of the grain at the flow meter (harvest lag time), and the time interval between the start of a harvest run and the point where flow meter output stabilizes to reflect crop input is unique to each harvester and crop. The software controlling the data logging could mathematically analyze the patterns of increasing initial grain flow to determine both the harvest lag time and the time period during which data are deleted at the start of each harvest run. During these initial 60 s, the time between the start of the harvest run and the point of inflection (maximum slope) of the curve describing the increase in grain flow at the sensor equates to the harvest lag time. By mathematical analysis of this increase in the grain flow at the sensor, this harvest lag time can be detected for each individual harvester and crop. Similarly, the determination of the peak of the curve (where the slope is zero) may be used to initialize logging in subsequent harvest runs. These adaptations to the software controlling logging enable data screening at the point of harvest using thresholds that are uniquely adapted to each harvester and crop.

The second major source of error in readings is the unpredictable interruptions in harvesting or sudden changes in forward speed. These must be detected and used to stop data logging immediately with further logging controlled as at the start of a harvest run. Normally, forward speed is very stable, and in the case of some harvesters, automatically controlled through sensing the load on the threshing mechanism (e.g. 'Cruisecontrol' from Massey Ferguson). The detection of sudden changes in forward speed, or changes which are not taking place under the harvester's automatic control system, could be used to delete associated yield data. This system must discriminate between 'normal' fluctuations in harvester speed and sudden changes. We recommend that the operator be given the opportunity to determine acceleration and deceleration thresholds that are used to block data logging immediately. The quantity of data deleted by these rigorous screening strategies may appear alarmingly high. By conventional mapping standards, however, the frequency of data logging in the remaining time is also very high. We strongly recommend that the logging systems are designed to produce valid data sets with invalid or doubtful data strictly eliminated. Some invalid data may remain, but the ratio of valid to erroneous data is greatly increased. Such screening strategies have the added advantage that data sets are reduced and the capacity of chipcards, etc. is extended in terms of harvesting time.

In practice, the selection of appropriate gridding techniques centers on inverse distance techniques and kriging. The data logging regime used on the harvester may simplify the data processing at this stage; the logging of data that have approximately the same lag distance in all directions facilitates normal search methods at the gridding stage. Thus, data would be logged in relation to distance rather than time, with the interval equal to the cutting width. Thus, with a 5 m cutting width, data are logged at a density of 400 ha^{-1}. With this strategy, the resultant grid is not affected by the direction of the harvester and, as with data screening, the potentially very large data sets are reduced in size.

Kriging is, statistically, the most valid gridding procedure. It is demanding in terms of computer time and memory, however, which is restrictive where such large data sets are concerned. Since many data are in the area over which yield is spatially dependent (i.e. within the semi-variogram range), we propose that simpler mathematically based techniques are appropriate. It is suggested that inverse distance technique could be adapted by putting a ceiling on the weightings so that data close to the grid point are averaged while data further away are weighted inverse to the distance. In this way, local averaging is combined with more distant searching. The ceiling on weighting used, and the weighting power given to more distant data requires investigation specific to each yield mapping system.

The selection of the gridding distance has a major effect in determining where small scale variation in crop yield are reflected in the grid values. Because of its use in the Massey Ferguson system, a standard of 10 x 10 m grid employing 100 nearest points for the production of yield maps on the farm has emerged in Europe. This has produced cartographically acceptable maps over a very wide range of crops.

Variations in the effective harvester cutting width and in crop moisture content are obvious sources of error in yield data. At present, the harvester operator inputs the cutting width into the data logger together with an estimate of the average moisture content. Technology is available to monitor these variables continuously at harvest. Radar instrumentation mounted on the table can detect the effective cutting width while a capacitance based moisture meter could be fitted to frequently sample the grain going to the clean grain auger. The monitoring of grain moisture content is particularly beneficial in volumetric based sensors since changes in grain moisture content also affect bulk density.

ACKNOWLEDGEMENTS

The authors acknowledge the wholehearted co-operation of the farmers on whose farms data reported here were collected: Peter and John Moore, Northumberland, England; Malcolm Forbes, Cleveland, England; and Hans De Neergaard, Vindum, Denmark.

REFERENCES

Anon. (1993). GPS World, October Edition.

Borgelt, S.C. 1993. Sensing and measuring technologies for site specific management. p. 141-157. *In* P.C. Robert et al. (ed.) Soil Specific Crop Management. ASA, Madison, WI.

Diekhans, N. 1985. Grundl. Landtechnik Bd. 35 (4): 111.

Matheron, G. (1971). The theory of regionalized variables and its applications. Cahiers du Centre de Morphologie Mathematique de Fontainbleau No. 5. France.

O'Callaghan, J.R. 1988. Engineering applications and developments. In Towards an Agro-Industrial Future. p. 45-50. *In* Proc. of the 6th Royal Show International Symposium. pp. The Royal Agric. Soc. of England, Stonleigh, Warwickshire, England.

Schnug, E., D.P.L. Murphy, and E.J. Evans. 1992. Evaluation of yield mapping techniques for use in combinable crops. p. 206-207. *In* Proc. of the Second Congress of the European Society for Agronomy.

Schnug, E., D.P. Murphy, E. Evans, S. Haneklaus, and J. Lamp. 1993. Yield mapping and application of yield maps to computer aided local resource management. p. 87-93. *In* P. C. Robert et al. (ed.) Soil Specific Crop Management. ASA, Madison, WI.

Searcy, S.W., J.K. Schueller, Y.H. Bae, S.C. Borgelt, and B.A. Stout. 1989. Mapping of spatially variable yield during grain combining. Trans. ASAE 32(3): 826-829.

Webster, R., and M.A. Oliver. 1990. Statistical methods in soil and land resource survey. Oxford University Press, Oxford, England. 316pp.

Contact addresses related to the technologies discussed

Claas Verttriebsgesellschaft mbH, P.O. 1140, D 4834, Harsewinkel, Germany. Tel: +49 5247 12549Fax: +49 5247 12434

Mr. J. Claydon, Claydon Yieldometer Ltd., Gaines Hall, Wickhambrook, Newmarket, Suffolk, England. Tel: +49 440 820327Fax: +49 440 820642

Dronningborg Industries, Udbyhojvej 113, P.O. Box 80, DK 8900, Randers, Denmark. Tel: +45 86 425855 Fax: +45 86 407010

RDS Technology Ltd., Stroud Road, Nailsworth, Glos. GL6 0BE, England. Tel: +44 453 834084 Fax: +44 453 835521.

4 Yield Measurement and Field Mapping With an Integrated GPS System

M. Eliason
D. Heaney
T. Goddard
M. Green
C. McKenzie
D. Penney

Alberta Agriculture
Food and Rural Development
Edmonton, Alberta, Canada

H. Gehue
G. Lachapelle
M. E. Cannon

Department of Geomatics Engineering
The University of Calgary
Calgary, Alberta, Canada

Instantaneous yield measurement using a commercial yield monitor integrated with two clear/access (C/A) code receivers is discussed in relation to establishing a Geographic Information System (GIS) database for a precision farming system. Yield measurements coupled with Global Positioning Satellite (GPS) positioning data obtained at the sub-meter level of accuracy are compared for several wheat (*Triticum aestivum* L.) fields in Alberta, Canada. Field maps containing site specific information are developed to form the basis for variable rate application of field inputs. Yield and position accuracy are compared against data obtained through more conventional means. The extension of this work through the development of a real-time DGPS navigation system for variable rate applications is also discussed.

INTRODUCTION

Producers have long recognized that field management practices based on a whole field approach leave many areas within the field with less than optimal production potentials. Field inputs managed under a best guess estimate of overall field averages either leave areas with below optimal inputs or growth potentials or cause an excess of inputs which reduce economic returns. Producer attempts to manage spatial variations are often intuitive. Management decisions may be based on a previous knowledge of the variability or direct observation during the field operation. Equipment operators often make equipment adjustments based on operating conditions. Equipment travel speed and tillage depth adjustments can be a spatial variability management decision. These parameters can be quickly and easily adjusted many times over the course of a field operation. Often, however, the operator is limited by ability to respond quickly to the variation and by the ability to remember when to do so.

Producers may also observe and desire to account for minor field variations but do not because of the effort required to do so. For example, some equipment settings are not adjustable "on the fly". Adjustments of this nature decrease the likelihood of the operator to stop and make the adjustment needed to address a minor spatial variability. So to, as equipment becomes larger, the desire to respond to spatial variability within the equipment operating width also becomes increasingly apparent.

In an attempt to address spatial variability management practices, a joint precision farming project was undertaken by Alberta Agriculture Food and Rural Development, the University of Calgary and the University of Alberta. Four sites within Alberta (Fig. 4–1) were chosen whereby yield, salinity, fertility and soil information data was gathered together with DGPS positioning information. The intent of the project is to better understand and develop control algorithms to manage soil and crop inputs.

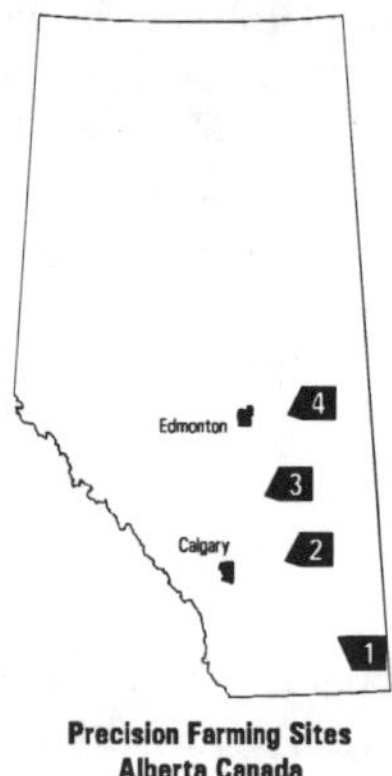

Fig. 4–1. Precision Farming Project Sites: 1) Bow Island, 2) Hussar, 3) Stettler, 4) Mundare.

MEASUREMENT SYSTEMS

Field data measurements were simultaneously recorded with position information to provide location references. Three–dimensional DGPS data were gathered with two 10–channel C/A code narrow correlator spacing NovAtel GPSCard™ sensors and Model 501 antennas (Fenton et al., 1991; Van Dierendonck et al.; 1992). Position accuracy to the sub-meter level using a robust carrier phase smoothing of the code approach was obtained (Gehue et al., 1994). Salinity data was acquired with an EM conductivity meter and previously described by Lachapelle et al. (1992a, 1993) and Cannon et al. (1994). Fertility and soil information was gathered via grid sampling and analyzed by the Soil and Feed Testing Laboratory of Alberta Agriculture. Yield information was gathered with an Ag Leader Technology™ Yield Monitor 2000™. GRASS (v.4.1, US Army Corps of Engineers - CERL) routines were used to surface the information layers.

YIELD MEASUREMENT

One goal of the project was to measure and use yield as an input parameter for subsequent management decisions. Yield mapping, combined with other spatial data, can be an important factor for site specific decision making (Schueller & Bae, 1987). Yield measurements can provide a feedback effect of the application inputs, allowing refinement of the application plan for future years (Borgelt & Sudduth, 1992). Reconstructing grain yield with a grain flowmeter, however, represents a challenge since the components of a combine redistribute grain and impose transport delays in such a manner that grain flow measurements do not accurately represent grain yields at the current combine location (Borgelt & Sudduth 1992). Searcy et al. (1989) modeled combinc transportation delays as a lumped-parameter system. Delays varied with grain flow rate and whether the combine was entering or leaving the crop. Borgelt and Sudduth (1992) described currently available yield sensor technology and equipment.

Yield measurement equipment which is easily adaptable to current model combines is not widely available. For the project, the Ag Leader Technology™ Yield Monitor 2000™ was chosen because of the availability and adaptability of the sensor to the various combine models used in the project. The three combine models currently used by project producers are: a John Deere 9600, an AGCO R72 and a Case/IH 1680. The Ag Leader Technology™ Yield Monitor 2000™ (Fig. 4–2) is a force-plate based sensor which, when installed into the clean grain elevator of the combine senses grain mass flow rate. Elevator speed, ground speed, header height and yield sensor signals were fed into a cab mounted digital readout which processed, displayed and stored data. Yield information from the digital readout was fed into a laptop computer at a 1 Hz interval. GPS information was combined with yield data for each reading.

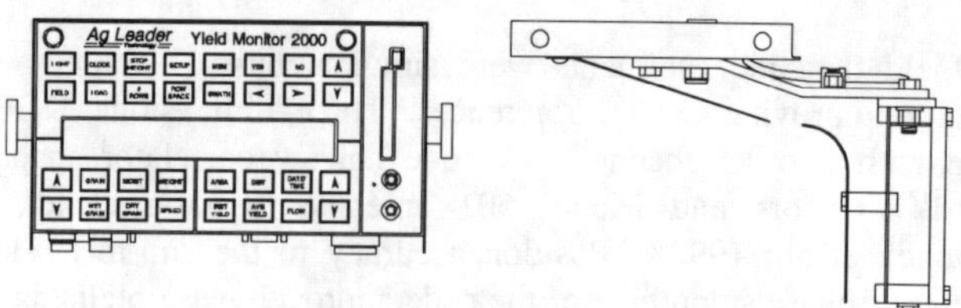

Fig. 4–2. Ag Leader Technology™ Yield Monitor 2000™.

YIELD RESULTS

The 1993 Alberta harvest season was characterized by cool damp weather with numerous harvest delays. Yield sensor equipment was obtained and installed just prior to the harvesting operations. Even though the harvest was delayed, insufficient time precluded extensive yield sensor calibration. Initial in-field calibrations of the yield sensors showed discrepancies from known values. Sensor calibration factors were adjusted initially and then left unaltered for the duration of the harvest operations. Final total yields as harvested from the sites showed discrepancies from the yield sensor values. To adjust yield data, total yields that were harvested from the sites were compared to the total yields as calculated from the yield monitor. Differences in the two values were the sum of the errors in the sensor systems and the errors in sensor calibration. Yield map results were scaled accordingly and as a result, represent at best, only an indication of yield variations within the field.

Figure 4–3 shows the combine trajectories for the Hussar site. This site was combined on September 20 and 21, 1993. Two combines were used to harvest the 81 ha site but only one was fitted with yield and GPS recording equipment. As a result, trajectories and yield data represent every other round of field information. Data was collected at a 1 Hz sample rate which, for a 6 km/h travel speed, corresponds to a data sample every 1.6 m.

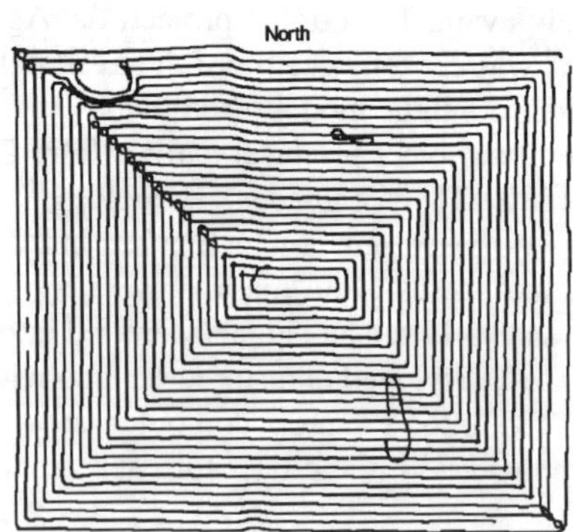

Fig. 4–3. Combine Grid Pattern for the Hussar Site.

Figure 4–4 shows yield results for the Hussar site. Variations in yield are evident as are data anomalies. Obvious data anomalies require modification. For example the high value data anomaly which occurs in the NE corner was likely caused by high grain moisture contents which returned a false reading from the yield sensor. The ridge in the SE corner, which runs into the middle of the image, coincides with a 90°corner in the swath. Data anomalies must be sorted to either accept, reject or alter data based on a quality control prescription. This prescription may be a personal judgement based on familiarity of the event or a numerical algorithm attempting to define the event. During testing, several data anomaly causing events were noted:

1. Yield data was collected when header heights were within a prescribed range. Combine operators did not always remember or desire to lift the header (shut-off data collection) in non-yielding areas (ie. headlands, while travelling over non-cropped or already harvested areas or while travelling to the unloading area).
2. The Hussar site was cut with a 7.6 m windrower prior to combining. Windrow depositions on 90° corners were uneven with heavy windrows occurring at the corners and light windrows occurring just after the corner.
3. Yield sensor readings were affected by the combine grain flow transportation delays. Transportation delays were evident when entering or leaving the crop, when slowing or stopping for combining problems, and when cornering.
4. Yield sensor readings assumed a constant cutting width was used. Variations in cutter bar utilizations occurred when detouring obstacles and/or when "straightening the row".
5. Yield and DGPS data were collected on a constant time based interval. This resulted in variable location based data densities. When travel speeds were reduced data densities were higher than desired.
6. A non-centre-line mounting location of the GPS antenna required a location offset calculation. This calculation was cumbersome when travel directions were "back-and-forth".

Data anomalies were subjected to various quality control prescriptions. Combine transport delays were assumed constant while either entering or leaving the crop and adjusted by a constant time delay. Cornering anomalies were adjusted by numerically testing GPS position and heading parameters. This allowed adjustments for combine corner looping and detouring around obstacles or non-cropped areas. Obvious non-yielding trajectory paths which were recorded as a result of operator error were edited manually. Cutterbar utilization anomalies which occurred while harvesting around obvious non-cropped areas were edited manually. Figure 4-5 shows the combine trajectories for the Stettler site. This site contained many irregular trajectories and cutterbar utilizations were highly variable. Yield variations at this site were not adjusted for cutterbar anomalies.

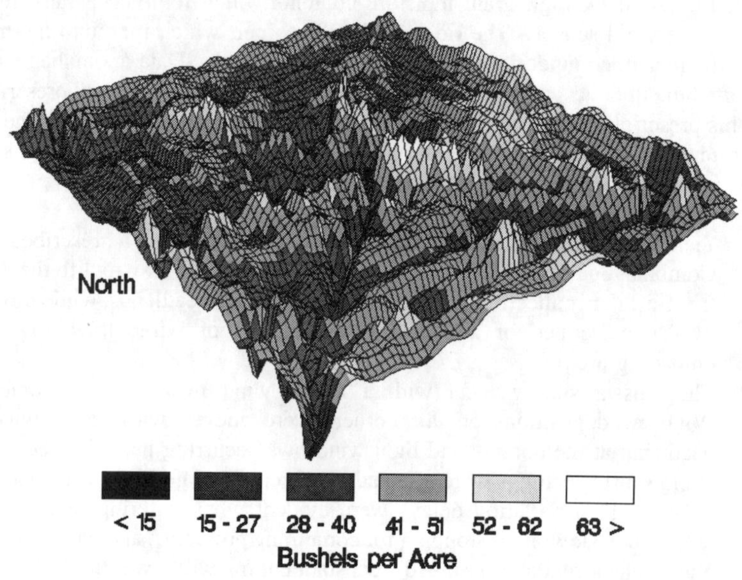

Fig. 4–4. Yield Map of the Hussar Site.

Fig. 4–5. Combine Trajectories for the Stettler Site.

GRASS routines were used to surface yield data onto position coordinates. Various smoothing effects were used. Binning or grouping yield values smoothed the yield map. Figure 4–6 shows grouped yield values for the Hussar site. Data densities affected smoothing routines. The high number of yield values collected along the trajectories as compared to the distance between the trajectories resulted in a skewing of yield values. This was especially noticeable for the Hussar site where data was collected on every other windrow. Data "along the trajectory" was dedensifed to more closely match "between the trajectory" distances. Dedensifying algorithms were either time or location based. Time based dedensifying prescriptions reduced data and only slightly reduced yield map skewing. The position based dedensifying prescription resulted in smoother and more uniform maps.

Collection of yield and GPS data required large computer memory storage requirements. Raw data files were recorded at a rate of about 3.5 megabytes per hr for the 1 Hz data sampling rate. Manipulation of the large data files was at times cumbersome. The need to dedensify data would imply that a position based or a ground speed based data recording trigger would optimize computer resources. This may be offset by the desire to subject data to real time quality control prescriptions. Real time accounting of combine grain flow transportation delays and cutter bar utilizations would be desirable. Post mission monitoring of combine position relative to a similar adjacent trajectory position may provide a measure of cutter bar utilization. To use this method in real time, however, would likely require a considerable amount of computing power. A cutting width sensing unit mounted on the cutter bar might be a more practical approach to real time measurement of cutter bar utilization. Accurate measurements of both grain flow transportation delays and cutter bar utilizations must be made for yield measurements to be meaningful at the sub-metre level of position accuracy.

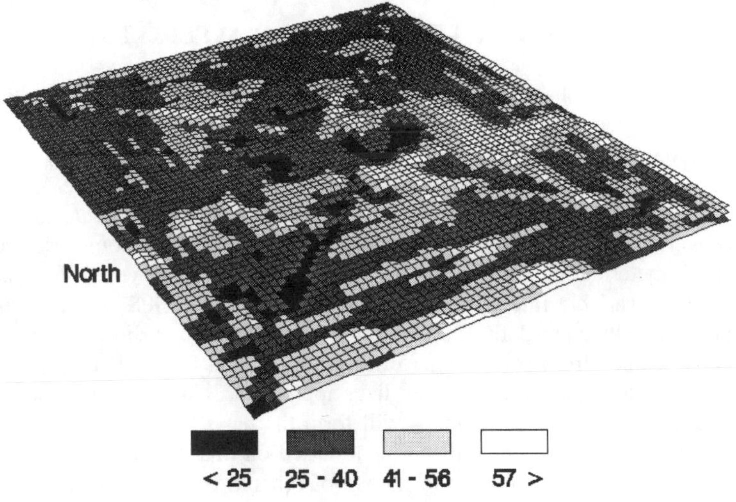

Fig. 4–6. Grouped yield values for the Hussar Site.

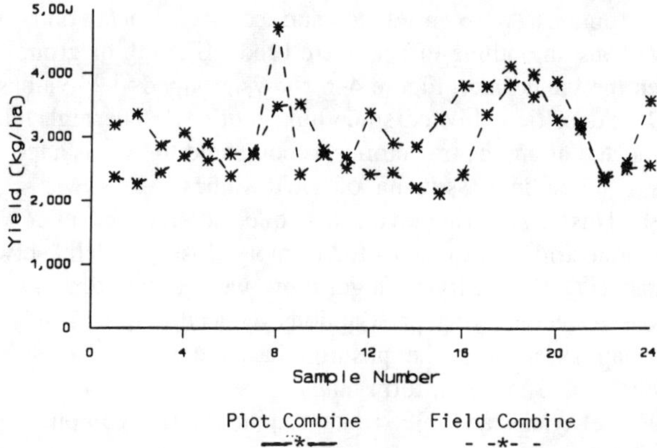

Fig. 4–7. Comparison of yield monitor and plot combine yields at various locations.

In an attempt to verify yield results, several transects at the Hussar site were harvested with a plot combine. Figure 4–7 compares yield results for various locations along a transect. Yield monitor values were corrected for calibration errors but do not contain smoothing effects. Yield values appeared similar but a statistical comparison has not been completed.

Yield data can be draped over other layers of information. Yield over topography, fertility or salinity provide numerous data interaction possibilities. GRASS routines provide a convenient means of draping and manipulating information layers. Various interaction effects will be studied over the duration of the project.

VARIABLE RATE FERTILIZER APPLICATION

The next phase of the precision farming project is to develop an on-the-fly variable rate fertilizing system. Variable rate fertilizer application is not new technology. Variable rates may be achieved through simple travel speed adjustments for machines lacking ground driven meters to variable adjustment electric control units for ground driven metering systems. What is lacking, however, is a prescription map needed to define fertility boundaries and to control metering devices independent of direct operator control. Operator unassisted variable rate fertilizer application requires DGPS operating in real time along with a pre-defined prescription map. Figure 4–8 shows a prescription map developed from soil sample fertility information for the Hussar site.

For the upcoming 1994 fertilizer application season, soil sample fertility information and 1993 yield maps will form the basis for the prescription map. Variable rate applications will be made based on this map using an air seeder and real time DGPS. As further layers of reliable field data are obtained and input algorithms developed, prescription maps will be adjusted accordingly. The prescription map can either be raster based, such that the pixel value of the

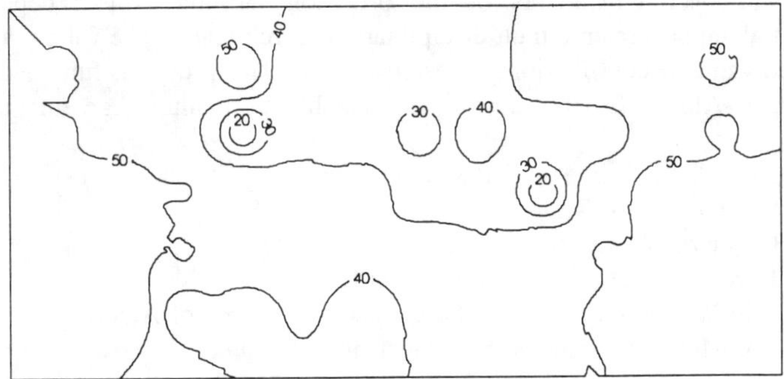

Fig. 4–8. Fertility prescription map for the Hussar Site.

screen dictates the action, or vector-based so the zone is determined and the appropriate response then taken. Initial attempts will use a raster-based map so polygon searching will not have to be performed. As variable rate technology develops, raster-based screen map/pixel resolutions may limit the variable rate control resolution. Initial attempts at pre-defined variable rate fertilizer application will focus on variable rate control along the trajectory path. Variable rate control for individual delivery meters located across the implement width is within the realm of real time DGPS capabilities. Variable rate control across the implement width may be attempted later in the project.

A further realistic development of real time DGPS is the ability to assist operator navigation. Real time DGPS coupled with a prescription map containing predefined trajectories based on implement width, can be used to assist operator navigation. Deviations from a current DGPS calculated position to the desired position can be displayed on a screen based navigational guide. The operator would be prompted to correct navigational deviations thereby reducing skips and overlaps. Development of a real time DGPS navigational aid is ongoing with a experimental prototype expected later this year.

CONCLUSIONS

This paper outlined early results from the precision farming project. Yield and soil information were collected together with GPS positioning data to form a GIS data base. Yield maps were derived and subjected to various data quality control prescriptions to resolve data anomalies. Several data anomalies caused events which were experienced such as operator recording errors, combine grain flow transportation delays and cutterbar utilization variations. Quality control prescriptions were developed for some of the anomalies. Field maps were developed to study and address field spatial variations. The next phase of the project is to use field information maps together with real time DGPS capabilities for use in fertilizer applications.

ACKNOWLEDGMENTS

The accomplishments of the precision farming project depend on substantial support from a multidisciplinary research team. The following have provided valuable contributions in results to date. This project is funded through the Canada-Alberta Environmentally Sustainable Agriculture Agreement.

Neil Clark, Technologist, AAFRD
Sheilah Nolan, Research Agronomist, AAFRD
Germar Lohstraeter, Technologist, AAFRD
Karen Skarberg, Agronomist, AAFRD
Tim Martin, CPU Applications Specialist, Univ. of Alberta
Tor Melgard, Graduate Student, Univ. of Calgary

REFERENCES

Borgelt, S.C., and K.A. Sudduth. 1992. Grain flow monitoring for in-field yield mapping. International Summer Meeting of The Am. Soc. Agric. Eng., Charlotte, NC. 21-24 June. ASAE Paper 921022.

Cannon, M.E., R.C. Mckenzie, and G. Lachapelle. 1994. Soil salinity mapping with electromagnetic induction and satellite-based navigation methods. Can. J. of Soil Sci. 74(3):335–343.

Fenton, P., W. Falkenberg, T. Ford, K. Ng, and A.J. Van Dierendonck. 1991. NovAtel's GPS Receiver, the high performance OEM sensor of the future. Proc. of GPS91, The Institute of Navigation, Alexandria, VA, pp. 49-58.

Gehue, H., G. Lachapelle, M.E. Cannon, T.W. Goddard, and D.C. Penney. 1994. GPS system integration and field approaches in precision farming. Proc. of ION National Technical Meeting, San Diego, CA, 24-26 January.

Lachapelle, G., M.E. Cannon, B. Townsend, and R.C. Mckenzie. 1992a. Mapping soil salinity with satellite-based positioning methods. The 38th Meeting of the Canadian Soil Science, Edmonton, Alberta, 9-13 Aug.

Lachapelle, G., R.C. Mckenzie, M.E. Cannon, B. Townsend, and N.F. Clark 1993. Mapping soil salinity with global positioning and electromagnetic induction methods. Proc. of the 30th Annual Alberta Soil Science Workshop, Edmonton, Alberta, 23 Feb., pp. 167-170.

Schueller, J.K., and Y.H. Bae. 1987. Spatially-attributed automatic combine data acquisition. Computers and Electronics in Agriculture 2:119-127.

Searcy, S.W., J.K. Schueller, Y.H. Bae, S.C. Borgelt, and B.A. Stout. 1989. Mapping of spatially variable yield during grain combining. Trans. ASAE 32(3):826-829.

Van Dierendonck, A.J., P. Fenton, and T. Ford. 1992. Theory and performance of narrow correlator spacing in a GPS receiver. Vol. 39, No. 3, The Inst. of Navigation, Alexandria, VA, pp. 265-283.

5 Yield Mapping of Potato[1]

Stephen L. Rawlins
USDA/Agricultural Research Service
Prosser, Washington

Gaylon S. Campbell
Dep. of Crops and Soil Science
Washington State University
Pullman, Washington

Ronald H. Campbell
HarvestMaster, Inc.
Logan, Utah

John R. Hess
Partner, Hess Farms, Inc.
Ashton, Idaho

The goal of the research supporting the development of this technology includes the integration of scientific knowledge into decision-support packages for farm managers. A recent publication from the Office of Technology Assessment publication (OTA, 1993) effectively states the justification for this research:

> "Computerized farm-management systems include the land-based or remote sensors, robotics and controls, image analysis, geographical information systems, and telecommunications linkages packaged into decision-support systems or embodied in intelligent farm equipment. Such systems will be increasingly important to the farmer's ability to increase yields, control costs, and respond to environmental concerns."

Continuous yield maps can help to determine precise requirements for management practices and inputs for each square meter of a field. Yield mapping technology is available for seed crops, but not for bulky products such as potato (*Solanum tuberosum* L.). Potato is a high-value crop with yield and quality highly sensitive to critical management decisions. Balancing profit against potential negative environmental impacts for this crop requires precise

[1]A portion of this work was funded by the U. S. Department of Energy Idaho Field Office Contract DE-AC07-76IDO1570.

application of irrigation water, fertilizer and pest and disease control practices. These factors make potato an ideal candidate for advanced information and management technology.

This paper describes development of potato yield mapping technology at two scales: one for a 1-row experimental harvester as part of an integrated agriculture experimental program at Prosser, WA; the other for a commercial machine operated by Hess Farms Inc. at Ashton, ID, which harvested 10 rows in a single pass. Only brief details are reported here of the complete design specifications and results from the Hess Farm system.

MATERIALS AND METHODS

Plot-Scale Harvester

The single-row harvester was custom built to measure yields from experimental plots. These plots provide data to parameterize and validate a potato crop simulator being developed to predict spatially distributed yield. As designed, this machine digs and transports potatoes over approximately 7 m of primary and secondary conveyer chains to a manual sacking unit. A treadway along the last 3 m of the secondary chain allows space for three or four people to stand while picking out rocks and debris.

To obtain continuous yield maps, we made two modifications to this harvester. First, we added a set of six, 10-cm diameter wheels, powered by a hydraulic motor, to the conveyer system to remove soil from the potatoes. Secondly, we replaced the sacking unit with the weighing conveyer belt shown schematically in Fig. 5–1. This 0.6 m-wide by 0.9 m-long belt transported potatoes from the harvester secondary chain to a weighed fruit bin. The lead end of this conveyer pivots on a bearing-mounted shaft; the tail end hangs from a 45-kg capacity S-shaped load cell (Model HM-308, HarvestMaster, Inc., Logan, UT). A lead counterweight balances most of the weight of the weighing conveyer, leaving only about 1 kg of weight hanging on the load cell when the belt is empty.

A fractional horsepower, 12-volt DC gear motor drives the belt at 0.2 m/s. Six magnets each were cemented onto the inside rim of one wheel of the harvester and one drum roller of the weighing conveyer to measure ground and belt speed. Signals sensed as these magnets pass by fixed proximity switches, as well as output from the load cell, are sent to a data logger (Model CR10, Campbell Scientific, Inc., Logan, UT). Output from this logger is displayed continuously on an electronic field notebook (HarvestMaster, Model HM1). Software to integrate these instruments and to provide the needed output, SPUDSIM, was designed and contributed by HarvestMaster.

The dimensions and speed of the weighing conveyer are based on an assumed maximum yield of 90 Mg/ha (9.0 kg/m^2). With a row width of 0.86 m and a harvester speed of 1.6 km/h (0.44 m/s), the rate of delivery of potatoes to the weighing conveyer is 3.5 kg/s. We measured the area occupied by average–sized potatoes to be about 0.004 m^2/kg. Therefore, 0.13 m^2 of belt surface must be provided to transport potatoes delivered by the harvester. This

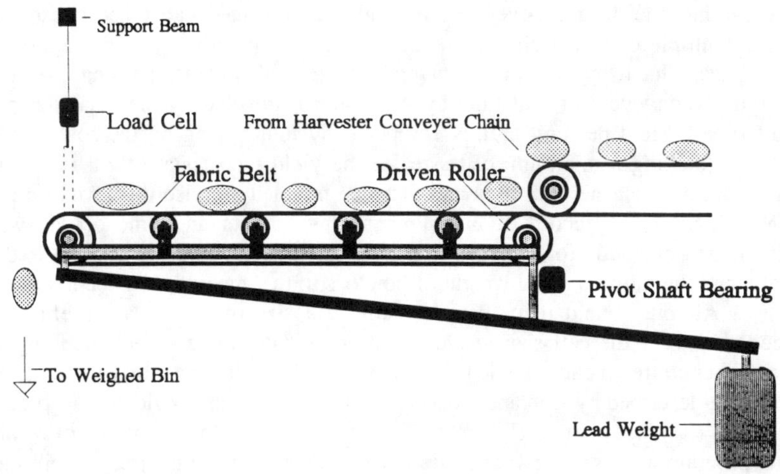

Fig. 5–1. Potato weighing conveyer. Potatoes transported on the 0.6-m wide by 0.9-m long belt are weighed with an electronic load cell which supports one end. The other end pivots on a bearing-mounted shaft. Most of the unloaded weight of the conveyer is counter balanced by a hanging lead weight. A 12-v DC motor drives the belt at 0.2 m per s.

requires a 0.6-m wide belt to move at a speed of 0.2 m/s. To obtain this belt speed, the 0.1-m diameter drive-roller must turn 40 rpm. With standard chain sprockets available, we were able to turn it at 43 rpm with a 90-rpm DC drive motor.

To make certain that neither row-length nor conveyer-weight data were recorded while the harvester turned around at the end of each row, a micro-switch was installed to deactivate two circuits. One was the 12-v power circuit to the conveyer drive motor; the other was the data circuit from the proximity switch on the harvester wheel. The micro-switch was mounted on the digger lift mechanism so that these circuits were closed only when the digger spade was lowered into the ground. To prevent the harvester from continuing to run, piling potatoes on the stationary weighing conveyer belt, the tractor operator disengaged power to the harvester at the same time he lifted the digging spade at the end of each row.

The SPUDSIM software provides means for zeroing the tare weight of the conveyer, as well as recording and displaying essential parameters. Load-cell weight is measured eight times per s, and these values are then averaged and recorded for a fixed distance along the row. This average weight, together with the linear travel of the weighing conveyer belt for this distance, provides the necessary information to compute yield per unit area for each mapping-unit cell. The fixed distance along the row for averaging load cell weight was chosen to be 3.2 m in this case.

Because linear travel of the harvester was measured by counting wheel revolutions, the distance over which yield was averaged had to be chosen as some multiple of wheel circumference. Weight of potatoes per unit length of row is calculated by dividing the average weight of potatoes passing over the weighing conveyer per unit time by the distance traveled by the harvester per unit time. Yield per unit area is computed by multiplying by the row width, 0.86 m, making a minimum cell size for the yield map equal to 2.8 m^2. The time of day, yield per unit area and distance traveled are also displayed on the HM1 monitor. A record of bin number, row number and time of day was maintained manually for each bin to permit calculation of lateral position in the field, and the ratio of actual weight of potatoes to that measured by the load cell.

Although yield maps based on this cell size of 2.8 m^2 are useful, we needed to vary the cell size to block out areas with borders corresponding to experimental treatments. To do this, we used the Field Information System (FIS) software described by Han and Goering (1993), which was provided to us by Dr. S. Han (Washington State University, Prosser, WA). A blocking protocol within this program allows somewhat random sample data to be converted to soil map data (cell data) at any desired grid scale, using statistical techniques described in Han et al. (1992). The input file is a string of ASCII data, each element of which contains values for the x and y position coordinates, and z, the variable being mapped.

To correct for displacement of potatoes from the point they were dug to the point they were weighed, we buried potatoes marked with red paint in rows at several locations, and measured the distance to the point that they passed over the weighing conveyer. A series of 15 measurements revealed this average displacement to be 8.5 m.

The same marked potatoes were also used to obtain potato samples for quality analyses from known locations on the ground. Two marked potatoes were buried in the row at both the beginning and at the end of each 3-m section to be sampled. When the first set of marked potatoes appeared, we began catching the potatoes from the weighing conveyer in a pan, and stopped catching them when the second pair appeared.

Commercial-Scale Harvester

The machine used was a standard two-row harvester manufactured in 1989 by Logan Manufacturing, Idaho Falls, ID. An initial attempt was made to weigh the potatoes carried by truck loading boom by recording its angle and the pressure in the hydraulic rams supporting it. Fluid leakage, presumably through the valves of the hydraulic system, caused substantial zero drift. This, along with the low theoretical sensitivity resulting from the large tare weight of the loading boom, made this approach unworkable.

The approach used consisted of mounting a weighing table placed under the upper picking table chain. Three sets of rollers were removed and replaced by a 0.71-m wide by 0.91-m long table, covered with a 2.5-cm thick square of UMHW plastic, which supports the chain and the potatoes. The table pivoted on bearings at the lead end, and rested upon two, 113-kg capacity electronic, button load cells (model LGC 250, Omega Engineering, Inc., Stanford, CT) at the tail end. Load cell weight as well as ground speed and chain speed over the table were sensed and recorded by the same instrumentation using proximity switches as described above for the plot harvester. Weight of potatoes per unit length of row was also calculated as described above.

Calibration of the weighing table was performed by running a series of ten weights, totaling 230 kg, across it while the harvester operated at normal speed. The harvester picked up potatoes from ten, 0.91-m spaced rows. Eight of these rows, four from each side of the two it dug, were piled onto the center two rows by a separate machine. Yields were summed at intervals of 7.6 m, making a harvest cell area of 69.5 square meters, or 0.00695 ha. As for the Paterson site, we determined field position by dead reckoning -- from the position along the row and the row number, also as described above.

RESULTS AND DISCUSSION

Plot-Scale Harvester

Figure 5–2 shows spatially distributed potato yield for four replicated 0.4-ha plots at Paterson, WA. The cell dimensions are 1.72 m wide by 3 m long (2 rows wide by 10 ft. long). Although the average yield of all plots is nearly the same, varying only from 75.2 to 77.7 Mg/ha, the yield within and plot varies substantially. Space does not permit a complete description of the treatment variables on these plots, but the yield variation does not correspond to treatment variables. Low yields across all replications near row lengths of 20 and 75 m correspond to the wheel tracks of a linear-move irrigation system. Other areas of low yield, such as at both ends, can be explained by destructive sampling of plants throughout the season. But the reasons for the high yields in large areas of each plot are not apparent. Yield maps such as these pose the question, if parts of the field can produce 100 Mg/ha, why not all of the field? Knowing where the high and low yields are within a field makes it possible to target a sampling pattern to begin to answer this question.

This site was broken out of virgin rangeland in 1991, which required extensive land leveling in some areas. Probably some of the yield variation corresponds to areas of cut and fill resulting from land leveling. But the crop was watered nearly every day, so variations in water holding capacity resulting from land leveling, which we can do very little about, should not have been a factor. If the variation is caused by soil nutrient levels, on the other hand, we should be able to do something about it with variable rate application of fertilizer. Distinguishing between those sites where yield is constrained by factors that cannot be modified, as opposed to those where it can, is essential to prescribing optimum management. In the first case, inputs probably need to be reduced to match the yield-producing capacity of the soil. In the latter case, inputs probably need to be increased to eliminate the yield constraint.

Because we weighed the bins of potatoes harvested for each half row, we had 384 checks on the stability of the weighing-conveyer calibration curve for these four replications. We used the measured value for this factor to calculate

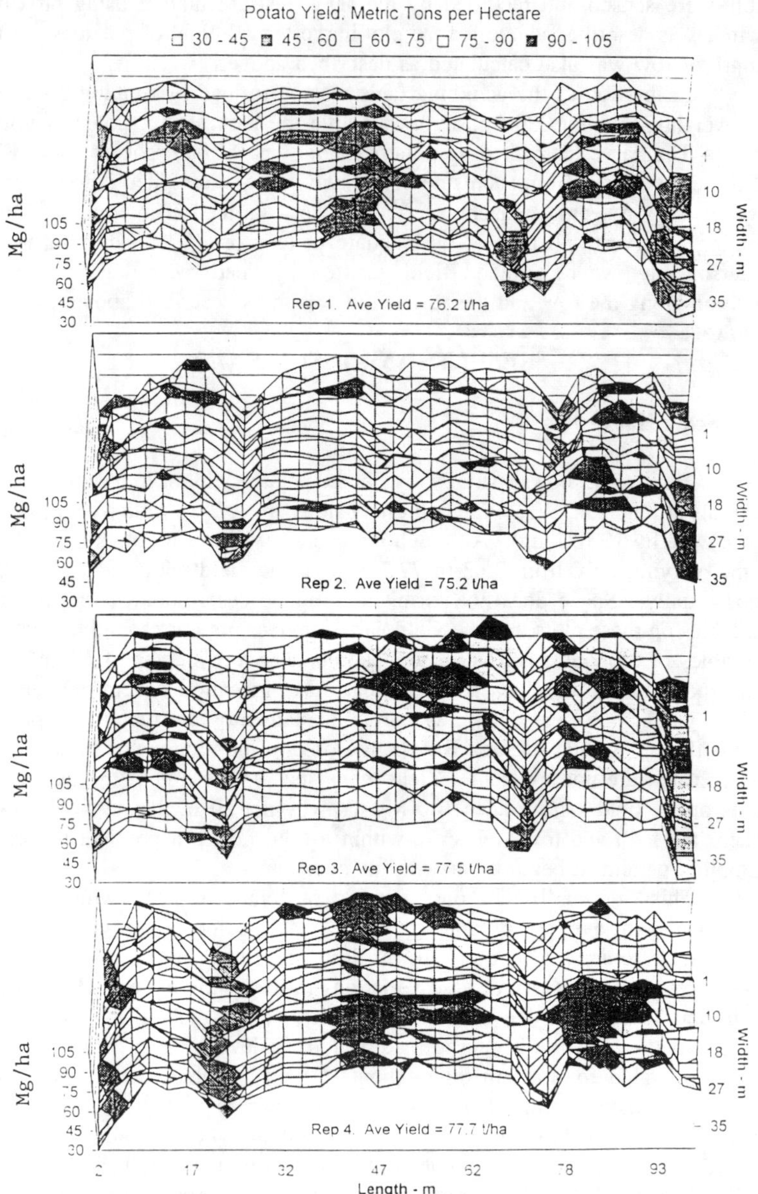

Fig. 5–2. Spatial distribution of potato yield for four experimental plot replications mapped on-the-go with a weighing conveyer platform. Low yields near 20 and 75 m in all replications correspond to the wheel tracks of a linear-move irrigation system.

the yield distribution along each half row, so drift in its stability did not affect our yield data.

Figure 5–3 shows the variation in this factor for averages of 12 bins each bin in the order in which they were weighed. Also plotted is the standard deviation for each of these averages. Although this factor remained reasonably stable for the first two thirds of the first plot harvested, something obvious happened from that point on. The standard deviation of the weighing-conveyer calibration factor for the first 48 bins was less than 5 percent. We assume this is the accuracy with which yields can be measured with this system. The problem from that point on appears to be related to faulty data for conveyer belt speed. We discovered that potato vines and other debris frequently wound around the shaft of the drum roller, and interfered with the magnets used to measure belt speed. In one case, near bin 288, all of the magnets but one were discovered to have been displaced to the point that they did not trip the proximity switch.

Commercial-Scale Harvester

Because of the pressure of harvest, we were not able to conduct extensive tests on the resolution and accuracy of the yield mapping equipment. During one test, the cumulative weight of potatoes indicated by the yield monitor was compared to the estimated weight of three truck loads. From experience, the trucks were known to hold about 15.5 Mg each, or 46.5 Mg total. The cumulative weight indicated by the yield monitor was 47.2 Mg -- well within error of estimation.

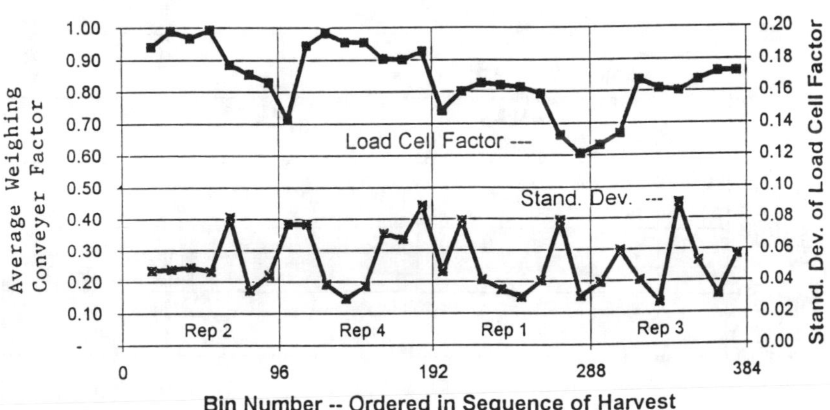

Fig. 5–3. Average and standard deviation of load cell factor for moving averages of 12 bins each in the sequence in which they were filled during harvest.

The data logger and load cells are capable of resolving changes in table weight of a few grams when the machine was not running or moving. The limitation for precision and accuracy, therefore, came from vibration and shock that occur in normal field operation. Figure 5–4 gives some idea of these sources of noise. The line shows raw yield measurements for one swath of 10 rows. The drop at about 390 m is at a sprinkler line, where the digger was pulled out of the ground. Some of the other low yields are probably associated with rock patches, but no record was kept to confirm this.

Considering points near the start of the transect, it appears that sample to sample variation is around 2 Mg/ha. This variation combines both the field variation and the yield monitor uncertainty, so the yield monitor must be at least this accurate. There are some individual points that change by as much as 6 Mg/ha compared to those around them. These may be from bumps, changes in tractor speed (there is a lag between the time potatoes are pickup up and the time they are weighed), or actual yield variations. Additional tests will be required to finally resolve questions about resolution and accuracy.

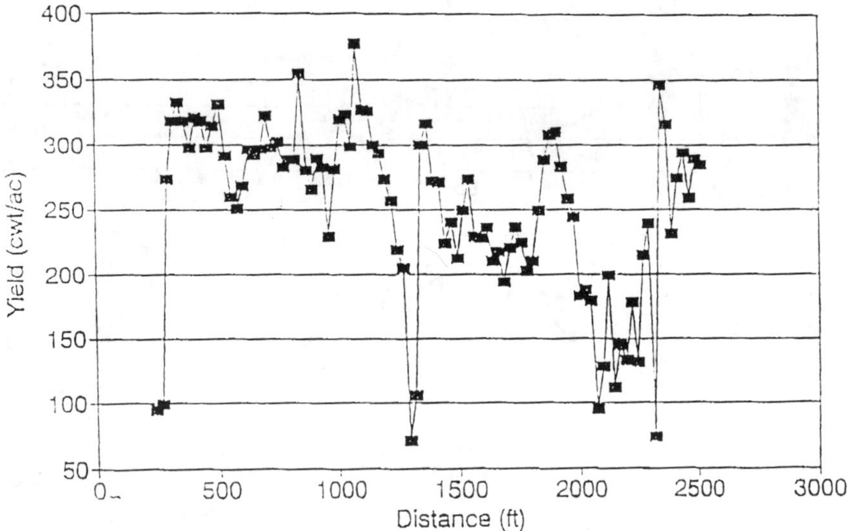

Fig. 5–4. Potato yield as a function of distance along one transect of Hess Farm.

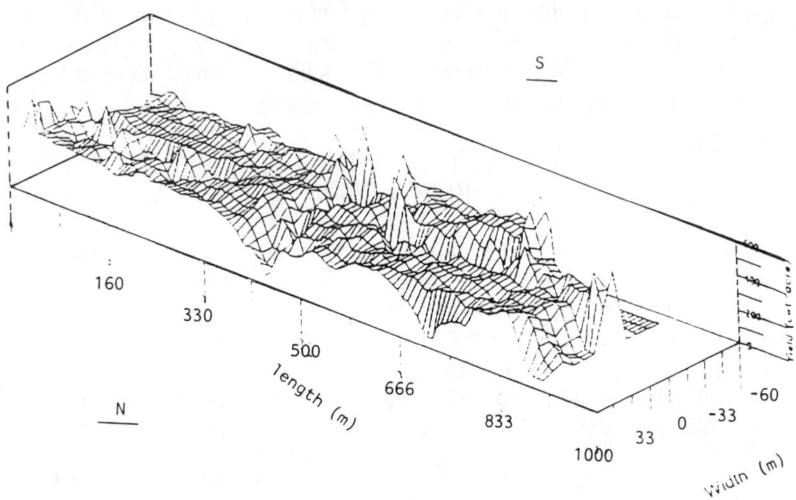

Fig. 5–5. Three-dimensional yield map for potato at Hess Farm.

Figure 5–5 is a three-dimensional plot of spatially distributed yield for the field harvested. The main features obvious on this plot are the sprinkler main line at about 360 m, where the digger is pulled out of the ground, and rock patches, which the digger also avoids. The largest rock patch is in the south west corner of the field (length increases toward the west; width increases toward the north). Other rock patches are indicated by low yield islands. There is an area of particularly high yield at about 600 m on the south side. A large area of low yield occurs on the south side between 75 and 150 m. The high yield spikes occur where the tractor stopped, but the harvester kept passing potatoes over the weigh table. They are generally accompanied by adjacent areas of low apparent yield. With the exception of these spots, yield over the remaining part of the field is surprisingly uniform.

CONCLUSIONS

Although additional tests are required, we have demonstrated technology that makes it possible to map yields of potato with both a plot scale and commercial scale harvester with accuracies of the order of five percent. The weighing conveyer technology can also be adapted for any crop that can be run over a conveyer, such as sugar beet. This new technology is important, because these crops have much higher value per acre, making excellent candidates for site specific management.

Improvements planned for further development include the use of GPS to locate the yield monitor and log terrain elevation. Also, a more reliable means for measuring conveyer belt speed for the plot harvester will be used.

Because product quality of potato and sugar beet (*Beta bulgaris* L.) is so important, maps of quality may be as important as maps of yield. Although we see no practical means to assess quality on the go, it may be possible to develop a mechanized means for collecting and packaging samples at specified grid points for laboratory evaluation.

REFERENCES

Han, S., M.D. Chan, C.E. Goering, and J.W. Hummel (1992). Blocking of spatial soil data for site-specific crop management. ASAE Paper 92–3608, ASAE, St. Joseph, MI.

Han, S., and C.E. Goering. 1993. FIS - A field information system for SSCM. ASAE Paper 933056, ASAE, St. Joseph, MI.

OTA. 1993. Preparing for an uncertain climate: Summary. Office of Technology Assessment, U.S. Congress, 63 pp. October 1993.

6 Sensing Grain Volumes on Individual Elevator Flights

John W. Hummel

Crop Protection Research Unit
USDA--Agricultural Research Service
Urbana, Illinois

Dohn W. Pfeiffer

Agricultural Engineering Department
University of Illinois at Urbana-Champaign
Urbana, Illinois

Norman R. Miller

Mechanical and Industrial Engineering Department
University of Illinois at Urbana-Champaign
Urbana, Illinois

Variations in characteristics such as soil type, fertility level, moisture level, etc. can directly affect crop production levels. Closely monitoring these farm field characteristics and applying only the necessary crop inputs could optimize yields in each portion of the field to maximize net returns and reduce adverse environmental impacts of the production system. Field navigation systems have been developed (Palmer 1988, Larson et al. 1988), and yield mapping based on field navigation systems is a subject of considerable interest to research scientists and production agriculture (Bae 1987, Peterson et al. 1989, Searcy et al. 1989, Tits et al. 1988, and Vansichen & de Baerdemaeker 1990).

Site specific crop yield information is needed to determine the desirability of using site specific nutrient application technology, and to evaluate the efficacy of the site specific management practices in use. Considerable effort has recently been placed on the development of sensors to measure grain flow rate through a combine as the crop is harvested in the field to determine instantaneous yield data (de Baerdemaeker et al. 1985, Wagner & Schrock 1987, Borgelt & Sudduth 1992, Pang & Zoerb 1990, Stafford et al. 1991, Pfeiffer et al. 1993).

OBJECTIVES

A sensor concept was sought that would eliminate the shortcomings of many of the current grain flow rate sensors. The overall objective was to design and test a sensor having the following characteristics: accuracy over a wide range of grain flow rates with no reduction in accuracy due to machine vibrations or field slopes, no impedance of grain flow or reduction in grain tank capacity, and minimal alteration to existing combine clean grain handling components.

SENSOR THEORY

Concept

The sensor concept selected for investigation was measurement of grain flow rate by measuring the height of the grain on individual flights of the clean grain elevator. A beam of light, projected across the clean grain elevator, might be used to sense the grain height on individual flights. Multiple diodes, located in a line opposite the light source, could sense the time periods when grain was moving past the sensor location and also provide information about the surface contour of the grain on each flight.

When the upper surface of the grain being transported upward by an elevator flight reaches the level of the light beam, light is blocked from striking the photodiodes. A dark time period corresponds to the length of time that the light beam is obstructed by the grain and flighting. A light time period occurs during the time when the light beam is not obstructed by the grain and flighting. By measuring the dark and light time periods, and knowing the spacing distance between adjacent flights, the speed of the elevator flighting can be determined and a grain height can be calculated for each flight.

Flow Rate Function

Development of the flow rate sensor began with measurement of the dark and light time periods for each flight passing the sensor. The dark time period was measured by counting the pulses, DC, from a frequency source driven at a known frequency, f_D, during the time that the light beam was blocked by the grain and flighting. The dark time period can be calculated as:

$$DT = \frac{DC}{f_D} \tag{1}$$

where DT - dark time period, s
DC - counter pulses
f_D - counter frequency, Hz.

The light time period was similarly measured by counting the pulses, LC, from a frequency source driven at a known frequency, f_L. The light time period counter was active during the time that the photodiodes received light from the light source, and the light time period can be calculated as:

$$LT = \frac{LC}{f_L} \tag{2}$$

where LT - light time period, s
LC - counter pulses
f_L - counter frequency, Hz.

The time period to complete a dark/light cycle, CT, is:

$$CT = DT + LT \tag{3}$$

where CT - cycle time, s.

The velocity of the elevator flighting can be calculated as:

$$V = \frac{L}{CT} \tag{4}$$

where V - velocity of flighting, m/s
L - distance between adjacent flights, m.

The distance between adjacent flights was the same for all flights except at the splice link, where a shorter or longer cycle distance may occur. Combining Eqns. 1, 2, and 3 with Eqn. 4 gives:

$$V = \frac{L}{\frac{DC}{f_D} + \frac{LC}{f_L}} \tag{5}$$

The grain height on each flight is determined by multiplying the dark period time by the linear velocity of the elevator flights. Thus:

$$h = \frac{L}{\frac{DC}{f_D} + \frac{LC}{f_L}} x \frac{DC}{f_D} \tag{6}$$

where h - grain height, m.

An expression for the grain carried on each flight was developed, assuming the grain flow rate was proportional to the height, width, depth, and density of grain on each flight. The mass of grain being carried by each flight is:

$$Grain/Flight = w \cdot d \cdot h \cdot \rho \tag{7}$$

where w - width of the elevator flighting, m
d - depth of the elevator flighting, m
ρ - density of the grain, kg/m³.

Substituting Eqn. 6 into Eqn. 7,

$$Grain/Flight = (w \cdot d \cdot \rho)\left[\frac{L \cdot \frac{DC}{f_D}}{\frac{DC}{f_D} + \frac{LC}{f_L}}\right] \tag{8}$$

The width, depth, and density of the grain on each flight was assumed to be constant. Since there is one flight moving past the sensor for each cycle time, the grain flow rate can be calculated as:

$$Grain\ Flow\ Rate = (w \cdot d \cdot \rho)\left[\frac{L \cdot \frac{DC}{f_D}}{\frac{DC}{f_D} + \frac{LC}{f_L}}\right]\left[\frac{1}{\frac{DC}{f_D} + \frac{LC}{f_L}}\right] \tag{9}$$

Combining terms and regrouping, the grain flow rate is:

$$Grain\ Flow\ Rate = (w \cdot d \cdot L \cdot \rho \cdot f_D)\left[\frac{DC}{\left(\frac{f_D}{f_L} \cdot LC + DC\right)^2}\right] \tag{10}$$

If the assumption of constant width, depth, surface profile and density of grain on each flight is correct, a linear relationship will be present between the grain flow rate and actual grain height on the elevator flights. If any of the other variables such as width, depth, surface profile, or density of the grain on the flight is varying, a polynomial regression will be necessary to develop the relationship between grain flow rate and the light and dark period counts.

SENSOR DESIGN

The sensor consisted of a light source, a signal conditioning circuit, and a data processing circuit. The light source and signal conditioning circuit were designed for mounting on the clean elevator (Fig. 6–1), and the data processing circuit was installed in the operator's cab of a combine.

The light source, a 12V quartz halogen automotive lamp, was located opposite the photodiodes on the outer ascending flight side of the elevator. Both elements of the dual-element lamp were energized. The light was projected between two adjustable plates used to focus and align the light beam onto the photodiodes. A tempered glass plate was mounted in the light source's aluminum base plate to exclude dust and debris from the light source enclosure.

The signal conditioning circuit was encased in a recess machined into two aluminum plates which were mounted between the ascending and descending elevator flights in the middle elevator wall. This circuit consisted of the four photodiodes, four operational amplifiers, and associated components (Pfeiffer 1993) to produce strong output signals to the data processing circuit. The four photodiodes were located behind a tempered glass on the ascending flight side face of the inner aluminum plate. A circuit board, supporting the electronic components, was located behind the photodiodes in the enclosure.

The glass plate covers for the light source and signal conditioning circuit were kept clear of dust and foreign matter by the ascending rubber flights and grain. The output signals from the four op-amps were sent through a six-wire shielded cable to the data processing circuit located in the combine cab.

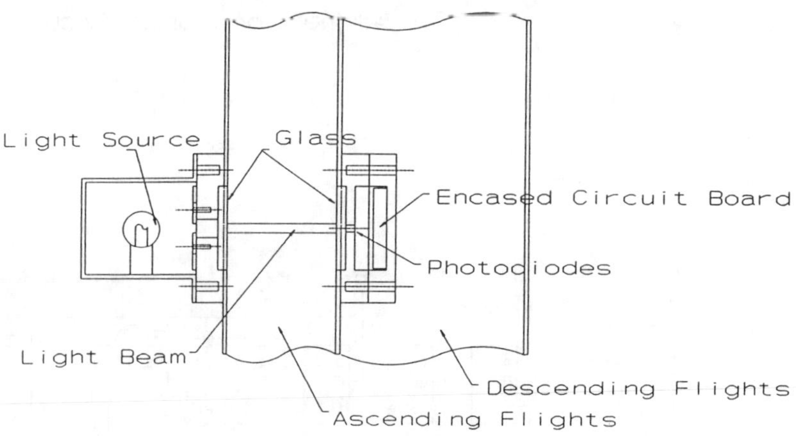

Fig. 6–1. Light source and signal conditioning circuit located in the clean grain elevator.

The data processing circuit (Fig. 6–2) processed the four sensor voltage outputs. Each of the four sensor data streams were routed into a quad comparator. A cutoff voltage of 0.6V was used for each comparator. When the line signal was greater than 0.6V, a zero voltage was outputted from the comparator with +5V outputted for a line voltage less than the 0.6V.

A single noise reduction filter was placed in each signal line after the comparator to remove voltage spikes present in the square wave output. The filter consisted of four D-type flip-flops and two triple input AND-gates placed in series with the line signal. The filter checked three consecutive flip-flop outputs. If a spike occurred with a period shorter than the time required for the spike to travel through three flip-flops, the spike was eliminated and the output voltage remained the same. This filter decreased the amount of noise in the signal lines to reduce counter errors.

After each input data stream had been filtered, it was passed through an AND-gate along with a clock pulse to achieve a counting pulse for the counters. The four dark period counters and light period counter operated using this principle.

A light status (LS) line was used by the data acquisition program to start and stop the light period counter. The LS line went high when all four dark period counters were counting, and went low after the first dark period counter stopped counting. The light period counter would stop, latch, and save the light period count while the LS line was high and began counting again when the line went low.

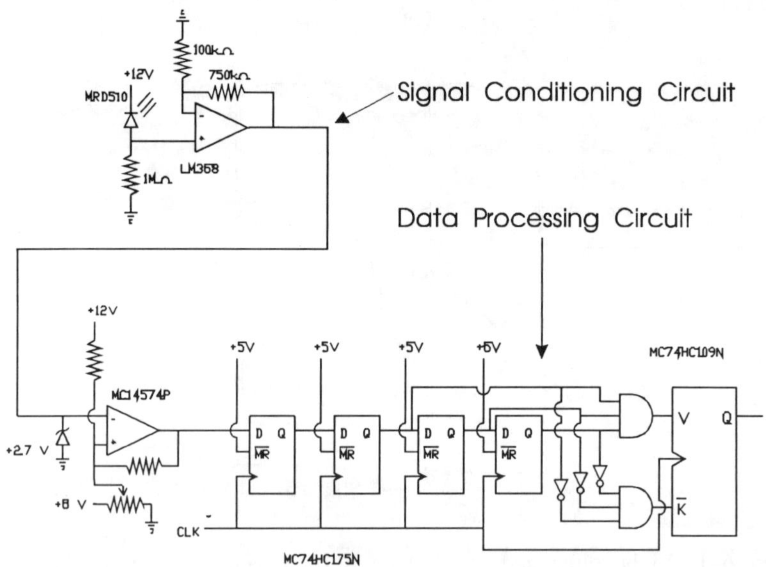

Fig. 6–2. One of four signal generation, signal conditioning and signal processing circuits used in the grain flow sensor.

A dark status (DS) line was used by the data acquisition program to signal that all the dark period counters were done counting so that the dark period counts could be latched and saved. The dark status line OR-gated the four signal lines. The dark status line was high when any one of the dark period counters was counting. The DS and LS lines simply carried the signals used to start and stop the counters.

EQUIPMENT

A laboratory test stand was constructed and equipped to collect and weigh total grain flow during a test. Pfeiffer (1993) conducted extensive tests of the grain flow sensor using the laboratory test stand.

For field testing, the grain flow rate sensor was installed in the clean grain elevator on a four-row Case-IH 1620 Axial-Flow[1] combine. The clean grain elevator of the combine contained forty-one equally spaced flights with a longer space at the connector link. At high idle engine speed, the clean grain elevator chain speed was 2.4 m/s. The average spacing between adjacent flights was 0.165 m. An instrumented weigh box (Pfeiffer 1993) was installed in the grain tank of the combine. The weigh box included a capacitance-type grain moisture sensor. The weigh box weighing system (Micro IV, MCS Corp., Owensboro KY) was used to record total weight and moisture content of the grain passing through the clean grain elevator during a test. The Micro IV unit was mounted in the cab of the combine under the overhead console. The sensor and light source were positioned 0.41 m below the upper clean grain elevator shaft in the grain bin extension section of the elevator.

A modification was made to the top of the clean grain elevator to facilitate calibration. The front semicircular face on the clean grain elevator discharge hopper was hinged just below the center of the grain tank auger to form a door which could manually be opened and closed. When the door was open, grain discharged through the opening into the grain tank, bypassing the grain tank auger and the weigh box. When the door was closed, the grain exited the clean grain elevator normally into the grain tank auger and to the weigh box. Closing and opening the door in the clean grain elevator hopper easily initiated and terminated flow to the weigh box for calibration purposes.

A Toshiba T1600 80286 laptop computer was used for all data acquisition during field testing. An external expansion slot (Won Under Co., Minneapolis MN) housed a PCL-830 counter/timer card (Advantech Co., Ltd.,). The computer, circuit, and light source were powered from the combine battery supply at all times, eliminating the need for an external power source. The QuickBASIC language was used for all data acquisition software developed to interface with the counter/timer card, and to collect the flow rate sensor data.

---

[1]Trade names are used solely for the purpose of providing specific information. Mention of a trade name, proprietary product, or specific equipment does not constitute a guarantee or warranty by the U.S. Department of Agriculture or the University of Illinois, and does not imply the approval of the named product to the exclusion of other products that may be suitable.

A simple switch was implemented to start and stop data acquisition by the computer during calibration. The switch, on an extension cord, was connected between the 5V source on the PCL-830 counter/timer board and one of the digital input lines. Using this simple addition to the main circuit, data acquisition by the computer began when the digital input line was +5V and terminated when the digital input line returned to zero volts. During calibration, the switch was used to remotely initiate and terminate data acquisition.

PROCEDURE

The sensor was calibrated by operating the combine at a constant ground speed, and collecting data over a range of grain flow rates. Each grain flow rate established a data point on a calibration curve. During calibration, the combine separator was activated and harvesting commenced. After an $\approx$ 12 s delay, grain began to flow past the flow rate sensor. The calibration door was initially open, allowing the grain to pass directly into the grain tank. When a constant grain flow rate was attained, the calibration door was closed, directing the grain into the grain tank auger and to the weigh box for measurement. Data acquisition by the computer was immediately initiated by engaging the extension cord switch. When $\approx$100 kg of corn was collected in the weigh box, the calibration door was manually opened and the switch was disengaged. The weight of grain accumulated in the weigh box was entered via the keyboard after data acquisition terminated. This procedure was repeated at each of six different grain flow rates to define sensor output over a range of grain flow rates.

A calibration curve was obtained by fitting a second order polynomial to the data. The calibration curve was called by the real time sensing portion of the program (Pfeiffer 1993) for flow rate calculations.

The accuracy of the sensor was tested by collecting grain flow rate data. A test plot was harvested, and the total quantity of grain sensed was recorded and displayed by the laptop computer. The sensor plot weight was then compared to the weigh box weight collected for the same sample. Calibration curves were developed at four different moisture levels as the corn crop dried down in the field. Real time sensing of corn plots at different moisture levels was also scheduled to test the sensor's ability to accurately measure the total quantity of grain harvested.

Additional tests of the sensor components were conducted after completion of the field tests. Some of these tests were made in the laboratory, and others were conducted on the combine, but under more controlled conditions, by introducing grain onto the cleaning shoe of the combine with the separator in operation. These tests were conducted after a commercially available impact grain flow rate sensor (Ag Leader 2000, Ag Leader Technology, Ames IA) had been installed in the combine. During installation of the commercial sensor, the length of the clean grain elevator was lengthened slightly, eliminating the long flight spacing at the elevator connecting link.

RESULTS AND DISCUSSION

Portions of a typical data set obtained from a single photodiode (Fig. 6–3) illustrate the data stream obtained from each of the sensor circuits. A positive dark period count for zero grain flow rate was expected, since the elevator flight excluded light from the photodiodes even when no grain was being transported. The dark period count obtained for the height of the elevator flights (at some constant combine operating condition, e.g., zero throughput, engine @ high idle) might be useful as a quick check of the sensor's calibration. The periodicity of the dark and light period counts at zero grain flow rate are indicative of the resolution of the sensor, since it was able to detect differences in the height of the elevator flights.

The light period counts are also shown (Fig. 6–3) with zero grain flow rate. For 40 flights, a light period count of ≈25 000 was recorded. For the 41st flight, however, the cycle distance was greater than for the other flights, resulting in a light period count of ≈35 000. This one light period count for each complete revolution of the elevator flight set was higher because more chain links were present between the two flights where the ends of the chain were joined.

Due to this unequal flight spacing, the data period counts from individual flights were averaged over one complete revolution of the flight set. Once data acquisition was initiated, 41-point averaging was delayed until 41 light and dark

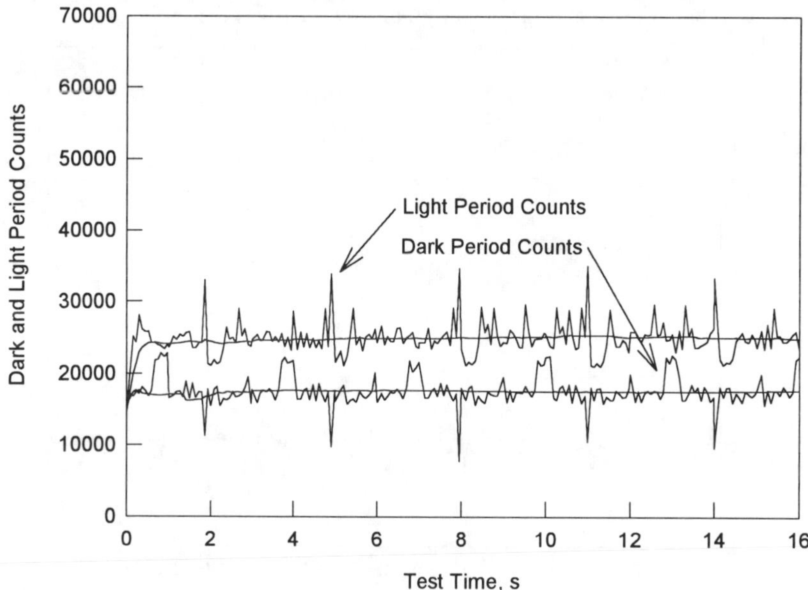

Fig. 6–3. Typical counts of clock cycles corresponding to light and dark periods sensed by one of the photodiodes during zero flow in the clean grain elevator of the CaseIH 1620 combine.

period counts had been accumulated. An average of the light and dark period counts, respectively, for the first 41 flights was calculated. When light and dark period counts for the 42nd flight were collected, average light and dark period counts for the 2nd flight through the 42nd flight were computed. This averaging technique continued until data acquisition was terminated and will be referred to as moving 41-point averaging. Results from the 41-point averaging are also plotted in Fig. 6–3, and show that much of the variability was eliminated using this data analysis method.

Calibration Curves

After running calibration tests, light and dark period counts for each of the four photodiodes were combined, using Eqn. 10, to calculate the instantaneous flow function for each flight. At zero flow rate, the height of the individual flights (Fig. 6–4) could be calculated. The variability in the rubber flighting material used on the elevator is evident. An artificially low value of height occurs for every 41st flight because a constant flight spacing was assumed. A mean instantaneous flow function value was calculated by averaging the instantaneous flow function values for the four photodiodes over the collected data set. The actual grain flow rate was calculated by dividing the plot weight by the test time. A coordinate pair on the calibration curve resulted for each of the four photodiodes at a given grain flow rate. The mean instantaneous flow function value was the abscissa and the actual grain flow rate was the ordinate.

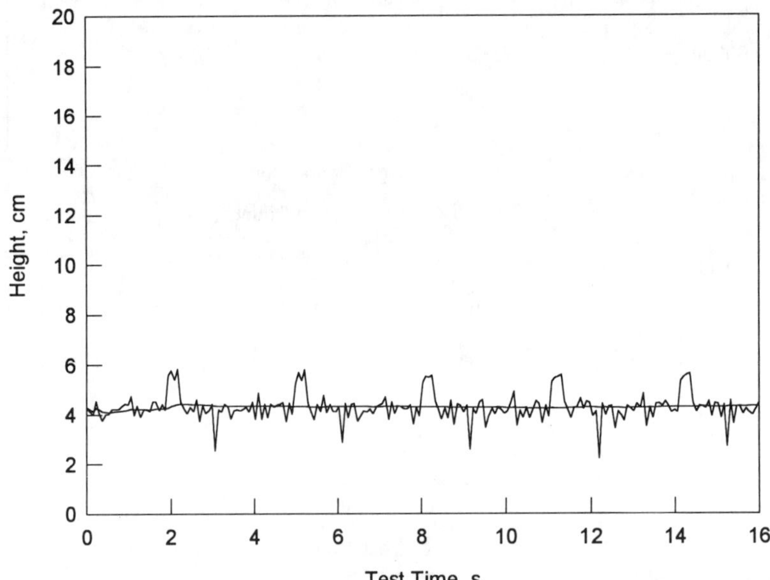

Fig. 6–4. Measured height of individual flights (zero flow) on the CaseIH 1620 combine's clean grain elevator using the grain volume sensor.

Flow Rate Function Results

A second order polynomial was fitted to data to establish calibration curves for the flow sensor (Fig. 6–5). Field calibration curves were established at corn moisture levels between 22% and 24% (WB). Data (Table 6-1) were collected from 18 m long plots (Field C). The overall average relative error in the recorded total harvested weights using the grain flow rate sensor was ≈ 2.65%. The weigh box weights were considered to contain no relative error in the weight measurements.

Table 6–1. Comparison of measured grain weights with grain weights calculated from measurements of grain height on individual elevator flights.

Plot Number	Moisture %	Weigh Box, kg	Sensed Weight†, kg	Absolute Error, kg	Relative Error, %
1	22.5	31.0	30.9	0.1	0.32
2	22.4	44.1	45.8	1.8	3.97
3	22.9	37.7	38.9	1.2	3.27
4	22.3	36.5	36.4	0.1	0.25
5	22.5	47.0	47.9	0.8	1.76
6	22.5	45.4	45.2	0.1	0.29
7	22.0	50.4	51.2	0.7	1.45
8	21.3	46.1	47.7	1.6	3.47
9	21.5	50.0	51.6	1.6	3.21
10	22.0	48.7	48.6	0.0	0.06
11	22.7	42.7	42.5	0.2	0.54
12	22.4	50.2	50.5	0.3	0.58
13	22.1	64.6	67.7	3.1	4.86
14	22.6	46.3	44.2	2.1	4.60
15	22.4	45.1	44.8	0.3	0.64
16	22.0	39.6	37.9	1.7	4.39
17	22.0	42.6	45.0	2.4	5.69
18	22.0	65.8	68.6	2.8	4.30
19	23.1	35.9	39.5	3.6	9.91
20	23.1	25.2	25.2	0.0	0.08
21	22.6	35.7	38.1	2.4	6.60
22	23.2	22.5	22.5	0.1	0.27
23	23.5	26.9	27.1	0.3	1.08
24	22.5	59.1	57.8	1.3	2.18

†Calibration curve relating grain height and grain weight developed from data collected on the same harvest date in the same field.

Table 6–2. Comparison of measured grain weights with grain weights calculated from measurements of grain height with the volume sensor.

Plot Number	Moisture, %	Weigh Box, kg	Sensed Weight[†], kg	Absolute Error, kg	Relative Error, %
1	23.3	158.7	156.5	2.2	1.41
2	23.2	185.0	181.9	3.1	1.69
3	24.4	180.1	185.2	5.2	2.87
4	23.1	175.1	173.0	2.2	1.24
5	23.5	182.5	175.4	7.1	3.87
6	22.5	181.7	178.0	3.6	2.00
7	24.8	171.6	167.0	4.7	2.73
8	23.0	193.7	191.9	1.8	0.92
9	23.9	159.5	156.0	3.5	2.19
10	23.1	192.0	199.7	7.7	4.00
11	24.2	211.8	212.7	0.9	0.42
12	22.7	150.6	153.1	2.5	1.66
13	23.5	158.2	160.6	7.6	4.52
14	23.6	190.6	195.2	4.6	2.41
15	24.9	173.8	177.4	3.7	2.11

[†] Calibration curve relating grain height and grain weight developed from data collected in a different field on a earlier harvest date.

At a later date, data were collected in a nearby field. Plot lengths were much longer in this field, allowing harvesting to continue until the weigh box was full. The calibration curve developed at an earlier date (Fig. 6–5, 22% moisture) was used. Comparisons in weigh box weight and the total quantity predicted by the grain flow rate sensor are shown in Table 6-2. The average error in measuring the total quantity of grain harvested compared to the weigh box sample weight was ≈2.3%. Overall results were very encouraging, since the calibration curve used for this test had been developed at an earlier date with different moisture content grain. The accuracy in measurement of total quantity of grain harvested and response to instantaneous grain flow to flow changes yielded quite accurate predicted weights.

Moisture effects on the calibration curve were difficult to assess due to the small span of moisture curves available. The data collected from three fields and four harvesting dates (Fig. 6–5) suggest that different calibration curves may be necessary as the handling properties of grain change from field-to-field, and as moisture conditions change.

When grain was being transported by the clean grain elevator, the fluctuations in the light and dark period counts increased (Fig. 6–6), as compared to fluctuations during zero flow (Fig. 6–3). This increased variability may have been due to instantaneous variations in the elevator speed as the combine was harvesting. Material feed rate through the combine varied, perhaps due to power demand spikes which may have produced instantaneous engine speed changes or due to the relative speeds of grain transport components, e.g., the clean grain elevator and the clean grain auger. Fluctuating count levels due to falling grain

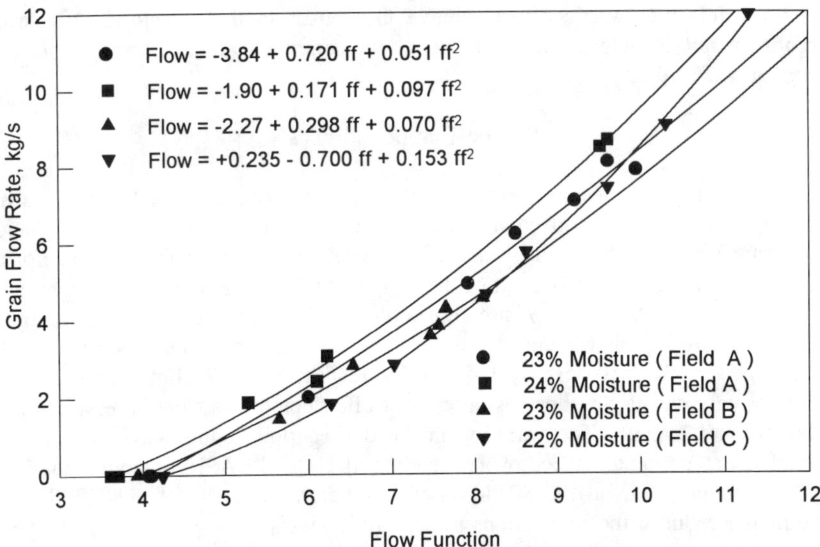

Fig. 6–5. Grain moisture effects on the flow function calibration curve for field test data. Data were obtained from different field research plots.

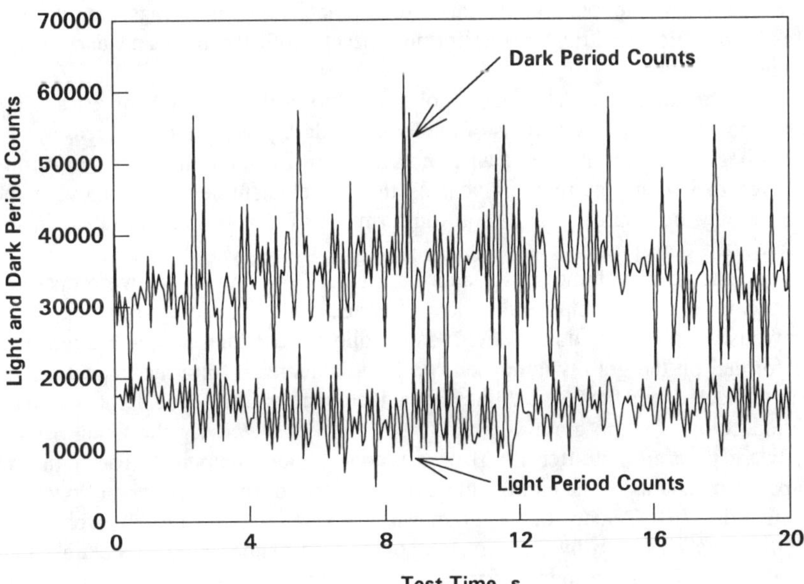

Fig. 6–6. Typical counts of clock cycles corresponding to light and dark periods sensed by one of the photodiodes during intermediate flow in the clean grain elevator of the CaseIH 1620 combine.

kernels can be discounted, since sensor location at the top of the clean grain elevator left little vertical space above the sensor in the elevator. Dust was probably not a factor, since little dust was generated at the harvest moisture content (22% W.B.).

Sensor Component Tests

During field testing, some concern about the source of the fluctuations in the light and dark period counts was raised. The data processing circuit and software were bench tested by using a 13.5 Hz square wave from a waveform generator to simulate the output of the signal conditioning circuit. The light and dark period counts (Fig. 6–7) are shown along with a 15-point moving average line. The light period counts are double the dark period counts since the clock frequencies were $1(10)^6$ Hz and $0.5(10)^6$ Hz, respectively. The light period counts did not vary during the short test interval, while fluctuations were observed in the dark period counts. The variation in the dark period counts was attributed to insufficient computer speed when running the QuickBASIC uncompiled data collection program on the 80286-based computer. The use of a 80486-based computer reduced the fluctuations to negligible levels.

Light and dark period data collected with zero flow in the clean grain elevator (Fig. 6–8) illustrate the improved data stream due to the uniform elevator flight spacing, as compared to the field test configuration data (Fig. 6–3). The data are from one photodiode, and a 15-point moving average line is also presented. Elevator flight nonuniformity affects both the light and dark period counts.

A portion of a data stream collected during the initiation of a test (Fig. 6–9) shows that the fluctuations of the light and dark period counts increase when grain begins to flow in the clean grain elevator. Elevator flight nonuniformity appears to become more prominent as flow level increases. Accuracy of the sensor may be improved if a more uniform set of flights were installed in the clean grain elevator.

The grain volume sensor and the impact flow rate sensor were operated simultaneously to compare the data streams. The impact sensor had been previously calibrated during field data collection, while no calibration was performed on the grain volume sensor. The data stream from the impact sensor was collected, via RS232 cabling, on a laptop computer. The light and dark period counts for the grain volume sensor were collected via the QuickBASIC software program (Pfeiffer 1993) on a second laptop computer. The light and dark period counts of the grain volume sensor were converted into grain flow rates using Eqn. 10. Density of the grain was 0.733 g/cm^3, and the elevator cross-section was 17.46 cm by 7.30 cm. A comparison of the two data streams (Fig. 6–10) shows similar responses for the two sensors. The impact sensor appears to have a low-flow threshold that is lower than that of the volume sensor. Significantly more data points were collected for the volume sensor, and the data stream illustrates the effect of the difference in the flighting material on the calculated flow rate. This variation could be reduced by installing a more uniform set of flighting in the elevator, or by modifying the data processing software to

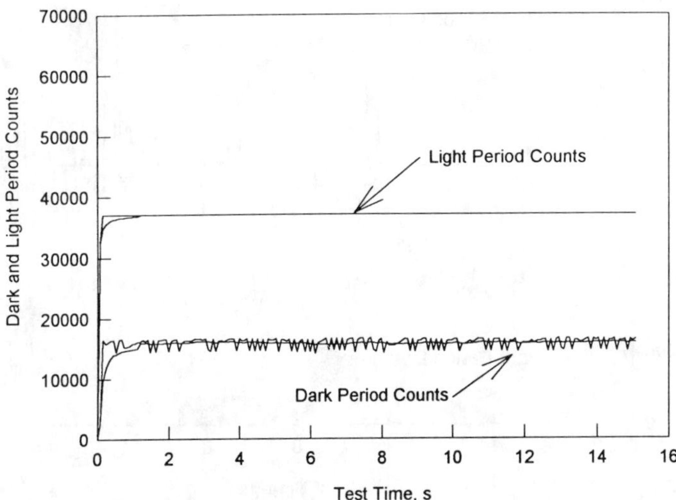

Fig. 6–7. Typical counts of clock cycles corresponding to light and dark periods sensed by one of the photodiodes while excited by a 13.5 Hz square wave from a waveform generator.

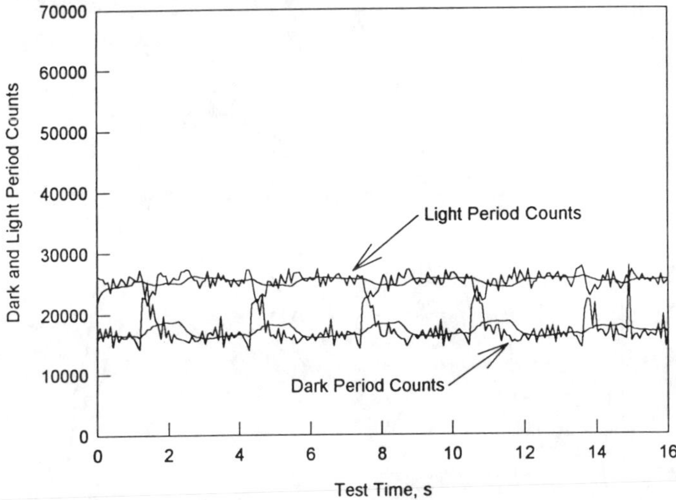

Fig. 6–8. Typical counts of clock cycles corresponding to light and dark periods sensed by one of the photodiodes during zero flow, after the clean grain elevator on the CaseIH 1620 combine had been lengthened and all flights were uniformly spaced.

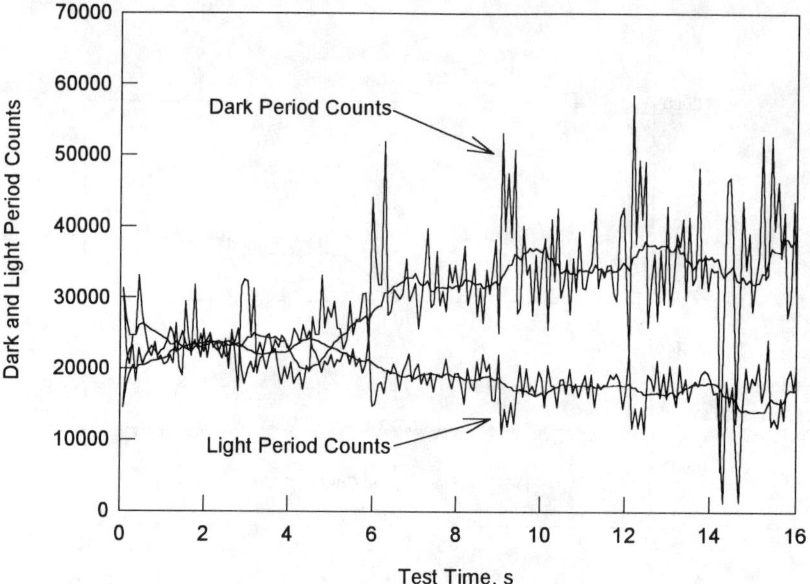

Fig. 6–9. Individual flight light and dark period counts of clock cycles obtained with an intermediate grain flow rate in the CaseIH 1620 combine clean grain elevator. The data illustrate the typical datastream obtained when grain flow is initiated.

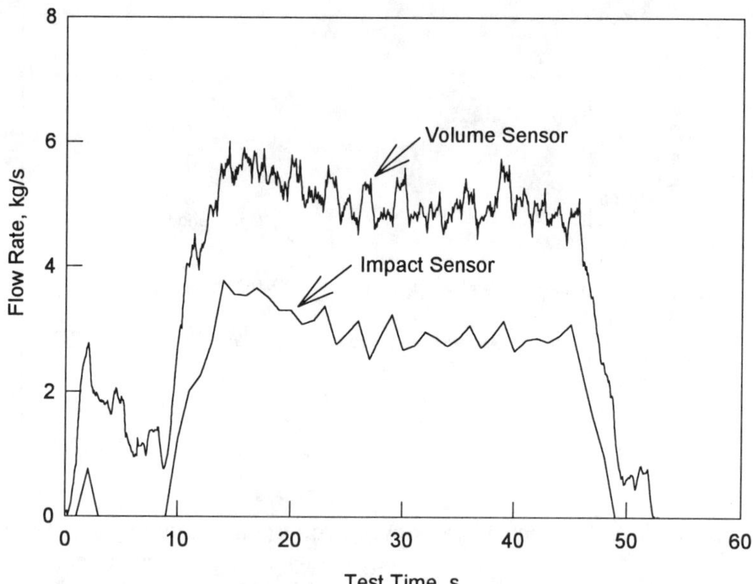

Fig. 6–10. Flow rate data streams from the impact and volume flow rate sensors.

include a flight-by-flight height correction calculation. The overall flow rate indicated by the volume sensor was greater than for the impact sensor, and calibration using weighed volumes of grain is required for accurate flow measurements with either sensor.

CONCLUSIONS

The grain flow rate through a combine can accurately be measured for the entire spectrum of flows possible in a combine by using photodiodes and a light source mounted in the clean grain elevator. Calibration of the sensor is required as the relation between measured grain height and actual grain flow rate is a function of the test weight and moisture content of the grain which introduces non-linear calibration curves. The grain flow rate sensor can accurately measure total quantity of corn harvested to within 3% in most field conditions. Nonuniformity of the combine's clean grain elevator flights, and nonuniform loading of the elevator flights appear to be the main contributors to the fluctuations in the light and dark period counts. The grain volume sensor appears to be able to detect and record the fluctuations in grain flow in the clean grain elevator as accurately as the commercially available impact sensor. The design of the sensing unit would permit retrofit into existing combines with minimal alteration, and results in no grain flow impedance or reduction of grain tank capacity.

ACKNOWLEDGMENT

The authors would like to thank Dennis King, Engineering Technician, for his assistance with the sensor electronics and computer interface, and Terry Holman, Biological Technician, for her participation in the data analysis phase of the research.

REFERENCES

Bae, Y.H. 1987. Determination of spatially variable yield maps. ASAE Paper 87-1533. ASAE, St. Joseph, MI.

Baerdemaeker, J. de, R. Delcroix, and P. Lindemans. 1985. Monitoring the grain flow on combines. Paper presented at Agrimation 1 Conference and Exposition, Feb. 1985, Chicago. ASAE Publ. 01–85.

Borgelt, S.C., and K.A. Sudduth. 1992. Grain flow monitoring for in-field yield mapping. ASAE Paper 92–1022. ASAE, St. Joseph, MI.

Larson, W.E., D.A. Tyler, and G.A. Nielsen. 1988. Field navigation using the Global Positioning System (GPS). ASAE Paper 88–1604. ASAE, St. Joseph, MI.

Palmer, R.J. 1988. Impact of navigation on agriculture. ASAE Paper 88–1602. ASAE, St. Joseph, MI.

Pang, S.N., and G.C. Zoerb. 1990. A grain flow sensor for yield mapping. ASAE Paper 90–1633. ASAE, St. Joseph, MI.

Peterson, C.L., J.C. Whitcraft, K.N. Hawley, and E.A. Dowding. 1989. Yield mapping winter wheat for improved crop management. ASAE Paper No. 89–7043. ASAE, St. Joseph, MI.

Pfeiffer, D.W. 1993. Real time corn flow rate sensor. Unpubl. MS Thesis, Library, Univ. of Illinois at Urbana-Champaign, Urbana, IL.

Pfeiffer, D.W., J.W. Hummel, and N.R. Miller. 1993. Real-Time Corn Yield Sensor. ASAE Paper 931013. ASAE, St. Joseph, MI.

Searcy, S.W., J.K. Schueller, Y.H. Bae, S.C. Borgelt, and B.A. Stout. 1989. Mapping of spatially variable yield during grain combining. Trans. ASAE, 32(3):826–829.

Stafford, J.V., B. Ambler, and M.P. Smith. 1991. Sensing and mapping grain yield variation. Proceedings of the 1991 Symposium on Automated Agriculture for the 21st Century, Chicago. Dec. 16–17, p. 356–365.

Tits, M., H. Delcourt, F. Vervaeke, and J. de Baerdemaeker. 1988. Yield maps and related field characteristics. Proceedings of the Eleventh International Congress on Agricultural Engineering, Dublin. Sept. 4–8, p. 2791–2796.

Vansichen, R., and J. De Baerdemaeker. 1990. A measurement technique for yield mapping of corn silage. Presented at the AgEng 90 Conference, Berlin, Oct. 24–26.

Wagner, L.E., and M.D. Schrock. 1987. Grain flow measurement with a pivoted auger. Trans. ASAE 30(6):1583–1586.

7 Perils of Monitoring Grain Yield On-The-Go

J. A. Lamb
J. L. Anderson

Dep. of Soil Science
University of Minnesota
Extension Service
St. Paul, Minnesota

G. L. Malzer
J. A. Vetch

Dep. of Soil Science
University of Minnesota
St. Paul, Minnesota

R. H. Dowdy
D. S. Onken
K. I. Ault

USDA-Agricultural Research Service
Dep. of Soil Science
University of Minnesota
St. Paul, Minnesota

With the advent of improved computer and sensor technology the adaption to agriculture has been proceeding quickly. One area of great interest to growers is the ability to measure grain yield across their production fields and use that information to make production decisions such as how much and where to spread fertilizer spread on the field. Since most fertilizer recommendations are based on some yield input, this information would be very useful. The idea of grain yield measurement sounds easy. Just hook a sensor to the combine, interface that with a computer and locating system, and off you go. As with all good ideas there are some perils (problems, pearls) which need to be overcome. The objective of this discussion is to identify perils which need to be taken into account and dealt with to make the best use of this technology.

Resolution

The first question that needs to be addressed by the potential user of this system is how large of an area do you want the yield determination to represent (resolution). This probably can be determined by:

1. The physical variability of field, such as whether the terrain is flat or hilly, and do soil textures change dramatically and in short distances.
2. The final use of the information you gather. For example, if a grower is going to use the yield to determine yield goal for fertilizer recommendations, factors such as - how fast can the applicator rates be changed, and how dense the soil sampling grid was - would be used to provide the additional information to the recommendations. The resolution may only need to be 61 to 91 m (200 to 300 ft) in this case. At the other extreme, a researcher would want a 1.5 to 3.0 m (5 to 10 ft) resolution to better understand the effect of variable soil properties or landscape on yield.

Timelags

The use of a production combine for continuous yield determination has characteristics which need to be identified and evaluated to get a useful yield map. Most important is being able to locate yield measurements to a point in the field. This requires adjustment for timelags in the machine itself. These lags include: i) the time to initially fill the combine with grain, ii) the time it takes the plant material to move from the edges of the header to the feeder chain of the combine, and iii) the time it takes to move from the feeder to the point of measurement (normally to the top of the clean grain elevator).

The time it takes to fill the combine will be dependent on the size of combine header, speed of travel, and size of threshing area. In our experience, this can be 15 s. Information gathered during that time is not of practical use because the grain flow through the machine is being established and the sensor measurements for yield changes are shadowed by the increase of grain flow through the machine. Yield estimates for the headland areas of the field will not be useful.

The second timelag occurs in the header of the combine. As an example, an ear of corn from the outside edge of a twelve-row corn head will spend more time in the header than an ear from a row in the middle. The net effect is that the grain in the middle will enter the combine earlier than the grain from the edge. This reduces the resolution of the grain yield data on the map. For newer farm production combines the distance between the feeder housing chain and edge can be 4.5 m (15 ft).

The time in the machine between the feeder housing and the sensor, normally located at the top of the clean grain elevator, can range from 10–15 s. Operating travel speed then factors into the resolution of the yield map and the ability to locate the yield on a map. Table 7-1 indicates the distance that a combine can travel during this lagtime of 10 or 15 s at five different speeds. It is not unusual for combines to harvest at speeds of 8.1 km hr^{-1} (5 mi hr^{-1}). This

Table 7–1. The effect of lagtime and combine travel speed on distance traveled.

Combine speed	Time in harvester	
	10 s	15 s
km hr^{-1} (mi hr^{-1})	------------ m (ft) ------------	
1.6 (1)	4.5 (14.7)	6.7 (22)
3.2 (2)	8.9 (29.3)	13.4 (44)
4.8 (3)	13.4 (44.0)	20.1 (66)
6.4 (4)	17.8 (58.7)	26.8 (88)
8.1 (5)	22.3 (73.3)	33.4 (110)

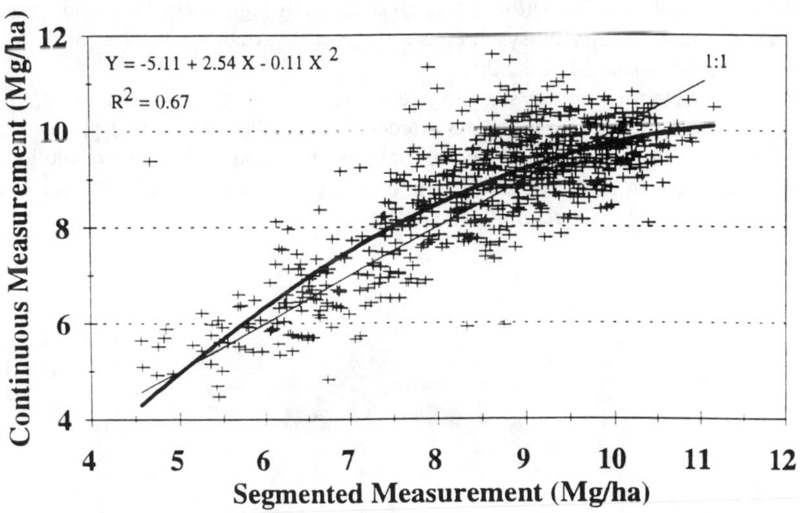

Fig. 7–1. Comparison of segmented measurement of corn grain yield by a research plot combine with continuous measurement of corn grain yield by an agricultural production combine in Sherburne County, Minnesota, Fall 1992.

means that the combine travels 22 to 33 m (73 to 110 ft) while the grain is being threshed and cleaned. Besides accounting for this lagtime, the grain is also being mixed. What went in at the same time will be mixed and take 10–15 s from the time the first grain goes by the sensor until the last grain is sensed.

Other Considerations

The use of electronic sensors requires the smoothing of the signal, data smoothing, and physical shock absorption to reduce the variability in the output. Vibration of the machine from the motor, threshing unit and the terrain it is traveling produces "noise" in the information. The noise can not always be filtered out.

In 1992 an experiment was conducted to compare the use of a plot combine which determined grain yield in 15.2 m (50 ft) segments with a agricultural combine equipped with a continuous grain yield measuring system (Fig. 7–1). The experiment was conducted on an eight ha (20 acre) irrigated corn field in Sherburne County Minnesota. Grain yields ranged from 4.4 to 11.2 Mg ha^{-1} (70 to 179 bu A^{-1}). At grain yields between 5.0 and 9.0 Mg ha^{-1} (80 to 144 bu A^{-1}) the continuous measurement over-estimates the grain yield determined by the segmented method. At grain yields greater than 9.0 Mg ha^{-1} (144 bu $A)^{-1}$ the continuous method under-estimates the grain yields. Another concern displayed by this information is the scatter of the data. The R^2 value is 0.67 which means the regression line fit to the data accounts for 67 percent of the variaction. For calibration of measurement devices used in research and perhaps agricultural production, although statistically significant, this is not good enough. The measurement system has a 16 percent variability which contributes to error besides natural variability.

New, more accurate, sensor systems are being developed. Hopefully, the new technology will minimize the effect of lagtime and noise and produce a yield map with more accuracy in the yield estimate and with better resolution. With increases in accuracy and resolution agriculture can use this information to more precisely manage crop production.

8 Yield Mapping of Wheat and Corn Using a Continuous, On-Combine Yield Monitor

Ron Campbell

HarvestMaster, Inc.
Logan, Utah

Both site specific management of crops, and on-farm testing of management practices require detailed and accurate knowledge of yield variations in the field. This information is best obtained using a standard harvester, so that routine use is possible with minimal expense. We designed and tested an accurate volumetric yield monitor which fits on a standard combine. Grain moisture was monitored simultaneously. Yield maps were produced for both dryland wheat and irrigated corn. Coefficients of variation for both crops were high. Yield varied by a factor of 4 over distances of several hundred feet, even in wheat stands that appeared quite uniform. These variations have a number of implications with regard to applicability of present crop models to management. We will present yield and moisture maps for wheat and yield maps for corn, and discuss some of the implications of the measured variations for both management and crop modeling.

SESSION I

SOIL RESOURCE VARIABILITY

9 Opportunities for Examining On-Farm Soil Variability

Donald R. Nielsen

Department of Land, Air and Water Resources
University of California
Davis, California

Ole Wendroth

Center of Agrolandscape and Land Use Research
Institute for Soil Research
Germany

M. B. Parlange

Department of Land, Air and Water Resources
University of California
Davis, California

During the past 20 yr, a global appreciation of environmental citizenship has risen with national and local legislation in many countries providing monies for research and information technology transfer to accelerate our ability to achieve sustainable agriculture. Today, optimizing inputs at the farm level, reducing agrochemical applications and maximizing agricultural production without soil and water degradation are common goals, and the topic of many international scientific and political conventions.

The objective of this presentation is to encourage the exploration of additional experimental methods and statistical tools already available in other disciplines to further improve our understanding of chemical, physical and biological processes that govern crop-soil interactions in the field. By observing and taking advantage of the spatial and temporal variability of our natural resources rather than ignoring them, we provide a glimpse into the future when our observational methods in agriculture and landscape ecology are more diagnostic and are an integral part of a fully developed field technology. A few examples will be given to illustrate when, where and how to better sample the behavior of an entire agricultural field or an ensemble of such fields.

Solute leaching, intercrop spacing, persistence of spatial field patterns and diagnosing variations in crop production are discussed in relation to spectral and cospectral analyses. State-space approaches are used to examine soil surface observations, to explain within field variations of crop nitrogen fixation and to prioritize soil attributes for soil specific crop management. Except to mention a couple of opportunities, geostatistical methods are not discussed. Suggestions for using split moving window techniques and fuzzy set analyses are presented for an ensemble of farms. Lidar observations illustrate new instrumentation for examining the variation of the land-atmosphere interface condition across a farm. And, educational reforms for B.S. and Ph.D. degrees in crop and soil sciences deemed appropriate for achieving sustainable agriculture with soil specific crop management are briefly mentioned.

Crop and soil scientists in partnership with farmers have made unexcelled achievements in crop production during the 20th century, and as we enter the next century, we are mindful of the delicate balance between food and fiber production and the quality of land and water resources in our global habitat. Our achievements in production were aided from laboratory and intensive small scale research efforts transferred to much larger sections of the landscape with the cooperation of farmers and agricultural professionals. Small, replicated plots established on sites believed to be "typical" or "representative" of a farmer's field or an agricultural landscape have provided their response to fertilizers, pesticides, irrigation as well as locations for testing attributes of potential cultivars. These small plots have also been used to suggest cropping and management alternatives. Assuming steady-state conditions, deterministic concepts and mass balance equations have usually been applied for relatively short time periods—minutes, days, weeks or for times no longer than a growing season. Attempts to assess the impacts of agricultural methods and weather events between and during several growing seasons have been made through long-term experimental plots managed for decades, and in a few cases for more than a century. These kinds of experiments remain effective today when one wishes to ascertain the effect of a particular treatment relative to crop or animal production. On the other hand, they do little for our improved understanding of how agricultural practices impact on the quality of water leaving a cultivated or rangeland region nor do they provide direct information regarding the all-too-often subtle changes in soil quality occurring on a farm or within a region. Indeed, with our more recent thoughts focused on sustainability, we have been gearing up to understand agriculturally-induced impacts on the quality of our environment and we have started to quantitatively analyze our efforts outside of those small plots and stretched our consideration across one or more farmer's fields.

We still have a tendency to rely solely on statistical methods developed > one-half century ago before we had anything but a slide rule, a primitive calculator, or a paper and pencil. For the entire century, we have asked the question, "Does the treatment cause a significant perturbation from the expected mean?" We have used analysis of variance and regression techniques designed to minimize the impact of spatial or temporal heterogeneity in field soils and until recently, have not even taken the time to record where we take an observation within an experimental plot or field. Some of our more progressive colleagues are turning to satellite data and geographic information systems to enhance our knowledge or appreciation of the behavior of landscape processes at specific locations. This knowledge cannot be effectively integrated with the vast majority of research activities being carried out now by agricultural scientists and extensionists because the results from the latter which rely on randomly sampled small plots are not easily transferable to other locations in space and time. Today, even hand-held personal computers provide computational opportunities that call for a major revision or perhaps a revolution of our experimental methods in crop and soil science.

We illustrate the above conjectures with two examples of research conducted by some of the best crop and soil scientists in the world who did an outstanding job with statistical tools limited to those taught and practiced at most agricultural research institutions. They also took advantage of the availability of the isotopes of carbon, oxygen, nitrogen and other elements that served as easily identifiable tracers to ascertain the fate of agrochemicals within the experimental field site. The first example is that conducted in Sweden. At a site just north of Uppsala, a six-yr randomized block design was replicated with four management treatments to assess their impact on the microbial ecology within the soil profile. The second example was conducted in the U.S. On a site at the University of California, Davis, a five-yr randomized block design was replicated with several irrigation-nitrogen fertilizer treatments to ascertain the eventual fate of the fertilizer—that in the crop, the soil profile, the groundwater and the atmosphere. The first ton of 99.99% pure ^{14}N in the form of ammonium sulfate ever produced was used to trace the fertilizer. In both of these examples, Sweden and U.S., the response of the harvested portion of the above-ground crop for each treated plot was easily measured and characterized statistically with analysis of variance and regression techniques. On the other hand, a quantitative mass balance of the nitrogen in the field plots was not achieved nor was the amount of nitrogen moving below the root zone into the water table satisfactorily estimated. Having spent > a million dollars for the latter experiment, the impact of agricultural production on the environment was inadequately assessed. Moreover, the environmental results from each of the localized experiments could not easily be transformed to other soils, other regions within each country or even to other locations of the same soil. Why not? There are two primary reasons. First, the experimental designs did not provide sampling strategies to quantify the spatial or temporal covariance structure of different observations within or between treatment plots. Second, no quantitative measures were taken to assure crop and soil properties outside the experimental plot area were indeed represented by the behavior of those inside the small plots. Hence, the intricacies of the chemical, physical and microbial processes occurring within the root zone of agricultural crops were not observed nor were there any guidelines developed from such experiments to know when, where and how to sample the behavior of the entire agricultural field or the impact of its cultivation on the quality of the surrounding region. This latter point needs amplification inasmuch as more than experimental design is involved.

One aspect of soil specific crop management is the quantity and rate at which agrochemicals and naturally occurring metabolites leave the soil profile to potentially pollute surface and ground water. The farmer needs to prevent leaching of valuable plant nutrients, soil-surface-applied agrochemicals, manure and organic wastes including crop residues into the profile, and at the same time in most irrigated regions of the world, leach excess salines from the profile beyond the recall of the crop root zone. This is a formidable task in view of the leaching properties of soil. At water saturation the soil hydraulic conductivity is 10 million times greater than that at air-dryness. In other words, a change of 7 percent soil water content yields a l0-fold change in the value of the hydraulic conductivity. Hence, small differences in soil water content across a seemingly uniform field soil gives rise to large differences at which water and its dissolved constituents move through the soil. Unfortunately, even the most uniform of field soils is heterogeneous. A few small plots currently used in a typical agronomic research field trial are inadequate to copy with the spatial variance structure of the soil's hydraulic conductivity and other properties.

Support for this latter statement can be seen in Fig. 9–1 where the concentration of nitrate nitrogen in the soil solution extracted from the 3-m depth of four plots treated identically from the U.S. experiment is plotted against time during the yr 1974. Each plot, initially at field capacity at the beginning of the cropping season, was cropped to corn, fertilized with 360 kg N/ha in the form of NH_4SO_4, and irrigated with water equal to 1/3 that amount lost by evapotranspiration on a biweekly schedule during the growing season. The N concentration in the soil solution ranged from < 1 mg/l to greater than 25 mg/l. No two plots exhibited the same behavior. For example, the N content below plot 17 remained < 5 mg/l during the early part of the yr and increased markedly to 20 mg/l midyr. On the other hand, the N content below plot 24 initially above 15 mg/l dropped to < 5 mg/l midyr and remained at nearly 10 mg/l for the remainder of the yr. Calculating an average value of N below the four plots is not meaningful, and particularly so when the more important concept of the rate of nitrogen leaving the rootzone is considered.

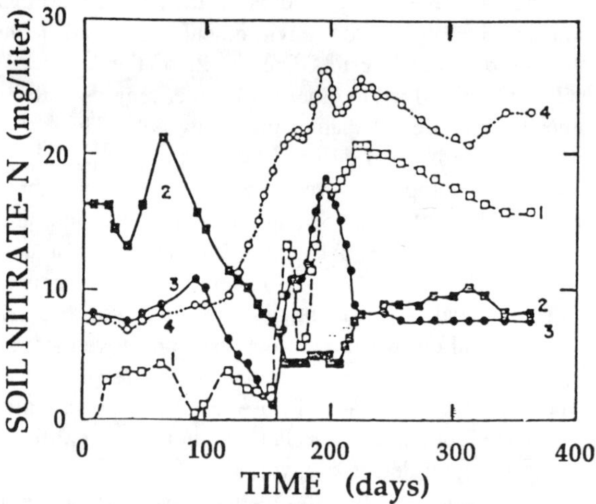

Fig. 9–1. Soil solution nitrate nitrogen concentration measured 3 m below the soil surface of four plots cropped to corn and fertilized and irrigated identically.

The rate of nitrogen leaving the rootzone J_N is obtained by multiplying the concentration of nitrogen in the soil solution C by the rate at which water is moving downward out of the rootzone J_w

$$J_N = J_w C \tag{1}$$

Recognizing that at any sampling location below any of the four plots both the N concentration and the soil water velocity each vary over one order of magnitude in space and time, the product J_N is indeed an unreliable estimate of the amount of N leaving the profile. Similar unreliable results were obtained from all other identically treated plots. The difficulty of not being able to interpret how much N is leaving the rootzone stems from the fact that the spatial and temporal variance and covariance structures of J_w and C were not measured nor quantified. This difficulty persists in virtually every sector of agricultural field research, and becomes worse with increased soil heterogeneity and when results from selected measurement sites are translated to average values across an entire field. Yet, sustainable production demands that we reliably estimate, effectively manage and control the constituents leaving the rootzone of an agricultural field. Moreover, managing the below ground effluent from a composite of agricultural fields in a given region or watershed is also a technology which has not yet been achieved nor generally taught in agricultural institutions of education and research. New experimental methods in conjunction with innovative statistical methods are required.

If we are to achieve soil specific crop management, prescription farming or sustainable agriculture, we need to develop methods to observe the attributes of crops and soils across the landscape and to be able to measure improvement or degradation of those natural resources without relying solely on the results from fertilizer trials and the like limited to a small fraction of the landscape. How do we help the farmer today? What research do we conduct to help farmers manage their fields? In the case of a fertilizer recommendation, what procedures do we presently follow? Do we assess the yield variation within the field that could be improved by non-uniform fertilizer applications? Do we estimate what fertilizer leaves his field through deep percolation or surface runoff? Are we, the farmer, and the farmer's neighbors content that the farmer's crop production is sustainable and environmentally compatible with the quality of the surrounding landscape? How do we cope with short term economic forces that provide no incentive to practice soil specific management of farms? Our answers are hopeful, but not confident.

A technology to directly assess the physical, chemical and biological properties of entire fields is now beginning to emerge but remains undeveloped and immature. Stochastic considerations are lacking, and procedures to ascertain subtle changes in soil quality or the long term consequences of present-day management are not available. Soil qualities presently described in soil mapping units are generally inadequate to develop alternative soil specific crop management schemes or reliable simulations of crop growth. And, too often, we judge the success of farming on the basis of this yr's crop yield. We note further that virtually all agricultural field research is based upon the idea of imposing treatments and measuring through an analysis of variance, the impact of a treatment relative to a crop or soil attribute. It is enlightening to hear the response of an agronomist who is asked, "What kind of field research would you conduct if you were not allowed to impose different treatments on replicated small plots?"

A BRIEF REVIEW

In classical statistics commonly used in the agricultural sciences, only the magnitudes of the observations are involved in the calculations while their coordinate values of space (or time) are frequently neglected. For example, observations of soil nitrogen taken in differently treated plots are expressed as

$$Z(i,j) = \mu + \sigma(j) + \varepsilon(i,j) \tag{2}$$

where Z is the ith observation of treatment j having an expected mean μ, a treatment deviation σ and an error ε. Usually, the probability distribution of the observations is assumed to be known without adequate experimental verification, and an analysis of variance conducted to ascertain the magnitude of a perturbation caused by a particular treatment.

An observation of soil nitrogen can also be considered in the form

$$Z(x) = \mu + \sigma(x) \tag{3}$$

where x is the spatial coordinate (in 1, 2 or 3 space dimensions or that of time). The value of μ across the landscape need not be constant and observations Z will reflect both local and regional variations as well as depending on sample size. This latter concept, being able to sample different attributes of the landscape over different scales of space and time, provides a panacea of new field research opportunity for crop and soil scientists. There has been already a remarkable amount of research in agriculture and agroecology using this regionalized variable approach to analyze anthropogenic activities or those occurring naturally within various kinds of ecosystems. Excellent texts are available to explain the assumptions upon which the theories and methods are based as well as the limitations of their application to a particular situation (e.g., Clark, 1979; Davis, 1973; Journel & Huijbregts, 1978; Shumway, 1988; and Webster & Burgess, 1983).

A concept fundamental to geostatistics and regionalized variable analysis is that of spatial dependence of soil or crop attributes. Intuitively, we do not expect field observations of soil properties to be necessarily spatially independent. We would expect measurements made close together to yield nearly equal values, and measurements made a small distance apart to yield values correlated with each other. We would also expect a partially repetitious behavior of soil observations as a result of cyclic tillage traffic and cropping patterns in cultivated fields, sequences of low and high topographical positions giving rise to cyclic locations of greater and lesser degrees of leaching, and sequences of soil mapping units not randomly located within a landscape owing to soil formation processes that are linked spatially to the coordinates of the soil surface. At the risk of being redundant, we describe and illustrate below the concepts of spatial dependence and probability distributions of observations which are often confused and confounded by soil and crop scientists.

A measure of the strength of the linear association between pairs of observations is useful in defining the separation distance between observations beyond which there is no correlation between pairs of values. This concept of spatial dependence or autocorrelation is helpful to answer the question, "How far apart should observations be taken, how large an area is represented by a single observation, or what size of observation is most appropriate?" Similar questions in time are also relevant. Consider a set of n observations $O(x)$ taken along a unidirectional transect across a field at a spacing h ($h = x_{i+1} - x_i$). Analogous to calculating the linear regression coefficient r for a set of measurements of crop yield y as a function of applied nitrogen fertilizer z, the spatial auto correlation coefficient r is calculated from

$$r(h) = \mathrm{cov}\left[O(x+h), O(x)\right]\left\{\mathrm{var}\left[O(x+h)\right]\mathrm{var}\left[O(x)\right]\right\}^{-1/2} \tag{4}$$

where cov is covariance and var is variance. When observations O are taken 1 unit apart, $r(1)$ is the value of the linear regression coefficient for $h = 1$ (lag 1) when values of $O(x + 1)$ are plotted against values of $O(x)$. In other words, nearest neighbors are plotted against each other. Similarly, when $h = 2$, $r(2)$ is the value of the coefficient when values of O are plotted against other values observed a distance of 2 units away.

The functional relation $r(h)$ has been expressed with several empirical formulae. One of the most commonly used expressions is

$$r(h) = \exp(-h/\lambda) \tag{5}$$

where λ is selected in order that the sum of the measured deviations of r from the above expression is zero. Note that the value of λ is equal to that distance h between measured values for which their correlation coefficient is l/e. λ is called the autocorrelation length or the scale of observation. A rather liberal interpretation of λ is that it represents the radial distance within a field characterized by a single observation.

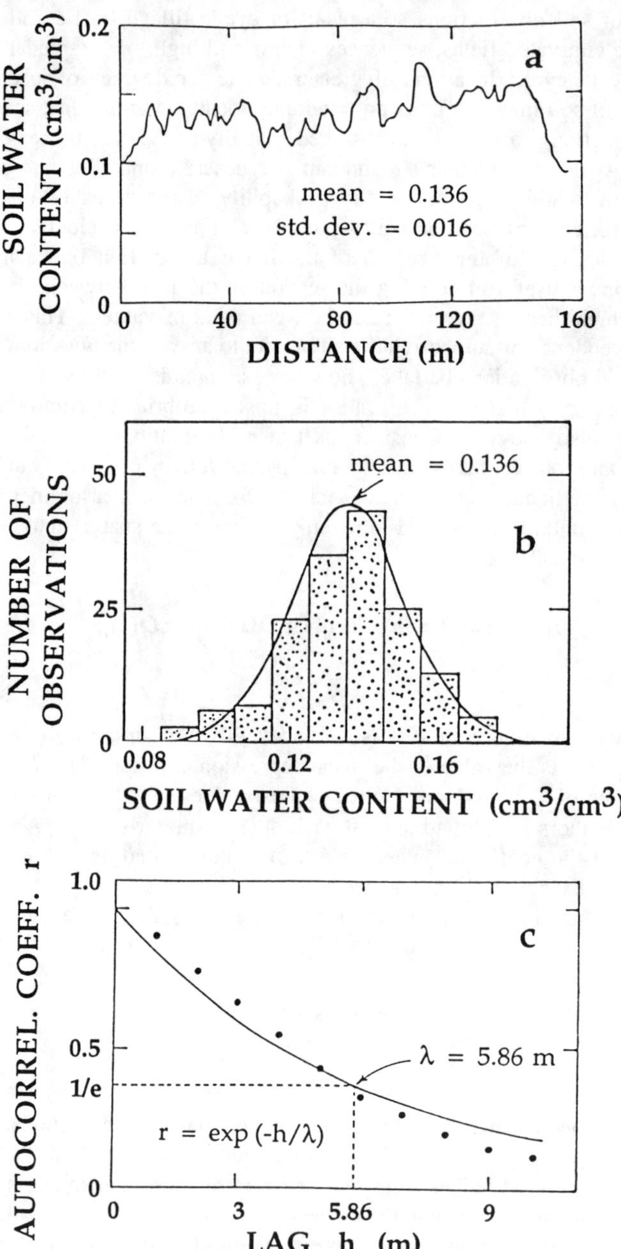

Fig. 9–2. (a.) Soil water content measured at 1-m intervals within a 160-m long transect; (b.) Histogram of values in a; (c.) Autocorrelogram of the values in a.

Figure 9–2a presents soil water content θ measured in a field at 1 m intervals with a neutron moisture meter at the 50 cm soil depth within a 160 m long transect. Neglecting the locations of the observations (n = 160), Fig. 9–2b is a histogram showing that the mean *m* and standard deviation *s* of the 160 observations are 0.136 and 0.0162 cm^3/cm^3, respectively. The height of each vertical bar represents the number of observations found in a range Δθ of 0.01 cm^3/cm^3. The smooth curve shown in the figure is for a normal probability density function

$$N = n\Delta\theta(2\pi s)^{-1/2}\exp\left[-(\theta-m)^2/2s^2\right] \tag{6}$$

where *N* is the number of observations of θ within θ ± Δθ/2. Having concluded that these observations are normally distributed, it is possible to ascertain the number of observations required to estimate the mean soil water content across the transect within prescribed levels of probability. On the other hand, the above formula says nothing about how far apart the observations should be taken. Figure 9–2c, the autocorrelogram of the 160 observations, shows that the autocorrelation length λ is about 6 m. Sampling at intervals < 6 m is somewhat unnecessary because the observations are related to each other. Sampling at intervals greater than 6 m does not allow meaningful interpolation between neighboring observations. It should be obvious that the functional relation between *r* and *h* depends upon the size of the sample, and that in general, the greater the sample size, the greater the value of the autocorrelation length λ. For any particular study, the investigator should consider the minimum distance between sampling locations (*h* = 1) in relation to the objectives of the experiment and the potential utility of the values of λ for each kind of observation.

The utility of using both of the above concepts in crop and soil science is great. Neither is a substitute for the other. The frequency distribution of θ given in Fig. 9–2b is approximately normal. Such a frequency distribution does not quantify the variability of the soil water content as regards their spatial arrangement but merely treats the values in terms of their magnitudes independent of their spatial position. The two concepts (frequency distribution and spatial variability) should not be confused. Using the first concept, it is possible to calculate the probability of an observation occurring within specified confidence intervals. This calculation, however, provides identical information for expected values for all locations. Different expected values for specific locations cannot be obtained. On the other hand, using the second concept, the autocorrelation function provides information about the separation distances within which observed values are related to each other. The spatial structure of a set of observations does not provide information about the frequency distribution of the observations. The fact that a set of observations manifests no spatial structure does not imply that the observations are normally distributed, and vice versa. The soil water content data in Fig. 9–2a constitute a set of observations being normally distributed, yet those observations separated by

<6 m are dependent on each other as expressed by l in Fig. 9–2c. The concept of randomness associated with the normal distribution should not lead to the conclusion that locations for field observations of soil water properties should be selected randomly.

The concept of spatial autocorrelation is easily extended to two kinds of observations (O_1 and O_2) to calculate the crosscorrelation coefficient rc

$$r_c(h) = cov[O_1(x+h),O_2(x)]\,(var[O_1(x+h)]\,var[O_2(x)]) \tag{7}$$

The utility of spatial or temporal cross-correlation should be obvious—it allows the investigator to consider the spacing and size of one set of observations with those of another. For example, at what distance should a tensiometer be placed in relation to a neutron meter access tube? Or, at what distance should a soil sample be taken in relation to the crown of a crop plant?

SPECTRAL AND COSPECTRAL ANALYSES

Above, the interpretation of the correlation coefficients r and r_c were applicable to observations compared more or less in near vicinity to each other (i.e., for h much, much < the width of the field being sampled). An opportunity to discern repetitious irregularities or cyclic patterns in soil or plant communities across a field exists with spectral analysis that utilizes the function $r\,(h)$, or with cospectral analysis that utilized the function $r_c(h)$. A spectral analysis identifies periodicities and can be calculated by

$$S(f) = 2\int_0^{\infty} r(h)\cos(2\pi f h)\,dh \tag{8}$$

where f is the frequency equal to $1/p$ and p is the spatial period. In a similar manner, $r_c(h)$ is used to partition the total covariance for two sets of observations across a field. A cospectral analysis is made by

$$Co(f) = 2\int_0^{\infty} \overline{r_c}(h)\cos(2\pi f h)\,dh \tag{9}$$

where

$$\overline{r_c}(h) = [r_c(h<0) + r_c(h>0)]2.$$

Spectral and cospectral analyses are potentially powerful tools for managing and increasing our knowledge of land resources. With them, we can spatially link observations of different physical, chemical, and biological phenomena. We can identify the existence and persistence of cyclic patterns across the landscape. In some cases, the cyclic behavior of soil attributes may be of more or equal importance than the average behavior. From a spectral analysis, some insights may be gained relative to the distances over which a meaningful average should be calculated. And, with spectral analyses, it is possible to filter out trends across a field to examine more closely local variations, or vice versa. The examples that follow illustrate the utility of spectral and cospectral analyses for improving field technology without dwelling on fundamental theoretical assumptions.

Solute Leaching

Gelhar et al. (1980) analyzed the solute distribution within a field by sampling the soil at 30 cm intervals to a depth of 22 m (See Fig. 9–3a). The spectrum $S(f)$ [equation (8)] shown in Fig. 9–3b has a peak at $f = 0.37\ m^{-1}$, which indicates a period of 2.7 m. Hence, the fluctuations of solute concentrations in Fig. 9–3a are indicative of annual cycles of leaching associated with irrigation in the desert environment of New Mexico. Disregarding the spatial distribution of the concentration value leads only to a standard deviation of about 0.5 mg/l — a classical parameter that adds little to our understanding of the leaching process.

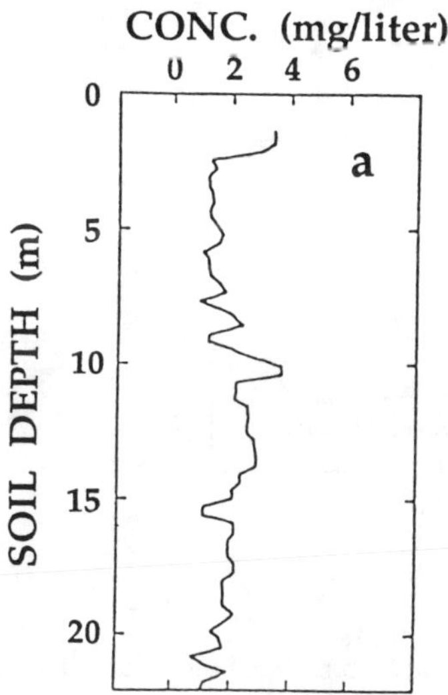

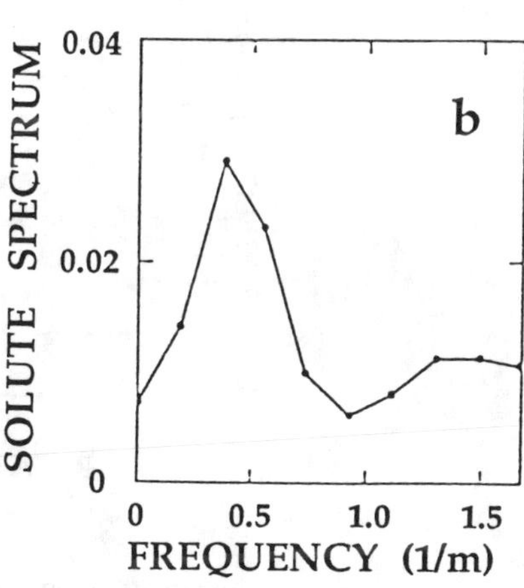

Fig. 9–3. (a.) Concentration distribution of solutes measured at 30-cm depth intervals; (b.) Spectral density function of the solute distribution of a.

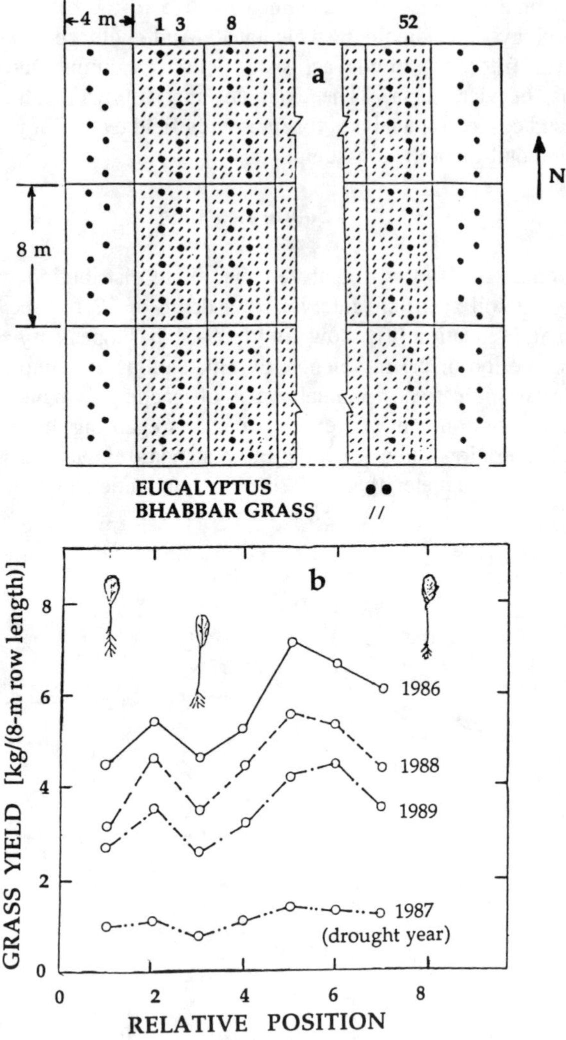

Fig. 9–4. (a.) Schematic diagram of rows of eucalyptus trees interplanted with rows of bhabbar grass; (b.) Average values of 23 observations of interplanted bhabbar grass yields at different positions from the tree rows of eucalyptus.

scales greater than 40 m. The coherency for this latter period clearly indicates that recharge has significantly altered the spatial pattern of soil water storage within local distances <40 m but has not affected the pattern at larger distances associated with the general elevation of the landscape. In other words, the hydrologic process of infiltration rather than that of evapotranspiration altered the spatial pattern of soil water storage.

Burrough et al. (1985) analyzing infra-red radiation emitted from a farm in the Flevopolders detected the impact of a single land reclamation procedure that persisted for 20 yr. The spectral analysis revealed a 24-m periodicity across the field that stemmed from plowing a winter wheat crop into the ground after an extremely wet yr. In the same manner, the persistence of soil compaction caused by farm implements and its impact on crops or soil properties could be ascertained with spectral and cospectral analyses of observations taken normal to the path of the farm implement traffic.

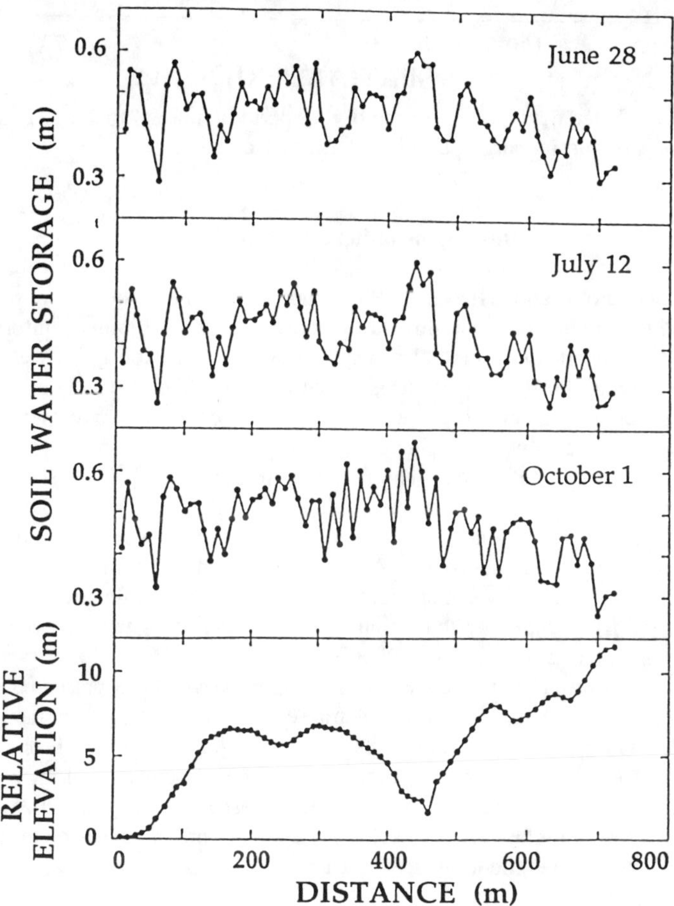

Fig. 9–5. Measured soil water storage in the 2-m soil profile and relative elevation.

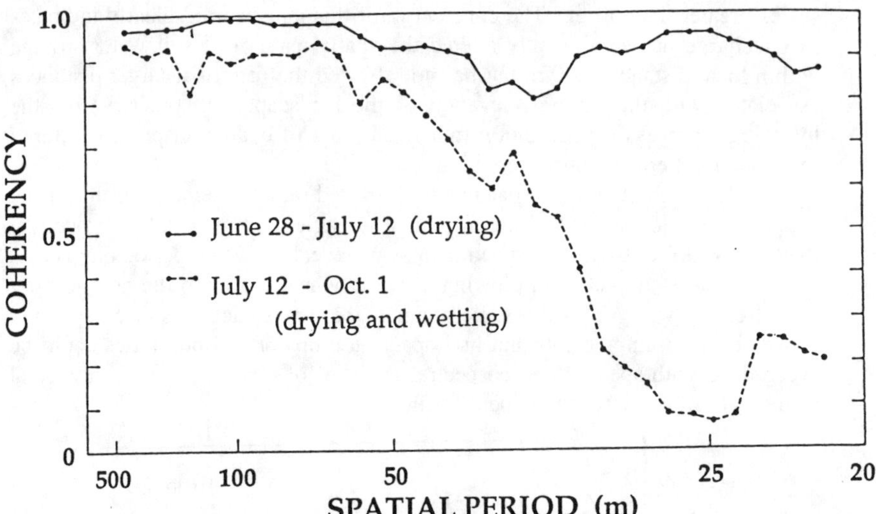

Fig. 9–6. Coherency spectra for the drying period (June 28 to July 12) and for the drying and recharge period (July 12 to October 1).

Diagnosing variations in crop production

Wendroth and Nielsen (1994) monitored some simply observable, integrative parameters in an almond orchard to derive preliminary information for field management. Almond tree trunk circumferences and gross yields were sampled along 300-m long transects in the orchard. Tree trunk circumference was used as a parameter integrating the growing conditions over several yrs, whereas nut yield was considered to integrate the dynamics of growing conditions within one season. The field was irrigated by sprinklers distributed regularly across the field. The farmer knew that the soil in his field was not uniform because it contained "sandy" and "cloddy" areas. Wendroth and Nielsen sought answers to the questions, i) Are the spatial trunk circumference and gross yield patterns affected by soil spatial variability and how are they related to each other?, and ii) Is there any indication that the irrigation system induced spatial variations amongst trees in the orchard?

Tree trunk circumference and gross almond yield are presented in Fig. 9–7 for two transects. The trees in transect 2 (though visually smaller and having smaller trunk circumferences than those in transect 1) had a larger average gross yield. Generally, high yields were present from 150 to 250 m distance in transect 1, while nearly the same transect section (100 to 250 m) in transect 2 showed relatively low yields. Trunk circumference was more strongly variable between neighboring trees in transect 1 compared with that in transect 2.

Trunk circumference and yield series for each of the transects were detrended via first order differencing

$$\Delta x_i = x_i - x_{i-1}. \tag{12}$$

By this procedure, small scale patterns which may have been obscured by an underlying large scale trend could be examined. As one can see from Fig. 9–8 in comparison with Fig. 9–7, the differencing removed the trends from the series. From the power spectra of trunk circumference differences (Fig. 9–9) and gross almond yield differences (Fig. 9–10), variability components at periods according to short cycles (every 2 to 3 trees) are apparent. The high variation between neighbor differences indicated by peaks at a frequency close to 0.5 (corresponding to every other neighboring pair of trees) reflects some kind of competition between trees. Further monitoring in future yrs could show whether these effects of competition are compensated or accumulated over different time intervals. Signals obtained for periods according to every third tree (see Fig. 9–9 for trunk circumference in transect 1, and Fig. 9–10 for yield in transect 2) indicate an inhomogeneous spatial distribution of irrigation water. These short cycle variance components were not apparent from signals obtained in power spectra for the original data.

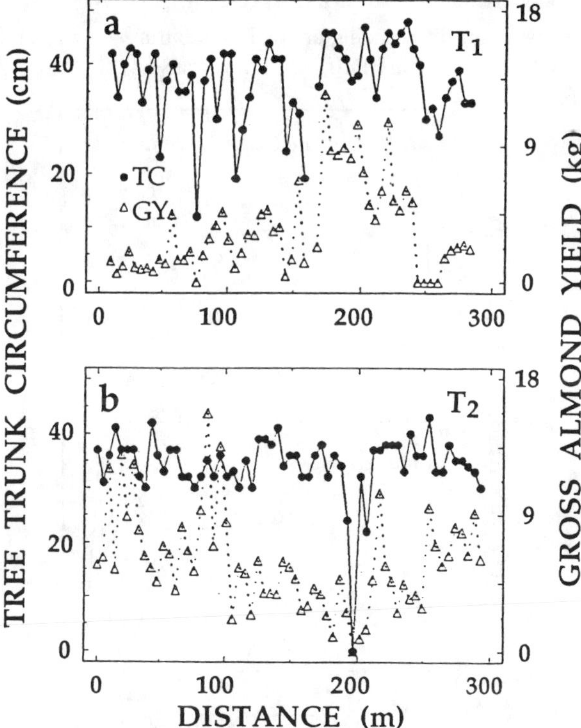

Fig. 9–7. Tree trunk circumference (TC) and gross almond yield (GY) along transects T_1 and T_2 in the south direction.

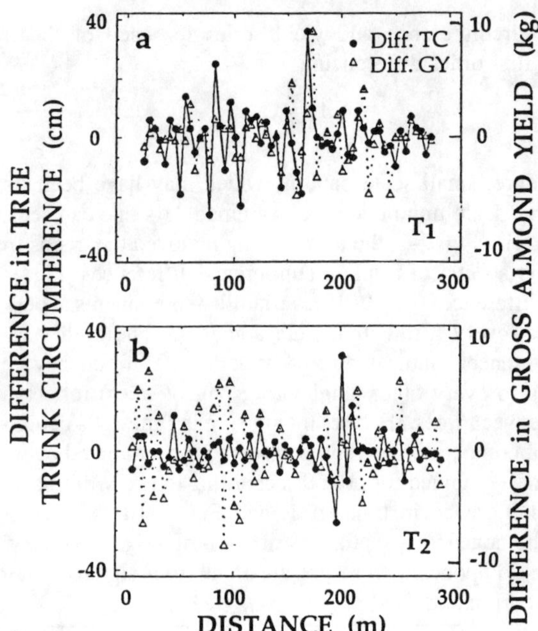

Fig. 9–8. First order differencing applied to tree trunk circumference (Diff. TC) and gross almond yield (Diff. GY) for transects T_1 and T_2 in the south direction.

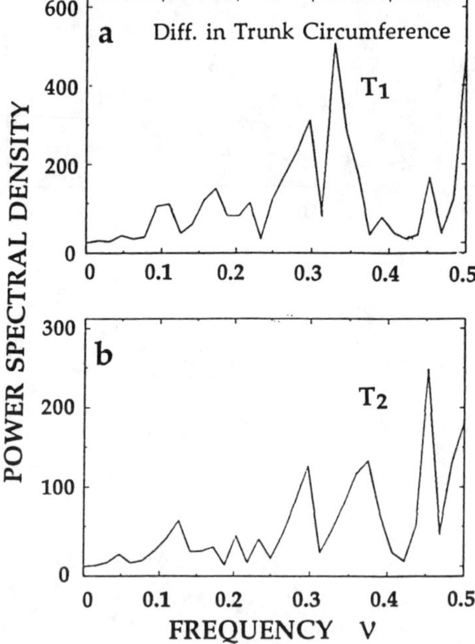

Fig. 9–9. Power spectra for neighbor differences of tree trunk circumference in transects T_1 and T_2.

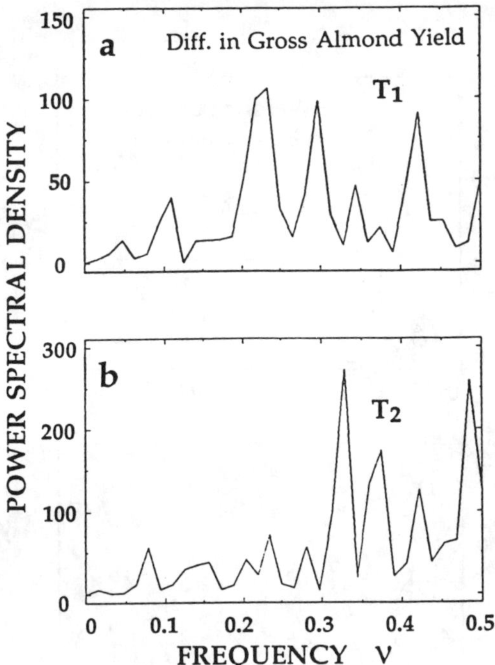

Fig. 9–10. Power spectra for neighbor differences of gross almond yield in transects T_1 and T_2.

Nut weights for the selected sampling locations across the transects together with almond gross yields are shown in Fig. 9–11. In transect 1, yields and nut weights are negatively correlated with r = -0.83 (n = 5). On the other hand, in transect 2 yields and nut weights are positively related with r = +0.58 (n = 8). The different gross yield versus nut weight relationships may indicate different production and reduction physiologies in transect 1 than those in transect 2. Probably, an early reduction of blossoms in transect 2 was not as distinct as it was in transect 1. Higher yields were produced with a high number of mainly medium size nuts in transect 2, whereas high nut weights in transect 1 did not compensate the low number of nuts which perhaps resulted from deficiency and reduction of blossoms earlier in the season.

Wendroth and Nielsen (1994) did not attempt to identify and measure deterministic details about the various on-going processes occurring in the field. They focused on the spatial pattern of some integrative parameters as indicators for spatial variability of growing conditions. The yield versus nut weight relation seemed to be appropriate for detecting spatial differences in growing conditions causing spatial differences in production physiology. For tree crops, identification of yield physiology over several yrs is required, and a larger number of nut-size samples than was used in their investigation. It should also be realized that the annual field operation of tree pruning causes physiological responses reflected in tree trunk circumference and nut yield. One could also

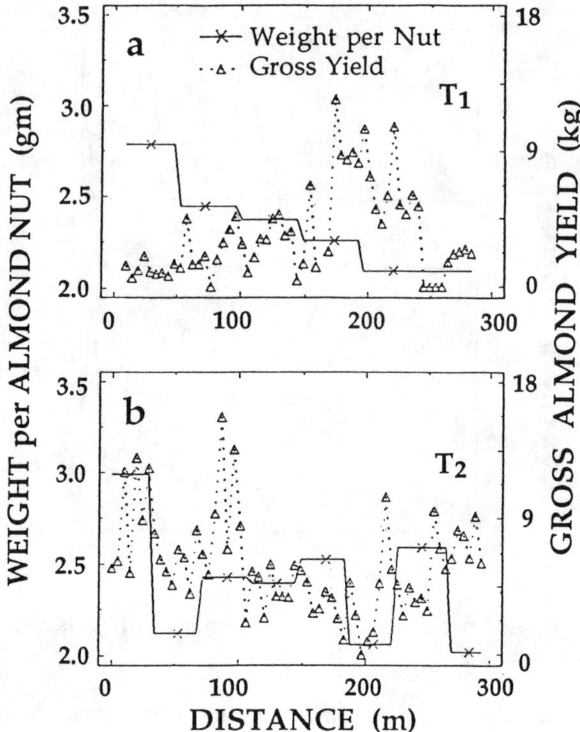

Fig. 9–11. Weight per almond nut and gross almond yield versus distance in the south direction for transects T_1 and T_2.

determine other processes underlying spatial yield variation, e.g. reactions of organic matter (Bhatti et al., 1991) and transport and retention of water and plant nutrients, which can be appropriately analyzed by on-site and site-specific sampling. Detecting various limiting effects at different locations and during different stages of growth would enhance field management decisions.

Spectral and cospectral analyses offer opportunities to examine the impacts of different cultural geometries on crop production as well as those related to environmental attributes. Similarly, spectral and cospectral analyses of microbial or biological responses to repetitive events (e.g. rainfall, irrigation, agrochemical application, cultivation etc.) could be evaluated at a particular location in a field with time rather than location being the variable of integration. The scales of time could range from minutes to yrs depending upon the problem or process being studied.

STATE-SPACE ANALYSIS

Thirty yr ago Kalman (1960) introduced the use of a linear estimation theory with stochastic regression coefficients to filter noisy electrical data in order to extract a clear signal or to predict ahead of the last observation. Kalman filtering is commonplace in electrical and mechanical engineering. Shumway and Stoffer (1982) extended the concept of Kalman filtering to smooth and forecast relatively short and non stationary economic series observed in time. Using this state-space model, a series of observations can be smoothed, missing observations can be estimated, and values outside the domain of observation can be predicted. For the purpose of prediction one can calculate confidence intervals to demonstrate the reliability of the forecasting. Reliability of the prediction will depend on the adequacy of the available model and number of predictions outside the domain of observations. Morkoc et al. (1985) utilized the state-space model to analyze spatial variations of surface soil water content and soil surface temperature as a dynamic bivariate system. Following the suggestion of Shumway and Stoffer (1982), Morkoc et al. described the variables of the observation model as

$$Y_i = M_i Z_i + v_i i = 1,2,3,\cdots \tag{13}$$

where M_i is a known q x p observation matrix which expresses the pattern and converts the unobserved stochastic vector Z_i into q x 1 observed series Y_i, and n_i is the observation noise or measurement error. Equation (13) written in matrix form for the bivariate system of Morkoc et al. is

$$\begin{pmatrix} W_i^0 \\ T_i^0 \end{pmatrix} = \begin{pmatrix} W_i \\ T_i \end{pmatrix} + \begin{pmatrix} V_{Wi} \\ v_{Ti} \end{pmatrix} \tag{14}$$

where W_i^o and T_i^o are the ith soil water content and the ith soil surface temperature observations, respectively. Equation (14) indicates that an observation of either soil variable consists of two parts. The observational noise or measurement errors may be generated either by errors in measuring W_i^o or T_i^o or by ignoring other variables which affect either soil water content or soil surface temperature. For example, soil temperature is not only affected by incoming solar radiation but also by mineralogical composition, microtopography, texture, structure and color of the soil surface. Hence, any omitted variables such as texture or color enter into the state-space system as observational noise or measurement error v_i.

Values of soil water content and surface temperature were modeled as a first-order multivariate process of the form

$$Z_i = \phi Z_{i-1} + \omega_i \tag{15}$$

where ϕ is a p x p matrix of state-space coefficients. Many other functions of multivariate processes equally applicable to soil specific crop management await future investigation. The matrix form of the above equation is

$$\begin{pmatrix} W_i \\ T_i \end{pmatrix} = \begin{pmatrix} \phi_{11} & \phi_{12} \\ \phi_{21} & \phi_{22} \end{pmatrix} \begin{pmatrix} W_{i-1} \\ T_{i-1} \end{pmatrix} + \begin{pmatrix} \omega_{Wi} \\ \omega_{Ti} \end{pmatrix} \tag{16}$$

Values of ϕ_{ij} and the error terms of the above equation can be calculated using recursive procedures (Shumway & Stoffer, 1982).

Details and the results of the state-space analysis of water content and temperature relations of a soil surface (without vegetation and having been nonuniformly sprinkler irrigated) are available (Morkoc et al., 1985). More directly relevant to the topic of this international conference is the application of equations similar to (13) and (15) to improve our understanding and management of biological processes as they occur in the field and to prioritize those soil attributes that contribute the most to crop quality and quantity.

Improved Understanding of Spatial Variability

For crop and soil scientists, particularly those conducting research related to crop production, the recent work of Wendroth et al. (1992) to determine the underlying process in a soil causing variation of plant yield of a nitrogen-fixing crop and a non-fixing crop is an example of the use of the above state-space analysis. The data analyzed by Wendroth et al. was part of a field study carried out at the FAO/IAEA Agricultural Laboratory experimental field in Seibersdorf, Austria. Two transects 96 m long, 1.8 m wide, and separated by a distance of 0.5 m were planted to ryegrass (*Lolium muldflorum*, L.) as the reference crop, and alfalfa (*Medicago sativa*, L.) as the nitrogen-fixing test crop, respectively. After germination, 20 kg/ha ^{15}N-labeled urea fertilizer with 4.8% ^{15}N atom excess were uniformly applied to both transects. Dry matter yield, total crop nitrogen content, ^{15}N atom excess, stone content and total soil nitrogen content were measured in the 63 plots along the transect.

Biological nitrogen fixation ($\%N_{fix}$) of the legume crop was estimated by the isotope dilution method using

$$\%N_{fix} = 100\ [1 - (\%^{15}N(fc) / \%^{15}N(nfc))] \tag{17}$$

where $\%^{15}N(fc)$ and $\%^{15}N(nfc)$ are $\%^{15}N$ atom excess values for the fixing and the nonfixing crop. The fraction of nitrogen in the shoot derived from fertilizer ($\%N_{fert}$) was calculated with

$$\%N_{fert} = 100 \left[\frac{\%\ ^{15}N\ atom\ excess\ in\ plant\ material}{\%\ ^{15}N\ atom\ excess\ in\ fertilizer} \right] \tag{18}$$

Nitrogen from the soil (N_{soil}) was calculated for ryegrass as the residual between N(nfc) and N_{fert} and for alfalfa as the residual between N(fc) and symbiotically

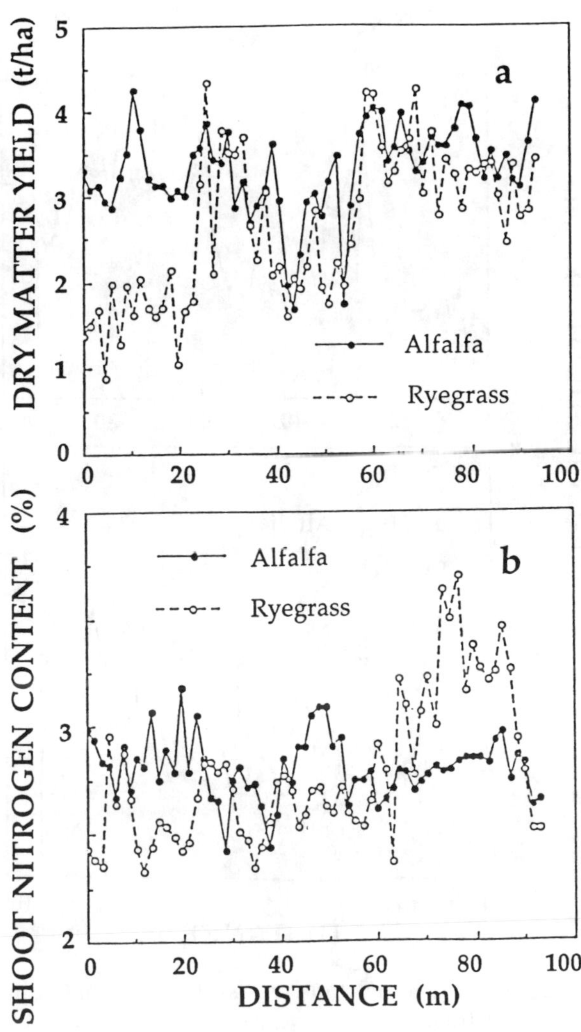

Fig. 9–12. (a.) Dry matter yield of alfalfa and ryegrass versus transect distance; (b.) Shoot nitrogen content of alfalfa and ryegrass versus transect distance.

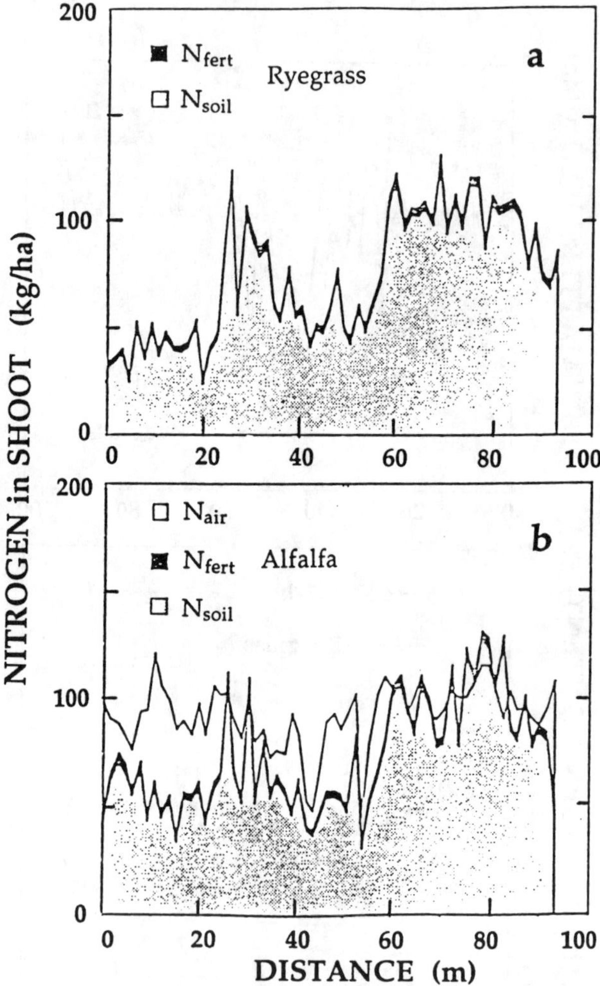

Fig. 9–13. (a.) Nitrogen in shoot of ryegrass derived from fertilizer (N_{fert}) and nitrogen derived from soil (N_{soil}) versus transect distance; (b.) Nitrogen in shoot of alfalfa from N_{fert}, N_{soil} and nitrogen derived from air (N_{air}) versus transect distance.

fixed nitrogen plus N_{fert}. Yields for both crops along the transects were described as multivariate autoregressive systems in state-space models (Shumway, 1988). Details will not be presented here but equations similar to (13) through (16) were used to analyze the data.

Yields of alfalfa and ryegrass along the transects plotted in Fig. 9–12a show a trend of increasing ryegrass yield from 0 to 93 m, which is not the case for alfalfa. Alfalfa yields were higher than ryegrass yields between 0 and 24 m distance but similar to ryegrass yields from 25 to 93 m. In Fig. 9–12b no similarities between both crops in behavior or trend of nitrogen contents can be observed along the transects.

Figures 9–13a and 9–13b show the contribution of different nitrogen sources to amount of nitrogen in shoots of ryegrass and alfalfa. In both crops a very large amount of total nitrogen is derived from soil and a very small amount of total nitrogen is derived from the fertilizer. For alfalfa, large amounts of nitrogen were derived from the atmosphere (N_{air}), especially in those regions where ryegrass yields were small (e.g. 0 to 24 m distance). Where ryegrass yields were large, only slight amounts of N_{air} were calculated for alfalfa. Linear regression of N_{soil} in ryegrass, N_{soil} in alfalfa, and N_{air} in alfalfa versus total nitrogen in soil gave small r^2-values of 0.152, 0.153, and 0.064, respectively.

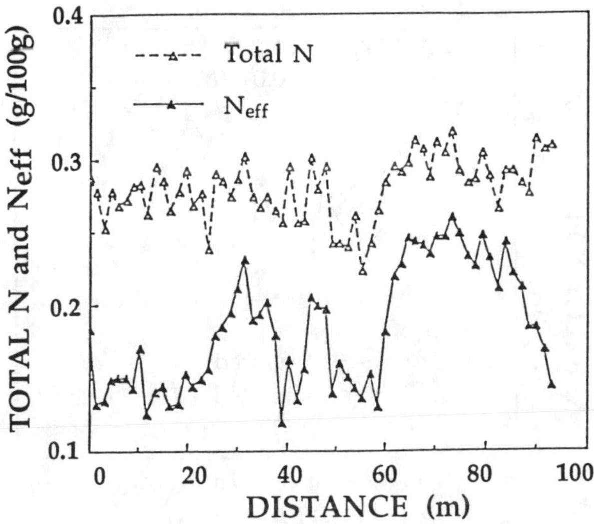

Fig. 9–14. Total soil nitrogen and effective soil nitrogen (N_{eff}) versus transect distance.

As is the custom worldwide, the researchers at Seibersdorf took samples of sieved soil (< 2 mm) for analyses of total soil N. Sieving the soil removed all the small stones having diameters > 2mm. A large volume of the soil profile was occupied by stones where the ryegrass yields were low (0 to 24 m), and a relatively small volume of stones were found where the ryegrass yields were large (60 to 90 m). The "effective" soil nitrogen content (N_{eff}) along the transect taking into account that volume of the profile occupied by stones is given in Fig. 9–14. Using this "effective" soil nitrogen content in the linear regression of N_{soil}in ryegrass, N_{soil} in alfalfa and N_{air} in alfalfa increased the above r^2 values of 0.152, 0.153 and 0.064 to r^2-values of 0.541, 0.470 and 0.393, respectively. A significant improvement, yet the regression against "effective" soil nitrogen content only accounted for 39 to 54% of the variance of nitrogen in the shoot. Why? Because the locations along the transect were ignored.

Variations along the transect are easily analyzed with a state-space model. Inasmuch as most of the nitrogen in the ryegrass was derived from soil, both ryegrass yield and N_{eff} along the transect were used to estimate the yield in a state-space model. The estimated and measured values of ryegrass yield fit the 1/1 line with r^2 equal to 0.884. Similarly, modeling the alfalfa yield with the state-space model using alfalfa yield, N_{eff} and N_{air} as input parameters along the transect gave an r^2 value of 0.972 for the 1/1 line. Figure 9–15 shows the results using the trivariate autoregressive state-space model for alfalfa yield along the transect. The equation in the figure is of the same form as (15).

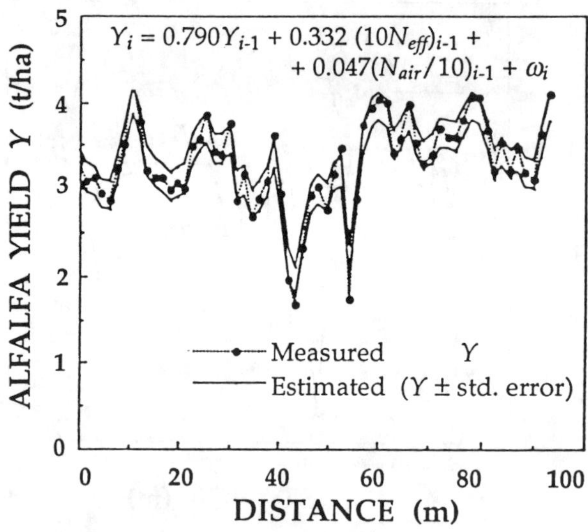

Fig. 9–15. Measured alfalfa yield Y and standard-error intervals of estimated Y along the transect using a first-order, trivariate state-space model.

Modeling these crop yields as bi- and trivariate autoregressive, the state-space model identified N_{eff} and N_{air} as underlying variables in that field soil. In this case, soil conditions causing yield variability were identified with an uncomplicated experimental design of only two transects under the same fertilizer treatment. This study showed that an improved interpretation of experimental results is easily achieved and that nitrogen fixing studies or other crop-soil interactions do not have to be limited to homogeneous soil conditions. In fact, a greater opportunity is potentially feasible when the heterogeneous soil conditions indigenous in a typical agricultural field are considered.

Prioritizing Soil Attributes for Soil Specific Management

Many yield-determining parameters can be easily accommodated in a state-space model. Efficient crop production achieved today stems from our knowledge to identify soil attributes that limit plant growth. That technology includes visual examination of plant parts to identify deficiencies or excesses of specific plant nutrients, field and laboratory tests of soil composition, observations of plant available water, using soil taxonomy to select soils best suited for a particular crop, development of cultivars best able to grow in a particular soil environment, etc. With that technology in mind, it has been rather easy to identify one or two of the most important soil attributes that could be mapped and managed for improving yields across a particular farm. Typically, those attributes were available soil nitrogen or some other macro (Hammond, 1993) or micro plant nutrient (Ascheman, 1993), relative position and slope of the terrain (Moore et al., 1993), and soil organic matter content (Macy, 1993). After identifying the first, second and perhaps third most important attributes, a rather long list of important secondary soil attributes are recognized, but their individual relative importance and priority across a particular farm are difficult to ascertain using typically small plot experimental designs involving complicated arrays of treatments.

If one uses state-space analysis, neither plots nor treatments are necessary to identify the relative priority of a dozen or more soil attributes. Crop plant and soil attributes can be measured on individual plants or clusters of plants and soil samples taken equally spaced along a transect from one end to the other of a farm. Having selected N primary and secondary attributes, the relative magnitudes of the coefficients f_i in (14) would reflect the priority of each attribute ($i = 1, 2, 3, \cdots, N$). With the more important attributes and their ranking identified, spatial patterns of their magnitudes could be estimated and analyzed geostatistically over the entire farm. Or, a more comprehensive small plot treatment experiment could be designed and analyzed to estimate the potential response of the crop to different treatments that alter the magnitudes of selected soil attributes contributing to crop yield.

State-space analysis offers countless opportunities to examine soil processes that control or mediate crop production across a farmer's field. It also can be used just as effectively to examine soil and plant processes along the vertical direction within a soil profile. Although not presented here, details of such state-space opportunities are discussed elsewhere (Nielsen et al., 1994).

GEOSTATISTICS

We note that several ideas regarding regionalized variable analysis were reported last yr at the first workshop on Soil Specific Crop Management (Robert et al., 1993). Most of them were related to geostatistics and opportunities to improve our assessment of soil and crop attributes within specified mapping units or site specific domains. Geostatistics was considered a valuable tool to spatially interpolate between point observations and calculations and to ascertain those interpolated values with a specified error using a minimum number of observations (Burrough, 1991). Mulla (1993) used kriging techniques to develop large scale maps showing spatial patterns in variability of selected soil properties and wheat yield. Mausbach et al. (1993) pointed out that applications of geostatistical techniques to mining are not much different than those to soil specific farming. We agree that the future of geostatistical techniques in soil specific farming is bright, and here we only wish to emphasize two topics.

Covariance Structures

The spatial covariance between soil and crop attributes as well as its transformation as a function of time remains largely unexplored in the agricultural sciences and their application to soil specific farming. Fundamental research is needed to even identify different scales of space and time to best investigate such covariance relationships. We choose soil erosion, a major impediment to sustainable agriculture, to illustrate the need for measuring spatial covariance structures. Soil erosion has long been studied using analysis of variance of observations from replicated plots carefully placed at different positions on the landscape e.g., the knoll, shoulder, back, foot or toe (Ruhe, 1960). On a rolling or hilly landscape, eroded material moves from higher landscape positions downhill and, as a result, regions within the field are "enriched" or "depleted".

Miller et al. (1988) used geostatistical analyses to evaluate the spatial variability of wheat yield and soil properties on complex hills under dry land agriculture conditions. They assumed that the variability of crop and soil properties was the result of erosion. Five 400-m long transects, each 50 m apart, were made over hills ranging in slope from 1 to 30%. Individual site descriptions and surface soil samples were taken every 20 m along each transect. Wheat (*Triticum aestivum* L. cv. Anza) was aerially sown at a rate of 135 kg/ha. A 1-m^2 sample of wheat was harvested at each of the 100 sites and the above ground biomass measured. Although a high correlation between percent slope and surface soil properties is generally expected, this was not the case. Neither were any simple relationships found between wheat growth and soil properties. Standard regression analysis, such as that given in Fig. 9–16a for above ground wheat biomass versus percentage clay in the soil surface, showed no relationship. On the other hand, when the spatial locations of these same observations were included as in Fig. 9–16b, a strong spatial dependency exists. Cross-variograms, calculated using geostatistical methods, showed well-defined dependencies with the average range of influence of both crop and soil properties roughly corresponding to the diameter of the hills. Cross-variograms are proving useful in determining spatial correlations of other field measured properties. For example, Nielsen and Alemi (1989) used cross-variograms to relate cotton yield to nematode infestation in a 40-ha field. They also examined the spatial covariance structure of nematode infestation at different times during the growing season and the competition between different kinds of nematodes within the field. Such relations were not evident using classical regression techniques.

Geostatistical analyses of soil and crop attributes are ideally suited to investigate their spatial covariance structure inasmuch as soil forming factors and erosion do not act uniformly but vary by position on the landscape.

Another Opportunity

Presently, there is a debate emerging amongst soil scientists regarding conventional and continuous methods of soil classification. Conventional classification methods establish a series of subdivisions which place individual soil profile descriptions into a hierarchical scheme such as that of Soil Taxonomy of the USDA (Soil Survey Staff, 1975). Such schemes prevalent in many countries, methodically constructed and developed during the past half century, rely on a modal (most frequently expected) soil profile to represent the land unit. The precision of any prediction using this classification is dependent on the homogeneity of the mapping units and hence on the within unit spatial variance that is typically not ascertained (Trangmar et al., 1985). And without a measure of the spatial variance structure within each mapping unit, little is known regarding the reliability of the modal soil profile to represent the mapping unit (McBratney and Webster, 1981). Data from measuring field soil properties generally exhibit both short and long range variations, are highly irregular and are multivariate. Various methods for obtaining optimal sampling strategies for mapping soil types according to hierarchical schemes based upon spatial distribution functions are available (e.g. Webster, 1973). On the other hand, the hierarchical classification of Soil Taxonomy does not account for the gradational nature of the soil continuum when crossing a soil mapping unit boundary. Hence, proponents of continuous methods of soil classification argue for a soil classification system which accounts for the continuous nature of soils in both vertical and horizontal directions. The method of fuzzy k-means analysis is a technique capable of accounting for the continuous nature of the soil (McBratney and DeGruijter, 1992) and allows an individual soil profile to belong totally, partially or not at all to a particular class (Zadeh, 1965). In fuzzy or continuous classification, soils being classified may well be a member, to a greater or lesser degree, of every class (Bezdek, 1981), and according to this definition, fuzzy classes are a generalization of discrete classifications (Odeh et al., 1990).

For either classification method — conventional or continuous — there remains no consensus as regards the choice of the state variables used to describe the soil profiles or a method to discern them. Here we suggest that those variables initially perceived as being descriptive of a soil (and its chemical, physical and biological properties) be observed and analyzed in a state-space model similar to (14). By examining the relative values of f_i, such a procedure would help identify the choice of the distance-dependent metrics or state variables as well as the development of functions of the state variables that could be used to describe soils within mapping units of either classification method. Covariance and cokriging of such state variables within the domain of a mapping unit of either classification method would be immediately helpful in soil specific crop management.

ENSEMBLE OF FARMS

Having made observations on individual farms as well as managing them using soil specific or site specific criteria, are there opportunities to consider observations of crops and soils that transcend an ensemble of farms? We believe so. The management of a farm is seldom environmentally independent of the neighboring farms. The quality and quantity of surface and subsurface waters, the fluxes of energy, water and gases at the land-atmosphere interface, populations of weeds and insects, distributions of plant and animal pathogens as well as those of soil fauna and higher organisms all are spatially dependent within and between farms. Opportunities exist to identify similarities and dissimilarities of soils being managed under different agricultural systems. What are measures of best management practices, and how can farmers better compare how each of them are succeeding as regards sustainable production and high quality environmental citizenship? We suggest answers are available if analyses are made of state variables of soil across ensembles of farms using split moving window techniques (e.g. Webster, 1973) and fuzzy set analysis (Triantafilis & McBratney, 1993).

Split Moving Window Technique

Recognizing that soil properties manifest both short and long range variations and are multivariate, Webster used a split moving window technique to locate boundaries between soil mapping units. He used the technique along a 6-km long transect of the upper Thames Valley sampled at 20-m intervals to a 1-m depth. Values of 27 properties of the soil profile were ascertained at each location. One property, the clay content is depicted in Fig. 9–16. Expecting a boundary somewhere in the middle of a selected portion of such data, he divided the sampling points in that portion into two groups, those on one side and those on the other. This effect was assessed by comparing the difference between the two groups with the variation within them. The greater the difference between the groups and the less the variation within them the more effective is the division. To identify the best position to divide the portion, he used Mahalanobis' generalized distance D calculated from

$$D^2 = (\overline{x_i} - \overline{x_2})' \, W^{-1} \, (\overline{x_1} - \overline{x_2}) \tag{19}$$

where $\overline{x_i}$ and $\overline{x_2}$ are the mean vectors of the principal component scores of the two halves of the window and **W** is the pooled within-halves variance-covariance matrix. If it is assumed that correlations between the principal components

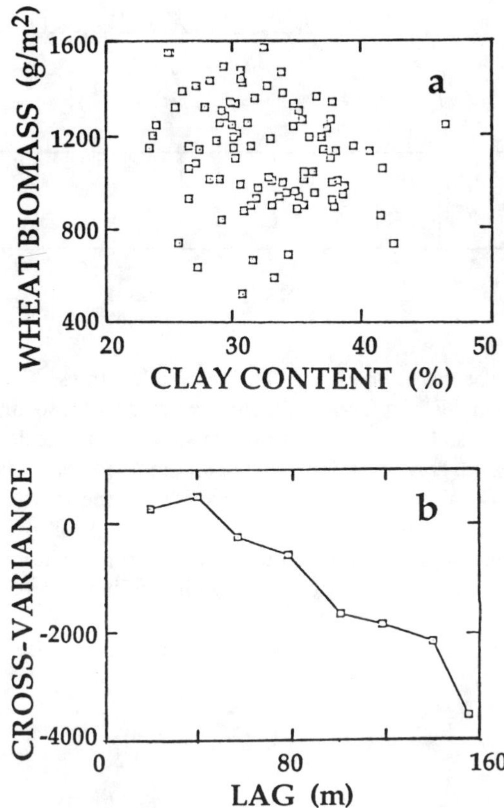

Fig. 9–16. (a.) Wheat biomass versus clay content; (b.) Cross-variogram of wheat biomass versus clay content.

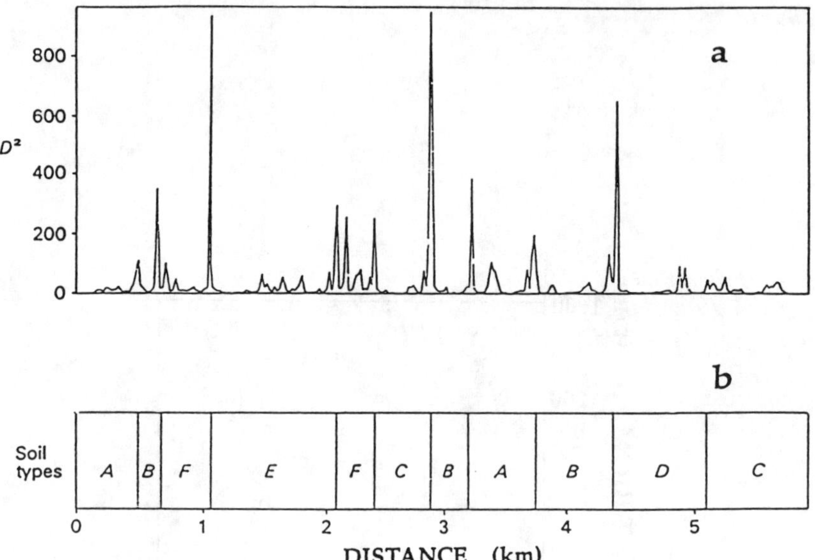

Fig. 9–17. The partition of the Thames Valley transect. (a.) Graph shows Mahalanobis' D^2 for a split window width of 280 m; (b.) Boundaries between soil types: *A* deep brown, more or less calcareous loam over limestone gravel, *B* greyish brown, mottled clay loam over clay, *C* gravelly calcareous brown loam over limestone gravel at about 0.5 m, *D* medium to heavy textured, dark grey, mottled soil over old alluvium, *E* darkgrey or greyish-brown mottled clay, river alluvium and *F* deep brown calcareous medium textured soil with mottling at depths over limestone.

within any portion of the transect are negligible, (19) reduces to

$$D^2 = \sum_{i=1}^{p} \frac{d_i^2}{s_i^2} \tag{20}$$

where d is the difference between means, s^2 the pooled within-halves variance and p the number of components. Values of D^2 for the Thames Valley transect are presented in Fig. 9–17a with soil types recognized in the field judged on soil taxonomy and airphoto evidence presented in Fig. 9–17b. The heights of the D^2 peaks leaves little doubt about the positions of the boundaries. The initial choice of the length of transect portion was based upon autocorrelation analysis inasmuch as the lag distance over which the correlation decays is related to the average distance between boundaries.

It should be obvious that the split moving window technique used to identify the boundaries between soil types *A* through *F* above could just as well have been applied to soils data to identify farms manifesting different levels of soil attributes associated with crop production management practices.

Fuzzy Set Analysis

Within the past decade we began to appreciate the opportunities in soil science afforded by fuzzy set analysis. In a recent report of Triantafilis and McBratney (1993), they applied fuzzy classification methods to interpret soil survey data. One outcome of their study was the identification of parcels of land deemed suitable or unsuitable for a number of land utilizations. We envision fuzzy or continuous classification methods to be increasingly used to assess the impact of soil specific crop management strategies within an ensemble of differently managed farms.

The challenge in front of us is to select the appropriate state variables indicative of soil quality and to include in our consideration the results from dynamic simulation models of crop growth (Bouma and Finke, 1993).

THE FUTURE

Soil specific crop management, the theme of this international workshop, can be achieved here and throughout the world if a bold, new step is taken by each of us to revolutionize our field research. The statistical examples above are only the beginning. Each day new and improved instrumentation and theoretical methodology become available. An example of each is mentioned below.

A Scanning Lidar Probe

A scanning lidar (light detection and ranging) probe based upon the techniques pioneered by Melfi et al. (1969) and Cooney (1970) has recently been built (Eichinger et al. (1994) and used to study land surface-atmosphere processes with high spatial and temporal resolution (Eichinger et al., 1993). This laser-radar system, a noninvasive technique which can measure water vapor concentration rapidly over relatively large areas or volumes with extremely high spatial resolution, is ideal to study the fluxes and associated turbulence of water vapor above a transpiring field crop. The system can be programmed to execute a number of different types of scanning strategies varying from single to multiple laser shots along a single line of sight to three dimensional scans.

An example of the ability of the probe to measure the concentration of water vapor over a transpiring crop of alfalfa is given in Fig. 9–18. The amount of variability within the 100-m by 100-m area of the field is striking considering that the field had been recently irrigated and laser leveled to ensure uniform water coverage. At the time of observation, the alfalfa was 80 cm tall, the field had been irrigated 5 days earlier and the field extended at least 1 km in the predominant upwind direction. The west side of the field is drier. Large water vapor concentrations and fluxes were observed on the north side of the field. The spatial covariance structure of these observations with those of soil water content were coherent throughout the day of measurements. The traditional hypothesis of a homogeneous evaporating surface of a well-irrigated crop is clearly incorrect. These results challenge some of the underlying assumptions of agronomic practices relative to water economy, crop production functions and leaching occurring within uniformly managed fields.

Modeling Mesocale Land-Atmosphere Interactions

The effects of land-use patterns on atmospheric convection, cloud formation and rainfall patterns are now being simulated with comprehensive meteorological models (Pielke et al., 1992). Eastman and Pielke (1994) used the Regional Atmospheric Modeling System (RAMS) with a fine resolution mesh of 3 km to explicitly resolve the microphysics occurring during a case day typical of a stagnant high pressure system centered over southeastern United States. On July 26, 1987 a substantial amount of Rayleigh convection was visible from satellite imagery throughout the morning hours. At roughly 13:00 EST the convection became more organized with a substantial system of cloud cells becoming apparent over the Atlanta, Georgia area. Eastman and Pielke hoped that the RAMS model could simulate this event and derived two different simulations.

For both simulations the model was homogeneously initialized with a sounding taken from Athens, Georgia, and the topography initialized with 30-second data. One simulation utilized the 176 vegetative data classes of the U.S. Geological Survey pared down to 18 classes to initialize the present-day surface land-use characteristics. For the other simulation the vegetation was set at a constant value that represented old growth forest — a condition crudely equivalent to land-use characteristics before civilization. The two simulations were integrated 12 hours forward in time, and their developing cloud fields compared. Figure 9–19 is an elevated view looking from the south with the city of Atlanta in the center. The x,y-plane in each graph manifests the vegetation distribution assumed for each simulation. Notice the large cell that develops in the variable vegetation simulation (Fig. 9–19a) compared to an absence of convective activity for the homogeneous old growth forest. These differences for the two simulations were also abundantly clear for surface accumulated rainfall. The impact of the variable vegetation is readily recognized and is attributed to the organization of mesoscale circulations that are dominant, especially when weak synoptic forcing is present. Opportunities and challenges await those of us who have responsibilities and abilities to develop and manageportions of the landscape that embrace different land-use schemes. We believe that site specific crop management alternatives for individual farms are today's mandate and that regionally specific crop management schemes are those of tomorrow.

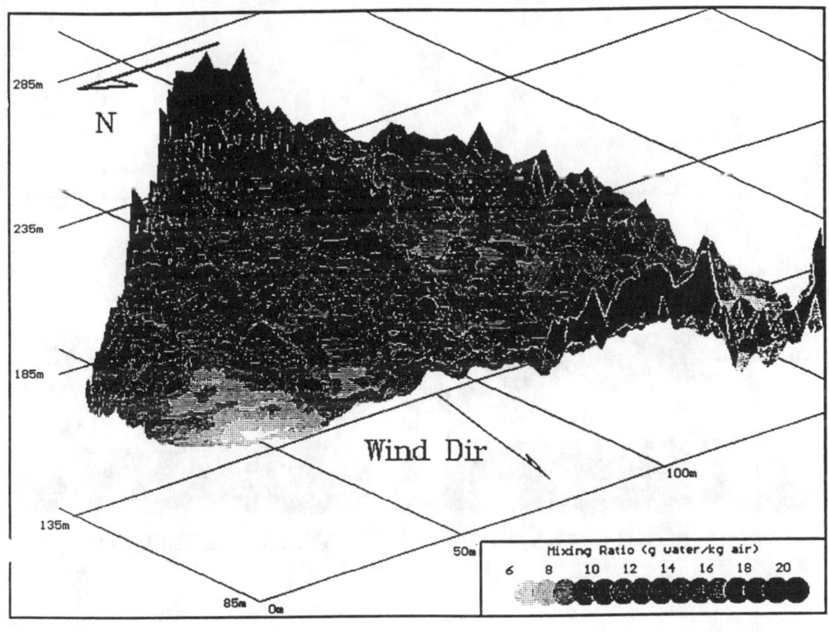

Fig. 9–18. Water vapor concentration at an elevation of 3.2 m over a uniform alfalfa field at 11:18 AM, June 28, 1990. The data were taken by a horizontal lidar scan over a period of 2 min.

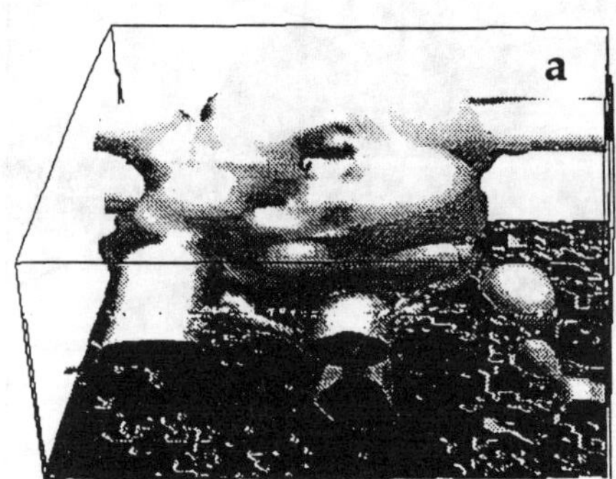

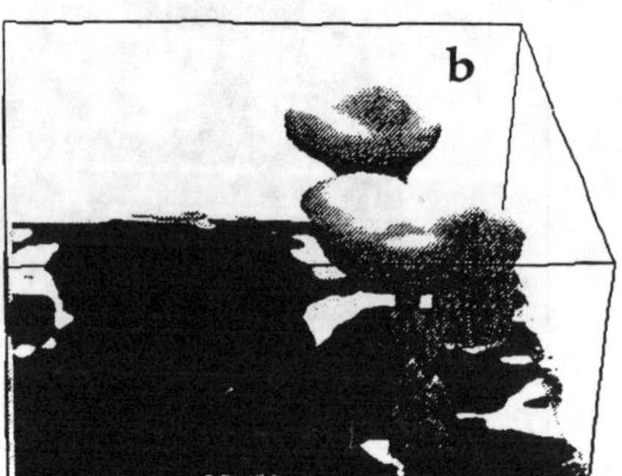

Fig. 9–19. Cloud cells at 2:00 PM simulated with RAMS in the vicinity of Atlanta, Georgia, July 26, 1987. **a.** today's land-use characteristics and **b.** an old growth forestry condition before civilization.

Experimentation on whole fields and ensembles of fields is the future — fewer, much fewer small "representative" plots are envisioned to help us manage the landscape. Sampling designs must be coupled with biological, chemical and physical parameter estimation methods and with crop growth models. Educational reforms of crop and soil science are also required at the B.S. and Ph.D. levels. Mathematics and physics in agricultural curricula should be taught and emphasized at the same level as biology and chemistry. If we can achieve site specific crop management, our next logical goal is to achieve a global population sufficiently educated to manage all of its continental resources without soil exhaustion and with sustained water quality.

REFERENCES

Ascheman, R.E. 1993. Some practical field applications. p. 79–86. *In* P.C. Robert, et al. (ed.) Proceedings of soil specific crop management: A workshop on research and development issues. Special Publ., Am. Soc. Agron., Crop Sci. Soc. Am. and Soil Sci. Soc. Am., Madison, WI

Bezdek, J.C. 1981. Pattern recognition with fuzzy objective function algorithms. Plenum Press, New York.

Bhatti, A.U., D.J. Mulla, and B.E. Frazier. 1991. Estimation of soil properties and wheat yields on complex eroded hills using geostatistics and thematic mapper images. Remote Sens. Environ. 37: 181–191.

Bouma, J., and P.A. Finke. 1993. Origin and nature of soil resource variability. p. 3–14. *In* P.C. Robert, et al. (ed.) Proceedings of Soil Specific Crop Management: A Workshop on Research and Development Issues. SSSA Spec. Publ. No. 28. Soil Sci. Soc. Am., Madison, WI.

Burrough, P.A., A.K. Bregt, M.J. deHeus, and E.G. Kloosterman. 1985. Complementary use of thermal imagery and spectral analysis of soil properties and wheat yields to reveal cyclic patterns in the Flevopolders. J. Soil Sci. 36: 141–152.

Burrough, P.A. 1991. Sampling designs for quantifying map unit composition. p. 89–127. *In* M.J. Mausbach, and L.P. Wilding (ed.) Spatial variabilities of soils and landforms. SSSA Spec. Publ. No. 28. Soil Sci. Soc. Am., Madison, WI.

Cooney, J.A. 1970. Remote measurements of atmospheric water vapor profiles using the Raman component of laser backscatter. J. Applied Meteorol., p. 182.

Clark, I. 1979. Practical geostatistics. Applied Science Pub. Ltd., London.

Davis, J.C. 1973. Statistics and data analysis in Geology. John Wiley & Sons, Inc., New York.

Eastman, J.L., and R.A. Pielke. 1994. SERON Modeling Study Report. Dep. Atmos. Sci., Colo. State Univ., Ft. Collins, CO, personal communication, March 10, 1994.

Eichinger, W.E., D.I. Cooper, D.E. Hof, D.B. Holtkamp, R.R. Karl, C.R. Quick, and J.J. Tiee. 1994. Development and application of a scanning, solar-blind, water Raman-lidar. Appl. Opt. (in press).

Eichinger, W.E., D.I. Cooper, M. Parlange, and G. Katul. 1993. The application of a scanning, water Raman-lidar as a probe of the atmospheric boundary layer. IEEE Trans. Geosci. Remote Sens. 31:70–79.

Gelhar, L.W., P.J. Wierenga, C.J. Duffy, K.R. Rehfeldt, R.B. Senn, M. Simonett, T.-C. Yeh, A.L. Gutjahr, W.R. Strong, and A. Bustamante. 1980. Irrigation return flow studies at San Acacia, New Mexico: Monitoring modeling and variability. Technical Progress Report No. H-3, New Mexico Inst. of Mining and Tech., Socorro.

Hammond, M.W. 1993. Cost analysis of variable fertility management of phosphorus and potassium for potato production in central Washington. *In* P.C. Robert, et al. (ed.) Proceedings of Soil Specific Crop Management: A Workshop on Research and Development Issues. SSSA Spec. Publ. 28. Soil Sci. Soc. Am., Madison, WI.

Journel, A.G., and Ch.J. Huijbregts. 1978. Mining Geostatistics. Acad. Press. New York.

Kachanoski, R.G., and E. de Jong. 1988. Scale dependence and the temporal persistence of spatial patterns of soil water storage. Water Resour. Res. 24:85–91.

Kalman, R.E. 1960. A new approach to linear filtering and predicting problems. Trans. ASME of Basic Eng. 8:35–45.

Macy, T.S. 1993. Macy Farms - Site specific experiences. p. 229–244. *In* P.C. Robert, et al. (ed.) Proceedings of Soil Specific Crop Management: A Workshop on Research and Development Issues. SSSA Spec. Publ. 28. Soil Sci. Soc. Am., Madison, WI.

Mausbach, M.J., D.J. Lytle and L.D. Spivey. 1993. Application of soil survey information to soil specific farming. p. 57–68. *In* P.C. Robert, et al. (ed.) Proceedings of Soil Specific Crop Management: A Workshop on Research and Development Issues. SSSA Spec. Publ. Soil Sci. Soc. Am., Madison, WI.

McBratney, A.B., and R. Webster. 1981. Spatial dependence and classification of soil along a transect in northeast Scotland. Geoderma 26: 63–82.

McBratney, A.B., and J.J. DeGruijter. 1992. A continuum approach to soil classification and mapping: Classification by modified fuzzy k-means with extragrades. J. Soil Sci. 43: 159–175.

Melfi, S., J.D. Lawrence, and M.P. McCormick. 1969. Observation of raman scattering by water-vapor in the atmosphere. Appl. Phys. Letters, p. 295.

Miller, M.P., M.J. Singer, and D.R. Nielsen. 1988. Spatial variability of wheat yield and soil properties on complex hills. Soil Sci. Soc. Am. J. 52:1133–1141.

Moore, I.D., P.E. Gessler, G.A. Nielsen and G.A. Peterson. 1993. Terrain analysis for soil specific crop management. p. 27-55. *In* P.C. Robert, et al. (ed.) Proceedings of Soil Specific Crop Management: A Workshop on Research and Development Issues. SSSA Spec. Publ. Soil Sci. Soc. Am., Madison, WI.

Morkoc, F., J.W. Biggar, D.R. Nielsen, and D.E. Rolston. 1985. Analysis of soil water content and temperature using state-space approach. Soil Sci.

Soc. Am. J. 49(4):798–803.

Mulla, D.J. 1993. Mapping and managing spatial patterns in soil fertility and crop yield. *In* P.C. Robert, et al. (ed.) Proceedings of Soil Specific Crop Management: A Workshop on Research and Development Issues. SSSA Spec. Publ. Soil Sci. Soc. Am., Madison, WI.

Nielsen, D.R., and H. Alemi. 1989. Statistical opportunities for analyzing spatial and temporal heterogeneity of field soils. p. 261–272. *In* M. Clarholm, and L. Bergström (ed.) Ecology of arable land - perspectives and challenges. Proc. of an International Symposium, 9-12 June 1987, Swedish University of Agricultural Sciences, Uppsala, Sweden. Kluwer Academic Publishers, Dordrecht.

Nielsen, D.R., G.G. Katul, Ole Wendroth, M.V. Folegatti, and M.B. Parlange. 1994. State-space approaches to estimate soil physical properties from field measurements. Trans. Int'l Congress Soil Sci., 15th (in press).

Odeh, I.O.A., A.B. McBratney, and D.J. Chittleborough. 1990. Design of optimal sample spacings for mapping soil using fuzzy k-means and regionalized variable theory. Geoderma 47: 93–122.

Pielke, R.A., W.R. Cotton, R.L. Walko, C.J. Tremback, M.E. Nicholls, M.D. Moran, D.A. Wesley, T.J. Lee, and J.H. Copeland. 1992. A comprehensive meteorological modeling system—RAMS. Meteor. Atmos. Phys.49: 69–91.

Robert, P.C., R.H. Rust and W.E. Larson. 1993. Proceedings of Soil Specific Crop Management: A Workshop on Research and Development Issues. SSSA Spec. Publ. Soil Sci. Soc. Am., Madison, WI.

Ruhe, R.V. 1960. Elements of the soil landscape. Trans. Int'l Congress Soil Sci., 7th (4): 165–170.

Samra, J.S., S.S. Grewal, and G. Singh. 1993. Modeling competition of paired columns of Eucalyptus on interplanted grass. Agroforestry Systems 21:177–190.

Shumway, R.H., and D.S. Stoffer. 1982. An approach to time series smoothing and forecasting using the EM algorithm. J. Time Ser. Anal. 3:253–264.

Shumway, R.H. 1988. Applied statistical time series analysis. Prentice-Hall, Englewood Cliffs, New Jersey.

Soil Survey Staff. 1975 Soil Taxonomy: A basic system of soil classification for making and interpreting soil surveys. USDA Agric. Handb. 436. U.S. Gov. Print. Office, Washington, DC.

Trangmar, B.B., R.S. Yost, and G. Uehara. 1985. Applications of geostatistics to spatial studies of soil properties. Advances in Agron. 38:45–94.

Triantafilis, J., and A.B. McBratney. 1993. Application of continuous methods of soil classification and land suitability assessment in the lower Namoi Valley. CSIRO Div. Soils Divisional Rep. No. 121, CSIRO, Australia. pp. 172.

Webster, R. 1973. Automatic soil-boundary location from transect data. Math. Geol. 5:27–37.

Webster, R., and T.M. Burgess. 1983. Spatial variation in soil and the role of kriging. Agric. Water Manage. 6:111–122.

Wendroth, O., A.M. Al-Omran, C. Kirda, K. Reichardt, and D.R. Nielsen. 1992. State-space approach to spatial variability of crop yield. Soil Sci. Soc. Am. J. 56:801–807.

Wendroth, O., and D.R. Nielsen. 1994. Opportunities of on-site monitoring in an almond orchard for sustainable field management. (submitted)

Zadeh, L.A. 1965. Fuzzy sets. Information and Control 8:338–353.

10 Yield and Nutrient Variability in Glacial Soils of Michigan

F. J. Pierce
D. D. Warncke
M. W. Everett

Dep. Crop and Soil Sciences
Michigan State University
East Lansing, Michigan

Site specific management (SSM) may have potential for production agriculture but evaluation of SSM is limited and results are mixed. This study assessed the variability of pH, P, K, Ca, Mg and corn (Zea mays L.) yield in fields at Adrian, Durand, and Plainwell, Michigan, sampled at 30.5m grid spacings, evaluated the potential for site specific fertilizer management, and calibrated the Harvest Yield® sensor. Each field averaged optimum pH and medium to high soil tests. Soil fertility within each field, however, ranged from deficient to excessive for most measured parameters. Fertilizer recommendations per hectare were similar for uniform versus SSM. Estimated cost of overfertilization by grid using uniform application was only a few dollars per hectare. Estimated yield loss from underfertilization by uniform fertilization, however, could exceed 2 Mg corn ha^{-1} at Durand. Grid spacing had little effect on field average fertility or fertilizer recommendations per hectare. Soils in no-tillage management had highly stratified fertility, with no consistent relation between the 0 to 5 cm and the 5 to 20 cm depths. Corn yield was variable but showed little spatial dependence, and was generally not correlated to soil fertility. Soil fertility was spatially dependent but spatial variation of the surface and subsurface were quite different for soils in no-tillage management. The absolute difference between the Harvest Yield® sensor and weigh wagon yields averaged over all transects was less than 5% at any location, although the grain flow calibration coefficient varied by combine. The site specific management of managed inputs and yield mapping at these sites appear to have potential benefits to corn production.

Site specific management (SSM) in agricultural production involves the variable management of soils, crops, and pests according to localized conditions within a field (Carr et al., 1991; Larson and Robert, 1991; Schueller, 1992). The basic steps in SSM include the quantitative assessment of variability within a field and the location specific management of that variability (Robert et al., 1993). The successful application of SSM requires the integration of a myriad of component technologies into spatially-variable crop production systems, many of which are only now emerging (Am. Soc. Agric. Eng., 1991; Schueller, 1992).

While many advances have been made in the last decade in technologies to quantify spatial variation and to spatially control application of crop production inputs, the evaluation of SSM has been limited. Field studies have evaluated only certain aspects of SSM over short duration, such as variable rate fertilizer application, and have generally evaluated SSM in terms of yield response or profitability to variable inputs with mixed results (Forcella, 1993; Hammond, 1993; Wibawa et al., 1993; Wollenhaupt and Buchholz, 1993; Wollenhaupt et al., 1993). While the impacts of SSM on yield and profitability are important, SSM may have significant environmental and social impacts, but these are difficult to assess and are not documented.

We are in the first phase of a 3-yr study, initiated in spring of 1993, to evaluate SSM soil and crop conditions in Michigan. This paper provides an assessment of the variability of pH, P, K, Ca, Mg and corn (Zea mays L.) yield in fields at Adrian, Durand, and Plainwell, Michigan, sampled at 30.5m grid spacings, evaluated the potential for site specific fertilizer management, and calibrated the Harvest Yield® sensor.

MATERIALS AND METHODS

Evaluation Sites

Three farm fields in southern Michigan were selected for study. Each field was cropped to corn in 1993. The 'Adrian' site is a 10 ha field located northeast of Adrian, Michigan (46°47'30"N 84°8'30"W). The field is managed in a corn-soybean (Glycine max L.)-wheat (Triticum aestivum L.) rotation under no-tillage. The soils were formed in loamy glacial till on a till plain and consist of a Blount loam (3 to 7% slope, Aeric Ochraqualf, fine, illitic, mesic) on the southern 65% of the field and a Morley loam (3 to 15% slope, Typic Hapludalf, fine, illitic, mesic) on north end of the field. Drainage is poor in both soils and the Morley soil is erosive. The 'Durand' site consists of a 16 ha section of a larger field located 6 km south of Durand, Michigan (47°47'30"N 83°52'30"W). The field is managed in a corn-soybean rotation under no-tillage, although corn was grown in 1992 and 1993 to accommodate our study. This 16 ha field contains 10 soil map units, with the dominant soils being Udollic Ochraqualfs, (fine-loamy, mixed, mesic) including the Conover loam (35%), the Metamora sandy loam (14%), and the Macomb loam (10%), mixed with the Walkill loam (Thapto-Histic Fluvaquents, fine-loamy, mixed, nonacid, mesic), and the Breckenridge sandy loam (Mollic Haplaquents, coarse-loamy, mixed, nonacid, frigid) (10% each) and other minor soil map units. The 'Plainwell' site is a 22 ha field located 4 km southeast of Plainwell, Michigan (46°45'35"N 85°34'W), in the floodplain of the Kalamazoo River. The field is managed in continuous corn production under irrigation, although no irrigation was applied in 1993 due to the occurrence of sufficient rainfall. The field is fall chisel plowed. The soils include the Coloma loamy sand (50%) (Alfic Udipsamment, mixed, mesic), the Brady sandy loam (35%)(Aquollic Hapludalf, coarse-loamy, mixed, mesic), and the Bronson sandy loam (15%) (Aquic Hapludalf, coarse-loamy, mixed, mesic) formed in glacial outwash.

Soil Sampling and Analysis

Each field was sampled on a 30.5 m grid in May, 1993. At each grid point, soil samples were taken at the grid intersection and at four locations 1.5 m distance from the grid intersection in a cross pattern. Soils at the Adrian and Durand sites were sampled at 0 to 5 and 5 to 20 cm depth, since these sites were under no-tillage management. The 0 to 5 cm depth was sampled twice at each subsample position. The Plainwell site was sampled at 0 to 20 cm depth since it was fall chisel plowed annually. Soil samples at each site were composited, air-dried, ground, and analyzed at the Michigan State University (MSU) soil testing laboratory for pH, P, K, Ca and Mg.

Fertilizer Recommendations and Potential Yield Response

Fertilizer recommendations were calculated for each grid point and for the field average using MSU recommendations for corn (Christenson et al., 1992) for a yield goal of 8.15 Mg ha^{-1} for Adrian, 8.78 Mg ha^{-1} for Durand, and 11.29 Mg ha^{-1} for Plainwell. The potential corn yield response to site specific management (YR_{SSM}) was calculated as the difference between the yield goal (YG) and yield estimated (Ye) from the MSU fertilizer response curve.

Corn Harvest

Corn grain yield, grain moisture, and harvest plant population were measured in November, 1993, at a subset of 30.5 m grid points from two 9.14 m corn rows centered on the grid intersections. Grain yields are reported at 15.5% moisture content.

Yield Mapping Calibration

Each farmer cooperator's combine was equipped with a grain flow sensor (MicroTrak Harvest Yield monitor), a grain moisture sensor, and an ultrasonic speed sensor (MicroTrak). The signal from each device was captured using the FARMGIS control and data acquisition system (Farmer's Software Association, Ft. Collins, CO). This system consisted of a differential global positioning system (GPS), analog and digital signal acquisition capabilities, and a computer. The differential GPS correction was generated from a NAVSTAR GPS base station established at each site and broadcast to the FARMGIS system using an FM radio. The GPS data was recorded along with grain flow, grain moisture, and speed on a one s interval. Each field was divided in swaths, equivalent to one pass of the combine, and numerically labeled from end to end of the field. During each pass of the combine, the time of entering and exiting each swath was recorded by the operator to exactly position each swath in the GPS record. This swath registration greatly assisted in the construction of the harvest yield map.

The procedure to calibrate the grain flow sensor was as follows. The grain harvested in the swath adjacent to the swath containing the harvested grid samples was weighed in a weigh wagon at the end of the field. The weigh wagon grain yield was corrected to 15.5% grain moisture using a grain moisture measurement on a subsample from the weigh wagon. The total grain weight per swath and grain yield per ha were obtained from the grain flow monitor in the cab and recorded. The harvest grid samples were averaged for each swath and compared to the grain flow and weigh wagon measurements.

Statistical Analyses

Soil test and yield data were summarized on a field basis using standard statistical analyses. Each parameter was subjected to semivariance analysis for each of five isotropic and five anisotropic models for defining semivariograms - linear, linear/sill, spherical, exponential, and gaussian or hyperbolic. Geostatistical analyses were performed using a GS+: Version 2.1, a geostatistical analysis package for the environmental sciences (Gamma Design Software, 1993).

Results and Discussion

Soil fertility for the 30.5 m grid spacing was highly spatially variable in all three fields and very stratified in the Adrian and Durand fields which were in no-tillage management (Table 10–1). Field average pH at the three sites was 6.0 to 6.7 but pH ranged from 2.0 to 3.1 units within the field. Field average soil test phosphorus (P) was adequate to very high for Durand and Plainwell and low to medium for Adrian. For Plainwell, P ranged by a factor of 4 but was high for all grid sampling points. Phosphorus ranged from deficient to excessive for Durand and for the 0 to 5 cm depth for Adrian. The 5 to 20 cm depth for Adrian was very low to medium levels for P. There was little correspondence between P in the 0 to 5 cm and 5 to 20 cm depths at either location. Potassium (K) averaged high to very high but ranged from deficient to extremely high for all locations. K was highly stratified in the Adrian and Durand fields. Calcium (Ca) and magnesium (Mg) varied considerably, with Mg at deficient levels within a portion of the Durand field. Correlations between soil fertility attributes were low regardless of location (data not shown).

The effect of grid spacing on field average soil fertility was minimal at all locations. Even though the number of samples per unit area decreases in proportion to the inverse square of the grid spacing, the estimate of the field average for each soil fertility attribute remain approximately the same, as illustrated for Durand in Table 10–2.

Fertilizer recommendations for P and K on a per ha basis were similar when calculated for the field average and for each grid spacing for all three sites (Table 10–3). Thus, fertilizer usage per ha would remain the same or be slightly higher in SSM, but the distribution of fertilizer within the field would be spatially different. Fertilizer recommendations for the 5 to 20 cm depth were higher

Table 10–1. Statistical summary of soil test values for farm fields in three Michigan locations sampled in 1993 at 30.5 m grid spacing.

Parameter	Location	pH	P	K	Ca	Mg	pH	P	K	Ca	Mg
		Depth									
	Adrian (n=74)	0-5 cm					5-20 cm				
		mg kg^{-1}					mg kg^{-1}				
Mean		6.5	23	210	1185	241	6.6	8	123	1316	276
Std. Dev.		0.9	9	71	268	50	0.5	4	46	309	73
Maximum		8.0	70	393	1684	377	8.0	22	298	2190	552
Minimum		5.2	11	97	600	124	5.1	3	72	655	96
	Durand (n=165)	0-5 cm					5-20 cm				
Mean		6.6	50	173	994	143	6.0	35	97	840	143
Std. Dev.		0.9	25	43	605	88	0.7	21	31	537	116
Maximum		7.8	150	293	3350	492	8.0	100	208	2358	612
Minimum		5.2	10	75	115	14	4.9	7	31	153	16
	Plainwell (n=174)	0-20 cm									
Mean		6.7	124	121	491	133					
Std. Dev.		0.4	32	359	119	24					
Maximum		7.7	240	313	1053	209					
Minimum		5.7	62	71	267	84					

Table 10–2. Effect of grid size on field mean soil test for a farm field in Durand, Michigan, in 1993.

Grid Size	pH	P	K	Ca	Mg
m		----- mg kg^{-1}----			
		0-5 cm			
30.5	6.6	50	172	994	142
61.0	6.6	45	176	947	140
91.5	6.5	55	178	1007	141
122.0	6.8	47	156	1033	150
		0-20 cm			
30.5	6.6	43	135	922	143
61.0	6.6	33	99	851	140
91.5	6.5	39	100	758	127
122.0	6.8	36	100	803	159

higher than those for the surface 5 cm due to stratification of P and K in no-tillage. Fertilizer recommendations for no-tillage are based on the weighted average of stratified soil sample. Recommendations for the 0 to 20 cm combined depth would be the weighted average of the value for the 0 to 5 and 5 to 20 cm depths.

The value of site specific fertilizer application was calculated in terms of (1) the amount of fertilizer over-applied, calculated as the sum of the positive differences between uniform application and SSM; and (2) the potential yield increase associated with applying fertilizer by location in addition to that applied under a uniform application. The yield increase calculation assumes that areas with soil tests lower than the field average would increase yield in proportion to the fertilizer response curve. This assumption seems reasonable if the cause of lower soil test levels was higher than average crop removal. If fertilizers were applied rather uniformly to the field and erosion was not responsible for soil test differences, then soil test levels maps may be inversely proportional to yield. These calculations were done for Durand for the 30.5 m grid spacing for the 0 to 5 cm, 5 to 20 cm and the combined 0 to 20 cm depths for P and K fertilization (Table 10–4). The average over-application of P and K was 18 and 16 kg ha^{-1}, respectively, for the 0 to 20 cm depth soil test. This amounts to only few dollars per ha. The estimated yield response to proper fertilization of each 30.5 m grid location was 1.5 and 1.3 Mg corn ha^{-1} for P and K, respectively. This value increases to 2.3 Mg ha^{-1} if the yield of the maximum of fertilizing either P or K at each site is selected. At $40 per Mg corn, the potential for corn yield increase is considerably greater than fertilizer savings.

Table 10–3. Comparison of P and K fertilizer recommendations for uniform versus site specific management relative to grid sampling intensity for three farm fields in Michigan.

				Grid Size (m)			
Location	Fertilizer	Depth	Uniform	30.5	61.0	91.5	122.0
		cm	kg ha^{-1}				
Adrian	P_2O_5	0 - 5	58	63	58	53	49
		5 - 20	103	83	105	93	99
	K_2O	0 - 5	0	6	4	6	0
		5 - 20	45	81	67	110	49
Durand	P_2O_5	0 - 5	0	18	23	21	29
		5 - 20	19	37	41	38	42
	K_2O	0 - 5	0	8	14	23	28
		5 - 20	83	100	100	97	100
Plainwell	P_2O_5	0 - 20	83	32	27	36	37
	K_2O	0 - 20	83	24	25	20	48

Table 10–4. Potential corn yield response to site specific P and K fertilizer management for a 30.5 m grid spacing on a farm field in Durand, Michigan. Potential corn yield was calculated as the difference between yield goal (8.78 Mg ha^{-1}) and corn yield predicted from Michigan State University's fertilizer recommendation using soil test plus fertilizer recommendation based on field average soil test.

	Difference in Fertilizer Application Field - SSM)		Corn Yield Increase Due to SSM Application		Maximum of Corn Yield Increase Due to P or K SSM Fertilization
Parameter	P_2O_5	K_2O	P_2O_5	K_2O	
	-- kg ha^{-1} --		------------------ Mg ha^{-1} ------------------		
				0 - 5 cm depth	
Mean	17	8	1.2	0.3	1.4
Std. Dev.	25	23	1.7	1.0	1.8
Maximum	92	155	6.2	6.7	6.7
Minimum	0	0	0	0	0
				5 - 20 cm depth	
Mean	16	7	1.6	1.5	2.3
Std. Dev.	33	73	1.7	2.0	2.0
Maximum	21	93	5.4	8.0	8.0
Minimum	0	0	0	0	0
				0 - 20 cm depth	
Mean	18	16	1.5	1.3	2.3
Std. Dev.	30	59	1.7	1.9	2.0
Maximum	79	243	5.4	9.1	9.1
Minimum	0	0	0	0	0

Table 10–5. Statistical summary of corn yield, grain moisture and plant population for farm fields in three Michigan locations sampled in 1993 at 30.5 m grid spacing.

	Corn Grain Yield	Grain Moisture	Harvest Population
	Mg ha^{-1}	g kg^{-1}	plants ha^{-1}
		Adrian	
Mean	5.6	376	39,460
Std. Dev.	1.4	47	5,148
Maximum	9.4	497	49,513
Minimum	1.8	299	26,550
		Durand	
Mean	7.0	269	58,996
Std. Dev.	1.0	23	6,704
Maximum	9.6	337	74,367
Minimum	4.6	229	40,119
		Plainwell	
Mean	9.6	244	64,917
Std. Dev.	0.8	10	3,340
Maximum	11.3	279	71,040
Minimum	7.4	224	55,254

Table 10–6. Pearson correlation coefficient (R) between corn yield and soil test levels for three farm fields in Michigan in 1993.

Soil Test	Adrian	Durand	Plainwell
pH	0.22	0.07	0.43‡
P	-0.09	-0.06	0.20
K	-0.12	-0.07	-0.39†
Ca	0.11	0.13	0.07
Mg	0.04	0.02	0.06
Grain Moisture	-0.70†	-0.68†	-0.65‡
Plant Population	0.41†	0.61†	0.55‡

†R is significant at p=0.05
‡R is significant at p=0.01

Variation in corn yield, as determined at grid intersections, was quite large at Adrian (CV=25%), moderate at Durand (CS=14%), and low at Plainwell (CV=8%) (Table 10–5). Lower yields at each location was due, in part, to lower populations and were associated with higher grain moisture, both of which may be associated with plant stress at these specific locations (Table 10–6). There was little correlation of yield with soil fertility, with the exception of pH and K for Plainwell (Table 10–6). Plainwell soils are sandy and the site has an area of low pH at the southwest corner of the field that was cleared of forest within past 20 yr.

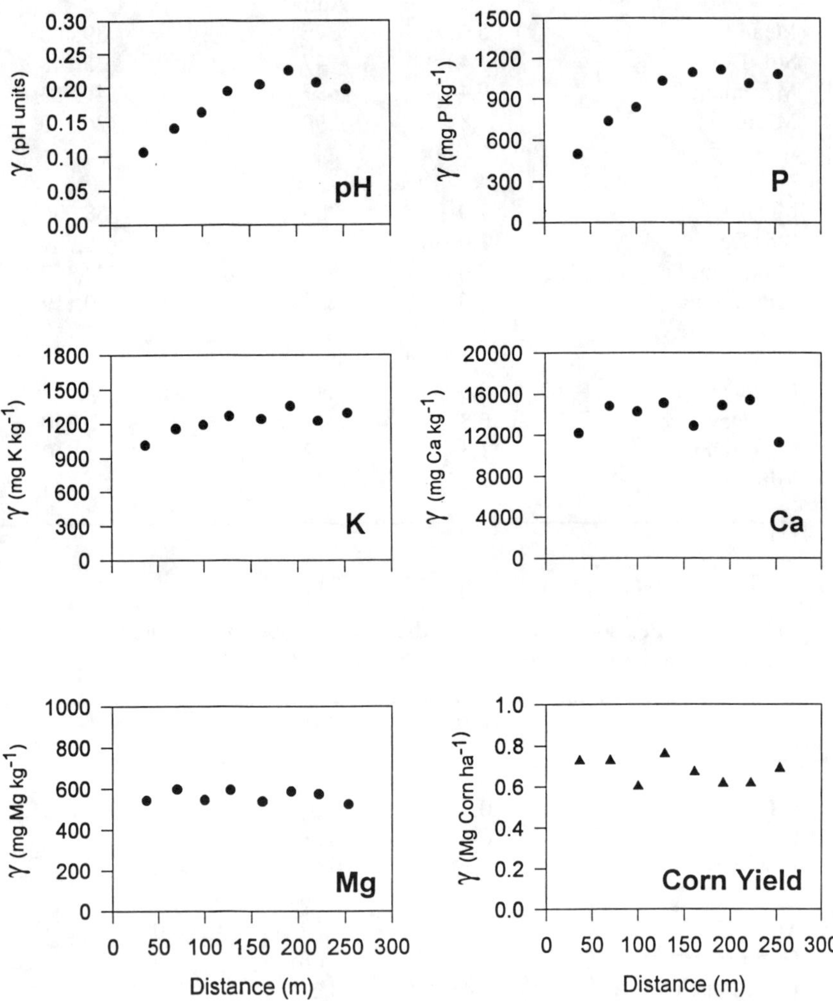

Fig. 10–1. Semivariograms for soil fertility of the 0 to 20 cm soil depth and 1993 corn grain yield for the Plainwell, Michigan, field.

Geostatistical analysis revealed a range of spatial dependence of soil tests and corn performance for the three locations. For Plainwell, pH and P were highly spatially dependent (Fig. 10–1). The spherical model fit the semivariance well, giving a range of 190 and 172 m and a nugget variance of 30 and 21% for pH and P, respectively (Table 10–7). The spherical model fit K reasonably well, but the nugget variance was high (69%) and the range was 157 m. There was little spatial dependence of Ca, Mg or corn yield. For Durand, spatial dependence was generally greater in the 5 to 20 cm depth than the 0 to 5 cm depth, with the exception of pH (Fig. 10–2). The linear model fit pH, P, Ca, and Mg at both depths, giving a range of 294 m for each parameter (Table 10–7). The spherical model fit K at both depths, giving a range of 115 and 174 m, respectively. Corn yield for Durand was somewhat spatially dependent (nugget variance 62%), with the isotropic semivariogram fit by the spherical model, giving a range of 231 m. Spatial analysis of the weighted average soil test for the 0 to 20 cm depth for Durand gave similar semivariograms for pH, Ca and Mg because the two depths had similar semivariograms, but mixed results resulted for P and K due to differences between depths (data not shown). For Adrian, spatial variability in pH was similar for both depths, with the spherical model giving a range of 95 and 105 m, respectively, for the 0 to 5 and 5 to 20 cm depths, and a nugget variance of 25% for both depths (Fig. 10–3, Table 10–7). The exponential model fit P and K well in the surface 5 cm, giving a range of 68 and 89 m and nugget variances of 28 and 31%, respectively. There was no spatial dependence of P or K in the 5 to 20 cm depth. The Ca semivariogram was best fit by the linear model in the 0 to 5 cm depth and the spherical model in the 5 to 20 cm depth, with nugget variances of 35 and 41%, and ranges of 255 and 175 m, respectively. The nugget variance was very low (< 4%) and the range was only 74 m for Mg in the 0 to 5 cm under the exponential model. The spherical model fit the Mg semivariance in the 5 to 20 cm depth, giving a range of 257 m and a nugget of 37%. Corn yield was not spatially dependent.

The geostatistical analysis showed that corn yield sampled on a 30.m grid spacing was not spatially dependent at two locations and only slightly dependent at Durand. Anisotropic analysis did not reveal much directional spatial dependence in corn yield. Soil pH was spatially dependent at all three locations and at both depths in soils under no-tillage. Soil test P and K were spatially dependent at all locations but not at both depths in soils under no-tillage at Adrian and Durand. Both Ca and Mg showed spatial dependence at Adrian and Durand at both depths but none at Plainwell. Where strong spatial dependence of pH, P or K exists, the prospects for creation of site specific fertilizer management zones are reasonable. This is confounded by the stratification of soil fertility, however, in which the spatial variation of the surface and subsurface are quite different. This may not be a problem with liming, since the spatial dependence of pH was similar for both depths and the weighted average 0 to 20 cm depth, as illustrated in the pH maps for Adrian.

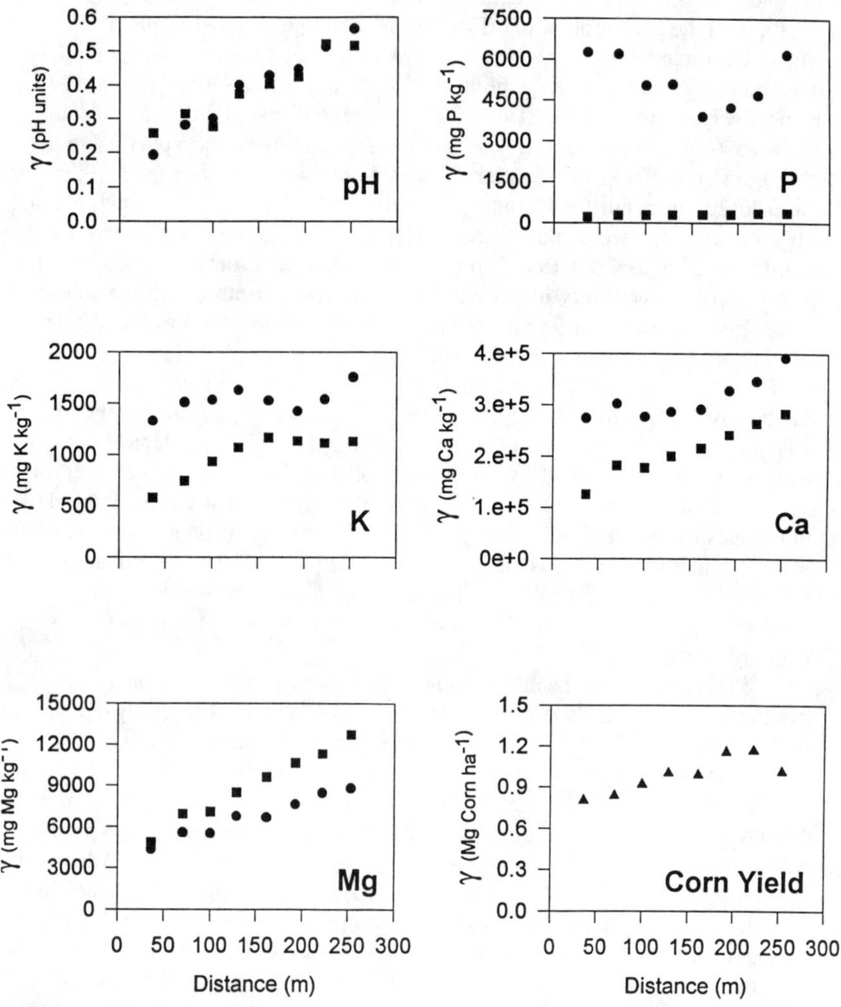

Fig. 10–2. Semivariograms for soil fertility of the 0 to 5 (●) and 5 to 20 (■) cm soil depths and 1993 corn grain yield for the Durand, Michigan, field.

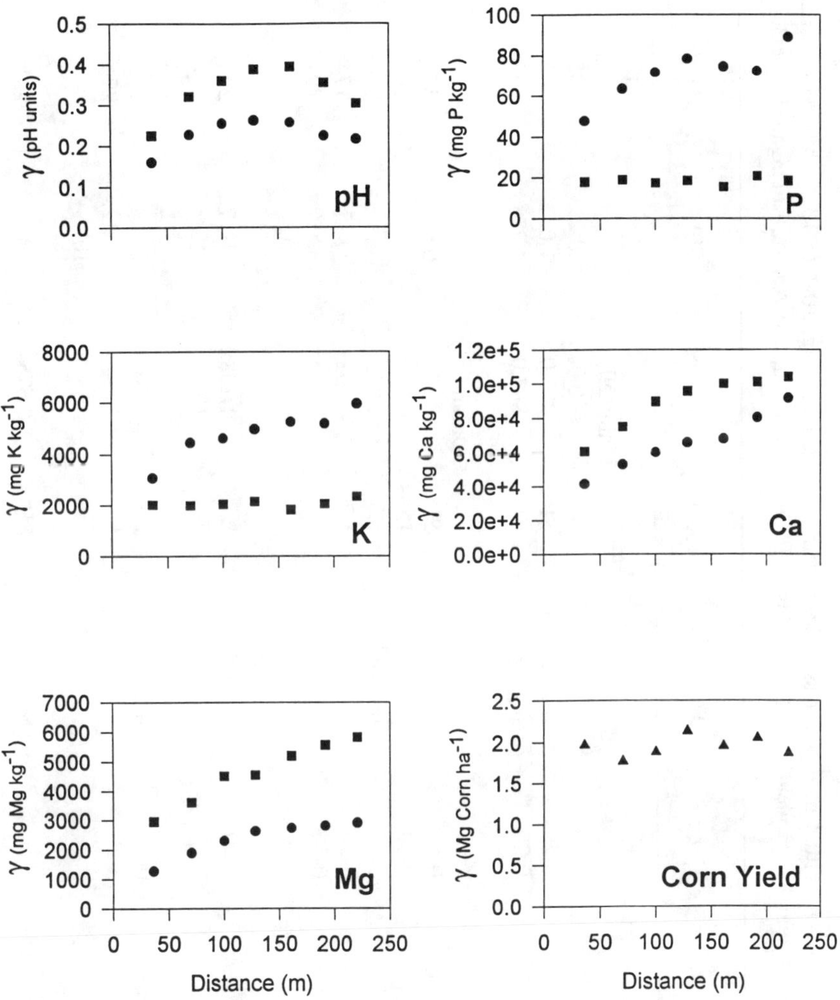

Fig. 10–3. Semivariograms for soil fertility of the 0 to 5 (●) and 5 to 20 (■) cm soil depths and 1993 corn grain yield for the Adrian, Michigan, field.

Table 10–7. Fitted models for significant semivariograms in Fig. 10-1 to 10-3 from geostatistical analysis of soil fertility and corn yield for a farm fields in Plainwell, Durand, and Adrian, Michigan, in 1993.

Parameter	Depth	Units	Model†	Nugget	Sill	Range	r^2
	cm					m	
				Plainwell			
pH	0-20		SP	0.06	0.21	190	0.96
P	0-20	mg kg^{-1}	SP	233.0	1,077	172	0.97
K	0-20	mg kg^{-1}	SP	887.0	1,278	157	0.85
				Durand			
pH	0-5		SP	0.127	0.601	354	0.98
	5-20		L	0.200	0.578	294	0.92
P	5-20	mg kg^{-1}	L	223	386	294	0.92
K	5-20	mg kg^{-1}	SP	302	1135	174	0.99
Ca	0-5	mg kg^{-1}	L	2.455×10^5	3.812×10^5	294	0.74
	5-20	mg kg^{-1}	L	1.138×10^5	3.117×10^5	294	0.97
Mg	0-5	mg kg^{-1}	L	3801	9733	294	0.97
	5-20	mg kg^{-1}	L	3934	1.412×10^4	294	0.97
Corn Yield		Mg ha^{-1}	SP	0.676	1.10	231	0.81

Table 10–7. Continued

Parameter	Depth	Units	Model†	Adrian Nugget	Sill	Range	r^2
pH	0-5		SP	0.078	0.318	95	0.77
	5-20		SP	0.089	0.360	105	0.75
P	0-5	mg kg^{-1}	EXP	23.6	84.09	68	0.85
K	0-5	mg kg^{-1}	EXP	1850	6018	93	0.93
Ca	0-5	mg kg^{-1}	L	3.346x 10^4	9.687x 10^4	255	0.97
	5-20	mg kg^{-1}	SP	4.190x 10^4	10.19x 10^4	175	0.99
Mg	0-5	mg kg^{-1}	EXP	112	3073	74	0.99
	5-20	mg kg^{-1}	SP	2190	5895	257	0.98

†Models include: L=Linear, SP=Spherical, G=Gausian, EXP=Exponential

‡ Anisotropic

Yield Monitor Calibration

The MicroTrak Harvest Yield® grain flow sensor was evaluated for all three farms using the farmer cooperator's combine. The average yields and standard deviations over all transects were similar between the Harvest Yield® grain flow sensor and weigh wagon adjusted to grain moisture of 15.5% (Table 10–8). The grid harvest yields were similar for Adrian but lower than the weigh wagon yields for Durand and Plainwell. The absolute difference between the sensor and weigh wagon yields averaged over all transects was < 5% at any location. The grain flow calibration coefficient varied by location, due to differences in combines and manufacturing differences in grain flow sensors. We varied the combine speed from slow to fast and found that the grain flow calibration was sensitive to combine speed and grain moisture.

Summary and Conclusions

Soil fertility was highly variable for three fields constituting a range of soil, climate, and management conditions for grain crop production in Michigan. Average values for each field showed optimum pH and average to high soil tests. Soil fertility, however, generally ranged from deficient to excessive for most measured parameters. Fertilizer recommendations on a ha basis were not different for uniform versus SSM, but the distribution of fertilizers within the field would vary considerably. Estimated cost of overfertilization by grid using uniform application was only a few dollars per ha. Estimated yield loss, however, from underfertilization by uniform fertilization could exceed 2 Mg corn ha^{-1} at Durand. Grid spacing had little effect on the field average fertility or fertilizer recommendation per ha. Soils in no-tillage management had highly stratified fertility with no consistent relation between the 0 to 5 cm and the 5 to 20 cm depths. Corn yield was highly variable and generally not correlated to soil fertility. Corn yield increased with plant population and declined with grain moisture content, indicating corn yields were lower where plants experienced stress that either reduced stand or delayed development. The geostatistical analysis showed that corn yield was not spatially dependent at two locations and only slightly dependent at Durand. Soil fertility was generally spatially dependent at all three locations but the spatial variation of the surface and subsurface were quite different for soils in no-tillage management. The absolute difference between the Harvest Yield® sensor and weigh wagon yields averaged over all transects was < 5% at any location, although the grain flow calibration coefficient varied by combine. The site specific management of managed inputs and yield mapping at these sites appear to have potential benefits. These data indicate, however, that soil physical properties or landscape, particularly in their effect on water relations, may be more important than fertility in explaining yield variation. The next phase of this project will evaluate the potential for SSM on these fields.

Table 10–8. Comparison of Harvest Yield® monitor, weigh wagon, and small plot yield estimates for corn from three farm fields in Michigan.

Location	Number of Transects	Harvest Yield® Grain Flow Sensor	Weigh Wagon	Grid Harvest	Absolute Difference [Sensor-Wagon]	Flow Calibration†
		----------	Mg ha^{-1}	----------	%	
Adrian	8	6.7∓.7	6.7∓.6	6.8∓.6	4.6∓2.6	65.8∓3.3
Durand	7	7.4∓.4	7.6∓.5	7.0∓.5	3.9∓2.5	74.9∓2.5
Plainwell	13	10.0∓.5	10.5∓.5	9.6∓.8	3.8∓3.0	168.4∓8.1

†The flow calibration is the factor used to convert grain flow to grain yield in the Harvest Yield® Grain Flow Sensor. The factory estimate of the flow calibration is 70.

Wollenhaupt, N.C., R.P. Wolkowski, and H.F. Reetz., Jr. 1993. Variable-rate fertilizer application: Update and economics. Paper presented at the Twenty Third North Central Extension-Industry Soil Fertility Conference, St. Louis, MO.

ACKNOWLEDGEMENT

The authors would like to acknowledge the farmer cooperators, John Anibal, Gordon Brighton, and Dan Klein, for their time and access to their fields and equipment. We acknowledge the technical assistance of Brian Long. We would also like to acknowledge Neil Havermale of Farmer's Software Association for the contribution of his FARMGIS data acquisition system and time in yield mapping and data analysis. This work was funded by the Michigan Agricultural Experiment Station and the Michigan Energy Conservation Project (MECP).

REFERENCES

American Society Agricultural Engineering. 1993. Automated agriculture for the 21st century. ASAE, St.Joseph, MI.

Carr, P.M., G.R. Carlson, J.S. Jacobsen, G.A. Nielsen, and E.O. Skogley. 1991. Farming soil, not fields: A strategy for increasing fertilizer profitability. J. Prod. Agric. 4:57–61.

Christenson, D.R., D.D. Warncke, M.L. Vitosh, L.W. Jacobs, and J.G. Dahl. 1992. Fertilizer recommendations for field crops in Michigan. Ext. Bull. E–550A. Coop. Ext. Serv., Michigan State Univ., East Lansing, MI.

Forcella, F. 1993. Value of managing within-field variability. p. 125–132. In P.C. Robert et al. (ed.). Proceedings of soil specific crop management: A workshop on research and development issues. ASA-CSSA-SSSA, Madison, WI.

Gamma Design Software. 1993. Geostatistical Analysis. Gamma Design Software, Plainwell, MI.

Hammond, M.W. 1993. Cost analysis of variable fertility management of phosphorus and potassium for potato production in Central Washington p. 213–228. In P.C. Robert et al. (eds). Proceedings of soil specific crop management: a workshop on research and development issues. ASA-CSSA-SSSA, Madison, WI.

Larson, W.E., and P.C. Robert. 1991. Farming by soil. p. 103–112. In R. Lal and F.J. Pierce (ed.) Soil management for sustainability. Soil Water Conserv. Soc., Ankeny, IA.

Robert, P.C. et al. (eds). 1993. Proceedings of soil specific crop management: a workshop on research and development issues. ASA-CSSA-SSSA, Madison, WI.

Schueller, J.K. 1992. A review and integrating analysis of spatially-variable control of crop production. Fertilizer Res. 33:1–34.

Wibawa, W.D., D.L. Dludlu, L.J. Swenson, D.G. Hopkins, and W.C. Dahnke. 1993. Variable fertilizer application based on yield goal, soil fertility, and soil map unit. J. Prod. Agric., 6:255–261.

Wollenhaupt, N.C. and D.D. Buchholz. 1993. Profitability of farming by soils p. 199–212. In P.C. Robert et al (ed.). Proceedings of soil specific crop management: A workshop on research and development issues. ASA-CSSA-SSSA, Madison, WI.

11 Site-Specific Yield Histories On a SE Coastal Plain Field

E. J. Sadler
W. J. Busscher

USDA-Agricultural Research Service
Florence, South Carolina

D. L. Karlen

USDA-Agricultural Research Service
Ames, Iowa

Site-specific farming research at Florence, SC, began in 1984 with a topographic survey and subsequent detailed soil mapping by local USDA-SCS staff. An area that was representative of the coastal plain soil types was planted to corn in 1985. Since then, five crops of corn (*Zea mays*), three of wheat (*Triticum aestivum*), three of soybean (*Glycine max*), and one of grain sorghum (*Sorghum bicolor*) have been grown. Harvested plots were located using surveying techniques, and the plot outlines were overlaid onto the soil map to determine the corresponding soil map unit. To date, over 3000 plots have been measured. Analysis of variance indicated that differences in mean yields were significant, but inspection suggested that intra-map unit variance was nearly as large as inter-map unit variance. Attempts to explain variation in yield using both statistical regression and mechanistic modeling were not successful. Geostatistical analysis produced the expected patterns of high and low yield, but yr-to-yr variation in mean yield masked underlying patterns. A method developed to normalize annual variability in mean yield, while accounting for shifts in location of sampled yields, produced composite maps of relative yield. These maps should be useful for setting target yields of various soil types, thus allowing calculation of fertilizer requirements. This research has provided much new knowledge about inherent variation expected for these soils, as well as having started a baseline from which to judge annual variability of yield for regional soils and crops. Interpretation of these results and extension of the information to make fertility and irrigation recommendations depends on the successful quantitative description of the causes of variation among soil types under regional climate. Despite problems encountered during this work, mechanistic simulation models appear to be the most likely tool to achieve this objective.

HISTORY AT FLORENCE

Field Studies

Site-specific farming research at Florence grew out of the erosion-productivity research topic in the early 1980's. At that time, the emphasis was on variation among soil types caused by historical erosion on sloping land in the Piedmont. The topography is more level in the Coastal Plain, and soil-to-soil variability results from soil genetic factors in addition to erosion. A predominant feature in this geographic area that remains unexplained is the Carolina Bay, which is a circular, shallow depression ringed by often inhospitable soils. The productivity of the bays is so poor that most are left to weeds. Whether the bays are left out of production or farmed, the reduction in productivity over the total area farmed is not trivial. Over 12% of the Florence location's experimental area are soils contained in or associated with Carolina Bays.

In 1984, the local USDA-SCS staff mapped the laboratory's experimental area, starting from a 15–m survey grid. Resolution was finer if changes in soil map unit were found between the grid points (USDA-SCS, 1986). At each auger hole, the classifiers also recorded the depth to the clay layer. This depth, at which most coastal plain soils change from >70% sand to >40% clay, is an easily-identified characteristic, and is used to distinguish among several soil types. The topographic map, the soils map, and the map of the depth to the clay layer were stored in computer format for later analysis (see Fig. 11–1 and Table 11-1 for information on soils).

In 1985, we planted corn on one representative 8-ha field, using uniform, conventional methods typical of local farmers. In similar fashion, five crops of corn, four of wheat, three of soybean, and one of grain sorghum have been grown on this area. For all crops, harvest plots were positioned using survey techniques, and the corresponding map units were identified from the soil map (e.g., Fig. 11–2). Over 3000 site-specific yields have been so obtained.

Statistical Analysis of Map Unit Means

Consistent with the erosion–productivity studies of the time, our early analyses focussed on the variations among map unit mean yields. Differences among map unit means for the first 5 yr were significant according to analysis of variance (Karlen et al. 1988; 1990), but later harvests were less conclusive. Tables 11-2 and 11-3 show soil map units, expected yields (from the USDA-SCS soil survey productivity rating), and measured mean yields for the cropping sequence.

Closer inspection of the yields for a map unit indicated that some inclusions were more productive than others. In some cases, the primary distinguishing characteristic between inclusions appeared to be the depth to clay. Except for a few soils, however, attempts to correlate yield to depth of clay were not successful (Fig. 11–3). Other possible causes of the variation included existence of an eluviated horizon, subsoil acidity, and non-uniform hardpan disruptionby in-row subsoiling. The interactions between these factors and depth

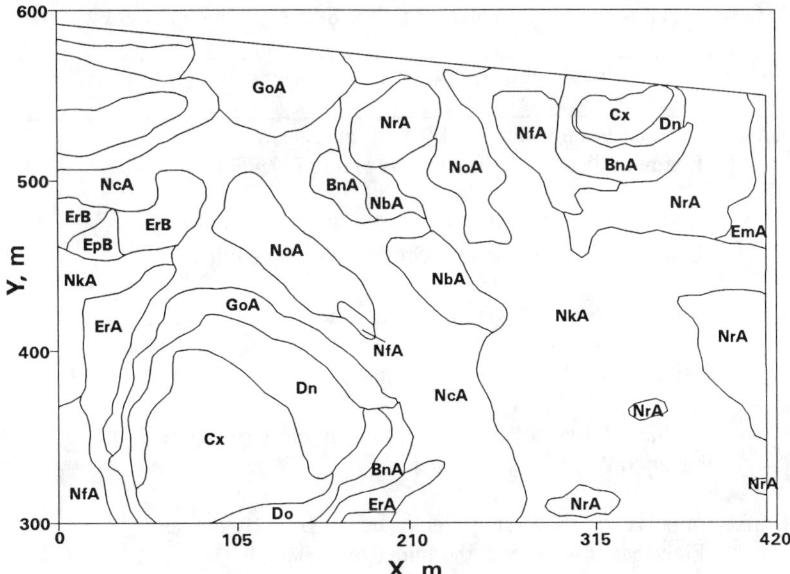

Fig. 11–1. Soil map of the fields used for site-specific farming research at the Florence, SC, USDA facility.

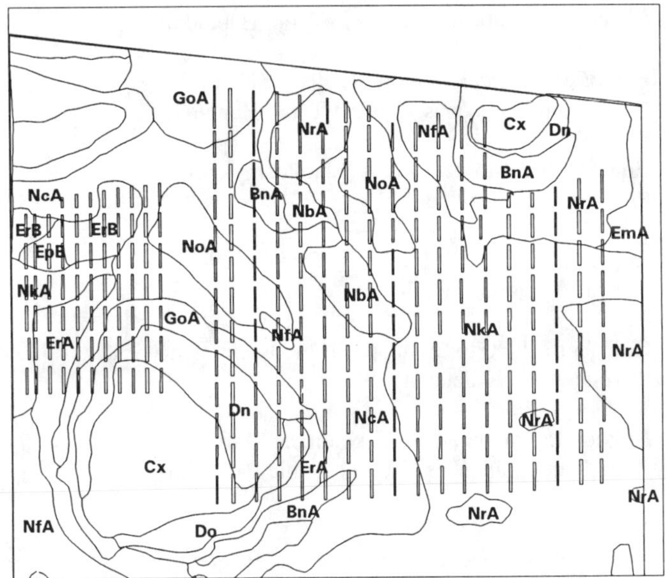

Fig. 11–2. Map of 1987 wheat harvest plots overlaid on the soils map.

Table 11–1. Proportionate distribution of soils within the 24-ha area at the Coastal Plains Research Center where the 8-ha experimental field was located.

Symbol	Soil classification	%
BnA	Bonneau Loamy fine sand (lfs), 0 to 2% slopes (Loamy, siliceous, thermic Grossarenic Paleudult†)	2.2
BoA	Bonneau Loamy sand (ls), 0 to 2% slopes, overwash (Loamy, siliceous, thermic Grossarenic Paleudult†)	0.2
Cx	Coxville Loam (Clayey, kaolinitic, thermic Typic Paleaquult)	4.5
Dn	Dunbar lfs (Clayey, kaolinitic, thermic Aeric Paleaquult)	2.2
Do	Dunbar lfs, overwash (Clayey, kaolinitic, thermic Aeric Paleaquult)	0.8
EmA	Emporia lfs, moderately thick surface, 0 to 2% slopes (Fine-loamy, siliceous, thermic Typic Hapludult)	1.7
EpA	Emporia lfs, thick surface, 0 to 2% slopes (Fine-loamy, siliceous, thermic Typic Hapludult)	2.5
EpB	Emporia lfs, thick surface, 2 to 4% slopes (Fine-loamy, siliceous, thermic Typic Hapludult)	0.2
ErA	Emporia fine sandy loam (fsl), 1 to 2% slopes (Fine-loamy, siliceous, thermic Typic Hapludult)	1.7
ErB	Emporia fsl, 2 to 4% slopes (Fine-loamy, siliceous, thermic Typic Hapludult)	6.0
ErD	Emporia fsl, 10 to 15% slopes (Fine-loamy, siliceous, thermic Typic Hapludult)	1.5
GoA	Goldsboro lfs, 0 to 2% slopes (Fine-loamy, siliceous, thermic Aquic Paleudult)	1.7
NbA	Noboco lfs, moderately thick surface, 0 to 2% slopes (Fine-loamy, siliceous, thermic Typic Paleudult)	1.7
NcA	Noboco lfs, thick surface, 0 to 2% slopes (Fine-loamy, siliceous, thermic Typic Paleudult)	7.9
NfA	Noboco fsl, 1 to 2% slopes (Fine-loamy, siliceous, thermic Typic Paleudult)	1.0

Table 11-1. Continued

NkA	Norfolk lfs, moderately thick surface, deep water table, 0 to 2% slopes (Fine-loamy, siliceous, thermic Typic Paleudult‡)	47.7
NnA	Norfolk lfs, moderately thick surface, very deep water table, 0 to 2% slopes (Fine-loamy, siliceous, thermic Typic Paleudult‡)	6.0
NoA	Norfolk lfs, thick surface, 0 to 2% slopes (Fine-loamy, siliceous, thermic Typic Paleudult‡)	4.5
NrA	Norfolk fsl, 1 to 2% slopes (Fine-loamy, siliceous, thermic Typic Paleudult‡)	5.2
W	Water	0.8

†Reclassified March 1990 to Loamy, siliceous, thermic Arenic Paleudult.
‡Reclassified March 1988 to Fine-loamy, siliceous, thermic Typic Kandiudult.

Table 11–2. Soil map units, productivity rating (USDA-SCS, 1986), and measured yields for the 5 corn seasons.

Soil	Productivity rating	Measured yield ± standard deviation				
		1985	1986	1988	1992	1993
	kg/ha			kg/ha		
BnA	5344			5104±1262	6543±1398	2232± 957
Cx	6916	3645± 963	728± 635	1364± 825	7802±1235	2456±1389
Dn	7230	3871± 921	163± 148	1192± 610	4889±2051	2751±1453
EpB	6287	7727± 978	2166± 254	3944± 604		
ErA	6916	7333± 738	2067± 349	3314± 722	6546± 405	2158± 251
ErB	6287	7686± 477	2187± 302	3775± 541		
GoA	7859	5097± 786	1329±1143	2102±1682	6275± 882	2230± 837
NbA	7230			4495± 548	8178± 480	3404± 681
NcA	7230	6038±1320	2228± 504	4160±1239	7497±1063	2629± 737
NfA	7230			4248± 560	7389± 909	1956± 628
NkA	6916	6806±1310	2738± 431	4471± 712	7609±1264	2262±1043
NoA	6916	8290± 57	2834± 489	4606± 671	7894± 995	2876± 819
NrA	6916		1496± 609	4378± 346	7209±1045	2452± 818
Mean		6319±1635	1865± 953	3510±1583	7310±1236	2481± 919
No. samples		130	145	331	256	209

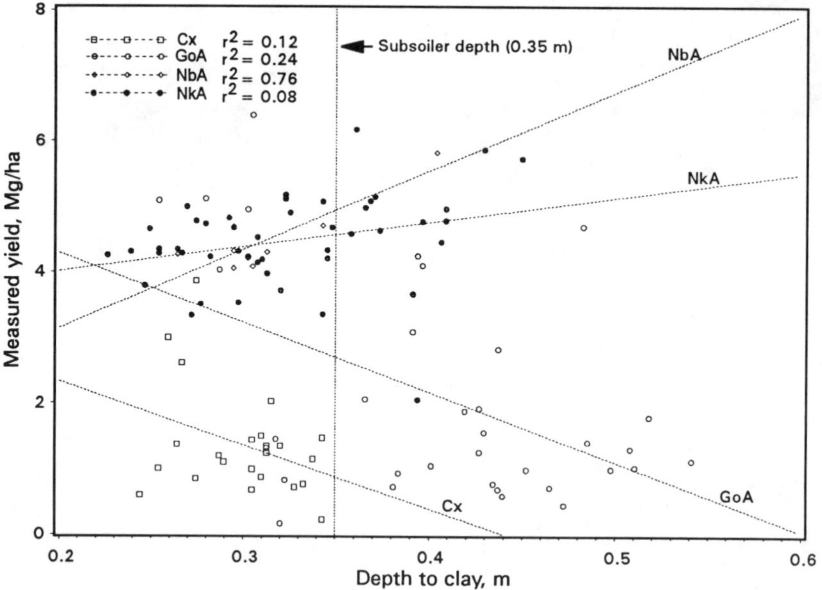

Fig. 11-3. Relationship of individual plot yields with depth to clay for the 1988 corn season.

Table 11–3. Expected yields (from soil survey productivity ratings), measured mean yields, and standard deviations for soybean and wheat grown on the various soil map units.

Soil	Soybean Measured yield ± standard deviation				Wheat Measured yield ± standard deviation			
	Rating	1989	1990	1991	Rating	1987	1989	1991
		kg/ha				kg/ha		
BnA	2021	1885±433	2497±117	1102±747		3815±1146	3808±713	1483±413
Cx	2695	2209±358	2491±301	1403±313	3368	3502± 299	4093±382	1431±432
Dn	3031	1827±313	2352±329	1796±156	3705	3715± 591	4133±485	.
Do	3031	1790±363	2316±271	1282±309	3705	.	3897±375	2112± 76
EpB	2021	.	.	.	3368	5053±1266	4711±230	.
ErA	2358	1867±229	1660±120	323±152	3705	5297± 844	4549±253	2706
ErB	2021	.	.	.	3368	6141± 195	4558±346	.
GoA	3031	1389±458	2545±301	1494±658	4042	4455± 810	3756±401	1486±298
NbA	3031	1918±341	2184±190	1783±356	4042	6229± 447	4298±438	2297±300
NcA	3031	1853±299	2368±224	1599±471	4042	5433± 734	4146±393	2053±355
NfA	3031	1711±281	2244±454	1072±510	4042	5261± 465	4049±424	2057±414
NkA	2695	2022±390	2000±460	912±654	4042	4529± 909	3961±764	1885±464
NoA	2695	1765±280	2353±319	1549±470	4042	5985± 446	4268±437	2027±373
NrA	2695	1794±373	2139±272	1376±608	4042	4816±1187	4026±637	2012±465
Mean		1841±387	2290±358	1395±606		4890±1052	4083±597	1967±445
No. samples		194	234	300		195	588	324

to clay were too complex to describe statistically. Describing the cause of this variation became the emphasis of the project. During this analysis, it became clear that yield variation within a map unit, or even within one inclusion of a map unit, could approach the variation among map units (Fig. 11–4).

Mechanistic Modeling of Map Unit Means

This emphasis on describing causes of the variation led to the acquiring, testing, and subsequent parameterization of the daily time step models of crop growth (Sadler et al., 1988): CERES-Maize (Jones & Kiniry, 1986), CERES-Wheat (Godwin et al., 1988), SORKAM (Rosenthal et al., 1989), CERES-Sorghum (Alagarswamy et al., 1988), and SOYGRO (Jones et al., 1989). Early data were more numerous and the yields were more varied for corn, so the initial effort concentrated on CERES-Maize. At that time, Versions 1.0 (Jones & Kiniry, 1986) and 2.0 (Ritchie et al., 1988) were both available. Sensitivity analyses indicated that combinations of parameters existed that would produce generally realistic yields, but attempts to match typical pedon descriptions to map unit mean yields were not successful (Fig. 11–5). In general, soils that did not have root-restricting horizons were adequately simulated, but soils with acidic subsoils or eluviated horizons were not. Algorithms were developed to produce estimates of rooting in horizons as a function of density, acidity, depth, and tillage. The accounting for root weighting in the CERES models, however, was not sufficient to prevent the simulated exploration by roots of these unexplored zones. Simulations were repeated for later versions (V2.10, Ritchie et al., 1989; V3.00 pre-release) with similar results, but this was expected because the rooting algorithm is common to all versions.

In addition, simulations during droughts indicated that the water balance was not simulated well because the model estimated too much infiltration during intense storms. For example, a 92-mm, 52-min duration storm occurred during the 1986 drought. The model simulated 22 mm of runoff, and 70 mm of infiltration. According to the model, that much infiltration was sufficient to carry the crop to the end of the season, and therefore simulated yields were about twice the measured ones. Qualitative observations of runoff, lower infiltration rates of local soils, lack of ponding except for a short time after the storm, and subsequent rapid onset of water stress all indicated that much more runoff occurred. Simulations using breakpoint rainfall and Green-Ampt methods resulted in a more likely outcome, which was about the reverse of the CERES result (Stone & Sadler, 1991). Reinsertion into CERES of lower rainfall totals, which forced infiltration to match the Green-Ampt results, produced simulated yields about halfway between the previously simulated and the measured yields. The authors of the models are addressing this rooting algorithm at this time.

Geostatistical Description of Spatial Yields

Concurrent with the later stages of the modeling effort, and following discussions during the first Site-Specific workshop (Sadler et al., 1992), a project was conducted to describe spatial variation using geostatistical methods (Sadler

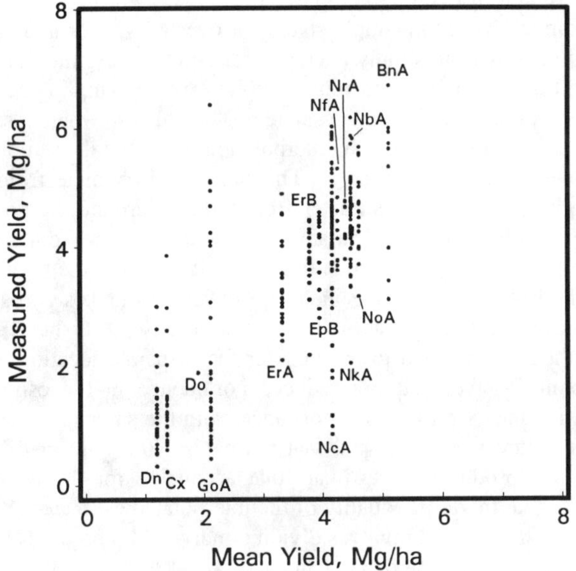

Fig. 11–4. Relationship of individual plot yields with the mean yield of the corresponding soil map unit for the 1988 corn season.

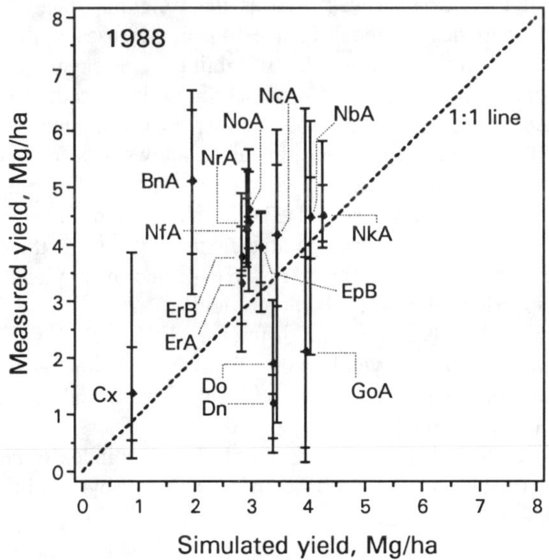

Fig. 11–5. Simulated and measured corn yield for the 1988 season.

& Busscher, 1992). Kriging of data from individual crop-yr produced yield maps (e.g., Figs. 11-6, 11-7). Typically, these maps were visually comparable, but clearly different from crop to crop (contrast Figs. 11-8, 11-9, and 11-10), and difficult to summarize into a single statement over all yrs for a single crop. If kriged interpolations were simply averaged for points across the field, the result would depend on the quality and reproducibility of the sampling scheme at the harvests. For example, consider a possible real-world case where a sample from a high-yielding yr is located near an interpolation point, but the sampling scheme skipped a low-yielding area nearby. The final yield estimate from that point would be high, as would the estimates from the area around the point. If data from a lower-yielding yr were available from within the area omitted from the high-yielding yr, one might conclude that the estimate from the high-yielding yr was biased higher because of the inclusion of high estimates across the zone. In fact, the kriged estimate was of less value because of distance from input data. A method was needed to account for the disparate locations of samples and to account for the resulting reduced confidence in the estimates when brought to a regular grid of varying distance from the samples.

A procedure was developed that normalized for yr-to-yr differences in mean yield, and produced an estimate that accounted for the location of the measurements and, therefore, quality of the interpolated estimates (Sadler et al., 1994). In brief, the method involves dividing individual plot yields by the mean yield for the yr, as suggested by Schnug et al. (1993), and then kriging the result. These normalized, or relative, yields can then be compared to other normalized yields on a standard grid. The problem of how to compute the average relative yield at an interpolation point was solved using the estimated variance of the kriged estimate, which is provided by the geostatistical software. The inverse of this variance was used as the weighting factor in a weighted average. The variance of the kriged estimate increases with distance from sample points, so by inversion, the weighting is strongest for information originating nearest the interpolated point. Using this procedure relieves one of the requirement to mask out areas that were sparsely sampled so that they will not overly influence the composite estimate. It allows data from multiple yrs to be aggregated into a single, objective yield map. The working map is of relative yield, which for our examples ranged ±0.4 from a mean of 1.0 (Figs. 11-8, 11-9, and 11-10). The only additional parameter needed beyond that required in a normal kriging process is an estimate of the expected mean yield, which conveniently corresponds to the farmer's target yield.

Sadler et al. (1994) examine differences between the weighted average yield map and individual yr yield maps. They also examine errors expected if, say, fertilization decisions were based on yield maps for individual yrs, relative yield maps, or the weighted-average-yield map. These decisions could be contrasted to decisions based on map unit productivity index ratings. Weighted average relative yields are shown for corn, wheat, and soybean in Figs 11-8, 11-9, and 11-10.

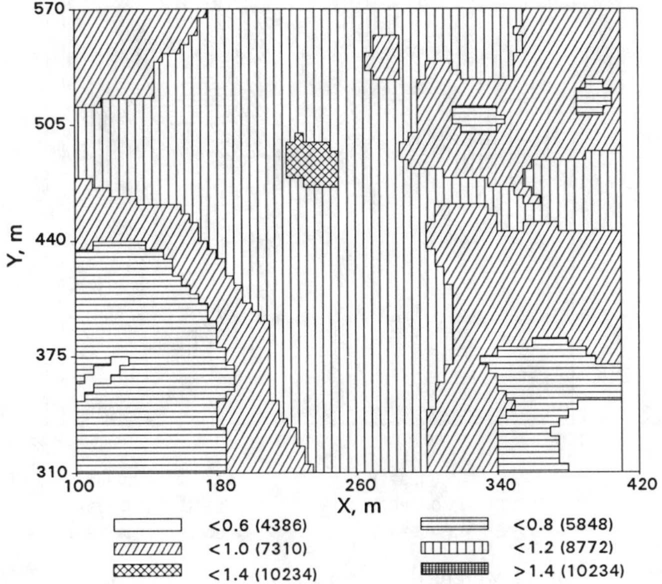

Fig. 11–6. Map of kriged corn yields from the 1992 season.

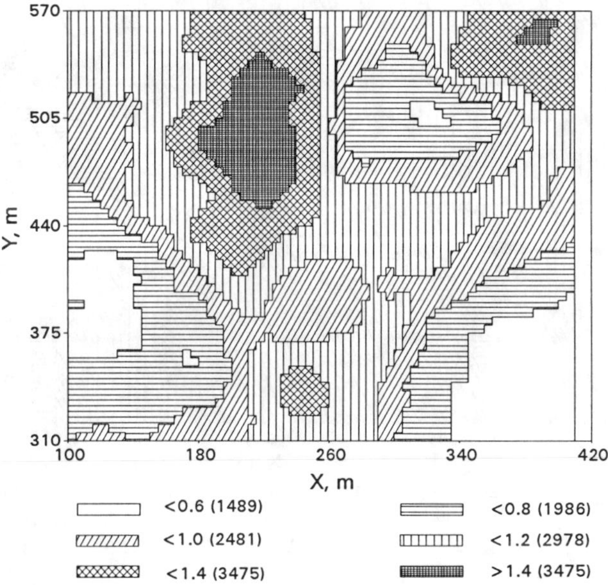

Fig. 11–7. Map of kriged corn yields from the 1993 season.

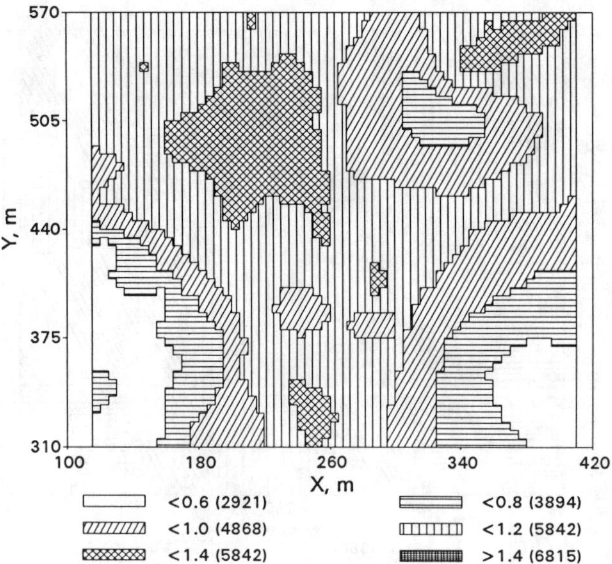

Fig. 11–8. Map of weighted mean corn yields, compiled from 3 seasons of yields on the larger field.

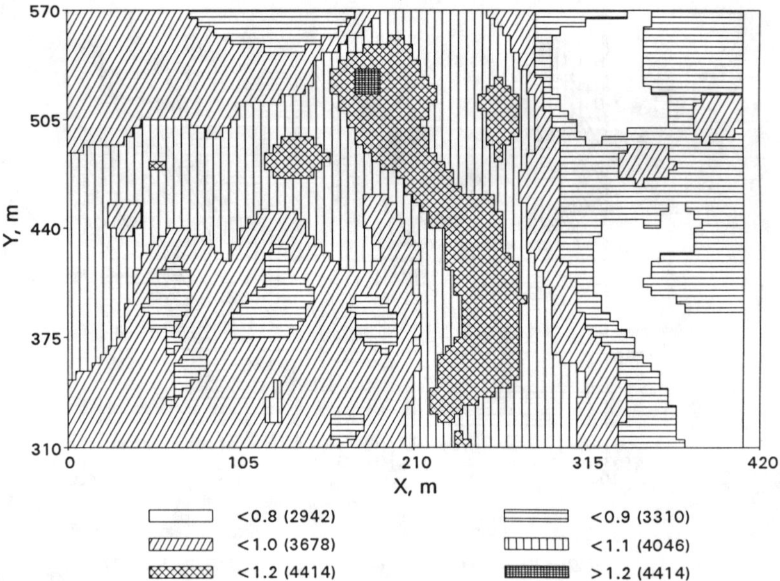

Fig. 11–9. Map of weighted mean wheat yields, compiled from 3 seasons of yields.

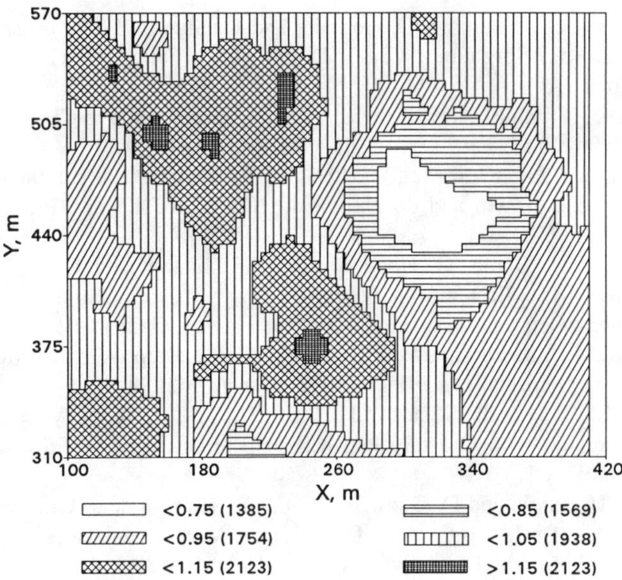

Fig. 11–10. Map of weighted mean soybean yields, compiled from 3 seasons of yields.

SUMMARY AND CONCLUSIONS

While the ongoing project will continue to provide insights into causes and effects of soil variability in the SE coastal plain, some conclusions have been obtained that will help direct future research at the location. First, the project has demonstrated the value of monitoring yields for multiple years. Second, it appears that limited progress can be made in quantitatively describing causes and effects of regional soil variability until mechanistic simulation models of crop growth can meet the water balance and can account for sub-optimal rooting conditions. Third, normalizing yield for annual variability shows some promise in improving utility of yield maps. Finally, geostatistical methods applied to normalized yield may allow composite summaries to be made of multiple-yr data.

ACKNOWLEDGEMENTS

The authors would like to thank the many persons who have worked on this project, including Dean Evans, Ernest Strickland, Bill Berti, Mike Boswell, Jimmie Vereen, Ron Schroll, Philip Ehlen, and Tosha Cain. Additional thanks to Dean Evans for creating the figures and Ellen Whitesides for producing the camera-ready copy.

REFERENCES

Alagarswamy, G., J. Ritchie, D. Godwin, and U. Singh. 1989. Auser's guide to CERES-Sorghum - V2.00 (Draft). Int'l. Fert. Dev. Center, Muscle Shoals, AL.

Godwin, D., J. Ritchie, and U. Singh. 1988. A user's guide to CERES-Wheat - V2.00. Int'l. Fert. Dev. Center, Muscle Shoals, AL.

Jones, C.A. and J.R. Kiniry (ed.). 1986. CERES-Maize: A simulation model of maize growth and development. Texas A&M Univ. Press, College Station, TX.

Jones, J.W., K.J. Boote, G. Hoogenboom, S.S. Jagtap, and G.G. Wilkerson. 1989. SOYGRO V.5.41: Soybean crop growth simulation model user's guide. Dep. of Agric. Eng., Univ. of Florida, Gainesville, FL.

Karlen, D.L., E.J. Sadler, and W.J. Busscher. 1988. Crop yield variation within a typical coastal plain field. Agron. Abstr. 80:278.

Karlen, D.L., E.J. Sadler, and W.J. Busscher. 1990. Crop yield variation associated with Coastal plain soil map units. Soil Sci. Soc. Amer. J. 54:859–865.

Ritchie, J., U. Singh, and D. Godwin. 1988. A user's guide to CERES-Maize - V2.00 (Draft). Int'l. Fert. Dev. Center, Muscle Shoals, AL.

Ritchie, J., U. Singh, D. Godwin, and L. Hunt. 1989. A user's guide to CERES-Maize - V2.10. Int'l. Fert. Dev. Center, Muscle Shoals, AL.

Rosenthal, W.D., R.L. Vanderlip, B.S. Jackson, and G.F. Arkin. 1989. SORKAM: a grain sorghum crop growth model. MP 1669, TAES Computer Software Documentation Series. Texas Agric. Exp. Stn., Texas A&M Univ., College Station, TX.

Sadler, E.J., and W.J. Busscher. 1992. Site-specific yield histories on a SE Coastal Plain field. Agron. Abstr. 84:312.

Sadler, E.J., D.L. Karlen, and W.J. Busscher. 1988. Modeling of soil variability effects on corn, wheat, and sorghum for SE Coastal Plain soils. Agron. Abstr. 80:284.

Sadler, E.J., D.E. Evans, W.J. Busscher, and D.L. Karlen. 1993. Yield variation across Coastal Plain soil mapping units. pp. 373-374. *In* P. C. Robert, et al. (ed.) Soil Specific Crop Management. ASA, CSSA, SSSA, Madison, WI.

Sadler, E.J., W.J. Busscher, and D.L. Karlen. 1994. Multi-year analysis of site-specific yields on a coastal plain field. (In preparation for submittal.)

Schnug, E., D. Murphy, E. Evans, S. Haneklaus, and J. Lamp. 1993. Yield mapping and application of yield maps to computer-aided local resource management. p. 87–93. *In* P.C. Robert, et al. (ed.) Soil Specific Crop Management. ASA, CSSA, SSSA, Madison, WI.

Stone, K.C., and E.J. Sadler. 1991. Runoff using Green-Ampt and SCS curve number procedures and its effect on the CERES-Maize model. ASAE Paper 91–2612.

USDA-SCS, 1986. Classification and correlation of the soils of Coastal Plains Research Center, ARS, Florence, South Carolina. South National Technical Center, Ft. Worth, TX.

12 Fertility Variability in the Minnesota River Valley Watershed in 1993 as Determined From Grid Testing Results on 52,000 Acres in Commercial Fields

Tom McGraw
Randall Hemb

Minnesota Crop Monitor
Buffalo, Minnesota

Evidence for substantial variability in soil test values for phosphorus (P), potassium (K), and zinc (Zn) in fields in south-central and southwestern Minnesota continues to grow. For example, an increasing number of fertilizer dealers, crop consultants, and farmers are reporting substantial yr-to-yr variation in soil test values for P and K. Sometimes, this variation can be explained by extreme variations in soil moisture content when soil samples were collected. In many situations, however, this variation is due to the fact that soil samples were not collected from the same location in the field in two consecutive yrs. This wide yr-to-yr fluctuations has convinced some producers to question the credibility of soil testing as a management tool.

EXPERIMENTAL PROCEDURE

The data presented in this paper were compiled from the files of Minnesota Crop Monitors. The data presented are the results of sampling more than 52 000 acres in southwestern and south-central Minnesota.

Fields were divided into grids. Each grid cell was ≈4.4 acres. The sample for each grid cell was a composite of 7 to 10 cores. The cores were collected by using an open-face hand probe. Sample depth was ≈6 in.

Soil samples were analyzed for P, K and Zn by Minnesota Valley Testing Laboratories, New Ulm, Minnesota. All data are reported as parts per million (ppm).

The analytical data were grouped or summarized in several ways. The data were first grouped into four geographical areas roughly defined in Fig. 12–1. The fields in each geographical area were then grouped according to size. Three size classifications (35 to 82 acres; 83 to 160 acres; 161+ acres were used. The data from 392 fields were summarized for this report (Fig. 12–2).

Minnesota

The Four Geographical Subset Areas of The Variability Study

1. **Renville and surrounding counties**
2. **Redwood and surrounding counties**
3. **Brown and the surrounding counties**
4. **Faribault and surrounding counties**

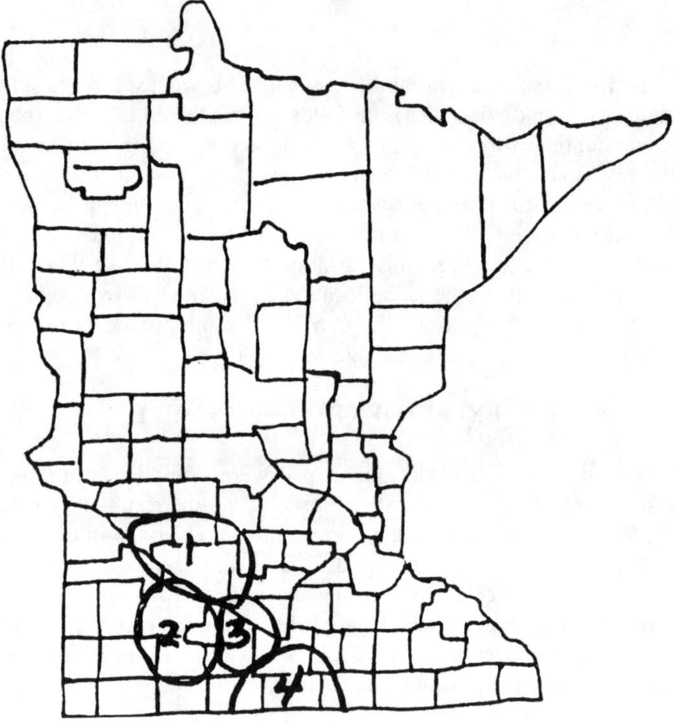

Fig. 12–1. The four geographical subset areas of the Variability Study.

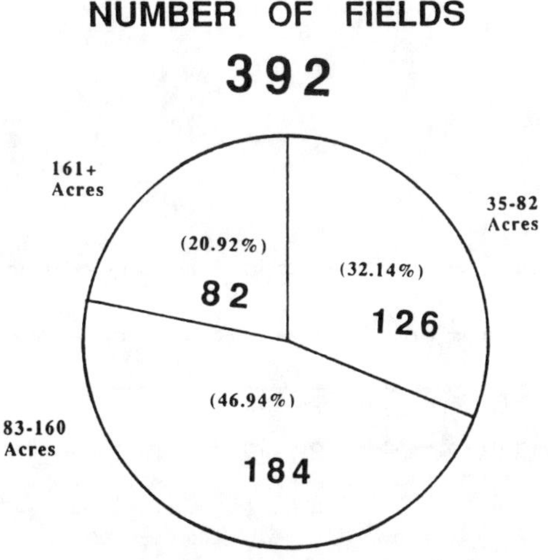

Fig. 12–2. Number of Fields.

RESULTS AND DISCUSSION

The highest percentage of samples collected was taken from fields that had a range of 83-160 acres (Fig. 12–2). Approximately 33% of the samples were taken from fields that had a size range of 35-82 acres. The smallest percentage of samples was taken from the large fields (161+ acres).

The results from fields in the Renville County area having a size of 35-82 acres are summarized in Table 12–1. These data illustrate the variability of soil test values for P, K and Zn that occurs in fields in central Minnesota. With P, for example, a high percentage of the samples taken from some fields have a high or very high soil test. In other fields, an approximately equal number of samples fall into each soil test category for P.

With K and Zn, a higher percentage of the samples have high or very high soil test values. In general, soils in the Renville County area of Minnesota tend to have high soil test levels for these two nutrients.

There was no relationship among the three nutrients measured. This is illustrated by the data from field No. 3. Twelve of the 17 samples had a very low level of Zn. Yet only one sample taken from this field had a very low soil test for P. This may reflect fertilization programs that were used in the past.

The data presented in Table 12–1 are summarized on a percentage basis in Table 12–2. For fields where soil tests for P were found in all five categories, the percentage was approximately the same in each of the categories. The higher percentage of samples of K and Zn in the high and very high categories is also shown in this table.

Table 12–1. Distribution of soil test values for P, K and Zn in 22 fields in Renville and adjacent counties in Minnesota in 1993.

Field	Samples	No. of P Categories					No. K Categories					No. Zn Categories				
	Field	VL	L	M	H	VH†	VL	L	M	H	VH	VL	L	M	H	VH
1	18	1	4	4	4	5		7	7	2	2		4	4	4	6
2	18	1	5	2	0	10		11	4	2	1	4	13	1		
3	17	1	2	2	5	7			12	4	1	12	2	3		
4	16	3	3	6	2	2		8	6	0	2		3	6	5	2
5	9	2	3	3	0	1		2	0	0	7		2	2	2	3
6	12	1	2	4	3	2		3	5	3	1		3	0	4	5
7	17	4	3	3	1	6		1	3	3	10		1	4	2	10
8	9	1	4	0	3	1		2	6	1			6	3		
9	13	3	2	1	2	5			4	4	5				5	8
10	19		6	12	0	1			3	1	15	1	8	4	4	2
11	18		2	4	2	10			5	5	8		2	5	1	10
12	18		7	3	6	2		5	11	1	1		3	8	1	6
13	18		5	7	4	2				1	17		1	4	6	7
14	15		3	3	1	8		3	2	4	6			4	4	7
15	16		1	7	1	7	1	0	5	1	9	1	3	6	1	5
16	12		1	4	3	4		3	2	1	6			3	2	7
17	10		1	4	1	4		2	4	1	3	2	2	3	1	2
18	6		1	1	3	1		3	2	1					3	3
19	18	6	6	5	1			13	2	2	1			8	5	5
20	9			5	4			2	3	2	2		1	2	6	
21	12		6	6				3	3	3	3	5	6	1		
22	15				1	14			2	0	13					15

†VL = very low, L = low, M = medium, H = high, VH = very high. Each sample represents ≈ 4.4 acres; samples collected at the center of the grid.

Table 12–2. The percentage of soil samples in each of five soil test categories when fields were sampled in Renville and adjacent counties in 1993.

Nutrient	No. of Categories	% of Fields	Soil Test Category				
			VL	L	M	H	VH
			- - - - - - - - - - -% - - - - - - - - - -				
Phosphorus	5	41	13	20	20	16	30
	4	41	-	20	34	16	30
	4	4	33	33	28	6	-
Potassium	5	5	6	-	32	6	56
	4	59	-	35	28	13	24
	3	23	-	-	32	17	51
	3	9	-	33	53	14	-
Zinc	5	14	9	30	30	13	18
	4	41	-	18	28	18	36
	3	9	-	-	36	28	36
	3	4	-	11	22	67	-
	3	14	45	45	10	-	-

The distribution of samples in each of the five soil test categories and the relationship to field size is shown in Fig. 12–3. The size of the field that was sampled did not appear to have a major effect on the distribution over soil test category in Renville and surrounding counties. This observation was consistent for each of the three nutrients (P, K, Zn). Likewise, there was a spread over four or five soil test categories for small (35–80 acres) as well as large (161+) acres.

The distribution and spread of samples in the five soil test categories for the Minnesota River Valley Watershed is shown in Fig. 12–4. For the entire 52,000 + acres, 27% to 31% of the samples had a low or very low level of soil test P. Samples from any given field fell into four or five of the soil test categories 70% of the time for small fields (35-80 acres) and 96% of the time for large fields (161+ acres). The observed range and distribution was somewhat expected when large fields were sampled. The observations from sampling the small fields were not expected.

In contrast to P, K soil tests were in the low and very low categories 14% of the time for small fields and 19% of the time for large fields. Similar percentages were observed with the soil tests for Zn.

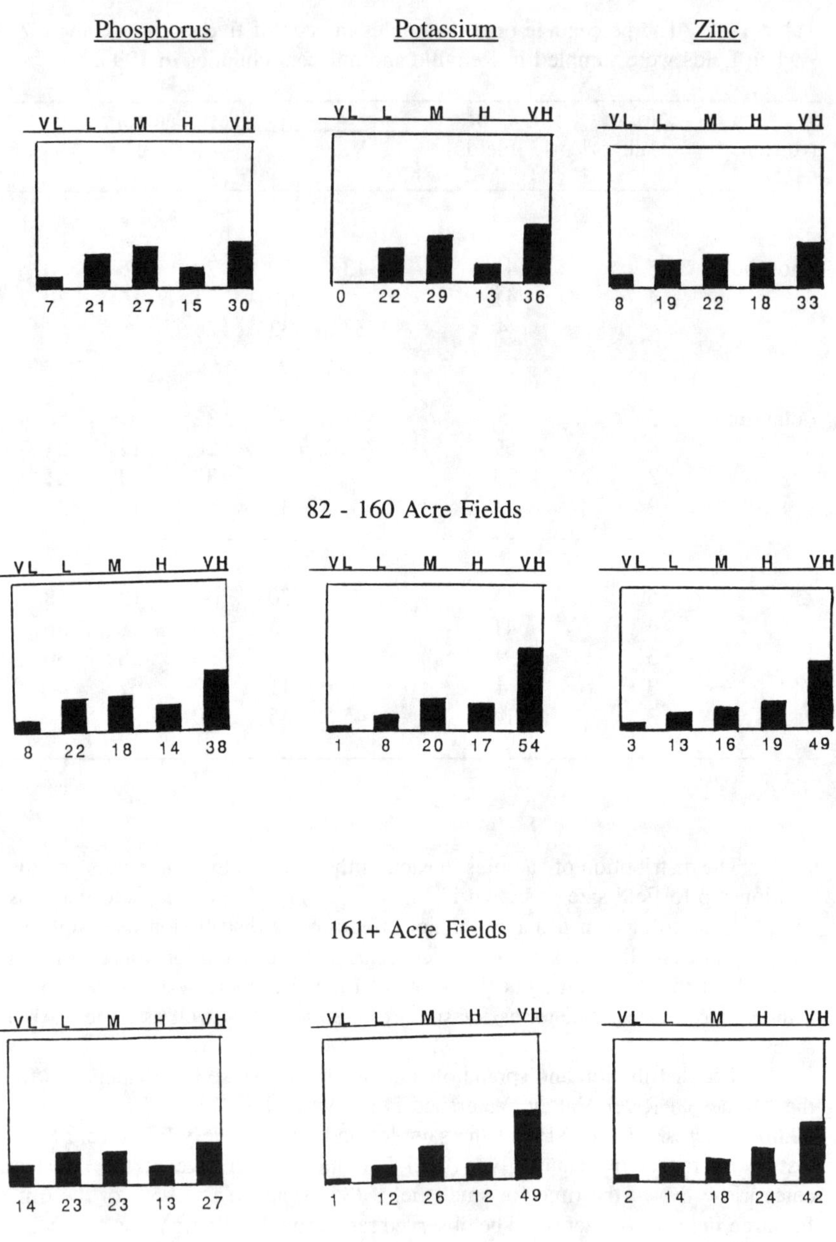

Fig. 12–3. Distribution of soil samples in five soil test categories and its relationship to field size in Renville and surrounding counties in 1993.

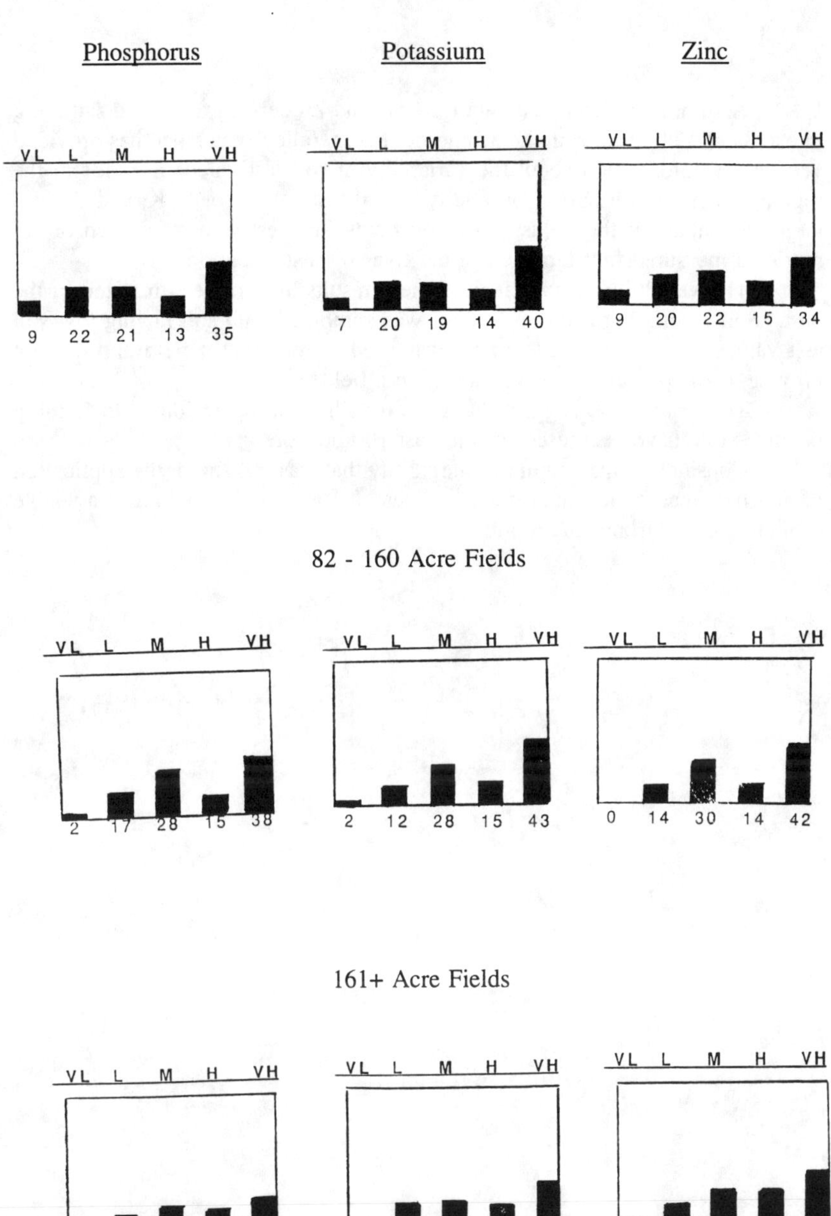

Fig. 12–4. Distribution of soil samples in five soil test categories and the relationship to field size in the Minnesota River Valley Watershed in 1993.

SUMMARY

A summary of soil test data for P, K and Zn derived from grid sampling in excess of 52 000 acres in the Minnesota River Valley Watershed has provided very useful information about the variability of soil test values in fields in the region. There is substantial variability in soil test values for P, K, and Zn in a large percentage of the fields. In many fields, soil test values for each of the nutrients measured fell into each of the five soil test categories.

The variability documented in these results has not been realized in the past. Soil having high soil test values was combined with soil having low soil test values. This combined sample was used to represent a field and did not provide a true picture of the variability in a field.

The results also suggest that the variability can be attributed to farming practices that have been used for the past 100 or more yrs.

Considering the amount of variability that was measured, the application of fertilizer based on an average soil test for a field is no longer a viable economic or environmental option.

13 Classical Statistical and Geostatistical Analysis of Soil Nitrate-N Spatial Variability

G. W. Hergert
R. B. Ferguson
C. A. Shapiro
E. J. Penas
F. B. Anderson

University of Nebraska
North Platte, Nebraska

Variable rate N application based on soil nitrate requires an optimum grid size to develop Kriged maps of N rate. The increased emphasis on preplant and pre-sidedress soil nitrate testing requires accurate estimates of sample numbers for determining the field average. Farmer fields from major corn production areas across Nebraska were sampled on a 30 m grid taking a single 5 cm core at each sampling point in the fall and spring. Samples were taken in 0.3 m increments to 1.2 m from areas 12 ha or less from 61 fields. Subsets of the original sets were generated for 60 and 90 m lag spacings. The top 1% high and bottom 1% low testing samples were reanalyzed as outliers. The high correlation (R^2=.97) between initial and rerun samples and a slope of 0.98 showed that lab variation was not the problem. Tests of normality showed that the data were usually log normally distributed, consequently the geometric mean of the LN distribution or the median of the normal distribution were better estimates of the field average. The CV of nitrate-N averaged 58% for the 30 m grid and the mean, median and CV for the 30, 60, and 90 m grid sets were not significantly different. Classical statistical analysis showed 12 to 20 cores per hectare were required to provide estimates of field nitrate content that were within ± 20% of the mean at confidence levels of 80% to 90%. Geostatistical analysis indicated the average range for sample independence was 90 m. For 10 to 12 ha fields a lag spacing of 75 to 90 m would provide reasonable field average accuracy and would provide a dense enough grid to develop Kriged maps for variable rate N application. Fall soil sampling provided similar results to spring sampling except on sandy soils.

INTRODUCTION

Soil sampling has always been an integral part of a profitable fertilizer program and guidelines for proper soil sampling have been defined since the mid-1940's (Cline, 1944; Peterson & Calvin, 1965; Sabbe & Marx, 1987). Soil scientists know how variable soils can be, so the concept of spatial variability is not new. Past guidelines for soil sampling recognized this variability and suggested dividing fields into different areas based on soil type, color, slope, salinity, erosion, drainage, or cropping history. Currently, there is renewed interest in soil sampling because of variable rate fertilizer application, although most of the application work has been done with P and K (Buchholz, 1991; Carr, et. al., 1991; Larsen & Robert, 1991; Mulla, 1993; Robert, et. al. 1991).

The focus of past soil sampling research has been to determine an adequate number of samples, *n*, to provide a reliable estimate of the mean, the most efficient sampling plan, and some measure of spatial variability. Peterson and Calvin (1965) defined the best sampling plan as one which gives the lowest sampling error at a given cost or the lowest cost at a given sampling error. Past research has shown that grid sampling almost always increases precision compared to random sampling due to the spatial correlation of values (Peck & Melsted, 1967; Peterson & Calvin, 1965; Sabbe & Marx, 1987). The underlying distributions of many soil parameters, including nitrate-N, are usually not normally distributed but are log normal (Reuss et al., 1977). Advances in the theory of regionalized variables enables estimation of the spatial dependence of soil properties regardless of the underlying distribution (Matheron, 1971). Geostatistical analysis provides 'statistically correct' contour maps of the soil parameter being measured. The Kriged maps can be used to develop variable rate fertilizer rates (Mulla, 1991).

In Nebraska there is a great concern about the influence of nitrogen on groundwater quality. Numerous N control and management areas exist (Hergert, 1987a; Ferguson & Peterson, 1991). Excessive N applications and leaching caused by poor irrigation management or excess rainfall have caused substantial increases in groundwater nitrate in several parts of Nebraska (Spaulding, et. al., 1978). In the western corn belt N recommendations can be improved significantly by using deep soil tests for residual nitrate to modify standard N recommendations (Hergert, 1987b; Penas, et. al., 1994). Spatial and temporal variability of nitrate-N in farmer fields needs to be quantified to develop optimum sampling plans for both uniform and variable rate N fertilization. Only limited work has been done in Nebraska during the last 30 yr in this area (Hooker, 1976; Reuss et al, 1977).

The objectives of this research were to: (i) determine the variability of nitrate and nitrate distributions within farmer's corn fields on selected benchmark soils across the state; (ii) determine seasonal variation in soil nitrate-N levels between fall and spring; (iii) determine the most appropriate number of samples required to estimate the mean nitrate-N value within prescribed confidence limits; and (iv) to utilize geostatistical analysis to determine optimum grid sample spacing for developing variable N fertilizer rates based on residual nitrate-N.

METHODS AND MATERIALS

Farmer's fields from major corn production areas across Nebraska were sampled from fall 1987 to spring 1991. Portions of fields generally < 12 ha were sampled using a 30 m lag as the standard spacing. Soil samples were taken in 0.3 cm increments from the soil surface to 1.2 m using a Giddings hydraulic soil probe with a core barrel diameter of 5 cm. Individual cores were bagged separately, air dried, ground, and then analyzed for nitrate-N. Fall sampling was done from early to late November and spring sampling was done from late March to mid April. Complete sets of fall versus spring samples were not taken during 1989 because winter precipitation was much below normal.

With a large number of samples there is always a possibility of laboratory error. No direct measurement of lab error was conducted, however, analyses were checked for outliers. The top 1% high testing samples and bottom 1% low testing samples of a site were resubmitted to the laboratory for analyses. After data screening, two additional data sets were developed for each site. In all fields, the northwest corner was designated as row 1 and column 1. This reference point was used to develop additional data sets representing lags of 60 or 90 m from the original 30 m lag data set.

The mean ($\overline{x}$), median, sample variance, and CV were determined using the Proc Univariate procedure (SAS Institute, 1988). The frequency distributions were viewed and compared to the Shapiro-Wilk *W*-test for both the raw data and a log normal transformed set. Geostatistical analysis was performed using GEO-EAS (USEPA, 1988).

A commonly used method to determine the number of samples required to estimate the mean of a soil parameter within a prescribed confidence interval was developed by Stein (1945):

$$n = \frac{(t^2)(s^2)}{d^2}$$

where t is the tabulated student t value for the desired alpha level and the degrees of freedom of the initial sample. s^2 is the variance of the initial sample and d is 1/2 of the width of the desired precision.

One limitation of the Stein equation is that it assumes an underlying normal distribution of the data, however, soil nitrate values are often distributed log normally (Reuss et al., 1977; Starr, et al., 1992). Parkin et al. (1988) suggested using the uniformly minimum variance unbiased estimator (UMVUE) method of Finney (1941) for log-normally distributed data as a less biased, higher accuracy method than the method of moments or maximum likelihood methods. This analysis was not performed due to time limitations but may be done at a later date.

Stein's formula was used with the current 30 m lag data set using a t value based on 80% and 90% confidence limits ($\alpha = .2$ and $\alpha = .1$). The value of d was calculated as ± 10% to ± 20% of the mean. The number of samples (n) calculated from the formula for this analyses should be interpreted with caution because of the uncertainty of the distributions.

RESULTS AND DISCUSSION

The field sites sampled during this project included irrigated and dryland sites of sandy (16 fields) and fine textured soils (45 fields) from major corn producing areas across Nebraska (Table 13–1).

Error Checking

A plot of the initial versus rerun nitrate values (data not shown) showed an excellent correlation ($R^2 = 0.97$) and the slope (0.98) was not significantly different than 1 indicating that lab variation was not the problem with apparent outliers. The sample reruns confirmed the natural large inherent variability in soil nitrate.

Descriptive Statistics

The Shapiro-Wilk test of normality for nitrate-N in 1.2 m (α=.01) showed 2 of the 61 sites were normally distributed, 42 were log normally distributed, and 17 were neither normal or log normal. The mean nitrate-N for the 30 m lag data sets ranged from 20 kg ha^{-1} to 757 kg ha^{-1} (Table 13–2). Overall there was a good indication that farmers have been practicing improved nitrogen management because of the fairly low residual nitrate levels (median=86 kg ha^{-1} in 1.2 m). The very high testing sites (38, 42, and 43) had a history of manure application. Site 40 also had a dairy manure history but did not test high. In most cases, the median value was lower than the mean confirming the log normal distribution. The geometric mean (87 kg ha^{-1}) was not significantly different than the median (86 kg ha^{-1}).

The coefficient of variation (CV) was high for most sites with the average CV over all fall and spring sites of 58% (Table 13–2). The complicating factor of a high CV is that more samples are required to obtain a given accuracy or predictability to be within a given amount of the true mean. The range in individual samples shows that there is a great deal of variability in soil nitrate from point to point in a field and it emphasizes the importance of taking an adequate number of samples.

Fall Versus Spring Sampling

A paired comparison of the 22 sites sampled during both the fall and spring (Table 13–1) showed no significant difference in the means (Table 13–3). The mean and median differed significantly, and the CV was higher for the fall sampling than the spring. This data confirms the fact that in most of the western corn belt, and especially in dry yr, very little change occurs in soil nitrate-N over winter. A comparison of the individual values among sites showed a somewhat different picture, (Table 13–1). Five of the locations changed < 10%, 11 showed a significant increase, and 6 showed a significant decrease in soil nitrate-N. The increase in soil nitrate-N may be attributed to late fall or very early spring mineralization between the two sampling times.

Table 13–1. Field sites sampled for the Nebraska soil nitrate spatial variability study.

	Sampled						Sampled				
Site	Fall	Sprg	County	Soil	Row X Col	Site	Fall	Sprg	County	Soil	Row X Col
1	y*	y	Scottsblf.	Tripp vfsl**	26 x 5	22	y	n	Logan	Hord sil	10 x 14
2	y	y	Scottsblf.	Mitchell sil	6 x 21	23	y	n	Lincoln	Valentine s	7 x 21
3	y	y	Wayne	Moody, Kenbec sicl	10 x 15	24	y	n	Lincoln	Hord sil	11 x 11
4	y	y	Dixon	Nora, Crofton, sil	17 x 8	27	y	y	Lincoln	Valentine s	10 x 14
5	y	n	Pierce	Thurman, Loup ls	12 x 14	28	y	y	Scottsblf.	Tripp vfsl	7 x 10
6	y	y	Lincoln	Caruso l	11 x 8	29	y	y	Scottsblf.	Tripp vfsl	7 x 10
7	y	y	Clay	Hastings sil	12 x 12	30	y	y	Dixon	Crofton sil	5 x 6
8	y	y	Clay	Hastings sil	12 x 12	31	y	y	Kearney	Holdrege sil	9 x 12
9	y	n	Hall	Wann l	7 x 12	32	y	y	Lincoln	Caruso l	11 x 9
10	y	y	Hall	Wann l	9 x 11	33	y	y	Butler	Butler sicl	9 x 11
11	y	y	Lincoln	Valentine ls	7 x 20	35	n	y	Hall	O'Neill sal	12 x 12
12	y	y	Lincoln	Cozad sil	20 x 7	36	y	y	Cedar	Moody sicl	5 x 6
13	y	y	Lincoln	Uly, Coly sil	9 x 16	37	y	y	Lincoln	Caruso l	11 x 9
14	y	y	Lincoln	Anselmo ls	7 x 20	38	y	y	Dawson	Cozad sil	7 x 21
16	y	n	Lincoln	Cozad, Hord sil	9 x 21	39	y	n	Dawson	Hord, Cozad sil	13 x 12
17	y	n	Dixon	Mdy, Thrm sil	14 x 8	40	n	y	Clay	Crete sil	12 x 12
18	y	n	Cedar	Crofton, Nora sil	8 x 14	41	n	y	Dixon	Nora, Crofton sil	17 x 8
19	y	n	Keith	Bayard sl	8 x 13	42	y	n	Saunders	Sharpsburg sicl	10 x 10
20	y	n	Scottsblf.	Alice vfsl	4 x 18	43	y	n	Saunders	Sharpsburg sicl	4 x 10
21	y	n	Scottsblf.	Keota sil	5 x 7						

*y = yes, n = no

**sil = silt loam; ls = loamy sand; sal = sandy loam; vfsl = very fine sandy loam; sicl = silty clay loam; l = loam; s = sand

Table 13–2. Soil nitrate parameters (kg nitrate-N in 1.2 m) from the sampling study.

Site	Range		CV	Low	Hi	Site	Range		CV	Low	Hi	Site	Range		CV	Low	Hi
	$\bar{x}$	Med					$\bar{x}$	Med					$\bar{x}$	Med			
1F	83	64	65%	25	345	12S	95	73	85%	34	626	30S	271	233	49%	30	1530
1S	88	74	56%	34	393	13F	87	71	73%	15	321	31F	150	127	70%	11	392
2F	44	40	32%	25	115	13S	102	81	70%	15	446	31S	88	74	66%	19	352
2S	44	39	42%	19	159	14F	118	108	36%	36	295	32F	83	72	48%	43	228
3F	45	43	37%	17	136	14S	56	50	43%	16	143	32S	100	92	37%	35	230
3S	60	59	32%	24	133	16F	106	96	46%	36	458	33F	44	41	34%	18	136
4F	134	99	79%	26	589	17F	75	63	57%	32	298	33S	46	45	29%	22	116
4S	131	103	70%	23	495	18F	108	77	76%	19	412	35S	26	24	37%	12	95
5F	69	56	59%	13	264	19F	109	82	72%	20	336	36F	53	50	18%	33	148
6F	52	44	57%	22	184	20F	113	77	81%	19	377	36S	46	45	37%	3	101
6S	59	54	39%	25	167	21F	101	77	60%	31	263	37F	123	101	81%	24	691
7F	49	35	79%	10	231	22F	24	20	47%	4	85	37S	139	125	64%	32	589
7S	56	48	51%	19	168	23F	38	24	52%	9	122	38F	314	227	84%	48	1465
8F	17	16	46%	7	60	24F	27	26	55%	8	124	38S	296	208	78%	57	1088
8S	22	21	31%	8	44	27F	27	24	48%	8	78	39F	166	138	72%	21	673
9F	43	39	37%	12	88	27S	30	28	41%	8	91	40S	43	21	134%	8	400
10F	56	46	74%	10	276	28F	109	59	95%	59	241	41S	241	196	86%	26	1509
10S	65	59	46%	20	200	28S	102	73	75%	36	461	42F	757	704	46%	161	1552
11F	60	57	37%	27	147	29F	96	47	91%	24	326	43F	178	159	54%	52	399
11S	40	38	36%	19	125	29S	114	74	76%	30	367						
12F	91	75	82%	25	473	30F	226	208	47%	17	1049						

Table 13–3. Comparison of fall versus spring nitrate statistics for the 30 m lag spacing.

	Mean		Median		CV	
	Fall	Spring	Fall	Spring	Fall	Spring
	--------kg Nitrate-N in 1.2 m--------					
Avg.	94	92	75	77	62	52
Range	17 - 314	22 - 296	16 - 227	21 - 253	18 - 95	29 - 88

Table 13–4. Soil nitrate-N parameters for the 61 sites sampled on different grid spacings.

Parameter	30 m lag	60 m lag	90 m lag
	--------kg Nitrate in 1.2 m--------		
n	111	30	14
mean	104	107	99
median	86	86	88
CV	58%	56%	52%
range	26-757	27-760	26-851

The decrease in soil nitrate on the sandy soils (sites 9, 11, 14, 28) most likely is due to leaching from over-winter precipitation. Losses can also occur on fine textured soil (sites 31 and 38) but are less common.

For most fine-textured soils in Nebraska, the data confirmed that fall sampling was a good estimator of residual nitrate-N. Taking fall samples allows farmers to plan their fertilizer program over the winter and gives them a wider time window for fertilizer N application.

Table 13–5. Core number to be within ± 10% or 20% of the mean at alpha levels of 0.1 or 0.2.

	specified accuracy													
	±10%		±20%			±10%		±20%			±10%		±20%	
	α = .1	α = .2	α = .1	α = .2		α = .1	α = .2	α = .1	α = .2		α = .1	α = .2	α = .1	α = .2
1F	113	69	28	17	12F	163	99	41	25	30F	116	70	29	18
1S	82	50	21	13	12S	207	126	52	31	30S	68	41	17	10
2F	28	17	7	5	13F	142	87	36	22	31F	135	81	34	20
2S	49	29	12	7	13S	133	81	33	20	31S	119	72	30	18
3F	38	23	10	6	14F	36	22	9	5	32F	63	38	16	10
3S	27	16	7	4	14S	50	30	13	8	32S	38	23	10	6
4F	168	102	42	26	16F	59	36	15	9	33F	32	19	8	5
4S	134	81	33	20	17F	89	54	22	14	33S	23	14	6	4
5F	95	57	24	14	18F	157	95	39	24	35S	38	23	10	6
6F	87	53	22	13	19F	139	85	35	21	36F	21	12	5	3
6S	41	25	10	6	20F	183	110	46	28	36S	29	17	7	4
7F	169	102	42	26	21F	82	49	21	12	37F	182	110	38	23
7S	69	42	17	11	22F	61	37	15	9	37S	111	67	28	17
8F	62	37	15	9	23F	74	45	19	11	38F	195	118	49	29
8S	26	16	6	4	24F	84	51	21	13	38S	167	101	42	25
9F	38	23	9	6	27F	62	38	16	9	39F	142	86	36	22
10F	151	91	38	23	27S	46	28	12	7	40S	489	296	122	74
10S	58	35	15	9	28F	254	154	64	38	41S	204	123	51	31
11F	38	23	10	6	28S	159	96	40	24	42F	59	35	15	9
11S	35	21	9	5	29F	233	141	58	35	43F	82	50	21	13
12F	163	99	41	25	29S	158	95	16	9	AVG	105	63	26	15

Table 13–6. Models and range (m) of semivariograms for the 61 sites.

Site	Model	Range	Site	Model	Range
1F	SPHR*	110	21F	LINR	122
1S	GAUS	115	22F	SPHR	50
2F	SPHR	90	23F	SPHR	52
2S	SPHR	90	24F	NONE	---
3F	SPHR	275	27F	SPHR	97
3S	SPHR	260	27S	SPHR	76
4F	SPHR	90	28F	GAUS	82
4S	SPHR	200	28S	SPHR	85
5F	SPHR	80	29F	SPHR	73
6F	SPHR	85	29S	SPHR	90
6S	SPHR	80	30F	SPHR	60
7F	SPHR	60	30S	SPHR	79
7S	SPHR	60	31F	GAUS	60
8F	SPHR	60	31S	SPHR	60
8S	GAUS	90	32F	GAUS	40
9F	SPHR	75	32S	GAUS	60
10F	SPHR	107	33F	SPHR	110
10S	SPHR	67	33S	SPHR	60
11F	SPHR	104	35S	SPHR	100
11S	SPHR	73	36F	SPHR	60
12F	NONE	---	36S	GAUS	100
12S	NONE	---	37F	SPHR	60
13F	SPHR	90	37S	SPHR	45
13S	SPHR	90	38F	GAUS	52
14F	SPHR	80	38S	GAUS	90
14S	SPHR	60	39F	SPHR	90
16F	SPHR	90	40S	GAUS	90
17F	SPHR	85	41S	SPHR	122
18F	SPHR	90	42F	GAUS	76
19F	SPHR	230	43F	SPHR	79
20F	GAUS	122			

*GAUS = gaussian, LINR = linear, SPHR = spherical, NONE = no spatial structure.

Number of Cores Required

The comparison of statistical parameters for the 60 and 90 m lag data sets and the 30 m set showed no statistical difference (Table 13–4). This is usually not the case as most studies show a decreasing CV as sample number increases (Peck & Melsted, 1967; Peterson & Calvin, 1965; Sabbe & Marx, 1987).

The Stein test showed that more samples than any producer might ever take would be required for a precision of ±10% at α=.1 (Table 13–5). Because of the predominance of log normal data, these numbers are skewed to the high side. The data indicate that in most instances at least 15-25 samples from an area 12 ha or smaller would be required for an acceptable accuracy. Sampling on a 60 or 90 m grid normally gave this many samples and would provide sufficient numbers for geostatistical analysis.

Geostatistical analysis

Most fields showed a definite spatial structure and different variogram models were developed (Table 13–6). There was no consistency in the nugget or sill values due to the extreme variability between sites. Values for the range of the semivariograms ranged from 40 to 275 m with the 25% to 75% quartile ranging from 60 to 90 m with a median of 85 m. The average value of the range over all sites was 93 m and was skewed due to 3 high values.

CONCLUSION

The geostatistical analysis coupled with the classical statistical analysis showed no significant differences in similar parameters for the 60 or 90 m lag spacing versus the 30 m lag spacing. This suggests that sampling distances of up to 90 m (~300 ft) can be used to determine nitrate-N content for 'field average' N recommendations if at least 20 samples are taken. For variable rate N recommendations based on residual nitrate this provides a reasonable number of samples (20–30) without being too costly. This sampling density conforms to current sampling guidelines in Nebraska (Penas et al., 1991) and is also similar to tabulated values from Table 13–4. The data point out the large variability inherent in a mobile nutrient like soil nitrate-N.

ACKNOWLEDGEMENTS

The authors acknowledge the financial support from the U.S. Geological Survey provided through the University of Nebraska Water Center and from the Nebraska Corn Development, Utilization and Marketing Board. Thanks to all the farmers who allowed us to sample their fields. We would like to thank Dr. Ken Frank and the staff at the UNL-Soil Testing Lab for their support in timely analysis of the thousands of samples we generated. Dave Dunn, Tonda Olson, and Bob Freicks helped take these soil samples for NEREC. Jim Petersen and Krystal Herrick at UNL-WCREC sampled fields in west central Nebraska, coordinated sampling operations, sample handling and storage, data management and analysis. They were the key to keeping the thousands of samples from this project organized.

REFERENCES

Buchholz, D.D. 1991. Missouri grid soil sampling project. p. 6–12. Proc. North Central Extension-Industry Soil Fertility Conf.

Carr, P.M., G.R. Carlson, J.S. Jacobsen, G.A. Nelson, and E.O. Skogley. 1991. Farming soils, not fields: A strategy for increasing fertilizer profitability. J. Prod. Agric. 4:57–61.

Cline, M.G. 1944. Principles of soil sampling. Soil Sci. 58:275–288.

Ferguson, R.B., and T. Peterson. 1991. Ground water quality research and demonstration projects in Nebraska. p. 107–113. Proc. North Central Extension-Industry Soil Fertility Conf.

Finney, D.J. 1941. On the distribution of a variate whose logarithm is normally distributed. J. Res. Stat. Soc. Suppl. 7:144–161.

Hergert, G.W. 1987a. Water quality issues in Nebraska. p. 87–92. Proc. North Central Extension-Industry Soil Fertility Conf.

Hergert, G.W. 1987b. Status of residual nitrate-nitrogen soil tests in the U. S.A. *In* J.R. Brown (ed.) Soil Testing: Sampling, correlation, calibration and interpretation. Soil Sci. Soc. Am. Spec. Publ. 21. Madison, WI.

Hooker, M.L. 1976. Sampling intensities required to estimate available N and P in five Nebraska soil types. M.S. Thesis Univ. of Nebraska.

Larson, W.E., and P.C. Robert. 1991. Farming by soil. *In* R. Lal and F. Pierce (ed.) Soil Management for Sustainability. Soil and Water Cons. Soc.

Matheron, G. 1971. The theory of regionalized variables and its application. Cah. Cent. Morphol. Math. Fontainebleau 5. Centre de Geostatistique.

Mulla, D.J. 1991. Using Geostatistics and GIS to manage spatial patterns in soil fertility. p. 336–345. *In* G. Kranzler (ed.) Proc. Automated Agriculture for the 21st Century. ASAE, St. Joseph, MI.

Mulla, D.J. 1993. Mapping and managing spatial patterns in soil fertility and crop yield. p.15–26. *In* P.C. Robert et al. (ed.) Soil Specific Crop Management. ASAE, Madison, WI.

Parkin, T.B., J.J. Meisinger, S.T. Chester, J.L. Starr, and J.A. Robinson. 1988. Evaluation of statistical estimation methods for log normally distributed variables. Soil Sci. Soc. Am. J. 52:323–329.

Peck, T.R., and S.W. Melsted. 1967. Soil Testing and Plant Analysis Part I: Soil Testing. SSSA Spec. Pub. #2. p. 25–36.

Penas, E.J., R.B. Ferguson, K.D. Frank, G.W. Hergert, and R.A. Wiese. 1991. Guidelines for soil sampling. Univ. of Nebraska. NebGuide G91–1000.

Penas, E.J., G.W. Hergert, and R.B. Ferguson. 1994. Fertilizer suggestions for corn. Univ. of Nebraska NebGuide G74–174 (Revised).

Peterson, R.G., and L.D. Calvin. 1965. Sampling. p. 54–72. *In* C.A. Black, et al. (ed.) Methods of Soil Analysis I. ASA. Monograph #9, Madison, WI.

Reuss, J.O., P.N. Soltanpour, and A.E. Ludwick. 1977. Sampling distributions of nitrates in irrigated fields. Agron. J. 69:588–592.

Robert P.C., W.H. Thompson, and D. Fairchild. 1991. Soil specific anhydrous ammonia management systems. p. 418–426. *In* G. Kranzler (ed.) Proc. Automated Agriculture for the 21st Century. ASAE, St. Joseph, MI.

Sabbe, W.E., and D.B. Marx. 1987. Soil sampling: Spatial and temporal variability. p. 1–14. *In* J.R. Brown (ed.) Soil Testing: Sampling, Correlation, Calibration, and Interpretation. Soil Sci. Soc. Am. Spec. Publ. 21, Madison, WI.

SAS Institute. 1988. SAS Users Guide. Statistics Version 6.03 ed. SAS Inst., Cary, N.C.

Spaulding, R.F., J.R. Gormly, B.H. Curtiss, and M.E. Exner. 1978. Nonpoint nitrate contamination of groundwater in Merrick County, Nebraska. Ground Water 16:86-95.

Starr, J.L., T.B. Parkin, and J.J. Meisinger. 1992. Sample size consideration in the determination of soil nitrate. Soil. Sci. Soc. Am. J. 56:1824–1830.

Stein, C. 1945. A two sample test for a linear hypothesis whose power is independent of the variance. Ann. Math. Stat. 16:243–258.

United States Energy Protection Agency. 1988. GEO-EAS (Geostatistical Environmental Assessment Software). Environ. Monitoring Laboratory EPA/600/4–88/033. Las Vegas, NV.

14 Managing Risk and Variability With Fuzzy Soil Interpretations

M. D. Mays
C. S. Holzhey

USDA-Soil Conservation Service,
Lincoln, Nebraska

I. Bogardi

Dep. of Civil Engineering
University of Nebraska-Lincoln
Lincoln, Nebraska

A. Bardossy

Dep. Civil Engineering
Institute of Hydrology and Water Resources
University of Karlsruhe
Karlsruhe, Germany

The risk a decision maker takes when using soils data to implement a practice or make an interpretive decision is dependent on the quality of the data, the tools for decisionmaking, the amount of pertinent data, how well the soil forming processes are understood, and how well such processes can be conveyed in an understandable way. Fuzzy logic is a tool that can be used to characterize uncertainty in soils information so that a risk-based method of soil interpretations can be implied. A methodology is presented to demonstrate how fuzzy soil interpretations provide a realistic approach to decisionmaking for risk-based soil interpretations. Map units from a soil survey are used to demonstrate the appropriateness of soils for septic tank filter fields and tillage.

INTRODUCTION

The use of Fuzzy Logic as a tool for assessing uncertainty in soil science has been slow to evolve despite studies pointing to its usefulness (Burrough, 1989; Paetzold, 1987; Mays et al. 1990; and Bogardi, et al., 1994). Much of the lack of use by many soil scientists can be related to a discomfort and an unfamiliarity with the engineering risk approach, the somewhat cumbersome mathematical computations, and the lack of a widespread computer friendly approach to soil interpretations. Fuzzy logic as a tool can be used in a "computer friendly environment" to express the uncertainty in data used to formulate decisions (Kosko & Isaka, 1993). Decisions about soils evolve into recommendations of seeding rates, fertilizer applications, pesticide applications, etc.

In many cases, the quality of data used to make recommendations is not known, even though it is crucial to the kind or level of accuracy with which a decision is made. If estimated data must be used because measured data for the site are not available, then the decision made should reflect data used to arrive at the decision. There is a need to develop a system of soil interpretations that will allow the incorporation of all data elements used to arrive at a decision. The purpose of this paper is to provide information on how fuzzy logic can be used as a tool to manage risk and variability in soil interpretations.

Risk

There is a need to express soil interpretations in terms of the risk assumed by a particular decision. Such an approach allows users to base their decision on the risk they are willing to assume with a given practice. Identification of risk enables the user to avoid, reduce, or otherwise manage these uncertainties that can be expressed as risk (Wilson & Crouch, 1987). Such risk may be expressed in the form of "inconvenience" and "discomfort." A consideration is given to the consequence of choices and is reflected in the final decision. This paper will provide the theoretical risk framework that will be made applicable in a computer program that is under development at the Soil Conservation Service National Soil Survey Center. The challenge for risk assessment in soil interpretations is to give the best estimate and the range of uncertainty associated with the practice. The advantage of mathematical risk assessment is that it openly considers uncertainties, so that the user can take advantage of such knowledge through risk management in an informed way. The disadvantage of a risk assessment approach to soil interpretations is that the process is time consuming and somewhat tedious. It may be necessary, however, in order to gain public confidence and consent for what we are doing in soil survey interpretations (Mays, 1994).

Since the estimates used in risk analysis have major uncertainties, it is important that cost estimates are applied to practices needed to reduce the risk. Such cost estimates are used to determine the feasibility of such risk reduction measures (an important issue for the user). The user does not want to waste resources on minimizing risk or on improvements that do not reduce risk beyond a required level.

Risk analysis in general includes three elements: i) load or exposure, ii) capacity or threshold, and iii) the failure event when load is larger than capacity. As we have described risk analysis in this paper, we have added a fourth step, the consequence of failure. Often the first three elements are used to define engineering risk. Details of the methodology are described in an internal report on "Risk Assessment for Soil Interpretations" (Bogardi & Bardossy, 1993).

Consequence

The consequence resulting from the application of a practice is the effect of carrying out a given practice as it relates to the cost or misgiving. Non-monetary consequences are those misgivings that are not related to financial failure but are an inconvenience or a discomfort to the user (individual or private consequence) or to others (social or public consequence). Social or public is used to describe those consequences that are perceived or would cause inconvenience or discomfort to others. Monetary (economic) consequences are the operation, installation, and maintenance costs associated with a practice. The monetary consequence is based on the actual value of property affected.

The addition of consequences to a risk analysis allows one to apply a cost factor to weigh in the decision of whether or not additional measures must be applied in order to meet a satisfactory level versus an acceptable risk taken. A continuum allows one to gauge the level of inconvenience or discomfort that may be associated with the degree of practice application. The user is in the best position to make the decision of whether or not the practice is feasible in proportion to cost.

MATERIAL AND METHODS

Fuzzy set techniques and multi-criterion decision making were combined in accordance with methods described by Bogardi and Bardossy (1983 and 1993) and Bogardi et al. (1994). The method involves the establishing of criteria which are used to calculate the ratings for each soil or map unit and is compared against standard criteria for that rating.

Criteria used to make ratings were adapted from the National Soil Survey Handbook (Soil Conservation Service, 1993). A panel of soil scientists from the National Soil Survey Center and the Wahoo, Nebraska Soil Survey Office was assembled to evaluate the criteria and to set the optimal values for each. Where supporting data were not available through field notes, laboratory data or other means, "best estimates" were made by the panel.

An optimal data set (standard) is described by a fuzzy set membership function (Bogardi et al. (1994). For example, the optimal set for the property permeability in rating a soil for tillage (Table 14–1 and Fig. 14–1) is described by: i) the largest possible interval (0 to 50 cm/hr) within which the rating would still be appropriate for the property, and ii) an interval (0.5 to 15 cm/hr) which can be considered to be an excellent rating. The actual set (Table 14–4) represents the measured or estimated data used in the rating.

The distance between the optimal and actual sets of properties is measured by the property evaluation measure (PEM). The closer the PEM is to 1.00 the closer the value for the actual set is to the value for the optimal set (Fig. 14–1 and 14–6). The PEM is used to indicate the degree of membership or partial membership the measured values (actual set) have to a set of standard values (optimal set). It reflects the degree to which the actual set of data has membership in the optimal set of data. The interaction between soil properties can be accounted for using a pairwise approach. (Bogardi et al. 1994).

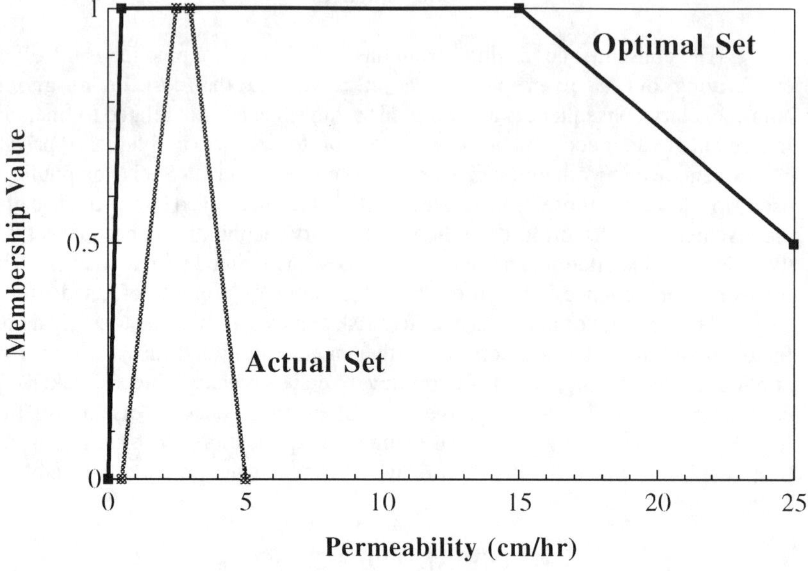

Fig. 14–1. Relationship of actual to optimal property values for permeability in rating Yutan-Judson complex, 5 to 11 percent slopes for tillage.

Map units often have more than one component. At least three possible ways of evaluating map units are recognized: i) the interpretations may be made based on the dominant soil in the map unit, ii) the interpretations may be made based on the most limiting soil in the map unit (a conservative approach), iii) the interpretations may be made based on the combination of soils in the map unit (the overall rating). For this study we have chosen to use the overall ratings.

In order to illustrate the use of the Fuzzy Logic program, we have chosen the Yutan-Judson complex, 5 to 11 percent slopes map unit (YUD) in Saunders County, Nebraska. The YUD map unit is comprised of 67 percent Yutan and similar soils, and 33 percent Judson and similar soils. Soil in this map unit developed on uplands in loess parent material. The family particle size class is fine.

Data for YUD were taken from field descriptions and records for use, when available. Using the Fuzzy Logic program, the distance between the actual data representing the map unit (YUD) and the optimal data for the practices septic tank absorption field and tillage were calculated (illustrated for the Yutan component in Tables 14–1, 14–2, 14–3, and 14–4). The distances are based on programming metric used to obtain a rating evaluation measure over data influencing a rating (Bardossy, et al., 1985). Different properties were weighted according to expert judgement. The rating for a map unit is represented by the fuzzy mean (R) from the composite program. For purposes of comparison with conventional ratings, we have defined the following categories:

Table 14–1. Definition of optimal sets of properties used in ratings for tillage.

Property	Acceptable Set		Appropriate Set		
	Min	Max	Min	Max	Weight
Permeability (cm/hr)	0	50	.5	15	10
Salinity (ds/m)	0	20	0	0	10
SAR	0	70	0	1	9
Texture	0	10	0	4	9
Organic Matter (%)	0	40	2	4	6
3-10 in. fractions (%)	0	90	0	3	6
> 10 in. fractions (%)	0	70	0	1	6
Ponding	0	10	0	0	6
Water table (cm)	2	900	90	800	5
Flooding	0	10	0	2	5
Wind Erodibility Group	1	9	8	9	5

Table 14–2. Definition of actual sets of properties for Yutan used in ratings for septic tank absorption fields.

Variable	Most likely set		Possible set	
	Min	Max	Min	Max
Total subsidence (cm)	0	0	0	0
Flooding	0	0	0	0
Depth to hard bedrock (cm)	1524	1600	1016	1778
Depth to cemented pan (cm)	500	550	180	900
Ponding	0	0	0	0
Depth to high water table (cm)	500	550	180	900
Permeability (24-69", cm/h)	3.8	3.8	1.5	5
Slope (%)	6	7	4	12
3-10 in. fractions (%)	0	0	0	5
> 10 in. fractions (%)	0	0	0	0

Table 14–3. Definition of optimal sets of properties used in rating septic tank absorption fields.

Property	Acceptable Set Min	Acceptable Set Max	Appropriate Set Min	Appropriate Set Max	Weigh
t					
Total subsidence (cm)	0	76	0	0	5
Flooding	0	10	0	0	10
Depth to hard bedrock (cm)	0	2000	1000	2000	10
Depth to cemented pan (cm)	0	2000	200	1000	8
Ponding	0	10	0	0	7
Depth to high water table	0	2000	200	2000	8
Permeability (61-175cm, cm/hr)	0	25	5	15	8
Slope (%)	0	50	2	2	5
3-10 in. fractions (%)	0	90	0	0	5
> 10 in. fractions (%)	0	70	0	0	5

Table 14–4. Definition of actual sets of properties for Yutan used in ratings for tillage.

Variable	Most likely set Min	Most likely set Max	Possible set Min	Possible set Max
Permeability (cm/hr)	2.5	3.8	.5	5
Salinity (ds/m)	0	0	0	0
SAR	0	0	0	0
Texture surface 30cm	5	5	3	6
Organic Matter (%)	1	1	.5	2
3-10 in. fractions (%)	0	0	0	1
> 10 in. fractions (%)	0	0	0	1
Ponding	0	0	0	0
Water table (cm)	180	200	180	900
Flooding	0	0	0	0
Wind Erodibility Group	7	7	6	8

R = 0.0 - 0.3 severe
R = .3 - .6 moderate
R = .6 - 1.0 slight

RESULTS AND CONCLUSIONS

Use of the fuzzy logic procedure to evaluate soil properties provides a reasonable approach for addressing uncertainty in soil interpretations. Using derived PEM's one may derive a fuzzy rating for a practice such as tillage or septic tank filter field. The fuzzy ratings take into consideration the full range of possible values for each rating criteria, thus allowing the contribution to the overall rating in accordance with expert opinion.

The ten criteria used to rate soils for septic tank filter fields were ascertained to not contribute equally toward the rating, therefore, the weighted value for each property was used for the relative importance. Properties such as permeability and slope had partial membership, that is their actual value was not completely within the optimal values and had rating < 1 for those criteria.

Using the conventional ratings for the YUD map unit for septic tank absorption field, the rating is moderate with limitations due to permeability and slope. The field soil scientist evidently decided to over ride the limitation on permeability because field conditions supported slight limitation due to permeability for the Judson soil. The fuzzy means for this map unit (YUD) is .80. Using the comparative scale, the rating would be slight. The plot of overall data for the YUD map unit (Fig. 14–2) shows a fuzzy rating ranging from .60 to 1.0 (Table 14–5). The fuzzy rating may be interpreted to show a possibility rating of about .80 or 80% slight. The fuzzy rating takes into account much more of the knowledge that the field soil scientists have at their disposal. Figures 14-3 and 14-4 show some of the membership functions of individual property ratings used in the overall rating (Fig. 14–2).

Table 14–5. Fuzzy means (R) of rating evaluation measures compared to field ratings.

Ratings	Fuzzy Mean	Fuzzy Rating	Fuzzy Rating Interpreted	Field Rating
Septic tank absorption field	.80	.60-1.0	.80 slight	moderate
Tillage	.88	.77-1.0	.88 well suited	suited
Remarks:	0.0 - 0.3	severe rating or poorly suited		
	0.3 - 0.6	moderate rating or suited		
	0.6 - 1.0	slight rating or well suited		

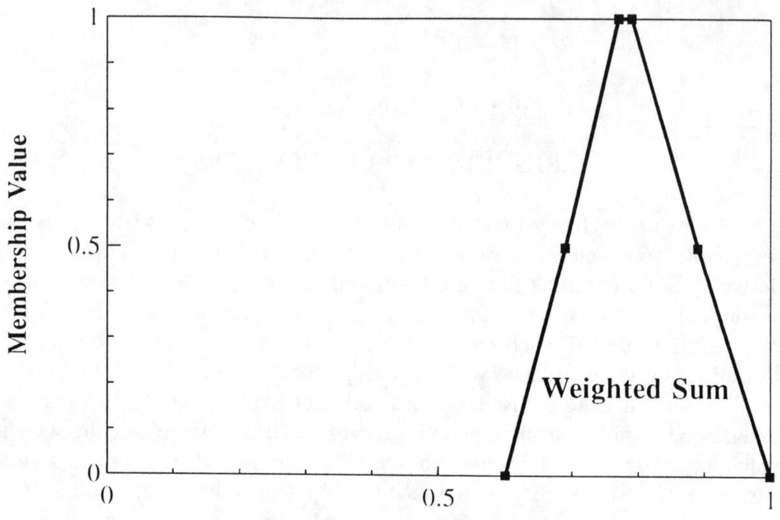

Fig. 14–2. Overall membership functions of properties for Yutan-Judson complex, 5 to 11 percent slopes for septic tank absorption fields.

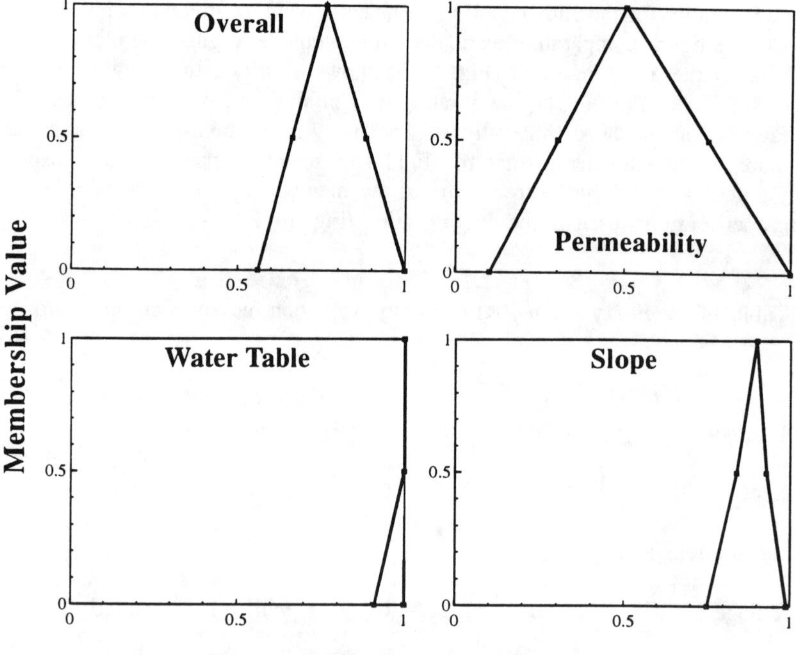

Fig. 14–3. Membership functions of selected soil properties for Yutan soils for septic tank absorption fields.

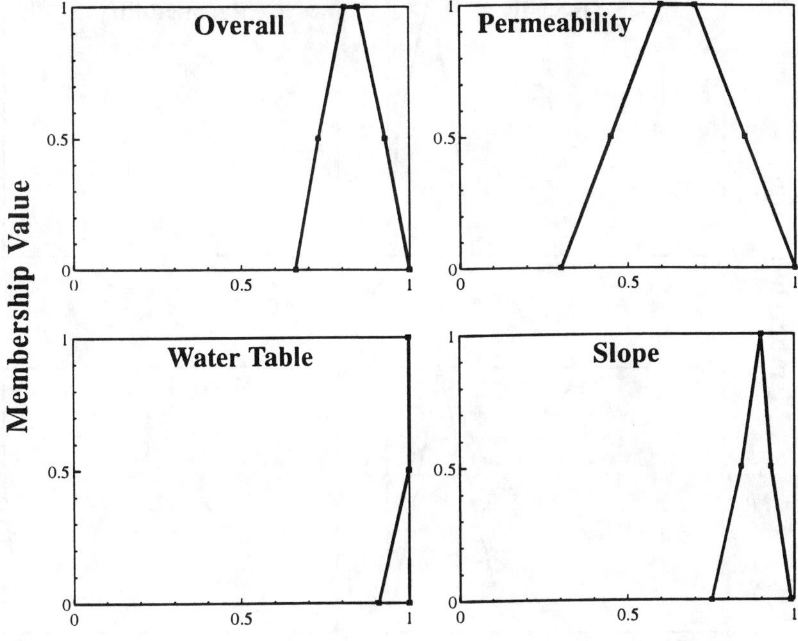

Fig. 14–4. Membership functions of selected soil properties for Judson soils for septic tank absorption fields.

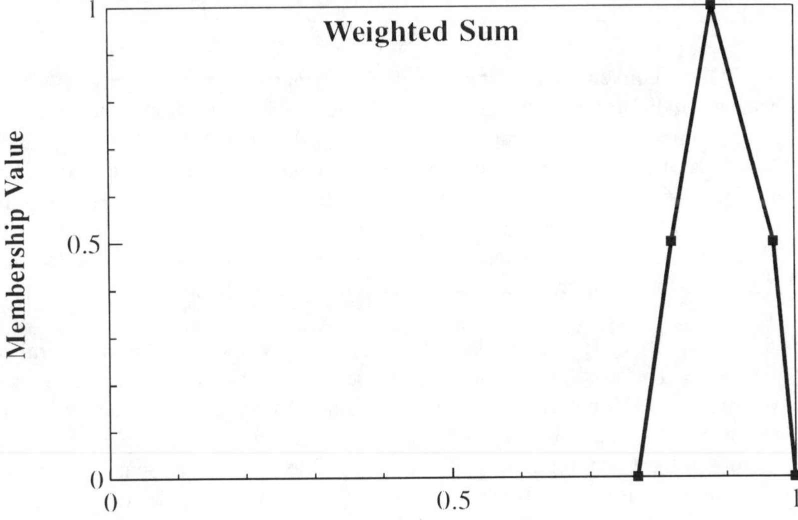

Fig. 14–5. Overall membership functions of properties for Yutan-Judson complex, 5 to 11 percent slopes for tillage.

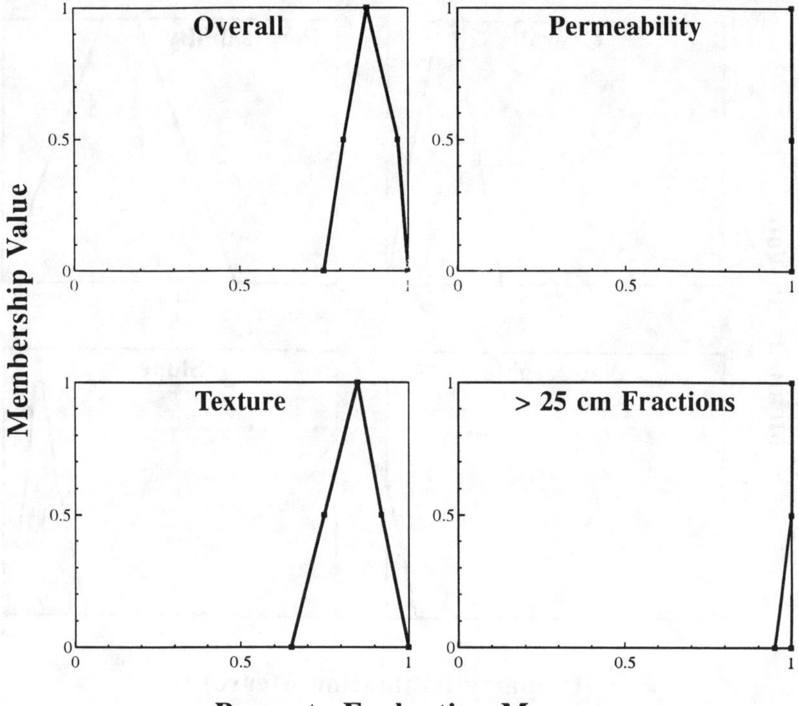

Fig. 14-6. Membership functions of selected soil properties for Yutan soils for tillage.

The conventional rating for the YUD map unit for tillage was "suited" because of limitations due to texture (Table 14–5). The fuzzy means for the tillage rating is .88 and ranges from .77 to .1.0 (Fig. 14–5). The membership functions of selected soil properties used in making the fuzzy rating are shown in Fig. 14–6 and 14–7. As with the septic tank absorption field fuzzy rating, one may interpret the rating to show a possibility rating that ranges from .77 to 1.0. This range may be interpreted to mean that the rating is about .88 or 88% slight. Again the fuzzy rating more closely reflects what is actually experienced in the field.

The fuzzy rating possibilities can be broadly interpreted as a general risk by: risk = 1 - possibility. One should be very cautious, however, in making assumptions or comparisons using these data before a detailed analysis is done. According to Bogardi and Bardossy (1994) in an internal report, the fuzzy vectors in the fuzzy logic program are applicable to the development of the risk assessment phase for soil interpretations that is currently under development at the National Soil Survey Center. Since risk is relative, one cannot fully interpret the meaning and the significance of these data until a number of soils from Saunders County, Nebraska covering a broad range are evaluated. Based on our knowledge of the YUD map unit, however, we expect the risk to be small for

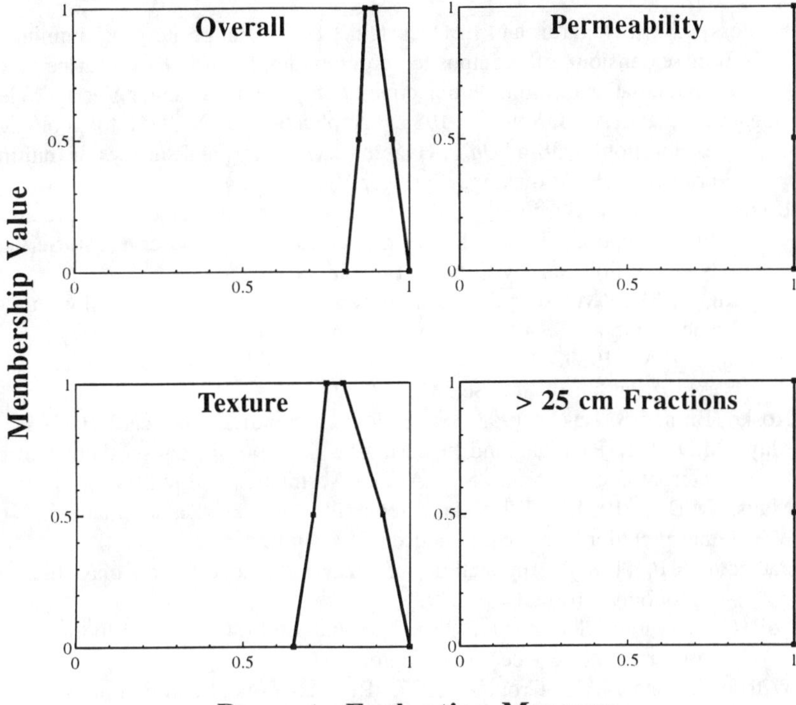

Fig. 14–7. Membership functions of selected soil properties for Judson soils for tillage.

septic tank absorption filter fields and for tillage. The use of the YUD map unit with the Fuzzy Logic program and the membership functions that are applicable to a risk assessment approach offer great prospect for development of a risk based system of soil interpretations. The fuzzy ratings reflect the uncertainty in data and can be transformed into a risk assessment defined by the consequences of the misgivings.

RECOMMENDATIONS FOR FUTURE DEVELOPMENTS

1. Programming of the risk phase of the Fuzzy Logic program.

2. Incorporation of factors that will directly reflect the quality of data in the final rating.

3. Refinement and further research on the interaction of soil properties and their effect on the final rating.

REFERENCES

Bardossy, A., I. Bogardi, and L. Duckstein. 1985. Composite programming as an extension of composite programming. *In* P. Serafine (ed.) Mathematics of multiple objective optimization. Springer Verlag, Wien.

Bogardi, I., and A. Bardossy. 1983. Application of MCDM to geological exploration. p. 38-47. *In* P. Hansen (ed.) Essays and surveys of multiple criteria decision making. Springer-Verlag. New York.

Bogardi, I., and A. Bardossy. 1993. Risk assessment for soil interpretation. Final Report, phase III. Fuzzy Logic, Soil Conservation Service, National Soil Survey Center. Lincoln, NE.

Bogardi, I., M.D. Mays, and A. Bardossy. 1994. Soil interpretations under uncertainty. SSSA, Madison, WI. (in review)

Burrough, P.A. 1989. Fuzzy mathematical methods of soil survey and land evaluation. J. of Soil Sci. 40:477–492.

Kosko, B., and S. Isaka. 1993. Fuzzy logic. Scientific America 269:32–41.

Mays, M.D., R.L. Paetzold, and M.J. Mausbach. 1990. Fuzzy set theory in soil interpretations. Amer. Soc. Agron. Abstracts. pp. 297-298.

Mays, M.D. 1994. Why data reliability and risk assessment in soil interpretations. SSSA. Madison, WI. (in review)

Paetzold, R.F. 1987. Applications of fuzzy set theory to soil classification. Agronomy Abstracts. p. 229.

Soil Conservation Service. 1993. National Soil Survey Handbook. Soil Conservation Service. Washington, D.C.

Wilson, R., and E.A.C. Crouch. 1987. Risk assessment and comparisons: An introduction. Science. 236:267–270.

15 Methods to Characterize Soil Resource Variability in Space and Time

J. Bouma

Dep. Soil Science and Geology
Agricultural University
Wageningen, Netherlands

Site specific soil management requires us to take a new look at methods to gather and interpret soil information. Definition of so-called *representative soil profiles* for mapping units on even detailed soil maps (1:10000 and larger) cannot be used to adequately represent spatial variability patterns on the field scale which are often gradual rather than abrupt, while considerable variation may occur within single mapping units. Moreover, the type of soil characteristics being observed or measured in regular soil surveys is not necessarily relevant for specific applications in the area of site specific management. The latter activity is strongly focused on soil behavior: how much fertilizer or biocide should be applied at what time to result in high production, coupled with low leaching or runoff of agrochemicals.

There are two ways in which soil resource variability can be characterized: (i) intensive monitoring of water and agrochemical-content of soils during the yr and of associated crop development, and (ii) dynamic simulation modeling of water and chemical fluxes as well as crop growth. The first procedure is time consuming and costly and can only cover conditions as they occur over a by-necessity-limited-period of time. The second procedure is much more flexible as it allows characterization for different conditions as long as these can be expressed by realistic parameters in model equations. The latter statement is crucial: there are examples of complicated deterministic models being used without adequate basic data, yielding irrelevant results. The obvious compromise is to use models with a degree of complexity that is in equilibrium with available data, while field monitoring data over a limited period of time are available to calibrate and validate the models. Lack of directly measured basic data to feed the simulation models can be compensated for by deriving pedotransfer functions which relate existing soil data to unavailable data, needed for modeling (e.g. Bouma, 1989).

A review will be presented in which the space and time dimension of soil data will be discussed in relation to site specific management, as explored in our work.

THE SPACE DIMENSION

Point observations can be interpolated to areas of land using geostatistical techniques which are widely accessible through user-friendly software. The question then is: what to observe at every point and how many points to observe. The first question has been the topic of much soil research resulting in rigid protocols for soil profile descriptions. Many of the features being described have little direct relation with parameters used for modeling. For instance, how to interpret a soil structure description in terms of: *weak fine subangular blocky*, when a simulation model for solute movement and crop growth needs parameters for hydraulic conductivity and moisture retention? The key to this problem is a so-called *functional analysis*, translating pedological features and parameters into parameters needed for simulation. Pedogenic soil horizons combine a number of soil parameters into characteristic combinations. Such horizons may be different from a soil-forming point of view but they may act identically when considered from a functional point of view. Wosten et al (1985) and Finke et al (1992) combined different pedological horizons to form functional horizons, each with characteristic hydraulic properties. The advantage of using such functional horizons as *carriers* of either soil chemical or physical information, is that soil horizons have a lateral continuity. Thickness of and depth to certain functional horizons are ideal continuous parameters to be used in interpolation techniques such as kriging. Moreover, combining soil horizons into functional horizons implies that a lower number of soil mapping units will occur, which, in turn, facilitates calculations.

The next question to be considered deals with the interpolation of point data to areal data. In land areas with a relatively low internal variability, a low number of data points may be sufficient for interpolation purposes, which may use classical statistics or geostatistics. More complex areas, on the contrary, require more data points. Use of geostatistics allows estimates of the optimal number of samples and of the accuracy of results obtained. Work of Stein et al (1991) has demonstrated that different landforms within an area of land can be used to divide the area into subareas, resulting in better estimates within each subarea as compared with a procedure where observation points are scattered randomly over the entire area. Figure 15–1 demonstrates spatial variability of the soil moisture capacity, calculated with a simulation model, for four major soil units in a sandy area in the Netherlands. Semi-variograms are characteristically different (Fig. 15–1a). These variograms have been used next to calculate graphs showing precision as a function of observation density (Fig. 15–1b). These graphs in Fig. 15–1 demonstrate that increasing the number of observations has a quite different effect in different soil units: precision is always increased but the increase is characteristically different for different soil units. Precision can never be lower than 10 mm for Plaggepts and 7 mm for Humaquepts and Haplaquods. As lines become steeper in Fig. 15–1b, increasing

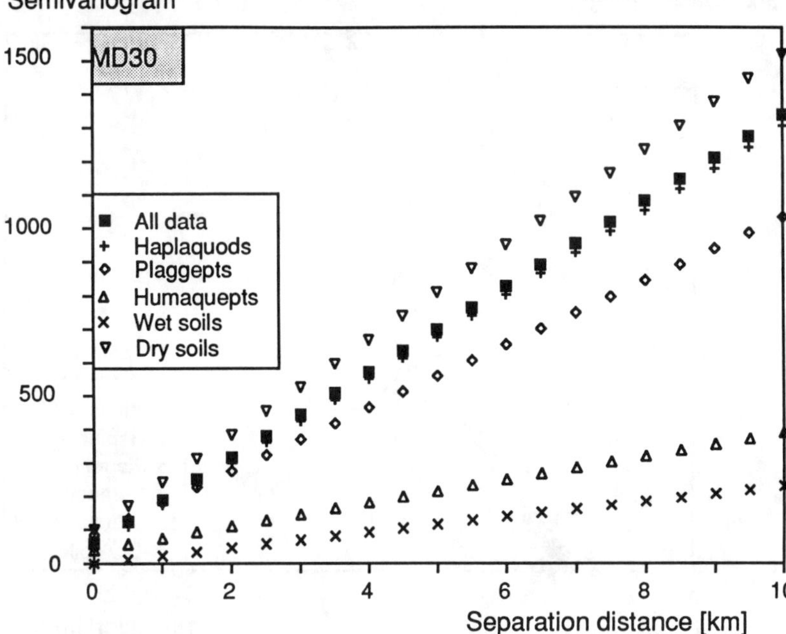

Fig. 15–1a. Variograms for the calculated average moisture supply capacity (based on climatic data for 30 yr) for three major soil series, occurring in the coversand area of the Netherlands. Also, variograms were made for all soils of the three types with relatively high and low water table levels. (After Stein et al, 1988).

boring density becomes less effective. One problem may be that the number of needed observations, according to a geostatistical analysis, is too high to be feasible from an operational point of view. In Dutch studies on farm level it was found that observations in a 50m by 50m grid was all that could be accomplished because of economical considerations. Still, the availability of a geostatistical analysis allows an estimate of the precision and error involved with such a relatively coarse grid.

THE TIME DIMENSION: LOOKING BACKWARDS

Long term monitoring of soil water or nutrient content yields a time series. Running a validated simulation model, however, adds an extra dimension because hypothetical conditions can be simulated for long periods. Hack-ten-Broeke, et al. (1993) used daily weather data for a 30–yr period to calculate the probability, for any given day during the yr, that water would flow from the rootzone into the subsoil. This provided a measure for potential pollution of the groundwater in the study being described. Interpolation techniques can not only be used for elementary point data, such as soil texture or organic matter content,

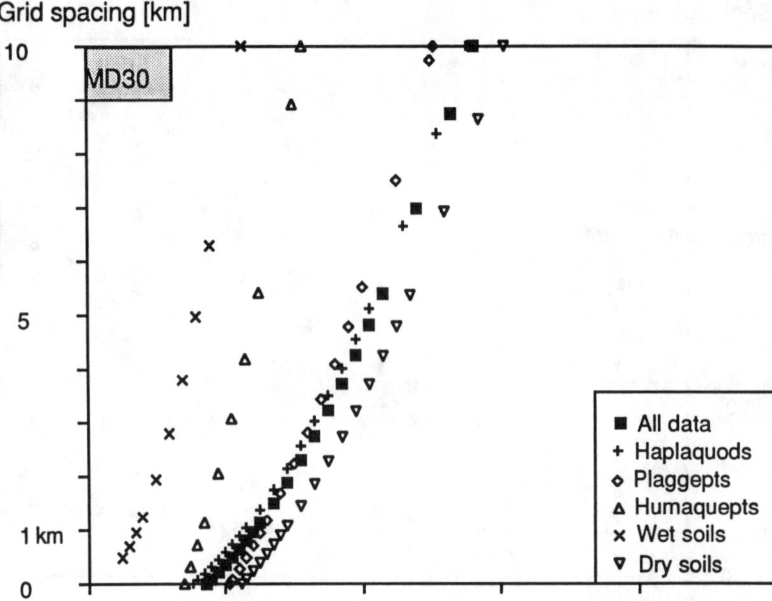

Fig. 15–1b. Precision as a function of grid spacings, as derived from data in Fig. 15–1a. These types of curves can be used to select a grid spacing (= boring density) which is associated with a pre-selected precision.

but also for calculated land qualities, such as the water supply capacity (discussed above), trafficability, aeration status, etc. which all include a time dimension (e.g. Stein et al, 1991).

By considering weather data for long periods, expressions can be obtained for any day in terms of probabilities of exceedance (e.g. Finke, 1993). In other words, the probability that trafficability on May 1 will be adequate is, for instance, 80%. This implies that there still is a 20% probability that trafficability is not adequate. Again, such data are based on historic weather records. Farmers will be particularly interested in future weather conditions for medium range periods, which are not based on historic records but on developments in a given yr. One possibility would be to run a validated simulation model on real time basis and use medium range weather forecasts as input. Thus, water content in soil would be predicted for such a period and farming practices could be adopted accordingly. Studies on using medium range weather forecasts have so far not been reported. Availability of validated simulation models (Wagenet & Hutson, 1989; Hansen et al 1990; Vereecken et al, 1991) allows such experiments now, however, and these should be of particular interest to site specific soil management.

CREATING NEW DATABASES

So far, well published techniques have been reviewed, dealing with point observations, interpolation, functional characterization and modeling. Even though continued field testing is necessary, modeling and interpolation techniques have great promise as future tools for site specific soil management, but operational procedures are still to be worked out and a major problem still is prediction of medium-term weather conditions. By now, it is reasonably clear which soil and crop data are needed to feed the models. Soil data, however, may vary considerably within a given soil type as a result of different management practices in the past (e.g. Bouma, 1992). Rather than run a model for observation points within farm fields, it would, in addition, be attractive for any farmer to know the possible range of properties that can occur within a given soil type as a function of soil management practices. Bouma (1994) has emphasized the need to define this possible range of soil properties. These ranges of properties are first to be identified in the field by interviewing farmers and extension agents. Next, observations are made in the field of the types of soil structure that are associated with these different types of management and measurements of hydraulic and chemical characteristics are to be made in the different soil horizons. Running simulation models, along the lines discussed above, would then provide expressions of dynamic soil behavior for each structure type within the soil type being distinguished,as a function of time. Figure 15–2 is included to tentatively visualize concepts being discussed. Note that each dot in Fig. 15–2 represents a particular type of structure that has been identified in the field within a mapping unit consisting of the soil series being considered and that has been characterized as discussed above. The lines between the dots represent particular types of management including an estimate of its duration. Results could be stored in a database, allowing estimates of soil behavior for soils of a particular soil series as a function of different types of management. Thus, options can be explored. Moreover, expressions are obtained for characteristic ranges of properties for each soil series. In Fig.15–2 three hypothetical soil series have been considered. Bouma (1994) has called the ranges of properties for each soil series, the *window of opportunity* for each soil type series, which is conceptually expressed in terms of soil quality in Fig. 15–2. Here we define soil quality as the degree to which a soil satisfies a well defined objective which is associated with a particular land use. Clearly, it will not be possible to define one single soil quality. Quality for the *agricultural production function* will be different from the *ecological function* and the *carrier function*, to mention only three functions. The schematic Fig. 15–2 indicates that the *window* is characteristically different for each soil type, in terms of its width and its range of quality. Expressions in terms of *windows of opportunity* provide the farmer with a framework for reference purposes. He obtains a means to define realistic options while the scheme allows him to avoid unrealistic plans. The scheme, in fact, illustrates that *anything cannot be done anywhere* and that each soil series has its limits. The scheme provides a framework indicating the natural boundaries of any soil series. As indicated, the scheme in Fig. 15–2 is still conceptual at this stage and will be further defined by research in our department

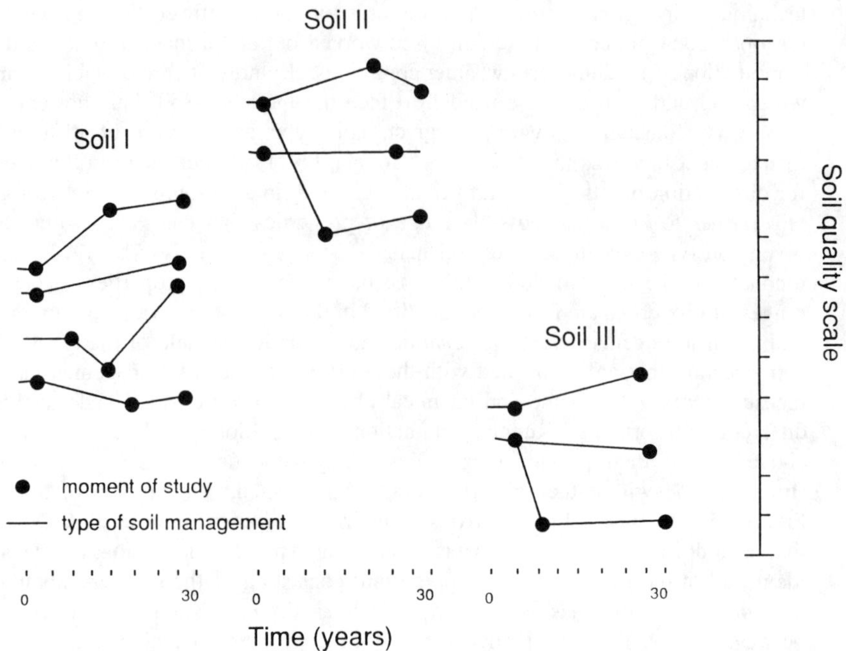

Fig. 15–2. Conceptual scheme which illustrates the possible ranges of soil qualities (*windows of opportunity*) which can be found in three soil series as a function of different types of soil management. Each dot is intended to represent results of a particular study site where the soil series occurs. The lines represent a particular type of management during a specified period. The three soil series have different soil qualities but management can influence quality significantly. The scheme is intended to visualize concepts used and does at this stage not represent real research results.

for major soil series in the Netherlands. Of course, when dealing with fields on farms, interaction between different soil types has to be considered as well and this creates new problems. Still, these can best be approached when data for the separate soils are available.

REFERENCES

Bouma, J. 1989. Using soil survey data for quantitative land evaluation. Advances in Soil Science. Vol. 9. B. A. Stewart (ed.) Springer-Verlag. p. 177–213. New York.

Bouma, J. 1992. Effect of soil structure, tillage and aggregation upon soil hydraulic properties. Advances in Soil Science. p. 1–37. *In* R. S. Wagenet, et al. (ed.) Special issue: Interacting processes in Soil Science. Lewis Publishers. Boca Rouge/Ann Arbor.

Bouma, J. 1994. Sustainable land use as a future focus for pedology? Guest Editorial Soil Sci. Soc. Am. J. 58:645–646.

Finke, P. A. 1993. Field scale variability of soil structure and its impact on crop growth and nitrate leaching in the analysis of fertilizing scenarios. Geoderma 60:89–109.

Finke, P. A., J. Bouma, and A. Stein. 1992. Measuring field variability of disturbed soils for simulation purposes. Soil Sci. Soc. Am. J. 56 (1):187–192.

Hack-ten-Broeke, M.J.D., H.A.J. van Lanen, and J.Bouma. 1993. The leaching potential as a land quality of two dutch soils under current and potential management conditions. Geoderma 60:73–89.

Hansen, S., H.E. Jensen, N.E. Nielsen, and H. Svendsen (1990). DAISY - A soil plant atmosphere system model. NPO-Research Report No. A 10. National Agency of Environmental Protection. Copenhagen, Denmark.

Stein, A., I.G. Staritsky, J. Bouma, A.C. van Eynsbergen, and A.K. Bregt. 1991. Simulation of moisture deficits and areal interpolation by universal co-kriging. Water Resourc. Res. 27(8):1963–1973.

Vereecken, H., M. Vanclooster, M. Swerts, and J. Diels. 1991. Simulating nitrogen behavior in soil cropped with winter wheat. Fert. Res. (27):233–243.

Wagenet, R.J., and J. Hutson. 1989. LEACHM: Leaching estimation and chemistry model - A process based model of water and solute movement, transformations, plant uptake and chemical reactions in the unsaturated zone. Continuum 2. Water Resources Institute, Cornell Univ. Ithaca, NY.

Wosten, J.H.M., J. Bouma, and G.H. Stoffelsen. 1985. The use of soil survey data for regional soil water simulation models. Soil Sci. Soc. Am. J. 49(5):1238–1245.

16 Soil-Terrain Modeling for Site-Specific Agricultural Management

J. C. Bell
C. A. Butler
J. A. Thompson

Dep. of Soil Science
University of Minnesota
St. Paul, Minnesota

Efforts towards site specific agricultural management recognize that the factors affecting both crop yields and environmental sensitivity vary in both space and time. In order to optimize the dual goals of economic sustainability and environmental protection, this variability needs to be considered to the extent that it is both economically and technically feasible. Defining and managing soil variability is crucial for site specific agricultural management because of the impact of soil properties on crop growth and the movement and accumulation of water within the landscape. The variability of soil properties per se are not of particular interest for site specific management, but rather the influence of these soil properties on land qualities such as soil moisture availability, soil nutrient status, and nitrate leaching potential (Bouma & Finke, 1993). These land qualities are frequently influenced by multiple soil and climatic characteristics as well as past management histories and current practices. The challenge for site specific management is to identify those soil properties affecting land quality that also vary across the landscape of interest.

Short-term temporal variability of soil properties, in the range of d to yr, is primarily driven by changes in weather conditions and management practices. For example, soil moisture, temperature, and fertility status are ephemeral properties and can vary considerably from yr to yr in response to weather conditions and agricultural management strategies. Other soil properties, such as soil profile morphology, texture, and subsoil structure, are less prone to short-term changes and reflect the long-term status of the soil resource. Our focus here is on defining the spatial variability of two soil properties that tend to be stable with time and are relevant for describing land quality in glaciated landscapes of the north-central United States; the depth of the A-horizon and depth to free carbonates.

There are three basic approaches that have been discussed in regard to mapping soil variability for site specific farming purposes. First, county (order II) soil surveys prepared by the national cooperative soil survey program document soil variability at scales typically ranging from 1:12 000 to 1:24 000. Soil surveys have the advantage of being available for many regions of the United States and use a consistent format and mapping methodology. Soil mapping units are defined by taxonomic classes and a considerable amount of soil attribute information is available on the properties of these classes (Mausbach et al., 1993). Soil surveys, however, were not created for the purpose of making recommendations for specific areas within a field (Kellogg, 1961). Consequently, the internal variability of soil mapping unit delineations is frequently too great, particularly with respect to soil nutrient status, for site specific management purposes (Holzhey et al, 1993). A second approach uses geostatistical interpolation techniques, such as kriging, to estimate the spatial distribution of soil properties from a network, usually a grid, of point samples (Mulla, 1993; Webster & Oliver, 1990). Geostatistical techniques have the capability to map large-scale spatial variability of both ephemeral and stable soil properties. The main disadvantage of geostatistical techniques, however, is that a relatively large number of soil samples must be collected and analyzed to create these maps. Additionally, the ephemeral nature of many measurements of soil nutrients will require periodic resampling to update soil nutrient status variables (Holzhey, 1993). The third approach takes advantage of relationships between temporally-stable soil properties and landscape features to estimate patterns of soil variability using soil-landscape models. This category includes correlating in-field variability with remote measurements of surface reflectance characteristics and topographic attributes defined by digital terrain analysis.

Quantitative models of soil-landscape relationships have been developed by numerous researchers (Troeh, 1964; Acton, 1965; Protz et al., 1968; Walker et al., 1968). At the time of their development, these models were of somewhat limited applicability because the spatial information required to apply these models to the landscape for practical purposes, such as quantitative terrain data, were not available. The relatively recent development of geographic information system (GIS) technology and digital terrain analysis provides the tools to quantitatively define landscape attributes and to simultaneously define the spatial patterns of multiple landscape attributes using digital overlay techniques. More recent works (Bell et al., 1992; Moore et al., 1993; Bell et al., 1994) demonstrate how statistical models of soil-landscape relationships can be used to map spatial patterns of soil properties.

Jenny (1941) theorized that the properties of a soil were a function of the interaction of several soil-forming factors (climate, topographic relief, parent material, biological activity, relative soil age, and others). If we assume that soil property spatial variability is a function of soil-forming factor spatial variability, then we need a means to depict and explain the spatial variability of these soil-forming factors. For our current research areas, relief (topography) is the soil-forming factor with the greatest variability and is assumed to be a primary factor contributing to soil variability. The spatial variability of topography can be depicted with a digital elevation model (DEM); a regular grid of elevation observations. Algorithms have been developed to estimate slope gradient, plan curvature, profile curvature, aspect direction, specific catchment area, and a variety of surface drainage proximity variables from DEMs (Bell et al., 1992; Moore et al., 1993; Bell et al., 1994). Several researchers (Klingebiel et al., 1987; Dikau, 1989; Bell et al., 1992; Moore et al., 1993) have found a high correlation between changes in these terrain variables and changes in soil drainage characteristics, A-horizon depth, organic matter content, extractable-P, pH, sand, silt, and soil taxonomic classes.

OBJECTIVES

This work was performed as an initial step to identify landscape level processes contributing to soil variability on a hillslope in west-central Minnesota, USA, and to examine the applicability of digital terrain analysis for landscape level studies of soil genesis. We will test the hypothesis that the spatial pattern of variability in the depth of the A-horizon and depth to carbonates can be explained by spatial patterns of terrain variability. The specific objectives of this research are to (i) define the statistical relationships between terrain attributes and the depth of the A-horizon and depth to carbonates, (ii) use these statistical relationships to map A-horizon depth and depth to carbonates for the 20-ha study site and (iii) based on the results, suggest the landscape processes that could be responsible for the genesis of A-horizon and carbonate depth (for future research) at this site.

METHODS

The 20-ha study site is located on the Alexandria glacial moraine of west-central Minnesota, USA, at 46°10'20" N, 95°56'15"W (Fig. 16–1). The hillslope that we studied is south to west facing with slope gradients ranging from 0 to 16% and ≈30 m of topographic relief. This site was selected because the hillslope is representative of soil-landscape conditions found in this part of the Alexandria moraine complex. The soil parent material is glacial till of late Wisconsinan age (10 000 to 14 000 ybp) from the Des Moines Lobe of the Laurentide ice sheet. The soils within the catena are classified as fine, mixed, frigid Typic Endoaquolls, Udic Argiaquolls, Pachic Haploborolls, and Typic Eutrochrepts. The climate is humid, continental with a mean annual precipitation of 597 mm including 1092 mm as snowfall and monthly mean temperature of 21.6°C in July and -14.5°C in January. The site is located in the forest-prairie ecotone with the native vegetation being a mixture of tall grass prairie and deciduous forests. After European settlement, the site was used for agricultural production and is currently in a mixture of grasses as part of the U.S. Department of Agriculture Conservation Reserve Program. Our overall methodology for determining statistical relationships between soil and terrain attributes and for creating predictive maps of soil attributes is illustrated in Fig. 16–2. This approach includes field descriptions of soil profiles along several transects, the calculation of terrain variables from a DEM, statistical modeling to define empirical relationships between soil and terrain attributes, and

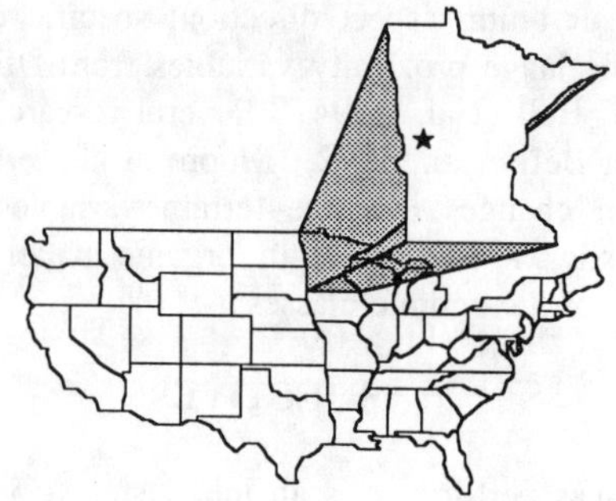

Fig. 16–1. Location of the study site (Dalton, MN).

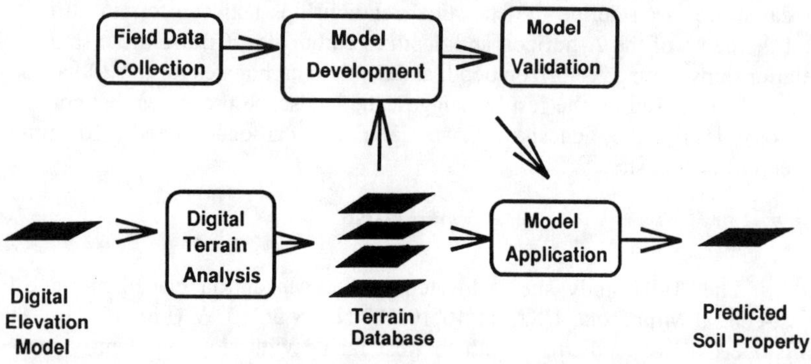

Fig. 16–2. Flowchart for soil-terrain modeling methodology.

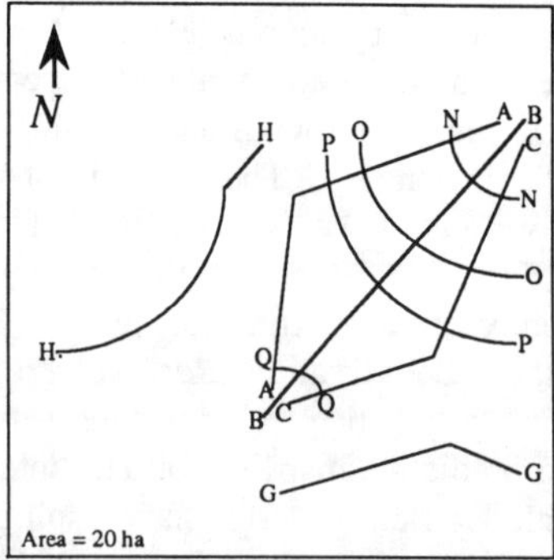

Fig. 16–3. Locations of the sampling transects for the Dalton site.

application of the statistical model to the DEM to map A-horizon and carbonate depth. The following sections describe the specific procedures used for each stage of this process.

FIELD SAMPLING

We described 191 soil profiles at 10– to 20–m intervals along a series of transects running both perpendicular and parallel to the direction of the maximum slope gradient (Fig. 16–3). This strategy provided continuous sampling across all hillslope positions plus repeated observations within the summit, shoulder, sideslope, toeslope, and depressional hillslope positions. Soil cores were extracted using a hydraulic, truck-mounted soil probe to depths of 1 to 2.5 m such that the bottom of the A-horizon and a horizon containing free carbonates was observed at each location. Standard nomenclature (Soil Survey Staff, 1951) was used to describe the morphology of all soil profiles. Depth to carbonates was defined as the depth to a soil horizon containing carbonate masses with a strong or violent reaction to 10% HCl. The geographic coordinates (universal transverse mercator (UTM)) of each soil profile description were recorded using a global positioning system (GPS) receiver with differential correction from a base-station at a known geographic location.

DIGITAL TERRAIN ANALYSIS

A ground survey was conducted on an approximate 10– to 15–m grid across the 20-ha site using a geodometer (electronic surveying device) to define terrain variability. The coordinates of the ≈1 500 field observations were transformed to UTM coordinates by referencing the field survey to two U.S. Geological Survey benchmarks in the vicinity of the study site. The measured elevations were interpolated into a regular grid using a universal kriging procedure (Royle et al., 1981) to create a 10–m DEM consisting of 1 978 points. Primary terrain attributes were calculated directly from the DEM whereas secondary terrain attributes were derived from combination of the primary attributes to characterize specific processes or landscape geometries. In order to estimate primary terrain attributes, we applied a second-order, central finite-difference scheme using a 3 x 3 moving window to calculate slope gradient (S), plan curvature (C_{plan}), and profile curvature (C_{pro}) (Moore et al., 1993). The algorithm described by Jenson and Domingue (1988) traces flow across the landscape based on elevation differences and was used to calculate flow directions, to determine drainage paths, and specific catchment area (A). For this hillslope, drainage paths were defined as grid cells receiving flow from > 100 upslope cells (1 ha area). This flow simulation assumes that the surface is impermeable and that water movement is solely a function of elevation gradients. While this assumption is unrealistic for this landscape, the algorithm can be used to separate regions of the landscape that will respond differently in regard to the collection or transportation of soil water based on topography. The elevation above nearest drainage path and horizontal proximity to nearest drainage path were calculated according to the procedure described by Bell et al. (1992). Four secondary terrain attributes (wetness index, stream power index, depression proximity index, accumulated flow index) were derived to separate landscape regions based on water flow characteristics. The wetness index (WI) identifies landscape positions where water is likely to accumulate and has been used to characterize the soil water content in landscapes (Moore et al., 1993). The stream power index (SPI) is a measure of the erosive power of overland flow.

The wetness index and stream power index are calculated as:

$$WI = \log(A / S) \qquad SPI = \log(A * S)$$

where A is the specific catchment area ($m^2\ m^{-1}$) and S is the slope gradient (%) (Moore et al., 1993). The drainage proximity index (DPI) and accumulated flow index (AFI) distinguish portions of the landscape that are likely to collect water based on hillslope geometry and are calculated as:

$$DPI = (E_{dr} / P_{dr}) * 100 \qquad AFI = (A_{dr} / E_{dr})$$

where E_{dr} is the elevation above the nearest drainage path, P_{dr} is the horizontal distance to the nearest drainage path, and A_{dr} is the specific catchment area (A) for the nearest drainage path. Computer programs to calculate primary and secondary terrain variables were written and implemented using the software tools of the Khoros image processing system (Khoros Group, 1991). The primary and secondary terrain attributes for each soil profile description were found based on the geographic coordinates of the soil observations in the field.

STATISTICAL MODELING

Correlations among all soil and terrain variables were calculated using Pearson's correlation coefficients. Logarithmic transformations were used for wetness index and stream power index to achieve near-normal distributions for these data. A split-sample approach was used to develop multivariate linear models to define relationships between soil and terrain attributes where 146 randomly selected observations were used to develop regression models and the remaining 45 were used to obtain an unbiased estimate of model performance. A stepwise procedure was used to select the optimal set of terrain variables for inclusion in the regressions. In order to be entered into or to remain in the model, we tested the hypothesis that the regression coefficient was equal to 0. Terrain variables were entered in a stepwise fashion if this hypothesis could be rejected at the 0.01 level of significance (Wilkinson, 1989). Once regression equations were developed for A-horizon thickness and depth to carbonates, these equations were applied to the 45 observations in the model validation set to evaluate the regression models.

RESULTS

A three-dimensional, wire-frame representation of the study site topography is shown in Fig. 16–4. Subsequent maps of terrain and soil attributes are draped over this three-dimensional surface to elucidate relationships between soil and terrain attributes and the land surface. For example, the derived map of slope gradient (Fig. 16–5a) indicates, as expected, that the highest slope gradients (light shades) occur on the sideslope positions with the lowest gradients (dark shades) at the summit and depressions. The high values for specific catchment area delineate the drainage pathways quite well (Fig. 16–5b). The diagonal striping on the hillslope is an artifact of the particular algorithm that was used which traces flow in an unrealistic manner in the uplands. The algorithm does, however, separate regions of the landscape receiving little, moderate, and high quantities of cumulative flow from upslope contributing areas. Proximity to the nearest drainage path (Fig. 16–5c) and elevation above the nearest drainage path (Fig. 16–5d), define the geometry of the landscape in relation to the drainage paths and separates low depressional areas (dark shades) from the uplands (light shades). Slope curvatures in the plan and profile directions were also calculated, but are not presented here because these two primary terrain variables were not significant in defining the regression models.

The wetness index (Fig. 16–6a) segregates regions of the landscapes where water is likely to collect based on the morphology of the land surface. High values were found in drainage paths and depressional areas with low values on drainage basin divides. The stream power index (Fig. 16–6b) separates areas of the landscape where high volumes or velocities of flowing water are likely to occur, such as drainage paths and steep sideslopes (high values), from gently

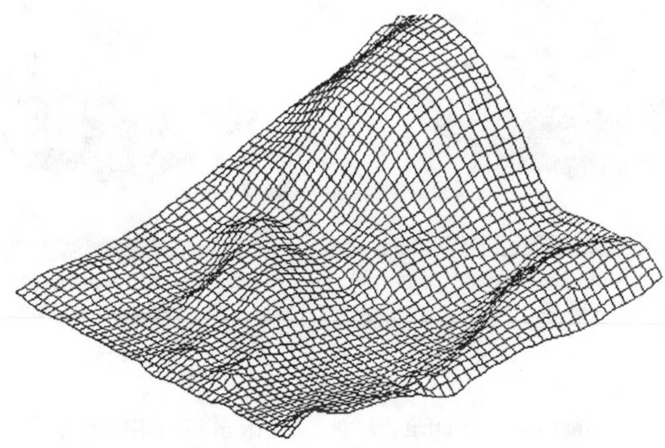

Fig. 16–4. Three-dimensional, wireframe view of the DEM for the Dalton site.

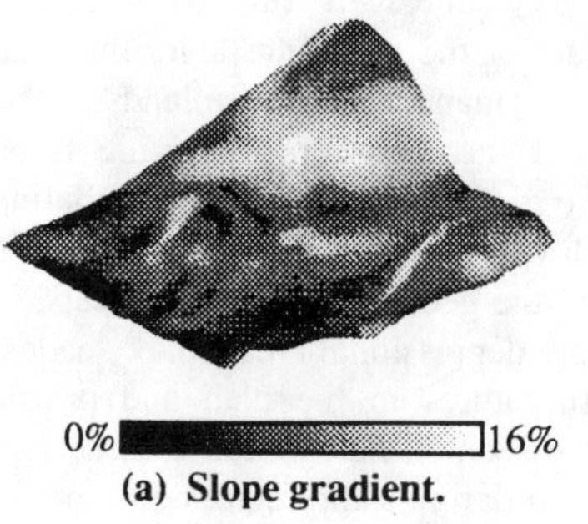

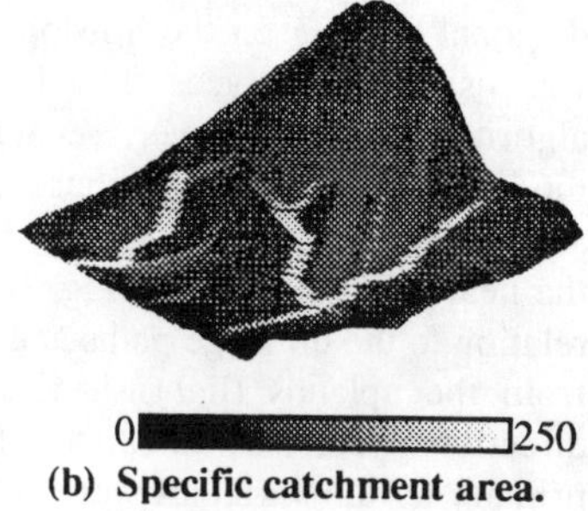

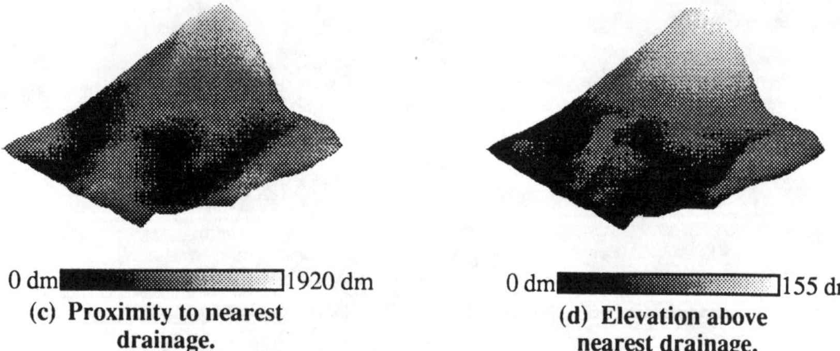

Fig. 16–5. Primary terrain attributes derived from the DEM. Gray-scale maps are draped over the terrain model (Fig. 16–4) for the Dalton site.

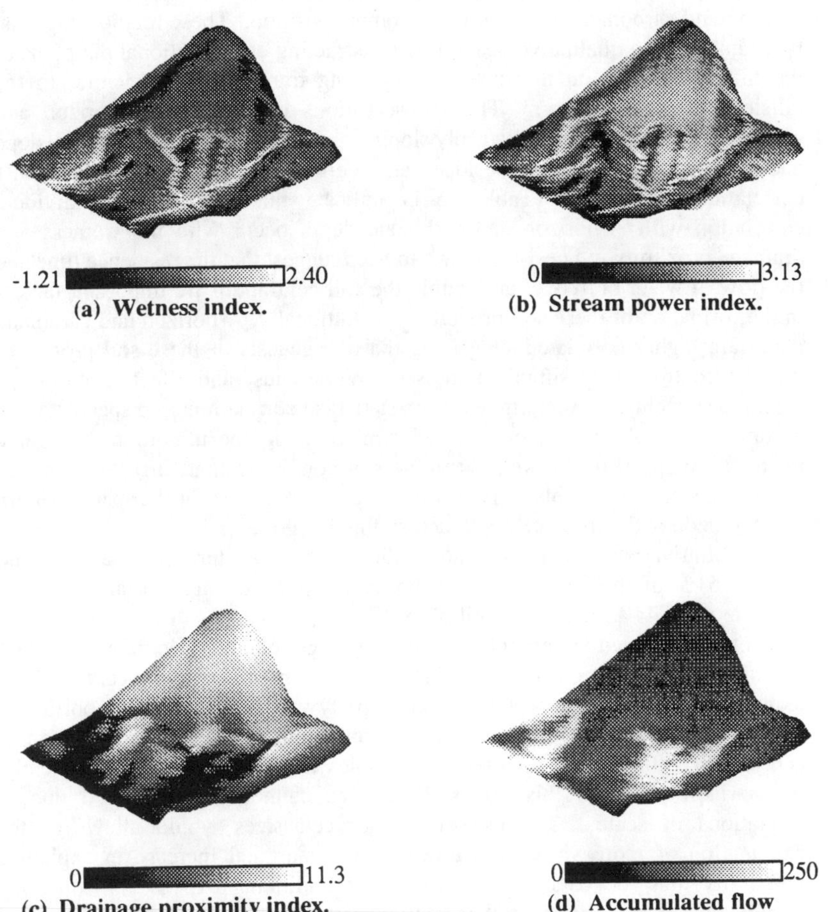

Fig. 16–6. Secondary terrain attributes derived from the DEM. Gray-scale maps are draped over the terrain model (Fig. 16–4) for the Dalton site.

sloping areas on short hillslopes and drainage divides that potentially could receive lower quantities of flow from upslope areas. The drainage proximity index (Fig. 16–6c) is essentially the mean slope gradient from any point on the landscape towards the nearest drainage path. Hence, this index provides another means of separating low, depressional areas on the landscape (low values) from the uplands (high values). The accumulated flow index (Fig. 16–6d) identifies depressional areas that could potentially accumulate water flow within the landscape (high values). These four secondary terrain indices are based on different combinations of the slope gradient, specific catchment area, elevation above the nearest drainage path, and horizontal distance to the nearest drainage path. Field sampling revealed consistent relationships between A-horizon and carbonate depth with topographic position. These relationships can be examined in a qualitative fashion by constructing cross-sectional diagrams of the hillslope based on the observations along transects perpendicular to the hillslope (Fig. 16–7a,b,c). These observations indicate that A-horizon and carbonate depth decreases on steeply sloping and convex portions of the hillslope and increase in concave, toeslope, and depressional positions. Pearson's correlation coefficients (Table 16–1) indicate that the highest individual correlation with A-horizon and carbonate depth occur with the wetness and drainage proximity indices. These two indices suggest that the residence time and the flow of water both over and within the soil continuum are important factors in the formation of these soil properties. Additionally, A-horizon and carbonate depth are highly correlated suggesting that the genesis of these soil properties may be controlled by similar processes. While cause-and-effect relationships cannot be conclusively determined from statistical correlations, we speculate that a combination of: (i) erosion and sedimentation, (ii) the inhibition of organic matter decomposition by wet, anaerobic soil conditions, and (iii) the extent of vertical leaching of soluble soil constituents are responsible for the spatial pattern of A-horizon and carbonate depth across this hillslope.

Multivariate linear regression indicated that two terrain variables could explain 51% of the variability in A-horizon depth and three terrain variables could explain 44% of the variability in depth to carbonates (Table 16–2). The percent of explained variability is within the range found by Moore et al. (1993) in a similar study, who speculated that it may be unrealistic to expect that these techniques could explain > 70% of soil property variability. A certain portion of the variability cannot be explained due to limitations of scale. For example, the cell size used to calculate the terrain variables was 10 x 10 m. Consequently, the variability within this 10- x 10-m area cannot be considered due to limitations of scale. The use of smaller cell sizes would allow for the explanation of more variability, however, the marginal increase in explained variability must be balanced against (i) the additional expense of collecting terrain data at a higher spatial resolution, (ii) the purpose for which the soil property information is required, and (iii) the spatial scale of the pedogenic processes of interest. The optimal spatial scale for studying pedogenic processes at the landscape level is unknown and probably varies among different climatic-physiographic regions.

Table 16–1. Partial correlation matrix of soil and terrain attributes.

AHD	0.91	-0.36	0.50	0.49	0.37	-0.33	-0.33	-0.57	-0.26	0.61
	DTC	-0.35	0.43	0.43	0.42	-0.33	-0.36	-0.59	-0.25	0.54
		A	-0.35	-0.25	-0.20	0.17	0.43	0.71	0.54	-0.38
			Pd	0.95	0.32	-0.23	-0.23	-0.43	-0.14	0.75
				Ed	0.35	-0.25	-0.21	-0.37	-0.05	0.83
					S	0.15	-0.09	-0.43	0.39	0.54
						Cpro	0.40	0.31	0.44	-0.29
							Cplan	0.61	0.57	-0.31
								WI	0.65	-0.55
									SPI	-0.07
										DPI

The regression analysis identified the wetness index and depression proximity index as the best predictors of A-horizon depth and carbonate depth. The wetness index was the terrain variable accounting for the largest portion of variability in A-horizon and carbonate depth followed by the depression proximity index and accumulated flow index. These indices distinguish topographic positions where hydrologic processes on the hillslope are different, suggesting that hillslope hydrology is a major factor influencing variability in the A-horizon and carbonate depth for the soil continuum at this site.

A comparison of the predicted vs. observed A-horizon and carbonate depths (Fig. 16–8a,b) indicates the regression predicts the A-horizon carbonate depth within 20 cm of the observed depth for ≈70% of the model validation observations. The largest residuals were underestimates associated with the three deepest observations of A-horizon and carbonate depths. If these observations were considered to be outliers and were eliminated from the analysis, then the R^2 would improve to 0.63 for A-horizon depth and to 0.48 for depth to carbonates. This also suggests that a non-linear model may be appropriate in this situation. We did not pursue further refinements, however, in curve fitting because our objective was to find the optimal set of terrain variables that could be used to predict these soil properties.

The wetness, drainage proximity, and accumulated flow indices were estimated for each 10- x 10-m grid cell for the study site and combinations of these indices were used to predict spatial patterns of A-horizon and carbonate depth using the regression equations described in Table 16–2 on a cell-by-cell basis. This procedure creates a digital map of A-horizon and carbonate depths based on spatial patterns of topography and provides a means to extrapolate point observations to the three-dimensional landscape and hence generate a spatial model of A-horizon and carbonate depths for the soil continuum. We draped the predicted maps of A-horizon and carbonate depths over the three-dimensional view of the site (Fig. 16–4) to help visualize the predicted spatial pattern of these soil properties in relation to the shape of the land surface (Fig. 16–9a,b). As expected, the spatial patterns for A-horizon and carbonate depth

Table 16–2. Coefficient and R^2 values for the regression equations relating A-horizon depth and depth to carbonates to terrain variables.

Terrain Variable	A-horizon Depth	Depth to Carbonates
Intercept	-89.04	-122.15
Wetness Index	-24.25	-39.19
Depression Proximity Index	6.30	8.55
Accumulated Flow Index	--	0.26
R^{2*}	0.51	0.44

*R^2 is an unbiased estimate determined from the validation data set.

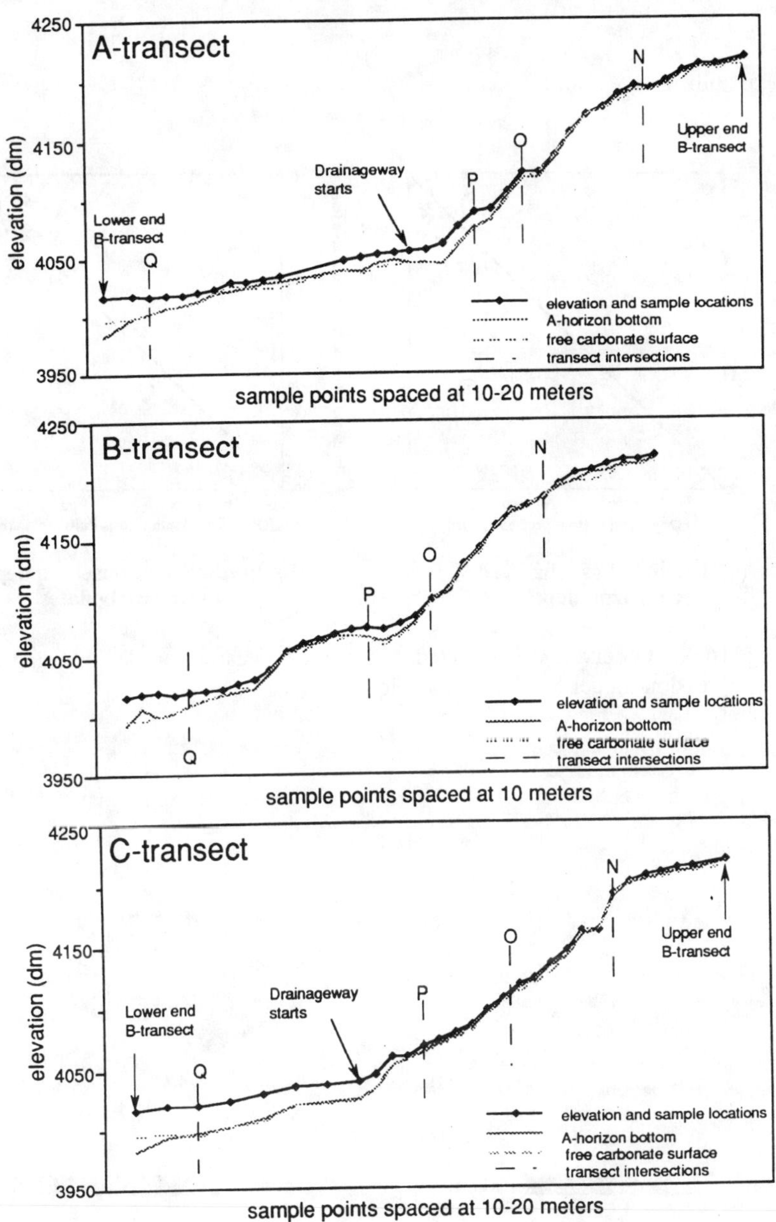

Fig. 16–7. Cross-sectional view of A-horizon depth relative to elevation for sampling transects A, B, and C at the Dalton site.

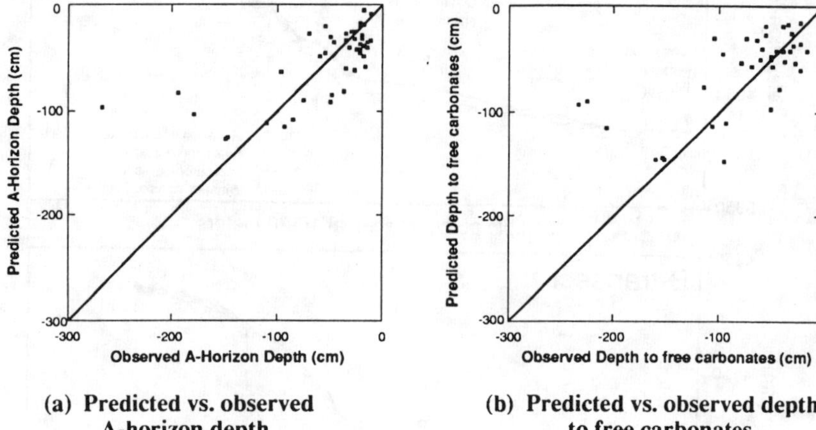

(a) Predicted vs. observed A-horizon depth.

(b) Predicted vs. observed depth to free carbonates.

Fig. 16–8. Observed vs. Predicted A-horizon and carbonate depths using the regression model for the Dalton site.

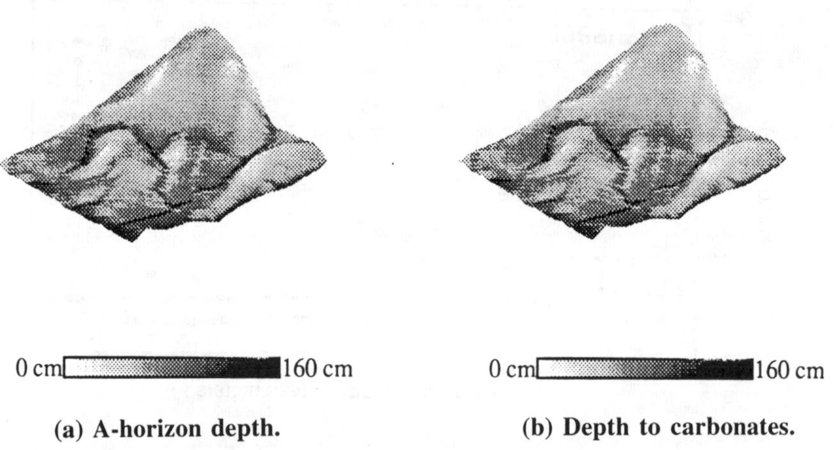

(a) A-horizon depth.

(b) Depth to carbonates.

Fig. 16–9. Predicted A-horizon and carbonate depths. Gray-scale maps are draped over the terrain model (Fig. 16–4) for the Dalton site.

are very similar. The estimated depths for both soil properties is greatest (dark shades) in the drainage paths and depression areas where water is likely to accumulate and move downward through the soil. The shallowest depth (light shades) for these soil properties are predicted to occur on steep and convex slopes in the uplands. These soil-terrain relationships agree with those observed on the sampling transects (Fig. 16–7). There is always the possibility that these statistical correlations are fortuitous in nature and may not be related to physical processes leading to soil genesis. In this case, the statistical correlations and estimated spatial patterns confirm our intuitive understanding of the landscape processes that could account for soil genesis in this setting. For example, the secondary terrain indices separate areas of the landscape that we would expect to have different hydrologic processes (erosion vs. deposition, lateral water flow vs. infiltration and vertical leaching, etc.) which would lead to the development of soils with different properties. The predicted locations of deep A-horizons are associated with deposition landscape positions or potentially wet positions where organic matter is likely to accumulate, Shallow A-horizons are predicted for steeper slopes and convex, water-spreading slopes. Similarly, deeper depths to carbonates are predicted in depressional areas and toeslopes where more vertical movement of water through the soil profile is expected and shallower depth to carbonates on sideslopes and convex slopes where more lateral movement of water would be expected. These relationships suggest a depression focused recharge situation where the groundwater is primarily recharged from water that collects in depressional areas from either surface or interflow (Richardson et al., 1992). We would expect that the relationships between soil properties and topography would be quite different in landscapes with flowthrough or groundwater discharge hydrology.

DISCUSSION

Digital terrain analysis and soil-terrain modeling have applications for site specific agricultural management in landscapes where soil characteristics change with topography. In this example we modeled A-horizon depth and depth to carbonates; however, spatial patterns of other soil properties that are related to topography could be modeled using a similar approach. Of particular interest would be the retention of available moisture within the rooting zone and seasonal fluctuations of saturated soil conditions that could be inferred from the depth to and presence of redoximorphic features within the soil profile. For site specific agricultural management applications, spatial patterns of A-horizon depth and depth to carbonates per se are of little interest. The spatial patterns of these properties could be used, however, to define common management regions within a field where we would expect soil conditions and processes to be similar. For example, if we assume that the predominant source of carbonates in the Alexandria moraine is the calcareous till, then the depth to carbonates could be considered as a surrogate for landscape differences in soil leaching intensity. Soil-terrain modeling could be used to identify portions of the landscape where water may move predominantly in the lateral, as opposed to the vertical, direction. This information could be used to adjust nutrient and pesticide application rates in portions of the field where the potential for direct leaching to the groundwater would be higher and, therefore, minimize adverse environmental impacts. Likewise, A-horizon depth, for some landscapes, is correlated with available moisture storage, soil nutrient status, and areas subject to erosion or depositional processes.

The relevance of the soil-terrain model developed in this study to other hillslopes within Alexandria glacial moraine is unknown at this time and will be the subject of future work. Previous studies (Bell, 1990; Hoosbeek & Bryant, 1992) suggest that it is difficult to extrapolate statistical correlations to other sites because the calibration site may not be completely representative of soil and terrain conditions for the broader geographic area. Statistical correlations most certainly could not be extrapolated to other climatic-physiographic regions where the soil-terrain relationships are likely to change or where other combinations of soil-forming factors may be of more importance. We view the identification of statistical relationships between soil and terrain attributes as one of the first steps towards developing process-based models of soil genesis at the landscape level. Statistical correlations, when combined with our existing knowledge of soil genesis, suggest possible mechanisms for the formation of specific soil properties.

This research, in combination with other work (McBratney et al., 1992; Moore et al., 1993; Bell et al., 1994; McSweeney et al., 1994), suggests a new paradigm for describing the spatial distribution of soils at the landscape level based on the spatial extent of soil horizons and defines the soil continuum in terms of stratigraphy, rather than as discrete pedons and polypedons (map units). Soil mapping units, as depicted in soil surveys, are conceptual constructs that were invented to help us convey information about the spatial variability of the soil continuum when pen and paper were the only means to convey this information. These constructs have served well, however, geographic information system technology and high-speed computing capabilities present new and innovative opportunities to both depict and understand the spatial dimension of soils and landscapes. We now have the ability to depict the continuously varying nature of the terrain quantitatively as a three-dimensional surface (Fig. 16–4). Consequently, to the extent that changes in the terrain are associated with changes of the soil continuum, we can model and depict the spatial variability of soil horizons in reference to the topographic surface (Fig. 16–9a,b).

Continuing advances in GPS and related technologies will make the creation of high-resolution digital terrain models cost effective for agricultural areas (Tyler, 1993). In the future, these technologies will make it possible to collect the surface elevation observations needed to create high-resolution digital terrain models during planting or harvesting field traverses. Because field observations of soil conditions are needed for this approach, accurate georeferencing of soil samples and soil profile descriptions taken as part of soil resource inventory activities will increase the database needed for developing soil-terrain models. This approach also provides a structure for viewing agricultural management comprehensively from a watershed perspective by defining hydrologic linkages among landscape elements. While soil-terrain modeling is a relatively new approach and will require further development, it provides soil scientist and agricultural managers a potentially valuable tool to make decisions for site specific agricultural management.

SUMMARY

1. Topographic variables derived from a 10-meter DEM can explain 51% of the variability in A-horizon depth and 44% of the variability in depth to carbonates for a hillslope in western Minnesota.

2. Spatial patterns of A-horizon depth and depth-to-carbonates can be mapped using statistical relationships with terrain variables.

3. Soil-terrain modeling techniques may be a cost-effective means of mapping soil variability and defining common management areas for site specific agricultural management for certain landscapes.

ACKNOWLEDGEMENTS

We wish to gratefully acknowledge the United States Department of Agriculture Soil Conservation Service, the United States Army Corps of Engineers Waterways Experiment Station, and the Graduate School of the University of Minnesota for their financial support of this project. Additionally we wish to acknowledge John Ladwig for his assistance with computer programming; Warren Lynn, Mike Leiser, Craig Prink, and Don DeMartelaere of the Soil Conservation Service, and David Ward, Gavin Patterson, and Jim Engstrom for their assistance in developing the digital elevation model.

REFERENCES

Acton, D.F. 1965. The relationship of patterns and gradient of slopes to soils. Can. J. Soil Sci. 45:96–101.

Bell, J.C. 1990. A GIS-based soil-landscape modeling approach to predict soil drainage class. Ph.D. thesis. The Pennsylvania State Univ. University Park, PA.

Bell, J.C., R.L. Cunningham, and M.W. Havens. 1992. Calibration and validation of a soil-landscape model for predicting soil drainage class. Soil Sci. Soc. Am. J. 56:1860–1866.

Bell, J.C., R.L. Cunningham, and M.W. Havens. 1994. Soil drainage class probability mapping using a soil-landscape model. Soil Sci. Soc. Am. J. 58:464-470.

Bouma, J., and P.A. Finke. 1993. Origin and nature of soil resource variability. pp. 3–14. *In* P. C. Robert, et al. (ed.) Proc. of soil specific crop management: A workshop on research and development issues. April 14–16, 1992. Am. Soc. Agron., Inc., Madison, WI.

Dikau, R. 1989. The application of a digital relief model to landform analysis in geomorphology. p. 51–77. *In* J. Raper (ed.) Three dimensional applications in geographic information systems. Taylor and Francis, New York.

Holzhey, C.S. 1993. Soil resources variability: Working group report. pp. 69–75. *In* P. C. Robert, et al. (ed.) Proc. of soil specific crop management: A workshop on research and development issues. April 14–16, 1992. Am. Soc. Agron., Inc., Madison, WI.

Hoosbeek, M.R., and R.B. Bryant. 1992. Towards the quantitative modeling of pedogenesis - a review. Geoderma 55:183–210.

Jenny, H. 1941. Factors of Soil Formation - A System of Quantitative Pedology. McGraw-Hill, New York.

Kellogg, C.E. 1961. Soil interpretation in the soil survey. In service publ. USDA-Soil Conservation Service, Washington, DC.

Khoros Group. 1991. Khoros users manual. Dep. of Elec. Computer Eng., Univ. of New Mexico. Albuquerque, NM.

Klingbiel, A.A., E.H. Horvath, D.G. Moore, and W.U. Reybold. 1987. Use of slope, aspect and elevation maps derived from digital elevation model data in making soil surveys. p. 77–90. *In* W.U. Reybold, and G.W. Petersen (ed.) Soil survey techniques. SSSA Spec. Publ. 20.

Mausbach, M.J., D.J. Lytle, and L.D. Spivey. 1993. Application of soil survey information to soil specific farming. p. 57–68. *In* P. C. Robert, et al. (ed.) Proc. of soil specific crop management: A workshop on research and development issues. April 14–16, 1992. ASA, Madison, WI.

McBratney, A.B., J.J. De Gruijter, and J.B. Brus. 1992. Spatial prediction and mapping of continuous soil classes. Geoderma, 54:39–64.

McSweeney, K.M., P.E. Gessler, B.Slater, R.D. Hammer, J.C. Bell, and G.W. Peterson. 1994. Towards a new framework for modelling the soil-landscape continuum. Chapt. 8. Factors of soil formation: A fiftieth anniversary perspective. SSSA Special Publ. 33.

Moore, I.D., P.E. Gessler, G.A. Nielsen, and G.A. Peterson. 1993. Soil attribute prediction using terrain analysis. Soil Sci. Soc. Am. J. 57:443–452.

Mulla, D.J. 1993. Mapping and managing spatial patterns in soil fertility and crop yield. p. 15–26. *In* P. C. Robert, et al. (ed.) Proc. of soil specific crop management: A workshop on research and development issues. April 14–16, 1992. Am. Soc. Agron., Madison, WI.

Protz, R., E. Presant, and R. Arnold. 1968. Establishment of the modal profile and measurement of variability within a soil landform unit. Can. J. Soil Sci. 48:7–19.

Richardson, J.L., L.P. Wilding, and R.B. Daniels. 1992. Recharge and discharge of groundwater in aquic conditions illustrated with flownet analysis. Geoderma 53:65–78.

Royle, A.G., F.L. Clausen, and P. Frederiksen. 1981. Practical universal kriging and automatic contouring. Geoprocessing 1:377–394.

Soil Survey Staff. 1951. Soil survey manual. Agric. Handb. No. 18. U.S. Gov. Print. Office, Washington, DC.

Tyler, D.A. 1993. Positioning technology (GPS). pp. 159–166. *In* P. C. Robert, et al. (ed.) Proc. of soil specific crop management: A workshop on research and development issues. April 14–16, 1992. Am. Soc. Agron., Madison, WI.

Troeh, F.R. 1964. Landform parameters correlated to soil drainage. Soil Sci. Soc. Am. J. 28:808–812.

Walker, P.H., G.F. Hall, and R. Protz. 1968. Relation between landform parameters and soil properties. Soil Sci. Soc. Am. Proceedings 32:102–104.

R. Webster, and M.A. Oliver. 1990. Statistical methods in soil and land resource survey. Oxford University Press, New York.

Wilkinson, L. 1989. SYSTAT: A system for statistics. SYSTAT, Inc. Evanston, IL, USA. Bell et al.

[illegible]

17 Terrain Modelling as a Basis for Optimal Agroecological Land Management Using Dynamic Simulation

J. Verhagen
P. Verburg
M. Sybesma
J. Bouma

Dep. of Soil Science and Geology
Agricultural University Wageningen
6700 AA Wageningen
The Netherlands

Current agricultural practices can be important sources of environmental pollution. Crop surpluses and low net return rates combined with new environmental legislation make the need for research for sustainable agricultural systems more apparent. Economic and environmental decision tools and new evaluation methods are needed.

Physical land evaluation is an important step in farming systems analysis; it includes a description of the biophysical properties of the soil (Bouma & van Lanen, 1986). Soil water movement is regarded as a key characteristic for both crop growth and leaching of agrochemicals. With the use of a dynamic deterministic simulation model areas with comparable water stress and leaching potential are indicated. Validated simulation models provide the opportunity to create risk or response maps that can be used to evaluate the effects of current and potential agricultural management activities. Simulation models can also be incorporated in decision support systems to define rules and evaluate conclusions derived from the systems.

The study area is a 110 ha experimental farm in the Wieringermeer polder in the northwestern part of the Netherlands (Fig. 17–1). The soils in this former tidal flat are strongly layered with textures ranging from sand to clay loam. Soils were classified as Typic Udifluvents (Soil Survey Staff 1975).

Resulting from the dynamic sedimentation environment soils vary strongly in both horizontal and vertical direction. The soils in the study area are described using the functional layer concept, which distinguishes soil horizons with identical hydrological properties, as determined by considering static and dynamic criteria. Spatial and temporal variations are characterized by interpolating results obtained from a deterministic, mechanistic simulation model applied for point data (SWAT-Modules, Vanclooster et al., 1993).

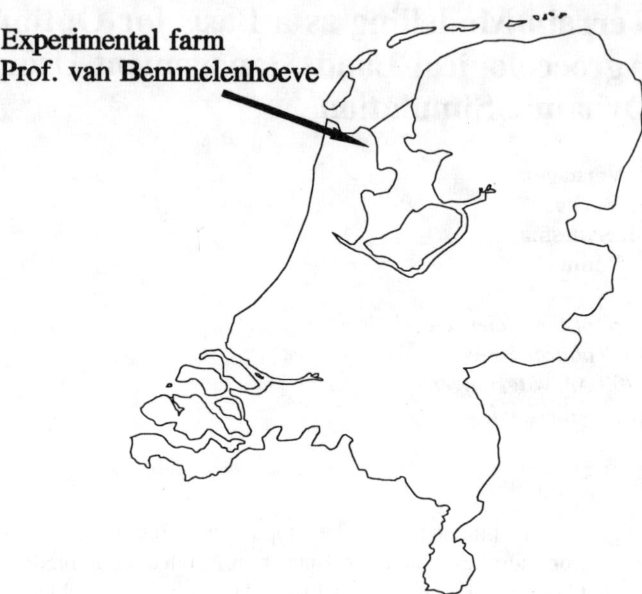

Fig. 17–1. Location of the study area.

MATERIALS AND METHODS

Functional layers

During a soil survey the soil pedon is described in terms of horizons as carriers of physical information. Texture, structure and organic matter largely determine the physical characteristics of a soil. In a standard soil survey, using auger holes to 120 cm depth, layers with comparable texture, structure and organic matter content can be identified; these layers may be combined to form functional layers.

A functional layer is a layer having significantly different hydrological properties as compared with other layers and a low intrinsic variability. Functional layers should be easily recognizable in an auger core (Wösten et al., 1986, Finke & Bosma, 1993). Two ways to arrive at functional groups are distinguished (Fig. 17–2). In the Netherlands there is a standard soil physical data set containing 40 different soil types (Wösten et al., 1987). This set can be used to make the conversion from pedogenic soil horizons to functional layers using soil textures and organic matter content (Fig. 17–2, right part). Because of the complex nature of the soil profile in the study area, the standard data set is not used and a separate soil physical data set is created, thus taking the second more complicated path (Fig. 17–2, left part).

The topsoil was treated as a separate group because of its high temporal variability. The study concentrated on deriving a set of functional layers for the subsoil.

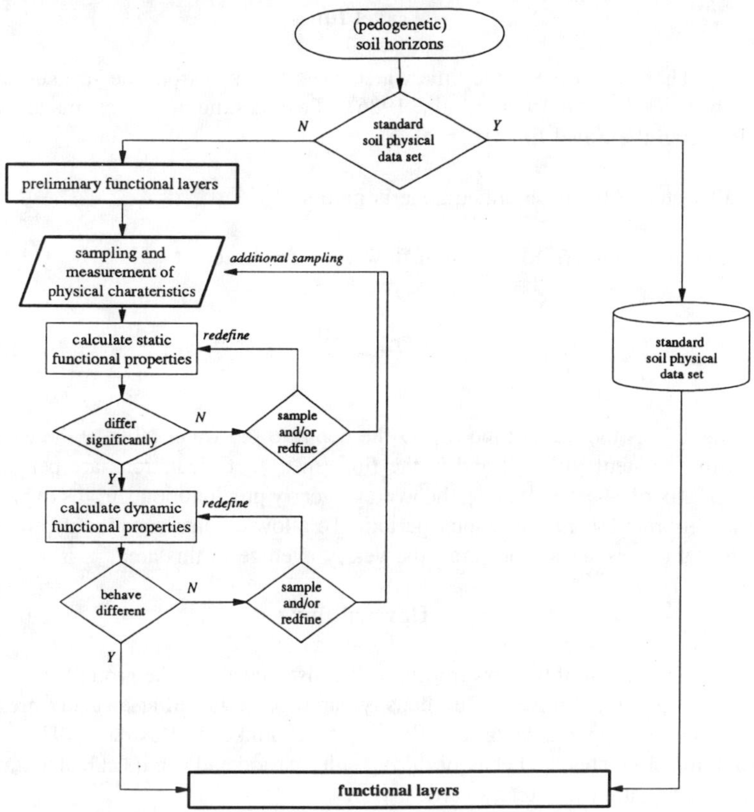

Fig. 17–2. Flowchart illustrating procedure to distinguish functional layers.

Preliminary functional layers are determined during the field survey and the physical characteristics are determined. Hydrological properties of the preliminary functional layers are compared using static, (Wösten et al. 1986, Finke & Bosma, 1993), and dynamic functional properties. The latter are land qualities acquired through dynamic simulation. Layers having different functional properties will behave differently under identical boundary conditions. The following static and dynamic functional properties are distinguished:

- Travel time (static): the time needed for infiltration from the soil surface to a defined water table.
- Height of upward flux (static): distance above the groundwater table to which a defined upward flux density can be maintained at a steady pressure head on the upper side of the flow system.
- Simulated flux (dynamic): simulated water movement under given boundary conditions.

Travel time

The travel time is the time water takes to travel from the soil surface to the water table (Wösten et al., 1986). Two assumptions are made when calculation the travel time:

1. all water in the unsaturated zone is mobile

2. flow is characterized by piston flow with unit gradient

$$T = \frac{D \times \theta}{N} \tag{1}$$

In which T is the travel time [d], D the depth to the water table, θ the average moisture content [m^3 m^{-3}] and N the flux [m d^{-1}]. Calculations are performed with a flux of 1.4×10^{-3} [m d^{-1}], the average yearly precipitation surplus expressed as a daily rate for the wet winter period. To allow comparisons the depth to the water table was set to 150 [cm], the yearly average in this area.

Upward flux

The height of the upward flux is the distance above the groundwater table to which a defined upward flux density can be maintained at a steady pressure head on the upper side of the flow system (Finke & Bosma, 1993). This condition simulates the behavior during a dry period and can be calculated using Darcy's equation for steady vertical flow.

$$q = -K(1 + \frac{dh}{dz}) \tag{2}$$

In which dh dz^{-1} is the vertical gradient of the soil matrix potential [kPa cm^{-1}], K the hydraulic conductivity [m d^{-1}] and q the flux density [m d^{-1}]. The upward flux is obtained by integration of equation 2.

$$z_n = -h_n \int_0^z \frac{dh}{(1 + \frac{q}{K})} \tag{3}$$

Where z_n is the vertical distance [m]. The flux density was set to 0.002 [m d^{-1}] and an upper boundary condition with a pressure head of -50 kPa. To allow comparisons calculations are made with all functional layers scaled to 150 cm.

Simulated flux

Water movement is simulated to assess the dynamic behavior of the samples under dry and wet conditions. To simulate fluxes through a soil profile the SWAT-modules, developed at the Institute for Land and Water Management (Vanclooster et al., 1993), were used. To simulate water extraction under natural conditions crop growth was included. The 1975/1976 growing season was chosen because it included an extremely dry summer. Calculations were made for homogeneous profiles using the hydraulic characteristics of the functional layers, the functional layers being scaled to functional profiles of 150 cm depth. Results of simulated fluxes are compared for two depths: 40 cm and 100 cm. First a short description of the used modules is given, next an overview of boundary conditions and crop parameters.

Water movement

The water balance module, S-WAT (Simulating Water Transport), is an adapted version of the deterministic mechanical water balance model SWATRER (Dierckx et al., 1986). Vertical unsaturated water flow in a soil is quantified using the Richards equation which is solved using a finite difference numerical solution technique. For this purpose the soil is divided into several compartments with a preset thickness of 1 cm. A functional soil layer usually consists of several compartments. Water transport can be calculated with the given boundary conditions of the soil water system.

For cropped soils the water balance has to consider crop transpiration and interception. Water uptake by plant roots is calculated as a function of the potential evapotranspiration rate and a sink term variable that is a function of the pressure head in the root zone (Feddes et al., 1987). At very high and low pressures the water uptake is reduced linearly with a dimensionless factor that is related to the pressure head.

Crop growth

The crop growth model is the simple and universal crop growth simulator (SUCROS87) which was originally developed at the Centre for Agrobiological Research in the Netherlands (Spitters et al. 1988). SUCROS87 is a mechanistic model that simulates dry matter accumulation from climatic data (radiation and temperature) and plant phenologic parameters. The rate of phenological development is mainly determined by temperature. Dry matter production is distributed over the various plant organs according to fixed, crop specific, distribution factors as a function of the development stage. Potential crop growth is limited when water stress occurs. Nutrient supply is assumed optimal. Soil water stress is described with a trapezoidal function. Above wilting point (pF4.2) and below field capacity (pF2) its value is set to zero indicating maximum reduction. The maximum value (1), is defined at soil water contents below 25% of field capacity or above wilting point. The intermediate trajectory is assumed to be linear. Root growth is modeled as a function of the development

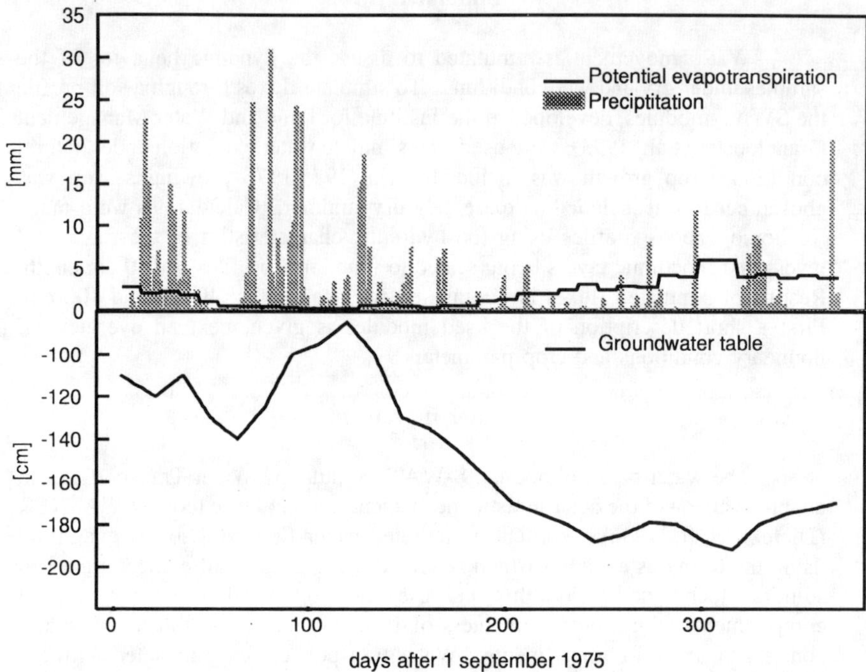

Fig. 17–3. Upper and lower boundary conditions that were used for simulations of water flow and crop growth.

stage using two state variables; root depth and root density. To calculate root density development, the accumulated root weight is translated into an enlarged root length that is divided into the different compartments using a weighing factor (sum of the weighing factors should be one).

Boundary conditions

Upper boundary conditions were obtained from a nearby climatic station, de Kooy. The lower boundary condition was derived from field monitoring of the water table level (Fig. 17–3).

Crop parameters

The growth of winter wheat was simulated. Because of the early planting date in September, a period with no management disturbances can be modeled, thus avoiding large temporal changes in the hydraulic characteristics of the topsoil. Crop development started at the end of March/early April, ± d 210. The development of the storage organ (the grains) started in June, d 275. Harvesting date was August 31, 1976.

Crop parameters were obtained from a standard set estimated from field trails under Dutch conditions (Spitters et al., 1988). Directly after planting the root length was set at 3.5 cm. The weighing factor for root development was set high in the upper 40 cm (0.70) gradually decreasing to zero at maximum rooting depth (85 cm).

Soil physical measurements

Both hydraulic conductivity and moisture retention data are used to compute the static and dynamic functional properties (Wösten et al. 1986). To determine the physical properties, laboratory and field experiments are necessary. The suction crust infiltrometer was used to measure the hydraulic conductivity near and at saturation (Booltink et al., 1991). This method can be applied both in the laboratory and the field. Unsaturated conductivities at lower pressure head can be determined using the one-step outflow (Kool et al., 1985). The water retention data can be obtained using a hanging water column equipped with pressure cells for the higher pressures. The hydraulic conductivity and moisture retention functions are described using the van Genuchten parameters (Van Genuchten, 1980).

Clustering of functional groups

To identify whether the static functional properties differ significantly a Multiple Analyses of Variance was preformed. The calculated travel time and upward flux groupings are defined and tested for a 99% confidence interval using Roy's largest root test to handle the dispersion. This test is a generalization of the F-test on multivariate data sets (Morrison, 1976). The dynamic functional property, flux, of the grouped samples at the depths of 40 cm and 100 cm is compared graphically. Average hydraulic functions are determined by geometrically averaging the grouped functions (Wösten et al., 1987).

Spatial and temporal variation

Spatial dependency of a soil property is expressed in the (semi-) variogram. The variogram expresses the degree of spatial dependence or semivariance between point observations as a function of the distance between points.

To arrive at spatial averages ordinary kriging was used. Kriging in this study is regarded as a regression procedure, the best linear unbiased predictor (BLUP) of a variable in an unvisited location depends on the observations and the semivariance values (Stein & Corsten, 1991). In this study the simulation results for the augering points were used for spatial interpolation. Stein et al. (1991) showed that this procedure, *calculate first interpolate later*, yielded better results than when soil data were interpolated towards points followed by simulation on these points. Interpolation of simulated point data forms the basis for the construction of a terrain model.

All soil profiles were characterised as a sequence of functional layers. Boundary conditions and crop parameters are used for calculation of the dynamic behaviour of the functional layers. Surface topography was used to translate groundwater levels to other locations.

Interpolated results were compared graphically for two depths at different points in time, thus providing data on the dynamic behavior of the soils in the area. Fluxes during and after a rain shower and during a dry period were compared in this study. The development of the grain [kg dry matter ha^{-1}] was monitored as an indicator of water stress levels.

RESULTS AND DISCUSSION

Functional layers

After a preliminary soil survey an impression of the different soil layers in the study area was obtained. The subsoils are complex fine layered with alternating sandy and clayey lenses of varying thicknesses. A preliminary subdivision was made (Table 17–1).

Table 17–1. Preliminary functional layers.

Layer	Description
1	sandy layer with few loamy/silty lenses
2	sandy layer with many loamy/silty lenses
3	silty/clayer layer with few sandy lenses

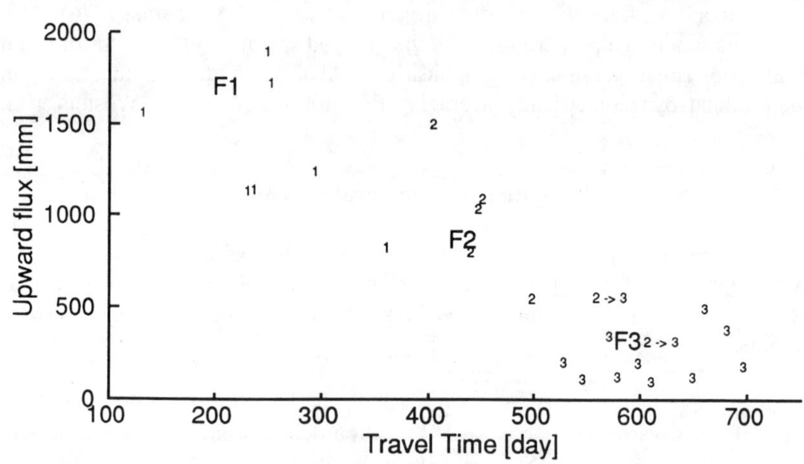

Fig. 17–4. Scatter plot of the functional groups based on the calculated two static functional properties, upward flux and travel time. Results for the measured and geometrically averaged hydrologic functions are plotted.

The layers were sampled and their hydraulic characteristics determined. The functional properties, travel time and upward flux, were calculated. Both properties were plotted (Fig. 17–4) and three groups could be identified. The groups were tested using Roy's largest eigenvalue. On basis of the calculated static properties two samples were moved from functional group 2 to functional group 3. After regrouping the preliminary groups the resulting three clusters were significantly different (Table 17–2).

Final testing for the subsoils was done by graphically comparing the behavior of the hydraulic functions with regard to dynamic fluxes at 100 cm depth under preset boundary conditions (Fig. 17–5). Based on the static and graphical analyses it was concluded that the three functional layers for the subsoil should be separated during the soil survey. The final grouping allowed geometrically averaged hydraulic functions to be determined for simulation purposes for each functional layer.

Two functional layers for the topsoil were identified; a sandy to sandy loam and a loamy to clay loam layer. The physical characteristic of the sandy to sandy loamy layer was assumed to be equal to those of the sandy to sandy loamy topsoil layer. The loamy to clay loamy topsoil was characterised using the parameters from the 1993 study (Finke & Bosma, 1993). Geometrically averaged hydraulic functions for the functional layers are shown in Fig. 17–6 and Table 17–3.

Table 17-2. Mean values and standard deviations (sd) of the functional properties.

Functional layer	mean travel time [days]	sd	mean height of upward flux [mm]	sd	n	Roy's largest eigenvalues
preliminary groups						
F1	252.3	69.3	1358.4	379,7	7	
F2	491.8	107.1	833.6	403.7	7	46.47
F3	617.6	56.8	236.8	171.8	10	
corrected groups						
F1	252.3	69.3	1358.4	379.7	7	
F2	434.0	50.1	993.2	354.4	5	94.62
F3	620.7	52.2	269.7	181.0	12	

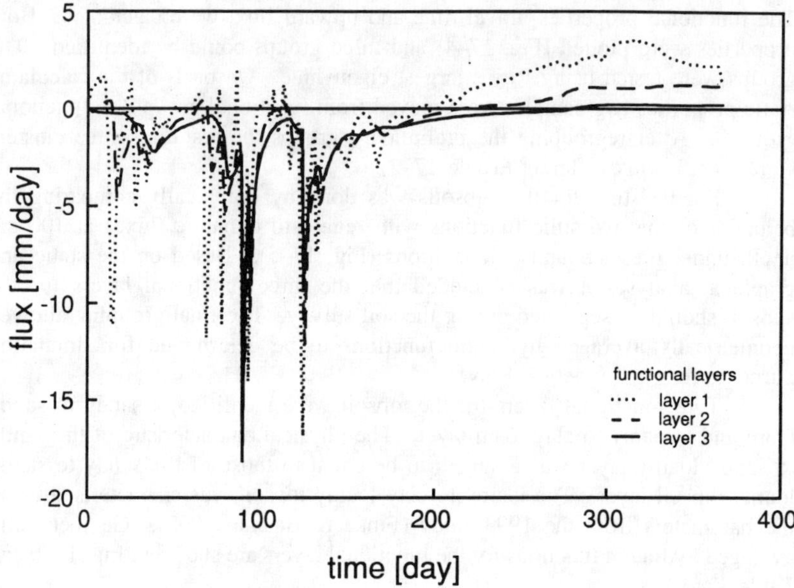

Fig. 17–5. Dynamic simulated flux [mm d^{-1}], below the rooting zone (100 cm depth) for the three identified functional layers for the subsoil.

Table 17-3. Description of the functional layers.

Functional layer	clay%	bulk dens [kg/dm3]	Θ-sat	K-sat	α	n	γ	Θ-res
			Van Genuchten parameters					
F1	0-8	1.48	0.42	183	0.03096	1.44236	-0.01715	0.00
F2	8-11	1.21	0.55	128.2	0.04852	1.20918	-0.25169	0.00
F3	11-23	1.08	0.60	7.3	0.03060	1.19300	-0.29187	0.00
F4	8-23	1.44	0.44	265	0.11300	1.14700	9.19400	0.00

Note: F1 topsoil and subsoil, F4 topsoil layer (after Finke and Bosma, 1993)

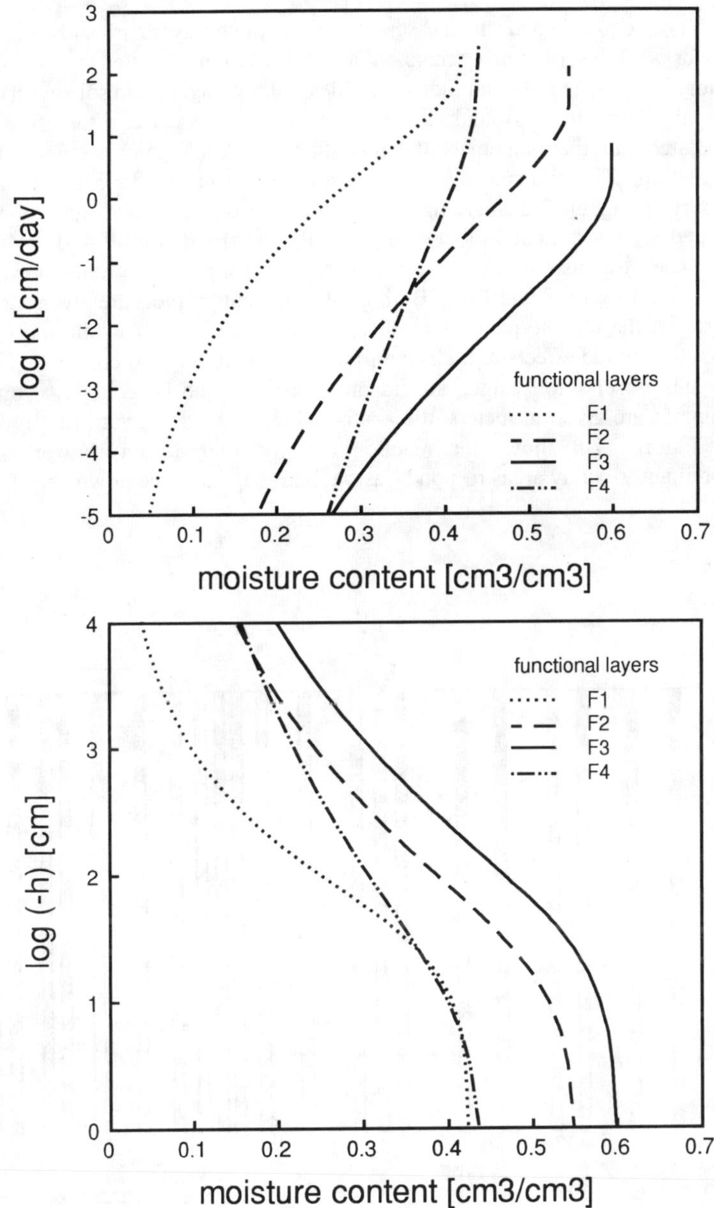

Fig. 17–6. Geometrically averaged hydraulic conductivity and moisture retention curves for all four functional layers.

Spatial and temporal variation

The sequence and thicknesses of the functional layers are shown in Fig. 17–7 for a series of soil borings in a cross section. Boundaries between functional layers are not continuous variables as they might be in soils with clear pedogenic horizons. Boundaries of functional layers were therefore not interpolated but the outcomes of calculations for each point were used for interpolation. The grid spacing of the survey was for practical reasons set at 50 meters resulting in 420 augerings. Soil water fluxes and crop growth were simulated with a validated model (Finke & Bosma, 1993) for all 420 profiles.

The response of the soil water regime following a rain shower is displayed in Fig. 17–8 and 17-9. Both 2.5-D and contour plots are given for two depths. At the top the patterns at 40 cm depth are displayed while the lower graphs display the effects at the same time interval for the 100 cm depth. Axis scale and observation angle are identical for all 2.5-D graphs; x and y coordinates are given in meters, the z-axis indicates the downward flux in [mm d^{-1}]. Figure 17-8 shows the reaction directly after a rain shower. The predominantly sandy areas respond fast at both depths. The downward fluxes

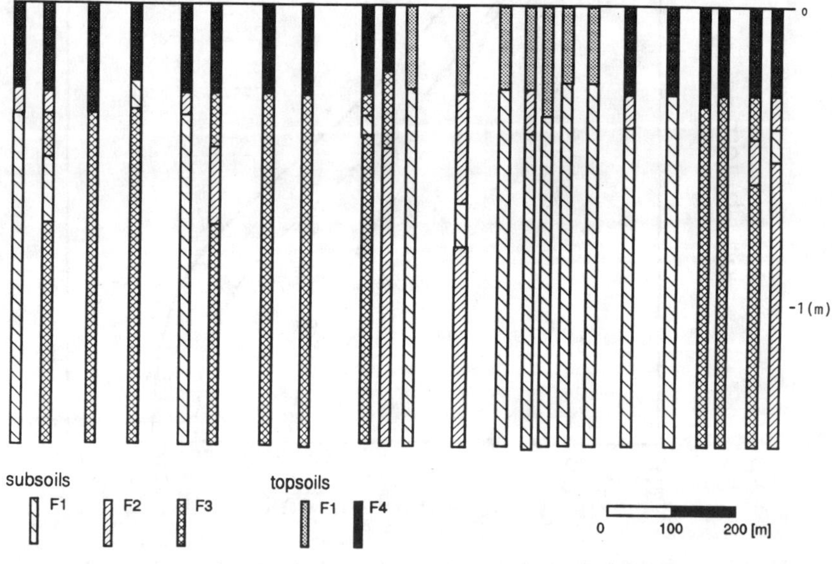

Fig. 17–7. Cross section of a series of soil borings that are described in terms of functional layers.

are high compared with the more clayey areas which show no clear reaction. A few days later (Fig. 17–9) the drainage front of the loamy and clayey area has caught up with the drainage front of the sandy area at a depth of 40 cm. The total downward flux has reached a level of 10 to 12 [mm d^{-1}] and no clear spatial structure remains. At 100 cm depth the different reaction rates are prominent; again the sandy area can be recognised by the relative high fluxes (≥7 [mm d^{-1}]) but also the predominant loamy versus clayey areas can be separated. The clayey areas yield very low downward fluxes (≤3 [mm d^{-1}]) as the loamy area display an intermediate range (3 ≤ flux ≥ 7 [mm d^{-1}]).

When taking the downward flux as an indication for sensitivity to leaching of agrochemicals (Hack-ten Broeke et al., 1993) the sandy areas provide a higher risk, under the given boundary conditions, of leaching chemicals followed by the loamy and clayey areas.

When looking at the upward flux as a measure for supplying water under dry conditions similar patterns are revealed (Fig. 17–10). It is shown that the sandy areas can support a substantial higher upward flux than the loamy and clayey areas.

The summer of 1976 was extremely dry, water storage in the profile was exhausted rapidly. Water necessary for crop development had to be supplied via the upward flux from the groundwater table. Simulated grain development is considered by looking at three dates, two at the initial stage (Fig. 17–11) and one at the ripening stage (Fig. 17–12). In Fig. 17–11 the grain production [kg dry matter ha^{-1}] at d 280, a few d after the grains were formed, and d 290 are plotted. Note that the scales of the z-axis, [kg ha^{-1}], are different. Although the initial grain production hardly displays any spatial structure, reduced production as result of water stress already becomes evident after ten days. As expected, the higher upward flux in the predominantly sandy areas resulted in significantly higher yields. A week before harvesting (Fig. 17–12) the loamy and clayey areas can also be identified.

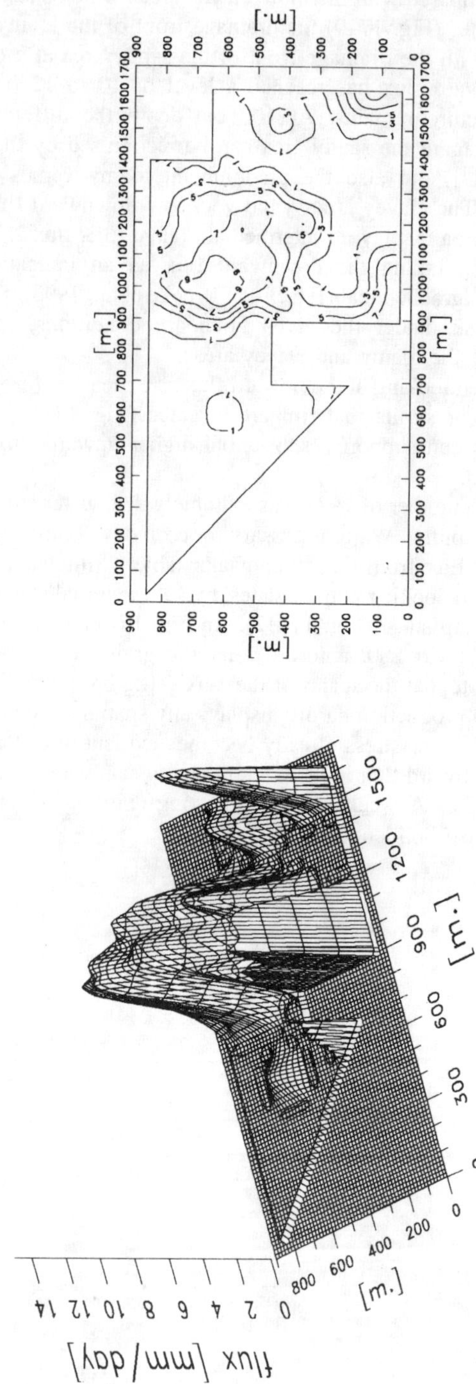

Fig. 17-8. The downward flux [mm d^{-1}] following a rain shower. At the top the patterns at 40 cm depth are displayed while the lower graphs display the effects at the same time interval for the 100 cm depth. Both a 2.5-D and contour plot are presented.

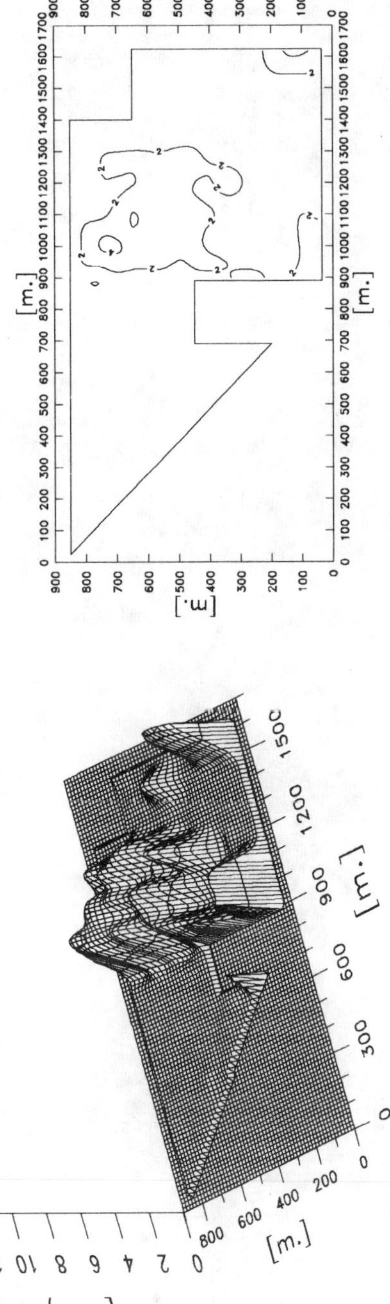

Fig. 17–8. Continued

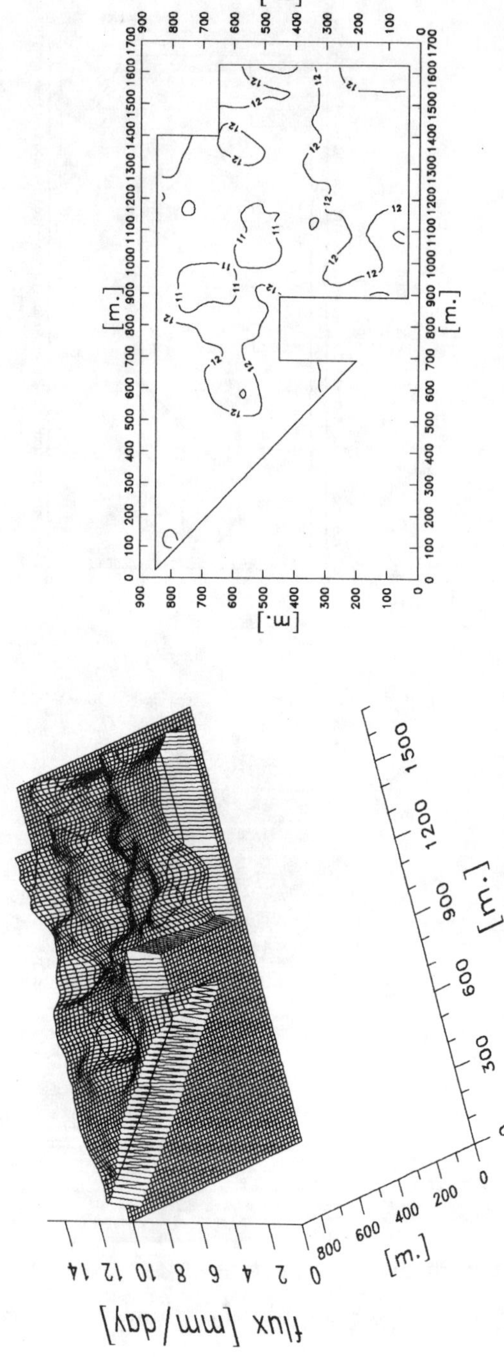

Fig. 17–9. The downward flux [mm d^{-1}] two days after a rain shower. At the top the patterns at 40 cm depth are displayed while the lower graphs display the effects at the same time interval for the 100 cm depth. Both a 2.5-D and contour plot are presented.

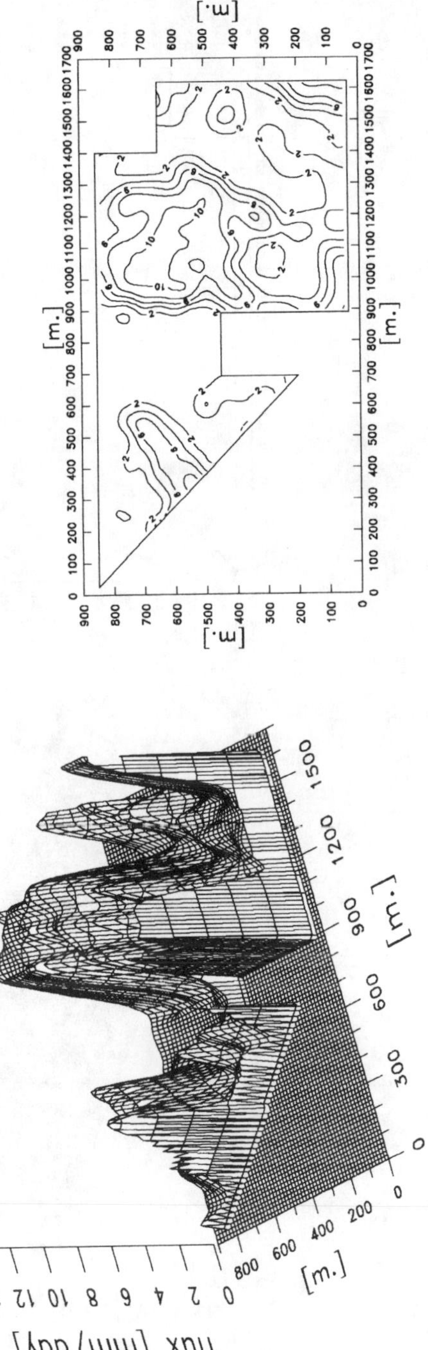

Fig. 17–9. Continued.

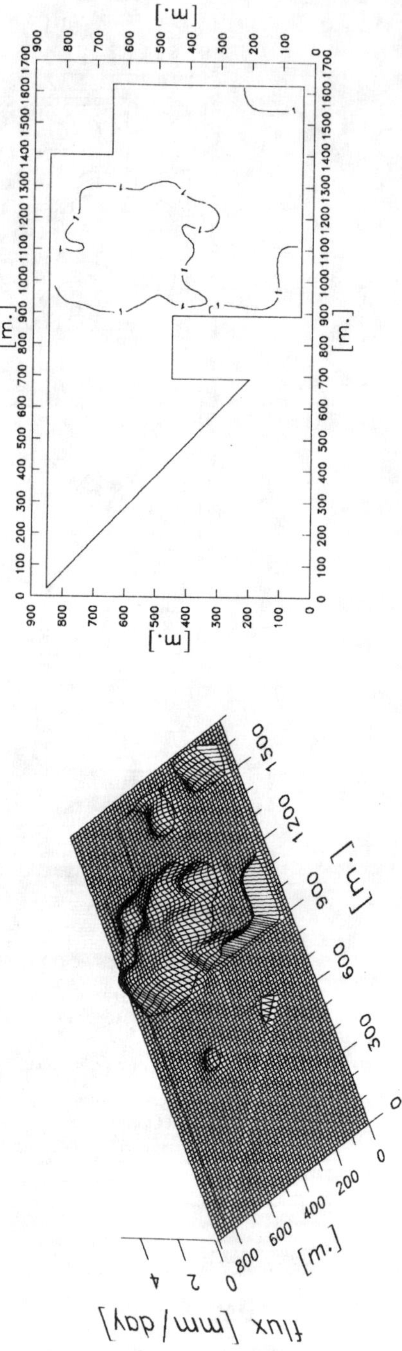

Fig. 17–10. The upward flux [mm d^{-1}] during a dry period. At the top the patterns at 40 cm depth are displayed while the lower graphs display the effects at the same time interval for the 100 cm depth. Both a 2.5-D and contour plot are presented.

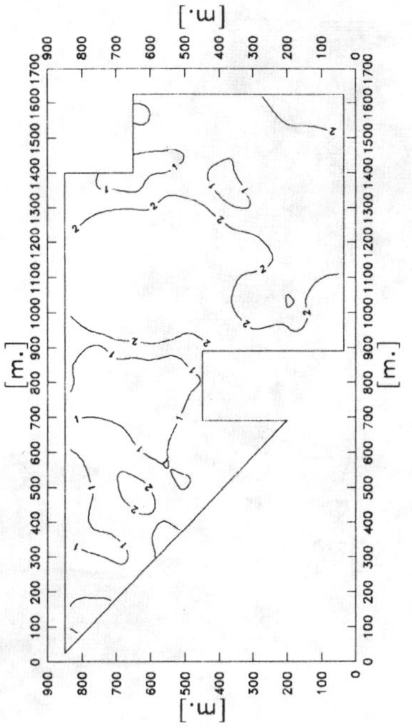

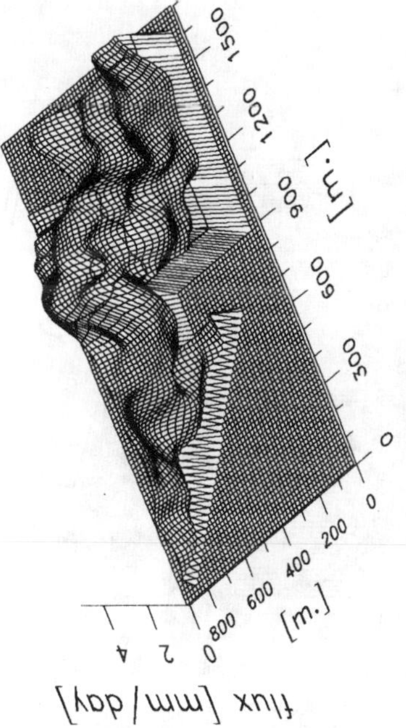

Fig. 17–10. Continued.

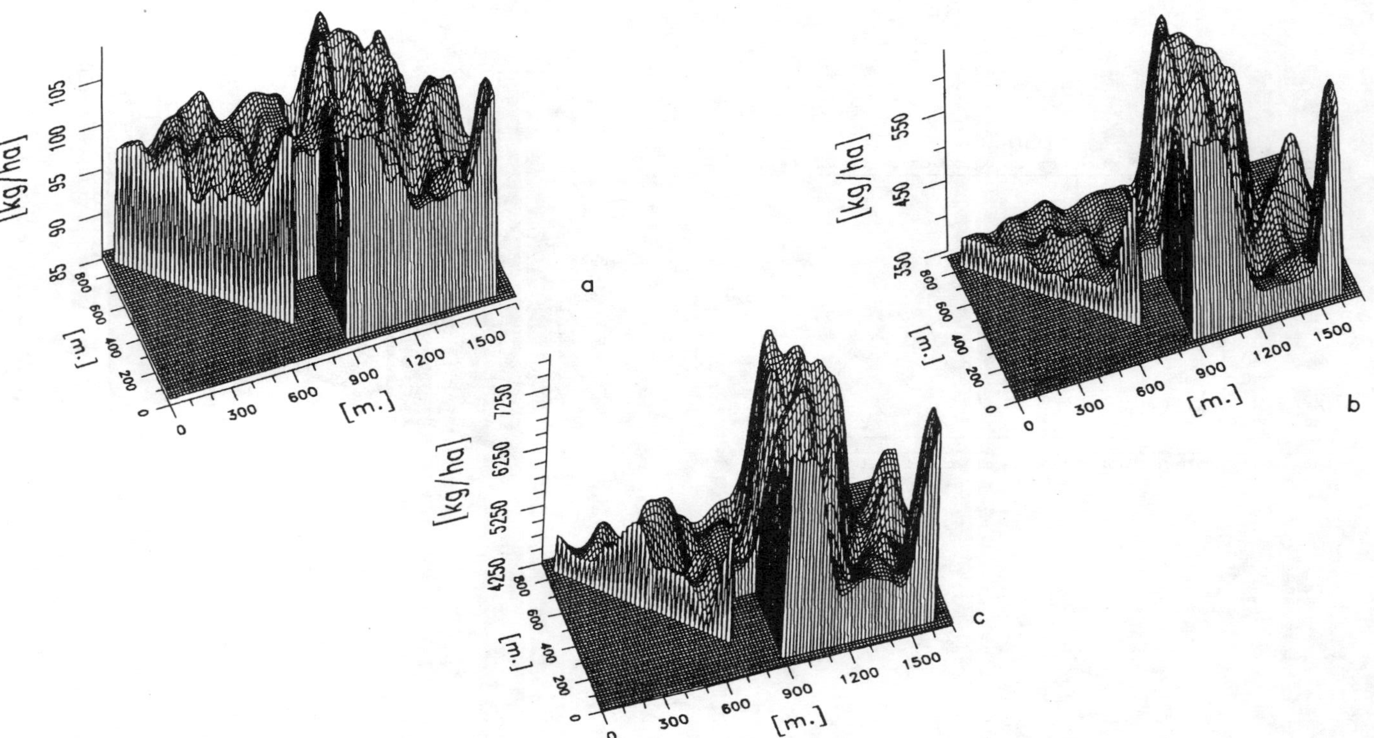

Fig. 17–11. Simulated grain production [kg dry matter ha^{-1}] at three dates; the intial stage at d 280 (a), ten days later (b) and a week before harvest (c). High yields are found in sandy areas with high upward flows.

Discussion

Complex soil profiles can be described in terms of functional layers. Hydraulic functions are used as basic input for deterministic mechanistic simulation models. Simulation models can be used to create risk or response maps. Areas with high leaching potential can be indicated for the study area. Fertilization regimes or its timing can be modified to reflect locally high leaching potential. Water limited crop production was modelled to detect areas with potentially different water stress levels. Indicated areas can be managed accordingly by irrigation or by raising of the groundwater table. Thus, model results provide relevant input data for site specific management practices. Additional modules for simulation of leaching of solutes and nitrate are already operational (SWAT-modules, Vanclooster et al., 1993). Simulation models can be used to not only evaluate current but also to explore the effects of potential management practices. The effects on crop production and the environment can thus be quantified and localized in space and time. Such techniques can become important tools to characterize and guide site specific management practices on a real time basis.

REFERENCES

Booltink, H.W.G., J. Bouma, and D. Giménez. 1991. A suction crust infiltrometer for measuring hydraulic conductivity of unsaturated soil near saturation. Soil Sci. Soc. of Am. J. 5:566–568.

Bouma, J., and H.A.J. van Lanen. 1986. Transfer functions and threshold values: from soil characteristics to land qualities. pp. 106–110. *In* K.J. Beek, et al. (ed.). Quantified Land Evaluation Procedures. ITC Publ. No. 6, Enschede.

Dierckx, J., C. Belmans, and P. Pauwels. 1986. SWATRER: a computer package for modelling the field water balance. Reference manual 2, Lab. Soil Water Eng., Leuven Belgium.

Feddes, R.A., P.J. Kowalik, and H. Zaradny. 1987. Simulation of field water use and crop yield simulation monographs. PUDOC, Wageningen, 188 pp.

Finke, P.A., and W.J.P. Bosma. 1993. Obtaining basic simulation data for a heterogeneous field with stratified marine soils. Hydrological Processes 7:63–75. J. Wiley and Sons, Chicester, England.

Hack-ten Broeke, M.J.D., H.A.J. van Lanen, and J. Bouma. 1993. The leaching potential as a land quality of two Dutch soils under current and potential management conditions. Geoderma, 60:73–88.

Kool, J.B., J.C. Parker, and M.Th. van Genuchten. 1985. Determining soil hydraulic properties from one step outflow experiments by parameter estimation: I. Theory and numerical experiments. Soil Sci. Soc. Am. J. 49:1348–1354.

Morrison, D.F. 1976. Multivariate Statistical methods, second edition. McGraw-Hill Kogaskusha, Tokyo.

Soil Survey Staff. 1975. Soil Taxonomy: A basic system of soil classification for making and interpretating soil surveys. USDA-SCS Agric. Handb. 436. U.S. Gov. Print. Office, Washington DC.

Spitters, C.J.T, H. van Keulen, and D.W.G. Kraalingen. 1988. A simple and universal crop growth simulation model, SUCROS87. *In* R. Rabbinge, et al. (ed.). Simulation and system management in crop protection. Simulation Monographs, PUDOC Wageningen, The Netherlands. p 147–181.

Stein, A., and L.C.A. Corsten. 1991. Universal kriging and cokriging as a regression procedure. Biometrics 47:575–587.

Stein, A., I.G. Staritsky, J. Bouma, A.C. van Eijnsbergen, and A.K. Bregt, 1991. Simulation of moisture deficits and areal interpolation by universal cokriging. Water Resourc. Res. 27(8):1963–1973.

Van Genuchten, M. Th. 1980. A closed form equation for predicting the hydraulic conductivity of unsaturated soils. Soil Sci. Soc. Am. J. 44:892–898.

Vanclooster, M., J. Diels, P. Viaene, and J. Feyen. 1993. The SWAT-Modules: theory and input requirement. Institute for Land and Water Management, Katholieke Universiteit Leuven, Belgium.

Wösten, J.H.M., M.H. Bannink, J.J. de Gruijter, and J. Bouma. 1986. A procedure to identify different groups of hydraulic-conductivity and moisture-retention curves for soil horizons. J. Hydrol. 86(1986):133–145.

Wösten, J.H.M., M.H. Bannink, and J. Beuving. 1987. Water retention and hydraulic conductivity characteristics of top- and sub-soils in the Netherlands: The Staring database. Report 1932, Soil Survey Inst. Wageningen, The Netherlands (in Dutch).

18 Quality Of Field Management Maps

D. S. Long
G. R. Carlson

Northern Agricultural Research Center
Montana State Univ.
Havre, Montana

S. D. DeGloria

Dep. of Soil, Crop, and Atmospheric Sciences
Cornell University
Ithaca, New York

To implement site–specific crop management, mapping methods have been developed for dividing fields into smaller areas that are relatively uniform in soil qualities that determine productivity. The resulting field areas would constitute management units for site–specific application of fertilizer, seeds, and tillage.

Established methods of deriving field management maps involve use of standard 1:20,000 scale soil surveys (Robert 1989), color infrared aerial photographs (Milfred & Kiefer 1976), and maps of kriged point data (Mulla 1991). Few studies, however, have been reported that document the reliability of management maps derived from these available methods.

Karlen et al. (1990) quantified yield variation for nineteen soil map units, classified as Ultisols representing seven soil series, in an 8 ha southeastern Coastal Plain field. A detailed soil map of the field was obtained from observing soils at 15 m intervals. Corn and wheat yield variation within an individual map unit was nearly as large as variation among soil map units. Yield variability was too extreme to suggest usefulness of soil maps for site-specific crop management.

Long et al. (1989) related digital brightness values in video-scanned color infrared aerial photographs with wheat yields in a 0.5 ha field in Montana. They found that dark-brightness values in cropped fields were more consistently related with wheat yield than dark-brightness values in crop-free fields. Further, dark values in cropped fields had significantly higher yields than light values. This suggested the usefulness of aerial photographs of cropped fields for expressing spatial patterns in crop variability and for selecting uniform parts of fields for crop experimentation.

Mulla (1989) proposed sampling soil variables on a grid and interpolating these variables with a geostatistical procedure termed kriging. Soil variables to be kriged would be ones whose spatial variation was related to patterns in plant available nutrients and water. Class maps were formed by dividing the frequency distribution of the kriged values of a soil variable into classes indicating low, medium, and high levels. Management maps for applying variable rates of fertilizer were developed from the class maps in two ways. The first way involved assigning a fertilizer recommendation to each class of a single class map based on potential crop yield and soil test nutrient levels (Mulla 1992). The second way involved combining class maps of two soil variables in a geographic information system (GIS) followed by assignment of fertilizer recommendations to each resulting class (Mulla & Annondale 1990, Mulla 1991).

The quality of management maps derived from soil survey interpretation, aerial photographic interpretation, and maps of kriged data needs to be determined to improve mapping methodology for site-specific crop management. The objective of the present study was to assess accuracy and precision of field management maps derived from: (i) interpreting a soil survey map, (ii) interpreting an aerial photograph, (iii) overlaying class-values of kriged point data in a GIS, (iv) combining values of kriged point data in a regression equation. The above mapping techniques were evaluated using cereal grain yield and soil variables sampled from a field in northern Montana.

MATERIALS AND METHODS

The study site is one 0.53 ha research field (40 m by 134 m) located on the Northern Agricultural Research Center near Havre, Montana, USA. The climate is semi-arid with about 30 cm of annual rainfall. The field includes Aridic Argiborolls belonging to the Telstad-Joplin soil association and Aridic Haploborolls belonging to the Fort Benton-Kremlin soil association. Telstad-Joplin soils are formed from glacial till whereas Fort Benton-Kremlin soils are formed from alluvium. In 1990, the entire field was planted to spring wheat (*Triticum aestivum* L.) in accordance with a 6 by 144 (864) plot uniformity trial design (Fig. 18–1). Grain yield was harvested from each plot using a plot harvester.

Developing Management Maps

A soil map, aerial photograph, and maps of kriged data were separately used to divide the field into areas that were expected to be relatively homogenous in grain yield. The resulting field divisions from each method were then depicted on maps. Each map constituted a management map for showing important differences in productivity.

The **soil survey method** involved mapping soils in the field at 1:660 scale. To expedite the survey, soil map units were defined to be an individual soil series, a soil series phase, or a soil series taxadjunct. A field management map was formed from grouping soil map units based on field observations and similar water-related characteristics noted in National Cooperative Soil Survey (NCSS) descriptions. The soil characteristics included such water-related attributes as available water capacity, runoff, depth to calcic horizon, permeability, and surface texture.

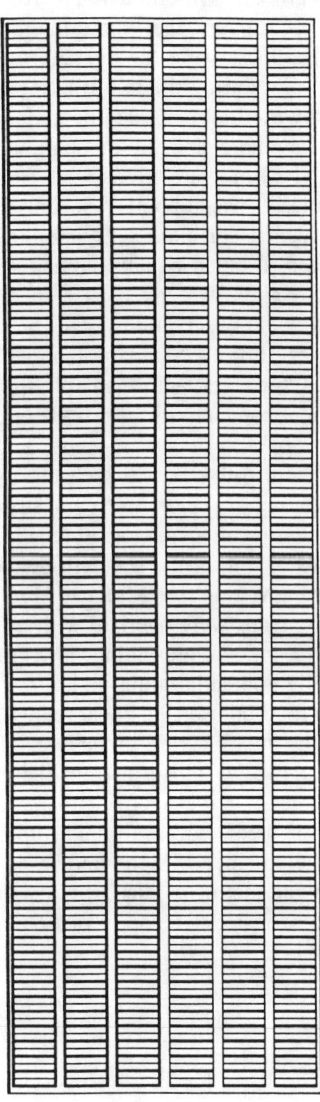

Fig. 18–1. Research field subdivided into 6 by 144 grid of 864 plots in accordance with uniformity trial design.

The **aerial photo interpretation method** entailed acquiring color infrared (CIR) photographs of the uniformity trial in mid-July when crop-growth differences were visible. This timing corresponded with the late heading, or flowering, stage of wheat growth. A photo-scale of 1:4,500 was chosen for single frame coverage of the field. A field management map was formed by delineating along sharp differences in photo tone that appeared between field areas differing in crop growth. Ground data on crop canopy height and relative biomass were used to support the photo-interpretations of crop growth. Measurements of dry biomass were the average of five electrical capacitance readings taken in each plot during the mid-tillering growth stage.

The **GIS overlay method** consisted of sampling soils in two north-south oriented transects 128 m long and 13 m apart, and five east-west oriented transects 40 m long and 22 m apart. Soils were sampled to a depth of 0.15 m at a total of 146 irregularly spaced points. Seventy two of the 146 points were located one to a plot which constituted 72 joint soil/yield observations. The remaining points were distributed equally among 18 plots. The mean of the values in each of these plots was taken as a measured observation and this constituted 18 joint soil/yield observations. This brought to 92 the total number of joint soil/yield observations. Standard laboratory methods were used in analyzing soil samples for nitrate-nitrogen, phosphorous, potassium, organic matter, pH, and electrical conductivity. Forward selection regression was used to identify the soil variables that were most correlated with grain yield. Unsampled values of these soil variables were then estimated to a 132 by 440 grid using ordinary kriging. The frequency distribution of each kriged soil variable was divided into low, moderate, and high classes based on cutoff values equalling the mean plus or minus one standard deviation and the mean plus or minus one half standard deviation (see Mulla 1991). The low, intermediate, and high values of two class maps were overlaid with a GIS to form a management map consisting of five units representing low, moderately low, moderate, moderately high, and high potential grain yield.

The **regression method** used the linear regression equation resulting from the above described forward selection procedure. This equation was used to predict grain yield to a 132 by 440 grid from the estimated values of two or more kriged soil variables. A management map was formed by dividing the frequency distribution of predicted grain yield into classes of low, moderately low, moderate, moderately high, and high yield based on the mean plus or minus one standard deviation and the mean plus or minus one half standard deviation. The GIS overlay and regression methods of combining the kriged data are illustrated in Fig. 18–2.

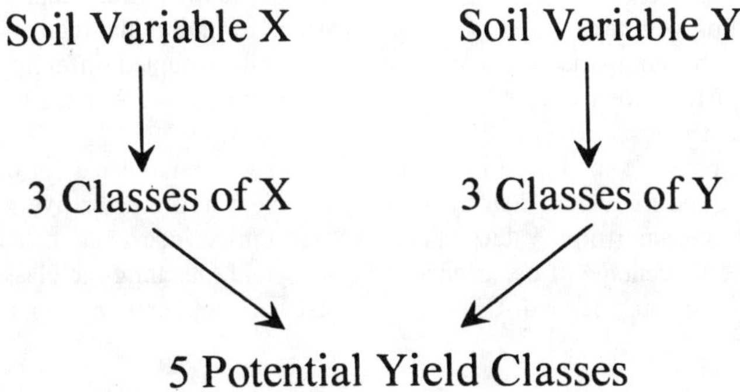

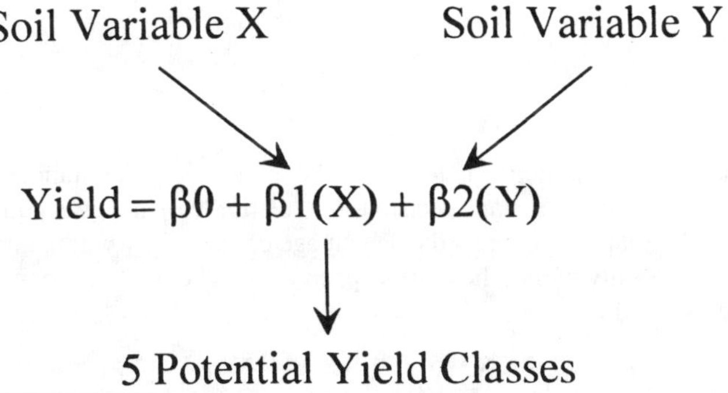

Fig. 18–2. Forming five potential yield classes with GIS overlay and regression methods of combining soil variable X and soil variable Y.

Evaluating Management Maps

Management maps were evaluated by comparing simple measures of accuracy and precision (Beckett & Burrough 1971, Marsman & DeGuijter 1986). These measures included the purity index and the average purity to describe map accuracy in terms of classification of grain yield, and relative variance to describe map homogeneity in terms of variability of grain yield.

For the comparisons, only quality measures for an entire map were used. The measures were calculated by averaging the values of all individual map units and weighing with the number of units. Measures for individual map units could not be used for comparisons because same map units occupied different areas on different management maps, and map unit numbers were unequal among management maps.

To measure map accuracy, a map legend was established for each map by dividing the frequency distribution of grain yield into four or five groups of low, moderate, and high values. The average purity index was calculated to express the percentage of occurrences of grain yield matching the classification criteria of the map legend. The purity of each map unit, $\overline{P_{jk}}$, in a map is expressed as

$$\overline{P_{jk}} = \sum_{i=1}^{N} \frac{p_{ijk}}{N} \tag{1}$$

where p_{ijk} takes on a value of **1** if the ith observation matches the class criterion of the jth map unit of the kth map and a value of **0** otherwise, and N is the number of observations within a map unit.

The average purity index, $\overline{P}$, is the arithmetic mean of the individual purities according to

$$\overline{P} = \frac{\sum_{i=1}^{N} \overline{P_{jk}}}{N} \tag{2}$$

where $\overline{P_{jk}}$ is the purity index from above, and N is the number of map units.

To measure map precision, the relative variance was computed from the pooled variance to express the percentage of variability within mapping units that is < variability within the total mapped area. The pooled variance, S_p^2, in a map is expressed as

$$S_p^2 = \frac{\sum (x_{1j} - \bar{x}_1)^2 + \sum (x_{2j} - \bar{x}_2)^2 + \ldots\ldots \sum (x_{ij} - \bar{x}_i)^2}{n_1 + n_2 + \ldots\ldots - n_t} \tag{3}$$

where $\Sigma(x_{ij} - \bar{x}_1)^2$ is the sums of squares for the ith management unit, n_i is the number of observations in the ith map unit, and n_t is the total number of units. Relative variance, R_{s2}, is expressed as

$$R_{S^2} = \frac{(S_t^2 - S_p^2)}{S_t^2} \qquad (4)$$

where S_p^2 is the pooled variance and S_t^2 is the total within field variance.

The 90% confidence interval was computed for the overall purity index of each management map (Snedecor & Cochran 1967). This interval for the purity index is based on the binomial distribution according to

$$\bar{x} \pm (1.64)\ \sigma \sqrt{n} \qquad (5)$$

where x is the decimal equivalent of the overall purity index, sigma is the standard deviation given by the square root of the product [(x)(1–x)], and n is the total number of observations. Differences in accuracy were judged to be significant if they exceeded 1.64 times the standard error of the 864 binary (1,0) observations.

The 90% confidence interval for the relative variance of each map was computed from

$$\frac{(n-1)\,S_P^2}{\chi^2_{(0.05,\,n-1)}} \le S_P^2 \le \frac{(n-1)\,S_P^2}{\chi^2_{(0.95,\,n-1)}} \qquad (6)$$

where n is the number of observations, S_p^2 is the pooled variance, and $\chi^2_{(0.05,n-1)}$ and $\chi^2_{(0.95,n-1)}$ are the critical values for the two-tail distribution of the chi square statistic. Untabulated critical values of the chi square statistic were derived from

$$\chi^2_{(\alpha,n-1)} = \frac{(\sqrt{2(n-1)}+z_\alpha)^2}{2} \qquad (7)$$

where $\chi^2_{(\alpha,n-1)}$ is the critical chi square value, n is the number of observations, z_α is the tabulated standard normal score for the desired level of probability, α. Differences in precision were judged to be significant at the 10% level of probability if regions of the corresponding confidence intervals were non-overlapping.

RESULTS AND DISCUSSION

Variability in Grain Yield

Observed values of grain yield averaged 1519 ± 242 kg ha^{-1} and ranged from a low of 1055 kg ha^{-1} to a high of 2291 kg ha^{-1}. Figure 18–3 shows the distribution of grain yield to mainly vary within 1000 kg ha^{-1} and 1500 kg ha^{-1} in the southern half of the field. In the northern half of the field, yield expresses a trend increasing from about 1000 kg ha^{-1} near the field center to about 2300 kg ha^{-1} in the northern part.

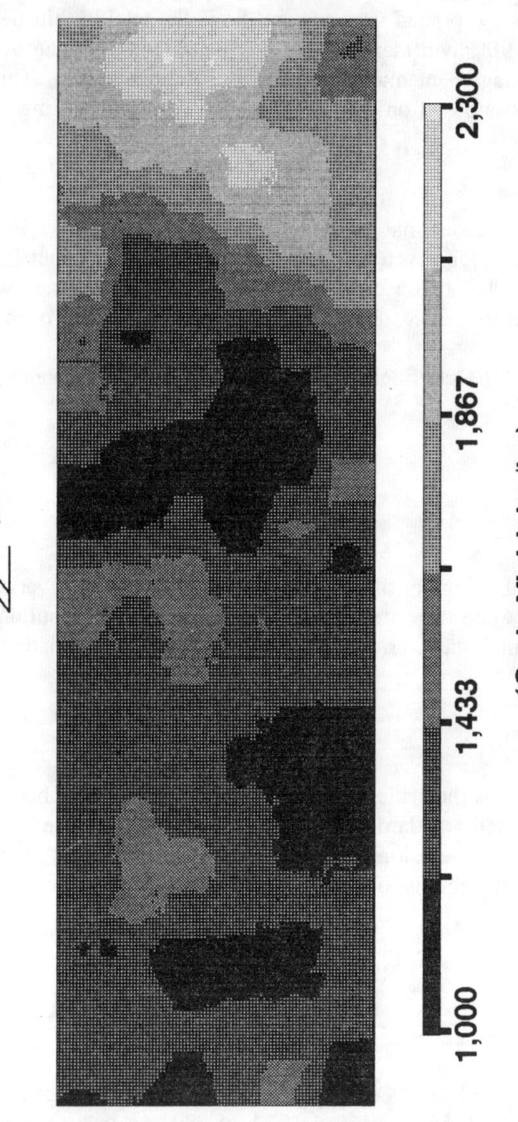

Fig. 18–3. Map of observed grain yield.

Field Management Maps

Soil Survey Method

The management map derived from the soil survey is shown in Fig. 18–4. Five management units, described in Table 18–1, were developed from interpreting water-related characteristics in field data and NCSS data.

Soils of the Telstad-deep phase formed a first management unit of high relative yield because of high available water capacity (AWC) that is suggested by field observations of the soils that show the A horizon to be dark colored and deep, and the Bt horizon to be fine textured, well developed and very deep. Very slow runoff is indicated by carbonates that are consistently below 51 cm

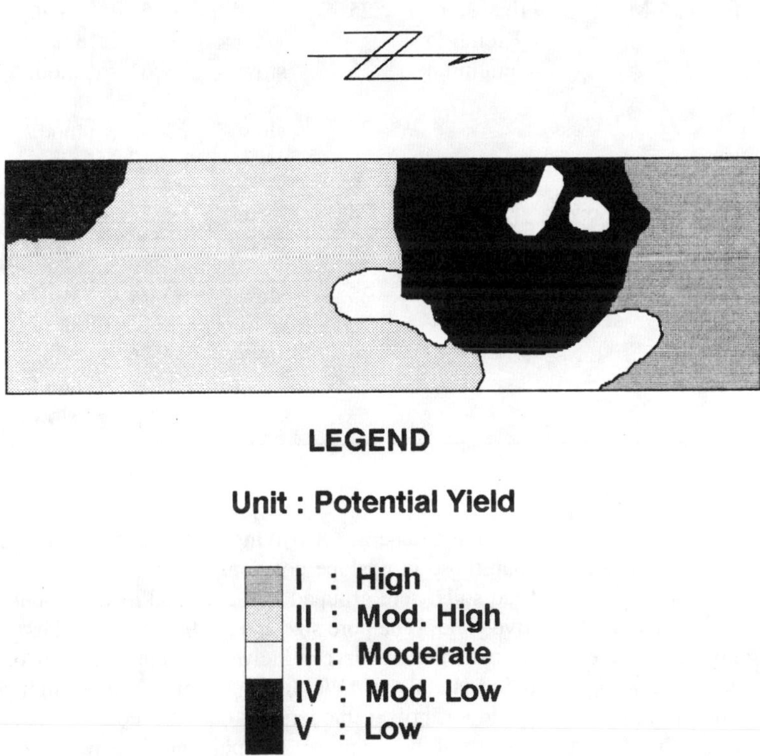

Fig. 18–4. Management map derived from the soil survey method.

Table 18–1. Data used in developing a management map with the soil survey method.

Map Unit	Potential Yield	Soil Series	AWC	Runoff	Depth to Bk	Perm.	Surf. Tex.
			-cm-		-cm-		
I	High	Telstad-deep	>28	v. slow	>51	slow	l
II	Mod. High	Telstad	28	slow	25–51	slow	cl
		Telstad-tax.	28	slow	28–44	slow	cl
		Scobey					
		Scobey-tax	28	slow	30–64	slow	cl
			28	slow	37–51	slow	cl
III	Mod.	Evanston	25	slow	20–51	mod.	l
		Kremlin					
		Kremlin-tax	25	slow	25–61	mod.	l
			25	slow	22–45	mod./ slow	l
IV	Mod. Low	Joplin	28	med.	<25	slow	cl
		Kevin	28	med.	<25	slow	cl
V	Low	Kenilworth	25	slow	30–71	slow	fsl
		Fort Benton					
			25	slow	38–76	mod. rapid/ slow	fsl

in the profile. These soils are situated in low lying terrain and thus are likely to receive more water than those in upslope positions.

Scobey and Telstad soils were grouped into a second management zone of moderately high relative yield. The pore size and configuration of these soils, judging from their clay loam texture, is conducive for high retention of soil water. Accordingly, NCSS data show AWC to be 28 cm. Further, high water retention is indicated by slow runoff, a moderately deep Bk horizon, and slow permeability. The taxadjuncts of these soils are morphologically similar and thus were likely to have similar AWC.

Evanston and Kremlin soils were grouped into a third management unit of moderate relative yield. AWC is reported in NCSS data to be 25 cm. The loam texture and moderate permeability indicate that water retention is likely to be somewhat < that of the Scobey and Telstad soils.

Joplin and Kevin soils formed a forth management zone of moderately low yield because of moderately low AWC. AWC is reported to be 28 cm but because of medium runoff the amount of water actually available is likely to be less. The increased runoff is also indicated by carbonates within 25 cm of the surface.

Fort Benton and Kenilworth soils formed a fifth management unit of low relative yield because of their low AWC that is reported to be 25 cm. Water retention is likely to be low because of moderately rapid permeability and fine sandy loam texture. Depth to carbonates is deeper than in more heavily textured soils. Thus, these soils are wetted to a greater depth because of low retention of water that is in turn related to the high sand content.

Aerial Method

The management map developed from the aerial method is illustrated in Fig. 18–5. Four management units were delineated based on contrasting photo tones appearing in the crop canopy on a CIR aerial photograph. Ground data on crop canopy height and biomass were used to support the interpretation of the crop differences. Table 18–2 shows that yield generally decreases with increasing photo tone and decreasing crop canopy height and relative crop biomass.

Table 18–2. Data used in developing a management map from aerial method.

Map Unit	Potential Yield	Photo Tone	Ground Reference Data	
			Mean Crop Height	Relative Crop Biomass†
			---cm---	
I	High	very dark	67.4 ± 2.8	202 ± 20
II	Mod. High	medium to dark	63.4 ± 2.5	174 ± 22
III	Mod. Low	medium to dark	59.3 ± 2.0	162 ± 23
IV	Low	light to medium	60.2 ± 2.3	149 ± 26

†measured by electrical capacitance.

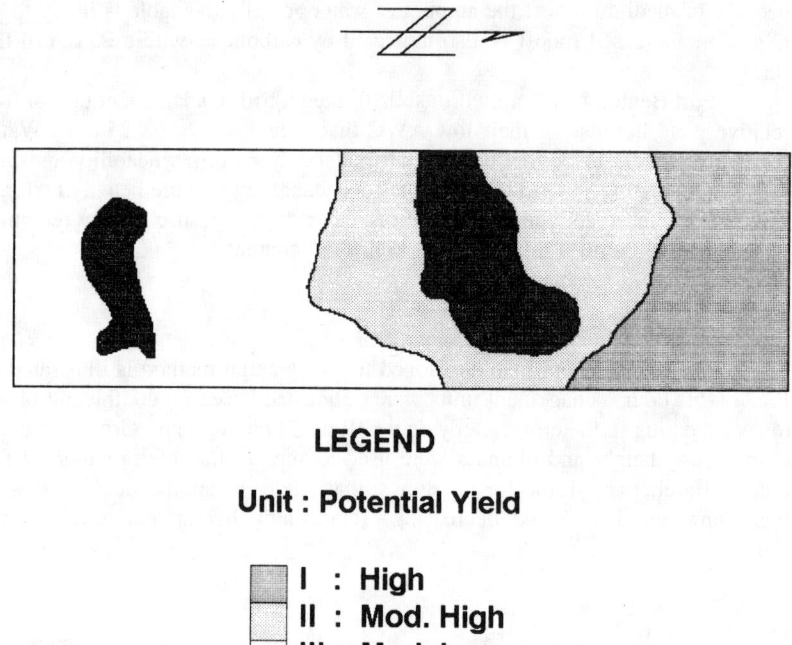

Fig. 18–5. Management map derived from the aerial photographic method.

Very dark toned areas of tall and vigorous crop growth formed a first management unit of high relative yield. These dark tones resulted from low reflectance owing to dense crop canopy, and dark colored, moist soils. Mean canopy height at 67 cm and mean relative biomass at 202 (by electrical capacitance) exceeded that of all other units.

Medium to dark toned areas with a mean crop height of 63 cm and mean relative biomass of 174 composed a second management unit of moderately high relative yield. The photo tones were lighter than that of the first unit because of decreasing crop canopy density and increasing soil surface exposure.

Medium to dark toned areas with a mean crop height of 59 cm and mean relative biomass of 162 formed a third management unit of moderately low relative yield. Though photo tones equalled that of the second unit, the crop growth characteristics were different.

Light to medium toned areas comprised a fourth management unit of low relative yield. The light tone was associated with low crop growth and sandy soils. Though mean crop height equalled that of the third unit, the mean relative biomass was different at 149 vs. 162.

GIS Overlay Method

pH and organic matter were identified as significant predictors of grain yield by the forward stepwise regression procedure. After semivariograms were computed for pH and organic matter, their values were estimated with kriging for an 132 by 440 grid. Two cutoff values for specifying the limits of low, moderate, and high classes in the frequency distribution of these variables were based on the mean plus or minus one half standard deviation. Figure 18–6 illustrates the GIS overlay method of forming five management units representing nine possible combinations of low, moderate, and high class levels of kriged pH and organic matter.

Figure 18–7 presents the management map derived from the GIS overlay method. Each management unit reflects the way pH and organic matter combine to affect grain yield. For instance, high yield of the first unit is characterized by high pH and organic matter whereas low relative yield of the fifth unit is characterized by low pH and organic matter. Moderately high relative yield is characterized by a combination of intermediate and high levels in the class values of these variables. In contrast, moderately low relative yield of the fourth management unit is characterized by a combination of intermediate and low class values. Moderate relative yield of the third unit is characterized by intermediate values.

OM \ pH	Low	Mod.	High
Low	Low	M. Low	Mod.
Mod.	M. Low	Mod.	M. High
High	Mod.	M. High	High

Fig. 18–6. Nine possible groupings of class levels in kriged pH and organic matter (OM). Terms within cells denote low, moderately low (M. Low), moderate (Mod.), moderately high (M. High), and high categories of potential yield.

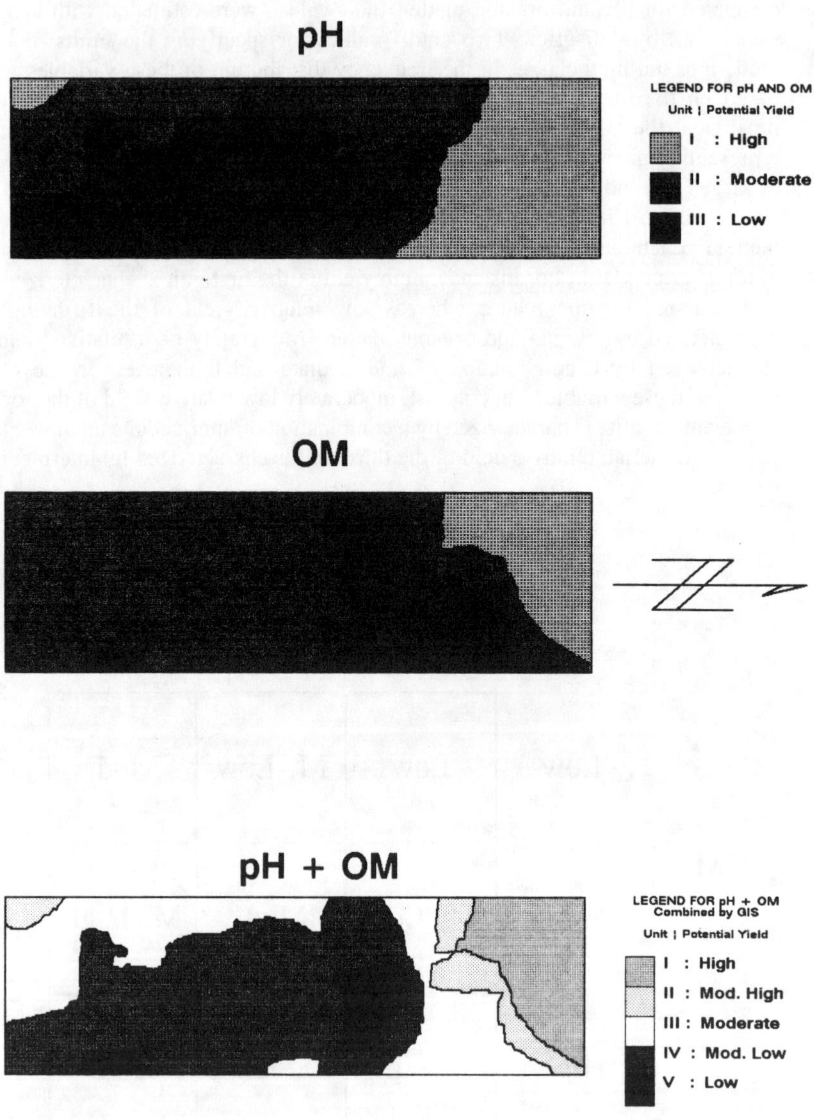

Fig. 18–7. Management map derived from GIS overlay method.

Regression Method

The stepwise regression equation that had identified pH and organic matter as significant predictors of grain yield is

$$Y = 3.37 + 0.36(X_1) + 0.04(X_2) \tag{8}$$

where Y is the natural log-transform of grain yield (kg ha^{-1}), X_1 is pH in 0 to 0.15 m depth, and X_2 is organic matter in 0 to 0.15 m depth. Grain yield was estimated with this equation using kriged values of pH and organic matter that had been interpolated to a 132 by 440 grid. Four cutoff values for specifying the limits of five classes in the frequency distribution of estimated grain yield were based on the mean plus or minus one standard deviation and the mean plus or minus one half standard deviation.

Figure 18–8 presents the management map derived from the regression equation, kriged values of pH and organic matter, and cutoff values that divide estimated grain yield in five classes. The five classes represent management units of low, moderately low, moderate, moderately high, and high relative grain yield. Conceptually, increases in pH and organic matter correspond with an increase in grain yield.

LEGEND FOR pH + OM
Combined by Regression

Unit ¦ Potential Yield

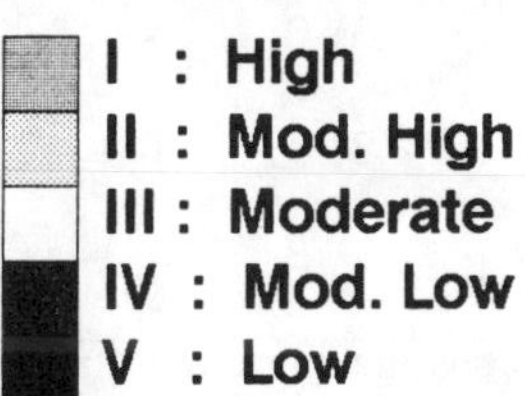

Fig. 18–8. Management map derived from the regression method.

Quality of Management Maps

Table 18–3 lists the measures of quality for different methods of developing a field management map. Accuracy is 60% for the aerial method, 42% for the regression method, 40% for the GIS overlay method, and 38% for the soil survey method. Precision is 67% for the aerial method, 60% for the regression method, 53% for the soil survey method, and 47% for the GIS overlay method.

The aerial method is significantly more accurate and precise than the other four methods as indicated by the 90% confidence intervals. Significant differences in accuracy among the other three methods were not detected. As indicated by the 90% confidence intervals, the four maps significantly differ in precision except for those made from the soil survey method and the GIS method.

To give some impression of differences between the mapping methods, each management map is visually compared with observed grain yield in Fig. 18–9. The map derived from the aerial method correlates well with the spatial patterns in yield over the whole field. This is the result that would be expected given the higher accuracy and precision of the map from the aerial method. The maps derived from soil survey and GIS overlay depict yield variation with some accuracy in the northern half of the field and with poor accuracy in the southern half of the field. The map derived from the regression method correlates with some accuracy over the entire field.

Table 18–3. Measures of quality for different methods of developing a field management map (values in parentheses are limits of 90% C.I.).

Method	Accuracy †	Precision ‡
	--------------------%	----------------------
Aerial	60 (57-63) A	67 (65–70) A
Regression	42 (39–45) B	60 (57-63) B
GIS Overlay	40 (37–43) B	47 (43–51) C
Soil Survey	38 (35–41) B	53 (49–56) C

† Percent of yield observations matching the map legend.

‡ Percent of yield variability in mapping units < variability in total mapped area.

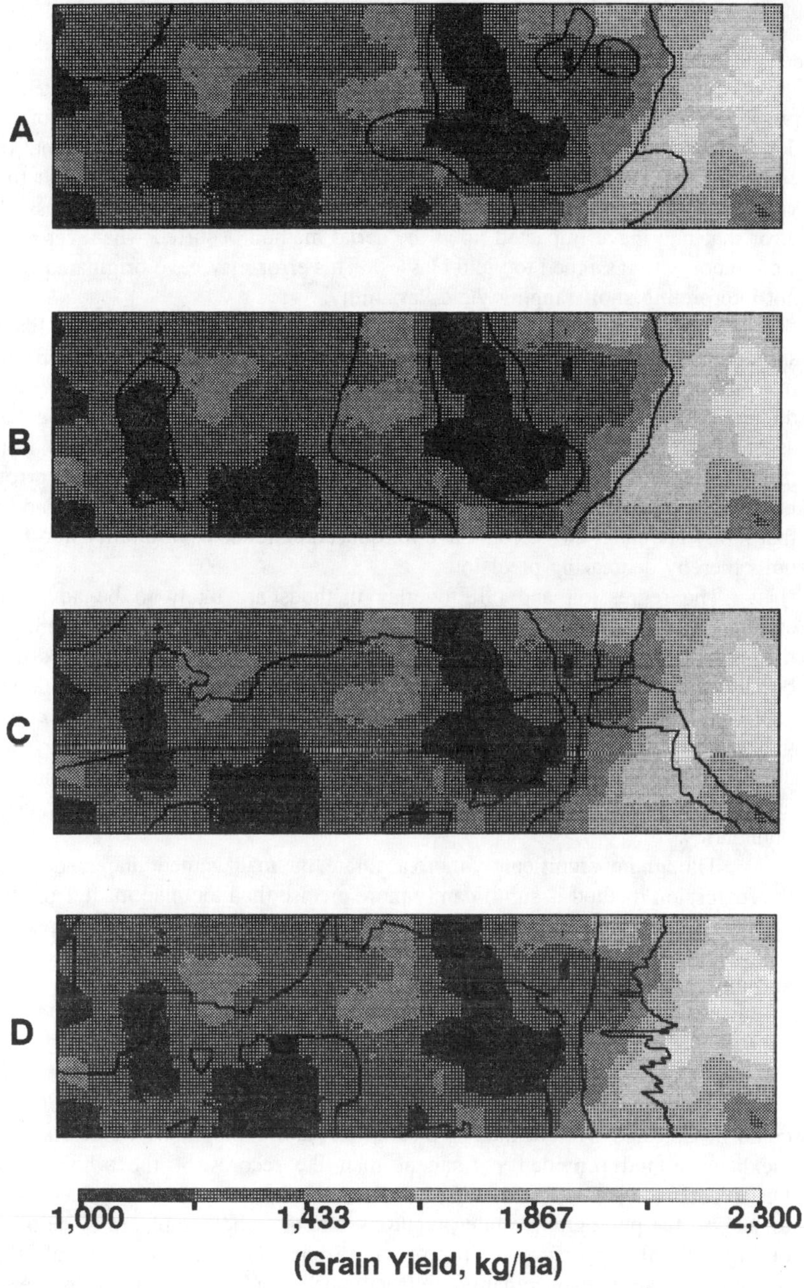

Fig. 18–9. Visual comparison of observed grain yield (background) with management maps (delineations) made from (A) soil survey, (B) aerial photograph, (C) GIS overlay, and (D) regression.

Assessment of Error

The management map made from the aerial method is more accurate and precise in terms of grain yield than either the other three maps because of the detailed information content of the imagery and availability of ground data on crop growth. This map, however, is based on crop growth of the same yr that yield data were taken and this may have favored the method. Further, a possible error that may have impacted upon the aerial method occurred whenever plots were incorrectly assigned to yield classes. This error may have originated from not interpreting short-ranging yield variability.

The low quality of the management map made from soil survey likely stems from (i) misinterpreting available water capacity for mapped soils, (ii) not mapping some soil variability that was important to yield, namely, impurities and differences within defined limits of mapped soils, and (iii) including other than assigned plots within a map unit. These errors possibly have sourced from inaccuracies in NCSS criteria and error in locating soil boundaries. All errors have led to misclassification of a plot resulting in a wrong purity index value for that plot. The latter two errors have introduced extraneous variability into map units thereby decreasing precision.

The regression and GIS overlay methods are likely to be adversely affected by sampling error arising from generalizing a small sample number to the entire field by means of kriging. Sampling error is inherent in kriged data because they are based on a subset of a population rather than on a population itself. In addition, the methods assume that yield classes, defined by cutoff values, coincide with important changes in crop variability. The relatively low accuracy of the management maps made from these methods suggests that this assumption is unrealistic and hence, leads to incorrect placement of class boundaries.

Though not significantly more accurate, the management map made from the regression method is significantly more precise than the map made from the GIS overlay method. By avoiding classification of kriged data, the regression method retains the original information content of the data used to estimate yield and this may account for the higher precision.

Cost/Benefit Relationships

The advantage of the aerial method was that crop and soil differences could be detected without intensive sampling, and photographs were relatively inexpensive and provided permanent map-like records of the whole field. Disadvantages were that ground reference data on crop growth were necessary to support the photographic interpretations, and that photographs must be taken during optimal appearance of crop or soil differences. Therefore, the question remains whether a management map from one yr would be useful in other years.

The advantage of the soil survey method was that interpretive data on NCSS soil series were available that could be used in grouping the soil map units into management units. The disadvantages, however, included large labor expenditure for intensive soil survey, difficulty in accurately delineating soil boundaries without aid of external topographic and soil features, and difficulty in classifying unusual soil types that were not described by NCSS standards. Further, standard procedures for designing mapping units for intensive soil survey were not available.

The advantage of the GIS and regression methods was use of kriging to estimate the unsampled values of a soil variable from a limited sample number. The disadvantages were high cost in labor and materials for sampling, sampling error due to generalizing a relatively small sample to a parent population, and the assumption that yield classes, defined by cutoff values, coincide with important changes in crop variability. The advantage of the regression method versus the GIS method was that values of kriged data could be used and this maintained the original information content of the data. Further, the regression method allowed for estimating potential yield from the continuous values of an unlimited number of kriged soil variables whereas the GIS method was limited to the class values of only two kriged soil variables. The regression method required that joint yield-soil observations were available for developing a regression equation and that yield was correlated with the observed soil variables.

SUMMARY AND CONCLUSIONS

This study investigated four mapping techniques of developing field management maps for site-specific crop management: (i) interpreting soil survey maps, (ii) interpreting aerial photographs, (iii) overlaying class maps of kriged point data in a GIS, and (iv) combining values of kriged point data in a regression equation. The methods were investigated in a 0.53 ha field that was planted to spring wheat in accordance with a 6 by 144 plot uniformity trial. Each method attempted to divide the field into smaller areas that were relatively uniform in crop yield.

The average purity index was used to express the percentage of occurrences of yield observations that matched the classification criteria of the map legend. Overall accuracy, or average purity index, was 38% for the soil survey method, 60% for the aerial method, 40% for the GIS overlay method, and 42% for the regression method. According to the 90% confidence interval, the aerial method was significantly more accurate than the other three methods.

The relative variance was used to measure the yield variation in a management map as a whole. Overall precision, or relative variance, was 67% for the aerial method, 60% for the regression method, 53% for the soil survey method, and 47% for the GIS overlay method. According to the 90% confidence interval, the aerial method and the regression method were significantly more precise than the GIS and soil survey methods.

To accurately divide fields into smaller areas that are relatively uniform in crop yield, the following should be included in future mapping strategies:

(1) Properly timed aerial photographs should be used together with ground reference data on crop growth,

(2) Values of kriged point data should be combined with the **regression** method rather than the **GIS overlay** method provided that joint soil/yield variables are available and these variables are correlated,

(3) Future research should verify whether management maps from one yr would be accurate and precise enough for use in other years. The conclusions of this study are mainly applicable to dryland cropland production systems. The results, however, also would pertain to crop production systems in other environments where differences in soil, topography, and precipitation leave noticeable effects on crop growth. Further work is needed to determine validity of the results in other locations with contrasting environmental conditions and crop management practices. Nevertheless, the results of this research indicate a need for further improvements in all types of mapping methodology for site-specific crop management.

ACKNOWLEDGEMENTS

Research funded by Montana Agric. Exp. Stn. Proj. No. 704 and Cornell Univ. Agric. Exp. Stn. Proj. No. 403. Computing resources provided by P. C. Robert, Dep. of Soil Science, Univ. of Minnesota.

REFERENCES

Beckett, P.H.T., and P.A. Burrough. 1971. The relation between cost and utility in soil survey. IV: Comparison of the utilities of soil maps produced by different survey procedures, and to different scales. J. Soil Sci. 22:466–480.

Karlen, D.L., E.J. Sadler, and W.J. Busscher. 1990. Crop yield variation associated with Coastal Plain soil map units. Soil Sci. Soc. Am. J. 54:859–865.

Long, D.S., G.A. Nielsen, and G.R. Carlson. 1989. Use of aerial photographs for improving layout of field research plots. Appl. Agric. Res. 4:97–100.

Marsman, A.B., and J.J. DeGuijter. 1986. Quality of soil maps: Comparison of survey methods in a sandy area. Soil Survey Papers No. 15. Stiboka, Wageningen.

Milfred, C.J., and R.W. Kieffer. 1976. Analysis of soil variability with repetitive aerial photographs. Soil Sci. Soc. Am. J. 40:553–557.

Mulla, D.J. 1989. Soil spatial variability and methods of analysis. p. 55–63. *In* C. Renard, et al. (ed.) Soil, crop, and water management systems for rainfed agriculture in the Sudano-Sahelian zone, ICRISTAT, Patancheru, India.

Mulla, D.J., and J.G. Annondale. 1990. Assessment of field-scale leaching patterns for management of nitrogen fertilizer application. p. 55–63. *In* K. Roth, et al. (ed.) Field-scale water and solute flux in soils. Birkhauser Verlag, Basel, Switzerland.

Mulla, D.J. 1991. Using geostatistics and GIS to manage spatial patterns in soil fertility. p. 336–345. *In* G. Kranzler (ed.) Proc. Automated Agric. 21st

Century. Am. Soc. Agric. Engineers, St. Joseph, MO.

Mulla, D.J. 1992. Mapping and managing spatial patterns in soil fertility and crop yield. p. 15–26. *In* P. C. Robert, et al. (ed.) Proc. soil specific crop management: A workshop on research and development issues, Minneapolis, MN. 14–16 April, ASA, CSSA, and SSSA, Madison, WI.

Robert, P.C. 1989. Land evaluation at farm level using soil survey information systems. p. 299–311. *In* J. Bouma, et al. (ed.) Land qualities in space and time. Proc. symposium ISSS. 22–26 Aug. 1988. Wageningen, Netherlands.

Snedecor, G.W., and W.G. Cochran. 1967. Statistical methods, The Iowa State University Press, Ames, IA.

19 Landscape Analysis Of Soil And Crop Data Using Regression

M. D. Tomer
J. L. Anderson
J. A. Lamb

Department of Soil Science
University of Minnesota
St. Paul, Minnesota

Geostatistical modeling of soil and crop variability requires that large numbers of samples be collected across the landscape, often at close intervals. Sampling costs are high, and intense sampling is generally impractical in a management context. Another problem with geostatistical analysis is the assumption of stationarity, which treats the data as a random field with some degree of correlation amongst neighboring sample points. The assumption applies poorly to soil properties and crop growth, which do not only vary in random fashion. There is also a systematic component of landscape variability. It is this systematic component, caused by trends in parent materials, hydrologic functioning, and soil development across the landscape, that is of greatest interest when soil or crop attributes are assessed on a site-specific basis. A better method of mapping variability of soils and crops would require less input data, and focus on systematic variability.

These concerns can be addressed through a regression approach to spatial modeling, using data collected from the land surface (e.g., topographic or remotely sensed data) to predict soil or crop attributes. The concept is not entirely new; soil scientists have traditionally used topographic information and aerial photographs in several ways. For example, hillslope partitioning (Ruhe & Walker, 1968) is one approach to explain topographic variation in soils or crops (Brubaker et al., 1993; Stone et al., 1985). Also, aerial photography has been used as an aid in soil mapping (Soil Survey Staff, 1951). Milfred and Kiefer (1976) showed that landscape patterns of loess thickness could be discerned using aerial photographs taken during crop growth. Other studies have applied regression analysis (or shown correlation) between surface-collected data and soil or crop parameters. Terrain modeling (Moore et al. 1991) has been used to map variation of soil properties within small drainage basins (Moore et al. 1993) where runoff is an important hydrologic component. Phototone data obtained by scanning aerial photographs (taken during bare soil conditions) has been related to soil organic matter (Robert, 1993; Zheng & Schreier, 1988), and (when photos were taken during crop growth) forage biomass (Everitt et al., 1987; Tucker, 1979).

Clearly, surface-collected data have potential to assess landscape variability of soils and crops. This paper provides examples of using surface-collected data and regression to model soil and crop variability across the landscape at a sand-plain location in east-central Minnesota.

SITE DESCRIPTION AND SCOPE OF WORK

The study was conducted at the Minnesota Management System Evaluation Area (MSEA) research site, near Princeton, Minnesota. The site is on the Anoka Sand Plain; topographic relief is < 3 m. Major research conducted at the site is designed to provide crop-management strategies that maintain productivity and protect ground water quality in sensitive environments. Details on the research design and some initial data from this major research are presented by Anderson et al. (1993) and Landon et al. (1993). The water-quality-monitoring component of the research is focused on sampling of ground water, rather than soil water stored above the water table. The project described here was designed to provide information on landscape variability at the site, to assist with interpretation of the monitoring data. Therefore projects were undertaken to describe variability of soil water storage, and crop growth and nitrogen uptake. The sandy soil has a high infiltration capacity, so runoff is negligible. We hypothesized that variability and lateral redistribution of soil water is caused by soil layering (both depositional and pedogenic), rather than runoff.

The soil at the site is a Zimmerman fine sand (mixed, frigid, Argic Udipsamment). A distinguishing feature of this soil is the presence of lamellae within 1.5 m of the surface. Lamellae are thin illuvial bands (usually < 5 cm thick) with small amounts of clay deposited by percolating water. Although the clay increase is slight (about 3%) lamellae are important to the hydrology of these soils. Water movement below lamellae occurs only upon their saturation, due to their smaller pores and abrupt boundaries. This effect, known as perching, helps lamellae retain soil water within upper soil layers. The result is improved crop productivity, and potentially slower chemical movement. Objectives of this research were: i) to determine the landscape distribution soil-water storage; ii) to correlate water storage with soil layering and topography, and iii) to determine if aerial photography can be used to map corn yield and nitrogen uptake. Patterns of corn yield and nitrogen uptake are interpreted by considering consistencies with soil-water modeling results.

METHODS

The presentation of this research is divided into sections describing topographic modeling of soil water storage, and modeling of corn yield and nitrogen uptake using aerial photography. These components of the study were conducted on separate research areas at the MSEA site. The soil-water monitoring project involved considerable disturbance. This study was, therefore, conducted on an uncropped hillslope (seeded with a grass mix) about 200 m east of the major research area (Fig. 19–1). The aerial photography study was conducted on the continuous corn research block (Fig. 19–1 shows relative locations of the two study areas).

In both study areas, a topographic survey was conducted at an approximate 4 m spacing, and a digital elevation model (DEM; 2 m cell size) was interpolated from the survey results. Slope and slope curvature (plan and profile) maps were calculated from the DEM's, using a 4 m cell size to reduce effects of microtopography and local interpolation errors.

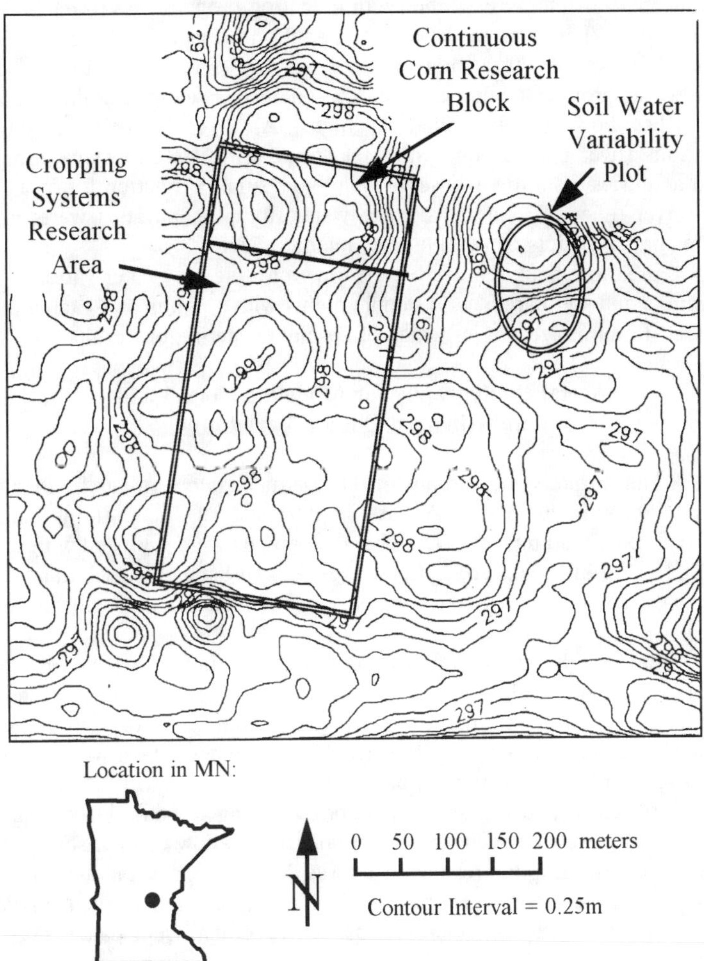

Fig. 19–1. Topographic map of the Minnesota MSEA site, located in Sherburne County (NE 1/4 of Section 18, T.35N., R.26E.). Relative locations of the soil-water variability plot and the continuous corn research block are shown

Topographic Modeling of Soil Water Variability

This study was based on a nested grid of 100 soil corings across a hillslope. Gaston et al. (1990) and Klich et al. (1990) provide examples of use of core surveys for investigating soil variability. The grid was hexagonal, and nested at distance intervals of 16, 8, 4, and 2 m between sampling locations. Each layer (soil horizon) observed to a 3-m depth was noted and sampled for analysis of particle-size distribution. Neutron-probe-access tubes were installed in each of the coreholes to allow soil-water monitoring. Soil-water profiles were measured in 79 of the access tubes with a neutron probe on seven dates during summer 1992. This was to determine spatial and temporal patterns of water storage. We expected that comparison of water profiles between wet and dry periods would provide the best characterization of soil-water variability. Rainfall during 1992, however, was relatively high. As a result, we observed high correlations (over 0.975) between water storage profiles collected during the wettest and driest conditions observed, and no spatial correlation for changes in storage over time. Therefore, average water storage conditions were used to develop spatial models of relative soil water storage.

Regression analyses were performed to predict water storage from topographic and soil parameters; sample points with a minimum spacing of 8 m were used to minimize the influence of spacial clustering on results.

Aerial Photography for Modeling Corn Yield and Nitrogen Uptake Variability

Aerial infrared photographs of the continuous corn research block were acquired at crop maturity (late August) in 1991 and 1992, and during bare soil conditions (post-seeding) in 1992. The photographs were digitized with a color scanner, to give infrared, red, and green reflectance bands for each image. The images were spatially coregistered, and averaged to a 2 m cell size (to match the DEM).

Grain and biomass samples were collected in 1991 and 1992 at 58 locations that could be plotted on the photographs. Nitrogen concentrations were determined (by Kjeldahl digestion) on grain, cob, and stover samples, to calculate total nitrogen uptake. Both yield and nitrogen uptake were expressed on a kilogram per hectare (kg/ha) basis.

Reflectance and elevation data corresponding to centroid cells at each harvest location were extracted from the imagery to allow comparison of image data with harvest sampling results. This was done by regression analysis, using harvest data as dependent variables, and reflectance as independent variables. Results were interpreted considering consistencies of landscape patterns between the two study sites.

RESULTS

Topographic Modeling of Soil Water Variability

Cross sections of soil layering and soil-water storage (Fig. 19–2) shows typical variability found, and provides information on the origin of this landscape. The finest sands are at the higher elevations, whereas coarser materials are found in low-lying (depressional) positions (Fig. 19–2b). This distribution is due to the aeolian (wind) forces that shaped the land after recession of the glacial melt waters that initially formed the sand plain. Depressions were formed by scouring winds, which would erode the deposits down to the water table or coarse alluvial deposits. The winds moved the fine sands into dunes, forming the low hills and ridges found today. This depositional variability has affected patterns of soil development. Lamellae development is restricted to the finer sands of the dune remnants. Lamellae are not found in depressional positions because the illuvial processes responsible for their development require stable movement of wetting fronts through the soil; movement of wetting fronts through coarse sands is generally unstable.

Lamellae are generally horizontal, and continuous in lateral extent (shown by thin layers of higher silt plus clay contents in Fig. 19–2c). Excavations at the site have indicated that this is particularly true of deeper lamellae (over 1.5 m below the surface), which are associated with a depositional discontinuity to coarser, alluvial sands. At upper-slope positions, lamellae become more strongly developed, but then terminate as they subcrop below the surface. (This is indicated in Fig. 19–2c, and has been observed in excavations). Perching of soil water by lamellae follows this trend, resulting in coincident "staircase" patterns of soil layering and soil water perching (Fig. 19–2d). This evidence supports the hypothesis that lamellae contribute to lateral water movement on a landscape scale.

The A horizon (topsoil) also contributes to landscape variability of soil water storage, with an effect contrasting that of lamellae (Fig. 19–3). The A horizon is thickest in lower landscape positions, due to recent accumulation of fines (by wind) and increased biomass production. This horizon is naturally thinner at higher topographic positions; recent tillage and wind erosion have probably increased this effect (Fig. 19–3a). The contrasting distribution of illuvial layering is also plotted (Fig. 19–3b). The topographic distribution of shallow soil-water storage (0 to 0.7 m) follows that of the A horizon thickness, while the distribution of sub-surface water storage (0.7 to 1.7 m) follows that of cumulative lamellae thickness (Fig. 19–3c, d). Data were separated into these depth intervals before regression modeling.

Water storage data were regressed as dependent variables against topographic parameters (elevation, slope, and slope curvature). The final models (Table 19–1) show elevation (cubic trend; all depth intervals), plan curvature (0 to 0.7 m depth interval), and slope (0.7 to 1.7 m and full profile) explain between 52 and 77% of the variance in water storage. When these models were applied to data from all 79 access tubes, spatial correlation was only evident at the 2 m sampling interval (by analysis of regression residuals).

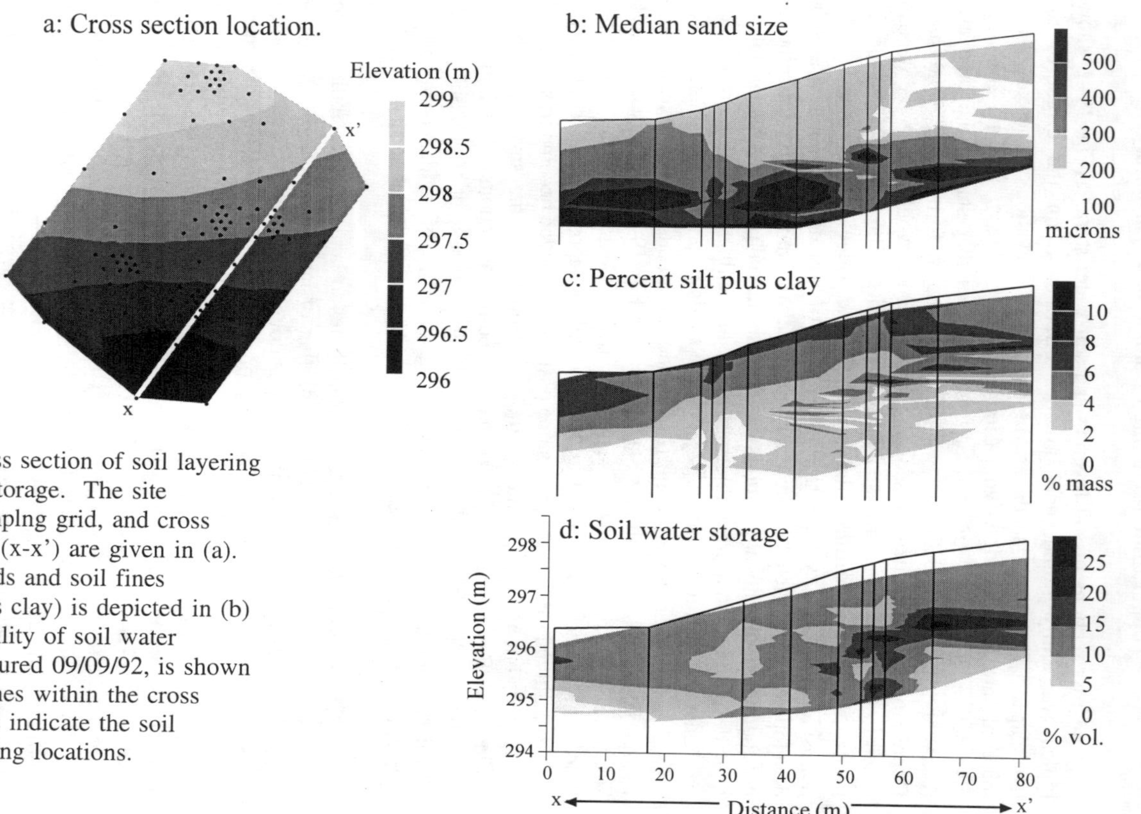

Fig. 19–2. Cross section of soil layering and soil-water storage. The site topography, samplng grid, and cross section location (x-x') are given in (a). Layering of sands and soil fines (percent silt plus clay) is depicted in (b) and (c). Variability of soil water storage, as measured 09/09/92, is shown in (d). Black lines within the cross section diagrams indicate the soil surface, and coring locations.

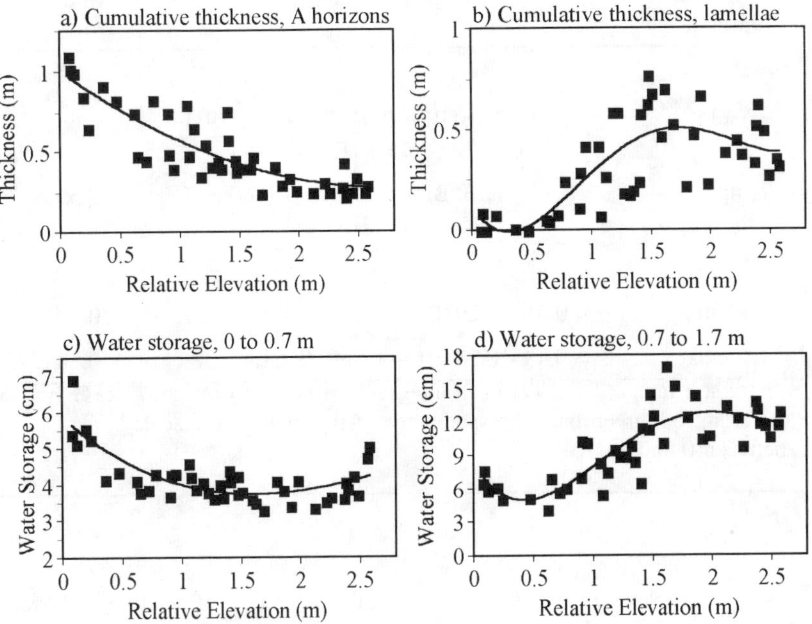

Fig. 19–3. Topographic distribution of soil horizons and soil-water storage. A-horizon thickness is greatest in topographically low areas, and decreases with increasing surface elevation (a). In contrast, lamellae are dominant in upslope areas (b). Soil-water storage follows this pattern: near-surface (0 to 0.7 m depth interval) water storage is greatest in topographically low areas (c). but deeper soils (0.7 to 1.7 m depth interval) are wettest in higher landscape positions (d). (Curves shown on the plots are for illustration.)

Table 19–1. Summary of regression models for prediction of relative water storage.

Depth Interval (m)	Regression model	Model R^2
0 to 0.7	$W = 2.57 + 1.42Z - 9.91Z^2 + 8.15Z^3 - 0.94PL$	0.52
0.7 to 1.7	$W = 0.31 + 1.35Z + 1.03Z^2 - 1.96Z^3 - 0.71SL$	0.77
0 to 1.7	$W = 0.49 + 1.04Z + 1.23Z^2 - 1.95Z^3 - 0.81SL$	0.72

W = relative water storage; Z = relative elevation; PL = plan curvature; SL = slope. All variables are scaled between 0 and 1, except for 0 to 0.7 m water storage, which was scaled (by a log-linear transform) between 0 and 3.

Table 19–2. Summary of regression models for prediction of corn yield and nitrogen uptake.

Dependent variable	Regression model	Model R^2
1991		
Yield	= 0.92 - 2.16*BI + 2.01*BR + 1.14*91I - 1.65*91R	0.64
N uptake	= 0.97 - 1.13*BI + 0.79*BG + 0.66*91I - 1.11*91R	0.59
1992		
Yield	= 0.51 - 0.20*BI + 1.67*92I - 1.45*92R	0.47
N uptake	= 0.43 - 0.25*BI + 1.62*92I - 1.49*92R	0.38

B, bare soil image; 91, 1991 crop image; 92, 1992 crop image; I, R, G, infrared, red, green bands, respectively. All variates were scaled between 0 and 1.

Regression results were used to calculate maps of relative soil water storage (Fig. 19–4) from the DEM, slope, and plan curvature maps. Output maps show relative values of soil water storage for the 0 to 0.7 m and 0.7 to 1.7 m depth intervals. Results provide maps of relative water storage capacity across the landscape that emphasize the contrasting distribution of surface and subsurface water storage. The maps further suggest that the steepest areas are likely to have minimal water storage, and possibly rapid groundwater recharge.

Aerial Photography for Modeling Corn Yield and Nitrogen Uptake Variability

Spatial patterns of the photographic imagery showed that crop reflectance was influenced more by soil conditions in 1991 than 1992. This is attributed to the cooler, wetter conditions of 1992 relative to 1991. As a consequence, harvest data were better predicted by reflectance data in 1991 than 1992. In 1991, 65% of yield variability and 59% of nitrogen uptake variability was captured by reflectance values from scanned photography. Whereas, only 47% of corn yield and 37% of nitrogen uptake variance could be similarly accounted in 1992 (Table 19–2). All regression models included a positive coefficient for infrared band crop reflectance, and a negative coefficient for red band crop reflectance. This is consistent with known patterns of plant growth, stress, and light reflectance: a photosynthetically-active plant will reflect infrared and absorb red light, while the opposite occurs with a stressed plant. Climatic conditions that create limited water stress in the plants are, therefore, best for discerning landscape patterns in crop growth. The difference in regression precision between

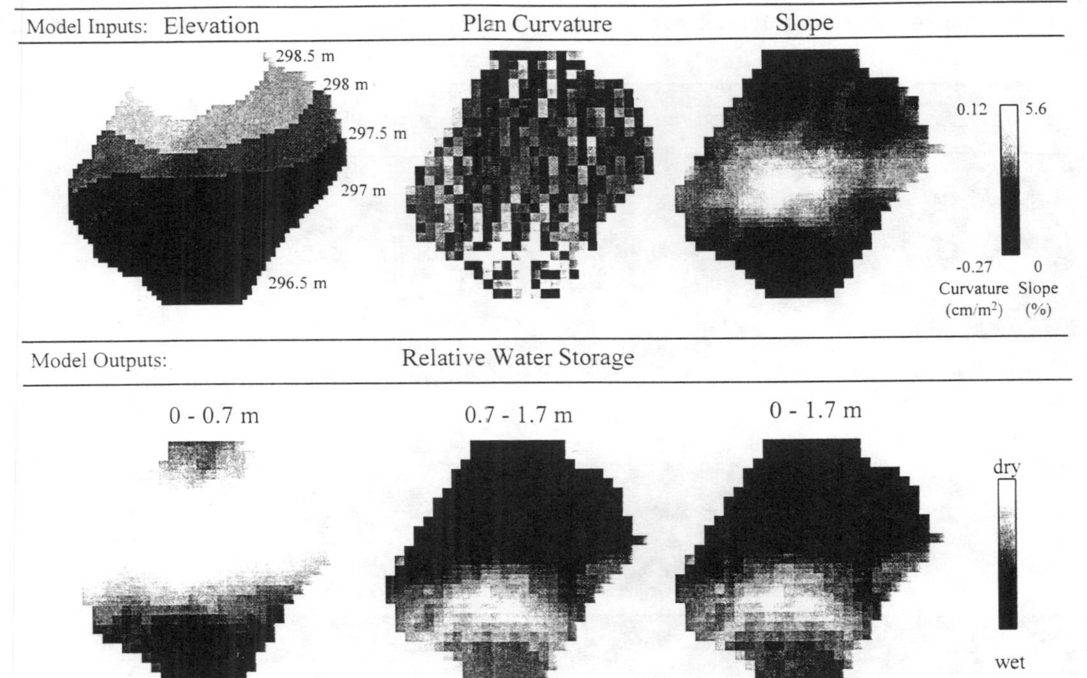

Fig. 19–4. Application of soil-water regression models to DEM data provides output maps of relative water storage for surface (0 to 0.7 m), subsurface (0.7 to 1.7 m) and full profile (0 to 1.7 m) depth intervals. The maps highlight the contrasting distribution of surface and sub-surface soil water, and suggest minimal water storage at the steepest slope position.

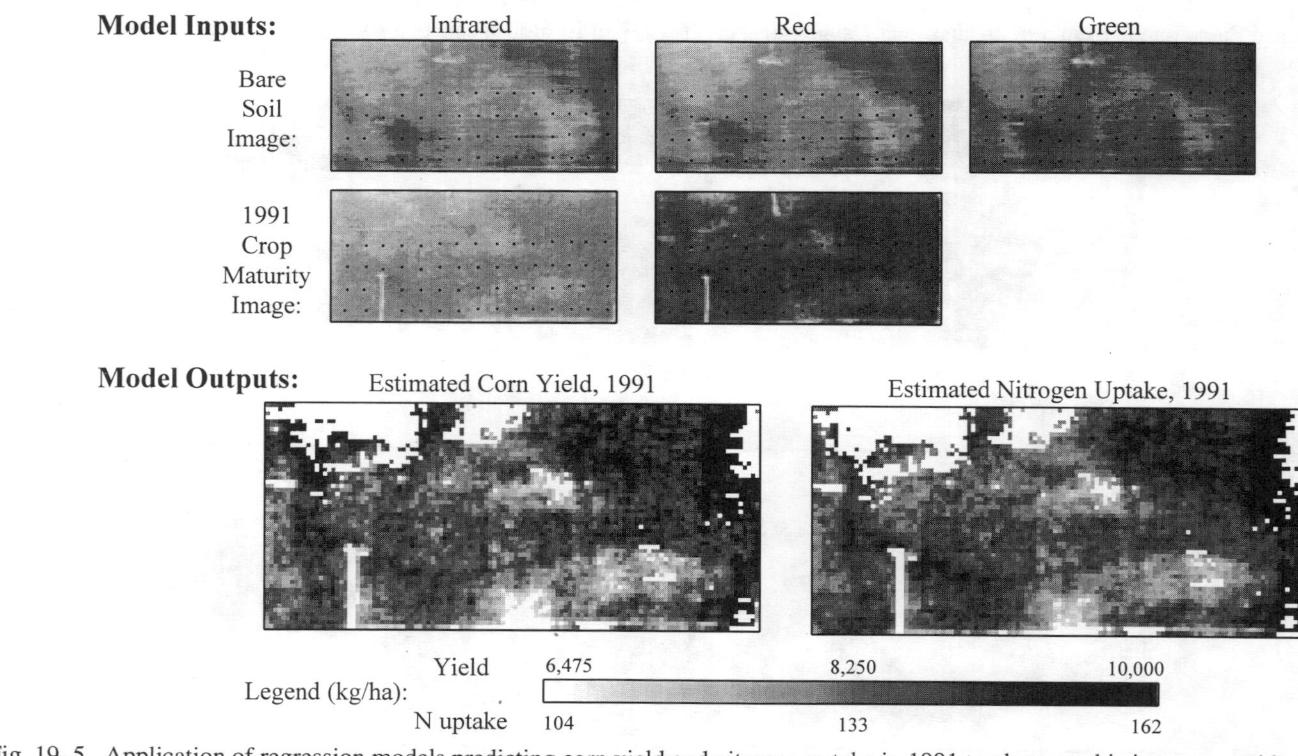

Fig. 19–5. Application of regression models predicting corn yield and nitrogen uptake in 1991 to photographic images provides output maps showing variability of corn growth in 1991. Greater yields and N uptakes generally occurred at topographically low position; the lowest productivity is predicted at upper slope positions (topographic map is given in Fig. 19–1.)

years emphasizes this point. In addition, all four regression models included a negative coefficient for infrared band reflectance from bare soil; this is related to soil organic matter, which is an efficient absorber of near-infrared wavelengths.

The regression models were also applied to provide maps of yield and nitrogen uptake (1991 results are given in Fig. 19–5). The maps also show areas with phototone values outside the range of regression data, where estimation of the dependent variable would require extrapolation of the models. The regression method had better interpolative capacity than kriging in all four cases; regression errors had lower absolute mean, lower standard deviation, and lower correlation with the measured variates than jackknife kriging errors. The maps show the lowest yields and N uptake are associated with steep, upper slopes. The highest yields and N uptakes are associated with lowland positions. This is consistent with the landscape distribution of surface soil-water storage (0 - 0.7 m depth interval) found in the hillslope study. DEM-derived data, however, could not explain > 30% of the variance in yield or N uptake.

CONCLUDING DISCUSSION

Distinct topographic trends in soil-water storage or crop growth would not necessarily be expected on a sand-plain landscape, since surface runoff is minimized by high infiltration. Topographic modeling of soil-water storage, however, was successful due to landscape patterns, aeolian deposition, and consequent soil development. Results of the photographic study show that infrared photography is a valuable tool for assessing variability of agricultural soils and associated crop growth. Interactions between crop growth, climate, and the landscape, however, must be understood before the utility and limitations of photographic survey in agriculture can be stated succinctly. Climatic conditions that induce some crop stress will provide the best understanding of inherent soil variability across the landscape. Results show the potential value of a regression approach to landscape modeling using surface-collected data. Further work is recommended to develop this approach for soil-specific-management applications.

ACKNOWLEDGMENTS

Thanks to Donald Kackman, Kurt Ault, and Dave Onken for their help with data collection, and to John Ladwig and Dr. Jay Bell for assistance with image preparation. This project was funded in part by the Water Resources Research Center (Grant no. USDI–14–08–0001–G2027). Other assistance was provided by the Management System Evaluation Area project, and by the University of Minnesota, through the Agricultural Experiment Station and a Graduate School Doctoral Dissertation Fellowship.

REFERENCES

Anderson, J.L., R.H. Dowdy, J.A. Lamb, G.N. Delin, R. Knighton, D. Clay, and B. Lowery. 1993. Northern cornbelt sand plains Management System Evaluation Area. p. 39–47. *In* Agricultural research to protect water quality Conf. Proc. Feb. 21–24, Minneapolis, MN. Soil Water Cons. Soc. Ankeny, IA.

Brubaker, S.C., A.J. Jones, D.T. Lewis, and K. Frank. 1993. Soil properties associated with landscape position. Soil Sci. Soc. Am. J. 57:235–239.

Everitt, J.H., D.E. Escobar, M.A. Alaniz, and M.A. Hussey. 1987. Drought-stress detection of buffelgrass with color-infrared aerial photography and computer-aided image processing. Photogram. Eng. Rem. Sens. 53: 1255–1258.

Gaston, L., P. Nkedi-Kizza, G. Sawka, and P.S.C. Rao. 1990. Spatial variability of morphological properties at a Florida flatwoods site. Soil Sci. Soc. Am. J. 54:527–533.

Klich, I., L.P. Wilding, and A.A. Pfordresher. 1990. Close-interval spatial variability of Udertic Paleustalfs in east central Texas. Soil Sci. Soc. Am. J. 54:489–494.

Landon, M.K., G.N. Delin, L. Gou, C.P. Regan, J.A. Lamb, J.L. Anderson, and R.H. Dowdy. 1993. Occurrence of agricultural chemicals in ground water at the Princeton, Minnesota Management System Evaluation Area. p. 434–438. *In* Agricultural Research to Protect Water Quality, Conf.Proc. Feb. 21–24, Minneapolis, MN. Soil Water Cons. Soc. Ankeny, IA.

Milfred, C.J., and R.W. Kiefer. 1976. Analysis of soil variability with repetitive aerial photography. Soil Sci. Soc. Am. J. 40:553–557.

Moore, I.D., R.B. Grayson, and A.R. Ladson. 1991. Digital terrain modeling: A review of hydrological, geomorphological, and biological applications. Hydrol. Processes. 5:3–30.

Moore, I.D., P.E. Gessler, G.A. Nielsen, and G.A. Peterson. 1993. Soil attribute prediction using terrain analysis. Soil Sci. Soc. Am. J. 57:443–452.

Robert, P. 1993. Characterization of soil conditions at the field level for soil specific management. Geoderma 60:57–72.

Ruhe, R.V., and P.H. Walker. 1968. Hillslope models and soil formation: I. Open systems. Trans. Int. Cong. Soil Sci. 9(4):551–560.

Soil Survey Staff. 1951. Soil Survey Manual. U.S. Dep. Agric. Handb. No, 18. Washington, DC.

Stone, J.R., J.W. Gilliam, D.K. Cassel, R.B. Daniels, L.A Nelson, and H.J. Kleiss. 1985. Effect of erosion and landscape position on the productivity of Piedmont soils. Soil Sci. Soc. Am. J. 49:987–991.

Tucker, C.T. 1979. Red and photographic infrared linear combinations for monitoring vegetation. Remote Sens. Env. 8:127–150.

Zheng, F., and H. Schreier. 1988. Quantification of soil patterns and field soil fertility using spectral reflection and digital processing of aerial photographs. Fertilizer Res. 16:15–30.

20 Sensitivity Of Computed Terrain Attributes To The Number And Pattern Of GPS-Derived Elevation Data

Damian J. Spangrud
John P. Wilson

Department of Earth Sciences
Montana State University
Bozeman, Montana

Gerald A. Nielsen
Jeffrey S. Jacobsen

Department of Plant and Soil Science
Montana State University
Bozeman, Montana

David A. Tyler

Department of Survey Engineering
University of Maine
Orono, Maine

Soil specific crop management requires precise knowledge of soil properties and soil-landscape processes (Bouma & Finke, 1993; Burrough, 1993; Larson & Robert, 1993; Mulla, 1993). Detailed soil maps at scales of 1:6000 or 1:8000 and spatially variable soil attribute data are needed to guide soil specific crop management in most landscapes (Moore et al., 1993a). Conventional soil survey maps, however, are produced at scales of 1:15,000 and larger and as such, these maps seldom delineate all of a field's variability (Fisher, 1991). Similarly, the range of soil attribute values reported for most mapping units is sufficiently large that these data cannot adequately represent soil attribute variation (Moore et al., 1993a).

Soil scientists have proposed a variety of pedotransfer (regression) and other geostatistical functions which use soil texture, organic matter, soil structure, and bulk density input data as cost-effective alternatives to expensive soil attribute data collection efforts during the past decade (e.g., Rawls et al., 1982; Mulla, 1991; Wagenet et al., 1991). Other groups have used terrain attributes derived from digital elevation models (DEMs) for this purpose. Klingebiel et al. (1987), for example, used slope, aspect, and elevation to increase the accuracy of mapped soil unit boundaries and thereby limit the within-unit variability, whereas Odeh et al. (1991) and Moore et al. (1993a, 1993b) have taken a slightly different approach and correlated soil properties with simple to measure primary and secondary terrain attributes that have physical meaning in an effort to improve soil attribute prediction.

Primary terrain attributes are calculated directly from elevation data and include variables such as slope gradient, aspect, specific catchment area, flowpath length, plan and profile curvature (Moore et al., 1991). Secondary attributes combine two or more primary attributes and are often used to characterize the spatial variability of specific processes occurring in the landscape. Moore et al. (1993a), for example, defined three compound indices (a wetness index, ω; a stream power index, Ω; and a sediment transport index, τ) and described their potential use in predicting the spatial distribution of soil properties as an aid to soil specific crop management.

These primary and secondary terrain attributes are usually computed from DEMs. A DEM is an ordered array of numbers that represents the spatial distribution of elevations above some arbitrary datum in a landscape. Moore et al. (1993a) review the various sources of DEMs and note that Global Positioning System (GPS) technology provides a rapid and relatively inexpensive way of obtaining data for the development of DEMs. This new technology offers important advantages in terms of scale and accuracy for soil specific farming applications given that the traditional sources of elevation data (e.g., 1:24000-scale USGS contour maps) and the 30 m DEMs derived from them with Z values rounded to the nearest meter offer data at too coarse a resolution for most site-specific farming applications. There is now a total of 24 GPS satellites in orbit and 5–10 of these satellites are visible to a receiver at any one time. A stationary receiver used in conjunction with a mobile receiver (in differential or kinematic mode) may provide X,Y,Z measurements down to centimeter accuracy (Tyler, 1993). The GPS operator determines the number and pattern (spread) of elevation data collected. Our research objective was to assess the impact of varying the number and pattern of GPS data on computed topographic attributes for a farm field in Montana.

METHODS AND DATA SOURCES

Elevation Data

Elevation data were collected at 6,284 different locations in 1991 with an Ashtech Sensor GPS receiver mounted on a pickup truck and an Ashtech P–12 GPS receiver operating in kinematic mode as part of a larger study examining potential nitrate contamination of ground water (Fig. 20–1a). These elevation data were converted to a regular 10 m grid with ANUDEM (Hutchinson, 1989) for subsequent analysis and display. This program takes irregular point or contour data and creates grid-based DEMs. ANUDEM automatically removes the spurious pits within user-defined tolerances, calculates stream lines and ridge lines

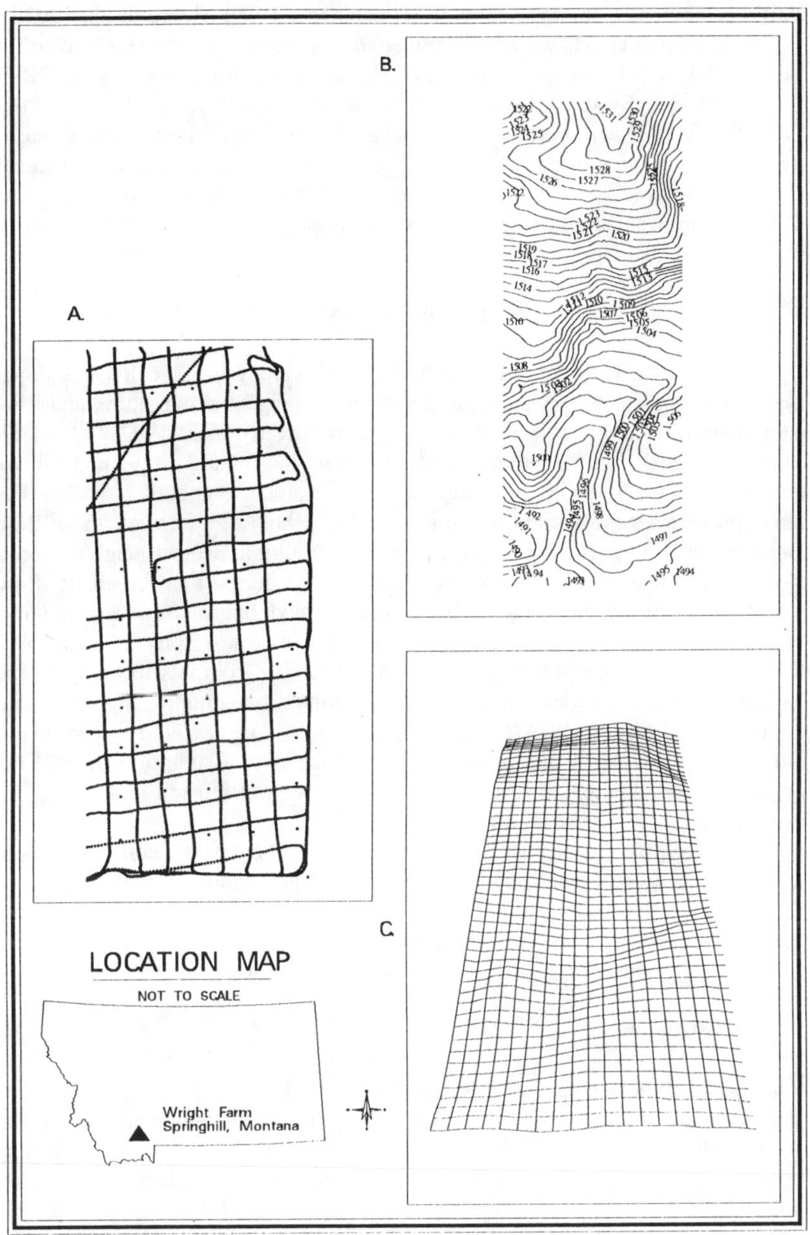

Fig. 20–1. Full GPS data set (a), contour map (b), and three-dimensional wire mesh diagram (c) for study area. Fig. 20–1 represents 6,284 X,Y,Z values obtained with a truck-mounted GPS receiver.

from points of locally maximum curvature on contour lines, and (most importantly) incorporates a drainage enforcement algorithm to maintain fidelity with a catchment's drainage network (Hutchinson, 1989; Moore et al., 1993d).

The 10 m DEMs were later converted to 1 m contour maps by converting the ANUDEM ASCII files consisting of X,Y,Z values to lattices in ARC/INFO (Environmental Systems Research Institute, Inc., Redlands, CA) using the LATTICECONTOUR command to convert these lattices into contour maps. Hence, the 1 m contour map reproduced in Fig. 20–1b was created in two steps: (1) all 6,284 X,Y,Z values were used by ANUDEM to create a 10 m grid; and (2) this grid was brought into ARC/INFO as a lattice and used to create a detailed contour map.

Terrain Analysis

TAPES-G is a grid-based method of terrain analysis that calculates slope, aspect, specific catchment area, profile, plan and tangential curvature, and flow path length for each cell in a square-grid DEM (Moore, 1992). Specific catchment areas can be estimated using either the D8 algorithm that allows drainage from a node to only one of eight nearest neighbors based on the direction of steepest descent, the quasi-random Rho8 algorithm, or the FRho8 algorithm that permits drainage from a node to multiple nearest neighbors on a slope-weighted basis. The Rho8 algorithm produces more realistic flow networks and the FRho8 algorithm permits the modeling of flow dispersion in upland areas, which is important in areas with convex topography (Moore, 1992; Moore et al., 1993d). Other programs use TAPES-G outputs to compute the spatial distribution of net radiation and minimum/maximum temperatures in complex topography (SRAD), spatially distributed wetness indices based on either a steady-state or quasi-dynamic subsurface flow assumption (DYNWET or WET), and the spatial distribution of soil loss and erosion and deposition potential in a catchment (EROS) (Moore, 1992).

Three primary attributes (elevation, slope gradient, and specific catchment area) and one secondary attribute (steady-state wetness index) were computed for this study. Elevation was estimated on a 10 m grid spacing with ANUDEM. The maximum slope or gradient, β (in degrees) was computed with a finite-difference algorithm in TAPES-G from the directional derivatives by:

$$\beta = \arctan\left[(f_x^2 + f_y^2)^{0.5}\right] \qquad (1)$$

Specific catchment area was calculated with the FRho8 algorithm that allows flow to be distributed to multiple nearest-neighbor nodes in upland areas above defined channels and uses the Rho8 or D8 algorithms below points of channel initiation. The points of channel initiation are designated (indirectly) by the user when they specify the maximum grading area in TAPES-G since this variable represents the minimum catchment area needed to initiate channel flow. A minimum drainage area of 20,000 m^2 (approximately 10% of the study area) was arbitrarily used to initiate channel flow in this study. The proportion of flow or upslope contributing area assigned to multiple downslope nearest neighbors above these channels was determined on a slope weighted basis using methods similar to those proposed by Freeman (1991) and Quinn et al. (1991), so that the fraction of catchment area passed to neighbor i is given by:

$$F_i = \text{Max}\ (0, \text{Slope}_i^{1.1})\ /\ \Sigma\ [\text{Max}(0, \text{Slope}_j^{1.1})] \tag{2}$$

where Slope is the slope from the current node to the nearest neighbor.

The compound topographic wetness index, $\ln(A_s/\tan\beta)$, has been used to characterize the spatial distribution of zones of surface saturation and soil water content in landscapes (e.g., O'Loughlin, 1986; Moore et al., 1988), to map forest soils (Skidmore et al., 1991), and to characterize the spatial variability of soil properties in a toposequence in Colorado (Moore et al., 1993a, 1993b). This particular version of the wetness index incorporates two important assumptions, that: (i) the gradient of the piezometric head, which dictates the direction of subsurface flow, is parallel to the gradient of the surface topography; and (ii) steady-state conditions apply, although Barling (1992) and Barling et al. (1994) recently proposed a quasi-dynamic version to overcome some of the limitations of this steady-state assumption.

The steady-state wetness index is used here to illustrate the effects of the number and pattern of elevation data on a hydrologically important compound topographic attribute. The value of this index increases with increasing specific catchment area and decreasing slope gradient, resulting in moderate values on hilltops (flat areas with low specific catchment area), high values in valleys (high specific catchment area and low slope) where water concentrates, and low values on steep hillslopes (high slope) where water drains more freely (Moore et al., 1993c).

Study Area and Evaluation Methods

The study area consists of a 20 ha (50 acre) farm field located at the base of the Bridger Mountains near the community of Springhill, Montana (T1N R6E, Section 8). It has a generally southerly aspect, moderately strong relief (43 m), and an average elevation of 1509 m (Fig. 20–1b and 20–1c). Minimum, maximum, and mean values for selected terrain indices are summarized in Table 20–1. A small intermittent stream runs through the field in a south-south-westerly direction (see bottom half of wire-mesh diagram). The soils are mostly fine-loamy Pachic and Udic Haploborolls and Agriborolls that have been farmed with a grain-fallow rotation for about 50 years.

Our primary goal was to determine how few GPS input data would be required to preserve key information about the computed terrain surfaces, since data collection, storage, and analysis may add additional costs to soil specific crop management applications. Elevation, slope gradient, specific catchment area, and the steady-state wetness index are treated as key information and the tests involve comparisons of various grids developed with partial GPS input data sets and the surfaces that were constructed from all 6,284 GPS data (Fig. 20–1a).

Table 20–1. Computed terrain indices, Springhill site, Montana.

	Minimum	Maximum	Mean
Slope (percent)	0.4	27.3	9.5
Specific catchment area (m^2 m^{-1})	10.0	886.0	222.4
Steady-state wetness index	5.7	15.4	8.5

Partial GPS data sets were chosen in ways that more or less matched those that a farm operator or consultant might have collected with a hand-held GPS receiver or a mobile receiver mounted on a vehicle.

GPS data sets that a farmer might have collected with a hand-held GPS receiver were obtained by dividing the field into 25 m, 50 m, and 100 m grid cells and randomly selecting a pre-determined number of X,Y,Z values in each grid cell (Table 20–2). This approach is equivalent to a stratified random areal sample design (Berry & Baker, 1968), although the random element disappears fairly quickly as more data are selected and the sample begins to mimic the routes travelled by the pickup truck and mobile GPS receiver (compare Fig. 20–2a, 20–2b, 20–2c, and 20–2d with Fig. 20–1a). This discovery may not be very important (practical) given that most people would switch from a hand-held receiver to one mounted on a truck if their goal was to collect > 200 sets of X,Y,Z data, although some of these data sets were retained for this study and used to demonstrate the impact of pattern on computed topographic attributes.

GPS data sets that a farmer might have collected with a truck-mounted GPS receiver were obtained from selected north-south and west-east truck routes in order to examine the impact of the number and pattern of GPS data on computed topographic attributes. A total of six data sets were compiled in this way. Three data sets were contructed by choosing every second north-south and/or west-east truck route and three data sets were constructed by choosing all north-south and/or west-east truck routes. The turns at the end of the truck routes were omitted from all six data sets. Figure 20–3 shows the pattern of GPS data sets and resulting contour maps with 1 m contour intervals. Between 1,120 and 4,817 X,Y,Z data sets were chosen with these options.

Each of the seventeen irregular X,Y,Z data sets was converted into regular 10 m grids with ANUDEM. The drainage enforcement option was used in ANUDEM to maintain fidelity with the stream network that was identified by the software itself. The resulting regular grids were transferred to TAPES-G to compute topographic attributes and to ARC/INFO to perform the statistical analysis and create the 1 m contour maps reproduced in Fig. 20–1 through 20–3. The TAPES-G attributes were also transferred to the GRID module in ARC/INFO where most of the statistical analysis was performed.

The elevation, slope gradient, specific catchment area, and steady-state wetness index surfaces (grids) were compared with equivalent surfaces developed from all the GPS data to evaluate the impact of the number and pattern of GPS data on computed topographic attributes. The contour maps in Figs. 20–1b, 20–2, and 20–3 allow visual comparisons and the statistics in Tables 20–3 and

Table 20–2. GPS data sets generated with stratified random areal sample simulating data from a hand-held GPS receiver.

Grid size (m)	No. of points selected	Max. desired no. of points	No. of points actually selected
100 m	1	32	29
100 m	2	64	58
100 m	4	128	116
50 m	1	120	111
50 m	2	240	221
50 m	4	480	441
50 m	16	1920	1747
25 m	1	435	391
25 m	4	1740	1535

20–4 provide additional details.

Mean absolute differences are reported to avoid the offset of positive and negative numbers. The root mean square error (RMSE) is commonly used to summarize the differences in values between two or more grids in GIS packages, and the Moran Index is used to measure how clustered or how randomly the differences in topographic attributes distribute spatially. Large mean absolute differences and RMSE values indicate large discrepancies between the sample grids and the grid constructed with all 6,284 GPS data points, whereas Moran Index values near zero indicate that the differences were randomly distibuted and values approaching +1 indicate positive spatial autocorrelation (i.e., that the differences are clustered). This approach relies on an implicit but important assumption: that the closer the sample surface is to the equivalent surface generated with the full GPS data set, the better the performance of the sample data set that is being assessed. This assumption and the resultant statistical measures match those used by Lee (1991) in his pioneering work comparing existing methods for building triangular irregular network models of terrain from gridded digital elevation models.

RESULTS

Figure 20–1b shows the contour map that was produced with ANUDEM, ARC/INFO, and all 6,284 GPS data points. Three topographic features are evident in this map and the accompanying three-dimensional wire mesh diagram (Fig. 20–1c): (i) the hilltop marking the northern boundary of the study area that is flanked by gentle side slopes to the west and south and very steep side slopes to the east; (ii) the shallow drainage (valley) that traverses the study area in a south-south-westerly direction in the bottom (southern) half of the study area; and (iii) the hilltop located in the southeast quadrant whose northern side slopes mark the southern margin of the valley identified as (ii). These topographic features were also observed in the field and they are used along with the statistical analysis of the differences in computed topographic attributes to assess the effectiveness of the different sample data sets.

Figures 20–2 and 20–3 show the GPS points that were actually selected and used to compute the topographic attributes as well as the contour maps that were produced with ANUDEM and ARC/INFO. The GPS points are so close together that they appear as lines in the left half of the figures. The contour maps were clipped to minimize edge effects and as a result, they cover a smaller portion of the study area than the GPS maps. The first three pairs of maps in Fig. 20–2 demonstrate the effect of selecting additional random points within 100 m grid cells. The first contour map was generated with only 29 GPS points and does not adequately delineate the western side slopes for the hilltop along the northern boundary nor the valley that traverses the southern half of the study area (Fig. 20–2a). Doubling the number of points selected in each grid cell (Fig. 20–2b) and doubling the number again (Fig. 20–2c) improved the delineation of the two hilltops and the valley. The third contour map is much better than the second map in defining the channel, in part because of the presence of a spurious pit near the southwest corner of the field in the second map (Fig. 20–2b). The contour map in Fig. 20–2d was produced when one X,Y,Z data point was chosen randomly from each 50 m cell. Both the total number of points selected (120 versus 128) and the resultant contour map are very similar to the best result that was achieved with 100 m cells (Fig. 20–2c). Finally, a comparison of these last two contour maps with the map produced from all 6,284 X,Y,Z data points indicates how few data were needed to generate a "reasonable" contour map of this particular field.

The contour maps produced with the linear samples tell a much different story. The first two contour maps reproduced in Fig. 20–3 were derived from the gridded DEMs that used every second and all of the north-south routes. Both contour maps show the major topographic features, although the second map captures much more of the fine-scale detail (Fig. 20–3a and 20–3b). The two contour maps derived from west-east sample data do much better in the northern half of the study area than they do in the southern half (compare Fig. 20–3c and 20–3d with Fig. 20–1b, and note the presence of spurious pits in the channel on both maps). This result was probably caused by the orientation of the truck routes relative to the channel and it illustrates the sensitivity of the surfaces produced with the orientation of the linear routes used to collect GPS data relative to the orientation of major topographic features (i.e., ridge and stream lines). The selection of every second north-south and west-east route captured 2,483 X,Y,Z data points (40% of the total points) and not surprisingly, this method produced a very similar contour map to the that produced with the full GPS data set (Fig. 20–3e).

Seventeen sets of elevation differences were computed by comparing the 11 grids from the stratified random area samples and the six grids computed from the linear samples with the grid generated with all 6,284 sets of X,Y,Z data. The 10 m by 10 m grid measured 63 rows by 24 columns, so that the total number of points is 1,512. A modest improvement in the fitted surface was achieved

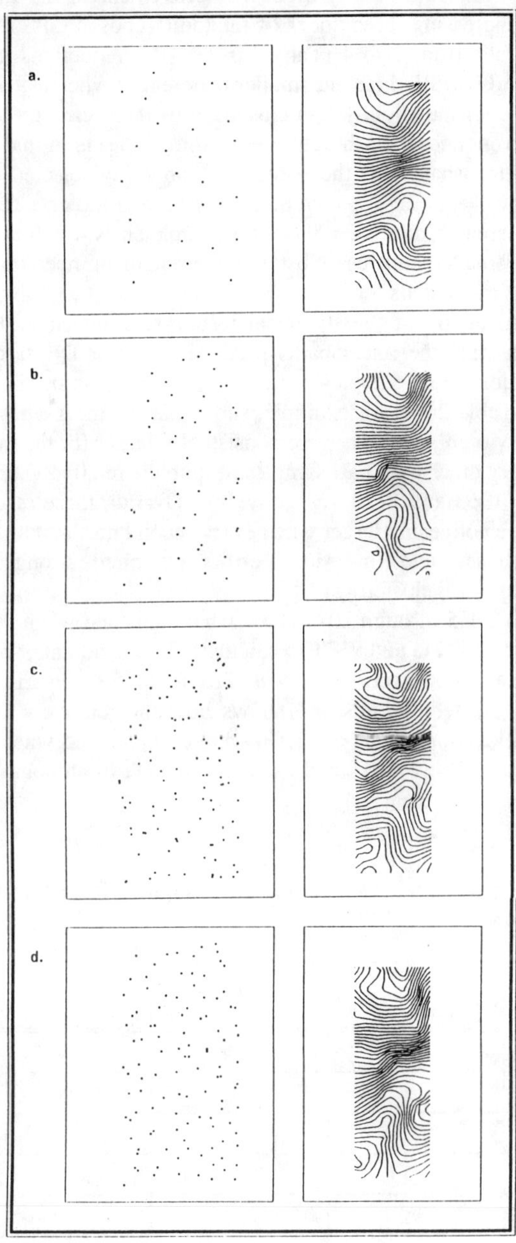

Fig. 20–2. GPS data sets and 1 m contour maps produced with 1 sample point per 100 m cell (a), 2 points per 100 m cell (b), 4 points per 100 m cell (c), and 1 point per 50 m cell (d).

number of points is 1,512. A modest improvement in the fitted surface was achieved by increasing the number of randomly chosen GPS data points in the 100 m grid cells from 1 to 4 (Table 20–3). The reductions in mean absolute differences and RMSEs indicate smaller differences whereas the declines in the Moran index indicate slightly less clustering of these errors. These trends are much more pronounced for the 25 m and 50 m data sets in that the values of all three indices fell sharply as the number of randomly selected GPS data points was increased (Table 20–3). Overall, these results indicate that the errors are strongly clustered (Moran's I > 0.60) for sample sizes < 200 and that the errors are typically smaller and less clustered if a small number of GPS points are selected from many cells.

A comparison of the statistical results reproduced in Tables 20–3 and 20–4 confirms that the linear paths produced inferior DEMs compared to the stratified random areal samples for roughly similar sample sizes. The results presented in Table 20–4, for example, show that the mean absolute differences, RMSEs, and Moran's I values were consistently larger for the every other north-south and every other west-east samples despite the relatively large sizes of these samples (n = 1363 and 1120, respectively). Overall, these results indicate that the errors were noticeably larger when every second north-south and/or west-east route was chosen and that these errors exhibited strong positive spatial autocorrelation (i.e., clustering).

Tables 20–5 through 20–7 show what happened when the elevation data were used in TAPES-G and WET to calculate slope gradients, specific catchment areas, and steady-state wetness indices. Table 20–5 summarizes the arithmetic means for the different grids and shows how the relatively small differences between the elevation grids (< 0.05%) were compounded when these data were input to TAPES-G and WET and used to calculate the topographic attributes.

Table 20–3. Spatial analysis of elevation differences from stratified random area sample points.

Grid size	No. points selected	Mean absolute elev. diff. (m)	RMSE (m)	Moran's Index
100/1	29	1.39	1.71	0.89
100/2	58	1.04	1.43	0.86
100/4	116	0.78	1.16	0.81
50/1	111	0.60	0.95	0.74
50/2	221	0.43	0.73	0.58
50/4	441	0.31	0.60	0.46
50/16	1747	0.16	0.40	0.19
25/1	391	0.27	0.54	0.39
25/4	1535	0.16	0.40	0.19

by increasing the number of randomly chosen GPS data points in the 100 m grid 100 m grid cells from 1 to 4 (Table 20–3). The reductions in mean absolute differences and RMSEs indicate smaller differences whereas the declines in the Moran index indicate slightly less clustering of these errors. These trends are much more pronounced for the 25 m and 50 m data sets in that the values of all three indices fell sharply as the number of randomly selected GPS data points was increased (Table 20–3). Overall, these results indicate that the errors are strongly clustered (Moran's I > 0.60) for sample sizes < 200 and that the errors are typically smaller and less clustered if a small number of GPS points are selected from many cells.

A comparison of the statistical results reproduced in Tables 20–3 and 20–4 confirms that the linear paths produced inferior DEMs compared to the stratified random areal samples for roughly similar sample sizes. The results presented in Table 20–4, for example, show that the mean absolute differences, RMSEs, and Moran's I values were consistently larger for the every other north-south and every other west-east samples despite the relatively large sizes of these samples (n = 1363 and 1120, respectively). Overall, these results indicate that the errors were noticeably larger when every second north-south and/or west-east route was chosen and that these errors exhibited strong positive spatial autocorrelation (i.e., clustering).

Tables 20–5 through 20–7 show what happened when the elevation data were used in TAPES-G and WET to calculate slope gradients, specific catchment areas, and steady-state wetness indices. Table 20–5 summarizes the arithmetic means for the different grids and shows how the relatively small differences between the elevation grids (< 0.05%) were compounded when these data were input to TAPES-G and WET and used to calculate the topographic attributes.

Table 20–3. Spatial analysis of elevation differences from stratified random area sample points.

Grid size	No. points selected	Mean absolute elev. diff. (m)	RMSE (m)	Moran's Index
100/1	29	1.39	1.71	0.89
100/2	58	1.04	1.43	0.86
100/4	116	0.78	1.16	0.81
50/1	111	0.60	0.95	0.74
50/2	221	0.43	0.73	0.58
50/4	441	0.31	0.60	0.46
50/16	1747	0.16	0.40	0.19
25/1	391	0.27	0.54	0.39
25/4	1535	0.16	0.40	0.19

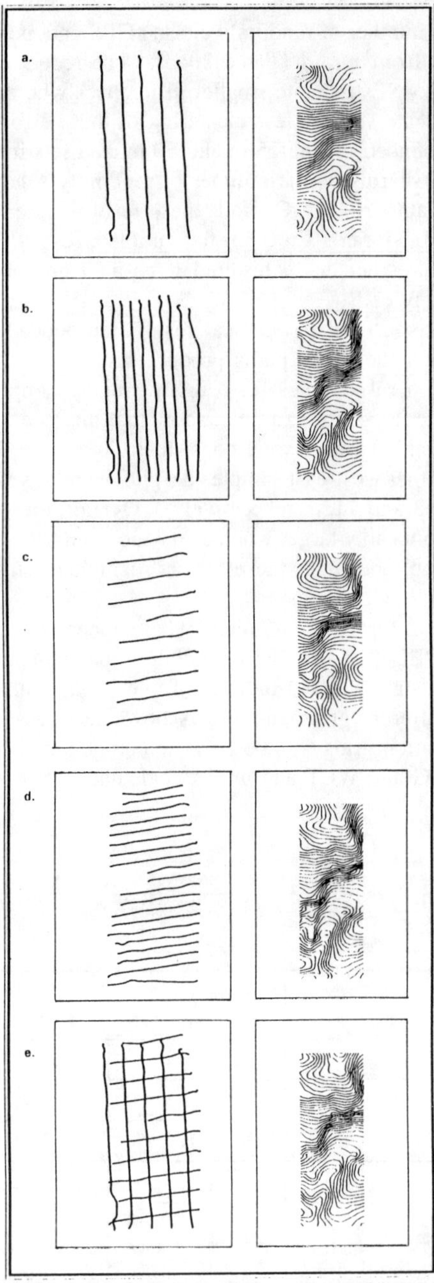

Fig. 20–3. GPS data sets and 1 m contour maps produced with every second north-south truck route (a), every north-south route (b), every second west-east route (c), every west-east route (d), and every second north-south and west-east truck route (e).

Table 20–4. Spatial analysis of elevation differences from linear routes simulating data from a truck-mounted receiver.

Linear routes	No. points selected	Mean absolute elev. diff. (m)	RMSE (m)	Moran's Index
0-1-0 NS	1363	0.35	1.20	0.81
All NS	2682	0.21	0.51	0.42
0-1-0 WE	1120	0.62	0.96	0.72
All WE	2135	0.22	0.48	0.34
0-1-0 NSWE	2483	0.35	0.66	0.60
All NSWE	4817	0.05	0.23	0.11

Table 20–5. Arithmetic means for selected topographic attributes.

Sample	Elevation (m)	Slope gradient (percent)	Specific catchment. area (m^2 m^{-1})	Wetness Index
100/1	1508.4	7.1	306.9	7.53
100/2	1508.9	7.6	263.5	7.33
100/4	1508.9	8.3	297.8	7.12
50/1	1509.0	8.6	238.8	7.09
50/2	1509.0	9.4	234.8	6.93
50/4	1509.0	9.9	243.5	6.78
50/16	1509.1	11.1	242.6	6.63
25.1	1509.1	9.9	238.8	6.79
25/4	1509.1	11.1	243.7	6.62
0-1-0 NS	1509.1	8.3	353.3	7.20
All NS	1509.0	10.0	240.2	6.77
0-1-0 WE	1509.2	9.1	281.2	6.95
All WE	1509.1	10.5	271.3	6.70
0-1-0 NSWE	1509.1	9.8	282.6	6.81
All NSWE	1509.1	10.8	255.9	6.67
All data	1509.1	9.5	222.5	6.76

Table 20–6. Spatial analysis of topographic attribute differences from stratified random samples.

Sample	Slope gradient			Specific catchment area			Wetness index		
	MAD	RMSE	Moran	MAD	RMSE	Moran	MAD	RMSE	Moran
100/1	3.48	4.14	0.78	333.6	1381.5	0.05	1.30	1.51	0.64
100/2	3.18	3.85	0.76	294.3	1262.2	0.05	1.13	1.44	0.64
100/4	2.68	3.54	0.72	299.5	1399.8	0.07	0.97	1.35	0.55
50/1	2.37	3.06	0.65	227.2	1041.9	0.01	0.83	1.17	0.49
50/2	2.03	2.72	0.57	234.6	1109.7	-0.09	0.71	1.11	0.37
50/4	1.97	2.65	0.53	131.2	738.6	-0.11	0.54	0.87	0.33
50/16	2.02	2.46	0.36	103.7	747.1	0.01	0.39	0.64	0.09
25/1	1.82	2.47	0.49	172.3	909.3	-0.12	0.55	0.90	0.24
25/4	2.06	2.42	0.35	98.6	694.9	0.07	0.41	0.65	0.16

The results summarized in Table 20–6 show how the size and pattern of the errors for the primary and secondary topographic attributes varied with the different stratified random area samples. The magnitude and clustering of the slope gradient errors decreased gradually as the cells decreased in size (i.e., the spread improved) and sample size was increased. The mean slope gradients for the sample grids ranged from 75% to 117% of the mean slope gradient (9.5 degrees) on the grid computed with the entire X,Y,Z GPS data set (Table 20–5). The specific catchment area means for the sample grids are all larger than that for the complete GPS data set and confirm that the sample elevation grids missed some of the fine-scale topographic variation (Table 20–5). The mean absolute differences (MADs) and RMSEs reported in Table 20–6 confirm this trend and indicate that there were substantial reductions in the magnitude of these errors as the number and spread of the sample data was increased. The spatial pattern of these errors was more or less random for all of the samples (as would be expected for a variable that accumulates flow from upslope cells). The relative magnitude and clustering of the steady-state wetness index errors expressed as percentages fell between those of the two primary attributes from which this compound index is calculated. The magnitude and level of clustering of the wetness index errors decreased as sample size and spread increased (Table 20–6).

Similar results were obtained with the linear samples, so that: (i) the magnitude and clustering of errors decreased as the number of truck routes was increased for slope gradient and the steady-state wetness index, and (ii) the magnitude of the errors decreased for specific catchment area although these errors were randomly distributed (Table 20–7).

Table 20–7. Spatial analysis of topographic attribute differences from linear samples.

Sample	Slope gradient			Specific catchment area			Wetness index		
	MAD	RMSE	Moran	MAD	RMSE	Moran	MAD	RMSE	Moran
0-1-0 NS	2.69	3.79	0.76	347.5	1443.8	0.14	0.99	1.42	0.58
All NS	1.67	2.31	0.47	163.4	941.0	-0.03	0.49	0.84	0.23
0-1-0 WE	2.48	3.28	0.68	259.8	1248.1	0.08	0.73	1.17	0.41
All WE	1.79	2.27	0.45	143.8	891.2	0.10	0.44	0.77	0.22
0-1-0 NSWE	2.01	2.78	0.57	242.7	1228.0	0.15	0.61	1.08	0.34
All NSWE	1.62	1.97	0.32	102.1	789.3	0.10	0.32	0.63	0.13

CONCLUSIONS

GPS technology is increasingly advocated as a cost-effective and accurate method for collecting the irregularly-spaced elevation data that are required as inputs to construct digital elevation models and compute primary and secondary terrain attributes. The grid surfaces generated from the 17 different sample data sets were compared with the surface developed with a DEM computed from all 6,284 GPS-derived X,Y,Z points in this study. Some of the results were shown graphically as elevation contours and spatial analyses of the differences in elevation, slope gradient, specific catchment area, and steady-state wetness index were used to summarize the performance of the different samples.

These statistical comparisons show that the number and pattern of the GPS input data will influence the DEM and terrain attributes that are computed. In particular, they show that: (i) the magnitude and clustering of the spatial distribution of these errors diminishes as sample size increases, (ii) the magnitude and clustering of the errors diminishes as the spread of the input data increases, and (iii) that seemingly small variations in elevation may result in large differences in the computed primary and secondary topographic attributes. This last observation has important implications for those who want to used these spatially-variable attributes to predict other environmental variables (e.g., Dikau, 1989; Moore et al., 1993a, 1993b) and those who want to use them as inputs to non-point source pollution and other kinds of environmental models (e.g., Panuska et al., 1991). Researchers, farmers, and consultants wanting to use GPS technology in order to build DEMs and perform terrain analysis should be aware of these differences and should organize their data collection efforts accordingly.

REFERENCES

Barling, R.D. 1992. Saturation zones and ephemeral gullies on arable land in southeastern Australia. Unpubl. Ph.D. dissert., Univ. of Melbourne, Melbourne, Australia. 339p.

Barling, R.D., I.D. Moore, and R.B. Grayson. 1994. A quasi-dynamic wetness index for characterizing the spatial distribution of zones of surface saturation and soil water content. Water Resour. Res. 30: in press.

Berry, B.J., and A.H. Baker. 1968. Geographic sampling. pp. 91–100. *In* B.J. Berry, et al. (ed.) Spatial analysis: A reader in statistical geography. Englewood Cliffs, NJ. Prentice-Hall.

Bouma, J., and P.A. Finke. 1993. Origin and nature of soil resource variability. pp. 3–13. *In* P.C. Robert, et al. (ed.) Soil Specific Crop Management. Madison, WI. Soil Sci. Soc. Am.

Burrough, P.A. 1993. Soil variability: A late 20th century view. Soils and Fertilizers 56(5):529–565.

Dikau, R. 1989. The application of digital relief models to landform analysis in geomorphology. pp. 51–77. *In* J. Raper (ed.) Three-dimensional applications in geographic information systems. New York, NY. Taylor and Francis.

Fisher, P.F. 1991. Modelling soil map-unit inclusions by Monte Carlo simulation. Int. J. Geogr. Info. Systems 5(2):193–208.

Freeman, G.T. 1991. Calculating catchment area with divergent flow based on a regular grid. Comp. and Geosci. 17:413–422.

Hutchinson, M.F. 1989. A new procedure for gridding elevation and stream line data with automatic removal of spurious pits. J. Hydrol. 106:211–232.

Klingebiel, A.A., E.H. Horvath, D.G. Moore, and W.U. Reybold. 1987. Use of slope, aspect, and elevation maps derived from digital elevation model data in making soil surveys. pp. 77–90. *In* W.U. Reybold, et al. (ed.) Soil Survey Techniques. Madison, WI. SSSA Spec. Publ. No. 20.

Larson, W.C., and P.C. Robert. 1993. Farming by soil. pp. 103–112. *In* R. Lal, et al. (ed.) Soil Management for Sustainability. Ankeny, IA. Soil and Water Conservation Society.

Lee, J. 1991. Comparison of existing methods for building triangular irregular network models of terrain from grid digital elevation models. Int. J. Geogr. Info. Systems 5(3):267–285

Moore, I.D. 1992. Terrain Analysis Programs for the Environmental Sciences: TAPES. Agric. Systems Info. Tech. 4(2):37–39.

Moore, I.D., G.J. Burch, and D.H. McKenzie. 1988. Topographic effects on the distribution of surface water and the location of ephemeral gullies. Trans. ASAE. 31:1383–1395.

Moore, I.D., P.E. Gessler, G.A. Nielsen, and G.A. Peterson. 1993a. Terrain analysis for soil specific crop management. pp. 27–56. *In* Robert, P.C., et al. (ed) Soil Specific Crop Management. SSSA. Madison, WI.

Moore, I.D., P.E. Gessler, G.A. Nielsen, and G.A. Peterson. 1993b. Soil attribute prediction using terrain analysis. SSSA J. 57:443–452.

Moore, I.D., R.B. Grayson, and A.R. Ladson. 1991. Digital terrain modeling: A review of hydrological, geomorphological, and biological applications. Hydrol. Proc. 5(1):3–30.

Moore, I.D., A. Lewis, and J.C. Gallant. 1993c. Terrain attributes: Estimation method and scale effects. p. 30–38. *In* Jakeman, A.K., et al. (ed) Modelling Change in Environmental Systems. Chichester. John Wiley & Sons.

Moore, I.D., A.K. Turner, J.P. Wilson, S.K. Jenson, and L.E. Band. 1993d. GIS and land surface-subsurface process modeling. p. 196–230. *In* Goodchild, M.F., et al. (ed.) Environmental Modeling with GIS. New York, NY: Oxford Univ. Press.

Mulla, D.J. 1991. Using geostatistics and GIS to manage spatial patterns in soil fertility. p. 336–345. *In* Kranzler, G. (ed.) Automated agriculture in the Twenty-first Century. St. Joseph, MI. Am. Soc. Agric. Engr.

Mulla, D.J. 1993. Mapping and managing spatial patterns in soil fertility and crop yield. p. 15–26. *In* Robert, P.C., et al. (eds), Soil Specific Crop Management. SSSA. Madison, WI.

Odeh, I.O.A., D.J. Chittleborough, and A.B. McBratney. 1991. Elucidation of soil-landfrom interrelationships by canonical ordination analysis. Geoderma 49(1):1–32.

O'Loughlin, E.M. 1986. Prediction of surface saturation zones in natural catchments by topographic analysis. Water Resour. Res. 22(5):794–804.

Panuska, J.C., I.D. Moore, and L.A. Kramer. 1991. Terrain analysis: Integration into the Agricultural Non-Point Source (AGNPS) pollution model. J. Soil Water Conserv. 46(1):59–64.

Quinn, P., K. Bevin, P. Chevallier, and O. Planchon. 1991. The prediction of hillslope flow paths for distributed hydrological modeling using digital terrain models. Hydrol. Proc. 5(1):59–79.

Rawls, W.J., D.J. Brakensiek, and K.E. Saxton. 1982. Estimating soil water properties. Trans. ASAE. 25: 1316–1320, 1328.

Skidmore, A.K., P.J. Ryan, W. Dawes, D. Short, and E. O'Loughlin. 1991. Use of an expert system to map forest soils from a geographical information system. Int. J. Geogr. Info. Systems 5(4):431–445.

Tyler, D.A. 1993. Positioning technology (GPS). p. 159–165. *In* P.C. Robert, et al. (ed.) Soil Specific Crop Management. SSSA. Madison, WI.

Wagenet, R.J., J. Bouma, and R.B. Grossman. 1991. Minimum data sets for use of soil survey information in soil interpretive models. p. 161–182. *In* M.J. Mausbach, et al. (ed.) Spatial Variabilities of Soils and Landforms. Madison, WI. SSSA Spec. Public. No. 28.

21 Evaluation Of Mapping Strategies For Variable Rate Applications

Wayne H. Thompson
Pierre C. Robert

Department of Soil Science
University of Minnesota
St. Paul, Minnesota

Farming systems have evolved from individually managed parcels to large uniformly managed tracts of land that are composites of smaller parcels. Each parcel has its own management history, accentuating variability within the large "modern" farming unit. Variable rate technology (VRT) addresses the issue of managing large fields more efficiently by accounting for variability that is either attributable to natural components of a heterogeneous landscape or the union of small individually managed parcels that form one large management tract.

Modeling landscape and nutrient variability is crucial for the development of optimal VRT mapping strategies. The most critical component of variable rate farming is the mapping base and strategy selected to generate the variable rate map. A variable rate map should subdivide a field into a number of homogeneous zones. Ideally, each zone will encompass a dominant characteristic that influences crop growth. This is not a trivial objective; climate, soil properties, topography, hydrology and management history all affect crop growth potentials. For example, differentials in aeration and the corresponding topographic variation across a landscape translate into hydrological gradients that, in turn, affect crop growth potentials (Bruce et al., 1990; Carr et al., 1991; Kleiss, 1970; Malo & Worcester, 1975; Miller et al., 1988; Spratt & McIver, 1972).

Catenary models (Ruhe, 1960) of landscape variability are the basis of soil surveys, but often conflict with spatial models describing management induced variability (Honeycutt et al., 1990a,b). Further, natural soil variability is confounded when management practices superimpose spatially independent variations in crop response patterns. Spatial modeling techniques based on local interpolation are sometimes used to approximate patterns of variability that do not follow landscape patterns, but require prohibitively large sample sets to accurately model variability (Burrough, 1986).

Simply subdividing a field into a number of small management units increases the probability of isolating and identifying homogeneous zones. This form of landscape partitioning can be accomplished using a number of approaches. The U.S. Soil Conservation Service (SCS) soil survey maps (1:20,000–24,000 scale) provide a catenary based approach that attempts to account for the natural chemical and physical soil characteristics in the mapping process, while a grid soil sampling and spatial statistics approach attempts to account for those factors not considered in the soil survey mapping method. Each method is based on the principle of partitioning the field into smaller subunits or management zones.

OBJECTIVES

The objective of this study was to evaluate two of the most common mapping strategies (the soil survey and grid sample methods) with the variable rate anhydrous ammonia application system in maize (*Zea mays* L.) relative to conventional nitrogen application methods. We do this by: (i) evaluating the most appropriate method, or combination of methods, that defines the observed variability of soils, residual nutrients and crop growth; and (ii) testing for variations in maize yields and economic performance between and within mapping strategies. Results from one site located in Southwestern Minnesota are presented.

MATERIALS AND METHODS

Site Description

The site is located in southwestern Minnesota on the eastern edge of the Coteau des Prairies. It is located on the lowland plain where soils formed over a shaley (calcareous) glacial till (Fig. 21–1, state map inset). The native vegetation was primarily tall grass prairie, varying with landscape drainage patterns. Drainage characteristics of this site range from very poorly drained to excessively well drained and eroded. Precipitation for the region is generally low (< 50 cm per yr^{-1}).

The 1:20,000 scale SCS soil survey map of the location (Soil Survey Staff, 1978) is presented in Fig. 21–1. The soil survey lists two soil consociations (similar soil types mapped as one unit) and two soil complexes (small inclusions of dissimilar soil types mapped as one unit) at this site, indicating the relative extent of soil variability. Soil variability attributable to the occurrence of soil inclusions is visible in Fig. 21–2 where a vertical aerial infrared photograph of the site is overlaid by contour lines (1 m intervals) from the digital elevation model (DEM) for the site.

The site's recent management history consists of a maize-soybean [*Glycine max* (L.) Merr.] rotation with moderately low, uniform fertilizer applications on maize. The preceding history (> 10 yrs prior to this study) includes manure applications across the east and north edges of the plot area, with occasional manure applications on the eroded convex landscape positions.

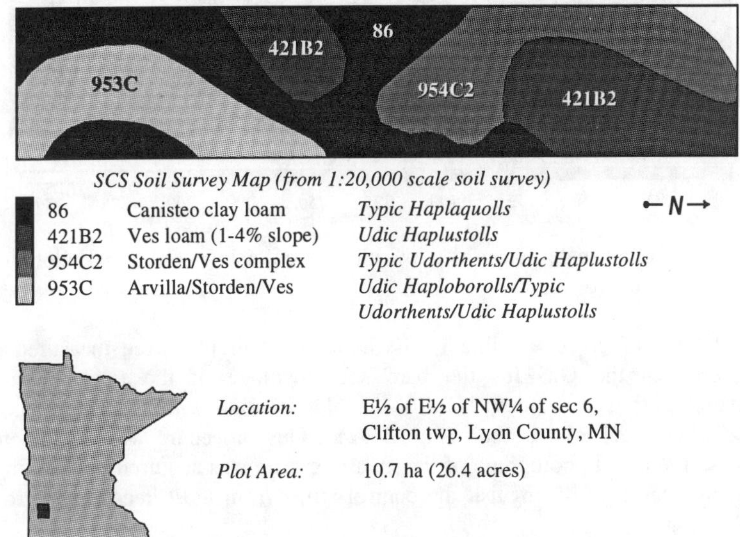

Fig. 21–1. The site is located in southwestern Minnesota on the eastern edge of the Coteau des Prairies. Soils formed over calcareous till.

Manure applications were most frequent over the areas close to the farmstead, located near the northeast corner of the plot area. Occasionally it was applied at the eroded area at the south end of the plot.

Low Altitude Aerial Photography

Two low altitude vertical aerial color infrared photographs were acquired for this site. The first photograph was taken in late March (1990) after snow melt, three to five d after a significant rainfall event (Fig. 21–2). Dark tones indicate poorly drained, concave landscape positions. Light tones correspond with eroded, convex landscape positions. Contour lines from the location's DEM are provided for reference. Small squares represent 61 m grid soil sample locations. A second photograph was acquired at early crop maturity, in mid to late September.

Scangal v.2.0 (Microsoft) was used in conjunction with a HP Scanjet II gray scale scanner (256 gray scale TIFF option) to digitize the vertical aerial photographs of bare soil and the maturing crop. Gray scale images of the maturing crop and bare soil were rescaled and warped with "rubber sheeting" to match layers using a PC-based raster geographic information system (GIS) [EPPL7, v.2.01; distributed by Land Management Information Center (LMIC), Minnesota Department of Administration, St. Paul]. Geo-referencing for the plot was accomplished using a laser theodolite (Geodometer 136, Type 571133064).

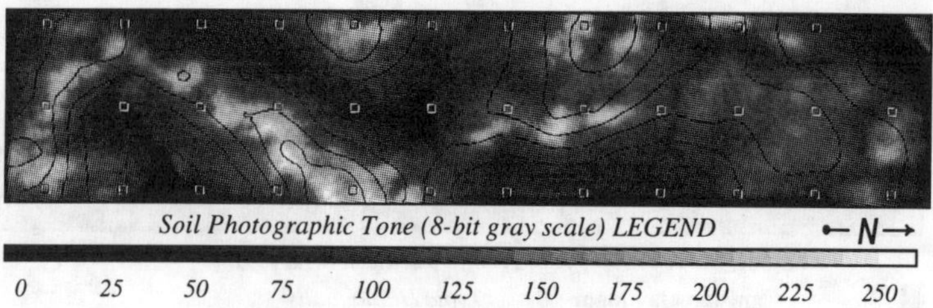

Fig. 21–2. Gray scale image of a low-altitude color infra-red aerial photograph of the site.

Coordinates of objects visible in the aerial photographs were measured and registered in the GIS for the bare soil coverage of the field. Ground measurements were used to identify the plot location within the field on the aerial photograph of bare soil (Fig. 21–2). This procedure was implemented because the aerial photograph of the maturing crop was acquired too late in the growing season to distinguish the control strips from those receiving nitrogen applications.

Digital Elevation Model

The theodolite was also used at this site to measure relative elevation using an effective sample spacing of ≈10 meters (Fig. 21–3). Approximately 280 measurements were kriged using simple kriging and imported into the GIS. The topography measurements were used to generate a digital elevation model (DEM) using ordinary kriging (Surfer, v.2.1; Golden Software). Slope gradient was calculated within the raster GIS (Fig. 21–4).

The site consists of a complex array of drainage patterns. Soils mapped as Canisteo (86) occur lower in the landscape with respect to the relative variations in elevation across the site. These areas are relatively higher in organic carbon (2.1–3.1%), have a thick A horizon (45 cm) and appear as zones with low photographic tone values (Fig. 21–2). Ves soils (421B2) appear as mottled zones in the image and are found higher on the landscape. Within these mottled zones, the areas with lower photographic tone values are generally small landscape concavities with limited surface drainage outlets. Soil complexes

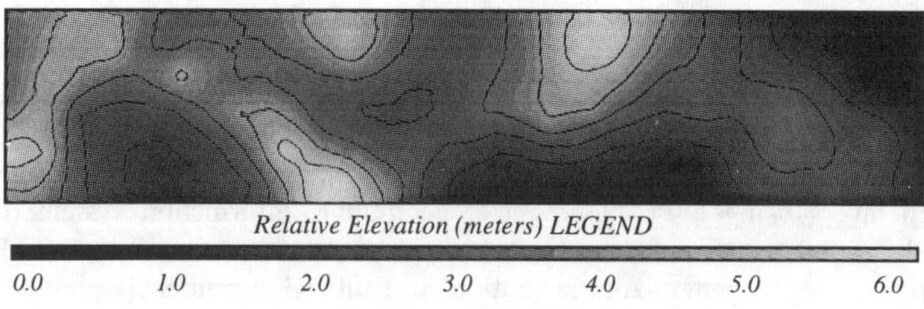

Fig. 21–3. Digital elevation model (DEM) of study area.

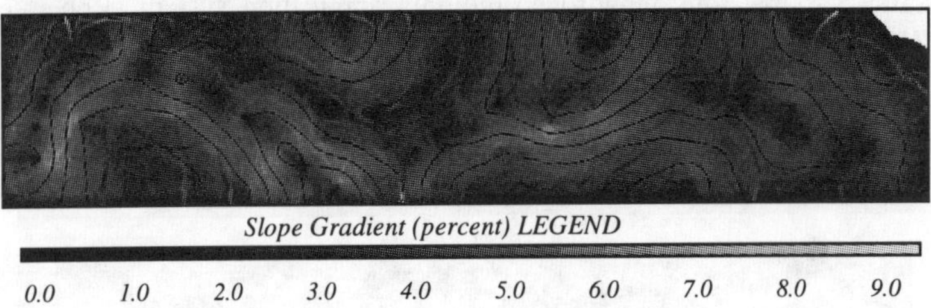

Fig. 21–4. Slope Gradient Map: derived from DEM of study area.

(953C and 954C2) occupy the highest landscape positions and are generally convex and highly eroded (Fig. 21–4). The associated soils appear as zones with the highest photographic tone values. Adjacent areas with slightly lower photographic tone values are usually associated with steeply sloping, highly eroded soils. This is clearly visible on the southern end of the plot within the Arvilla-Storden-Ves complex (953C).

The landscape of this site undulates with a gradual overall increase in elevation from north to south and east to west (Fig. 21–3). Again, the relatively unbroken zone of low photographic tone values (< 50, Fig. 21–2) represents the lowest landscape positions, the zones with continuous high photographic tone values (> 150) characterize the highest landscape positions and eroded steeply sloping sideslopes (Fig. 21–4), and mottled zones (50–150), the areas with complex discontinuities in photographic tone values, represent the undulating soils located at medium landscape positions with variable drainage characteristics.

Treatment Descriptions

Four treatments were incorporated into the experimental plan for each site. Each treatment represents a mapping strategy:

1) a control receiving no anhydrous ammonia;
2) conventional uniform anhydrous ammonia rate corresponding to the collaborating grower's management practice;
3) the grid sample method (variable rates by kriged intervals, Fig. 21–5a);
4) the soil survey method (variable rates by soil mapping unit, Fig. 21–5b).

Treatments were laid out as sixteen randomly ordered parallel strips (transects) across the longer dimension of the plot area (Fig. 21–5c). Each mapping strategy (treatment) was replicated four times as four individual transects. The width of the application toolbar (9.1 m) dictated the width of each transect (9.1 m). The overall plot width at all sites was 146 m, while plot length was 730 m. The length of each strip was maximized to increase the potential number of harvest sub samples per transect. Nitrogen was applied in the spring of 1990 one week before planting maize using a prototype of the Soil Teq Inc. variable rate anhydrous ammonia application system (Robert et al., 1992).

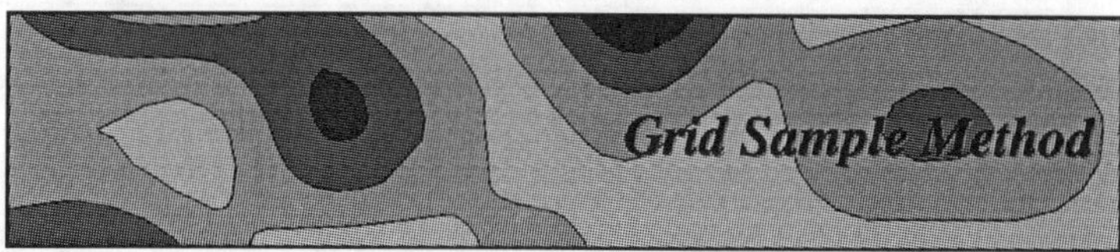

Fig. 21–5a. Variable Rate Map: Grid sample method.

Fig. 21–5b. Variable Rate Map: Soil survey method.

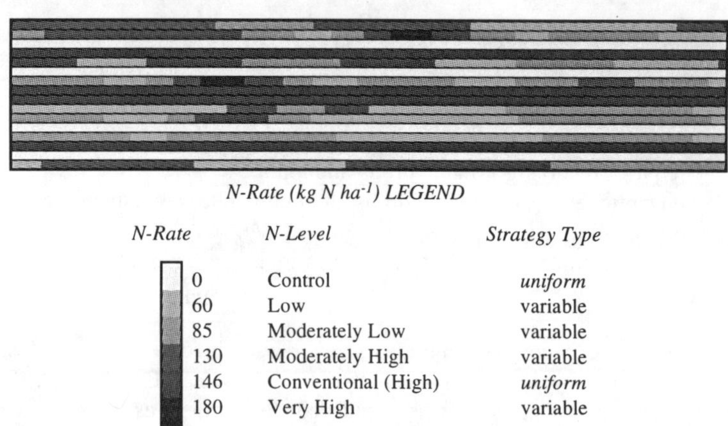

Fig. 21–5c. Experimental design for evaluating variable rate mapping strategies.

Variable Rate Treatments: Mapping Strategy Preparation

The average nitrogen application rate for each mapping unit was applied with the soil survey method, while the variable rates of the grid method were separated by kriged contours.

Soil Survey Method

This method uses the SCS soil survey mapping units as landscape partitions to separate individual management units. Mapping units are categorized by their different parent materials and inherent physical and chemical characteristics that are expressed through a variety of horizonation patterns. Outlined in Fig. 21–5b is the SCS soil survey map for this location. The principle underlying this method is that all related soil map units (soil types) are assumed to have a similar nutrient status and drainage characteristic for influencing plant growth. This presumption serves as the basis for determining the productivity potential of each soil (Anderson et al., 1992). Using this idea, we assumed relative uniformity across all individual polygons of individual mapping units from the 1:20,000 scale soil survey maps. The result provided composite (average) soil sample nutrient values for each mapping unit, regardless of the number of individual polygons or their spatial separation. An average nitrogen application rate was calculated (Rehm and Schmitt, 1989) for each soil mapping unit. This interpretation did not consider the potential impact of variable management histories in creating additional variability across the landscape, nor the spatial variability (dissimilar inclusions) within the 1:20,000 scale soil map unit polygons.

Grid Sample Method

The method is based on extensive soil sampling in a uniform grid pattern across the field. We selected a 61 meter (200 feet) grid interval to provide a manageable number of coordinate points for the relatively small rectangular plots. A nitrogen application rate was calculated (Rehm and Schmitt, 1989) for each grid sample location. Simple kriging was used to interpolate between the grid sample points (Surfer v.2.1; Golden Software). This software package assumes an isotropic linear semivariogram model. The kriging procedure provided a nitrogen rate contour map. The continuum of variable nitrogen rates expressed in a contour map was subdivided into five zones. The partitions were made at cut-off values derived from the standard deviations of the variable nitrogen rates across grid samples (McBratney et al., 1991). Variations in the nutrient status across the farming landscape resulting from non-uniform management practices are expressed in the variable rate map generated using the grid sample method.

Plot Harvest Procedures

Each of the sixteen transects was subdivided into twelve to sixteen harvest samples, varying with plot length. The control and conventional treatment transects were subdivided into uniformly sized grain yield samples while the variable rate treatment transects were subdivided into unequally sized (lengths) grain samples. The latter was done to avoid overlapping nitrogen rates within any grain yield sample.

Plots were harvested using the collaborators' harvest equipment. Grain yield samples were harvested with a conventional combine. They were passed directly to a weigh-wagon (accurate to within about one kilogram) where the sample weight was recorded and a small subsample of grain was collected to approximate the sample harvest grain moisture.

The harvest data was entered into a spreadsheet and subsequently transferred to the microcomputer GIS (EPPL7 v.2.0). The harvest information included: transect number, grain yield sub-sample identifier, uncorrected grain yield and harvest grain moisture. A spreadsheet (Excel v.4.0, Microsoft) was used as the database management system (DBMS) to support the GIS. PC-SAS (SAS, 1988) was used to perform statistical diagnostics and analyses.

Statistical Procedures

Classical statistical procedures, analysis of variance (ANOVA) were used to compare the performance of the different mapping strategies (outlined above) against the conventional uniform method of nitrogen fertilizer application. The control treatment (no nitrogen fertilizer applied) is the base line for weighing the relative importance of the application rates used in the different mapping strategies. The assumptions and methodology of ANOVA procedures are well documented (Snedecor and Cochran, 1980). Other less common procedures were used in the analysis to model spatial components of landscape variability: (i) analysis of covariance; and (ii) trend surface modeling.

The objective of including a concomitant variable, or covariate, is to strengthen the statistical basis for treatment comparisons by reducing the amount of unexplained variability within treatments. For example, internal variations within treatment classifications may follow predictable trends defined by a covariate. To use a covariate in this case, the variable in question must be linearly related to the covariate with a correlation coefficient (r) of at least 0.30 (Lane and Nelder, 1982). In contrast, if the data is stratified or partitioned by a covariate no assumption of linearity is required (Koch et al., 1982). Each of these approaches will reduce the random error (ANCOVA error term, e) by accounting for more of the variability within individual treatments.

Conventional trend surface modeling is based on the principles of regression analysis. It is a global approach to modeling variability using a general interpolator (Griffith and Amrhein, 1991). The 2-way model is used to characterize the extent of variation across samples, usually ignoring variation between adjacent samples. This implies that the long range trends of spatial data can be identified with a 2-way model, but the short range variation between adjacent samples is lost. Trend modeling can be applied to minimize unwanted long range spatial trends (non-stationarity) before applying a local interpolation procedure, such as kriging. The trend model is derived using uniformly dispersed samples across a surface. The validity of the interpolations and extrapolations between and beyond sample locations is measured through evaluation of the error terms associated with each of the coefficients in the model. Low order polynomials are used to describe the general long range trends across a surface. Highly complex polynomials are sometimes generated in an attempt to model the intricate variability of a surface, but can be difficult to interpret and usually provide unacceptably large or small extrapolated values at the surface margins (Burrough, 1986). Additionally, because of the nature of a least squares method of regression, the 2-way model estimates rarely fit the measured values. This limitation of the global model is a function of the spatial error associated with fitting a generalized response surface to spatial data. Outliers are especially problematic and may distort and invalidate the resulting model of a surface, if they are not properly weighted before modeling.

Landscape Models

Soil photographic tone (aerial photograph, Fig. 21–2) relative elevation (Fig. 21–3) and slope gradient (Fig. 21–4) values corresponding to each soil sample were extracted from the GIS layers and were used as independent predictors for estimating spatial variations in soil nutrients (Lane & Nelder, 1982).

Landscape Partitioning

Four conventional rectangular blocks were used in the plot design to minimize the impact of error associated with experimental plot management. The second objective was to reduce the random error term in the statistical model by properly subdividing treatments according to a dominant spatial component of the landscape that influences crop development. This was done to help clarify the degree to which treatment effects are controlled by variations (short and long range) in either soil nitrate or soil organic carbon (Koch et al., 1982; Oliver & Webster, 1989; McBratney et al., 1991). Tests for interactions between total carbon and soil nitrate were not evaluated for lack of sufficient numbers of harvest samples.

GIS layers representing a continuum of total carbon and soil nitrate variability were individually partitioned into four asymmetrical zones. Each of the five GIS layers (Fig. 21–8 and 21–9) representing soil and management variability were subdivided into four zones (classes), where each zone represents ~25% of the harvested area within the plot (Fig. 21–10 and 21–11) excluding buffer strips between treatment transects. The plot area was subdivided in this manner to maximize the probability of attaining equal numbers of harvest samples across treatments, blocks and landscape partitions.

The modal partition identifier (e.g. total carbon or soil nitrate) for each grain yield sample was selected for use as a corresponding classification variable in ANCOVA. The harvest data set was compiled with a DBMS in preparation for statistical analysis (all classification variables generated in the GIS were merged with harvest data). Each mapping strategy was included as four distinct transects (four replicates of each strategy, Fig. 21–5c) where each transect contained at least twelve grain yield subsamples.

Residual soil nitrate and carbon estimates were incorporated into the statistical model to evaluate maize grain yields and economic net returns across mapping strategies. Partitioned landscape components were used in ANCOVA as asymmetrical partitions to stratify landscape variability into relative classes.

RESULTS AND DISCUSSION

Landscape Models and Partitioning

The soil color and natural drainage patterns expressed in the photograph correlate well with a number of nutrients, specifically those most closely associated with soil texture and drainage (Milfred & Kiefer, 1976; Zheng & Schreier, 1988). This is illustrated in Fig. 21–6c by the relationship between soil photographic tone (8-bit pixel values) and soil organic carbon ($|r| > 0.70$). Relative elevation and slope gradient (Fig. 21–6a and 21–6b) and their interaction (Fig. 21–7) demonstrate similar correlations with soil carbon. Each of the three parameters serves as a strong predictor of total carbon variation ($|r| > 0.50$; $p < 0.0001$ for each model). Extrapolation of these relationships across a landscape provides a clear depiction of short range spatial variations in total carbon (Fig. 21–8). [Model (1): from linear relationship with soil photographic tone (8-bit gray scale, Fig. 21–6c); Model (2): from linear relationship with slope gradient and relative elevation (Fig. 21–7.); Model (3): 3-way relationship between relative elevation, slope gradient and soil photographic tone.]

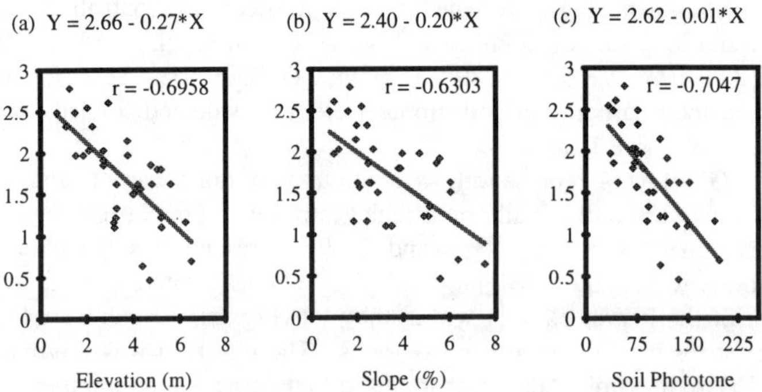

Fig. 21–6. Correlation of total carbon with relative elevation (a), slope gradient (b), and soil photographic tone (c).

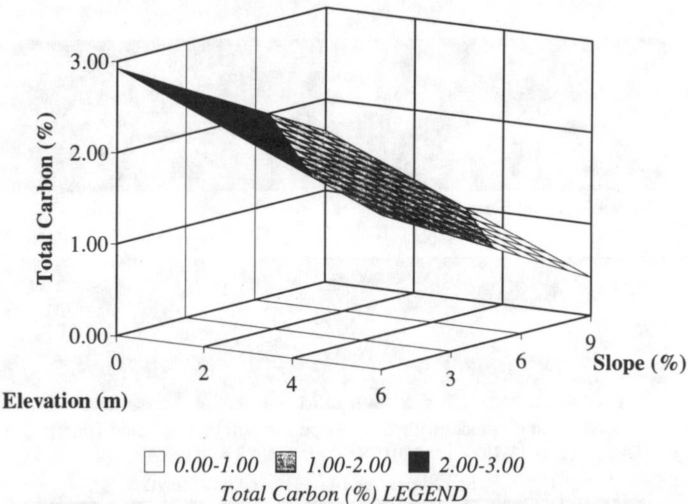

Fig. 21–7. Estimating total carbon using slope gradient and relative elevation from the DEM (Fig. 21–3).

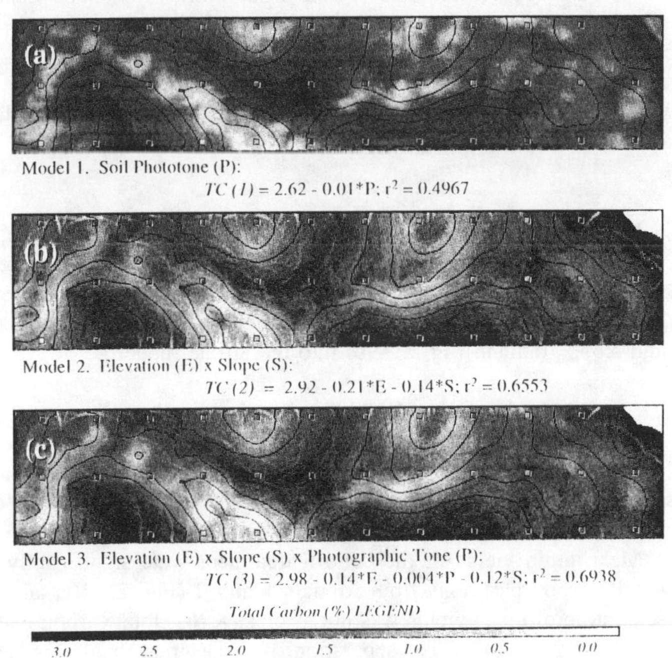

Fig. 21–8. Three approximations of soil carbon variability. The gray scale tones within the small squares represent measured values, while the surrounding tones represent total carbon estimates. Elevation contour lines (from Fig. 21–3) are provided for reference.

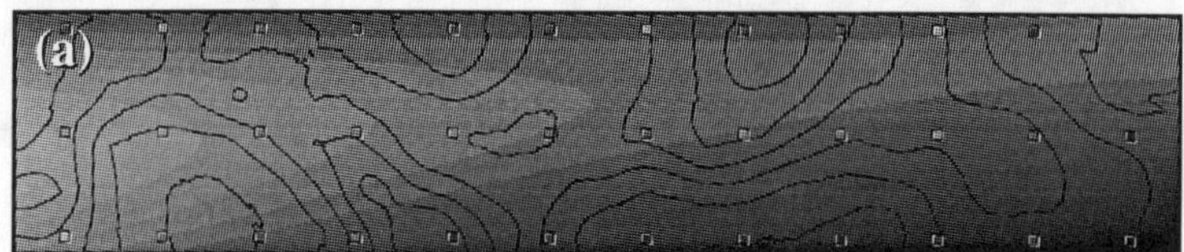

Model 1: Trend Surface:
$N(1) = 16.16 + 0.0002*X - 0.1069*Y + 0.0001*X*Y - 0.0000*X^2 + 0.0004*Y^2$

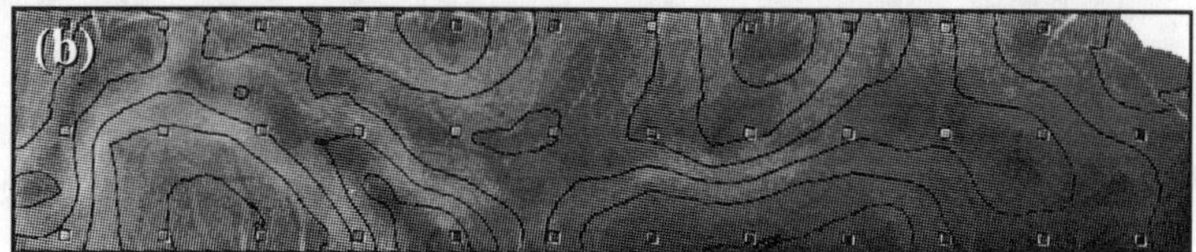

Model 2: Best Fit model with X, Y, Slope and Soil Photographic Tone:
$N(2) = 13.2821 + 0.0001*X*Y - 0.8546*S + 0.0190*P$

Soil Nitrate (mg-NO_3-N kg-soil^{-1}) LEGEND

30 25 20 15 10 5 0

Fig. 21–9. Two approximations of soil nitrate variability. The trend surface model (a) illustrates distinct long-range variability. The second model (b) provides an estimate of both long- and short-range variations where X (northing), Y (easting) interaction, soil photographic tone and slope correlate with variability (measured values correspond with tones within small squares.

Long range variation in soil nitrate due to isolated manure applications (Fig. 21–9a) is not visible in the aerial photograph (soil nitrate versus soil photographic tone: |r| » 0.30, data not shown). It was modeled by combining long range components (X, Y) and short range components (soil photographic tone and slope gradient, Fig. 21–9b) into the nitrate model.

Maize Grain Yields by Landscape Partition

Within treatment variations of maize grain yields were evaluated across landscape positions. Grain yields across partitions were evaluated by testing for differences between treatment means across partitions (Table 21–1).

Mean grain yields of the control treatment were significantly different across soil carbon partitions from Models 1 and 2 (Fig. 21–10a and 21–10b). This separation of mean yields was not as clear across total carbon partitions of Model 3 (Fig. 21–10c). Clear separation of mean grain yields for each of the treatments receiving nitrogen fertilizer was best accomplished with Model 3 (Fig. 21–10c). There is a clear increase in grain yield with estimated organic carbon with each method of modeling total carbon variability. No absolute differences between total carbon modeling methods are observed with this harvest data.

Table 21–1. Maize grain yields (Mg ha^{-1}) by landscape partition. Grain yield differences across partitions by treatment are presented (mean ± std dev). A least squares method was used to test for differences between means.

Total Carbon Partitions [Model 1: Soil Photographic Tone]

	Control	*Conventional*	*Grid Method*	*Soil Method*
1.5	4.91 ± 1.97 a	5.89 ± 1.42 a	5.93 ± 1.54 a b	6.33 ± 1.24 a
1.8	6.10 ± 0.95 b	6.45 ± 0.84 a b	6.61 ± 1.12 a b c	5.81 ± 1.02 a
2.0	6.09 ± 1.19 b	5.85 ± 1.94 a	6.95 ± 1.02 b c	7.19 ± 0.84 b
2.6	7.34 ± 0.86 c	6.83 ± 1.21 b	7.33 ± 1.39 b c	7.45 ± 0.95 b

Total Carbon Partitions [Model 2: Slope Gradient and Relative Elevation]

	Control	*Conventional*	*Grid Method*	*Soil Method*
1.3	4.59 ± 1.81 a	5.68 ± 1.40 a b	5.84 ± 1.81 a	5.90 ± 1.24 a b
1.7	5.82 ± 0.70 b	6.20 ± 1.20 a b c	6.63 ± 0.95 b	6.36 ± 1.12 a b c
2.0	6.50 ± 1.17 b c	6.55 ± 1.17 b c	7.02 ± 0.78 b	6.94 ± 0.63 b c
2.8	7.15 ± 0.70 c	7.22 ± 1.18 c	7.36 ± 1.44 b	7.57 ± 0.82 c

Total Carbon Partitions [Model 3: Soil Phototone, Slope Gradient and Relative Elevation]

	Control	*Conventional*	*Grid Method*	*Soil Method*
1.3	4.98 ± 1.80 a	5.63 ± 1.34 a b	5.73 ± 1.54 a	5.98 ± 1.33 a b
1.7	5.61 ± 1.10 a	6.22 ± 1.27 a b c	6.53 ± 1.17 b c	6.19 ± 1.10 a b c
2.0	6.64 ± 1.14 b	6.55 ± 1.22 b c d	7.23 ± 0.64 b c d	6.82 ± 0.66 b c
3.4	7.16 ± 0.75 b	7.16 ± 1.15 c d	7.38 ± 1.41 c d	7.60 ± 0.72 d

Soil Nitrate Partitions [Model 1: Trend Surface]

	Control	*Conventional*	*Grid Method*	*Soil Method*
13	5.55 ± 2.27 a	6.19 ± 1.53 a b	6.18 ± 1.38 a b	6.35 ± 1.08 a
14	6.20 ± 1.21 b	5.83 ± 1.56 a b	6.67 ± 1.05 a	6.77 ± 1.45 a
16	6.29 ± 1.49 b	6.46 ± 0.72 a b c	6.84 ± 1.43 a b	6.81 ± 0.95 a
30	6.39 ± 1.28 b	7.06 ± 0.87 b c	7.14 ± 1.40 b	6.92 ± 1.38 a

Soil Nitrate Partitions [Model 2: X, Y, Slope Gradient and Soil Photographic Tone]

	Control	*Conventional*	*Grid Method*	*Soil Method*
13	4.92 ± 2.15 a	5.43 ± 1.53 a	6.62 ± 1.70 a b	6.09 ± 1.00 a
14	6.16 ± 1.30 b	6.00 ± 1.39 a b	6.38 ± 1.13 a	6.74 ± 0.94 a
17	6.32 ± 1.19 b	6.81 ± 1.12 b	6.67 ± 1.05 a b	7.30 ± 1.29 a
28	6.91 ± 0.76 b	6.81 ± 0.90 b	7.19 ± 1.51 b	7.14 ± 1.28 a

mean grain yields across partitions (by treatment) with the same letter are not significantly different at p = 0.05.

Soil nitrate partitions separate mean grain yields of the control treatment from the low nitrate partition and the three partitions representing higher nitrate concentrations in both models (Fig. 21–11a and 21–11b). Short range variability brought out by Model 2 (Fig. 21–11b) increased the range in mean grain yields across partitions, relative to the mean grain yields using Model 2 partitions (Fig. 21–11a). Minimal separation of means is observed for treatments receiving nitrogen. Grain yield suppression due to excessively high nitrogen rates occurred within the conventional treatment at nitrate partitions representing low soil nitrate concentrations (Table 21–1 and 21–2).

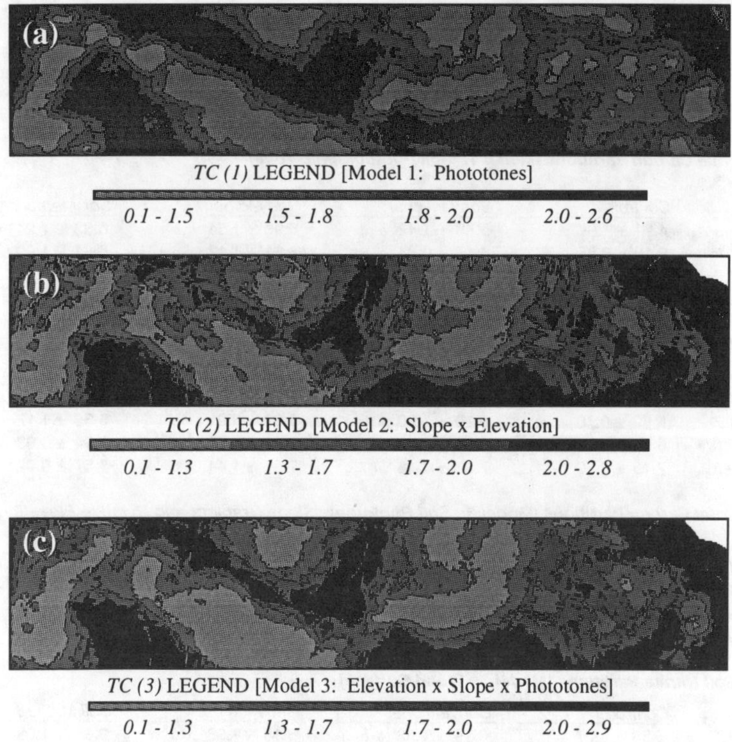

Fig. 21–10. Total carbon estimates were grouped into four classes (partitions), where each class represents 25% of the harvested area within the plot.

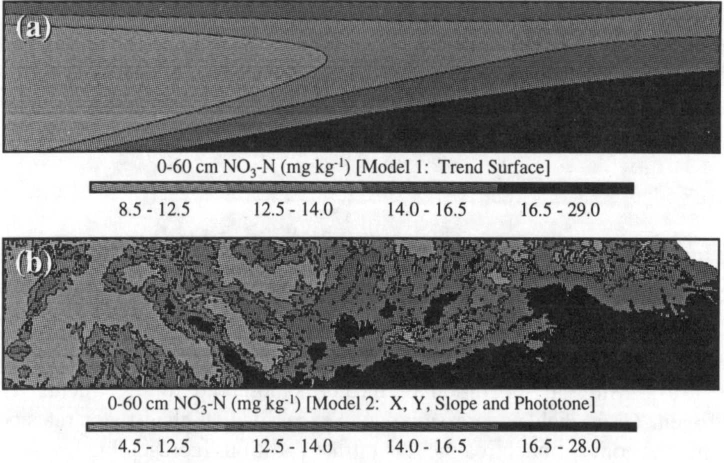

Fig. 21–11. Soil nitrate landscape partitions.

Table 21–2. Nitrogen application rates (kg N ha^{-1}) by treatment and landscape partition (mean ± std dev).

Total Carbon Partitions [Model 1: Soil Photographic Tone]

	1.5	1.8	2.0	2.6
Control	0 ± 0	0 ± 0	0 ± 0	0 ± 0
Conventional	146 ± 0	146 ± 0	146 ± 0	146 ± 0
Grid Method	100 ± 21	104 ± 26	97 ± 32	83 ± 27
Soil Method	78 ± 27	86 ± 29	104 ± 24	113 ± 27

Total Carbon Partitions [Model 2: Slope Gradient and Relative Elevation]

	1.3	1.7	2.0	2.8
Control	0 ± 0	0 ± 0	0 ± 0	0 ± 0
Conventional	146 ± 0	146 ± 0	146 ± 0	146 ± 0
Grid Method	100 ± 29	102 ± 32	97 ± 22	83 ± 23
Soil Method	75 ± 26	105 ± 28	104 ± 25	103 ± 29

Total Carbon Partitions [Model 3: Soil Phototone, Slope Gradient and Relative Elevation]

	1.3	1.7	2.0	3.4
Control	0 ± 0	0 ± 0	0 ± 0	0 ± 0
Conventional	146 ± 0	146 ± 0	146 ± 0	146 ± 0
Grid Method	97 ± 22	103 ± 35	101 ± 19	81 ± 25
Soil Method	72 ± 24	90 ± 29	108 ± 21	110 ± 26

Soil Nitrate Partitions [Model 1: Trend Surface]

	13	14	16	30
Control	0 ± 0	0 ± 0	0 ± 0	0 ± 0
Conventional	146 ± 0	146 ± 0	146 ± 0	146 ± 0
Grid Method	101 ± 28	98 ± 24	102 ± 30	79 ± 20
Soil Method	85 ± 32	106 ± 26	94 ± 32	100 ± 26

Soil Nitrate Partitions [Model 2: X, Y, Slope Gradient and Soil Photographic Tone]

	13	14	17	28
Control	0 ± 0	0 ± 0	0 ± 0	0 ± 0
Conventional	146 ± 0	146 ± 0	146 ± 0	146 ± 0
Grid Method	109 ± 30	91 ± 29	92 ± 26	89 ± 19
Soil Method	86 ± 31	100 ± 33	94 ± 27	103 ± 26

Differences in mean grain yields between treatments (mapping strategies) were evaluated by partition (Table 21–3). Differences were most apparent within the low total carbon zone. The differences diminished as the relative total carbon increases. No differences between treatments were observed at the highest total carbon partition. No consistent differences in mean grain yields were found between treatments using nitrate partitions.

Landscape aeration and hydrology gradients are expressed in the nutrient partitions because of the distinctive precipitation patterns found in southwestern Minnesota during the 1990 growing season. For example, rainfall in this region was limiting during pollenation and grain fill. As a result, crop growth over the well and excessively well drained soils, low in organic carbon was severely depressed (Fig. 21–2 zones with high photographic tone). In contrast, the moderately well drained and poorly drained soils, specifically within the concave landscape positions (Fig. 21–2, low photographic tone values, high in organic carbon), had sufficient plant available soil moisture to sustain adequate crop growth.

Table 21–3. Maize grain yields (Mg ha-1) by mapping strategy. Grain yield differences across treatments by partition are presented (mean ± std dev). A least squares method was used to test for differences between means.

Total Carbon Partitions [Model 1: Soil Photographic Tone]				
	1.5	*1.8*	*2.0*	*2.6*
Control	4.91 ± 1.97 a	6.10 ± 0.95 a b	6.09 ± 1.19 a	7.34 ± 0.86 a
Conventional	5.89 ± 1.42 a b	6.45 ± 0.84 a b	5.85 ± 1.94 a	6.83 ± 1.21 a
Grid Method	5.93 ± 1.54 a b	6.61 ± 1.12 a	6.95 ± 1.02 b	7.33 ± 1.39 a
Soil Method	6.33 ± 1.24 b	5.81 ± 1.02 b	7.19 ± 0.84 b	7.45 ± 0.95 a
Total Carbon Partitions [Model 2: Slope Gradient and Relative Elevation]				
	1.3	*1.7*	*2.0*	*2.8*
Control	4.59 ± 1.81 a	5.82 ± 0.70 a	6.50 ± 1.17 a	7.15 ± 0.70 a
Conventional	5.68 ± 1.40 b	6.20 ± 1.20 a b	6.55 ± 1.17 a	7.22 ± 1.18 a
Grid Method	5.84 ± 1.81 b	6.63 ± 0.95 b	7.02 ± 0.78 a	7.36 ± 1.44 a
Soil Method	5.90 ± 1.24 b	6.36 ± 1.12 a b	6.94 ± 0.63 a	7.57 ± 0.82 a
Total Carbon Partitions [Model 3: Soil Phototone, Slope Gradient and Relative Elevation]				
	1.3	*1.7*	*2.0*	*3.4*
Control	4.98 ± 1.80 a	5.61 ± 1.10 a	6.64 ± 1.14 a	7.16 ± 0.75 a
Conventional	5.63 ± 1.34 a b	6.22 ± 1.27 a b	6.55 ± 1.22 a	7.16 ± 1.15 a
Grid Method	5.73 ± 1.54 a b	6.53 ± 1.17 b	7.23 ± 0.64 b	7.38 ± 1.41 a
Soil Method	5.98 ± 1.33 b	6.19 ± 1.10 a b	6.82 ± 0.66 a b	7.60 ± 0.72 a
Soil Nitrate Partitions [Model 1: Trend Surface]				
	13	*14*	*16*	*30*
Control	5.55 ± 2.27 a	6.20 ± 1.21 a	6.29 ± 1.49 a	6.39 ± 1.28 a
Conventional	6.19 ± 1.53 a	5.83 ± 1.56 a	6.46 ± 0.72 a	7.06 ± 0.87 a
Grid Method	6.18 ± 1.38 a	6.67 ± 1.05 a	6.84 ± 1.43 a	7.14 ± 1.40 a
Soil Method	6.35 ± 1.08 a	6.77 ± 1.45 a	6.81 ± 0.95 a	6.92 ± 1.38 a
Soil Nitrate Partitions [Model 2: X, Y, Slope Gradient and Soil Photographic Tone]				
	13	*14*	*17*	*28*
Control	4.92 ± 2.15 a	6.16 ± 1.30 a	6.32 ± 1.19 a	6.91 ± 0.76 a
Conventional	5.43 ± 1.53 a b	6.00 ± 1.39 a	6.81 ± 1.12 a b	6.81 ± 0.90 a
Grid Method	6.62 ± 1.70 b	6.38 ± 1.13 a	6.67 ± 1.05 a b	7.19 ± 1.51 a
Soil Method	6.09 ± 1.00 a b	6.74 ± 0.94 a	7.30 ± 1.29 b	7.14 ± 1.28 a

mean grain yields across treatments (by partition) with the same letter are not significantly different at p = 0.05.

Economic Net Return by Landscape Partition

Economic analysis revealed statistically significant variability within treatments (Table 21–4). As with grain yields, the total carbon partitions proved reliable for separating economic returns across this location's landscape. No differences in economic return were identified between the control or conventional treatments versus variable rate mapping strategies, nor between variable rate mapping strategies (Table 21–5). This implies that the variable rate mapping strategies employed in this experiment did not accurately represent the cropping requirements for this site.

Table 21–4. Economic net returns (\$ ha^{-1}) by landscape partition. Net return differences across partitions by treatment are presented (mean ± std dev). A least squares method was used to test for differences between means.

Total Carbon Partitions [Model 1: Soil Photographic Tone]

	Control	*Conventional*	*Grid Method*	*Soil Method*
1.5	387 ± 155 a	423 ± 112 a	439 ± 123 a	477 ± 96 a
1.8	480 ± 75 b	468 ± 66 a b	492 ± 82 a b	433 ± 77 a
2.0	480 ± 93 b	421 ± 153 a b	520 ± 77 b	537 ± 67 b
2.6	578 ± 68 c	498 ± 95 b	554 ± 113 b	555 ± 75 b

Total Carbon Partitions [Model 2: Slope Gradient and Relative Elevation]

	Control	*Conventional*	*Grid Method*	*Soil Method*
1.3	362 ± 143 a	407 ± 110 a	433 ± 139 a	444 ± 96 a
1.7	459 ± 55 b	448 ± 95 a b	494 ± 74 a b	471 ± 87 a b
2.0	512 ± 92 b	476 ± 92 b	526 ± 64 b	518 ± 46 b
2.8	563 ± 55 c	529 ± 93 b	557 ± 116 b	567 ± 64 c

Total Carbon Partitions [Model 3: Soil Phototone, Slope Gradient and Relative Elevation]

	Control	*Conventional*	*Grid Method*	*Soil Method*
1.3	392 ± 141 a	403 ± 106 a	425 ± 122 a	451 ± 103 a
1.7	442 ± 87 a	450 ± 100 a b	486 ± 87 b	462 ± 87 a b
2.0	523 ± 90 b	476 ± 96 a b	541 ± 52 b	507 ± 50 b
3.4	564 ± 59 b	524 ± 90 b	559 ± 114 b	568 ± 58 c

Soil Nitrate Partitions [Model 1: Trend Surface]

	Control	*Conventional*	*Grid Method*	*Soil Method*
12.7	437 ± 178 a	447 ± 120 a	459 ± 109 a	476 ± 84 a
14.1	488 ± 95 a	419 ± 123 a	497 ± 84 a	505 ± 111 a
16.4	496 ± 117 a	469 ± 57 a	510 ± 110 a	510 ± 74 a
29.9	503 ± 101 a	516 ± 69 a	540 ± 111 a	518 ± 104 a

Soil Nitrate Partitions [Model 2: X, Y, Slope Gradient and Soil Photographic Tone]

	Control	*Conventional*	*Grid Method*	*Soil Method*
12.8	387 ± 169 a	388 ± 120 a	491 ± 131 a	455 ± 77 a
14.4	485 ± 102 b	433 ± 110 b	478 ± 88 a	503 ± 72 b
16.7	498 ± 94 b	496 ± 88 b	500 ± 86 a	548 ± 102 b
28.1	545 ± 60 b	496 ± 71 b	541 ± 121 a	535 ± 97 b

Mean Economic Returns across landscape partition classes within the same treatment with the same letter are not significantly different at p = 0.05

Table 21–5. Economic net returns ($ ha^{-1}) by mapping strategy. Net return differences across treatments by partition are presented (mean ± std dev). A least squares method was used to test for differences between means.

Total Carbon Partitions [Model 1: Soil Photographic Tone]

	1.5	1.8	2.0	2.6
Control	387 ± 155 a	480 ± 75 a	480 ± 93 a b	578 ± 68 a
Conventional	423 ± 112 a	468 ± 66 a	421 ± 153 a	498 ± 95 b
Grid Method	439 ± 123 a	492 ± 82 a	520 ± 77 b	554 ± 113 a b
Soil Method	477 ± 96 a	433 ± 77 a	537 ± 67 b	555 ± 75 a b

Total Carbon Partitions [Model 2: Slope Gradient and Relative Elevation]

	1.3	1.7	2.0	2.8
Control	362 ± 143 a	459 ± 55 a	512 ± 92 a	563 ± 55 a
Conventional	407 ± 110 a	448 ± 95 a	476 ± 92 a	529 ± 93 a
Grid Method	433 ± 139 a	494 ± 74 a	526 ± 64 a	557 ± 116 a
Soil Method	444 ± 96 a	471 ± 87 a	518 ± 46 a	567 ± 64 a

Total Carbon Partitions [Model 3: Phototone, Slope Gradient and Relative Elevation]

	1.3	1.7	2.0	3.4
Control	392 ± 141 a	442 ± 87 a	523 ± 90 a	564 ± 59 a
Conventional	403 ± 106 a	450 ± 100 a	476 ± 96 b	524 ± 90 a
Grid Method	425 ± 122 a	486 ± 87 a	541 ± 52 a	559 ± 114 a
Soil Method	451 ± 103 a	462 ± 87 a	507 ± 50 a b	568 ± 58 a

Soil Nitrate Partitions [Model 1: Trend Surface]

	13	14	16	30
Control	437 ± 178 a	488 ± 95 a	496 ± 117 a	503 ± 101 a
Conventional	447 ± 120 a	419 ± 123 b	469 ± 57 a	516 ± 69 a
Grid Method	459 ± 109 a	497 ± 84 a b	510 ± 110 a	540 ± 111 a
Soil Method	476 ± 84 a	505 ± 111 a b	510 ± 74 a	518 ± 104 a

Soil Nitrate Partitions [Model 2: X, Y, Slope Gradient and Soil Photographic Tone]

	13	14	17	28
Control	387 ± 169 a	485 ± 102 a	498 ± 94 a	545 ± 60 a
Conventional	388 ± 120 a	433 ± 110 a	496 ± 88 a	496 ± 71 a
Grid Method	491 ± 131 b	478 ± 88 a	500 ± 86 a	541 ± 121 a
Soil Method	455 ± 77 a b	503 ± 72 a	548 ± 102 a	535 ± 97 a

Mean Economic Returns across treatments within landscape partition class with the same letter are not significantly different at p = 0.10

CONCLUSIONS

This study indicates that statistically significant grain yield and economic profit variations exist within a farming landscape. Comparisons of the two variable rate mapping strategies used in this study with the uniform nitrogen application rate revealed distinct trends favoring variable rate applications, but minimal statistically significant economic benefits.

The experimental design incorporating predetermined variable rates imposes a restriction on statistical analysis. This restriction eliminates the possibility of accurately determining grain yield response variations and economically optimum fertilizer application rates. A better design is to subdivide the field into transects and apply a uniform, but different rate of fertilizer to each transect (Malzer et al., 1994). This modification facilitates crop response evaluation across all landscape positions.

Large differences in grain sample size diminish the reliability of comparisons across treatments. Increasing the number of grain samples and maintaining uniform sample dimensions would minimize this source of error.

The nutrient variability models demonstrated in this study are powerful tools for segmenting the landscape into zones of relative homogeneity. Specifically, DEM derived attributes and aerial imagery serve as excellent covariates and enhance the resolution of these spatial models. Additionally, this method requires fewer soil samples than local interpolation techniques, such as kriging or simple distance weighted interpolation.

ACKNOWLEDGMENTS

The combined efforts of Jim Schnyder (collaborating farmer; Marshall, MN), Leon Kitzer (local fertilizer distributor, Harvcst States; Marshall, MN), Leo Langer (crop consultant, CENTROL; Cottonwood, MN) and the Marshall High School Ag. program made it possible to complete this study. Special thanks to Dean Fairchild and Roger Knutson of Soil Teq, Inc. for their humor, technical skills and organizational assistance.

REFERENCES

Anderson, J.L., P.C. Robert, and R.H. Rust. 1992. Productivity factors and crop equivalent ratings for soils. AG-BU-2199. Revised 1992, MN Ext. Serv., St. Paul.

Bruce, R.R., W.M. Snyder, A.W. White, Jr., A.W. Thomas, and G.W. Langdate. 1990. Soil variables and interactions affecting prediction of crop yield pattern. Soil Sci. Soc. Am. J. 54:494–501.

Burrough, P.A. 1986. Chapter 8: Methods of spatial interpolation. p. 147–166. *In* Principles of geographical information systems for land resources assessment, Monograph on Soils and Resources Survey No. 12. Oxford Sci. Publ.

Carr, P.M., G.R. Carlson, J. S. Jacobsen, G. A. Neilsen, and E. O. Skogley. 1991. Farming soils, not fields: A strategy for increasing fertilizer profitability. J. Prod. Agric. 4:57–61.

Griffith, D.A., and C.G. Amrhein. 1991. Chapter 15, An introduction to trend surface models. p. 406–416. *In* Statistical Analysis for Geographers. Prentice-Hall, New Jersey.

Honeycutt, C.W., R.D. Heil, and C.V. Cole. 1990a. Climate and topographic relations of three Great Plains soils: I. Soil morphology. Soil Sci. Soc. Am. J. 54:469–475.

Honeycutt, C.W., R.D. Heil, and C.V. Cole. 1990b. Climate and topographic relations of three Great Plains soils: II. Carbon, nitrogen, and phosphorous. Soil Sci. Soc. Am. J. 54:476–483.

Kleiss, H.J. 1970. Hillslope sedimentation and soil formation in Northeastern Iowa. Soil Sci. Soc. Am. Proc. 34:287–290.

Koch, G.G., I.A. Amara, G.W. Davis, and D.G. Gillings. 1982. A review of some statistical methods for covariate analysis of categorical data. Biometrics 38:563–595.

Lane, P.W., and J.A. Nelder. 1982. Analysis of covariance and standardization as instances of prediction. Biometrics 38:613–621.

Malo, D.D., and B.K. Worcester. 1975. Soil fertility and crop responses at selected landscape positions. Agron. J. 67:397–401.

Malzer, G.L., T.J. Graff, P.C. Robert, D.R. Huggins, B.J. Fuchs, W.H. Thompson, G.L. Gaalaas, and T.W. Bruulsema. 1994. Site specific crop management: N-response by soil condition at Hildreth site 1993. *In* Field Research in Soil Science 1994. p. 172–176. (Soil Series #140) Misc. Publ. 83–1994. Minn. Agric. Exp. Stn. U of MN, St. Paul.

McBratney, A.B., G.A. Hart, and D. McGarry. 1991. The use of region partitioning to improve the representation of geostatistically mapped soil attributes. J. Soil Sci. 42:513–532.

Milfred, C.J., and R.W. Kieffer. 1976. Analysis of soil variability with repetitive aerial photography. Soil Sci. Soc. Am. J. 40:553–557.

Miller, M.P., M.J. Singer, and D. R. Neilsen. 1988. Spatial variability of wheat yields and soil properties on complex hills. Soil Sci. Soc. Am. J. 52:1133–1141.

Oliver, M.A., and R. Webster. 1989. A geostatistical basis for spatial weighting in multivariate classification. Mathematical Geology 21(1):15–35.

Rehm, G., and M. Schmitt. 1989. Fertilizing corn in Minnesota. AG-FO-3790. MN Ext. Serv., St. Paul.

Robert, P.C., W.H. Thompson, and D. Fairchild. 1992. Soil specific anhydrous ammonia management system. p. 418–426. *In* Automated Agriculture for the 21st Century, 1991 ASAE Proc., Chicago, IL

Ruhe, R.V. 1960. Elements of the soil landscape. 7th Int. Congress of Soil Sci., Madison, WI. 23:165–169.

SAS. 1988. Chapter 2, Introduction to analysis of variance procedures. *In* SAS/STAT User's Guide, Release 6.03 ed. SAS Institute, Cary, NC.

Snedecor, G.W., and W.G. Cochran. 1980. Chapter 20, Two-way tables with unequal numbers and proportions. p. 414–433. *In* Statistical Methods,

7th ed. The Iowa State University Press, Ames.

Soil Survey Staff, USDA-SCS. 1978. Soil Survey of Lyon County, MN. U.S. Gov. Printing Office.

Spratt, E.D., and R.N. McIver. 1972. Effects of topographical positions, soil test values, and fertilizer use on yields of wheat in a complex of black chernozemic and glaysolic soils. Can. J. Soil Sci. 52:53–58.

Zheng, F., and H. Schreier. 1988. Quantification of soil patterns and field soil fertility using spectral reflection and digital processing of aerial photographs. Fert. Research 16:15–30.

[illegible]

[illegible]

[illegible]

22 Continuous Models of Soil Variation For Continuous Soil Management

A. B. McBratney
B. M. Whelan

Department of Agricultural Chemistry and Soil Science
University of Sydney
Sydney, NSW 2006, Australia

A generalized spatial model for continuous soil variables is proposed to incorporate previously known and real–time sensed data. A particular form of the model is discussed whereby the known information is continuous soil class data constructed from fuzzy-set theory. Techniques required to obtain continuous soil class information from discontinuous soil maps are also discussed. The importance of utilizing accurate data in the system, and the consequences of inaccuracy on financial management decisions, is highlighted by fertilizer response–waste examples.

INTRODUCTION

The spatial variation of agronomically significant soil attributes has become a subject of importance to the farming and wider communities. Of concern is the accepted notion that agriculturally productive land can be treated as a relatively homogeneous resource. Such an assumption can lead to inappropriate input resource application and associated financial, environmental and social costs. Site-specific treatment (continuous soil management) is being proffered as the remedy to many of these environmental and resource-use efficiency problems (Larson & Robert 1991). This soil management system is based on matching input resource application and soil management techniques with present resource levels as they vary across a site.

Achieving this synchronization requires a base system to describe, and delineate suitable responses to, the inherent soil variation in the agricultural landscape. Such a system comprises a union of data acquisition operations, information processing and decision formulation procedures. Ensuring this system produces a meaningful and useful treatment of the soil variation, depends largely on the ability to accurately model the overall soil variation from the observed data.

HISTORICAL PERSPECTIVE

The premise that soil heterogeneity influences the productive potential of agricultural land can not be regarded as a new concept. Equally, the knowledge that measuring the degree of heterogeneity and using this as base data with which to manipulate management operations is long held.

Haines and Keen (1925) employed a dynamometer to record the continuous variation in draw-bar pull required during parallel transects of a field. They were effectively measuring the resistance of the soil to the plow, a factor influenced by the soil cohesion, plasticity and surface friction which indirectly reflects a wide range of other soil variables such as moisture content, organic matter levels and soil structure.

Figure 22–1 shows a continuous map produced by the study, with lines of equal force (isodynes) used to depict the variation in the field. Their conclusions as to the influential effect of this inherent soil heterogeneity on field experiments and crop growth are fundamental to soil-specific management.

Russell (1959) understood this effect of soil variability and speculated (with considerable elegance) on the extent to which mechanical and chemical changes could be imposed on the soil:

"Automation may yet invade the farm, commercial market garden and nursery even though an escapist reaction may keep it out of the private garden. An electronic controller programmed to respond to changing weather and soil conditions as revealed by self recording instruments, and to changing plant conditions recorded by photographic devices operating light cells, may yet by

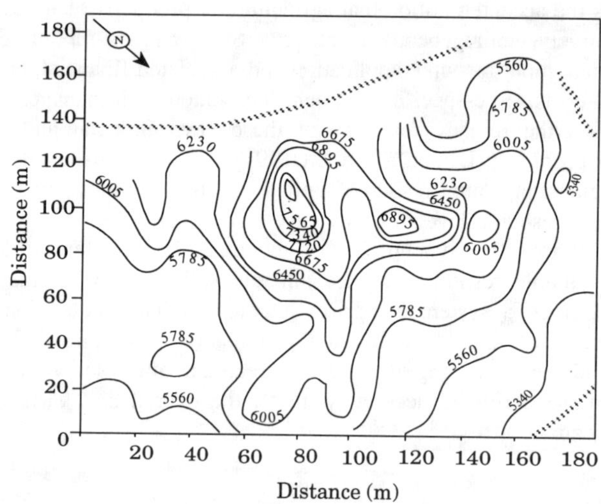

Fig. 22–1. Continuous map of draw bar pull (newtons) after Haines and Keen (1925).

remote control send the proper cultivating implement to the proper spot and direct its operations, select the proper chemical agent for each particular purpose and direct a discharge of the proper amount to the proper place; never itself making a mistake but being eternally watchful to correct those of any human intruder who thinks he knows better."

From a soil surveyor's perspective, McBratney (1984) alluded to the benefits of soil-specific management when discussing the future of geostatistical applications to soil survey:

"I imagine that geostatistical soil survey will become very inefficient and cumbersome if information about a great many very local areas of a much larger region is required i.e. where many thousands of samples have to be taken and analyzed. Technological advances in microchips and sensing and a growing need by farmers for more and more local information to optimize their operations leads me to the conclusion that very detailed surveys will be performed in the future using some sort of proximal sensing of soil in real-time, with perhaps some concomitant ameliorative procedure. The need for sampling will become unnecessary."

The technology is now available to tackle the operations suggested above (Borgelt 1993). The real challenge is to develop the ability to observe and respond to continuous soil variation in a way that is accurate and cost-effective.

A GENERALIZED SPATIAL MODEL

A generalized spatial model for continuous soil variables s at location x ($s(x)$) is proposed (1).

$$s(x) = f(x) + e(x) \tag{1}$$

where:

$f(x)$ = fitted model of $s(x)$
$e(x)$ = residuals of $s(x)$.

Further, $f(x)$ is described by (2)

$$f(x) = a(x) + z(x) \tag{2}$$

where:

$a(x)$ = auxiliary data
$z(x)$ = new real-time input data.

The models may be generalized linear models (McCullagh & Nelder 1989) or non-linear generalized additive models (Hastie & Tibshirina 1990), tree-based models or neural network models.

Auxiliary data would be interpolated in real-time from previously geo-referenced site knowledge e.g., digitized soil maps, digital elevation models (Odeh et.al., 1994), seasonal aerial infrared scans of the site or other site-specific data. New input data, providing good estimates of short-range variation, can be sourced from calibrated real-time sensing devices in use at the time of the soil management operation e.g., electromagnetic induction and tine-mounted nitrogen or moisture content sensors. The provision of the calibration equations and the optimal combination of auxiliary and new real-time input data offers a new challenge and paradigm for soil survey and land evaluation.

To begin with, a form of auxiliary data is required that will augment the continuous nature of the new input data. Fuzzy sets offer a way of providing this type of data.

FUZZY SETS

Quantitative interpretations of traditional soil maps assume the units are hard classes and the best predictor of any soil property is the class mean. This is based on the idea that the soil changes relatively sharply at the boundaries of map units.

Since the mid-1980's and the advent of Geographic Information Systems (GIS) (Burrough, 1986) existing soil maps have been digitized, so that the Boolean methods of GIS could be utilized. Around the same time, soil and other environmental scientists began questioning the limitations of what was essentially a hard-set approach and began investigating the potential of fuzzy sets. Fuzzy sets, formalized by Zadeh (1965), allow quantification of the degree of membership in given classes and as such are intuitively appealing for describing the continuous variation of soil across the landscape (McNeill & Freiberger, 1993).

In traditional soil management many of the logical operations are wasteful of information. The construction of fuzzy decision rules would allow much more information to be carried through to the final decision. This process has been developed on individual soil attributes for land evaluation [Burrough (1989); Tang et al.,(1991); Triantafilis & McBratney, (1993)] and has further implications for environmentally-related map operations in Geographic Information Systems [Burrough et al., (1992); Heuvelink & Burrough (1993)].

SOIL-MAP FUZZIFICATION ALGORITHM

A process for fuzzifying conventional soil maps can be described in 4 steps:

(1) Rasterise map (with c classes)

The rasterisation process should, as well as creating information on the class (primary class) at each pixel, also record for each pixel the nearest-neighbor (secondary) class, and should label pixels where n, (n>2), classes meet, as n-tuple points.

(2) Apply a "distance transform" to the image, .i.e, the distance (d) from the boundary of each map areal is estimated to produce a continuous surface with valleys at the boundaries.

(3) Apply a distance-membership function to the distance data. It seems intuitive that the membership on the boundary where two classes meet should be one half in each and the membership should increase away from the boundary. The rate of increase may be related to the maximum distance, d_{max}, from a boundary in the map areal, so that long thin units have steep membership gradients at their edges. Given d_{max}, the distance transform image and a function such as Fig. 22–2 which has the form of an exponential model rising to unity at

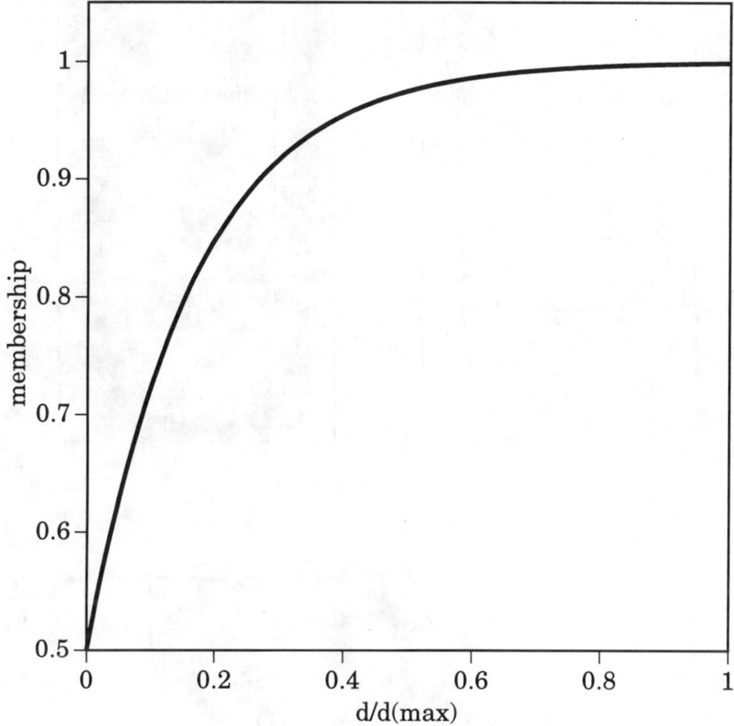

Fig. 22–2. Primary class membership as a function of relative distance to soil unit boundary.

at 0.95 d_{max}, the distance image can be transformed to membership in the primary class, membership in the secondary class can be found by subtraction of the membership from unity. The result is c maps or images which sum to unity pixelwise but which have zero membership in all but c-2 of the classes, these reflect memberships based on their distance from the boundary.

(4) There still remains a problem at places where > two classes meet. After the membership boundary model has been applied, the n-tuple points should be given memberships of $1/n$ in the n appropriate classes. A moving average filter should be applied to "spread out" the fuzziness around these points. It is suggested that a 3–by–3 moving average filter be applied to each class in neighborhoods where there is a non-zero membership in > two classes. Memberships are adjusted pixel-wise to assure they sum to unity. Repeat this filter r times, r should be about 5, but an objective way of deciding this needs to be found.

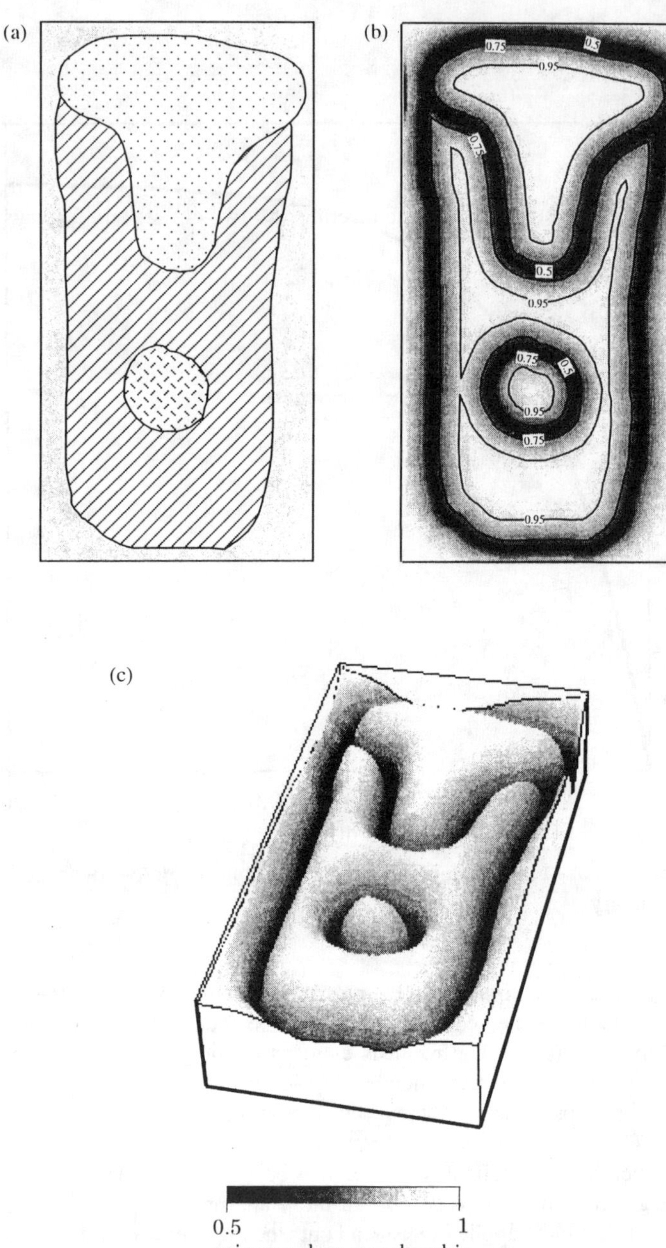

Fig. 22–3. Depicting the soil map fuzzification process: (1) hard-set soil class map; (b) fuzzy-set classification with primary class membership contours; (c) perspective diagram of the fuzzy soil map.

A Simple Example of the Algorithm

Figure 22–3(a) depicts three soil mapping units representing traditional hard-set classification. Following rasterisation and distance transformation of the "map" using a program developed in-house, the distance-membership function (Fig. 22–2) was applied to produce the fuzzy-soil map (Fig. 22–3(b)). Contours display the primary class membership.

This simple example avoids the meeting of > two classes and displays the primary class membership so that the basic procedure can be shown clearly. Secondary class membership can be calculated as above.

The continuous nature of soil variability can now be represented by a perspective diagram of the fuzzy-soil map (Fig. 22–3(c)). Such a map provides greatly increased subtlety in depicting changes in soil type, a feature that successful site-specific soil management demands.

USING THE FUZZY SOIL MAP

The beneficial use of fuzzy soil maps for continuous soil management operations can by demonstrated by considering the following. In traditional fertilizer management, the required applications may be made based on a mean site soil nutrient level (3).

$$s(x) = m_i(x) + e(x) \qquad (3)$$

where:

$m_i(x)$ = mean site soil nutrient level for soil class i

Using the fuzzy soil map, the required applications across the site can be calculated using the sum of all membership values across all classes (4).

$$s(x) = m_i(x) + e(x) \qquad (4)$$

where:

mi = membership values in the ith of k classes

As an example of this procedure the memberships in three sequential soil classes were calculated for a soil variable along a theoretical field transect using the membership function described by Fig. 22–2. A plot of the class membership along the transect is shown in Fig. 22–4.

Applying Equation (4) to the class memberships and soil variable data for the transect produces a continuous, smoothed trace of the soil variable levels. Figure 22–5 compares these results with the levels obtained using Equation 3 in conjunction with a hard-set soil map. The information obtained from the fuzzy soil map, when combined with real–time sensed data, may provide a more sensitive and accurate information base for fertilizer application decisions.

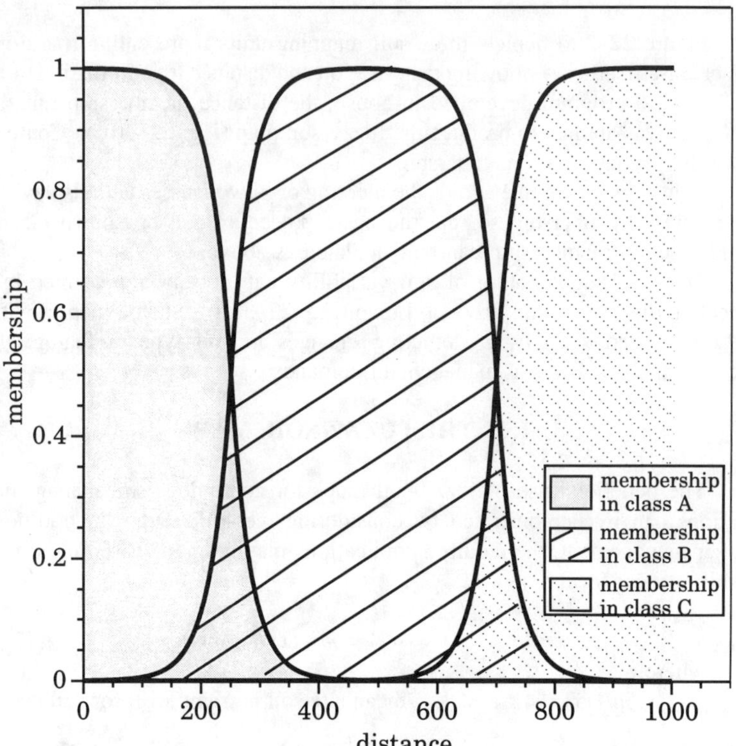

Fig. 22–4. Fuzzy soil-class membership for a theoretical field transect.

SOME ECONOMICS

Site–specific management appears to be a logical approach to utilizing fragile and valuable resources. The question of its acceptance will depend on the benefits displayed, and initially the economic benefit will be weighted much higher in any on–farm decision. Costing the social and environmental benefits is extremely difficult (Wollenhaupt & Buchholz 1993) but basic analysis of income–expenditure can be used to highlight any financial savings the system may provide.

The results from a simple one dimensional model are presented here, based on broadacre sorghum grown on lateritic red–brown earth (rhodic palexeralf). The initial parameters used were: sorghum @ \$150.00 tonne^{-1};[1] Single super phosphate (8.6% available phosphorous (P)) @ \$247.00 tonne^{-1} = \$2.87 (kg available P)$^{-1}$; maximum grain yield @ 2.5 tonne ha^{-1}.

[1]"The Land" February 10, 1994. Rural Press, Richmond, Australia.

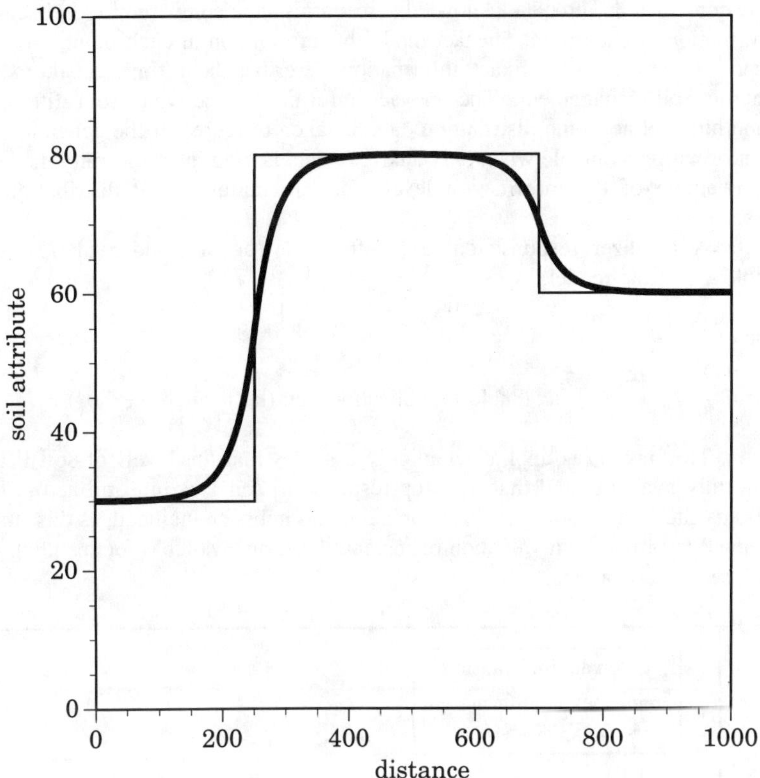

Fig. 22–5. Comparison of hard-set classification with fuzzy-set classification for soil attribute levels across a theoretical field transect.

The Model

The model tests the difference in financial return obtained from differential fertilization of a theoretical 1000 ha site using varied levels of knowledge of the initial soil P content. P distributions were generated using a first-order autoregressive model (5).

$$P s_i = b\ 10^{s_i} \qquad (5)$$

where:

$P s_i$ = soil phosphate level (kg ha^{-1})
b = median regulating coefficient
s_i = $a\ s_{i-1} + k\ h$
a = autoregressive parameter
k = coefficient
h = random sample from a normal distribution (N(0,1))

Five log–normally distributed populations of 1000 initial soil P levels were constructed. Three data knowledge scenarios were compared for each: exact information at each point, inexact model of information at each point, and the mean value of the site. Exact information describes the ultimate goal of site-specific soil management. The inexact information, derived from fitting a smoothing spline to the distribution data, is taken to represent the general level of information available when only auxiliary data is used in soil fertility models. An example of the information levels for one initial soil P distribution is presented as Fig. 22–6.

A fertilizer response curve (6) from Helyar & Godden (1977) was employed in the model.

$$y^R = 2500\,[1 - e^{-0.04\,(f + 8.8)}] \qquad (6)$$

where:

y^R = yield (kg ha^{-1})
f = available P application rate (kg available P ha^{-1})

The model, as applied, implicitly assumes that the levels of soil P are uniformly available and that the crop response to soil P is uniform across the 1000 ha site. The cost of fertilizer spreading has not been included, as this study is aimed at providing a valuation of increased returns available for spending on site-specific operations.

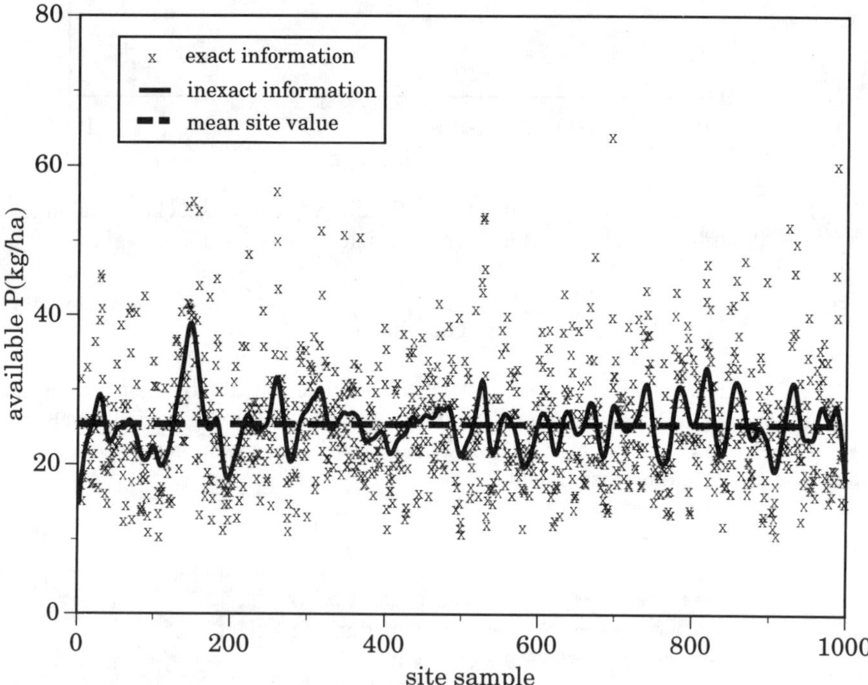

Fig. 22–6. Theoretical available P distribution and associated information knowledge levels.

Table 22–1. Simple economic analysis of information-based fertilizer programs. Autoregressive initial P distribution with a range of mean values.

	Available P distribution (kg/ha) min. = 4.45 med. = 10.52 mean = 11.04 max = 27.73			Available P distribution (kg/ha) min. = 10.24 med.= 24.21 mean = 25.40 max = 63.78			Available P distribution (kg/ha) min. = 12.90 med.= 30.52 mean = 32.03 max = 80.40		
	Exact information	Inexact information	Mean application	Exact information	Inexact information	Mean application	Exact information	Inexact information	Mean application
Total available P required (kg)	21630.4	21630.4	21630.4	8377.2	7377.9	7301.2	4340.85	1963.74	687.74
% site fertilized	100	100	100	82.6	97.8	100	60.6	58.9	100
Available P wastage (kg/ha)	—	1.16	1.36	—	1.66	2.05	—	0.53	0.29
Yield wastage (kg/ha)	—	23.6	27.7	—	58.95	69.97	—	75.56	92.34
Yield gain (kg/ha)	—	20.16	23.13	—	28.25	34.48	—	9.36	5.34
Return ($/ha)	312.88	309.03	308.29	350.94	344.45	342.82	362.53	357.91	359.16

Table 22–2. Simple economic analysis of information-based fertilizer programs. Autoregressive initial P distribution with a range of variance values.

	Available P distribution (kg/ha) min. = 16, med.= 25.11, mean = 11.04, max = 40			Available P distribution (kg/ha) min. = 10.24, med.= 24.21, mean = 25.40, max = 63.78			Available P distribution (kg/ha) min. = 3.45, med. = 21.91; mean = 25.40, max = 119.5		
	Exact information	Inexact information	Mean application	Exact information	Inexact information	Mean application	Exact information	Inexact information	Mean application
Total available P required (kg)	7385.4	7301.2	7301.3	8377.2	7377.9	7301.2	10756.4	7841.53	7301.25
% site fertilized	96.1	100	100	82.6	97.8	100	76.5	89.1	100
Available P wastage (kg/ha)	—	1.21	1.42	—	1.66	2.05	—	1.75	2.21
Yield wastage (kg/ha)	—	26.53	31.22	—	58.95	69.97	—	118.06	125.63
Yield gain (kg/ha)	—	21.27	24.64	—	28.25	34.48	—	28.94	6.88
Return ($/ha)	353.79	349.77	348.96	350.94	344.45	342.82	344.11	334.08	334.37

A desired P level for maximizing financial return (32.7 kg available P ha^{-1}) was obtained from marginal analysis using Eq. 6. Total available P required for the site, and the percentage of the site fertilized, was calculated by differencing the known and desired levels. When other than exact initial P fertility information was used, the residuals obtained from comparison with the exact levels were used to calculate points of under or over fertilization. Under-fertilized sites received a yield penalty based on (6) and over–fertilized sites had excess fertilizer costs tempered by increased yield achieved from the higher fertility.

Table 22–1 displays the results of applying the model to three "fields" with similar soil P distributions but differing mean values. Exact information produced an increase in returns over the less detailed information ranging from $3.38 ha to $8.11 ha at all mean P levels. Importantly, the gain in returns for more precise information appears to increase, and subsequently, decrease as the mean of the P distribution falls from that of the desired fertilizer level. This raises the possibility of a zone of mean P fertility that is best suited to specific fertilizer application.

By maintaining a constant mean P level and modifying the distribution variance, the influence of within-site variability was examined. Table 22–2 suggests that the more variable a field, under these conditions, the greater the return for detailed information (up to $10.03 ha).

Interestingly, the inexact modelled information provided a small gain in returns over the mean application in all but the highly variable field and the field with a mean soil P status close to the desired level. In a highly variable or well fertilized field, the site mean may be an economical estimate for constructing P fertilizer programs.

This simple economic model provides a basic insight into the improvement in returns offered by attention to detail in a P fertilizer application regime. Returns for exact information are greater than the $1.80 ha to $3.15 ha quoted by Wollenhaupt and Buchholz (1993) as the cost of operating a site-specific fertilizer program. Inexact modelling (using auxiliary data only) to the degree in this example would not support such a P fertilizer program. Improving auxiliary data using fuzzy soil maps and incorporating accurate real time sensed data appears imperative for successful site-specific management.

CONCLUSIONS

The success of a site-specific soil management depends on the accuracy with which the variation in significant soil parameters are monitored and adjusted. Fuzzy soil-maps will enable the true soil variation to be more precisely identified, thereby improving the detail available for fertilizer application decisions and increasing the probability and profitability of more site-specific management programs.

REFERENCES

Borgelt, S.C. 1993. Sensing and measurement technologies for site-specific management. p. 141–157. *In* P.C. Robert, et al. (ed.) Proceedings of soil-specific crop management: A workshop on research and development issues. American Society of Agronomy. Madison, WI.

Burrough, P.A. 1986. Principles of geographic information systems for land resources assessment. Oxford University Press.

Burrough, P.A. 1989. Fuzzy mathematical methods for soil survey and land evaluation. J. Soil Science 40:477–492.

Burrough, P.A., R.A. MacMillan, and W. van Deursen. 1992. Fuzzy classification methods for determining site suitability from soil profile observations and topography. J. Soil Sci. 43:193–210.

Haines, W.B., and B.A. Keen. 1925. Studies in soil cultivation. II. A test of soil uniformity by means of dynamometer and plough. J. Agric. Sci. 15:387–394.

Hastie, T.J., and R.J. Tibshirani. 1990. Generalized additive models. Chapman and Hall. London.

Helyar, K.R., and D.P. Godden. 1977. The biology and modelling of fertilizer response. J. Aust. Inst. Agric. Sci. 43:22–30.

Heuvelink, G.B.M., and P.A. Burrough. 1993. Error propagation in cartographic modelling using Boolean logic and continuous classification. International Journal of Geographical Information Systems. 7:231–246.

Larson, W.E., and P.C. Robert. 1991. Farming by soil. p. 103–112. *In* R. Lal, et al. (ed.) Soil management for sustainability. Soil Water Conservation Society, Ankeny, IA.

McBratney, A.B. 1984. Geostatistical Soil Survey. Ph.D thesis. The University of Aberdeen.

McCullagh, P., and Nelder. 1989. Generalized linear models. 2nd edition. Chapman and Hall. London.

McNeill, D., and P. Freiberger. 1993. Fuzzy Logic. Bookman Press. Melbourne.

Odeh, I.O.A., A.B. McBratney, and D.J. Chittleborough. 1994. Further results on the spatial prediction of soil properties from attributes derived from a digital elevation model: heterotopic cokriging and regression-kriging models. Geoderma. (submitted).

Russell, E.J. 1959. The world of the soil. Collins, London.

Tang, H.J., J. Debaveye, D. Ruan, and E. van Ranst. 1991. Land suitability classification based on fuzzy set theory. Podology. 51:277–290.

Triantafilis, J., and A.B. McBratney. 1993. Application of continuous methods of soil classification and land suitability assessment in the lower Namoi valley. Div. of Soils Divisional Rep. 121, CSIRO Australia, Melbourne.

Wollenhaupt, N.C. and D.D. Buchholz. 1993. Profitability of farming by soils. p. 199–212. *In* P.C. Robert, et al. (eds.) Proceedings of soil-specific crop management: A workshop on research and development issues. American Society of Agronomy. Madison, WI.

Zadeh, L.A. 1965. Fuzzy sets. Information and Control. 8:338–353.

23 Microrelief And Spatial Variability of Some Selected Soil Properties on an Agricultural Benchmark Site In Quebec, Canada

C. Wang
M. C. Nolin
J. Wu

Centre for Land and Biological Resources Research
Agriculture Canada
Ottawa, Canada

Questions about soil quality changes and sustainability of production in Canada have not been seriously addressed until recently. Baseline soil data and long-term soil and farm management information are needed for assessing soil quality changes and the sustainability of agricultural production.

In the process of establishing soil baseline data for a benchmark site at St. Lawrence Lowland, Québec, it was observed that the spatial variability of many important soil chemical and physical properties was closely related to relief or landscape position. Within this nearly level site, there is a broad, low ridge with side slopes of 1% running east-west near the middle of the field. Simple correlation analysis indicated that this microrelief feature has a profound affect on the spatial distribution patterns of organic C content, pH and soil moisture within the site. Organic C content was correlated (at 1% level) with soil moisture of both 0–15 cm and 15–30 cm depths. Organic C was, however, negatively correlated (at 1% level) with soil pH in 0.01 M $CaCl_2$ and water. The results of this study indicate the need to monitor soil quality on a landscape basis.

INTRODUCTION

A network of 23 benchmark sites across Canada was selected and sampled between 1989 and 1992 for monitoring and assessing soil quality change on representative soils and landforms under typical agricultural practices (Centre for Land and Biological Resources Research, 1992). As part of establishing the baseline data, each benchmark site has very detailed soil and topographic maps, analytical data of soil properties that were systematically collected, as well as the detailed farming history of the site. One of the benchmark sites which represents typical farmland on level clayey marine deposits in the St. Lawrence Lowlands near Montreal has demonstrated a close spatial relationship between microrelief and some agronomically important soil properties. The spatial distribution patterns of organic C, pH and soil moisture in the rooting zone were profoundly affected by minor changes in relief. The objectives of this paper are to report such spatial relationships, examine some of the probable explanations, and discuss the implications to soil sampling for monitoring and assessing soil quality.

MATERIAL AND METHODS

Site

The 85 x 500 m benchmark site is located near St-Antoine, Quebec on Montreal Lowland (site 18-QU on Fig. 23–1) and was mapped as Providence series. It belongs to Orthic Humic Gleysol subgroup of the Canadian system of soil classification (Agriculture Canada Expert Committee on Soil Survey, 1987). It is a Typic Endoaquoll in U.S. Soil Taxonomy (Soil Survey Staff, 1992). The parent material was clayey marine deposit of the Late Wisconsinan. The site had an overall 0.5% slope from north to south with a broad ridge (about 1.2% slope) running east-west in the middle of the site (Fig. 23–2). The site has been under arable cropping for > 100 yr. Since 1980 the field was cropped to a corn-wheat-soybean-barley rotation. A minimum tillage system was implemented in 1988.

Methods

The Ap horizon (mostly 25 to 30 cm thick) was sampled in early May 1989 before planting of soybean on a 30 m x 30 m grid leaving a 10 to 15 m buffer strip along the edges of the site. The relative elevation was measured on all sampling points with a level and metric rod. Soil samples were air-dried, ground and sieved for laboratory analyses according to standard procedures (Sheldrick, 1984). All laboratory analyses were done according to procedures presented in an analytical manual by Sheldrick (1984) as follows: soil reaction (pH) in 0.01M $CaCl_2$ and in H_2O, per procedures 84–001 and 84–002 respectively; total carbon by LECO induction, per procedure 84–013; particle size distribution by pipette method, per procedure 84–026. Soil moisture content was determined gravimetrically and expressed on an oven dry basis at 105°C. Two-dimensional contour maps and three-dimensional diagram were produced by a computer-based graphics package, SURFER (Golden Software Inc., 1990). Simple correlation coefficients (Table 23–1) were calculated by a statistical software, CoStat (CoHort Software, 1990).

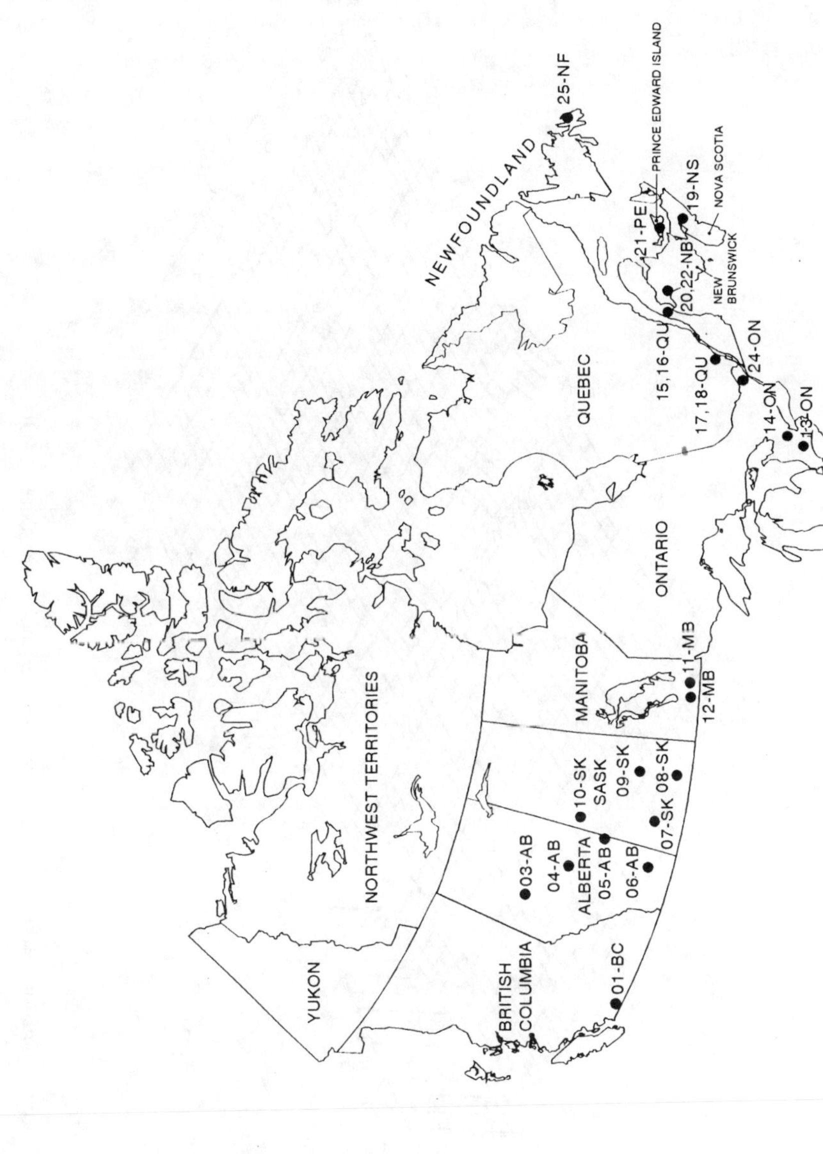

Fig. 23–1. General locations of the benchmark sites.

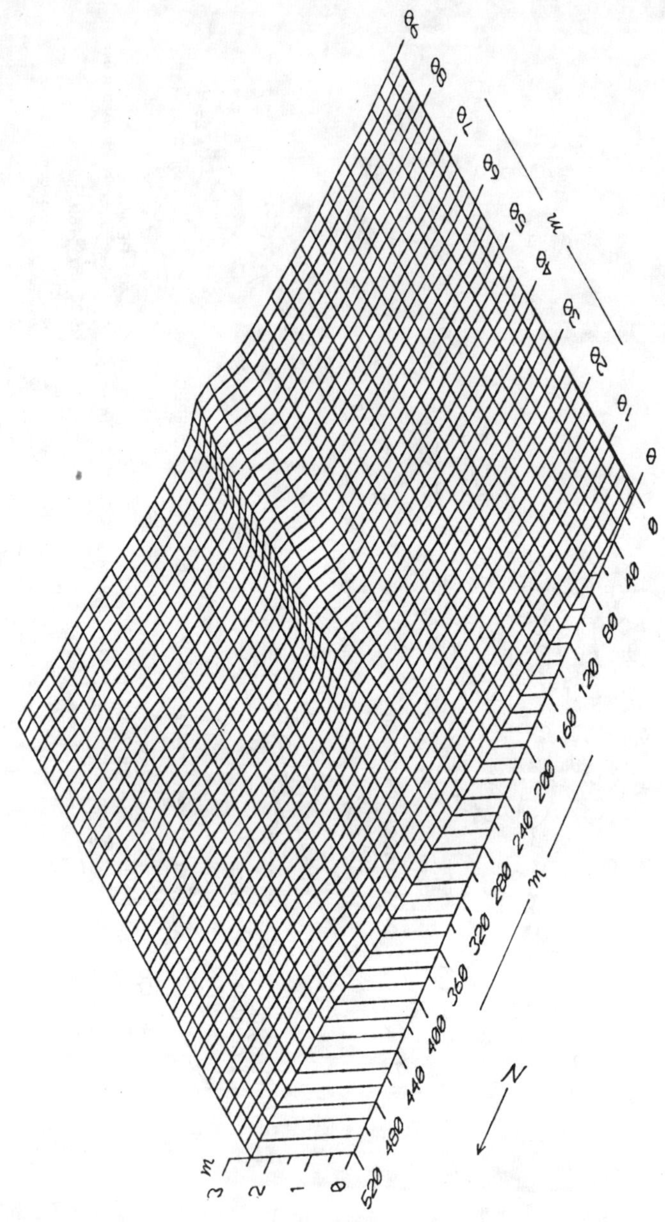

Fig. 23–2. Three dimensional topographic diagram of the studied site (site 18–QU).

Table 23–1. Simple correlation coefficients (r) of % Carbon, pH and moisture content.

	Carbon	pH $CaCl_2$	pH H_2O	Moist. 15
pH $CaCl_2$	-.63†			
pH H_2O	-.62†	.98†		
Moist. 15	.70†	-.28‡	-.26‡	
Moist. 30	.60†	-.14	-.09	.70†

† Significant at $P < 0.01$
‡ Significant at $0.05 < P < 0.1$

RESULTS

Carbon content ranged between 1.6 and 4.9. The highest carbon content was located near the center of the 350 m mark (Fig. 23–3). The carbon content decreased sharply in all directions to about 2.8% at the 520 m and 230 m marks. The carbon content below the 230 m mark was mostly about 2.5% decreasing gradually from east to west.

Values of pH (in $CaCl_2$) ranged between 5.2 and 6.5. The lowest pH was located near the east edge of the site at the 320 m mark (Fig. 23–4). It increased gradually westward to about 6.0 at the northwest edge of the site as well as at the west edge of the 260 m mark. Below 260 m, the pH increased gradually from 5.6 at the east edge of the site to 6.5 near the west end of the site. The pH in H_2O (Fig. 23–5) had a similar spatial trend except that the average value was about 0.5 unit higher than in $CaCl_2$.

Moisture content for both depths ranged between 22% and 30%. There were distinct differences on soil moisture distribution above and below the broad, low ridge for the 0–15 cm (Fig. 23–6) and 15–30 cm (Fig. 23–7) depths. Soil moisture was higher on the northern half of the site (above the ridge) for both 0–15 cm and 15–30 cm depths. For both depths, the highest moisture content (about 29%) was located near the middle of the 440 m mark while the moisture content was lower (25% or less) and relatively uniform below the ridge. There was an additional maximum (about 29%) of moisture content near the west edge of the field at the 300 m mark for the 15–30 cm depth.

Particle size distribution of Ap horizons varied within a relatively narrow range. Typically the surface soil (Ap) had 40–50% clay, 40–50% silt and < 15% sand.

Simple correlation coefficients between carbon, pH and soil moisture are shown in Table 23–1. Carbon was significantly correlated to soil moisture at both the 0–15 cm and 15–30 cm depths ($r = .70$ and $.60$, respectively). Carbon was, however, significantly negatively correlated to both pH in 0.01M $CaCl_2$ ($r = -.63$) and water ($r = -.62$).

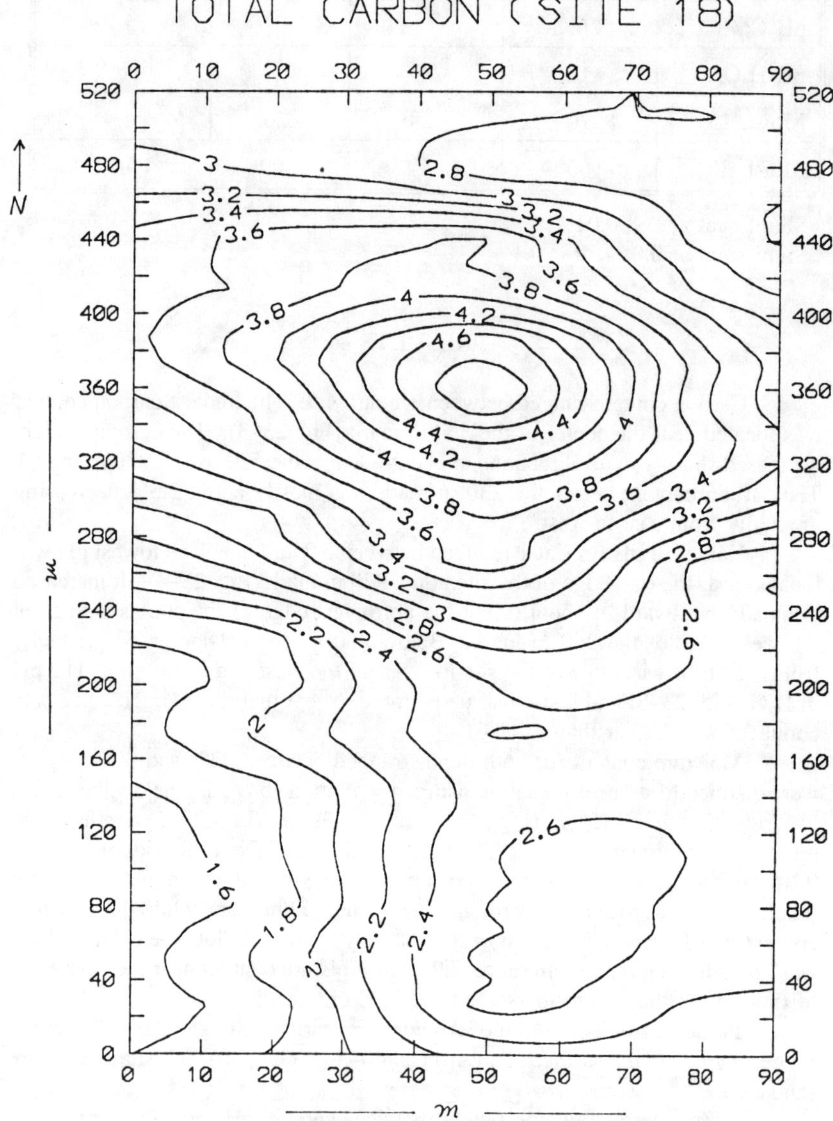

Fig. 23–3. Contour map of carbon content (%).

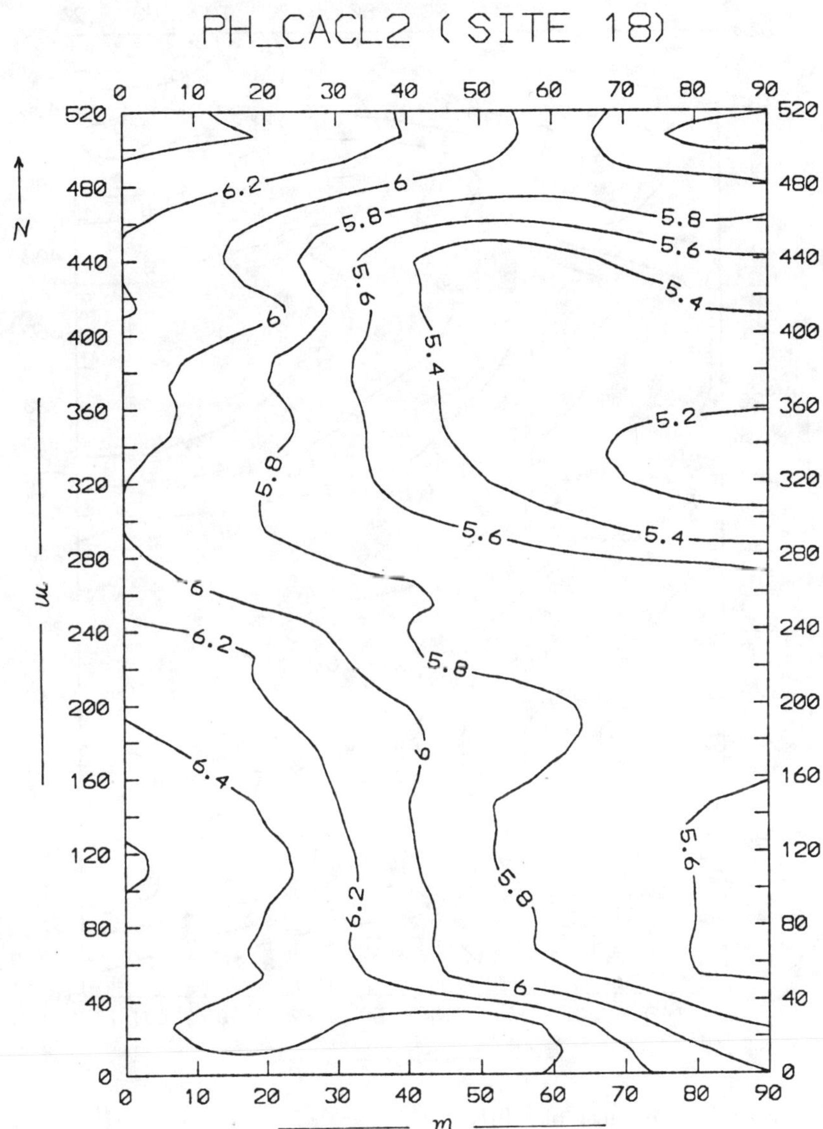

Fig. 23–4. Contour map of Ph ($CaCl_2$).

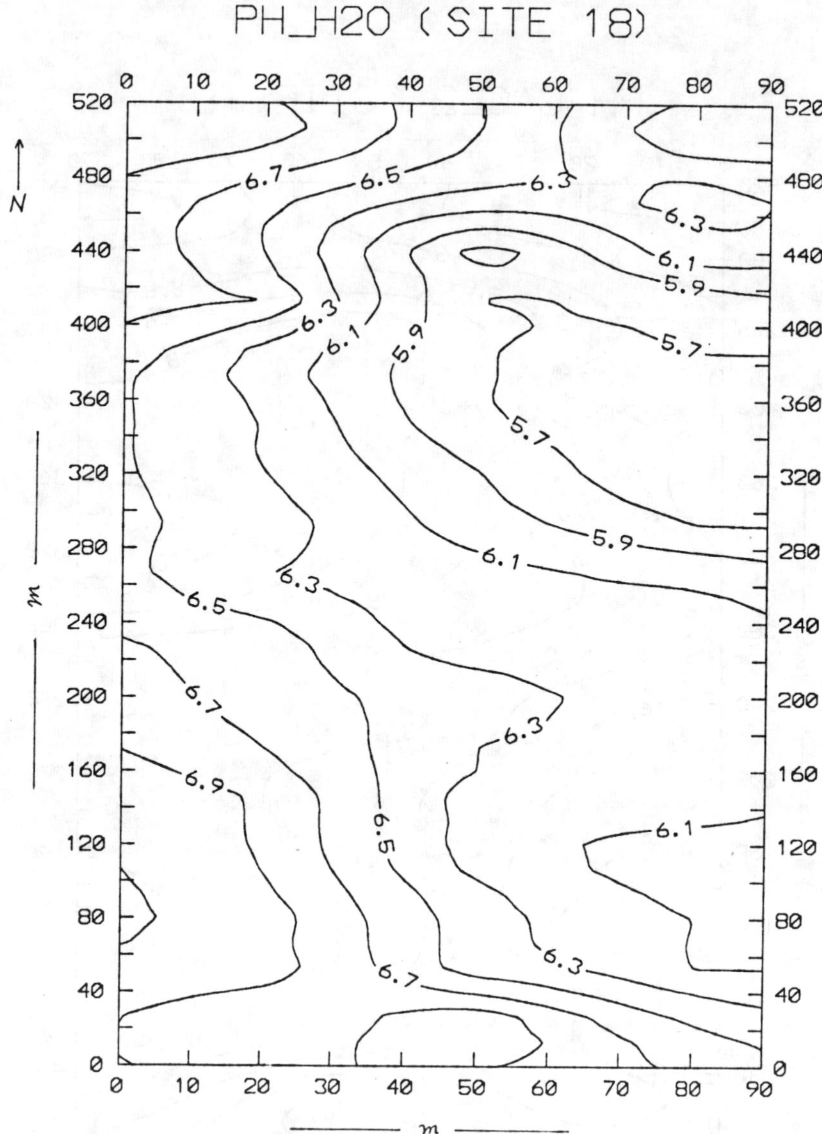

Fig. 23–5. Contour map of pH (H_2O).

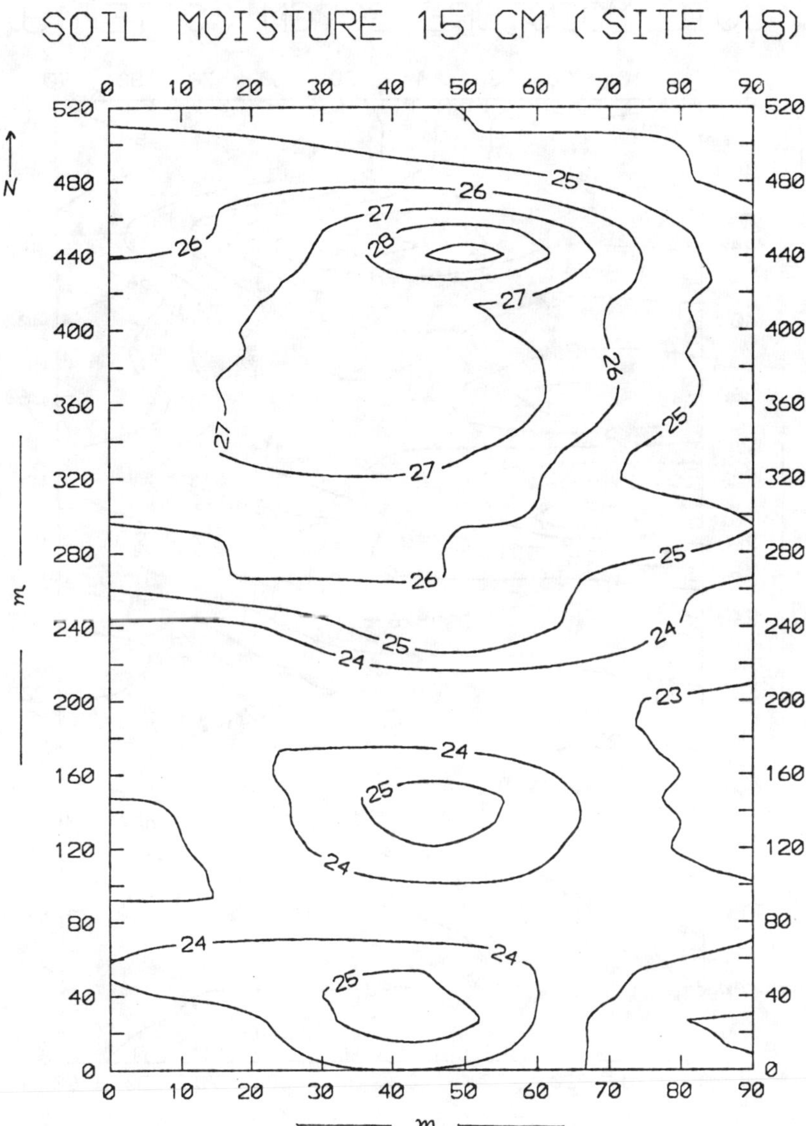

Fig. 23–6. Contour map of soil moisture (%) in 0–15 cm.

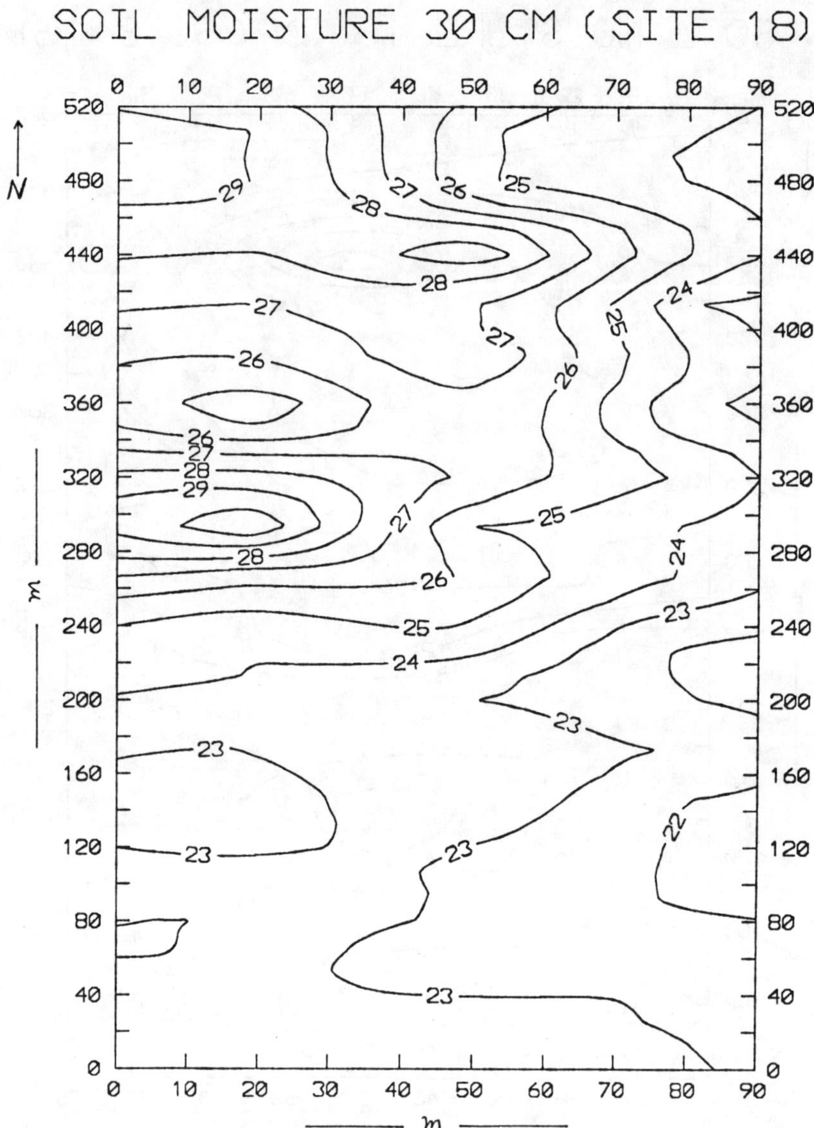

Fig. 23–7. Contour map of soil moisture (%) in 15–30 cm.

DISCUSSION

Microrelief is known to affect soil properties. Kachanoski et al. (1985) observed that A horizon thickness of some soils in Saskatchewan, Canada was significantly correlated to surface curvature in the direction of maximum surface gradient. Bocock et al. (1982) reported a close relationship between soil temperature and microrelief. Wang et al. (1991) demonstrated that microrelief affected soil reaction, drainage and the presence of imogolite in soil. All of these studies had a surface gradient greater than this benchmark site. The accumulation of carbon just above the low, broad ridge (compare Fig. 23–2 and 23–3) and the fact both soil pH and soil moisture were significantly correlated to carbon (Table 23–1) indicated that even very minor relief changes can have major influence on spatial distributions of agronomically important soil properties. This also indicates that in soil quality monitoring, relief (or landscape), even if of minor differences, has to be part of the overall sampling design. Landscape position and carbon content were significantly correlated with soil erosion and potato quality and yields in one of the benchmark sites in New Brunswick (Cao et al., 1994).

Due to the acidic nature of all samples of the Ap horizon, all carbon in Ap horizon is of organic forms. The accumulation of organic carbon just above the low ridge may be caused by the combination of three factors.

1. Relatively high moisture content above the ridge thus impeded aeration of this clayey soil which in turn slowed organic matter decomposition. The significant correlation between carbon and soil moisture (Table 23–1) supports this hypothesis.

2. Soil erosion and redistribution occurred along the slope and soil rich in organic carbon accumulated just above the ridge.

3. High carbon was correlated with lower pH (Table 23–1), and the lower pH generally resulted in lower biological activities hence the lower rate of organic matter decomposition.

The negative correlations between soil carbon and pH and between pH and soil moisture content at 0–15 cm (Table 23–1) suggested that organic acid produced from organic matter decomposition may have lowered soil pH (Deol et al., 1992) and that leaching may also be a factor for the negative correlation between soil moisture at 0–15 cm and pH values.

The northern half of the site had higher moisture content in May, in spite of higher elevation (Fig. 23–2), than the southern half for both 0–15 cm and 15–30 cm depths (Fig. 23–6 and 23–7). This reversal in the moisture distribution may be due to the low ridge near the middle of the site (Fig. 23–2).

The relatively narrow range of variation in particle size distribution is normal for a soil developed on marine deposits. It was assumed that the texture of the surface soil had minor influence on spatial variabilities of organic C, pH and soil moisture content.

In conclusion, the study has demonstrated that minor changes in relief can result in large spatial variations for some agronomically important soil properties. Relief (or landscape), therefore, is an important factor to be considered in sampling design for soil quality monitoring and evaluation.

REFERENCES

Agriculture Canada Expert Committee on Soil Survey. 1987. The Canadian System of Soil Classification. 2nd ed. Agric. Can. Publ. 1646. 164 pp.

Bocock, K.L., A.D. Bailey, and M. Hornung. 1982. Variation in soil temperature with microrelief and soil depth in a newly planted forest. J. Soil Sci., 33:55–62.

Cao, Y.Z., D.R. Coote, H.W. Rees, C. Wang, and T.L. Chow. 1994. Effects of intensive potato production on soil quality and yield at a benchmark site in New Brunswick. Soil and Tillage Res.

Centre for Land and Biological Resources Research. 1992. Benchmark sites for agricultural land in Canada. CLBRR Contri. No. 92–28. Research Branch, Agriculture Canada. Ottawa.

CoHort Software, 1990. CoStat, version 4.20. Berkeley, CA. U.S.A.

Deol, Y.S., C.M. Monreal, and D.W. Anderson. 1992. Effect of incorporation of wheat straw and nitrogen addition on the flux of soil gas (CH_4 and N_2O) at two moisture and temperature regimes. Soils and Crops Workshop. University of Saskatchewan, Saskatoon.

Golden Software, Inc. 1990. Surfer, Version 4. Golden, CO. U.S.A.

Kachanoski, R.G., D.E. Rolston, and E. De Jong. 1985. Spatial and spectral relationships of soil properties and microtopography. Soil Sci. Soc. of Amer. J., 49:804–812.

Sheldrick, B.H. (ed.). 1984. Analytical Methods Manual. Land Resource Res. Inst., Res. Branch, Agriculture Canada. Ottawa.

Soil Survey Staff. 1992. Keys to soil taxonomy, 5th ed. SMSS Technical Monograph No. 19. Blacksburg, VA. Pocahontas Press, Inc. 556 pp.

Wang, C., G.J. Ross, J.K. Torrance, and H. Kodama. 1991. The formation of Podzolic B Horizons and Pedogenic Imogolite as influenced by microrelief within a Pedon. Geoderma, 50:63–77.

24 Soil Properties Across A Landscape Continuum As Affected By Planting Wheel Traffic

M. J. Lindstrom
W. B. Voorhees

USDA-Agriculture Research Service
North Central Soil Conservation Research Laboratory
Morris, Minnesota

T. E. Schumacher

Department of Plant Science
South Dakota State University
Brookings, South Dakota

Soil compaction has been shown to reduce crop yields in the northern corn belt of the U.S.A. and Canada (Adams et al., 1960 and Raghaven et al., 1979). Modern row crop production can result in many trips over the field and as a result much of the field area is subjected to field traffic, some areas many times. Compactive force from field traffic is determined by the surface contact pressure of the tires while the depth that compaction extends into the soil is determined by axle load (Sohne, 1958). Controlled wheel-traffic studies by Voorhees et al. (1978) on a loam textured soil showed that normal row crop farming operations with equipment weight limited to 4.5 Mg per axle could result in soil compaction below 30 cm.

Compaction can influence crop production by changing bulk density, soil strength, aggregate size distribution, and pore size distribution. These changes in turn affect infiltration, drainage, water availability, aeration, root exploration, and nutrient uptake. Larson et al. (1980) reported that soil compactibility can be categorized based on type of clay and amount of organic matter. Larson et al. (1980) also showed that soil compaction increased with increased soil water content. Gupta and Allmaras (1987) found that the compression index increased with increasing clay content up to 33% and then leveled off showing that clay soils are more susceptible to compaction than sandy soils. The degree of soil compaction for a given axle-load is dependent on several soil properties such as clay, organic matter, and water content at the time of compaction. Differences in these soil properties are frequently associated with both landscape position and soil series in agricultural fields (Brubaker et al., 1993). Since these properties vary with soil series and landscape position, one might expect the effects of wheel-tracks on soil properties to vary across a field in a predictable pattern.

Raghaven et al. (1979) demonstrated that higher mechanical impedance from repeated wheel traffic passes with contact pressure of 62 kPa restricted depth of corn (*Zea mays* L.) root penetration and overall root density. In this same trial, Raghaven et al. (1979) reported a compaction-weather interaction. In a dry season, higher corn yields were obtained on slightly compacted soil than on non-compacted soil; whereas, in a wet season, corn yields decreased with increased cumulative ground contact pressure. Greater soil water availability in the slightly compacted soil during the dry season was the explanation extended for increased yields with soil compaction. Similar beneficial yield results with wheat (*Triticum aestivum* L.) and soybean [*Glycine max* (L.) Merr.] were experienced from moderate soil compaction by Voorhees (1987) during dry growing seasons, but during growing seasons of adequate soil moisture crop yields declined with compaction. Differences in landscape position may partially mimic the effect of climatic interactions on compaction since landscape positions in the foot and toeslopes frequently have higher moisture contents than the summit and backslope segments of the landscape within a given year.

Fausey and Dylla (1984) found that wheel track compaction from five passes with a tractor with a rear axle load of 3.5 Mg resulted in a significant corn yield reduction when fertility was limiting, but that yields were not depressed when soil fertility and soil moisture were not limiting. Fields with significant differences in soil fertility such as may occur in fields with significant eroded phases could respond with differences in the compaction-productivity relationship. Soil compaction is not a direct cause and effect process. Raghaven et al. (1990) categorized the effects of soil compaction into three main elements of machine, soil, and crop which reflect an action-reaction-output approach. The first element is subdivided into compaction and tillage machinery. The second element consists of soil characteristics and the third element reflects the water, air, and nutrients needs of the crop. Interactions between these elements and subelements make soil compaction across landscapes a complicated problem that has significant relevance to questions of site-specific farming.

The glacial morainic landscape of eastern South Dakota consists of moderately rolling hills with many closed undrained depressions. Well-drained soils and somewhat poorly drained to poorly drained soils are commonly located on the same farm or field. Soil moisture regimes will vary due to landscape position and drainage characteristics, but these soils are commonly tilled and planted at the same time because of their close proximity. Farmers working within these soil catenas should be aware of potential soil compaction problems and tillage interactions that may exist due to soil variability within a single field. Wheel-traffic compaction in these landscapes could warrant special consideration in site-specific farming systems (Voorhees et al., 1993).

This study was designed to investigate the effects of row crop planting wheel traffic and the interaction with tillage systems across a landscape continuum containing soils with differing soil moisture regimes due to landscape position and drainage characteristics on interrow soil properties and corn yields.

MATERIALS AND METHODS

This trial was initiated in the spring of 1990 across a landscape continuum on the Eastern South Dakota Soil and Water Research Farm near Brookings, SD on a Vienna-Lismore soil association. The Vienna soil series (fine-loamy, mixed Udic Haploboroll) is a deep, well-drained soil formed in silty material over loam glacial till with moderately slow permeability with a typical mollic depth of 23 cm. Vienna soils are nearly level to hilly on uplands having plane and slightly convex surfaces. Slope gradients typically are < 6%, but range from 0 to 15%. The Lismore soil series (fine-loamy, mixed Pachic Udic Haploboroll) is a deep, moderately well-drained soil formed in silty sediments over glacial till with moderate to moderately slow permeability and a typical mollic depth of 50 to 60 cm. Lismore soils are found on nearly level to gently sloping upland flats and in upland scales and are subject to a perched water table at depths of 90 to 120 cm over the winter and spring seasons.

The plot area had previously been uniformly cropped for at least 10 years. The entire plot area was moldboard plowed the fall of 1989. Nine plots were established to include three tillage systems and two crop rotations of continuous corn and corn-soybean in the spring of 1990. Two plots of corn-soybean were established in each tillage system so that both crops would be present each year. Plot width was 18.3 m (24–0.76 m rows). Plot length extended across the landscape continuum, a distance of 286 m.

The plots started in the summit position, proceeded through the shoulder and backslope positions, and concluded after passing through the footslope position. Soils encounterd in this transect were Vienna 1–3% or B slope (VB) at the summit, Vienna 3–5% or C slope (VC) on the shoulder and upper backslope, Vienna B slope on lower back slope, and Lismore 0–1% slope (LA) in the footslope position. The Vienna 3–5% slope (VC) located on the shoulder and upper backslope positions was commonly in the eroded phase. The Vienna 1–3% slope located on the lower backslope commonly had a greater mollic depth (43 cm) in the surface horizon, resultant from deposition of eroded soil material from positions higher in the landscape, and is differentiated from the Vienna 1–3% slope (VB) located at the summit position by designation as a deep phase Vienna (DVB).

The tillage systems were initiated during the 1990 crop season and included fall moldboard plow, spring disk (MP); fall chisel plow, spring disk (CP); and ridge till (RT). Moldboard plowing was done with a six bottom plow to a depth of 20 cm with 46 cm plow shares. Chisel plowing was done with a 4.5 m wide chisel plow equipped with 7.6 cm twisted shovels on 30 cm centers to a depth of 15 cm. Ridges were formed in the RT system with the second cultivation. Ridge heights were 10 cm for corn and 7.5 cm for soybean. Secondary spring tillage on the MP and CP plots was done with a 4.5 m wide tandem disk at a depth of 7.5 cm. Tillage wheel traffic over the plots was determined by tillage equipment width and did not necessarily coincide with planting wheel traffic. Soil moisture content during fall tillage was dry and probably did not contribute to compaction. Soil moisture content during spring secondary tillage was similar to soil moisture content at time of planting;

therefore, wheel traffic from the secondary tillage may present a conflicting variable not accounted for in this analysis. Additional wheel traffic associated with corn and soybean production such as spraying, fertilizing, and cultivation was confined to the initial wheel tracked interrow from the planting operation. Fertilizer nutrients were added to the plot area based on South Dakota State University soil test recommendations. Standard pre-and post-emergence herbicides were applied to the respective crops for weed control. Crop harvest was done with two row equipment with axle loads of < 2 Mg per axle. Although the frequency of wheel traffic was high, because of the dry soil conditions and the low axle loads present at harvest, this wheel traffic was not considered to be a conflicting variable.

Planting wheel track compaction was applied annually in 1990, 1991, and 1992 with a two wheel drive tractor having a total weight of 8.1 Mg, with 3.2 Mg on the front axle and 4.9 Mg on the rear axle. The rear axle was equipped with single tires. A ground contact pressure of 180 kPa was exerted by each rear tire. Plots were planted with a commercial 8-row planter giving two wheel tracked interrows for each eight planted rows. In the 24–row plots, three passes were required to plant the plots. Additional wheel track compaction was applied immediately after planting by shifting the tractor over one interrow area in the second set of eight planted rows. In the third set of planted rows, the original wheel tracked interrow were driven over two more times and then shifted over one interrow and tracked three times. Plot areas were set out to ensure that planting wheel traffic was applied in the same area each year. This sequence of wheel traffic resulted in non-tracked interrows (NWT), single wheel tracked interrows (WT), and interrows wheel tracked three times (WT3X). Yield comparisons were made on 2-crop rows with no wheel traffic on either side of the row, with single wheel traffic on both sides of the row, and wheel traffic three times on both sides of the rows.

Soil moisture content was measured at depth increments of 7.5 cm down to 30 cm for each soil series, tillage, and crop rotation sequence at the time of wheel traffic application in 1991 and 1992 from non-wheel tracked interrows. Soil core samples (5.1 by 5.0 cm) were taken at depths of 2.5–7.5 cm and 15.0–20.0 cm from all soil series, tillage, crop rotation and interrow wheel track variable one month after planting in 1991 for bulk density and macroporosity measurements. Penetrometer resistance measurements were taken at increments of 2.5 cm to a depth of 30 cm from the same combination of treatments. In 1992, soil core samples at these same depths and treatment combinations were taken from the previous years corn rotation treatments three weeks after planting. Penetrometer resistance measurements were taken from all treatments. Proctor density of the four soils were determined by the South Dakota Civil Engineering Laboratory using standardized procedures (ASTM, 1981). Proctor density is maximum bulk density for the bulk density-water content curve for a given energy level.

RESULTS AND DISCUSSION

Soil water content by 7.5 cm increments to 30 cm at the time of planting wheel traffic compaction for the four soils present in the landscape continuum are shown in Fig. 24–1. A Proctor density of 1.72 and 1.71 Mg m^{-3} at optimum water contents of 17.6 and 17.0% (g g^{-1}) were measured for the VB and VC soils. A Proctor density of 1.61 and 1.60 Mg m^{-3} at optimum water contents of 19.9 and 20.0% (g g^{-1}) were measured for the DVB and LA soils. The changes in Proctor density and optimum water content are a result of higher clay and organic matter contents for the DVB and LA soils. Soil water contents measured at the time of wheel traffic application showed that the DVB and LA soils had a higher water content than the VB and VC soils and were near or above the optimum water content for Proctor density at 0 to 7.5 cm. At depths greater than 7.5 cm, the water content was greater than the optimum. The water content in the 0 to 7.5 cm zone was near or less than the optimum in the VB and VC soils and generally slightly greater than the optimum below 7.5 cm. The MP treatment had a lower water content level than the CP or RT for all soils in the surface zone.

Table 24–1. Mean bulk density values by soils, tillage systems, and interrow compaction treatments for the two depth increments in 1991 and 1992.

	1991		1992	
	Depth (cm)			
	2.5 – 7.5	15.0 – 20.0	2.5 – 7.5	15.0 – 20.0
	Mg m^{-3}			
Soil				
VB	1.43a†	1.47a	1.31a	1.42a
VC	1.40a	1.48a	1.31a	1.42a
DVB	1.30b	1.36b	1.25a	1.36a
LA	1.26b	1.34b	1.24a	1.35a
Tillage				
MP	1.29a	1.38a	1.25a	1.33a
CP	1.31b	1.43a	1.21a	1.43b
RT	1.44b	1.44a	1.37b	1.4lab
Compaction				
NWT	1.22a	1.37a	1.16a	1.32a
WT	1.42b	1.46b	1.25a	1.37a
WT3X	1.39b	1.42ab	1.41b	1.47b

† Within columns and measured variable, means followed by the same letter are not significantly different at the 0.05 probability level.

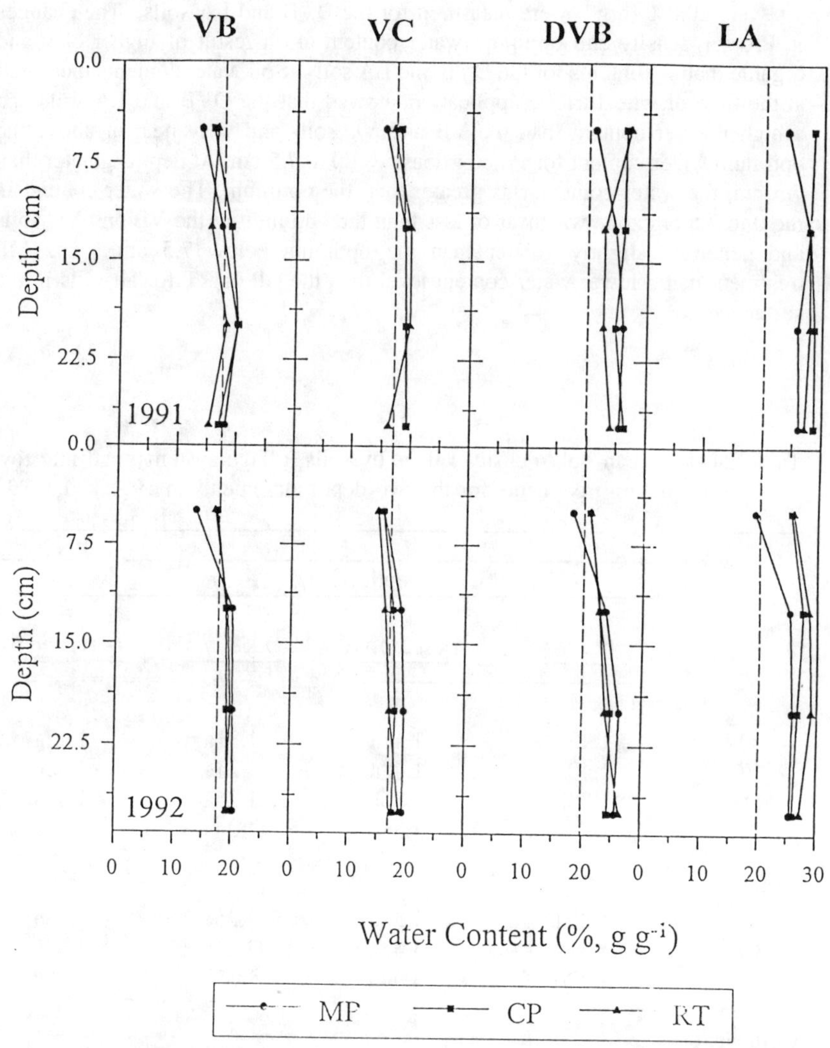

Fig. 24–1. Gravimetric soil water content of the non-tracked interrow for the four soils at time of planting wheel traffic by tillage systems in 1991 and 1992. Vertical dashed line represents the optimum soil water content for Proctor density.

Bulk density values by soils, tillage system, and interrow compaction treatment by depth increments for the two years are shown in Table 24–1. The VB and VC soil series had a higher bulk density than the DVB and LA soil series in 1991. These differences were not present in 1992, although similar trends were observed. The RT tillage system had a higher bulk density at 2.5–7.5 cm both years. Differences were not as conclusive at 15.0–20.0 cm. The effect of planting wheel track compaction on interrow bulk density was apparent with the WT3X compaction treatment resulting in a higher density than the NWT interrow at both depths and years. The WT compaction treatment also had a higher density than the NWT treatment in 1991.

Volume of macropores by soils, tillage systems, and interrow compaction treatments are shown in Table 24–2. Soil did not appear to have an effect on macropore volumes at 2.5–7.5 cm. Differences were observed at 15.0–20.0 cm in 1991, but not in 1992. The RT tillage system showed a reduced macropore volume at 2.5–7.5 cm both years and at 15.0–20.0 cm in 1992. Planting wheel traffic also reduced macropore volumes with both the WT and WT3X treatments being lower than the NWT for both depths and years.

Table 24–2. Mean macropore volumes by soils, tillage systems, and interrow compaction treatments for the two depth increments in 1991 and 1992.

	1991		1992	
	Depth (cm)			
	2.5 – 7.5	15.0 – 20.0	2.5 – 7.5	15.0 – 20.0
	%, v/v			
Soil				
VB	11.4a†	9.1a	15.4a	12.6a
VC	13.1a	8.6ab	12.9a	10.4a
DVB	14.1a	8.2ab	13.3a	9.5b
LA	11.6a	5.6b	13.7a	8.7b
Tillage				
MP	15.1a	8.8a	16.1a	13.3a
CP	14.5a	7.9a	16.2a	9.5b
RT	7.9b	7.0a	9.1b	8.2b
Compaction				
NWT	19.2a	10.6a	19.2a	13.1a
WT	9.0b	6.2b	11.5b	9.6b
WT3X	9.6b	6.8b	10.7b	8.4b

† Within columns and measured variable, means followed by the same letter are not significantly different at the 0.05 probability level.

Both bulk density and macropore volumes were influenced by tillage systems and wheel traffic. This influence, however, did not appear to be strongly associated with soils as interactions between soils and tillage systems or soils and compaction treatments were not observed. An increase in bulk density resulted in a decrease in macropore volumes (Table 24–1 and 24–2).

Penetrometer resistance readings by soils, tillage systems, and interrow compaction treatments for 1991 and 1992 are shown in Fig. 24–2 at 2.5 cm increments to 30 cm. Penetrometer resistance readings were significantly different for the individual soils, but this relationship was probably strongly related to moisture content of the soils at the time of measurement. The VB and VC soils were similar in soil strength in both 1991 and 1992 and were greater than the DVB which was also greater than the LA. The MP tillage treatment had lower resistance reading than the CP or RT. The RT tillage was greater than the CP to about 17.5 cm, then the values become similar.

The NWT interrow consistently had lower resistance values than the WT or WT3X compaction treatments. The WT and WT3X had similar values except between 5 to 10 cm where the WT3X resulted in greater resistance than WT. Differences, however, were small and the magnitude of all penetrometer resistance values suggest that soil strength was not a major concern.

Penetrometer resistance data for the VB and LA soils by tillage systems in 1991 and 1992 (Fig. 24–3 and 24–4) show a greater difference between the NWT and the WT or WT3X in 1992 indicating a cumulative effect of wheel traffic over years, but at this time we do not have adequate data to make this conclusion.

Corn yield by soils, tillage systems, and interrow compaction treatments are shown in Table 24–3. Soils showed significant yield differences for both years. Yield differences observed in 1991 were primarily what would be expected from landscape position effect on moisture regimes. The lower backslope position (DVB) had the highest yields followed by the summit position (VB); the shoulder and lower backslope position (VC) had a lower yield (Jones et al, 1989). The lower yields from the footslope (LA) was due to excess water stress during portions of the growing season. Similar yield variations were observed in 1992, with the LA soil experiencing excess moisture stress early in the growing season. Yields were also reduced in 1992 due to an early fall frost that killed the crop before maturity.

Tillage systems affected corn yields in 1991, but not in 1992. A soils by tillage system interaction was also present in 1991. The CP and RT tillage systems performed better than the MP system on the VB and VC soil series where moisture may have been limiting. Wheel traffic compaction at time of planting had no effect on yields in either year.

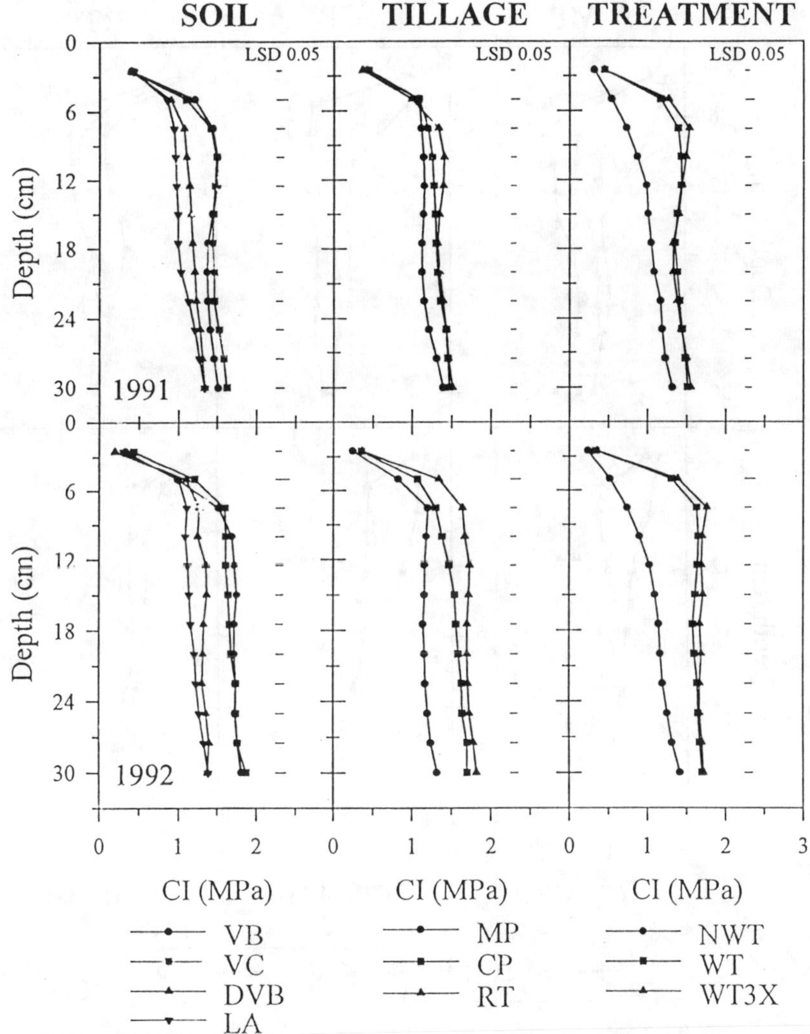

Fig. 24–2. Penetrometer resistance readings by soils, tillage systems, and interrow compaction treatments for 1991 and 1992.

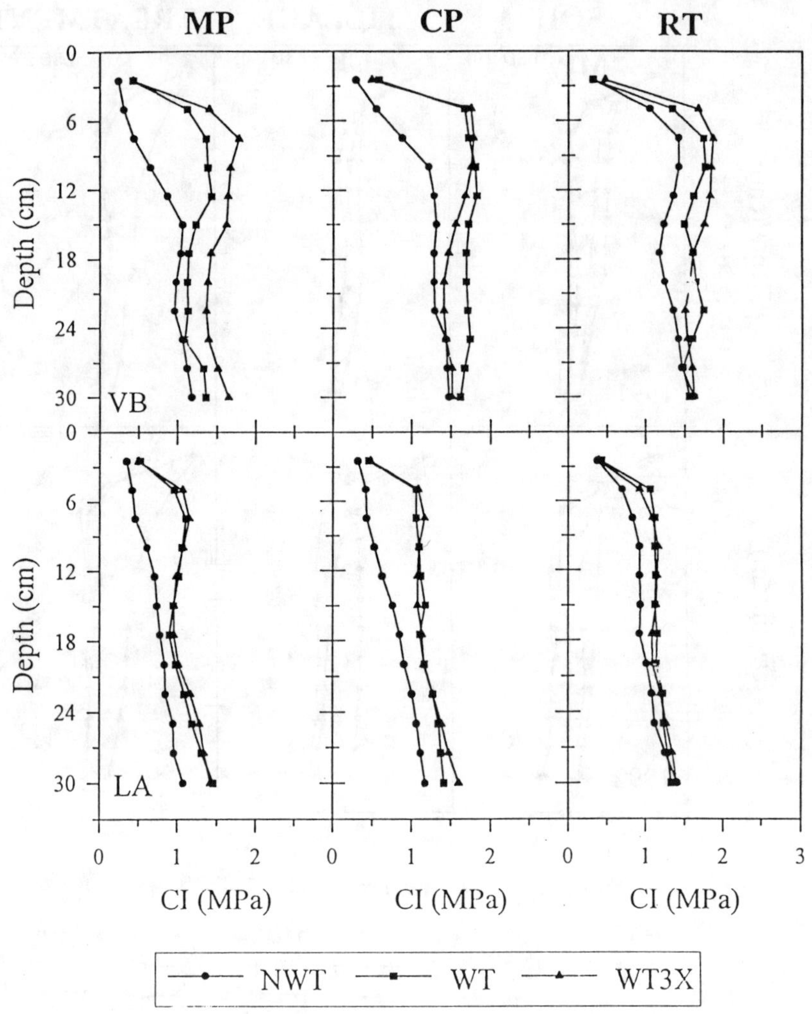

Fig. 24–3. Penetrometer resistance readings for the VB and LA soils by tillage systems and interrow compaction treatments in 1991.

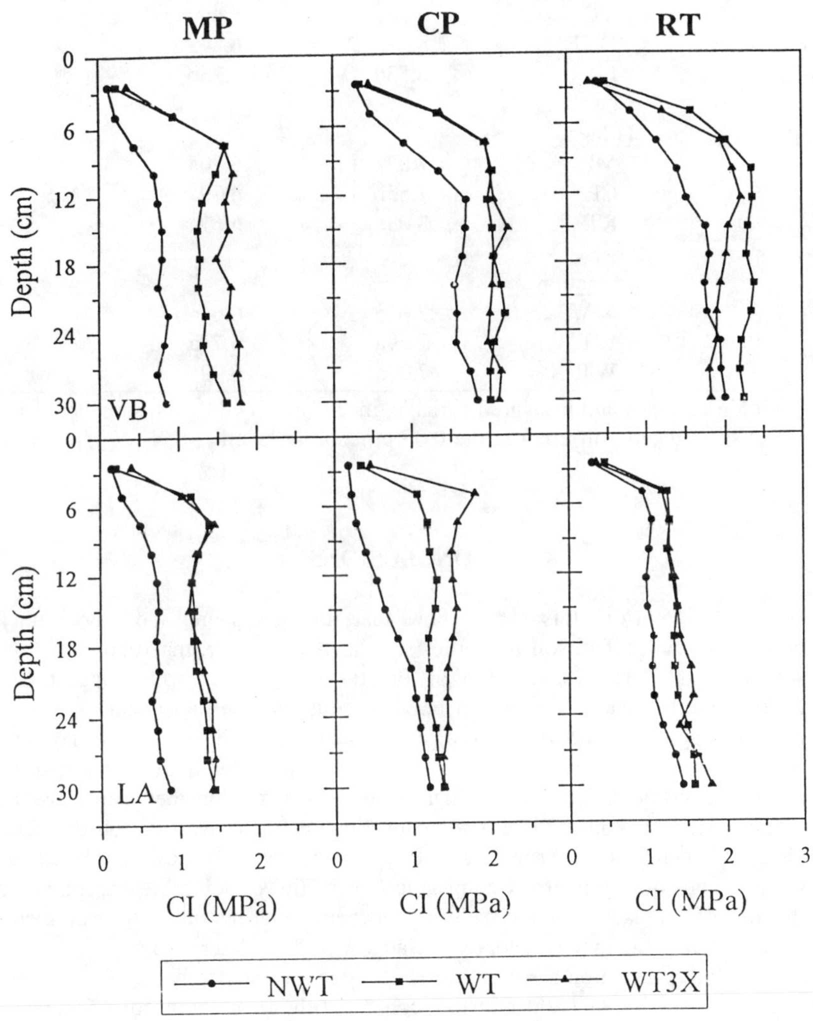

Fig. 24–4. Penetrometer resistance readings for the VB and LA soils by tillage systems and interrow compaction treatments in 1992.

Table 24–3. Mean corn yields by four soils, tillage systems, and interrow compaction treatments in 1991 and 1992.

	1991	1992
	Corn Yields	
	----------- Mg ha^{-1} ----------	
Soil		
VB	7.73b†	7.16a
VC	7.20c	6.40b
DVB	8.58a	6.84a
LA	6.53d	6.25b
Tillage		
MP	7.03a	6.70a
CP	7.55b	6.79a
RT	8.03c	6.61a
Compaction		
NWT	7.61a	6.64a
WT	7.69a	6.77a
WT3X	7.78a	6.99a

† Within columns and measured variable, means followed by the same letter are not significantly different at the 0.05 probability level.

CONCLUSIONS

Results from this study show that tillage systems and wheel track compaction will affect soil properties such as bulk density, macroporosity, and soil strength. The relationship of Proctor density and soils suggest that differences in surface compaction between soils may be dependent on annual moisture regimes. The existence of a compaction by soil interaction may only be apparent in years where soil moisture levels are near optimal for specific landscape positions and not for other positions. Interactions between soils and tillage systems or soils by interrow compaction treatments were not observed for either bulk density or macroporosity. Penetrometer data did show soils by tillage systems and soils by interrow compaction interactions. Soil moisture content at the time of compaction and at the time of penetrometer resistance measurements were probably the major factors associated with these interactions.

Soil compaction has been adequately demonstrated to be a major concern in crop production and soil moisture content at the time compactive forces are applied is an important parameter determining the degree of compaction. The increase in soil bulk density, decrease in macropore volumes, and increase in soil strength observed in this study suggest potential soil physical problems with planting wheel traffic. Corn yields, however, were not significantly affected by planting wheel traffic for the four soils or for the three tillage systems indicating that compactive forces applied when planting a crop with a medium sized tractor exerting 180 kPa on the rear axle (single pass or three passes) when traffic is confined to specific interrow areas were not a major row crop production concern. Differences in corn yield across soils suggests that inherent soil productivity differences should be the primary consideration of site-specific farming implementation, followed by the interaction on management and landscape position on soil properties.

REFERENCES

Adams, E.P., G.R. Blake, W.P. Martin, and D.H. Boelter. 1960. Influences of soil compaction on crop growth and development. Vol 1, pp. 607–615. Trans. 7th Int. Cong. Soil Sci., Madison, WI.

ASTM. 1981. Annual Book of ASTM Standard. American Society for Testing Materials. Philadelphia, PA.

Brubaker, S.C., A.J. Jones, D.T. Lewis, and K. Frank. 1993. Soil properties associated with landscape positions. Soil Sci. Soc. Am. J. 57:235–239.

Fausey, N.R., and A.S. Dylla. 1984. Effects of wheel traffic along one side of corn and soybean rows. Soil Tillage Res. 4:147–154.

Gupta, S.C., and R.R. Allmaras. 1987. Models to assess the susceptibility of soils to excessive compaction. Adv. Soil Sci. 6:65–100.

Jones, A.J., L.N. Mielke, C.A. Bartles, and C.A. Miller. 1989. Relationship of landscape position and properties to crop production. J. Soil Water Conserv. 44:328–332.

Larson, W.E., S.C. Gupta, and R.A. Useche. 1980. Compression of agricultural soils from eight soil orders. Soil Sci. Soc. Am. J. 44:450–457.

Raghaven, G.S.V., E. McKyes, F. Taylor, P. Richardson, and A. Watson. 1979. The relationship between machinery traffic and corn yield reductions in successive years. Trans. ASAE 22:1256–1259.

Raghaven, G.S.V., P. Alvo, and E. McKyes. 1990. Soil compaction in agriculture: A view towards managing the problem. Adv. Soil Sci. 9:1–36.

Sohne, W. 1958. Fundamentals of pressure distribution and soil compaction under tractor tires. Agric. Eng. 39:276–281, 290.

Voorhees, W.B. 1987. Assessment of soil susceptibility to compaction using soil and climatic data bases. Soil Tillage Res. 10:29–38.

Voorhees, W.B., R.R. Allmaras, and M.J. Lindstrom. 1993. Tillage considerations in managing soil variability. *In* Soil Specific Crop Management. ASA, CSSA, SSSA, Madison, WI.

Voorhees, W.B., V.A. Carlson, and C.G. Senst. 1978. Compaction and soil structure modification by wheel traffic in the northern Corn Belt. Soil Sci. Soc. Am. J. 42:344–349.

25 Spatial Corn Yield During Drought in the SE Coastal Plain

E. J. Sadler
P. J. Bauer
W. J. Busscher

USDA-Agricultural Research Service
Coastal Plains Research Center
Florence, South Carolina

Throughout the southeastern USA coastal plain, Carolina Bays are low, depressional areas that often have different soils and a wide variation in crop yield. Corn (*Zea mays*) appears to be the most susceptible to soil variation, especially during periods of drought. During 1993, a severe drought yr in this region, corn yields were measured at 209 sites within an 8-ha field where yield variation among soils had been evaluated for 12 crops. Site-specific effects of soil variation on crop phenology, biomass, and yield components were measured at 11 sites. Time-domain reflectometry (TDR) soil moisture probes were installed at eight of those sites, two within each of four map units. These were monitored from 40 days after planting until after maturity. Drought stress during vegetative growth caused severe leaf rolling in several areas, while other areas suffered no visually-apparent stress. This observation was supported by infrared thermometer measurements of canopy temperature (TC), which ranged from ambient air temperature (TA) to about 10°C higher. Rain following the period of drought reduced TC-TA to near zero for all soils, indicating that the stress was relieved. Plant height five days before the rain ranged from 0.48 to 1.34 m. Mid-silk leaf area index ranged from 1.15 to 2.56. Time of tasselling and black layer formation ranged over three weeks. Grain yield ranged from 104 to 318 g m^{-2} dry weight at the 11 primary sites and from 18.5 to 419.8 g m^{-2} in the entire field. Mean yield over 209 18-m^2 plots was 214.8 ± 79.3 g m^{-2} on a dry weight basis (2481 ± 916 kg ha^{-1} at 15.5% moisture). This high variation in crop yield presumably resulted in large differences in residual N since fertilizer applications were uniform across the entire field. We suggest that additional analysis and stochastic simulation are needed to estimate risk from such an occurrence and to develop environmentally safe N management practices for subsequent crops.

CONTEXT OF FIELD STUDY

This research was conducted during the 12th cropping season of a study designed to document spatial variability of crop yield within a representative coastal plain field (Karlen et al., 1990; Sadler et al., 1993; 1994a). The field, which includes 14 soil map units (USDA-SCS, 1986), had been cultivated since 1985 using uniform conventional techniques. These included offset disking to incorporate weeds and crop residues, in-row subsoiling for row crops, and field cultivation to prepare the seedbed for planting. The field studies were initiated to document inherent variation among and within soil map units and to provide data that could be compared with the results of mechanistic computer simulation models. Preliminary results from the simulation studies suggested that the soil water balance was not being adequately described (Stone & Sadler, 1991). Field observations of grain fill occurring after models simulated crop maturity during 1992 indicated that crop phenology was also not correctly simulated. These observations suggested that more site-specific data were needed to calibrate the simulation models to coastal plain soils.

Our objectives were to evaluate in detail corn response to differences in soil characteristics. Paired samples within four soil map units were chosen to allow comparison within and among soil map units. Soil properties and plant growth characteristics were measured so that known weaknesses in simulation models could be addressed. Where possible, sampling strategies were designed to allow both soil-to-soil comparisons and geostatistical analysis.

FIELD STUDY - 1993

The field was disked on 9 and 23 March. Granular fertilizer was broadcast 30 March at an application rate of 17–40–121–15 kg ha^{-1} N-P-K-S. On 8 April, metolachlor herbicide [Dual[1] 8E, 2-chloro-N-(2-ethyl-6-methylphenyl)-N-(2-methoxy-1-methylethyl) acetamide, Ciba-Geigy Corp., Greensboro, NC] was sprayed at an application rate of 2.8 kg ha^{-1} and incorporated. On 9 April, 'Pioneer Brand 3165' corn was planted using a KMC (Kelley Manufacturing Co., Tifton, GA) in-row subsoil unit with Case - IH (Chicago, IL) Model 800 planters. Broadleaf weeds were controlled on 11 May with 2,4-D [(2,4-dichlorophenoxy) acetic acid] herbicide (Weedar, Rhône-Poulenc Ag Company, Research Triangle Park, NC) at an application rate of 0.5 kg ha^{-1}. Sidedress N fertilizer was banded on both sides of the crop row on 28 May, at an application rate of 112 kg N ha^{-1}. Corn was harvested from 16 to 24 September using a plot combine (GWC, Inc., Nevada, IA).

---

[1] Mention of trade names is for information purposes only. No endorsement implied by the USDA-ARS.

REPRESENTATIVE SITES

Four soil map units were selected as being representative of the range of soils within the field. These included:

1) the predominant map unit, Norfolk loamy fine sand (NkA; moderately thick surface, deep water table, 0 to 2% slopes; Fine-loamy, siliceous, thermic Typic Paleudult; reclassified 3/88 to Fine-loamy, siliceous, thermic Typic Kandiudult),
2) a historically low-producing Coxville loam (Cx; Clayey, kaolinitic, thermic Typic Paleaquult),
3) a local inclusion that under-performed expectations for Goldsboro loamy fine sand (GoA; 0 to 2% slopes; Fine-loamy, siliceous, thermic Aquic Paleudult), and
4) Bonneau loamy fine sand (BnA; 0 to 2% slopes; originally mapped as Loamy, siliceous, thermic Grossarenic Paleudult; reclassified 3/90 to Loamy, siliceous, thermic Arenic Paluedult) that produced high yields in spite of having the lowest productivity rating.

Two sites were chosen for each map unit. The positions of these sites within the larger field are shown in Fig. 25–1. Site-specific characteristics that were determined after selection and installation of the TDR probes suggested that two of the sites were not as representative as desired. Site #1, which was to represent GoA, was placed by error on the boundary between GoA and Dunbar (Dn;

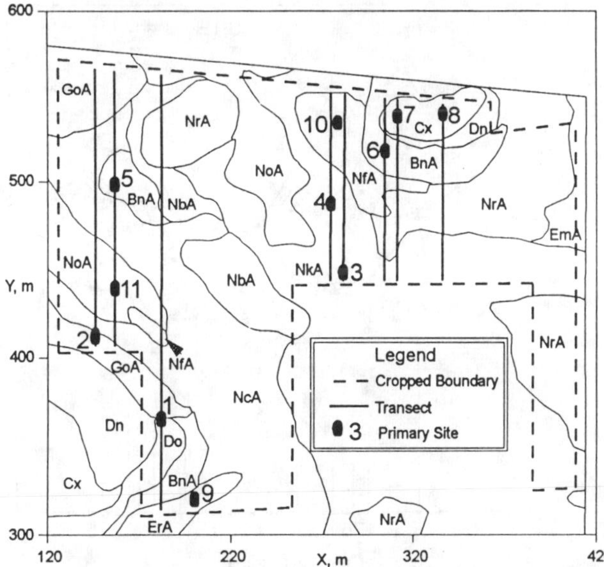

Fig. 25–1. Site plan for 1993 corn study. Locations for 8 TDR sites plus 3 additional representative sites are indicated, as are transect locations for spatial variation measurements during the season.

Clayey, kaolinitic, thermic Aeric Paleaquult), but the difference between these two inclusions is < between typical pedons of the two soils. Site #8, which was the second Cx location, was determined to encroach on an area recently disturbed during land forming around a drainage culvert. The differences between the two Cx sites in horizon depths and textures were small. The differences in strength that might exist between them because of the previous disruption may be significant.

Each of these eight sites was instrumented with time-domain reflectometry (TDR) probes to monitor soil moisture to a depth of 1 m. Probe depths were selected to represent the soil horizons at the site. A TDR (1502B, Tektronix, Inc., Beaverton, OR), laptop computer, switching devices, battery, and required cabling were assembled onto a 2-wheel hand truck (Sadler & Busscher, 1993) that was used to make the soil moisture measurements. The TDR traces from the 5–6 probes at each site were automatically obtained and reduced using the computer programs of Baker and Allmaras (1990). The season's soil water balances for these eight sites are shown in Fig. 25–2.

The most obvious difference among the soil types occurred because of the 46-mm rain on 12 June. As can be seen in Fig. 25–2, site #8 showed a greater increase in profile soil water than average (about 2x), and sites #1 and #5 showed a smaller increase (about 0.5x). After correcting for evaporation on days between the measurements, the best estimates of infiltration ranged from 21 and 33 mm for sites #1 and #5 to 98 mm for site #8. The latter was > twice the rainfall amount, indicating a large run-on to the site. Site #7, at 50 mm, also had a net increase larger than the rainfall amount. The remaining four sites

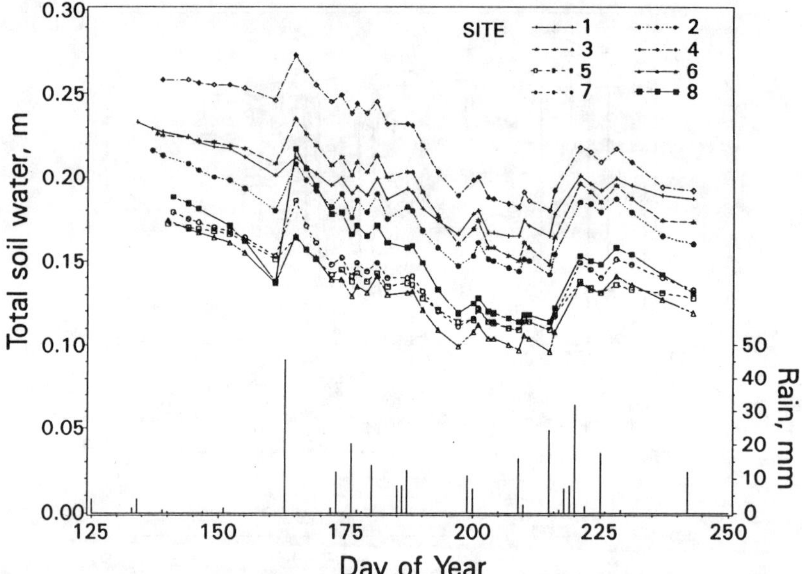

Fig. 25–2. Soil water balance at the 8 original representative sites. Rainfall is indicated on the right axis.

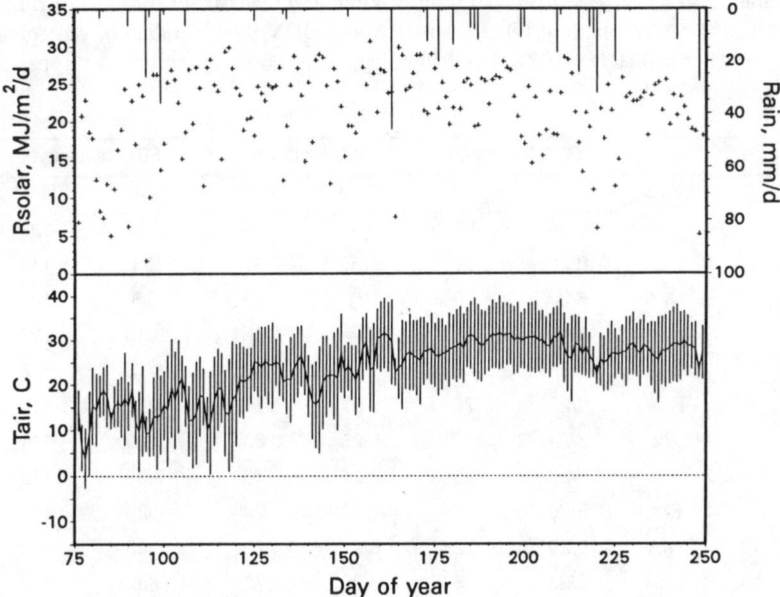

Fig. 25–3. Daily air temperature extremes, daily solar radiation, and daily rainfall during 1993.

ranged from 38 to 41 mm. Subsequent evaporation from the eight sites showed faster losses from site #8 and slower losses from sites #1 and #5.

By the time of this rain, at which time the corn was approaching the tasseling stage, the drought (Fig. 25–3) had caused substantial differences in plant height. Corn in three areas of the field was taller and more developed than the rest of the field. When emerging tassels confirmed the advanced development, we added these sites to the study and collected plant samples. These included an Emporia fine sandy loam (ErA; 1 to 2% slopes; Fine-loamy, siliceous, thermic Typic Hapludult), a Noboco fine sandy loam (NfA; 1 to 2% slopes; Fine-loamy, siliceous, thermic Typic Paleudult), and a Norfolk with a thicker surface horizon (NoA; thick surface, 0 to 2% slopes; Fine-loamy, siliceous, thermic Typic Paleudult; also reclassified 3/88 as Kandiudult). The Noboco is similar to the Norfolk, but has a higher water table during part of the yr. These three additional sites are identified as Nos. 9, 10, and 11, respectively, on Fig. 25–1.

Measurements of phenology, biomass, leaf area, and yield components at various growth stages during the season were made at these 11 sites. The phenology measurements were made by repeatedly rating the 10 plants in either direction adjacent to the TDR site (20 plants marked for three additional sites) for tasseling, silk emergence, and black layer. When 50% of the plants had reached these developmental stages, the dates were recorded. The data are reported in Table 25–1.

Table 25–1. Dates of 50% tasselling, silking, and black layer for the 11 primary sites. Date of planting (DAP) was 9 April (DOY 99), and date of emergence was 18 April (DOY 108). No differences were observed among soil types for emergence date.

Site		50% tasselling		50% silking		50% blacklayer	
#	Soil	Date	DAP	Date	DAP	Date	DAP
1	GoA	6/28	80	7/2	84	8/10	123
2	GoA	6/26	78	6/30	82	8/1	114
3	NkA	6/26	78	7/1	83	8/8	121
4	NkA	6/29	81	7/4	86	8/10	123
5	BnA	6/28	80	7/1	83	8/15	128
6	BnA	6/23	75	6/27	79	8/3	116
7	Cx	6/24	76	6/28	80	7/29	111
8	Cx	6/23	75	6/24	76	7/30	112
9	ErA	6/20	72	6/22	74	7/29	111
10	NfA	6/20	72	6/21	73	7/28	110
11	NoA	6/23	75	6/25	77	8/6	119
min/max			0.889		0.849		0.859

On 7 June, plant height, leaf area, and biomass were measured on 1-m samples of crop row. At this time, height was determined to the top of plant tissue directly above the stem (Table 25–2). When 50% of the plants had silks emerged, the same measurements were taken on 2-m samples of row. Height at this time was determined to the bottom of the tassel (Table 25–3). Leaf area was measured both times with a leaf area meter (Model Li 3100, LiCor, Lincoln, NE). Biomass was determined by weight of tissue after drying for three days at 70°C. At the 50% silk-emergence sample, potential kernel number for each ear was determined by multiplying the number of rows by the number of potential kernels in the rows (Table 25–4). Two 3.05-m rows were hand harvested to determine yield components at the time of the final combine harvest. A mechanical sheller was used to thresh the grain. Kernel number and dry weight were determined by counting and weighing subsamples (ranging from 340 to 754 kernels) of the shelled grain (Table 25–5).

ADDITIONAL SPATIAL MEASUREMENTS

Corn growth during 1993 was extremely variable because of the drought, so additional measurements were made in an attempt to document what appeared to be an unusual occurrence. The first was simply to record, on 7 June, plant height on 10-m intervals along the rows that included the eight sites (Fig. 25–1 shows transects). Results from the transect including site #1 are shown in Fig. 25–4. To further document the conditions at the primary sites, heights of all individual plants being monitored for phenology were recorded on 23 June. The variation within and among sites is shown in Table 25–6.

Table 25–2. Corn growth characteristics measured on 6/7/93.

Site			Plants	sd	Height	sd	LAI	sd	Weight	sd
#	Soil	n[†]	# m^{-2}		m		m^2 m^{-2}		g m^{-2}	
1	GoA	3	4.6	0.9	0.682	0.013	0.3783	0.0220	60.6	4.8
2	GoA	3	4.2	0.2	0.911	0.136	0.6580	0.3508	92.4	26.1
3	NkA	3	3.7	0.9	0.612	0.045	0.1577	0.0341	42.8	6.9
4	NkA	2	4.7	1.2	0.608	0.052	0.2037	0.1282	47.4	13.8
5	BnA	3	3.9	0.4	0.625	0.031	0.1747	0.0711	41.5	2.6
6	BnA	1	5.2	-	0.873	-	0.4380	-	80.1	-
7	Cx	1	6.0	-	0.821	-	0.4407	-	80.1	-
8	Cx	1	3.4	-	1.228	-	1.2232	-	130.9	-
	min/max		0.567		0.495		0.129		0.317	

[†]n refers to number of 1-m samples taken.

Table 25–3. Corn growth characteristics measured at silking, from a single 2m sample.

Site #	Soil	Harvest Date	Height m	Leaf area $m^2 m^{-2}$	Dry Weight $g m^{-2}$
1	GoA	7/2	0.897	0.8537	---
2	GoA	6/30	1.313	0.8073	211.4
3	NkA	7/1	1.001	0.7521	---
4	NkA	7/9	1.270	0.6367	---
5	BnA	7/1	1.018	1.1361	---
6	BnA	6/28	1.075	1.0532	241.1
7	Cx	6/28	1.326	0.9459	280.3
8	Cx	6/28	1.655	1.3910	353.9
9	ErA	6/28	1.278	1.6717	288.8
10	NfA	6/24	1.486	1.2714	198.2
11	NoA	6/24	1.482	0.9981	288.8
min/max			0.542	0.381	0.560

Table 25–4. Potential kernels per plant determined at silking. Means and standard deviations computed from individual plant measurements.

Site #	Soil	rows ear^{-1} Mean	sd	kernels row^{-1} Mean	sd	kernels ear^{-1} Mean	sd
1	GoA	14.4	1.6	44.8	2.9	640.8	56.3
2	GoA	17.3	2.7	37.3	6.2	643.1	143.0
3	NkA	14.5	1.4	50.9	2.1	736.5	62.7
4	NkA	14.8	1.8	42.2	4.8	623.9	115.2
5	BnA	14.4	1.3	49.9	4.2	717.4	80.3
6	BnA	15.0	2.7	39.4	4.1	587.4	93.7
7	Cx	14.5	1.1	39.6	4.4	573.8	70.0
8	Cx	15.5	1.1	41.7	7.6	648.4	139.7
9	ErA	13.2	1.2	41.9	4.3	550.3	62.2
10	NfA	15.5	1.4	43.2	5.0	669.9	105.6
11	NoA	13.5	2.1	43.4	5.2	590.3	133.3
min/max		0.763		0.733		0.747	

Table 25–5. Yield components taken at harvest. Means and standard deviations are calculated from two 3.04-m samples per soil type. Plot yield is in dry weight; spatial yield is from the four nearest harvest plots and is also expressed here as dry weight.

Site		Plant density	sd	Ear density	sd	Ker-nel #	sd	Kernel wt	sd	Plot yield	sd	Spatial yield	sd
#	Soil	# m^{-2}		# m^{-2}		# ear^{-1}		g $kernel^{-1}$		g m^{-2}		g m^{-2}	
1	GoA	4.9	0.30	3.7	0.91	231.1	63.1	0.189	0.040	173.8	118.2	139.5	29.1
2	GoA	6.7	0.30	4.5	0.30	175.2	17.1	0.156	0.064	128.4	71.0	129.3	35.6
3	NkA	6.0	1.22	5.4	0.30	242.7	32.4	0.232	0.023	301.6	27.2	255.6	52.7
4	NkA	5.2	1.22	3.0	1.22	195.3	7.4	0.181	0.018	103.6	28.4	154.5	61.2
5	BnA	6.2	0.30	6.0	0.61	222.8	18.1	0.210	0.038	287.9	102.2	257.1	31.0
6	BnA	5.6	0.61	5.2	0.61	159.4	32.6	0.151	0.006	125.1	35.4	138.2	92.4
7	Cx	5.4	1.52	4.9	1.52	153.7	22.9	0.165	0.011	126.5	48.6	128.3	75.5
8	Cx	5.6	0.00	4.5	0.91	378.2	75.2	0.190	0.009	317.8	14.7	327.4	81.3
9	ErA	4.7	0.61	4.3	0.61	233.3	40.6	0.189	0.021	228.5	24.4	200.2	23.2
10	NfA	4.9	0.91	4.5	0.30	369.0	86.8	0.146	0.010	248.3	91.2	180.4	67.0
11	NoA	4.9	0.30	4.9	0.30	389.7	9.2	0.145	0.014	277.7	8.1	222.0	63.0
min/max		0.701		0.500		0.394		0.625		0.326		0.392	

Table 25–6. Variation in plant height, measured to the topmost leaf tip, on 23 June, 1993. Areas sampled were used for 50% silking and black layer determination.

#	Soil	No.	Mean height	Std. dev.	C.V.	Min	Max	min/max
		-	m	m	%	m	m	min/max
1	GoA	19	1.00	0.23	23.26	0.51	1.37	0.372
2	GoA	19	1.14	0.11	10.01	0.86	1.30	0.662
3	NkA	20	1.23	0.13	10.52	0.86	1.42	0.606
4	NkA	20	1.07	0.18	16.58	0.69	1.35	0.511
5	BnA	20	1.18	0.18	15.18	0.84	1.65	0.509
6	BnA	22	1.29	0.14	11.18	0.94	1.60	0.588
7	Cx	20	1.34	0.15	11.23	1.07	1.60	0.669
8	Cx	22	1.72	0.16	9.21	1.37	2.08	0.659
9	ErA	21	1.69	0.14	8.50	1.45	1.93	0.751
10	NfA	20	1.85	0.22	11.80	1.27	2.11	0.602
11	NoA	19	1.47	0.12	8.02	1.24	1.65	0.752
	min/max		0.541	0.478	0.345	0.352	0.616	0.242

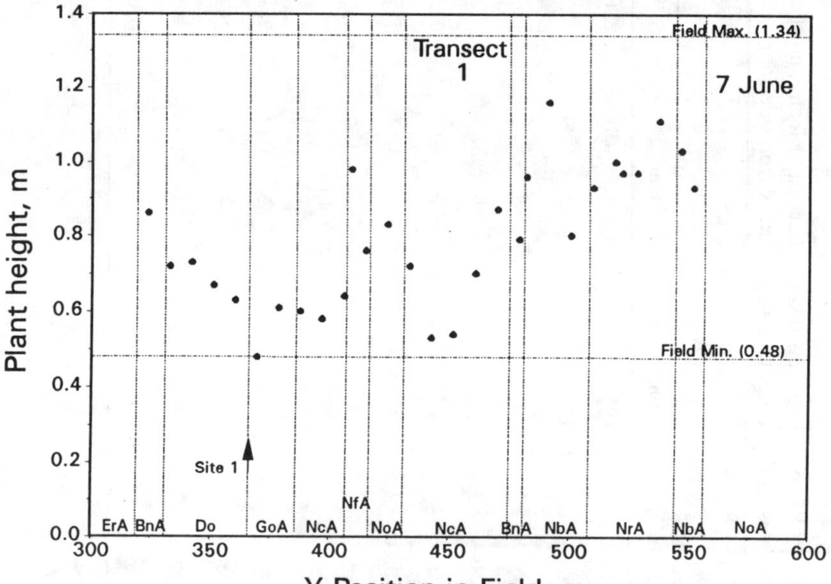

Fig. 25–4. Plant heights measured on 10-m spacings during the season. Transect corresponds to number 1 on Fig. 25–1.

The second measurement was also obtained opportunistically. On 10 June, the more-stressed areas in the field were visually distinct from others that were not apparently stressed. The plants in stressed areas were as short as 0.48 m; the non-stressed plants were up to 1.34 m tall. Other signs included severe leaf rolling, with the associated straightening, to the point that many plants did not extend > about 15 cm into the space between the rows. The stressed plants also had a blue-gray cast. In contrast, the areas with the tallest plants showed no visually-apparent stress: no rolling, normal leaf position, and typical green coloring.

Attempts on such short notice to locate a thermal scanner to image the field were not successful, so we made infrared thermometer (IRT; Everest Interscience, Tustin, CA; model 4000, 4° field of view) measurements on the same transects. To do this, an IRT was connected to a datalogger (CR21X, Campbell Scientific, Logan, UT) mounted on a platform strapped to the operator's waist. The operator walked along the row at a steady pace, pointing the IRT forward and down at a 45° angle above the row. The datalogger recorded the temperature at 1-s intervals. A manual switch allowed the operator to start and stop at known locations. Assuming the pace was steady (average was 1.4 m s^{-1}) allowed computation of location from the time of the individual measurements. The first such measurements were made on 10 June during the most severe stress period. After the 46-mm rain on 12 June, measurements were made on 14, 17, and 19 June.

Data from transects 1 and 5 are shown in Fig. 25–5 and 25–6. Transect 1 includes site #1, which appeared to be the area with the most severe stress. Transect 5 includes site #11 (NoA), which never showed visible signs of stress. The timing of the four sets of measurements shows the severity of the drought stress before the rain, near total relief of stress afterwards, and the development of drought conditions nearly as severe as before. The variation in the canopy temperature increased with drought stress. This was suggested by Aston and Van Bavel (1972) as an early indicator of the need for irrigation. Cloudiness in the humid Southeast may prevent successful implementation of field-scale variation in IRT readings as an irrigation scheduler. It may be possible, however, to isolate an indicator area, for which a limited transect might be made during cloud-free periods. The first 40 m of transect 5 might be such an area, because it includes the GoA, NcA, and NoA soils, which span the range of response observed in the field.

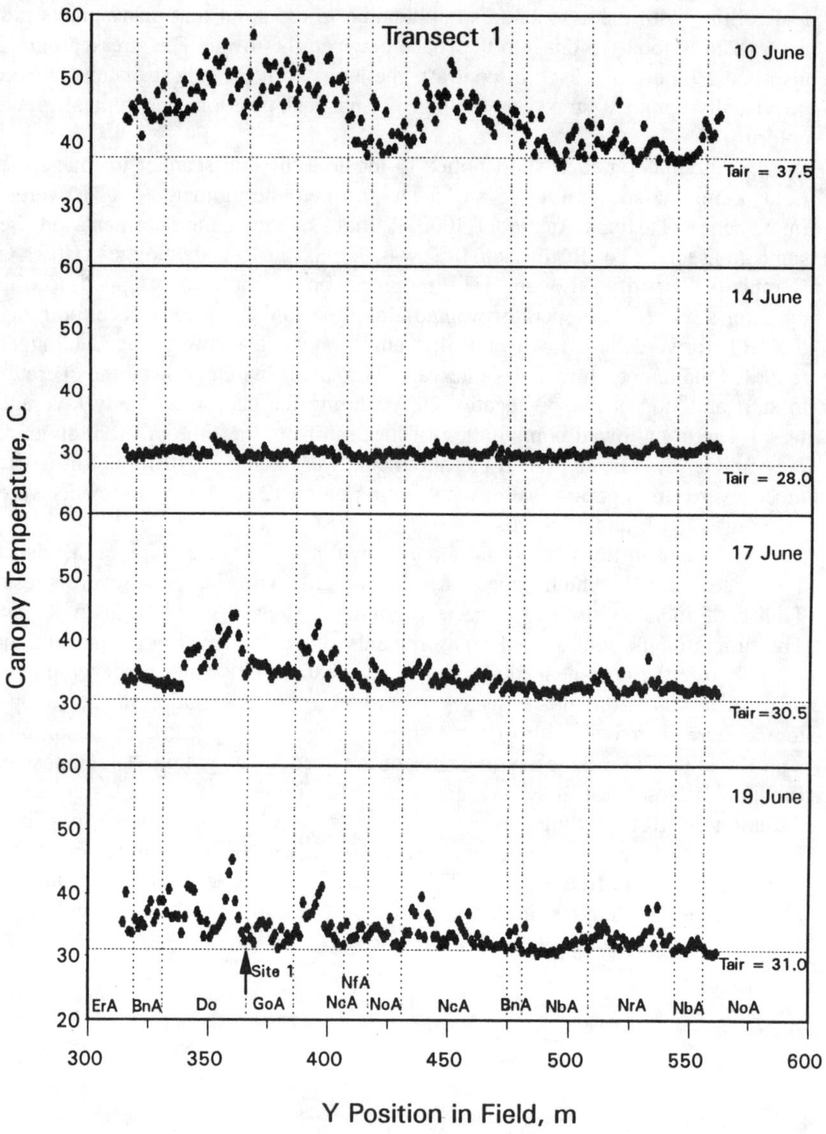

Fig. 25–5. Crop temperature measured with an IRT as a function of distance along transect #1 for one date before and for 3 dates after a 46-mm rain. The area around site #1 was apparently the most stressed in the field.

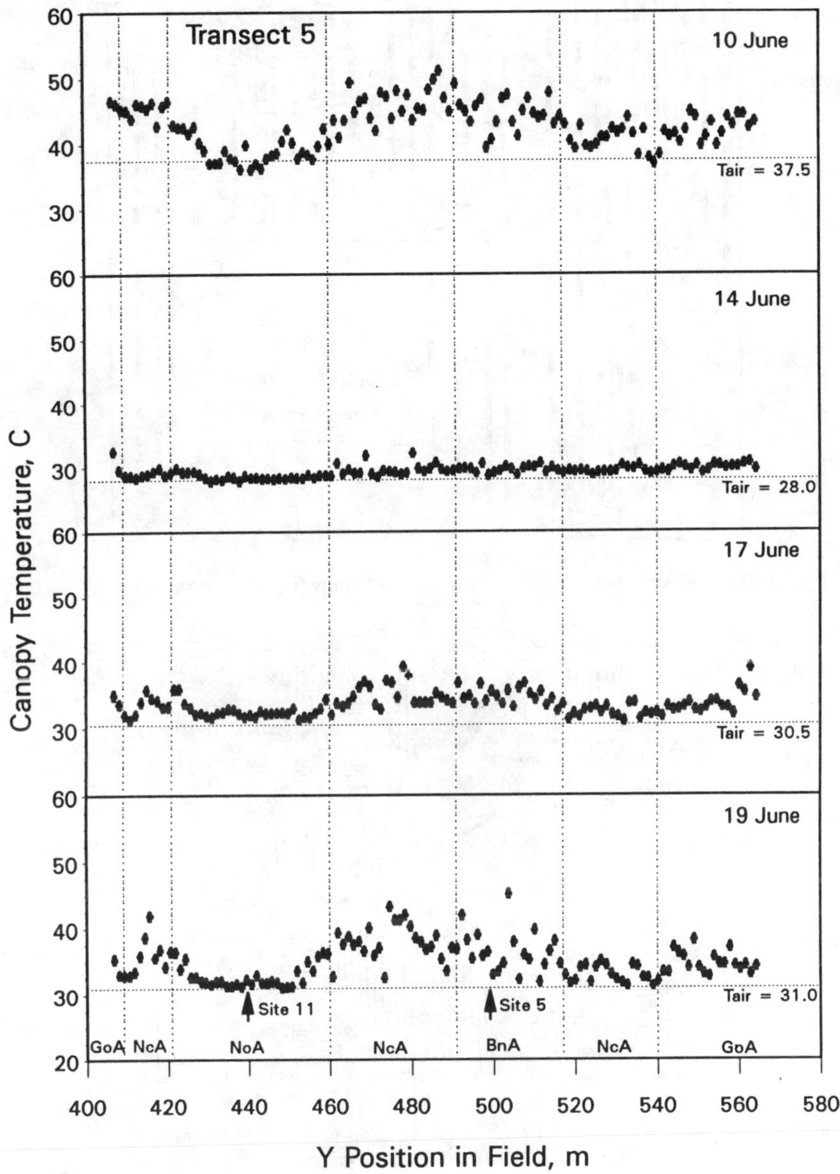

Fig. 25–6. Crop temperature measured with an IRT as a function of distance along transect #5 for one date before and for 3 dates after a 46-mm rain. The NoA site noted was never observed to be stressed.

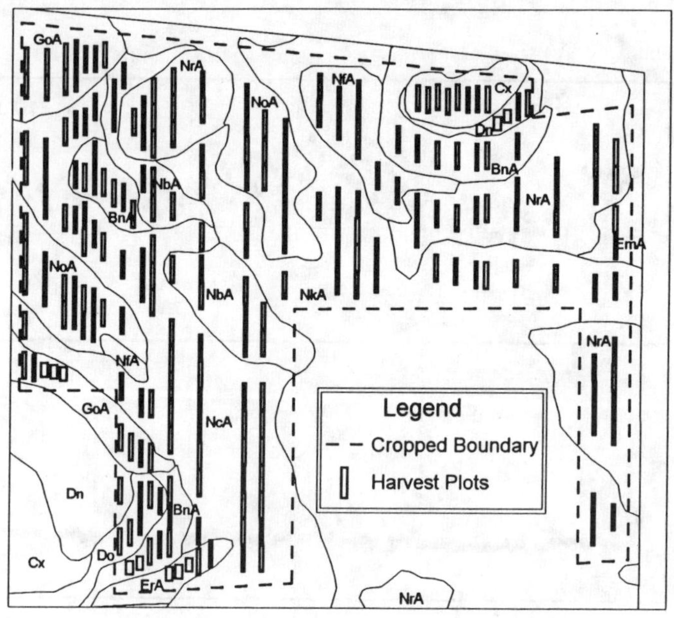

Fig. 25–7. Harvest plots overlaid on the soil map for 1993 corn harvest.

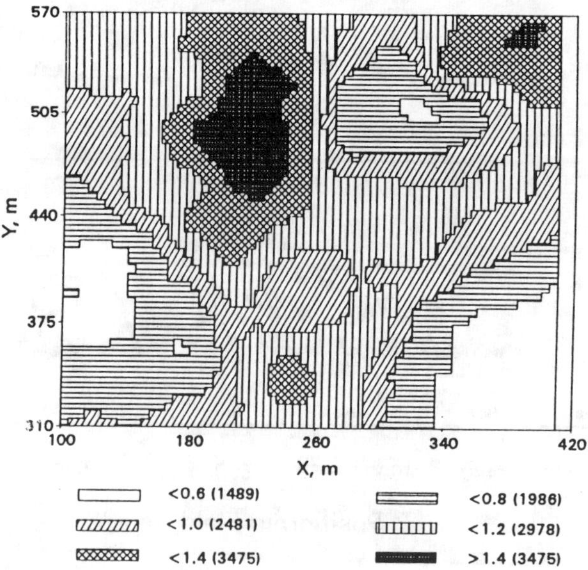

Fig. 25–8. Relative corn yield, computed as harvest plot yield divided by the field mean, kriged and mapped over the sampling area.

SPATIAL YIELD SAMPLING

Site-specific harvest plots were obtained in a manner similar to other yrs. Individual plots (18 m^2) were planned based on CAD drawings of the field, and then were flagged in the field. As mentioned above, the field plot combine was used to harvest the corn. Plot yields were determined and attributed to the corresponding soil type from the map (Fig. 25–7). Yields from harvest plots proximal to the 11 representative sites are reported with data from the sites in Table 25–5. The relative yield map of the entire field is shown in Fig. 25–8. It was produced by dividing all yield data by the mean yield for the field and then kriging (Sadler & Busscher, 1992; Sadler et al., 1994a).

SMALL- VS. LARGE-SCALE VARIABILITY

Variability in this and other fields exists on scales ranging from the small (plant-to-plant in a row; <1 m) to the large (field sections; >100 m). Accounting for the large-scale variability was the original intent of soil mapping. In this field, as observed in 1993 (Fig. 25–9) and before (Karlen et al., 1990), variation within map units approaches the magnitude of variation among map units. Yields between separate inclusions of the same map unit are often markedly different. For these reasons, sampling strategies used in the project have evolved from those that allow primarily map-unit comparisons to those that also allow strictly spatial comparisons (i.e., kriging) as well. Both techniques require coordinate locations of yield plots, which have always been determined for this project.

The 1993 study allowed an examination of variation in several physiological and physical characteristics of the crop, both in space and across soil map units. The four paired samples of soil types allowed direct comparison of soil water balance and various physiological characteristics, although between only two sites per unit. Where possible, in all tables, standard deviations and the ratio between the minimum and maximum measurement are provided so that the reader may judge how much variation existed, both in the samples and among soil types. Table 25–6 in particular shows variation in plant height both within and among samples of about 20 plants. In general, variation among map units appears larger than that within map units, but the magnitudes are so similar that they reduce confidence in comparisons among map units. The transects of plant height (Fig. 25–4) and canopy temperature (Fig. 25–5 and 25–6) also support the need for spatial analysis. Comparison of the soils map (Fig. 25–1) and the kriged yield map (Fig. 25–8) suggests some, but not total, correspondence.

Several preliminary conclusions can be derived from the 1993 results. First, under drought stress, large differences in most measurable parameters can be obtained, both within and among map units. The water balance showed measurable differences in rainfall/runoff partitioning for single storms, as well as noticeable variation in rate of water use. The IRT measurements documented the variation in canopy temperature in space, the recovery after rain, and the subsequent reoccurrence of stress. This confirms the recommendation by Aston and van Bavel (1972) that increased spatial variation in canopy temperature may

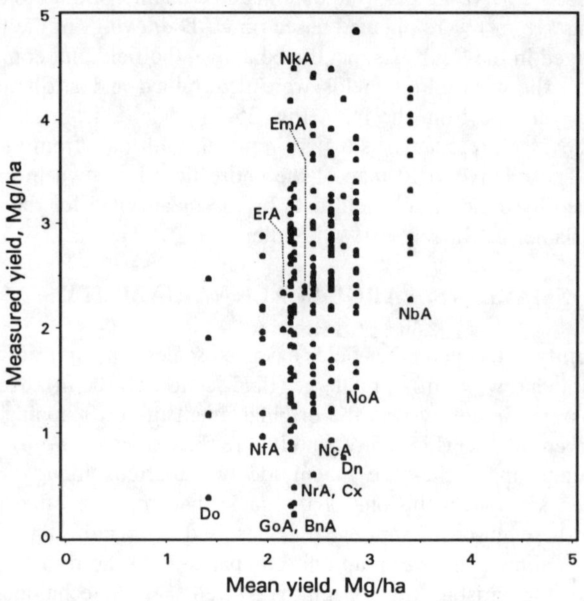

Fig. 25–9.Individual harvest plot yields plotted against the mean yield for each map unit. Plot area was 18 m^2 for each of 209 plots. Mean yield was 2481 ± 916 kg ha^{-1}, and the range was from 214 to 4849 kg ha^{-1}, all expressed at 15.5% moisture.

SUMMARY AND CONCLUSIONS

be a useful early indicator of water stress.

Variation in timing of tasselling, silking, and black layer was about 15% of the maximum. Stress appeared to delay tasseling and silking, but no clear-cut result obtained for time of maturity. If stress delayed silking, but shortened the grainfilling period, as suggested by Shaw (1988), differences among soils in timing of maturity might be masked. Although differences in timing were relatively < differences in other parameters, most simulation models rely primarily on air temperature to drive phenology. Because air temperature would be the same across soil types, models are not likely to account for variation in timing of maturity. Variation in LAI and biomass during vegetative growth on 7 June was very large - 87% and 68% of the maximum, respectively. Variation in height on that day was about 50% of the maximum. By mid-silk, variation in LAI had dropped to 62% and variation in biomass had dropped to 44% of the maximum. Variation in potential kernel number was about 25% of the maximum observed. By harvest, the actual kernel number was, as expected, more variable - about 61% of the maximum. Minimum kernel weight and ear number per unit ground area were 38% and 30% of the maximum. Final yield at the 11 primary sites varied from 125 to 318 g m^{-2} dry weight; for this, the range is 67% of the maximum. Yield over the entire field (209 plots) averaged 2481 ± 916 kg ha^{-1} and ranged from 214 to 4849 kg ha^{-1}, all expressed at 15.5% water content.

Such large differences in grain yield indicate that large differences in residual N may exist across this field. Although one could not have foreseen the severity of the drought, an operator will have to deal with potential leaching if fertilizer application for subsequent crops do not account for the residuals. Additional analysis and stochastic simulation could help estimate risk from such an occurrence.

REFERENCES

Aston, A.R., and C.H.M. van Bavel. 1972. Soil surface water depletion and leaf temperature. Agron. J. 64:368–373.

Baker, J.M., and R.R. Allmaras. 1990. System for automating and multiplexing soil moisture measurement by time-domain reflectometry. Soil Sci. Soc. Amer. J. 54:1–6.

Karlen, D.L., E.J. Sadler, and W.J. Busscher. 1990. Crop yield variation associated with Coastal Plain soil map units. Soil Sci. Soc. Amer. J. 54:859–865.

Sadler, E.J., and W.J. Busscher. 1992. Site-specific yield histories on a SE Coastal Plain field. Agron. Abstr. 84:312.

Sadler, E.J., and W.J. Busscher. 1993. Soil water content and water use for conventional and conservation tillage in the SE Coastal Plain. Agron. Abstr. 85:327.

Sadler, E.J., W.J. Busscher, and D.L. Karlen. 1994a. Site-specific yield histories on a SE Coastal Plain field. Site-Specific Management for Agricultural Systems. Proc. 2nd Workshop. ASA/CSSA/SSSA, Madison, WI. (This Proceedings)

Sadler, E.J., W.J. Busscher, and D.L. Karlen. 1994b. Multi-year analysis of site-specific yields on a Coastal Plain field. (in preparation).

Sadler, E.J., D.E. Evans, W.J. Busscher, and D.L. Karlen. 1993. Yield variation across Coastal Plain soil mapping units. pp. 373–374. *In* Robert, P.C., et al. (eds). Soil Specific Crop Management. ASA/CSSA/SSSA Madison, WI.

Shaw, R.H. 1988. Climate requirement. pp. 609–638. *In* Sprague, G. F., et al. (eds). Corn and corn improvement. Agronomy No. 18. American Society of Agronomy, Madison, WI.

Stone, K.C., and E.J. Sadler. 1991. Runoff using Green-Ampt and SCS curve number procedures and its effect on the CERES-Maize model. ASAE Paper No. 91–2612.

USDA-SCS, 1986. Classification and correlation of the soils of Coastal Plains Research Center, ARS, Florence, South Carolina. South National Technical Center, Ft. Worth, TX.

26 Yield Mapping By Electromagnetic Induction

D. B. Jaynes
T. S. Colvin
J. Ambuel

USDA-Agricultural Research Service
National Soil Tilth Lab
Ames, Iowa

Traditional farming has treated the field as the smallest management unit. Even in fields of limited extent, however, the soils within will normally vary in their properties from point to point. These variations can profoundly affect the soil's response to management such as tillage and chemical inputs. To optimize the impact of chemical inputs, it is best to tailor the application rate to the specific requirements of a location, rather than to apply a uniform application that reflects the mean requirement of the whole field. This strategy may lead to more "bang for the buck", but may also reduce off-site impacts of chemical applications since vulnerable areas within a field could be treated differentially.

The concept of farming the soil is an old one (Goering, 1993). It has only been recently, however, that serious consideration has been given to the concept, as emerging technology has overcome some of the earlier hurdles to implementation. Several strategies exist for soil specific farming. These include on-board sensors that can measure some indicator of chemical need, such as soil nitrate concentration (or a surrogate measure such as organic carbon), and allow for real time adjustments of fertilizer application rate. A second strategy is to map the soil variations or capabilities across the field using some relational reference frame. This map can then be stored in the memory of a smart applicator which, when connected to a real-time position sensor, can vary the application rate based on soil capability and some preprogrammed response strategy.

For this second strategy to work, methods must be available to accurately position equipment under variable weather conditions and to build a soil capability map on which to base application rates. The emergence and refinement of real-time and accurate global positioning systems (GPS) has virtually achieved the first requirement. Real-time positioning is now possible with accuracies on the order of meters. The second requirement of developing capability or requirement maps has not seen as much progress. Numerous studies have explored using soil survey maps for this purpose, but within map unit variability of the soil physical and chemical properties that determine yield limit this approach. Also, converting the maps to a field scale which usually involves an enlargement of the original map scale (1:15840), results in poorly positioned soil unit boundaries, compromising their usefulness. Lastly, soil survey maps have a minimum soil unit size (1.2 ha) below which delineations are not represented. Thus, soils of contrasting characteristics but limited extent are not represented by these maps.

Ideally, what is needed is an accurate, fast, inexpensive method of producing soil capability maps at a level of resolution that is comparable to the scale of chemical application (boom width or even distance between individual applicators). One possible method of quickly and cheaply measuring soil properties is non-contacting electromagnetic induction (EM). EM responds to the electrical properties of soil which are determined by a complicated interplay of soil clay content, water content, and salinity. Because bulk electrical conductivity is a function of numerous soil properties, EM has been successfully used to measure not only soil salinity (Rhoades & Corwin, 1981), but also soil water content (Kachanoski et al, 1988), soil clay content (Williams & Hoey, 1987), soil cation exchange capacity and exchangeable Ca and Mg (McBride, et al., 1990), and atrazine partition coefficients (Jaynes, et al., in review).

Crop yields are influenced by a wide range of intrinsic and extrinsic properties of which nutrient and water stress are the most important. Where fertilizer management has eliminated most nutrient stress, yield variability across fields is controlled mostly by spatial variations in water stress, either droughtiness or excess moisture. Areas of droughtiness or excess moisture should also have differing electrical properties since we would expect these areas to have different clay, water, and salt contents. Thus, a mapping of the electrical properties as provided by EM measurements may provide a mapping of yield variability as influenced by spatial patterns of potential water stress. Since EM meters are non-contacting, they could be used to map a field very quickly if linked to a positioning system such as GPS.

In this study, we investigate the correlation between measurements made with an EM meter linked to a GPS system and measured corn and soybean yields. If successful, this approach would allow for the delineation of high and low yielding areas within fields as affected by water stress.

METHODS

Three fields, known as the Bassett, Baker, and Thompson fields, were chosen for this initial study. All of the fields were located within 15 km of Ames, IA. The predominant soil association in this area is Clarion-Nicollet-Canisteo, with the Clarion soils being well drained, the Nicollet soils somewhat poorly drained, and the Canisteo soils poorly drained. In addition to the major soils in the association, important soils found within these fields include Webster and Harps, both poorly drained, and Okoboji, a very poorly drained soil occupying closed depressions within the landscape.

The Bassett field was surveyed by EM on 27 April 1992 before field work had started for the year. Measurements were made in early spring to minimize effects of crop transpiration and variable evaporation on soil profile water content. The 31 ha field was planted to corn in 1992 and soybean in 1993 and spatial yield patterns measured at harvest both years. The second two fields adjoined each other and were each ≈15 ha. EM surveys of each were conducted 3 d after the Bassett field survey. Although yield data have been collected for these fields for several years, only yields from the 1992 and 1993 growing seasons were analyzed for this study. In 1992, the Baker field was planted to soybean and the Thompson field to corn. In 1993, the crops were switched with corn planted on the Baker field and soybean on the Thompson field.

EM Measurements

Electromagnetic induction measurements were made with an EM38 induction meter (Geonics Limited, Ontario, Canada). (Mention of commercial products is solely to provide specific information and does not constitute endorsement by the USDA over other products not mentioned). Readings were made with the meter in the vertical dipole orientation, 20 cm above the soil surface. The EM38 integrates electrical conductivity measurements over a distance approximately equal to its length of 1 m and over a depth of ≈3 m, although the measurement is mostly influenced by properties in the 0 to 1–m depth increment (McNeill, 1980b). A vertical rather than horizontal orientation was used so that slight variations in the spacing between the meter and soil surface would have minimal effect on the readings (McNeill, 1980b). The meter produces a sinusoidal magnetic field that induces electrical currents in the soil which in turn create a secondary magnetic field that is measured by the meter (McNeill, 1980a). The apparent electrical conductivity is measured in mS/m.

The meter was attached to the center of a 3–m long, 5– x 15–cm (2"x6") wooden boom. A bicycle wheel was attached to the end of the boom and the front of the boom was hitched to a five wheel all-terrain vehicle. A long wooden boom was required because the meter is very responsive to metal components and electrical engine noise. For the same reason, duct tape was used to secure the EM38 meter to the boom. A cable from the meter attached to a data logger (Polycorder 516C-A, Omnidata International Inc., Logan UT) on the vehicle. The data logger converted the analog signal from the meter to a digital signal and transmitted the value every 3 s to the GPS computer which combined and stored the measurements with GPS location information.

GPS Measurements

The GPS system used has been described in Ambuel et al., (1991). It consists of a mobile and base GPS receiver, two radios, RF modems and computer. The mobile receiver, radio, modem, and computer were secured to the back of the vehicle. The base station GPS receiver, radio, modem, and computer were placed at a known global position within half a mile of the field. In this configuration, the GPS system can be used as a real-time positioning system. For this study, however, all data was recorded within the computer and processed at a later date. The data contained the vehicle's position by latitude and longitude and the EM meter reading at that point. The vehicle was driven at ≈8 km/h along straight transects. With the three second recording interval and this speed, the distance between readings was about 6 m. Eight or nine equally spaced transects were driven across each field so that the entire width was covered. Field edges had to be avoided when barbed wire fences were present (Bassett field), because of the meter's sensitivity to metal objects.

Yield Measurements

Yields measurements were made with the field-plot combine described by Colvin (1990). This system consists of a weigh tank and moisture meter. Harvesting was conducted along transects. An approximate distance of 20 m was driven on the Bassett field (15 m on Baker & Thompson) while grain was loaded into the weight tank. At the end of the measured distance, the tank was weighed, the moisture content measured, and the grain dumped into the grain wagon below. No positioning system was attached to the combine. Distances along the transects were determined by counting wheel rotations, while the transects themselves were surveyed after harvest. Again eight or nine transects spaced across each field were harvested. No attempt was made to co-align harvest and EM survey transects. Three rows of corn were harvested in each transect and four rows of soybean. Yields for each section of a transect were converted to Mg/ha for reporting.

Computations

Positions corresponding to individual EM readings were corrected for the distance between the GPS and the EM meter. Contour maps of EM and yield were created for each field using Surfer software (Golden Software, Inc., Golden, CO). Values on a regular grid (~16 m spacing) were interpolated from the measured data using a kriging interpolator provided with the Surfer software and a quadrant search pattern for nearest neighbors. A gray scale map of yield was then produced from the gridded data. Contours of equal apparent conductivity were also created from the interpolated grid of EM values with the Surfer software. These contours were then overlain on the yield maps to allow qualitative comparison between the two coverages.

Comparing data collected at different locations (and measured with different sample sizes) across a field is difficult. Overlaying maps at the same scale provides qualitative comparisons of spatial data but does not allow for a numerical measure of the correlation between the two. Since the gridded data for both yield and EM were on the same grid, this data could be compared quantitatively. Correlation maps were created by computing the statistic:

$$r_{i,j} = (EM_{i,j} - \langle EM \rangle)(YI_{i,j} - \langle YI \rangle)/(s_{EM}{}^2 s_{YI}{}^2)^{1/2}$$

where $EM_{i,j}$ and $YI_{i,j}$ are the EM and yield values at the i,j grid location, $\langle EM \rangle$ and $\langle YI \rangle$ are the mean values and $s_{EM}{}^2$ and $s_{YI}{}^2$ the variances. The traditional correlation coefficient can be found by summing $r_{i,j}$ over all grid points and dividing by the total number of grid points. In addition, the area of the fields contributing to negative or positive correlations can be computed giving an indication of the spatial extent of the correlation across a field.

RESULTS

EM Measurements

The EM measurements varied smoothly across the fields but varied over a relatively large range. Apparent electrical conductivity ranged from a low of 10.0 mS/m to a high of 65.0 mS/m. The mean EM value (and standard deviation) was 40.3 (8.0) mS/m for the Bassett field, 35.1 (9.8) mS/m for the Thompson field, and 31.6 (9.6) mS/m for the Baker field.

The EM measurements were not random but showed distinct patterns in each field (Fig. 26–1 to 26–3). In the Bassett field (Fig. 26–1), high EM values (> 50 mS/m) were found in the southeast quadrant and in central and north central areas of the field. High EM values tended to be found in local depressions or drainageways. Low EM values (< 30 mS/m) were found throughout the field and were associated with locations higher in elevation. These observations supported the hypothesis that EM identified areas where we would expect different degrees of water stress.

Similar patterns of EM measurements were found for the Thompson and Baker fields. In the Baker field (Fig. 26–2) high EM values were measured in a large area west of mid-field and along the southeast side. Low EM values were measured along the north edge, east central, and southeast side. The lowest EM values corresponded to eroded hilltop locations. In the Thompson field (Fig. 26–3), high EM values were found in the southwest corner and east central locations while low values were measured in the southeast corner and west central locations.

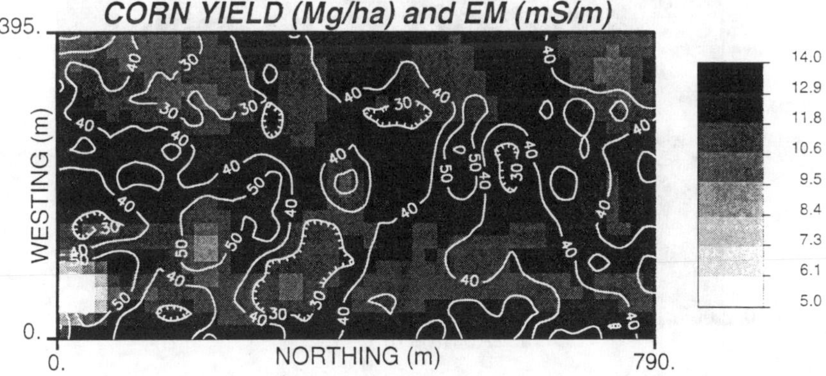

Fig. 26–1. Bassett field corn yield for 1992 in gray-scale and EM contours.

1992 Yields

Yields within all fields were variable. The growing season of 1992 started and ended drier than average with a very wet July. Average corn yields were 11.8 Mg/ha in the Bassett field and 10.7 Mg/ha in the Thompson field. Soybean yield in the Baker field was 3.13 Mg/ha. Overall, yields were very good in 1992, with the Iowa average being the highest on record for both corn and soybean.

From the yield maps shown in Fig. 26–1 to 26–3, it is clear that there were areas within all three fields where yields were well below and above the field average. Comparison between the EM and yield maps indicate that EM may be correlated with yield, but the correlation varies between fields. In the Bassett field (Fig. 26–1), high yields are associated with moderate EM readings and low yields with both low and high EM readings. Thus, there was no correlation between yield and EM with an r of -0.09. Approximately 44% of the field area had negative correlations between yield and EM (Fig 26–7a) with distinct areas within the field contributing to either a positive or negative correlation. If crops in the areas differentiated by EM mapping experienced water stress, stress from both too much and too little water must have reduced yields, and thus, resulted in poor correlations.

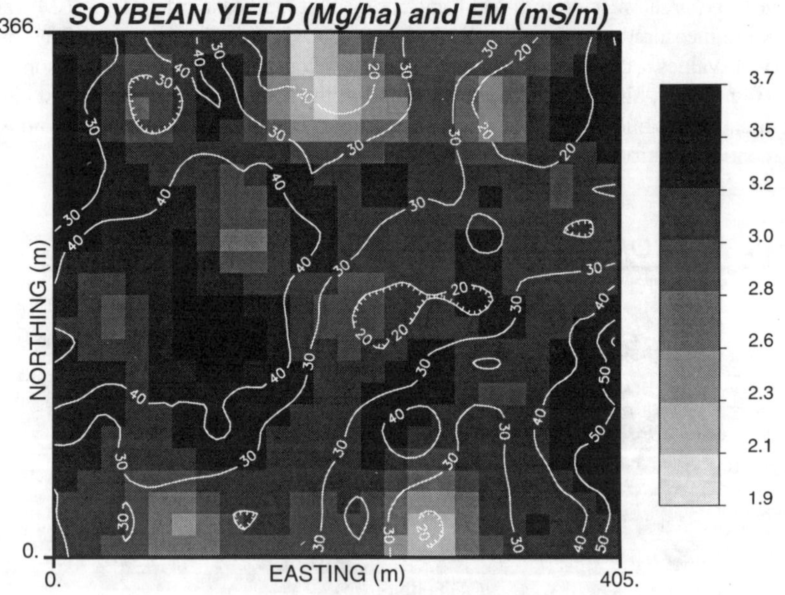

Fig. 26–2. Baker field soybean yield for 1992 in gray-scale and EM contours.

In the Thompson field a much stronger field-wide correlation was found between corn yield and EM values. The field-wide correlation was -0.55 with 76% of the field area contributing to the negative correlation (Fig. 26–8b). No consistent correlation was found between low EM readings and yield as in the Bassett field whereas low yields were clustered in areas with high EM values. Thus, this supported our hypothesis that yields were reduced by excessive moisture in these areas.

In contrast, the soybean yields in the Baker field showed a strong positive correlation with EM values. The field wide correlation was 0.45 with 68% of the field area contributing to the positive correlation. Assuming that high EM readings indicate wetter locations, soybean may have been more drought stressed in 1992 than corn leading to relatively higher yields in the wetter (high EM locations) and lower yields in the drier (low EM) areas (Fig. 26–3 and 26–8a).

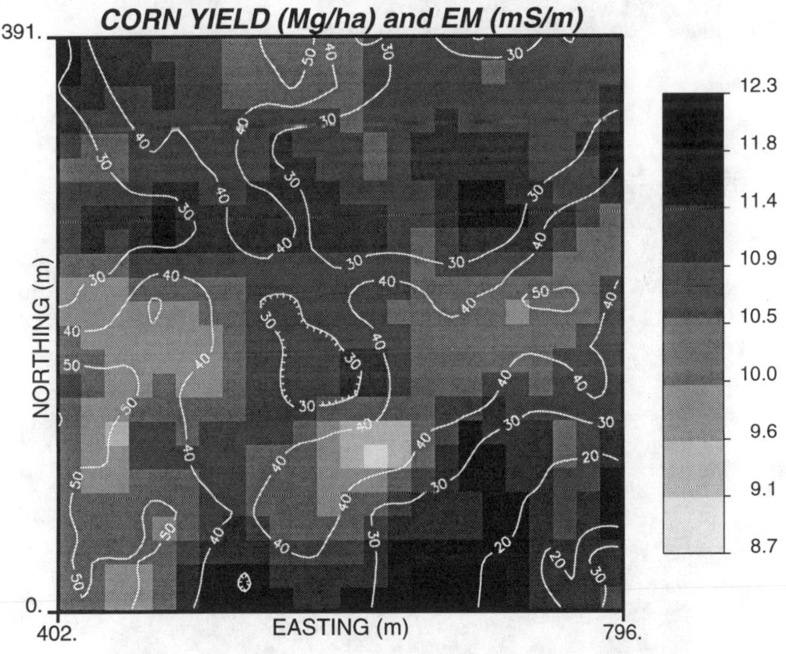

Fig. 26–3. Thompson field corn yield for 1992 in gray-scale and EM contours.

1993 Yields

The yr 1993 was extremely wet and corn and soybean yields were drastically reduced both state-wide and locally. Depressional areas were often filled with standing water for several days intermittently throughout the spring and summer leading to little or no yield in these locations. Figure 26–4 to 26–6 show the EM readings superimposed on the gray-scale representation of yield forthe three fields. Average yields were 1.94 and 2.39 Mg/ha for soybean in the Bassett and Thompson fields and 5.27 Mg/ha for corn in the Baker field.

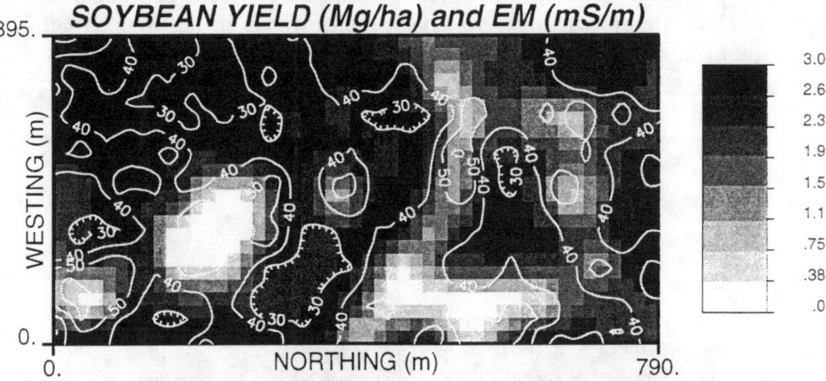

Fig. 26–4. Bassett field soybean yield for 1993 in gray-scale and EM contours.

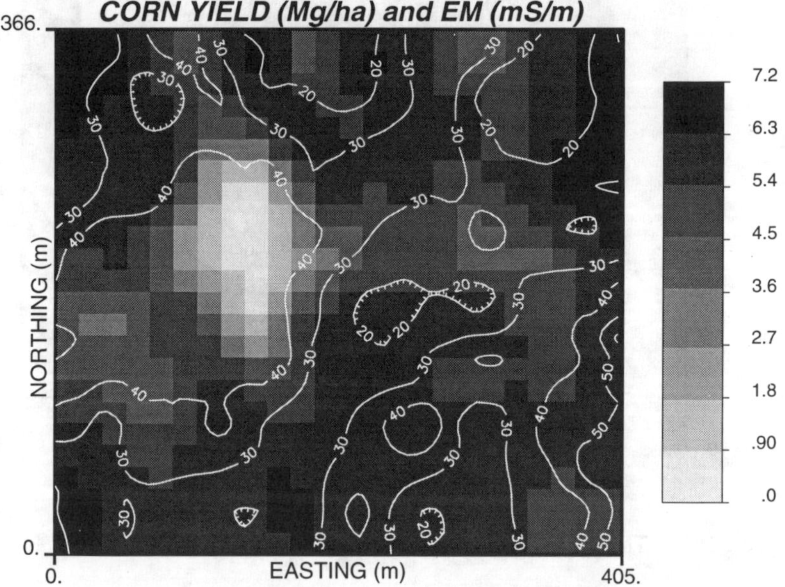

Fig. 26–5. Baker field corn yield in 1993 in gray-scale and EM contours.

Within the Bassett field (Fig. 26–4) large areas produced no yields. All of these locations are associated with high EM readings. Lower yields were also associated with some of the low EM readings which may have been due to the eroded nature of the locations rather than moisture stress. Overall for the Bassett field, nearly 80% of the field area contributed to a negative correlation of -0.63 (Fig. 26–7b).

In the Baker field, the two large areas of low corn yields (<4 Mg/ha) were located in the west central portion of the field (Fig. 26–5). These two areas are clearly associated with high EM values as indicated when overlaying the two maps. A second area of high EM values along the southeast side of the field does not correspond to an area of lower yields in 1993. These positive and negative correlations are clearly shown in Fig. 26–8c. For areas with low EM values, yields were average or above average, in contrast to 1992 where they were low. This is consistent with the hypothesis that low EM values correspond to droughty soils which produced low yields with lower rainfall but average yields when high rainfall removes drought stress. Thus, the field-wide correlation between yield and EM reversed in 1993 to -0.50 with ≈69% of the field contributing to the negative correlation.

Soybean yields in the Thompson field were below 1 Mg/ha in three large areas of the field (Fig. 26–6). Correlation between yields and EM values appears to be very good (Fig. 26–8d) with low yields again associated with high EM readings. The correlation map shows that most of the field area (86%) contributed to a the negative correlation of -0.73.

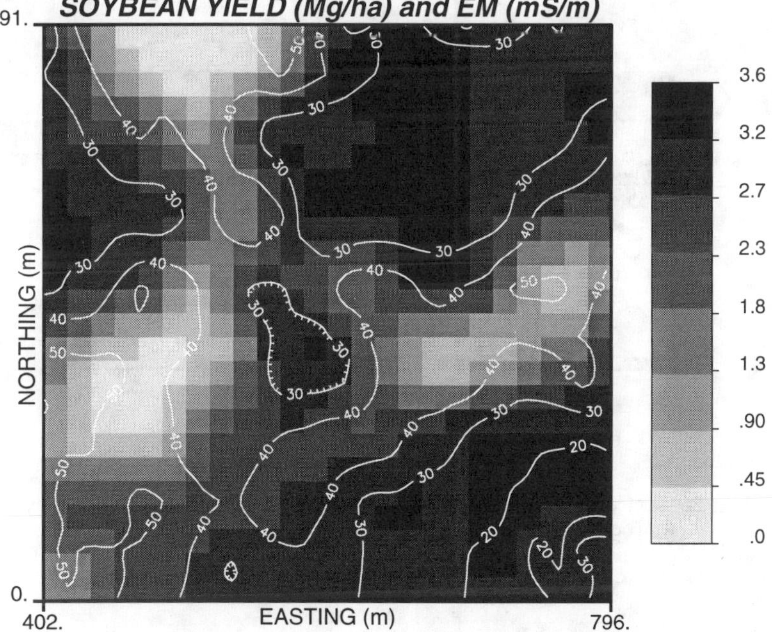

Fig. 26–6. Thompson field soybean yield for 1993 in gray-scale and EM contours.

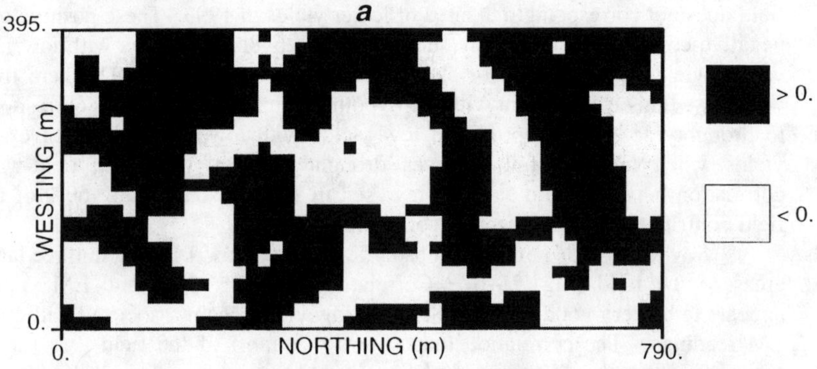

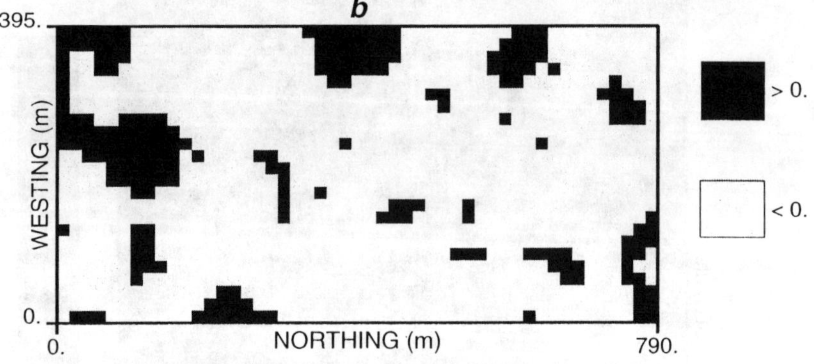

Fig. 26–7. Correlation map for Bassett field between EM and a) 1992 yield and b) 1993 yield.

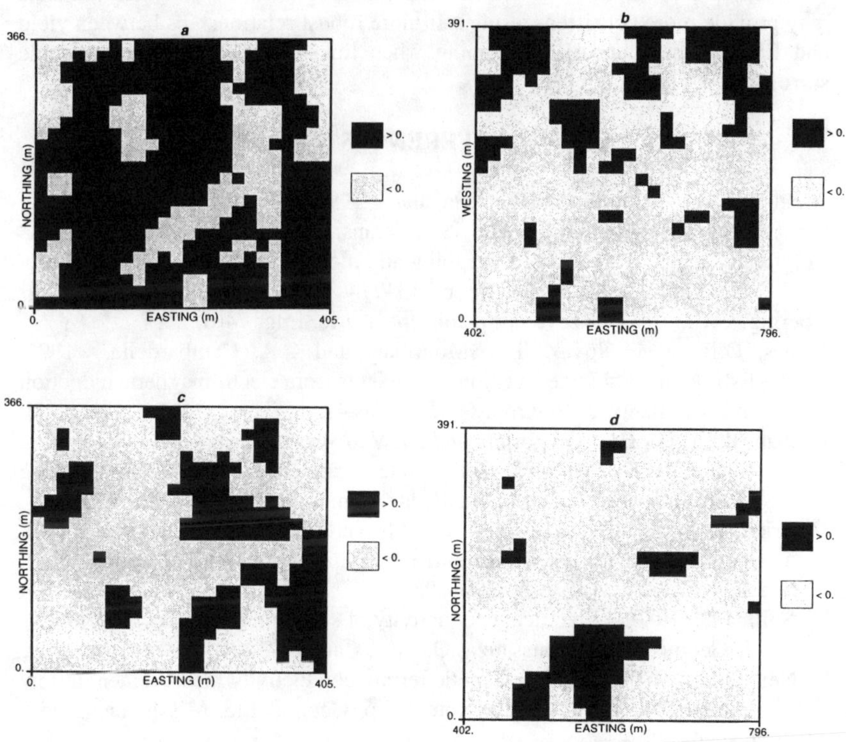

Fig. 26–8. Correlation map between EM and 1992 yield in a) Baker field and b) Thompson field, and 1993 yield in c) Baker field and d) Thompson field.

SUMMARY

EM measurements are correlated to yield, but not consistently from yr-to-yr. In the wet yr of 1993, both corn and soybean yields had strong negative correlations with EM readings in all three fields. We hypothesize that excess soil moisture was the dominant cause of low yields and these areas were clearly identified by EM mapping. In the more average rainfall yr of 1992, a simple correlation between EM and yield resulted in no correlation in one field, negative correlation in a second, and positive correlation in the third. We interpret these results to indicate that both excessive moisture and droughtiness were causing reduced yields especially in the Bassett field.

Variable correlations between fields may also be due to differences in crop response (corn vs. soybean) to low/high water stress, differences in degree of erosion between fields, and differences in management (eg. fertilizer management). A more complicated relationship between EM and yield rather than a simple linear one or a relationship that incorporates yearly climate data may provide more consistent results. If more robust relationships between yield and EM can be found, EM mapping when linked to GPS may be a viable surrogate for yield mapping.

REFERENCES

Ambuel, J., T.S. Colvin, and K. Jeyapalan. 1991. Satellite based positioning system for farm equipment. SAE Trans., 100–2:323–329.

Colvin, T.S. 1990. Automated weighing and moisture sampling for a field-plot combine. Appl. Eng. Agric. 6:713–714.

Goering, C.C. 1993. Recycling a concept. Agric. Eng. 74(6):25.

Jaynes, D.B., J.M. Novak, T.B. Moorman, and C.A. Cambardella. 1995. Estimating herbicide partition coefficients from electromagnetic induction measurements. J. Environ. Qual. 24:36-41.

Kachanoski, R.G., E.G. Gregorich, and I.J. Van Wesenbeeck. 1988. Estimating spatial variations of soil water content using noncontacting electromagnetic inductive methods. Can. J. Soil Sci. 68:715–722.

McBride, R.A., A.M. Gordon, and S.C. Shrive. 1990. Estimating forest soil quality from terrain measurements of apparent electrical conductivity. Soil Sci. Soc. Am. J. 54:290–293.

McNeill, J.D. 1980a. Electrical conductivity of soils and rocks. Tech. Note TN-5, Geonics Ltd. Mississauga, Ontario, Canada.

McNeill, J.D. 1980b. Electromagnetic terrain conductivity measurement at low induction numbers. Tech. Note TN-6. Geonics Ltd. Mississauga, Ont., Canada.

Rhoades, J.D., and D.L. Corwin. 1981. Determining soil electrical conductivity-depth relations using an inductive electromagnetic soil conductivity meter. Soil Sci. Soc. Am. J. 45:2552–260.

Williams, B.G., and D. Hoey. 1987. The use of electromagnetic induction to detect the spatial variability of the salt and clay contents of soils. Aust. J. Soil Res. 25:21–27.

SESSION II

MANAGING VARIABILITY

27 Managing Spatially Variable Weed Populations

D. A. Mortensen
G. A. Johnson
Dawn Y. Wyse
Alex R. Martin

Dep. of Agronomy
University of Nebraska
Lincoln, Nebraska

Currently, spatial variation in weed type and infestation level are overcome by using higher inputs to manage for a mean field condition. Because of the patchy nature of weed distributions, often weed populations in large areas of fields are below economic thresholds and don't warrant control. An alternative approach involves characterizing spatial variation in weed populations, and managing with rather than overcoming it. Field studies designed to describe the numerical and spatial distribution of weed populations revealed that weed population distributions are best described by the negative binomial model. Spatial maps constructed from the same data suggest a high degree of spatial aggregation for individual and multi-species assemblages. While aggregation is observed in most fields, the agronomic relevance of these spatial patterns is poorly understood. Results of spatial analyses of weed populations in 12 Nebraska farm fields indicate that postemergence herbicide applications could be reduced by 71% and 94%, respectively, for broadleaf and grass weeds if herbicides were applied to existing populations. These reductions reflect the large areas of farm fields completely free of weeds. Spatial maps of weed seedling populations have also been used to simulate the spatial distribution of yield loss and of control decision rules. The purpose of these simulations was to determine the extent to which agronomic decisions are influenced by spatially variable weed populations. The number of unique weed seedling combinations far exceeded the number of % yield loss and weed control decision conditions, though significant spatial variation in % yield loss and weed control decision rules was observed. In conducting these analyses, aspects of weed population spatial variation were identified needing further study. First, most studies have been conducted in a reduced tillage corn-soybean rotation, thus spatial patterns in other crops and tillage systems have not been studied. Second, little is known about the spatial stability of weed populations through time. Finally, the link between spatial variability and agronomic decisions must be understood more fully. This paper uses spatial pattern data to determine the potential benefits of such information.

INTRODUCTION

The annual value of corn and soybean losses due to weeds in the United States is estimated at > 1.9 billion dollars (Bridges et al., 1992). Farmers have reduced the risk of crop losses from weed competition by increasing herbicide usage. A 1988 survey indicated that herbicides were used on 96% of the corn and 100% of the soybean acreage in the north central states (USDA-ERS, 1989), up from 57% and 49% in 1966 (Swanson & Dahl, 1989). Herbicide use accounts for ≈80% of the pesticides used in corn and soybean.

Pest management specialists have looked to economic threshold principles as an alternative management approach, and have experienced limited success in implementing threshold theory. Adoption has been constrained by concerns over risk of pest control failure, seed rain from uncontrolled weeds, and the fact that threshold densities are often so low that fields would be sprayed in most yrs. In two recent studies, Bauer and Mortensen (1992) and Mortensen et al. (1993a) demonstrated that when economic thresholds are implemented herbicide spray frequencies can be reduced for some weed species (short seedbank longevity), while others like velvetleaf (*Abutilon theophrasti*) require spraying or some other therapeutic control measure in most yrs. An alternative method of accomplishing herbicide reduction would be to design a cropping system that takes advantage of the unique aspects of individual farmsteads, crops (phenologies, cultural practices, and interference) and pest biology to accomplish the pest management objectives. Alternatively (or in addition), herbicides could be applied only where weeds occur. Cousens (1987) recognized the value of this approach when he wrote "even when herbicide spraying is to be on a whole-field basis, recognition of the patchy nature of weed distributions can be important in applying thresholds. If weeds have an aggregated distribution, such as the negative binomial, estimates of yield loss using the mean field density may differ markedly from estimates calculated on a quadrat by quadrat basis".

Weed seedlings are spatially aggregated in the field (Dessaint et al., 1991; Mortensen et al., 1993b; Navas, 1991; Thornton et al., 1990; VanGroenendael, 1988; and Wiles et al., 1992). An aggregated population is one in which individual plants occur in clumps or patches of varying density and size with few or no plants occurring between clumps (Cottam et al., 1957). Based on the aggregate nature of weed seedling populations, herbicide cost and environmental impact may be reduced if herbicide is applied only where weeds are present or exceed some predetermined threshold. Several intermittent herbicide application systems have been or are being developed as a way of applying herbicide to weed patches or individual plants (Duff, 1993; Felton et al., 1991; Mortensen et al., 1993c; Thompson et al., 1991; VonBargen et al., 1992; and Woebbecke et al., 1994). The purpose of this paper is to describe the relationship between spatially variable weed populations and resulting weed management decisions. In addition, spatial population distribution maps were used to simulate the potential for herbicide use reduction if intermittent herbicide application was commercially feasible.

MATERIALS AND METHODS

In 1993, weed seedling populations were sampled in seven corn and five soybean fields prior to the first cultivation or postemergence herbicide application. In each field, a preemergence herbicide application (farmers choice) was made in a 38 cm band over the row. This method of application provided an opportunity to study the effect of herbicide treated and untreated areas on the spatial arrangement of weed seedlings in the fields. Weed seedling populations were sampled just prior to the first cultivation or postemergence herbicide application (for corn, May 15 through June 4, and for soybeans, June 11-June 22). Scouting is commonly performed at this time to determine the need for and type of postemergence herbicide. In each field, a 4–ha area was sampled in a 7–m grid to facilitate geostatistical analysis. A total of 812 X,Y intersection points were sampled in each field. At each intersection an inter–row and an intra–row sample was taken. The first quadrat was used to assess the inter–row weed seedling population in a 0.76 m^2 area (herbicide-free). The second sample was used to assess the intra–row weed (herbicide treated). Weed density by species was recorded at all sampling points for both quadrats.

Potential for Herbicide Use Reduction Through Intermittent Application

Geostatistical methods were used to quantify the spatial dependence of weed seedling populations. For this analysis, composite grass and composite broadleaf weed populations were studied since current herbicide treatments target groups of weeds rather than individual species. Indicator kriging was used to estimate the conditional probability of an unsampled location being above or below a density threshold (Issaks & Srivastava, 1989). Indicator kriging is based on a binary transformation of weed density estimates, i.e., above or below a set point, in this case a weed density threshold (Rossi et al., 1992). Indicator kriged probability estimate maps were converted to probability estimates (binary maps) using Bayes classification rule (Fig. 27–1a and Fig. 27–1b) (Solow, 1986).

Estimated threshold levels of 0, 1, or 3 broadleaf weeds and 0, 3, or 5 grass weeds per 0.76 m^2 were chosen for the analysis. In this study a range of thresholds was used to estimate the effect of threshold level on the proportion of field area above and below those thresholds. The range of thresholds used in this paper is valid for grasses and small seeded broadleaf species but is high for larger seeded broadleaf plants (Bauer & Mortensen et al., 1991, Coble and Mortensen, 1992). The weed free area or 0 plant threshold, however, was absolutely weed-free. Geostatistical analysis was performed using GS+.[1]

[1]Gamma Design Software, P.O. Box 201, 457 East Bridge St., Plainwell, MI USA 49080.

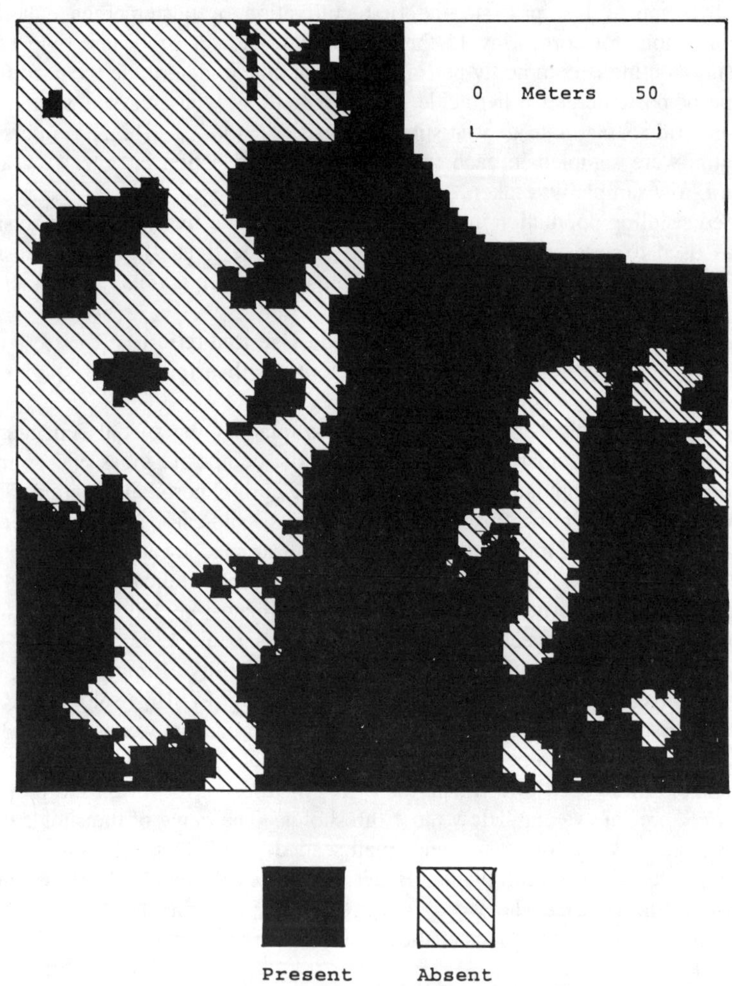

Fig. 27–1a. Presence–absence map of inter–row composite broadleaf weeds in corn (Field #3). Spatial maps were derived using indicator kriging estimates.

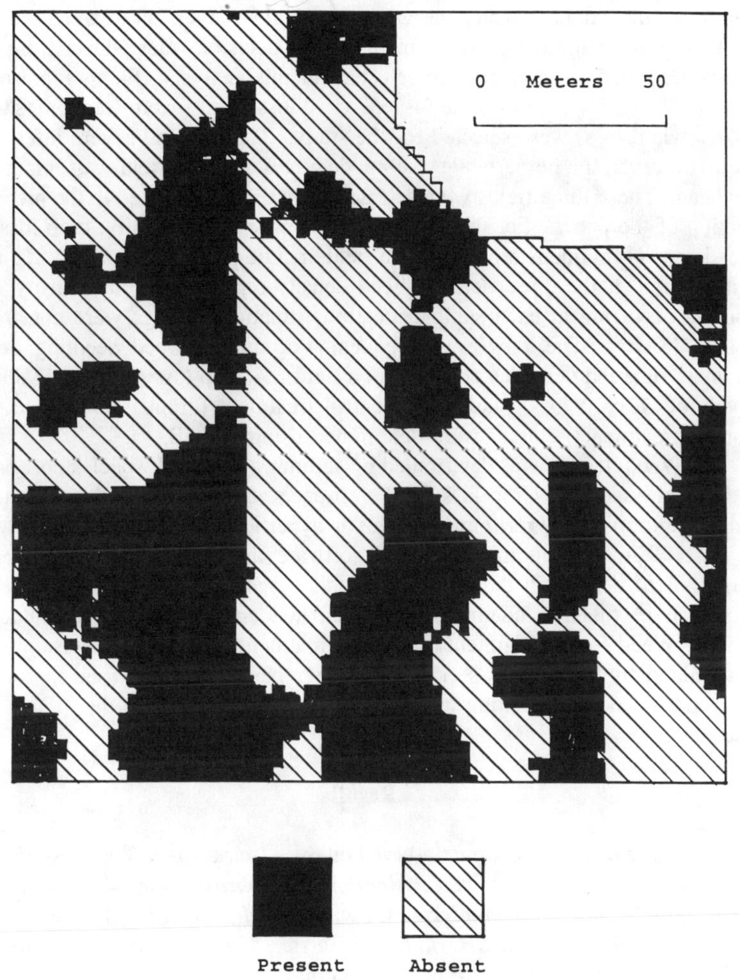

Fig. 27–1b. Presence–absence map of inter–row composite grass weeds in corn (Field #3). Spatial maps were derived using indicator kriging estimates.

Indicator kriging probabilities were plotted based on Bayes classification rule. The binary spatial maps were then digitized using Compeyes[2] and imported into Image-Pro Plus[3] where the field area above and below the threshold was determined.

Spatial Variation in Weed Density, Crop Loss, and Weed Control Decisions

This analysis was conducted to characterize the spatial arrangement of weed density, crop loss, and weed management decision rules. Secondly, to determine the extent to which variation in spatial arrangement of weed seedlings influences the variation in weed management decision rules. Data from three fields described previously were selected for this analysis. Fields representing low (<4% yield loss), moderate (4-26% yield loss), and severe weed infestations (>26% yield loss) were selected for the study. The low and severe infestations were observed in corn, while the moderately infested field was planted to soybean. These three fields were selected from the 12 analyzed in the preceding section of the paper. For this analysis, composite weed density, crop loss, and weed control regimens representing a field area of 49 m^2 were posted 812 times within each of the three fields. Weed density represented total density of all species occurring at each sample location. Spatial postings were constructed using GSLIB. The range and frequency of unique weed seedling density (number of weed densities observed in a field), % crop loss (number of discrete levels of % crop loss), and weed control treatment (number of unique weed control treatments) classes were determined using GSLIB.

Weed density was determined by summing across all species observed in individual quadrats. Crop loss (in the absence of weed control) was estimated using the crop loss algorithm (rectangular hyperbola) in NebHERB (Mortensen et al., 1992) an economic threshold software application used by Nebraska farmers (for details on the crop loss function see Coble & Mortensen, 1992). For each of the 812 quadrats, the anticipated crop loss was calculated and posted. Weed control regimens (herbicide treatments or no-spray decisions) were determined from the same seedling data set. NebHERB was used to compute % crop loss and the resulting weed control decision for each of the 812 seedling data sets.

Results

Ten weed species were observed on the 12 farm sites. The most common broadleaf weed species were *Abutilon theophrasti* Medik., *Amaranthus retroflexus* L., and *Solanum nigrum* L., while *Sorghum bicolor* (L.) Moench and *Setaria sp.* (which include *Setaria faberi* Herrm., *Setaria viridis* (L.) Beauv., and *Setaria lutescens* Hubb.) were the most common grass weed species.

---

[2]Digital Vision, Inc., 270 Bridges St., Dedham, MA USA 02026.

[3]Media Cybernetic, 8484 Georgia Ave., Silver Spring, MD USA 20910.

Potential for Herbicide Use Reduction Through Intermittent Application

Inter–row Weed Population

The mean density of composite broadleaf weed species ranged from 1.3 to 51.2 seedlings per linear meter of row with an average of 11.1 seedlings across all fields (Table 27–1). Mean density of composite grass species was lower and ranged from 0.1 to 15.7 with an average of 3.8 seedlings per linear meter of row. Broadleaf weed density was exceptionally high in fields 4, 8, 9, 11, and 12 while grass weed density was high in fields 7 and 12. In all fields, weed distribution was non-uniform. Composite broadleaf and grass weed density maps, derived through geostatistical analysis, showed isolated aggregates of high density with areas between aggregates free or nearly free of weeds (Fig. 27–1a, 27–1b, and Fig. 27–2a, 27–2b). Estimates of mean broadleaf and grass weed density were influenced by these dense aggregates. This suggests that threshold values derived through estimates of a field population may need to be adjusted to account for the aggregated nature of weed populations. In addition, the frequency distribution of seedling populations was highly skewed with a majority of quadrats containing no weeds (Fig. 27–3).

Table 27–1. Sample means for composite grass and broadleaf weed population in the inter–row and intra–row area†.

Field	Inter–row Area		Intra–row Area	
	Broadleaf	Grass	Broadleaf	Grass
	--------- Seedling density per linear meter of row ---------			
1	1.3	6.1	-	-
2	-	-	0.5	1.9
3	1.8	3.5	0.4	0.1
4	9.7	0.1	6.1	0.1
5	2.4	0.2	0.2	0.1
6	-	-	0.5	0.1
7	4.9	10.6	0.2	0.1
8	9.1	0.4	3.1	0.1
9	18.1	1.4	5.1	0.1
10	1.9	0.3	0.1	0.1
11	10.6	0.1	0.2	0.1
12	51.2	15.7	4.3	1.4
AVERAGE	11.1	3.8	2.1	0.4

†Density per 0.76 m^2 and 0.38 m^2 for inter–row and intra–row data, respectively.

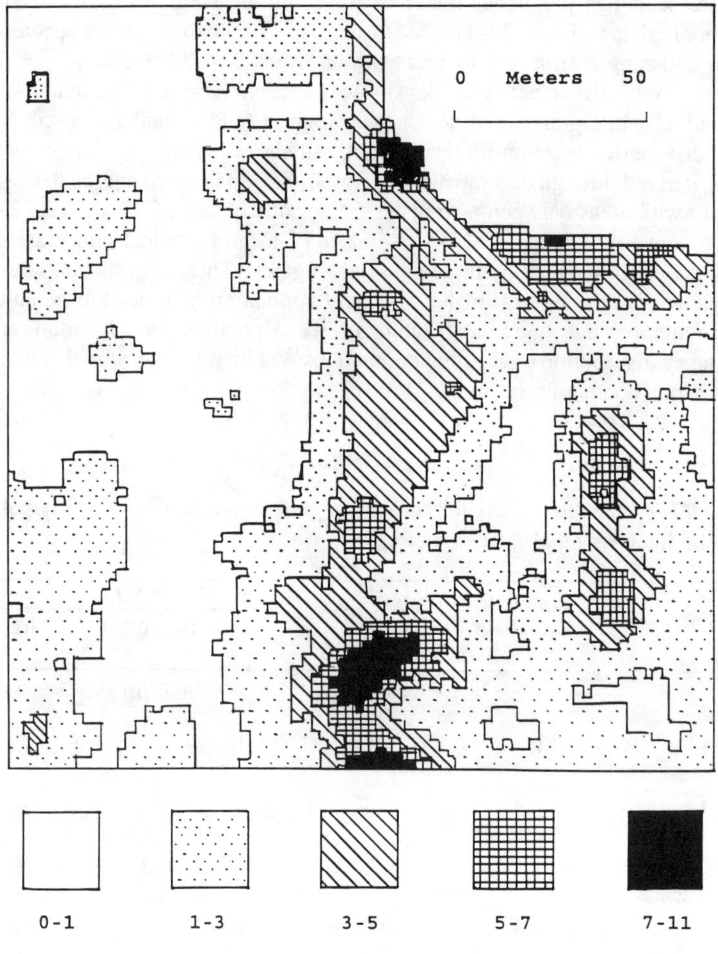

Fig. 27–2a. Presence–absence map of inter–row composite broadleaf weeds in corn (Field #3). Spatial maps were derived using punctual kriging estimates.

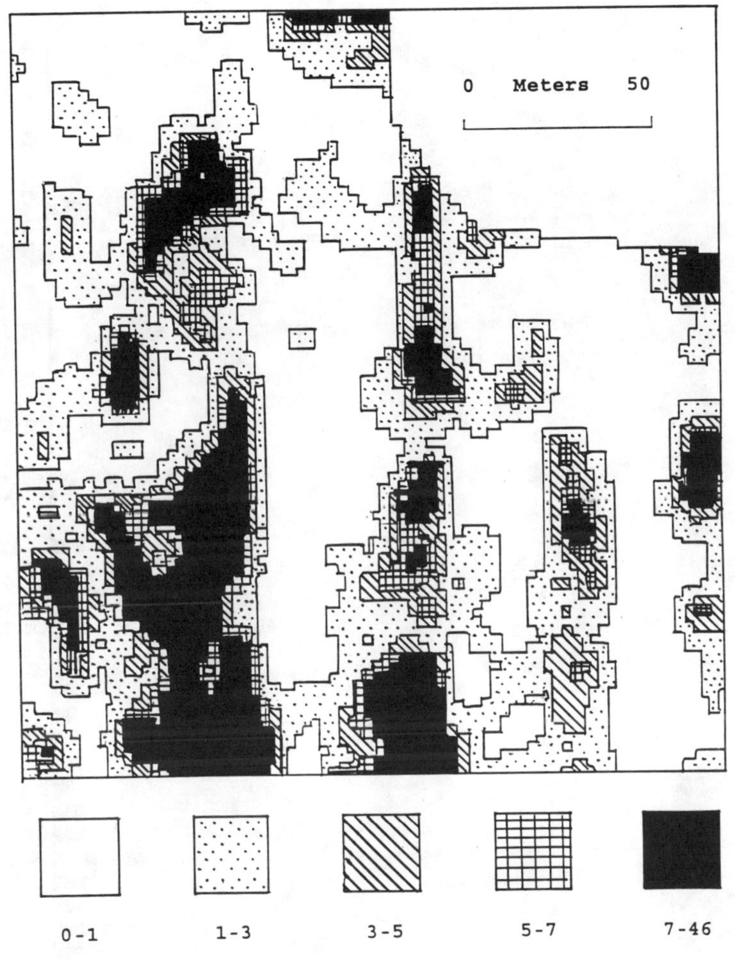

Weed density (m^{-1} of crop row)

Fig. 27–2b. Presence–absence map of inter–row composite grass weeds in corn (Field #3). Spatial maps were derived using punctual kriging estimates.

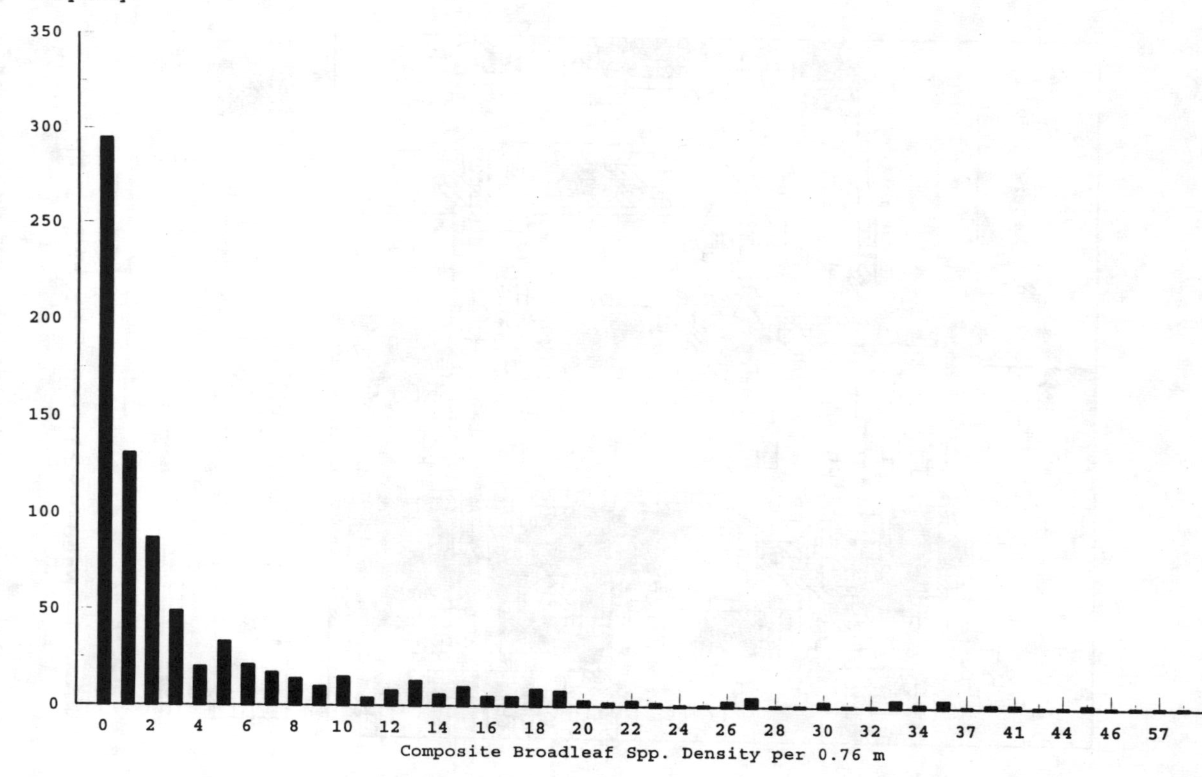

Fig. 27–3. Frequency distribution of composite broadleaf spp. count data in field 7.

The relationship between weed seedling density and spatial aggregation was tested using simple linear regression. Weed seedling density (broadleaf and grass) was regressed against weed-free area (zero threshold). A negative linear relationship ($r^2 = 0.41$) existed between seedling mean density and percent weed free area. On average, 30% of the inter–row sample area in the 12 fields surveyed was free of broadleaf weeds and 70% free of grass weeds when a threshold of 0 weeds per $0.76m^2$ was used (Table 27–2). As threshold levels increased, the proportion of area below the threshold also increased for both composite broadleaf and grass weed populations. In 6 of 10 fields sampled, >50% of the field area was below a threshold of 1 broadleaf weed per linear meter of row. When the threshold was increased to 3 broadleaf weeds per linear meter of row, > 50% of the field area was below the threshold in 7 of 10 fields. On average, 52% and 67% of the study area was below a threshold of 1 and 3 broadleaf weeds per linear meter of row. For grass weeds, >50% of the field area was below thresholds of 3 and 5 plants per linear meter of row in 9 of 10 fields sampled with an average of 85 and 89% of the study area below thresholds of 3 and 5 grass weeds per linear meter of row.

Table 27–2. Percent area below a specified threshold level for inter–row broadleaf and grass weed populations.

		Broadleaf Population			Grass Population		
		Threshold values (Number–meter of row)					
Farm	Crop	0	1	3	0	3	5
		% Area below threshold					
1	Corn	55	82	97	35	68	75
2	Corn	-	-	-	-	-	-
3	Corn	34	69	93	68	85	92
4	Corn	4	15	31	99	99	99
5	Corn	20	70	93	97	99	99
6	Corn	-	-	-	-	-	-
7	Corn	42	55	79	57	75	81
8	Soybeans	29	46	65	97	99	99
9	Soybeans	<1	<1	2	70	96	99
10	Soybeans	65	84	95	97	99	99
11	Soybeans	45	72	79	98	98	99
12	Soybeans	1	24	34	6	32	45
Average	-	30	52	67	72	85	89

Intra–row Weed Population

On average 71 and 94% of the sampled area was free of broadleaf and grass weeds using a zero population density threshold (Table 27–3). The mean intra–row grass weed density was low (0.39 plants per linear meter of row) due in part to low initial population levels (reflected in the low densities observed in the inter–row space) and the added suppression from the preemergence herbicide application (Table 27–1). Most farmers applied a preemergence grass herbicide alone or in combination with a broadleaf herbicide. Density maps of intra–row weed populations showed areas of high density with few or no weeds between these areas (data not shown). Intra–row weed seedling density (broadleaf and grass) was highly correlated ($r^2 = 0.92$) with weed-free area. The use of herbicides reduced the variability in mean seedling density and increased aggregation resulting in a stronger linear relationship than that of the untreated inter–row data.

Table 27–3. Percent area below a specified threshold level for intra–row broadleaf and grass weed populations.

		Broadleaf Population			Grass Population		
		Threshold values (Number–meter of row)					
Farm	Crop	0	1	3	0	3	5
		% Area below threshold					
1	Corn	-	-	-	-	-	-
2	Corn	88	98	99	59	94	98
3	Corn	87	99	99	99	99	99
4	Corn	13	30	55	99	99	99
5	Corn	97	99	99	99	99	99
6	Corn	88	98	99	99	99	99
7	Corn	98	99	99	99	99	99
8	Soybeans	57	71	80	98	99	99
9	Soybeans	<1	8	27	98	99	99
10	Soybeans	99	99	99	99	99	99
11	Soybeans	99	99	99	99	99	99
12	Soybeans	58	71	80	81	85	97
Average	-	71	79	85	94	98	99

Spatial Variation in Weed Density, Crop Loss, and Weed Control Decisions

The three agricultural fields chosen for this analysis represented low, moderate, and severe weed infestations. Weed infestation classifications were based on % weed free area, where a low infestation had greater than 88% weed free area, moderate 48-88%, and severe 0-47%. Of the 12 fields analyzed in this way approximately half of the corn fields had low and half had moderate infestations (one severely infested). In contrast, a broader range of infestations were observed in soybean, where a third of the fields had low, a third moderate, and a third severe weed infestations.

The frequency distributions computed for the 812 samples of weed population density, % crop loss, and weed management decision rules were highly skewed at low weed densities. Eighty-nine percent (89%) of the low, 49% of the moderately, and 18% of the severely infested fields required no weed control inputs (Figs. 18-4 and 18-5). While the field with the low infestation had the greatest area requiring no weed management inputs, the range of weed management decision rules was similar for each of the three fields (Table 27–4). A total of nine weed management decision rules were recommended in the low and moderate fields, while only seven were recommended for the severely infested field. This similarity in the total number of treatments recommended (7-9 treatment recommendations for 812 weed population estimates) reflected the nature of the weed infestations in each field. Selection of a weed management decision recommendation is based on the efficacy of the weed control practice, cost, weed species composition and density, where weed species and density influence both treatment efficacy and % crop loss. If weed density is high, in this field a large number of quadrats contained between 10 and 40 plants per linear meter of row, and if the species composition is relatively uniform, the same treatment recommendation will frequently occur. For example, in the severely infested field Aatrex 4L was recommended 21%, Marksman (3pt) was recommended 41%, 2,4-D amine 16% of the time.

Table 27–4. The number of composite seedling density, % crop loss and weed management decision rules observed in the low, moderate, and severe weed infested fields.

Infestation Level	Seedling Density	Crop Loss	Weed Management Decision Rule
Low	10	20	9
Moderate	35	39	9
Severe	41	43	7

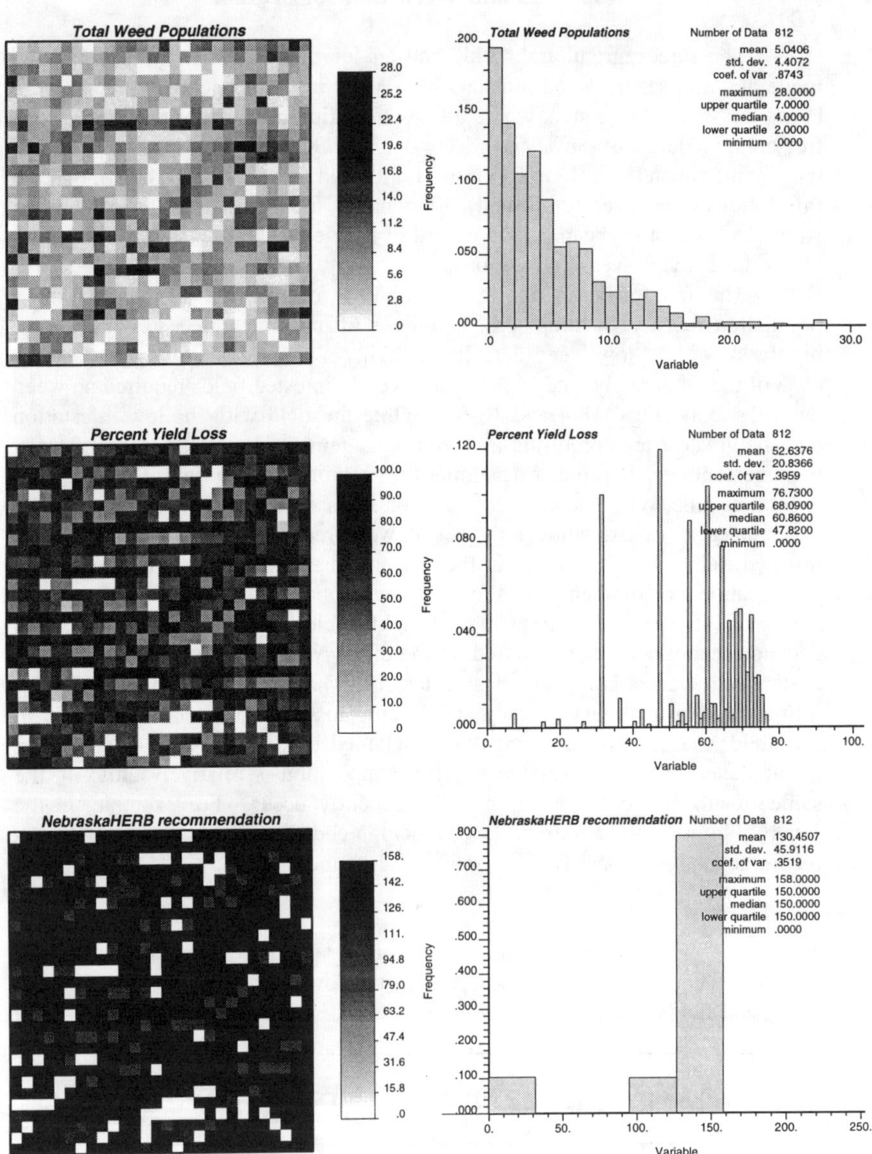

Fig. 27–4. Spatial postings and frequency distribution for composite weed density, % crop loss, weed management decision rules in a corn field with a low weed infestation.

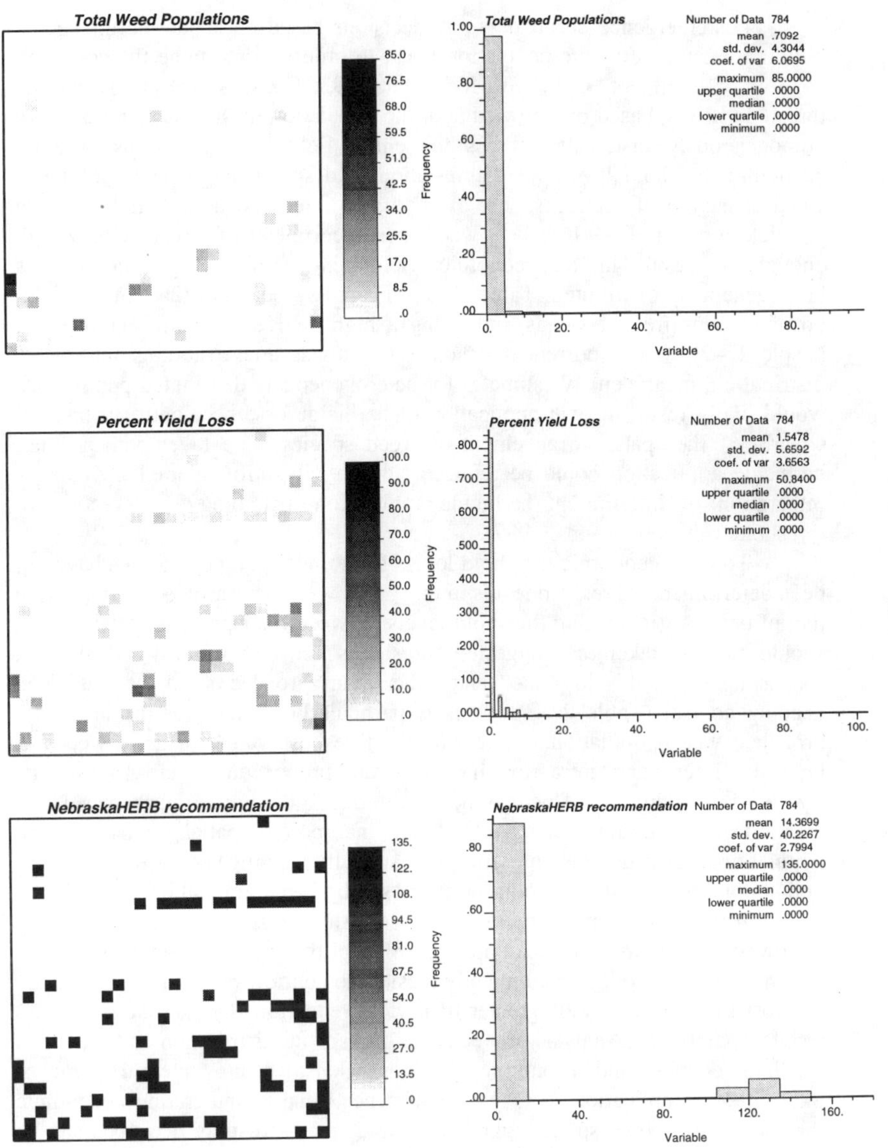

Fig. 27–5. Spatial postings and frequency distribution for composite weed density, % crop loss, weed management decision rules in a corn field with a high weed infestation.

DISCUSSION

Postemergence herbicide spray decisions based on economic thresholds require an estimate of the population mean in order to determine the costs and benefits of treatment (Coble & Mortensen, 1992; Cousens, 1987). Economic threshold levels, based on knowledge of the population mean, assume weeds are homogeneously distributed across the entire field. Relationships between economic threshold level, weed infestation, and spatial aggregation influence weed management decisions. For example, the inter–row area of field 7 had a population mean of 4.9 broadleaf weeds per linear meter of row, a density high enough to result in an economic yield loss - if the population was homogeneously distributed (Table 27–1). Forty-two percent (42%) of the field, however, was free of weeds, indicating a high degree of spatial aggregation (Table 27–2). Since current threshold estimates assume a homogenous weed distribution, mean density estimates for heterogeneously distributed populations would likely result in over-application of herbicide or other control measures. Clearly, if the spatial arrangement of weed species were taken into account herbicide application could be reduced substantially through implementation–integration of intermittent herbicide application technology with economic threshold values (Cousens, 1987).

To date, economic threshold levels for composite weed species have not been determined. Threshold levels in this paper were arbitrarily set to determine the influence of increasing threshold levels on weed management decisions and should not be taken as suggested thresholds for composite weed species. Increasing thresholds to some value other than zero decreased the field area needing control (Table 27–3). This is particularly evident for the inter–row broadleaf weed population. Increasing the threshold value for grass weeds in both the inter– and intra–row increased the proportion of area below the threshold only slightly. This may have been due to the initially large weed-free area. Ideally, economic thresholds should incorporate spatial trends in weed populations, i.e., in the presence of spatially aggregated weed seedling populations the threshold should be increased by some amount.

Analysis of spatial structure can also influence spray–no spray decisions on a whole field basis. For example, if >80% of the area is below an economic threshold, it may not be economically feasible to control weeds in that field, i.e., the cost of control may be greater than yield reductions by present and future weed populations. Analysis of weed seedling spatial distribution and the mean seedling density and economic threshold densities are integral parts of intermittent application. Presence–absence maps integrating economic thresholds, degree of spatial aggregation, and spray–no-spray decisions can be constructed and used as a decision rule for intermittent application systems. Intermittent applicators can use these maps in conjunction with spatial positioning devices to apply herbicide only where needed. Although this seems like a reasonable approach, the cost of such analysis and implementation has yet to be determined on a field-scale basis. Field sampling required to describe the weed seedling populations is a significant limiting factor in implementation of this technology. It is clear that sampling must be seen as a continuing process in which data are recorded and used to aid in less-intensive subsampling in subsequent yrs. The extent of subsampling will, in large part, depend upon the spatial stability of seedling patterns in time.

The range of treatment options was smaller than expected. This was particularly true for the severely infested field (Fig. 27–5). The 7-9 recommendations made for each of the fields analyzed could be further reduced by sacrificing net gain (most ideal weed management input) for a smaller number of treatments. This could be accomplished by placing a bounds on the net gain decision in order to identify some 3-5 treatment decisions. Decreasing the range of outcomes would make the spray–mechanical control process more manageable. Where instead of changing treatments every third or fourth cell (49 m^2), perhaps the change is only required every 10 or 15 cells.

The use of real–time plant sensing represents a promising alternative to historic spatial maps. Real–time data acquisition systems do not rely on accurate assessment of weed seedling distribution. Rather, spray–no-spray decisions are made by sensing individual plants at the time of application. Herbicide use efficiency of such systems is greatest when weeds are spatially aggregated and population densities are relatively low. The present study shows that weeds are indeed spatially aggregated and that intermittent herbicide applicators, based on real-time data acquisition, are a critical first-step in implementation of this technology on a field scale basis. The use of real-time sensing could reduce herbicide use by an average of 30 to 72 % if in-row plant species discrimination were possible. Intermittent herbicide application represents a significant change in herbicide application technology that addresses economic, environmental and agronomic concerns. Technology is, or soon will be, available to apply herbicides only when weeds are present. This approach, whose value was suggested as early as 1987 by Cousens, lacks basic information on the spatial and temporal dynamics of weed populations in agricultural fields. This paper shows that herbicide use can be substantially reduced if information on spatial weed distribution is used as a basis for intermittent herbicide application or refinement of economic thresholds.

ACKNOWLEDGMENTS

The authors wish to thank David Marx, comments and discussion, and Thomas Harvill for programming support. The authors would also like to thank the following farmers for their cooperation and input: Randy Bergman, Dale Hansen, Matt Hansen, Ron Larson, Robert and Lyle Probst, Everett Wilkins, and Dwayne Wollenburg.

REFERENCES

Bauer, T.A., and D.A. Mortensen. 1992. Comparison of economic and economic optimum thresholds for two annual weeds in soybeans. Weed Technol. 6:228–235.

Bridges, D.C. 1992. Crop losses due to weeds in the United States - 1992. Weed Sci. Soc. Amer., Champaign, IL, 403 pages.

Coble, H.D., and D.A. Mortensen. 1992. The threshold concept and its application to weed science. Weed Technol. 6:191–195.

Cottam, G., J.T. Curtis, and A.J. Cantana Jr. 1957. Some sampling characteristics of a series of aggregated populations. Ecology 38:610–621.

Cousens, R. 1987. Theory and reality of weed control thresholds. Plant Protection Quarterly 2:13–20.

Dessaint, F., R. Chadoeuf, and G. Barralis. 1991. Spatial pattern analysis of weed seeds in cultivated soil seedbank. J. Applied Ecol. 28:721–730.

Duff, P. 1993. Detectspray system. Proc. of the Weed Science Society of America, 33:45.

Felton, W.L., A.F. Doss, P.G. Nash, and K.R. McCloy. 1991. To selectively spot spray weeds. Am. Soc. of Agric. Eng. Symp. 11–91:427–432.

Isaaks, E.H., and R.M. Srivastava. 1989. An Introduction to Applied Geostatistics. Oxford University Press, New York, 562 pages.

Mortensen, D.A., A.R. Martin, and F.W. Roeth. 1992. NebHERB: An economic threshold-based weed management program. Univ. Neb. Coop. Ext. Serv.

Mortensen, D.A., A.R. Martin, T.E. Harvill, and T.A. Bauer. 1993a. The influence of rotational diversity on economic optimum thresholds in soybean. In the 8th Eur. Weed Res. Symp. p. 815–823.

Mortensen, D.A., G.A. Johnson, and L.J. Young. 1993b. Weed distribution in agricultural fields. pp. 113–124. *In* Soil Specific Crop Management, P.C. Robert, et al. (ed.) Agron. Soc. of Am.

Mortensen, D.A., G.E. Meyer, and K. Vonbargen. 1993c. Plant sensing and mapping for intermittent herbicide application. Reviews in Weed Science. (in press)

Navas, M.L. 1991. Using plant population biology in weed research: a strategy to improve weed management. Weed Research 31:171–179.

Nelson, T. 1993. SprayVision, a selective sprayer technology developed in North America. Proc. of the Weed Science Society of America, 33:43.

Rossi, R.E., D.J. Mulla, A.G. Journal, and E.H. Franz. 1992. Geostatistical tools for modeling and interpreting ecological spatial dependence. Ecological Monog. 62:277–314.

Solow, A.R. 1986. Mapping by simple indicator kriging. Mathematical Geology 18:335–352.

Swanson, J.A., and D.C. Dahl. 1989. The U.S. pesticide industry: Usage, trends, and market development. Staff Paper P89–5. Dep. of Agric. Econ., Univ. Minnesota, St. Paul.

Thompson, J.F., J.V. Stafford, and P.C.H. Miller. 1991. Potential of automatic weed detection and selective herbicide application. Crop Protection 10:254–259.

Thornton, P.K., R.H. Fawcett, J. B. Dent, and T. J. Perkins. 1990. Spatial weed distribution and economic thresholds for weed control. Crop Protection 9:337–342.

VanGroenendael, J.M. 1988. Patchy distribution of weeds and some implication for modeling population dynamics: A short literature review. Weed Research 28:437–441.

Von Bargen, K., G.E. Meyer, D.A. Mortensen, S.J. Merritt, and D.M. Woebbecke. 1992. Red-near infrared reflectance sensor system for detecting plants. Optics in Agric. and Forestry 1836:231–238.

Wiles, L.J., G.W. Oliver, A.C. York, H.J. Gold, and G.G. Wilkerson. 1992. Spatial distribution of broadleaf weeds in North Carolina soybean (Glycine max) fields. Weed Science 40:554–557.

Woebbecke, D.M., G.E. Meyer, K. Von Bargen, and D.A. Mortensen. 1994. Plant species identification, size, and enumeration using machine vision techniques on near-binary images. Mid Cent. Agric. Eng. Paper 93 ___. (in press)

28 Yield Variability Within a Long-Term Corn Management Study: Implications for Precision Farming

D. R. Huggins
R. D. Alderfer

University of Minnesota
Southwest Experiment Station
Lamberton, Minnesota

Precision farming (PF) is an emerging farm management system that seeks to optimize management on small portions of a field according to specific site characteristics. This farming approach has intuitive appeal as management strategies can be tailored to small field areas, thereby improving resource efficiency and decreasing potentially adverse environmental impacts. Technological advancements have provided agriculture with tools enabling sophisticated, site-specific management (i.e. computer software and hardware, geographical information systems, global positioning systems, variable rate fertilizer, herbicide, and seed equipment, and real-time field condition assessment). Crop management systems that incorporate and optimize the use of new PF tools, however, lag far behind the capabilities of the current technology.

Long-term field experiments conducted on small plot areas could be used to assess the potential viability of a PF management practice. When designing agricultural trials, care is taken to identify uniform areas with minimal soil and site variation. The density of experimental plots vary, but can range up to 500 plots per ha. (200 per acre). These small experimental areas can be considered fundamental management units for precision farming where inputs would not be varied. If the experiment is conducted over a long time period, environmental-treatment interactions specific for the study location are expressed. This site-specific information can be used to evaluate if an optimal treatment can be prescribed for a given site. If the optimal treatment changes from yr-to-yr due to unpredictable environmental-treatment interactions, then the practice is suspect for precision farming management.

This paper reviews a long-term nitrogen (N) management study for corn (*Zea mays* L.) conducted near Lamberton, MN. Our objectives are to evaluate yr-to-yr variation in corn yield response to N management; identify sources of corn yield variation; and assess the potential for precision N management across southwestern Minnesota landscapes.

EXPERIMENTAL

Treatment Description and Field Methods

The field study was located in southwest Minnesota on the University of Minnesota, Southwest Experiment Station near Lamberton. The study was initiated in 1960 on a tile-drained Normania loam (Fine-loamy, mixed, mesic Aquic Haplustolls) and has been planted to continuous corn through 1993. Treatments consisted of 4 N rates, 0, 45, 90, 180 kg N/ha, (0, 40, 80, 160 lbs N/ac), 2 N fertilizer forms (ammonium nitrate and urea), 3 application timings (fall, spring, and sidedress), and two fall fertilizer placements (fall incorporated and non-incorporated N). The 18 treatments were arranged in a randomized block design with four replications.

The fertilizer treatments have been applied annually to the same plot area, 6.1 x 23.6 m, (20 x 77 ft) for 34 yr. During the 1976 drought, treatments were applied but no grain yields were collected. Data for individual plots are missing for 1960, 1961, 1963, 1964, 1966, and 1968, however, treatment means are available for these years. Following corn harvest and stalk chopping, fall N treatments were broadcast and incorporated with a moldboard plow, except for treatments where N was not incorporated. Treatments with non-incorporated N were moldboard plowed prior to the surface application of fall N. All other treatments were fall moldboard plowed to a depth of 30 cm (12 in). Spring applied N was broadcast and incorporated during secondary tillage operations in late April or early May. Corn was planted in 76 cm (30 in) rows, and starter fertilizer was banded 5 cm (2 in) to the side and 5 cm (2 in) below the seed across all treatments. Rates of starter fertilizer have varied from yr-to-yr ranging from 0 to 22 kg N/ha (20 lbs/ac), with an average of 112 kg/ha (100 lbs/ac) of 14-41-15 applied. Treatments with sidedressed N were broadcast and incorporated during the first cultivation in June.

Statistical Analyses

Analysis of variance was used to assess treatment effects (SAS, 1988). Multiple regression analyses were conducted with grain yield as the dependent variable and independent variables consisting of weather and management data collected since 1960 at the Southwest Experiment Station. To assess within field variation, regression analyses were conducted on plot means as well as individual plots for yrs that had complete information for each data set.

RESULTS AND DISCUSSION

Treatment Effects on Yield

Average long-term corn yields showed a significant response to N rate and timing (Table 28–1). No significant difference between urea and ammonium nitrate fertilizer forms were expressed for grain yield. Maximum grain yields were obtained with applied N rates of 180 kg N/ha (160 lbs N/ac) across N

Table 28–1. Average corn yields (1960 through 1993) as influence by N rate, form, timing, and placement.

	Nitrogen timing and placement							
	Fall				Spring		Sidedressed	
N form	Surface		Plowdown		Incorporated			
and rate†	Yield	SE‡	Yield	SE	Yield	SE	Yield	SE
$kg\ ha^{-1}$	$Mg\ ha^{-1}$							
Urea								
61	5.51	0.18	5.63	0.18	5.75	0.18	6.05	0.18
106	nd§	nd	6.42	0.20	6.77	0.19	7.00	0.18
195	nd	nd	7.05	0.19	nd	nd	nd	nd
Ammonium nitrate								
61	5.43	0.18	5.19	0.18	5.79	0.18	6.06	0.17
106	nd	nd	6.38	0.18	6.63	0.19	6.49	0.19
195	nd	nd	6.94	0.18	nd	nd	7.29	0.19

$LSD_{0.05}$ = 0.31 $Mg\ ha^{-1}$.
Check plots (16 $kg\ N\ ha^{-1}$) averaged 3.98 $Mg\ ha^{-1}$ with an SE of 0.16 $Mg\ ha^{-1}$.
†N rates include 16 $kg\ N\ ha^{-1}$ applied as starter fertilizer.
‡SE = standard error.
§nd = no data.

and form. The difference in N rate between 90 (80) and 180 kg N/ha (160 lbs N/ac) is large, however, and the optimal N rate for grain yield lies between these extremes. The 90 kg N/ha (80 lbs N/ac) treatment with urea shows the effect of application timing on grain yield (Table 28–1). Long-term average grain yields significantly increased from 6.42 (103 bu/ac) to 7.00 Mg/ha (112 bu/ac) as N application timing was delayed from fall to sidedressed N.

Grain Yield Variation

Large variations occurred in yr-to-yr and within yr yield response to treatments (Fig. 28–1). Treatment differences within a yr ranged from <2 Mg/ha (32 bu/ac) in 1963 and 1977 to greater than 7 Mg/ha (112 bu/ac) in 1986. Large variations in yield also occurred from yr-to-yr. In 1963 and 1977, all treatments produced greater grain yields than treatments in the previous or following yr. During these years, yr-to-yr climatic and management conditions had a greater effect on yield than N treatments.

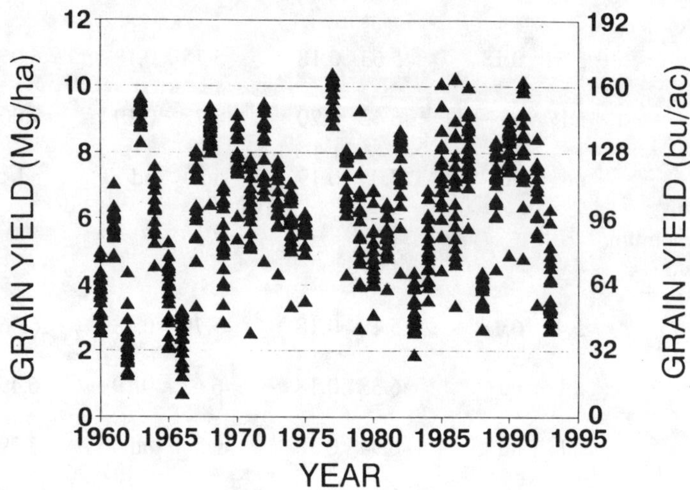

Fig. 28–1. Corn grain yield variation by yr and treatment for 1960–1993.

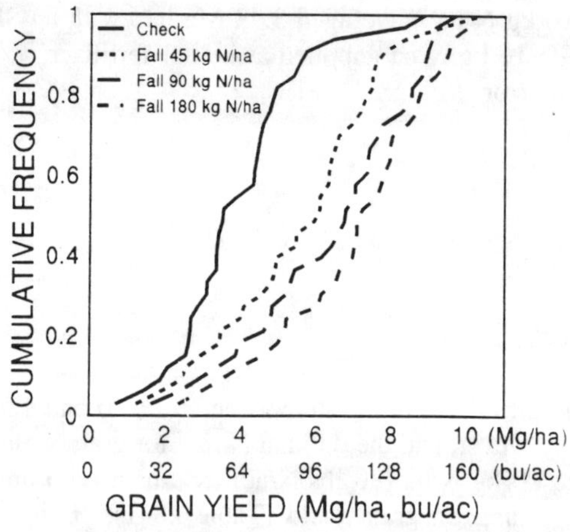

Fig. 28–2. Cumulative frequency of grain yield for selected treatments (1960–1993).

Cumulative frequency distributions depict the range in grain yield that occurred for selected treatments from 1960 to 1993 (Fig. 28-2). Grain yield ranged from <2.3 Mg/ha (37 bu/ac) to greater than 8 Mg/ha (128 bu/ac) for any given treatment including the check. The large yr-to-yr variation in grain yield that occurred for a given treatment makes it difficult to prescribe accurate N recommendations or predict the effects of N management on grain yield.

Regression Analyses

Multiple regression models using weather, management, and treatment variables were developed to increase understanding of grain yield variation. Treatment variables with significant model parameters were N rate and timing, and the regression model including quadratic terms had an R^2 of 0.15. The R^2 indicated that treatment variables only explained 15% of the variation in yield. Regression models using weather, management, and treatment variables increased the R^2 to 0.82. Significant model variables and positive (+) or negative (-) correlations with yield were: average monthly temperature for April (+), May (+), July (-), and August (-); average monthly rainfall for May (-), June (-), and July (+); available soil moisture for the previous fall (+); date of planting (-); N rate (+) and N timing (positively correlated with delayed time of application). Other variables that were included in the regression analysis but not significant were: N form, growing degree d, stress index, leaching index, annual precipitation, and date of killing frost. Another regression model that included weather, management, and treatment variables was developed using treatment means in order to assess variation associated with the site. The R^2 for the model using treatment means increased to 0.92.

Interpretation of the regression analyses are as follows: 15% of the yield variation was explained by N treatments; 67% of the yield variation was explained by climatic and other management variables; 10% of the yield variation was explained by site variability; and 8% of the variation was unexplained. Therefore, climatic variables had the greatest effect on yr-to-yr yield variation and could seriously diminish the capability of managing yield variability through PF.

The large, relatively unpredictable and uncontrollable source of yield variation from climatic variables has significant implications for PF. Variables such as yield and components of N use efficiency that interact strongly with climate will be difficult to predict across a landscape for use in formulating management recommendations. Also the ability to manage the yield variability through PF will be limited if climate is the major source of the variability. This situation will be particularly evident if optimal management strategies are not consistent or predictable from yr-to-yr. In addition, interpretation of experiments designed to evaluate N use efficiency across the landscape will require intensive measurement of weather and site variable interaction. One conclusion from this analysis is that greater success in PF will be initially achieved through emphasis on factors that have greater variation predictability (i.e. organic matter, P, K, pH,) as compared to highly variable, unpredictable factors (i.e. weather, yield). Thus, PF practices should emphasize variables with high spatial variability but low temporal variability.

Maximum Yields

Maximum yields of corn, irrespective of treatment, ranged from 3.2 (51 bu/ac) to 10.2 Mg/ha (163 bu/ac) during the time period of the experiment (Fig. 28-3). Contrary to Minnesota Agricultural Statistics that reported an annual increase in corn yield of 0.14 Mg/ha (2.19 bu/ac) from 1950-1991 (Baker, 1992), grain yield within the study has not increased over time. The consequences of long-term monoculture corn have probably limited grain yield. No one treatment consistently resulted in the greatest yield, an indication of climatic-soil interaction effects on treatment outcomes. Fall applications of N produced the greatest yield in 10 of 33 yr, spring applications in 7 of 33 yr, and sidedressed applications in 16 of 33 yrs. Nitrogen rates required for maximum yield also ranged from 90 (80) to 180 kg N/ha (160 lbs/ac).

The number of years within defined yield classes was almost uniform across the range of maximum yields (Fig. 28-4). If yield goal was based on average yield than a large number of years would occur with yields significantly below or above the yield goal. The uniform yield distribution makes it difficult to establish realistic yield goals for N rate recommendations or other agronomic recommendations based on yield projections.

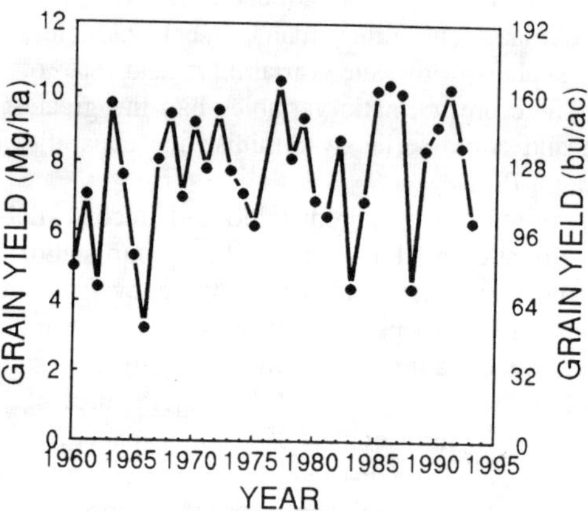

Fig. 28–3. Maximum grain yield irrespective of N treatment for 1960-1993.

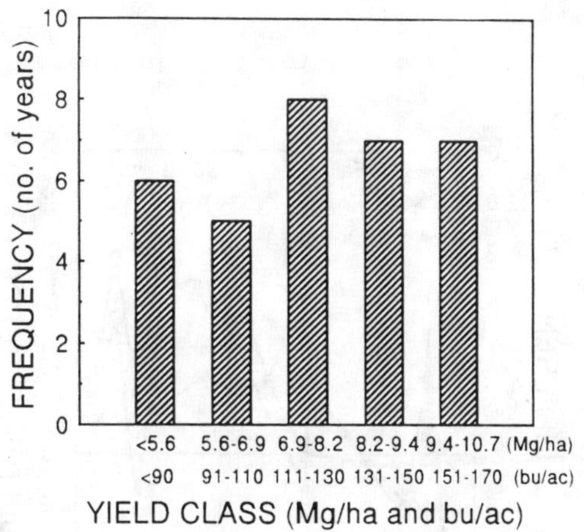

Fig. 28–4. Frequency of years that occurred within five maximum yield classes.

Soil Moisture and Treatment Relations

In the Northern Corn Belt of Southwest Minnesota, fall applied N comprises ≈80% of the N applied to corn. Fall applications are favored over spring and sidedressed N because fall N prices are typically 10% less, conflicts with other field operations are minimized, and weather and field conditions are often more favorable for N application than in the spring. Disadvantages of fall applied N are associated with increased time (6 to 9 mo) for N losses as compared to spring or sidedressed N. Losses of N can also be high for sidedressed N if rates exceed crop requirements and result in high soil residual levels of nitrate after harvest (Huggins et al., 1994).

It was hypothesized that increased soil moisture levels during the fall could be related to N loss and would be reflected in the yield response of the following corn crop. To test this hypothesis, yield differences between fall, spring, and sidedress applied N were related to available soil moisture conditions of the previous fall (Fig. 28–5). The years 1981 to 1993 were selected, as extremes in fall soil moisture occurred over this time period. Negative yield differences indicate greater yields occurred for fall N applications as compared to spring or sidedressed N, or greater yields resulted from spring N as compared to sidedressed N. Years with low available soil moisture in the fall tended to favor fall or spring applications of N. As available soil moisture increased to greater than 14 cm (5.5 in), however, a distinct yield advantage for spring and sidedress applied N as compared to fall applied N was expressed. Differences in yield were greatest during the 1993 season where sidedress and spring treatments were 2.5 (40) and 1.75 Mg/ha (28 bu/ac) greater, respectively, than N fall treatments.

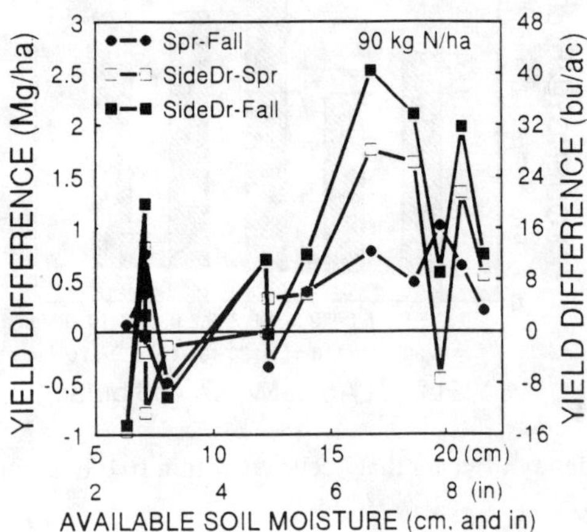

Fig. 28–5. Grain yield difference between N timings as related to available soil moisture the previous fall.

Differences in soil moisture regimes across the landscape must be evaluated to put soil moisture data from the long-term study into a landscape perspective. The long-term study is located on a Normania loam, a moderately well-drained soil that is commonly positioned in a catena with Ves loam, a well-drained soil, and Webster clay loam, a poorly-drained soil. Soil moisture data collected weekly in 1992 across a catena consisting of Storden-Ves loam, Normania loam, and Canisteo clay loam (poorly-drained soil) were compared to soil moisture data from the long-term corn study (Fig. 28–6). Seasonal soil moisture regimes for Normania loam at the two different sites were similar during 1992. In contrast, the soil moisture regime for the Storden-Ves loam was consistently 15-20 cm (6-9 in) of moisture drier, and the Canisteo clay loam 10-15 cm (3-6 in) moisture wetter than the Normania loam in the upper 150 cm (5 feet) of soil. Furthermore, soil moisture regimes of Normania loam in the long-term study during years of either extremely dry (1988) or moist (1993) conditions were bounded by the moisture regimes expressed by well drained Storden-Ves loam and poorly drained Canisteo clay loam. The extremes in soil moisture regimes across the landscape (Fig. 28–6) and the relationship of yield to N timing and fall moisture conditions (Fig. 28–5) indicates that management decisions regarding N timing could be improved through monitoring fall soil moisture.

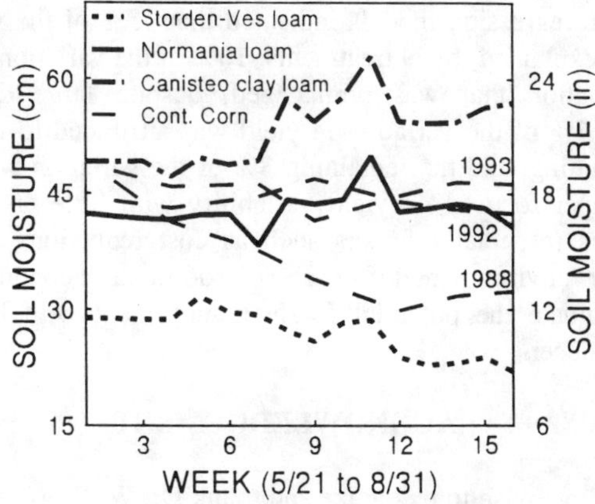

Fig. 28–6. Seasonal soil moisture regimes (15 to 150 cm deep) across a toposequence.

Currently, Best Management Practices for N in Southwestern and West-central Minnesota recommend spring or fall applications of N (Randall and Schmitt, 1993). Recommendations for fall N, however, include delaying application timing until soil temperatures are below 10°C (50°F) at a 15 cm (6 in) depth, and use of ammoniacal forms of N to reduce the susceptibility of N loss. The long-term study indicated that differences in soil moisture regime could be used to further refine N timing recommendations across the landscape. Fall applications of N may result in greater N use efficiency on well-drained soils than on moderately or poorly-drained soils. Spring and sidedress applied N may result in greater N use efficiency for moderately well-drained and poorly-drained soil. Monitoring fall soil moisture conditions across the landscape could be a useful PF approach to aid N timing recommendations.

CONCLUSIONS

We have contended that long-term management-oriented research plots are similar to fundamental management units for PF. Therefore, long-term field studies can be useful for assessing the viability of PF management practices. Multiple regression models of long-term corn N management trials near Lamberton, MN, showed that yr-to-yr climatic variations far outweighed N management treatment effects on grain yield. The unpredictable nature of weather effects on grain yield variability make it difficult to prescribe PF recommendations based on yield goals. The magnitude of climatic effects on yield variation also describes the limits of managing yield variation with PF. Management of variables such as N that are strongly effected by climatic variables will be more difficult to use with PF than managing variables that are more stable with respect to climate. Long-term results indicated that differences in soil moisture regime could be a promising avenue of research to further refine N timing recommendations for corn across the landscape.

Multiple regression models indicated that 15% of the variation in yield was managed (explained) by N treatments; 10% of the variation in yield was due to spatial variation that was unmanaged, despite large differences in N management; 67% of the variation in yield was attributed to weather variables and date of planting; and the remaining 8% of the variation was unexplained.

This paper focused on yield variability with little or no discussion of producer profits (especially factors such as cost reductions and output price uncertainty) or environmental impacts. Additional economic analyses are planned to evaluate the potential for N management with PF to financially benefit the producer.

ACKNOWLEDGEMENT

The authors wish to recognize and thank Dr. W. W. Nelson, former Head and soil scientist at the Southwest Experiment Station, for his foresight and perseverance in continuing this long-term experiment.

REFERENCES

Baker, D.G. 1992. The positive and negative effects of the recent "benign climate". Field Research in Soil Science 1992. Misc. Publ. 75–1992. MN Agric. Exp. Stn. pp. 12–19.

Huggins, D.R., D.J. Fuchs, J.A. Staricka, and G.L. Malzer. 1994. Long-term N management effects on corn production and nitrate leaching potential at the Southwest Experiment Station, Lamberton, MN. 1994. *In* Agricultural research to protect water quality. Proc. of the Conference, Vol. 2, Poster paper presentations. Soil and Water Conserv. Society, Ankeny, IA. pp. 657–661.

Randall, G.W., and M.A. Schmitt. 1993. Best Management Practices for nitrogen use in southwestern and west-central Minnesota. Univ. of MN Ext. Serv., AG-FO-6128-C.

SAS Institute. 1988. SAS/STAT user's guide. Release 6.03 ed. SAS Inst., Cary, NC.

29 Comparison of Variable Rate to Single Rate Nitrogen Fertilizer Application: Corn Production and Residual Soil NO_3–N

N. R. Kitchen
D. F. Hughes

Soil and Atmospheric Sciences
University of Missouri
Columbia, Missouri

K. A. Sudduth
USDA-Agricultural Research Service
University of Missouri
Columbia, Missouri

S. J. Birrell

Agricultural Engineering
University of Missouri
Columbia, Missouri

The questions most often asked by farmers when discussing variable rate (VR) fertilizer inputs are "How much will it cost?" and "How will it affect my yields?" Many farmers and custom application services share a reluctance to invest in the time and technology required for variable fertilizer application since studies to answer these questions are few and results have been inconsistent.

The magnitude of soil and landscape variability within fields is itself inherently variable. It should surprise no one that improvements in production or savings in fertilizer costs from VR strategies will also vary. A further complication is that fertilizer response research has historically been conducted on small plots. Fertilizer recommendations were then developed by combining results from a number of state- or region-wide fertilizer response studies into simplified "universal" recommendation equations. Consequently, fertilizer recommendations largely ignore response differences due to soil variability. In contrast, VR strategies for agricultural inputs are based directly on field-scale variability. While many nutrient management principles learned from small plot research are still applicable for VR management, sensitivity of the recommendations to changes in agricultural inputs for the individual field may be lacking, due to the general nature of the recommendations.

Though great technological strides have been made in the areas of positioning systems, yield monitoring and mapping and on-the-go sensors, many agronomic questions are unresolved. Of highest importance is the need to define and prove decision criteria that can be used to guide VR fertilizer applications. Credibility and subsequent adoption of variable fertilizer strategies will follow as both economic and environmental benefits are documented.

In Missouri, we are conducting field-scale studies, evaluating the impact of VR fertilizer applications on production and the environment in a desire to establish field-proven decision criteria for profitable precision input management. The majority of our work has been focused on VR nitrogen (N) applications. As a fertilizer input, N is the nutrient that most often will have an impact on production. Because of its soluble nature, N is also a fertilizer nutrient prone to off-field movement, and thus, can negatively impact ground and surface water. For N fertilizer in Missouri, expected yield has the greatest impact on rate recommendations. Previous VR nutrient studies have been mostly based on intensive soil sampling or potential production by soil unit (Mulla et al., 1992; Carr et al., 1991). The use of yield maps to predict potential productivity as the basis of varying fertilizer inputs has previously been suggested (Colvin et al., 1991), but few studies have been reported.

This report will present the results of several studies that address the following research questions: i) Can variable rate technology for the application of N fertilizer sustain or improve production while reducing the risk of potential NO_3 leaching?; and ii) Can the previous years' corn or soybean yield maps be used to predict yield potential and corresponding VR N fertilizer maps?

METHODS

During the 1992 and 1993 growing seasons, VR N studies were conducted in Missouri on three claypan soil sites and on one alluvial soil site. The claypan soil studies were located in north-central Missouri, near Centralia, and the alluvial soil study was located near Oran in the Bootheel region of the state. Single rate (SR) (or conventional rate) and VR treatments were established side-by-side through fields with various soils as described in Table 29–1 (See Appendix Table 29–1 for English units).

For the claypan soil studies (studies 1–3), the SR application treatment of N fertilizer was determined using University of Missouri recommendations (Buchholz, 1992), a field-average soil test and the historical "best yield" from these soils as expected yield (or yield goal). The VR N treatments were determined for different management units (MU), and the expected yield for each MU was based upon the relative yield differences observed from the previous one or two yr corn or soybean yield maps. Thus, variation in expected yield based upon the yield maps was the main factor that caused variation in N fertilizer recommendations. MUs were assigned letters such that in alphabetical order expected yields decreased (e.g. for study 1 "A"=most productive and "C"=least productive). The SR treatments corresponded with MU "A". Using this approach, the expected benefits from VR fertilizer inputs compared to SR inputs on MUs other than MU A would be fertilizer cost savings and a reduction in the potential for off-site N losses (e.g. leaching or runoff).

For the alluvial soil study (study 4), the SR treatment for N, phosphorus (P), and potassium (K), fertilizers were determined using a field average soil test and the historical field-average yield for expected yield. MUs for the VR treatment were determined from a previous yr corn yield map and the results from a 30–M grid soil sampling (at the intersections) of the field. Expected yield and accompanying fertilizer inputs for the VR treatments were determined so that the expected yield when weighted by the area of MUs was approximately equal to the expected yield for the SR treatment. Thus, total fertilizer inputs for VR were also the same as for SR. With this approach, the expected benefits from VR compared to SR would be increased yields for the more productive areas of the field and a reduction in the potential for off-site nutrient losses for the less productive areas of the field.

Nitrogen fertilizer as urea-ammonium-nitrate (UAN) solution was pre-plant broadcast applied along with herbicides and incorporated in studies 1 and 2. Planting occurred the same d for studies 1 and 2. UAN fertilizer was side-dressed 3 weeks following planting for studies 3 and 4. Planting population was 54 000 seed–ha (22 000 seed/A) for studies 1–3 and 69 000 seed/ha (28 000 seed/A) for study 4. The alluvial soil study was irrigated with a center pivot system.

Yields were measured using a combine instrumented with a continuous grain flow sensor which allowed for mapping of yield variability (Birrell et al., 1993). Apparent unrecovered fertilizer N was estimated for studies 1-3 by sampling and analyzing N uptake in the above ground portion of the crop at harvest (Bremmer & Mulvany, 1982), subtracting the crop N uptake in plants grown in a non-fertilized area of each MU, and calculating the amount of fertilizer N not accounted for in the above ground portion of the crop. Representative soil samples were taken at 15–cm increments to a depth of 75 cm following harvest and analyzed for NO_3-N (Keeney & Nelson, 1982; Lachet Chemicals Inc., 1988).

Fertilizer treatments were implemented in a "variant" strip-plot experimental design with sub-units (MUs) non-randomized within randomized strips (fertilizer applications)(Cochran & Cox, 1957). Analysis of variance F-tests were performed accordingly. Variable rate was compared to SR within each MU by a single-degree of freedom orthogonal contrast. If no simple effect of fertilizer application within MUs was found, then main effects due to fertilizer application (averaged over MU) were examined. The experimental arrangement provided no valid statistical test for the main effect of MU since it was a non-randomized factor. Inferences to MU were based solely on means. Studies 1 and 3 had two replications and study 4 had three replications. Study 2 was un-replicated, and therefore, only means could be provided.

RESULTS AND DISCUSSION

Grain Yields

The amount of variability occurring within a given field is unique for that field. Consequently, variations in inputs for different MUs in our studies were unique for each field (see Table 29–1). Also, yr-to-yr climate variability had pronounced effects on our results.

Studies 1-3 (claypan soils)

Excellent corn yields were obtained in much of north-central Missouri in 1992 because of good soil moisture and favorable temperatures during seed fill. The expected yield of 9410 kg/ha (150 bu/A) for SR in study 1 was obtained or exceeded in landscape positions corresponding to MUs A and B (Fig. 29–1). For SR, expected yield and N inputs were the same for all MUs. Actual yield from MU C (eroded side-slope area) of SR was significantly less than expected yield by 1066 kg/ha (17 bu/A). For VR treatments, yield exceeded expected yield by 627, 501, and 1191 kg/ha (10, 8, and 19 bu/A) for units A, B, and C, respectively. VR of N did not significantly reduce yield when compared to SR within any single MU, but averaged over all MUs did significantly reduce yield by 380 kg/ha (6 bu/A). Thus, N was more limiting than we expected it would be and VR N application did not optimize production in 1992. Increased N limitation was a consequence of the better-than-usual climate and soil moisture conditions during seed fill.

For studies 1–3 we chose a high expected yield value for the SR treatment, based on corn production from the best yrs for this area. Whether it be best yrs or the perceived best areas of the field, or both, most farmers are likely to use expected yield values that exceed their long-term, whole-field yield average. This is because N fertilizer costs are low relative to the potential risks associated with N deficiency. Relative to SR, VR treatments exhibited reduced rates, and therefore included more risk of being N-deficient in favorable-climate yrs. The high expected yield chosen for SR application meant that there was no opportunity to evaluate it as an under-application when compared to VR application. For non-irrigated claypan soils in this area, an expected yield of 6270 to 8150 kg/ha (100 to 130 bu/A) would be more representative of the mean of better yrs and whole fields. Had we used an expected yield from within this range for SR, and based the VR application on expected productivity relative to that, N input for VR would have exceeded N input for the SR application in at least one MU area. Using this approach, yield would have improved with VR over SR, based on the type of results obtained from study 1.

In 1993, studies 2 and 3 were adjacent to each other and differed only by timing of N fertilizer (pre-plant vs side-dressed, respectively) and corn variety. Statistics are not available for study 2, but yield values reflect the mean of multiple MU areas for treatments A and B. With both studies 2 and 3, yield was relatively constant over fertilizer application treatments and MUs (Fig. 29–1). For most yrs, the primary non-nutrient plant-limiting factor on the claypan soil

Table 29–1. Description of study locations and treatments.

Study #	Site	Yr	Fertilizer Treatment	Management Unit	Soil Series	% of Area	Expected Yield	Fertilizer Applied N	Fertilizer Applied Other
							kg/ha	kg/ha	kg/ha
1	Centralia	1992	SR	A,B,C	(same as VR)	100	9410	190	-
			VR	A	Adco silt loam & Mexico silty clay loam, overwash	26	9410	190	-
			VR	B	Mexico silty clay loam	44	8030	162	-
			VR	C	Mexico silty clay loam, eroded	30	6580	130	-
2	Centralia	1993	SR	A,B,C	(same as VR)	100	9410	190	-
			VR	A	Adco silt loam & Mexico silty clay loam, overwash	55	9410	190	-
			VR	B	Mexico silty clay loam	22	7520	157	-
			VR	C	Mexico silty clay loam, eroded	23	5640	106	-
3	Centralia	1993	SR	A,B	(same as VR)	100	9410	190	-
			VR	A	Mexico silty clay loam, overwash	45	9410	190	-
			VR	B	Mexico silty clay loam	55	8030	157	-
4	Oran	1993	SR	A,B,C,D	(same as VR)	100	9530	213	K_2O: 62
			VR	A	Farrenburg fine sandy loam	28	10,970	241	K_2O: 84 P_2O_5: 75
			VR	B	Lilbourn fine sandy loam	16	10,350	218	K_2O: 84
			VR	C	Malden/Clana loamy fine sand	38	9090	207	K_2O: 56
			VR	D	Malden loamy fine sand	18	7210	185	K_2O: 56

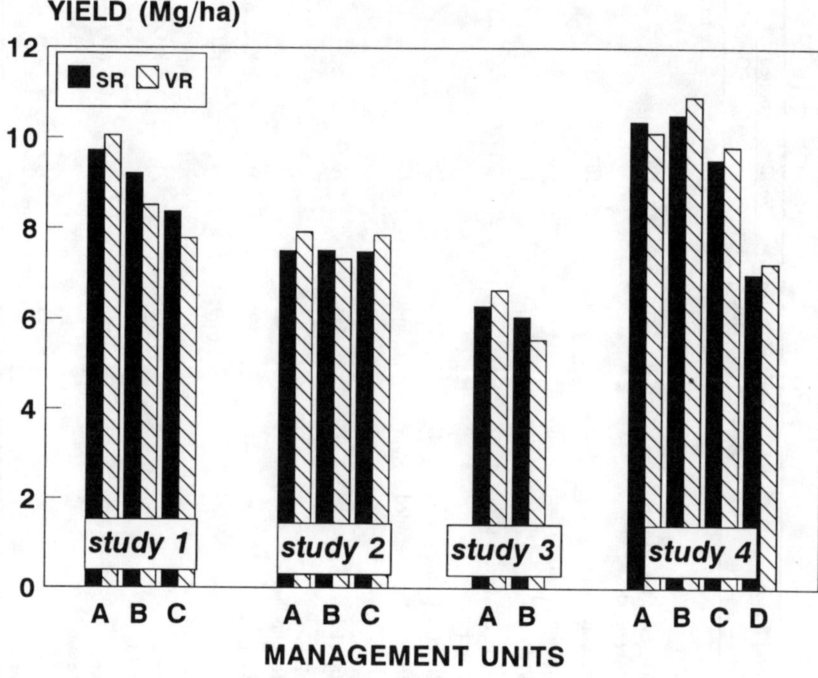

Fig. 29–1. Corn yield from four Missouri studies comparing the impact of variable rate (VR) to single rate (SR) N fertilizer applications over different management units within fields.

landscape is water. Water was not limiting in the 1993 growing season (157 cm (62 inches) annual precipitation). Prolonged wet conditions during several 1-2 week periods during the summer produced conditions ideal for N loss from the soil by denitrification. The anoxic condition of the soil appeared to also have other adverse affects on plant growth. This was especially evident in study 3 where some undetermined wet-environment-by-hybrid interaction occurred that resulted in premature senescence. Seed fill was poor and resulted in lower yields than in the adjacent study 2.

Study 4 (alluvial soils)

Study 4 was located on an alluvial soil site in the southeast corner of the state. With irrigation, average corn yield in this region of the state is about 30-50% higher than the state average (Missouri Farm Facts, 1993). Yields were not significantly different by N applications (VR vs SR) and ranged from 9530 to 10 910 kg/ha (152 to 174 bu/A) for MUs A,B, and C in study 4 (Fig. 29–1). Even with irrigation, yield from the sandier soils of MU D was less than that for the other MUs by 3010 kg/ha (48 bu/A). Irrigating to prevent water stress in MU D would result in over watering in other MUs (personal communication with cooperator). Even though N, P, and K fertilizer rates were varied according to expected yield and soil test for the VR treatment, there was no impact on yield.

Using Yield Maps for Predicting Expected Yield

The strength of yield mapping for determining VR applications is that it integrates soil, landscape, crop, and climate factors together into an expression of relative productivity. If yield variation patterns within fields changed from yr-to-yr and from crop-to-crop, then yield mapping would offer little guidance for developing VR application strategies. Although we only began collecting yield variability data in 1991, some productivity trends have been expressed over both yr and crops. For example, Fig. 29–2 from the Centralia claypan soil site shows relative yield along a 180-M transect with summit (6-42 M) to toe-slope (160-180 M) landscape positions over three yr. Relative yield changes for 1991 corn and 1992 soybean were similar with lowest relative yields occurring both yrs on an eroded side-slope. With water non-limiting in 1993, yield was less depressed on the side-slope.

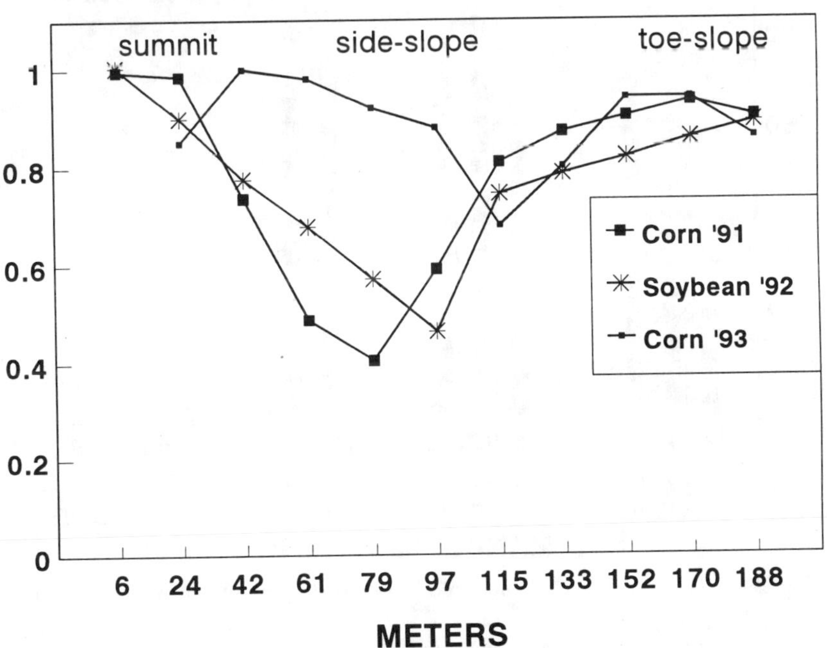

Fig. 29–2. Relative yield along a single claypan soil transect over three yr.

While yield response did not precisely match expected yield for MUs within these studies, patterns of yield were similar to a previous year's yield maps for studies 1 and 4. We recognize that defining different management units from yield mapping in order to develop plans for varying fertilizer rates should be based on more than one or two yr of data (as with these studies). The process of identifying relative differences in productivity within fields needs to be based on an accumulation of as many yr of mapped yields as possible.

Unrecovered N and Soil NO_3-N

In studies 1-3, the apparent fertilizer N recovery efficiency was measured and then used to calculate the amount of unrecovered N in the above ground portion of the plant (Fig. 29–3). In all three claypan soil studies, unrecovered N decreased with VR in the least productive soils. For study 1, the difference in VR and SR at MU C (eroded side-slope) was about 38 kg N/ha (34 lbs N/A). About 60% of the unrecovered N in SR at this landscape position could be accounted for in the greater post-harvest soil NO_3-N (Fig. 29–4). In MU C, the SR treatment received 60 more kg N/ha (54 lbs N/A) than the VR application

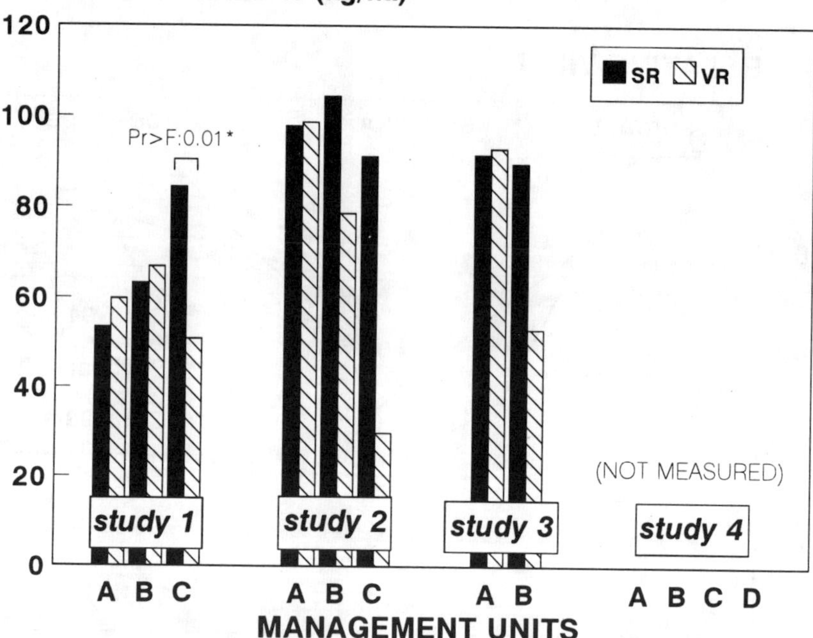

Fig. 29–3. Unrecovered N from three Missouri studies comparing the impact of variable rate (VR) to single rate (SR) N fertilizer applications over different management units within fields. * = a significant orthogonal comparison between SR and VR at that management unit.

which resulted in about 22 kg additional NO_3-N/ha (20 lbs/A) in the soil following harvest. While we would expect some of this extra NO_3-N to be available for the following year's crop, we also could expect water leached from the SR treatment root zone soil to have a higher NO_3 concentration than the VR in MU C.

Lower unrecovered N in studies 2 and 3 with VR (Fig. 29–3) did not result in a reduction of soil NO_3-N following harvest (Fig. 29–4). The wet yr likely caused significant losses through denitrification. With normal crop uptake of N and seed fill in study 2, little NO_3-N was left in the root zone at crop maturity. In study 3, the wet yr also caused early crop death, and thus less N uptake, leaving ≈45 kg/ha (40 lb/A) more NO_3-N in the soil than in study 2. Fertilizer application (VR vs SR) and MU did not affect post-harvest soil NO_3-N in either study.

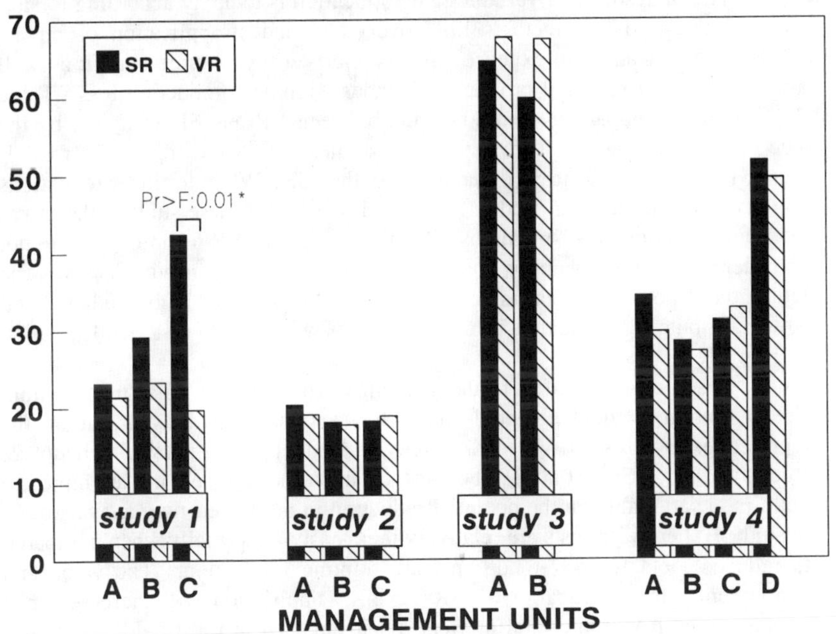

Fig. 29–4. Post-harvest soil NO_3-N from four Missouri studies comparing the impact of variable rate (VR) to single rate (SR) N fertilizer applications over different management units within fields. * = a significant orthogonal comparison between SR and VR at that management unit.

In study 4, soil NO_3-N after harvest was greatest in MU D (Fig. 29–4). Like yield, soil NO_3-N was not different between VR and SR in this MU, and yet the SR treatment had received about 15% more N than VR (Table 29–1). The soil in MU D was very sandy. We expected that fertilizer N would leach as NO_3 very quickly in this MU when rain or irrigation exceeded water holding capacity. The water stress on the crop in this MU (supported with visual observations) and resultant yield reduction was likely the reason for elevated soil NO_3-N. Since there was a rate difference between VR and SR, either more NO_3 leached beyond the sampling zone for SR, or the rate difference between VR and SR was too subtle to detect a difference in soil NO_3-N. Whether the higher NO_3-N after harvest was from over-fertilization or from some other contribution (e.g. late-season N mineralization) did not change the conclusion that N was not likely the limiting factor for crop growth in this area. Nitrogen fertilizer probably could have been reduced more in the VR treatment MU D without decreasing yields.

Comparison of Benefits From Fertilizer Application Strategies

The philosophy of variable-rate application is to apply according to site-specific needs so that there is little over- and under- application of inputs. "Over-" implies an input expense with no short-term economic benefit as well as possible environmental problems. "Under-" implies production loss. Table 29–2 offers a comparison of net benefits between VR and SR weighted by the size of MUs as listed in Table 29–1. In studies 1-3, N fertilizer costs for VR were \$10 to \$12/ha (\$4 to \$5 dollars/A) less than SR. While fertilizer reductions slightly lowered yields in study 1, averaged over all of these studies, there was not much economic loss or gain by VR. Extra costs for VR technology were not considered. Environmentally, the only direct measured gain with VR was the reduction of potentially leachable NO_3-N in MU C of study 1. In studies 2 and 3, we attribute the reduction in unrecovered N with VR to less denitrifciation than in SR.

Based on our studies to date, savings in N fertilizer costs using within field variation in yield potential and VR N application will be less than the reported extra costs of VR data collection and application (Wollenhaupt & Buchholz, 1992; Carr et al., 1991). At the same time, we have not attempted to put a monetary value on the potential reduction in nitrate leaching as was found in study 1. Fertilizer costs are relatively inexpensive, especially when compared to reduced yield returns if nutrients are limiting production. The value and consequently the practicality of VR fertilizer application will increase when implementing it within a total nutrient management plan for a field.

Table 29–2. Returns from VR and SR applications.

Study #	Site	Yr	Fertilizer Treatment	Fertilizer Cost[1]	Expected Yield Weighted by % Area	Yield Difference from Expected Yield[2]	Grain Return - Fertilizer Cost[3]	Environmental Return With VR Compared to SR
				- \$/ha -	---------- kg/ha ----------		- \$/ha -	
1	Centralia	1992	SR	84.00	9410	-315	811.40	
			VR	70.50	7960	750	768.00	reduced potential NO_3-N leaching in MU C
2	Centralia	1993	SR	84.00	9410	-1820	663.20	
			VR	72.10	8150	-565	675.05	none
3	Centralia	1993	SR	84.00	9410	-3260	521.20	
			VR	75.80	8650	-2510	529.30	none
4	Oran	1993	SR	110.20	9530	0	828.40	
			VR	124.70	9470	65	813.90	possible reduction in NO_3-N leaching in MU D

[1] Using: \$0.44/kg N; \$0.57/kg P_2O_5; \$0.26/lb K_2O.
[2] Actual yields weighted and summed by management unit and fertilizer application treatments when significantly different.
[3] Corn price: \$0.04/kg.

SUMMARY

Four studies have been conducted comparing the effects of yield-map derived VR N fertilizer application to whole-field SR N fertilizer application. These studies were conducted on both claypan and alluvial soil sites in Missouri. Previous year's yield maps had shown 30 to 50% variations in corn and soybean yields within these fields. Grain production, unrecovered N in the corn crop, and post-harvest soil NO_3-N were measured.

Nitrogen was more limiting for corn in 1992 than in a typical yr, and thus yield was slightly decreased. The same rates in a more typical yr may have resulted in some greatly over-applied areas with the SR treatment. Yield otherwise was not helped or hurt by VR N applications. Application of N using "best yrs" or "best areas of the field" as a basis for expected yield can result in over-application of N and increased residual soil NO_3-N following harvest (MU C of study 1). Using yield mapping as a predictive tool for expected yields improved our ability to accurately apply N fertilizer. Including more yr of mapped yields in our dataset will enhance our ability to predict productivity differences within fields.

There are several challenging steps to implementing a VR program. Collection and management of relevant soil and crop production data is one step. A second, and just as critical step, will be the process of developing appropriate decision criteria as a basis for variable applications. This on-going work will help us achieve these goals.

ACKNOWLEDGMENTS

The authors wish to acknowledge the farmer cooperators on these studies, Don Collins of Centralia and Bill Holmes of Oran. We further recognize that others have made significant contributions to this research work, namely Steve Borgelt, Larry Mueller, Scott Drummond, and Bill Wilson.

REFERENCES

Birrell, S.J., K.A. Sudduth, and S.C. Borgelt. 1993. Crop yield and soil nutrient mapping. ASAE Paper #93–1556. ASAE, St. Joseph, MI.

Bremmer, J.M., and C.S. Mulvany. 1982. Nitrogen-total. *In* A.L. Page (ed.) Methods of soil analysis, part 2, chemical and microbiological properties. 2nd ed. Agronomy (part 2) 9:595–624. Am. Soc. Agron., Madison, WI.

Buchholz, D.D. 1992. Soil test interpretations and recommendations handbook. Univ. of Missouri, Dep. of Agronomy.

Carr, P.M., G.R. Carlson, J.S. Jacobsen, G.A. Nielsen, and E.O. Skogley. 1991. Farming soils, not fields: A strategy for increasing fertilizer profitability. J. Prod. Agric. 4:57–61.

Cochran, W.G., and G.M. Cox. 1957. Experimental designs. 2nd ed. John Wiley and Sons, Inc., New York.

Colvin, T.S., D.L. Karlen, and N. Tischer. 1991. Yield variability within fields in central Iowa. p. 366–372. *In* Automated agriculture for the 21st century. Proc. Chicago, IL. 16-17 December 1991. ASAE, St. Joseph, MI.

Keeney, D.R., and D.W. Nelson. 1982. Nitrogen-inorganic forms. *In* A.L. Page (ed.) Methods of soil analysis, part 2, chemical and microbiological properties. 2nd ed. Agronomy (part 2) 9:643–698. ASA, Madison, WI.

Lachet Chemicals, Inc. 1988. QuikChem method no. 12–107–04–1–A. Nitrate and nitrite from soil extracts. Mequon, WI.

Missouri Farm Facts. 1993. Missouri Dep. of Agric. and U.S. Dep. of Agric.

Mulla, D.J., A.U. Bhatti, M.W. Hammond, and J.A. Benson. 1992. A comparison of winter wheat yield and quality under uniform versus spatially variable fertilizer management. Agric. Ecosyst. Eviron. 38:301–311.

Wollenhaupt, N.C., and D.D. Buchholz. 1992. Profitability of farming by soils. p. 199–211. *In* P. C. Robert, et al (ed.) Soil specific crop management. Proc. Minneapolis, MN. 14–16 April, 1992. ASA. Madison, WI.

Appendix Table 29–1. Description of study locations and treatments in English units.

Study #	Site	Yr	Fertilizer Treatment	Management Unit	Soil Series	% of Area	Expected Yield(bu/A)	Fertilizer Applied N(lb/A)	Fertilizer Applied Other
1	Centralia	1992	SR	A,B,C	(same as VR)	100	150	170	-
			VR	A	Adco silt loam & Mexico silty clay loam, overwash	26	150	170	-
			VR	B	Mexico silty clay loam	44	128	145	-
			VR	C	Mexico silty clay loam, eroded	30	105	116	-
2	Centralia	1993	SR	A,B,C		100	150	170	-
			VR	A	Adco silt loam & Mexico silty clay loam, overwash	55	150	170	-
			VR	B	Mexico silty clay loam	22	120	140	-
			VR	C	Mexico silty clay loam, eroded	23	90	95	-
3	Centralia	1993	SR	A,B		100	150	170	-
			VR	A	Mexico silty clay loam, overwash	45	150	170	-
			VR	B	Mexico silty clay loam	55	128	140	-
4	Oran	1993	SR	A,B,C,D		100	152	190	K_2O: 55
			VR	A	Farrenburg fine sandy loam	28	175	215	K_2O: 75 P_2O_5: 67
			VR	B	Lilbourn fine sandy loam	16	165	195	K_2O: 75
			VR	C	Malden/Clana loamy fine sand	38	145	185	K_2O: 50
			VR	D	Malden loamy fine sand	18	115	165	K_2O: 50

Appendix Table 29–2. Returns from VR and SR applications in English units.

Study #	Site	Yr	Fertilizer Treatment	Fertilizer Cost[1]	Expected Yield Weighted by % Area	Yield Difference from Expected Yield[2]	Grain Return - Fertilizer Cost[3]	Environmental Return Compared to SR
				- \$/acre -	---------- bu/acre	----------	- \$/acre -	
1	Centralia	1992	SR	34.00	150	-5	328.50	
			VR	28.55	127	12	318.20	reduced potential NO_3-N leaching in MU C
2	Centralia	1993	SR	34.00	150	-29	268.50	
			VR	29.20	130	-9	273.30	none
3	Centralia	1993	SR	34.00	150	-52	211.00	
			VR	30.70	138	-40	214.30	none
4	Oran	1993	SR	44.60	152	0	335.40	
			VR	50.50	151	1	329.50	possible reduction in NO_3-N leaching in MU D

[1] Using: \$0.20/lb N; \$0.26/lb P_2O_5; \$0.12/lb K_2O.

[2] Actual yields weighted and summed by management unit and fertilizer application treatments when significantly different.

[3] Corn price: \$2.50/bushel.

30 Managing Spatial Variability with Furrow Irrigation to Increase Nitrogen Use Efficiency

R. B. Ferguson
J. E. Cahoon

University of Nebraska
UNL-SREC
Clay Center, Nebraska

G. W. Hergert

University of Nebraska
North Platte, Nebraska

T. A. Peterson
C. A. Gotway
A. H. Hartford

University of Nebraska
Lincoln, Nebraska

Increasing levels of nitrate in groundwater have been observed in some river valleys in Nebraska since the mid–1950's. These valleys are characterized by relatively shallow, coarse-textured soils overlying shallow aquifers. Much of the land is irrigated and cropped to continuous corn. Ground water in parts of the central Platte river valley have shown steady increases in NO_3-N concentration, in some areas on the order of 1 ppm yr^{-1} (Engberg & Spalding, 1978). In the most recent statewide evaluation of ground water nitrate and pesticide levels, Exner and Spalding (1990) found that over half of the wells in the state testing higher than 10 ppm NO_3-N were in irrigated river valleys hydrogeologically subject to contamination. They found that more than 20% of the wells sampled contained in excess of 10 ppm NO_3-N, and that both the concentration of nitrate in ground water and the areal extent of contamination had increased since earlier surveys.

More recent investigations in Nebraska have focused on upland soils developed for irrigation overlying much deeper aquifers, generally at least 15 m below the surface. Ground water at this depth is generally of high quality across Nebraska, but incidences of wells testing in excess of 10 ppm NO_3-N are increasingly frequent. Exner and Spalding (1990) found a significant number of wells exceeding 10 ppm NO_3-N in southern Phelps and southwestern Kearney counties, an irrigated corn producing area of fine-textured soils with depths to groundwater between 15 and 30 m. Spalding and Kitchen (1988) found that nitrate had moved down at least 18 m in 15 yr under furrow-irrigated research plots on a silt loam soil receiving 224 kg N ha^{-1} annually. With additional sampling and analysis, Bobier (1992) estimated the downward rate of movement of nitrate-N at this site to be 0.6–0.8 m yr^{-1}. This situation is representative of much of the furrow-irrigated cropland in Nebraska, and suggests that ground water NO_3-N concentrations in deeper aquifers across the state will continue to increase as nitrate in transit in the intermediate vadose zone reaches the aquifer. This situation may also be representative of regional concerns in the Midwest where corn is produced.

Inherent inefficiencies associated with furrow irrigation tend to accelerate nitrate-N leaching in many situations. A typical furrow irrigation infiltration profile is illustrated in Fig. 30–1. An irrigated silt loam soil sampled in 1988 in Hamilton county, Nebraska (Watts et al., 1990) contained over 1000 kg NO_3-N ha^{-1} to a depth of 24 m at the upstream end of the field and over 2700 kg NO_3-N ha^{-1} at the downstream end of the field. The length of the field was ≈800 m. Due to the application time required to adequately water the lower end of many fields, excessive infiltration and nitrate leaching occur at the upstream end of the field. Root zone and intermediate vadose zone samples collected at Mid-Nebraska Water Quality Demonstration Project sites show a similar influence of furrow irrigation on accumulated nitrate-N (Ferguson, et al., 1992). Figure 30–2 illustrates differences between nitrate-N accumulations in the intermediate vadose zone at two Mid-Nebraska Project sites in 1993. These trends are typical of many furrow-irrigated fields sampled in Nebraska. Nitrate-N in the root zone and intermediate vadose zone near the upstream end of the field is often less than near the downstream end, due to greater leaching or denitrification potential near the upstream end of the field.

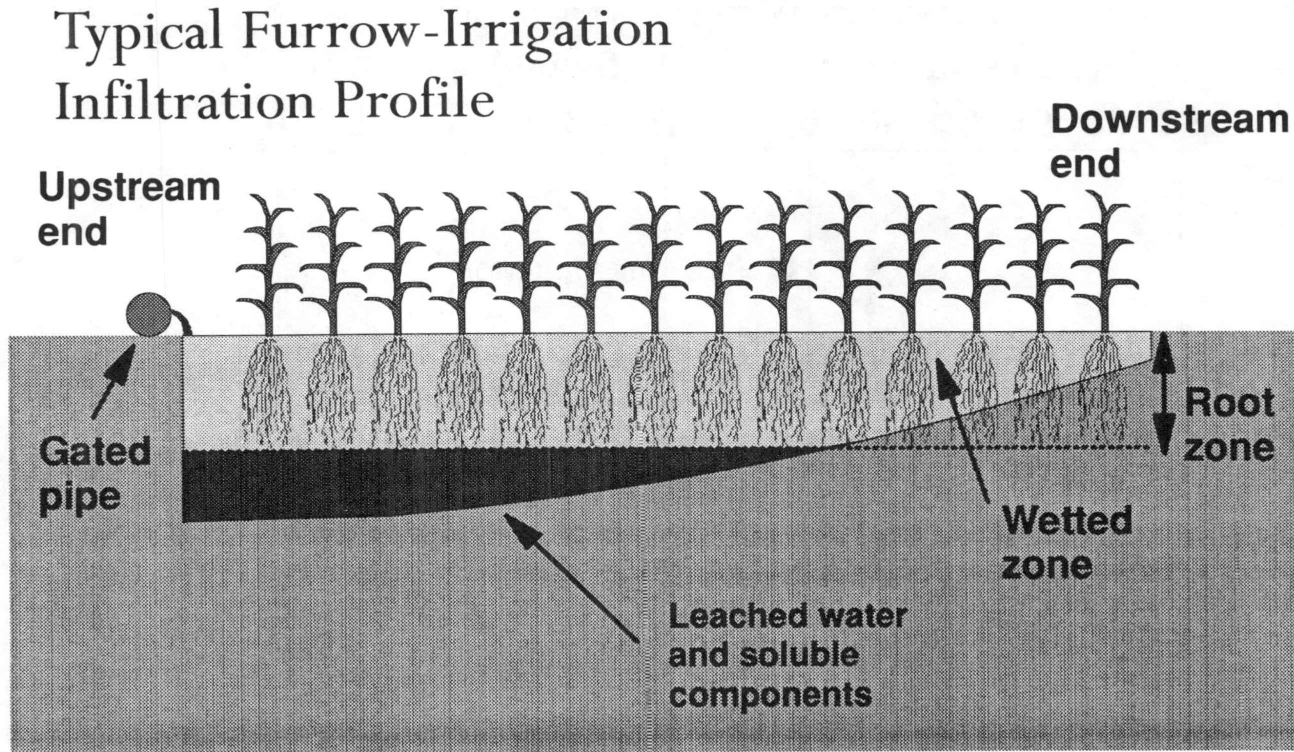

Fig. 30–1. Typical furrow-irrigation infiltration profile.

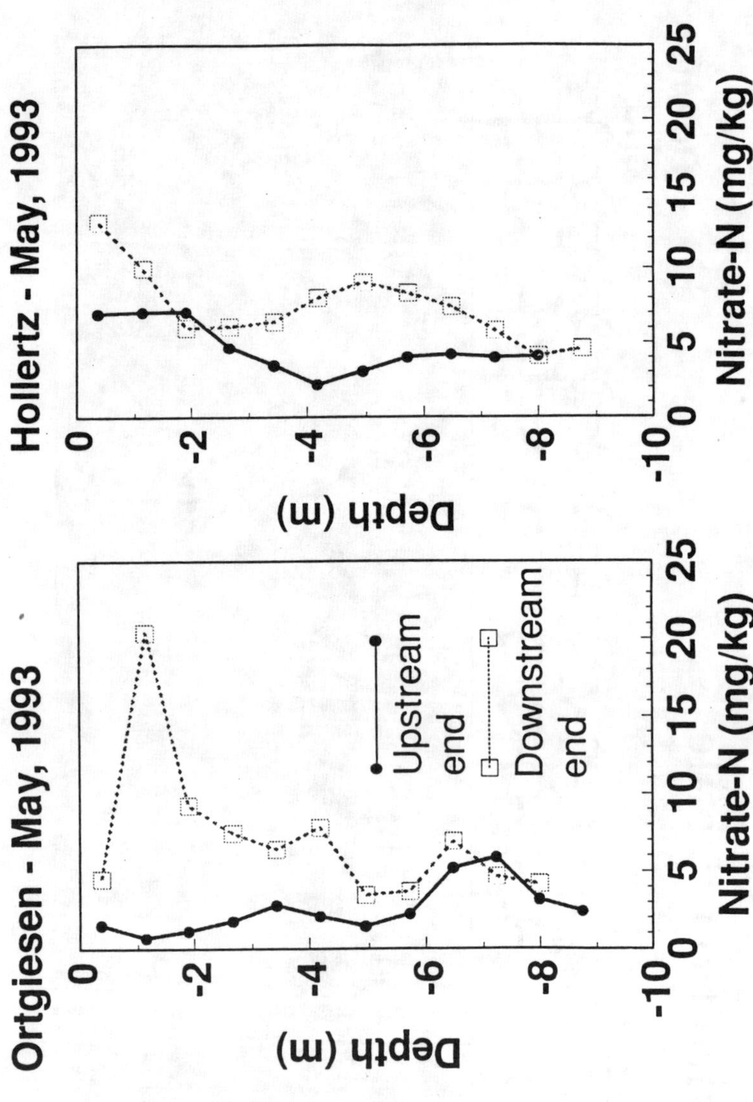

Fig. 30–2. Nitrate-N accumulation in the intermediate vadose zone at two Mid-Nebraska Water Quality Demonstration Project sites, May, 1993.

VARIABLE NITROGEN RATE DEMONSTRATIONS

In Nebraska, ≈1.7 million ha of irrigated corn are produced annually with nearly half under furrow irrigation. Variable rate technology has recently been adapted to anhydrous ammonia application, and has the potential to increase N use efficiency and reduce nitrate leached to ground water in irrigated corn production. Of the N fertilizer applied in 1989 in Nebraska, ≈75% was in the form of anhydrous ammonia (1989 Nebraska Agricultural Statistics).

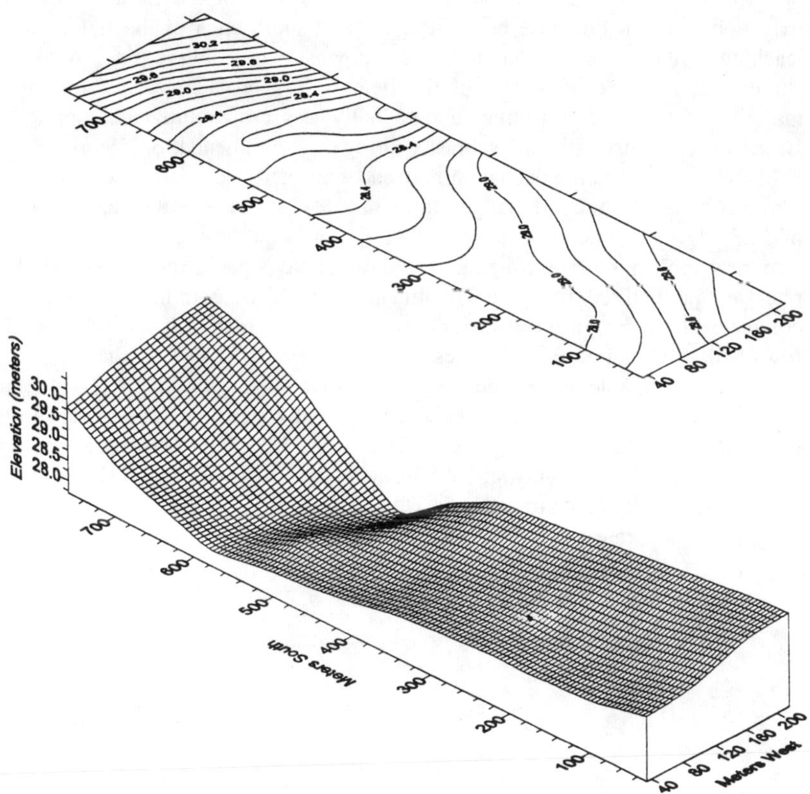

Fig. 30–3. Field topography, Ruhter variable N rate demonstration site, Adams County, 1992.

Limited use of a variable rate anhydrous ammonia applicator began in Nebraska in 1992, primarily on a demonstration basis to become familiar with the equipment. A study was conducted on a site in Adams county, in the south-central area of the state, to compare variable versus fixed rate applications. The majority of the field was a Hord silt loam (fine-silty, mixed, mesic, Cumulic Haplustoll) with Butler silt loam (fine, montmorillonitic, mesic Abruptic Argiaquoll) in the low, central area of the field (Fig. 30–3). The field was 760 m in length, and because it sloped from both ends towards the low area was irrigated from gated pipe at both ends. Surface water drained from the low central area to the east off of the field.

Soil samples were collected on a square grid with 61 m spacing in March, 1992. Individual cores were collected to a depth of 1.2 m in 0.3 m increments and analyzed for NO_3-N. The resulting map of residual NO_3-N is shown in Fig. 30–4. The accumulation of residual nitrate-N in the central portion of the field is likely related to both irrigation and soils. Since the opportunity time for infiltration is greater at the end of the field nearest the irrigation pipe (in this case both the north and south ends of the field), more leaching is likely to occur than in the area downstream from the pipe. Also, the Butler soil in the central area of the field is less productive than the Hord majority of the field, resulting in potentially less crop removal of nitrate-N. Based on the University of Nebraska nitrogen recommendation algorithm for corn, which used soil residual nitrate and expected yield, the recommended fertilizer N rate for the field ranged from 34 to 240 kg ha^{-1}, for an expected yield of 11.3 Mg ha^{-1}. The cooperator applied 34 kg N ha^{-1} at planting, resulting in sidedress application rates ranging from 0 to 206 kg N ha^{-1}. Besides the variable rate treatment, three fixed rate treatments were also used in a randomized, complete block design with six replications applied to field length strips eight rows wide. The three fixed N rates were 140, 202 and 240 kg N ha^{-1}.

Two variable N rate demonstration sites were used in 1993 - one in Lincoln county at the West Central Research and Extension Center near North Platte, and the other in York county. Data from the Lincoln county site is not presented. The York county site was located on a furrow-irrigated, Hastings silt loam (fine, montmorillonitic, mesic Udic Argiustoll). The field was ≈370 m long, with moderate slope (Fig. 30–5).

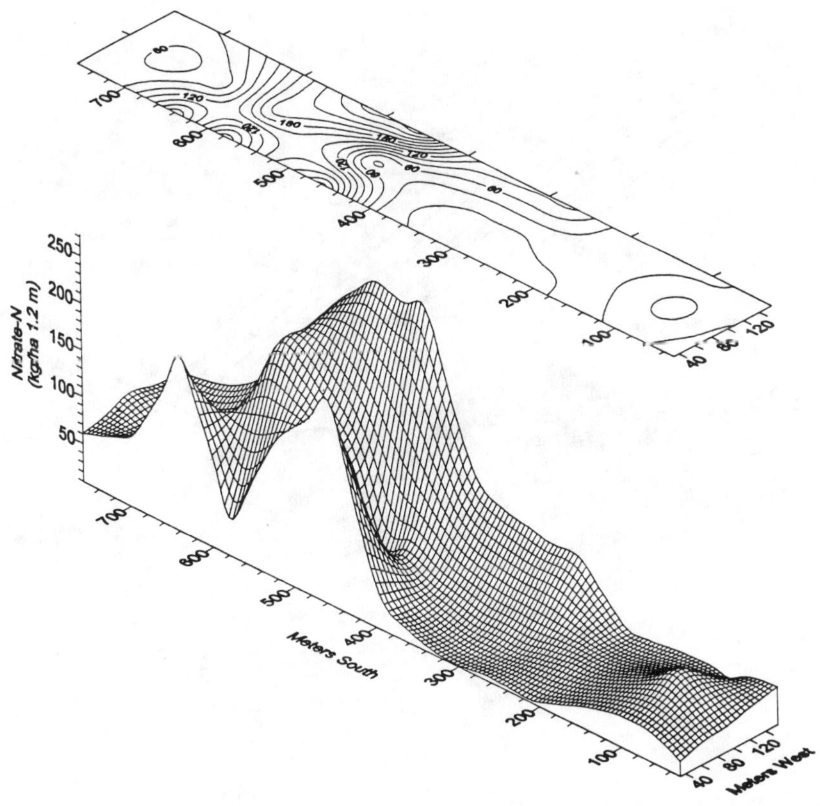

Fig. 30–4. Residual nitrate-N to 1.2 m, Ruhter variable N rate demonstration site, Adams County, March, 1992.

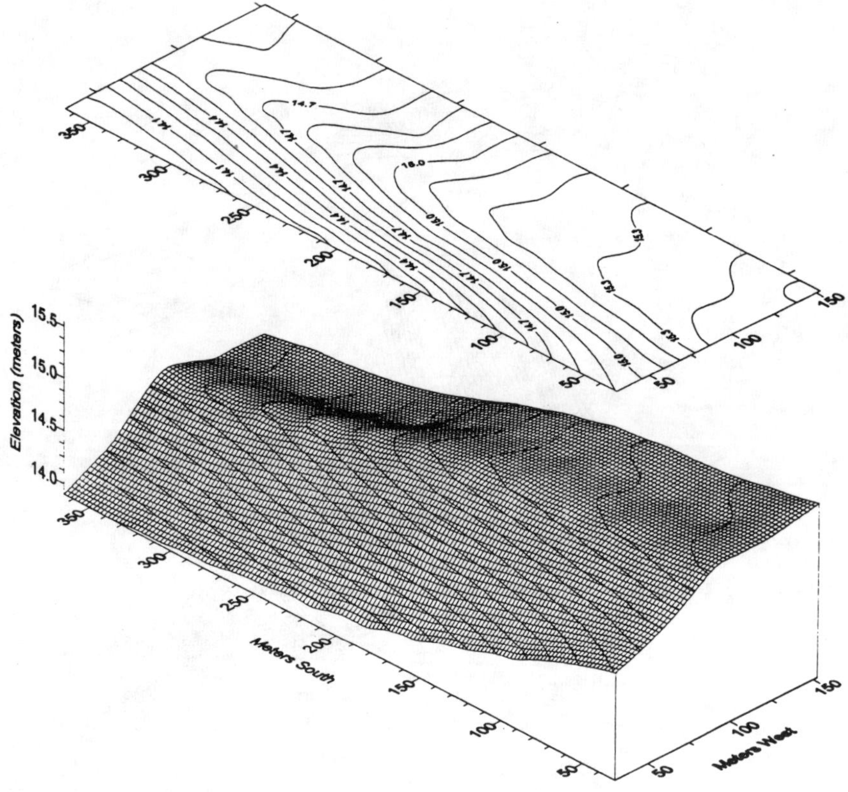

Fig. 30–5. Field topography, Stahr variable N rate demonstration site, York County, 1993.

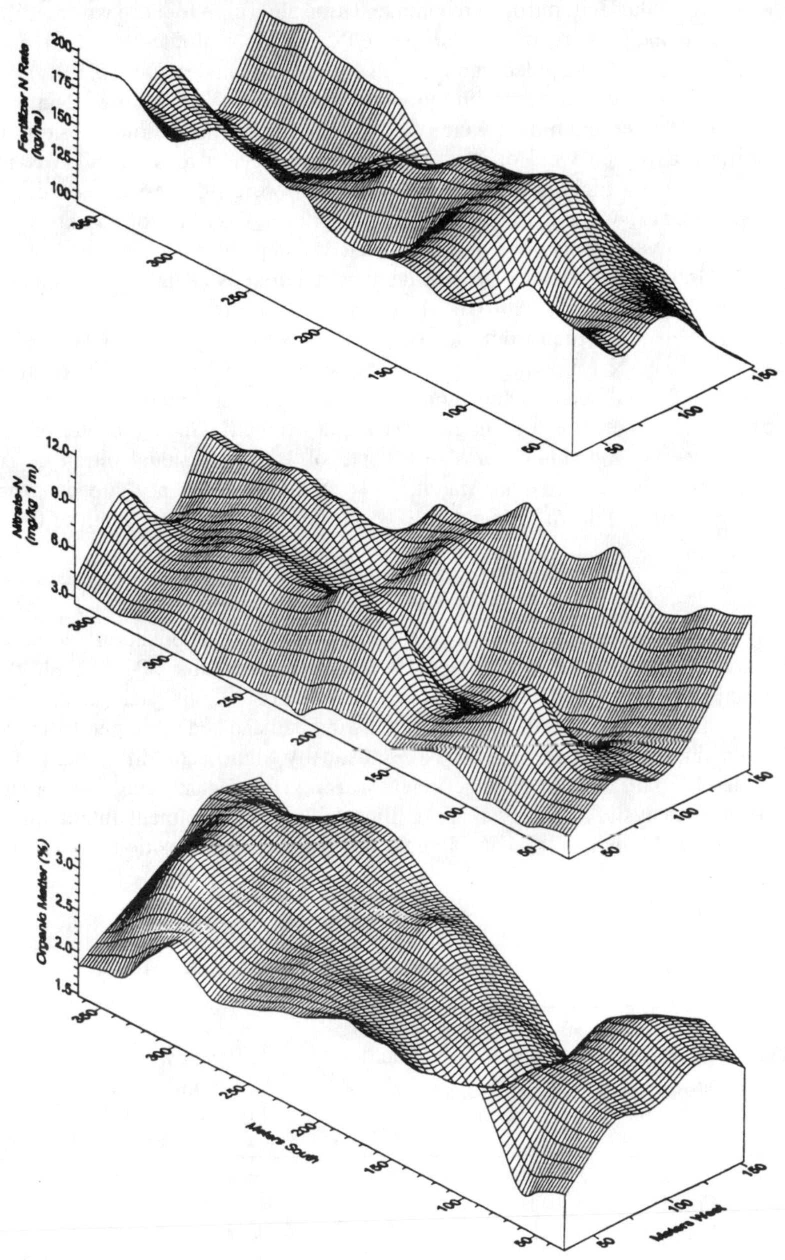

Fig. 30–6. Organic matter content, residual nitrate-N, and fertilizer N rate, Stahr variable N rate demonstration site, York County, April, 1993.

Since the UNL nitrogen recommendation algorithm for corn was modified in 1992 to include organic matter as well as residual nitrate-N and expected yield, the field was sampled and mapped for both organic matter (0.2 m depth) and nitrate-N (0.9 m depth) with individual cores at 30.5 m spacing. The maps of organic matter and nitrate were used with the recommendation algorithm for a uniform expected yield of 11.8 Mg ha^{-1}, to produce a recommended N rate map (Fig.30–6). Field length strips 16 rows wide were sidedressed according to two treatments - variable rate approach or a uniform, fixed rate of 155 kg N ha^{-1}, applied to seven replications in a randomized, complete block design.

Grain was harvested at both demonstration sites using a yield mapping combine in cooperation with Dr. Mark Schrock from Kansas State University. Yield was also measured from each strip using a weigh wagon. Soil samples for residual nitrate-N following harvest were collected in March 1993, from the variable and fixed, recommended rate strips at the Adams county site. Sampling was again conducted on a 61 m grid spacing down the field,at the center of each strip, collecting individual cores to a depth of 1.2 m. Residual nitrate N soil samples will be collected in March, 1994 at the York county site on a grid spacing of 30.5 m down the field at the center of each strip to a depth of 0.9 m.

Demonstration Results

Average N rates, grain yield, N use efficiency and application rate ranges for demonstration sites in 1992 and 1993 are given in Table 30–1. Yield data in Table 30–1 was collected from the weigh wagon. Grain yield at the York county site (1993) was reduced because of wind damage and some denitrification N loss due to standing water. No statistically significant differences were detected in grain yield or N use efficiency, using strip mean yields and classical statistical analysis. There was a significant block by treatment interaction in 1992 for yield, indicating that, as might be expected, the position in the field

Table 30–1. Variable N rate demonstration summary, 1992 - 1993.

Year	Treatment	Average N Rate (kg ha^{-1})	Yield (Mg ha^{-1})	Efficiency (kg grain kg N^{-1})	N Rate Range (kg ha^{-1})
1992	Fixed - 62	140	9.54	68.1	-----
	Fixed	202	9.92	49.1	-----
	Fixed + 39	241	9.73	40.4	-----
	Variable	205	9.86	48.1	34 - 241
1993	Fixed	155	8.85	57.1	-----
	Variable	144	8.73	60.6	123 - 185

influenced yield response by treatment. Geostatistical analysis of yield and residual nitrate-N data from the 1992 Adams county site and 1993 York county site is still being conducted. Several factors have combined to complicate geostatistical analysis of the data. The large size of the yield data set has resulted in computer hardware and software constraints (over 7 000 yield points were mapped). Also, the recorded locations of yield (mapped using post-processed, differential GPS) did not directly coincide with the center of each strip. Finally, preplant residual nitrate-N, grain yield, and post-harvest residual nitrate-N were not recorded on the same grid spacing. Figure 30–7 illustrates the strip locations, yield measurement locations, and post-harvest residual N locations. In order to facilitate data analysis, yield measurement points were moved to the center of each strip, and integrated within cells to reduce the dataset size. The resulting averaged, aligned dataset is shown in Fig. 30–8.

The size of the yield dataset has not allowed the use of the interactive geostatistical package, GEOEAS, for analysis. Instead, preliminary geostatistical analysis of yield has been conducted using a non-interactive package. The resulting semivariogram for yield is shown in Fig. 30–9. Although considerable scatter is present in the semivariogram, a reasonable estimate results from the use of a spherical model, with a nugget of 550, a sill of 900, and a range of 700 ft (213 m). Geostatistical analysis to separate the treatment structure from the analysis, and incorporate post-harvest residual nitrate-N, is still in progress. No geostatistical analysis has been conducted to date on data collected at the York county site in 1993.

Some useful information has resulted from the 1992 and 1993 variable N rate demonstrations:

1. It is evident that variable N application will not always result in more efficient N use or greater yield at less cost. There will likely be fields at which a fixed rate approach will be as efficient as a variable rate approach. It will be important to develop procedures to identify fields or cropping situations where variable rate N application will be potentially beneficial.

2. All variables need to be measured on the same grid structure.

3. Large datasets collected with yield mapping combines can complicate data manipulation and analysis.

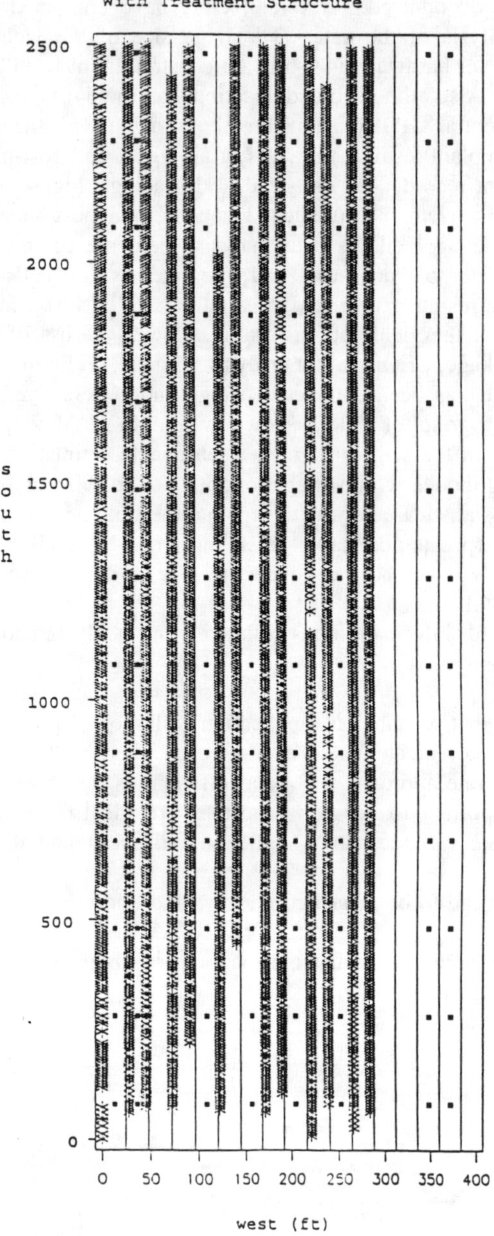

Fig. 30–7. Ruhter variable N rate demonstration site, Adams County, 1992. Treatment strips are 8 rows wide; yield measurements are marked X; post-harvest residual nitrate-N sample points marked with a square.

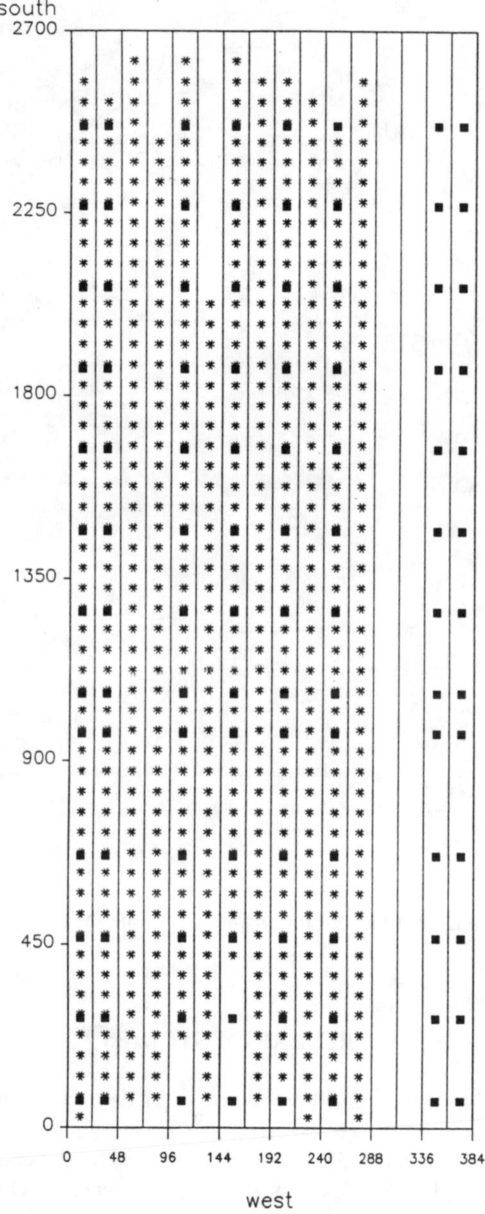

Fig. 30–8. Ruhter variable N rate demonstration, Adams County, 1992. Averaged, aligned yield points at center of treatment strips. Post-harvest residual nitrate-N sample points marked with a square.

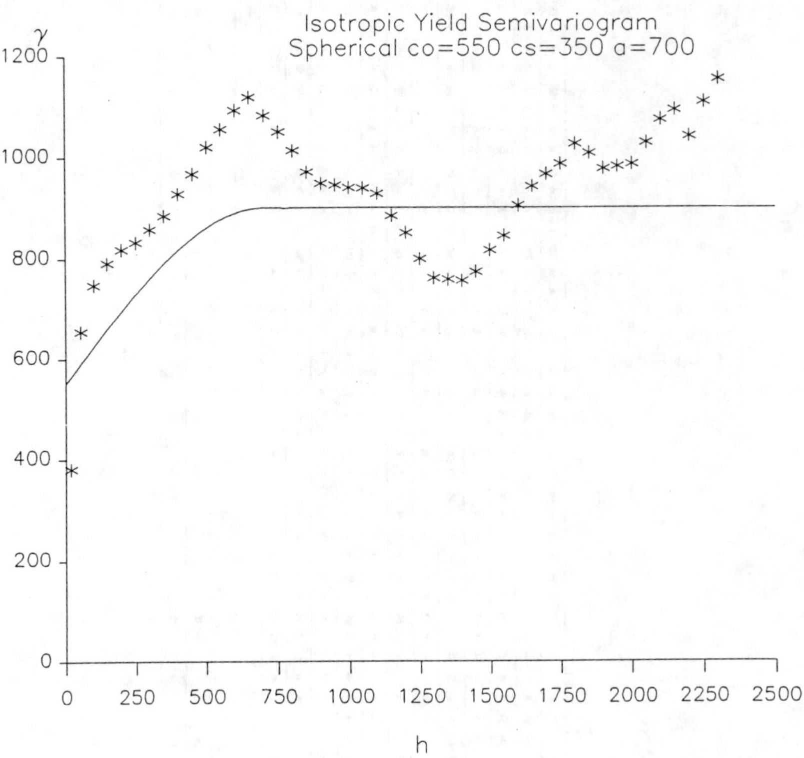

Fig. 30–9. Yield semivariogram, Ruhter variable N rate demonstration site, Adams county, 1992.

RESEARCH STUDIES

With experience gained from two yr of demonstrations, and questions resulting from that work, two research projects were initiated in the fall of 1993 to investigate the potential for variable rate nitrogen application to improve N use efficiency, and reduce N loss to the environment. One study is being conducted under sprinkler irrigation at sites in the central Platte valley and the Sandhills of Nebraska, and will not be discussed in detail in this paper. The other is being conducted at three furrow-irrigated sites in south-central and west-central Nebraska. Questions being addressed in both studies include:

1. What grid sampling density is necessary to accurately predict residual soil nitrate-N levels for a field?
2. Is it possible to not have to grid sample fields every yr for nitrate-N; can yield maps, sampling selected points, or leaching models estimate accurately the N requirement map for the field?
3. How does variable N rate application compare to fixed N rate application in terms of grain yield, N use efficiency, soil residual N and leaching losses?

Based on preliminary geostatistical analysis from the 1992 Adams county demonstration site, it appeared that a grid spacing of less than 61 m was necessary to more accurately estimate the nugget effect and produce accurate, kriged maps. Consequently, an alternating grid design is being used on the research sites, including several smaller cross grids with spacings of 0.76, 1.5, 2.3, 3.7 and 5.3 m in all four primary directions. This design (Fig. 30–10) results in smaller lag spacings which may more accurately predict the nugget effect without the prohibitive time and expense to sample the entire field on a denser grid.

Semivariograms resulting from this grid design for mean nitrate-N (mg kg^{-1}) to 0.9 m and P (mg kg^{-1}) to 0.2 m are shown in Fig. 30–11 and 30–12. The closer grid spacings in the sampling design result in more samples at smaller lags, defining the semivariogram at short distances (Fig. 30–11). Figure 30–13 illustrates the kriged map of mean nitrate-N for the field. The field was found to have a strong directional anisotropy for phosphorus and zinc. Although the field has been cropped to corn and soybeans for at least the previous 25 yr, the patterns of P were indicative of manure application (Fig. 30–14). In consulting with the cooperator, we found that there is evidence of a past farmstead in the north part of the field, as he occasionally turns up debris with tillage. It is likely that a livestock confinement lot was associated with the farmstead, resulting in accumulations of P and Zn which are evident 30 or more years later. Interestingly, these patterns are not evident for organic matter, nitrate-N, or pH.

Furrow-irrigation component

Additionally, the furrow-irrigation study is evaluating how the spatial trend of water infiltration and leaching inherent to furrow irrigation can be accounted and managed for. Ideally, infiltration functions and intake opportunity times would be measured on the same grid interval as other soil and crop parameters, and used to develop infiltration estimates using volume balance techniques. Time and labor constraints, however, make this approach not feasible in this project. Consequently, information will be collected to gain some indication of the variability in infiltration that results from variable soil factors, and to indicate the variability in infiltration resulting from opportunity time differences from the upstream to the downstream portion of the field.

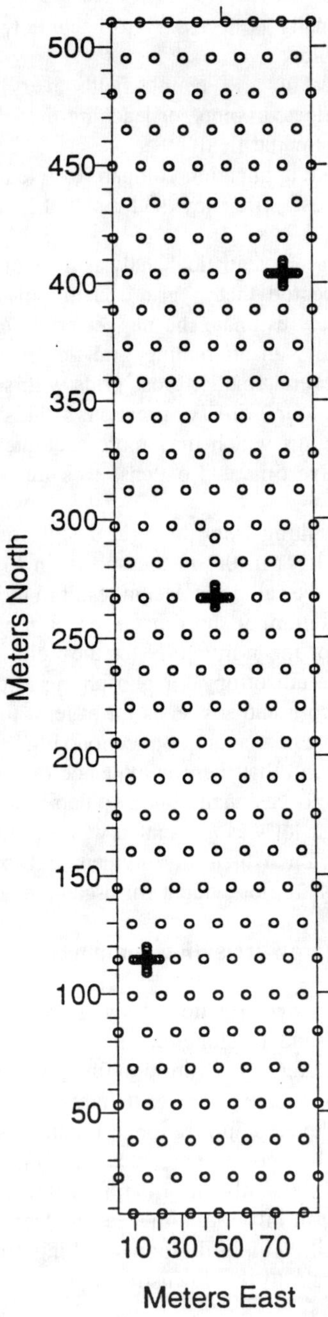

Fig. 30–10. Grid sampling design, furrow-irrigation site 9470, Clay County, fall, 1993.

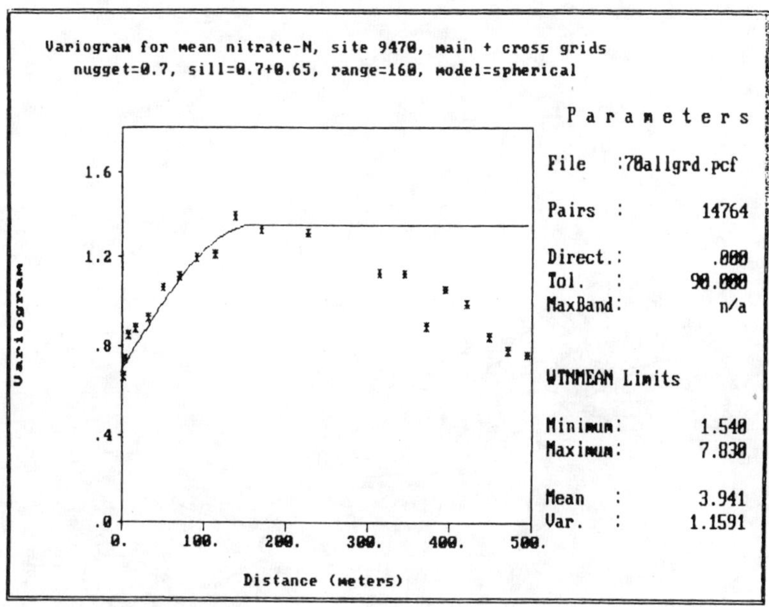

Fig. 30–11. Semivariogram for mean soil nitrate-N, furrow-irrigation site 9470, Clay County, fall, 1993.

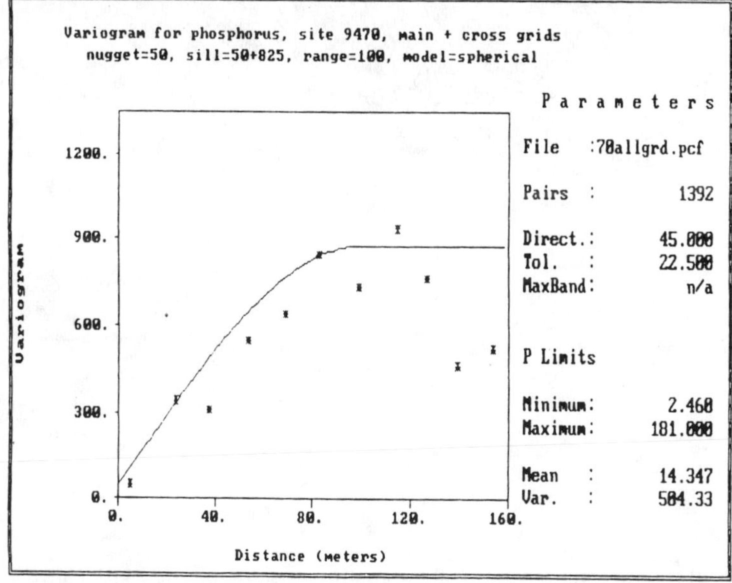

Fig. 30–12. Semivariogram for soil phosphorus, furrow-irrigation site 9470, Clay County, fall, 1993.

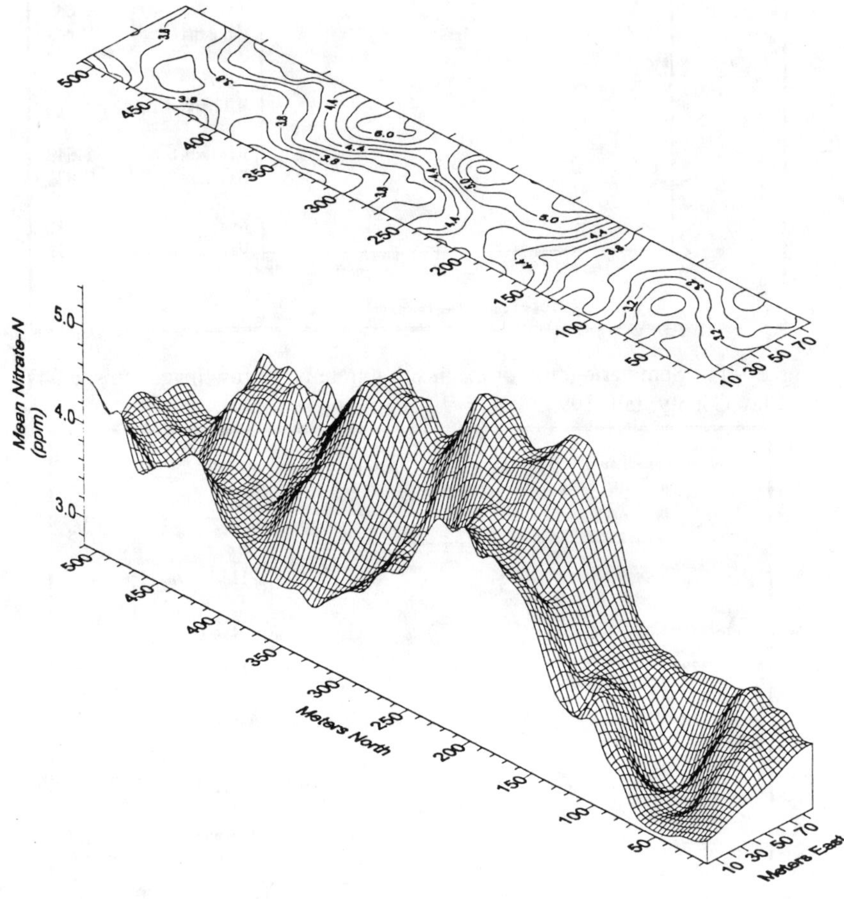

Fig. 30–13. Soil nitrate-N distribution, furrow-irrigation site 9470, Clay County, fall, 1993.

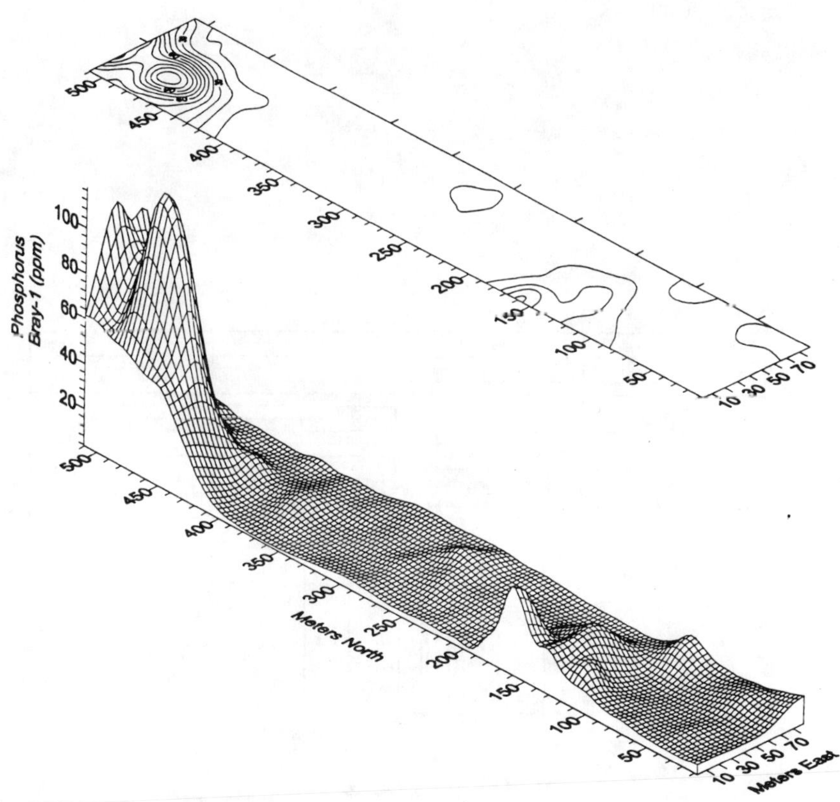

Fig. 30–14. Soil phosphorus distribution, furrow-irrigation site 9470, Clay County, fall, 1993.

A combination of sorptivity tests (Clothier & White, 1981; Green et al., 1986) and volume balance techniques will be used to establish the following relationships at 12 point in the field:

$$F_1 = st^{1/2}$$

$$F_2 = a(t - t^*)$$

where s = the sorptivity (cm/min$^{1/2}$),
t = opportunity time (min)
a = steady state infiltration rate (cm min^{-1})
F_1 = cumulative infiltration during the transient stage (cm)
F_2 = cumulative infiltration during the steady state stage (cm)
t^* = time at which infiltration changes from transient to steady state (min).

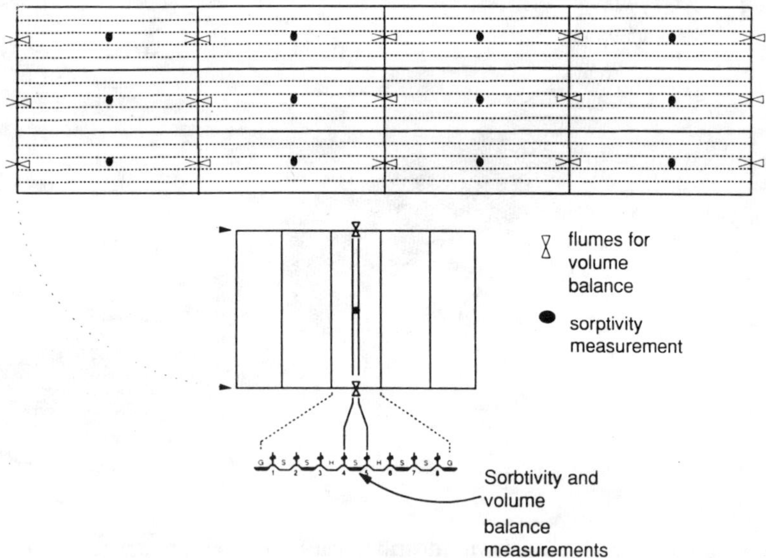

Fig. 30–15. Sorptivity and volume balance measurement locations, furrow-irrigation study sites, 1994.

Figure 30–15 illustrates the planned arrangement of sorptivity and furrow flume measurements in the field, in quarters of the furrow length. Sorptivity tests will be performed prior to the irrigation event. Near the end of the irrigation event, irrigation flumes will be measured to establish the steady state infiltration rate for each quarter length section. Advance and recession will be tracked on the observation furrow. This information will allow the estimation of infiltrated depth at each sorptivity measurement location. Using volume balance techniques, an infiltration profile for each observed furrow (similar to Figure 30–1) can be developed, from which infiltration at any point down the furrow can be estimated.

SUMMARY

Only preliminary investigations of variable rate nitrogen application have occurred in Nebraska to date. The primary focus thus far has been on gaining experience with application equipment and software, evaluating soil sampling grid designs, and developing the experimental protocols for more rigorous research studies. Considerable range in the required N fertilizer rate has been found in some fields, associated with soil and irrigation management properties. No consistent advantage has been observed to date for variable rate N application in yield or N use efficiency. Current studies are addressing questions of what fields–soils–cropping systems are most likely to benefit from variable rate N application; what sampling grid density is necessary; what are the options to annual grid sampling for N; and under what conditions can variable rate application reduce the potential for N loss to the environment.

REFERENCES

Bobier, M.W. 1992. Agrichemical movement in the intermediate vadose zone at the South Central Research and Extension Center near Clay Center, NE. M.S. Thesis, Dep. of Geology, Univ. of Nebraska.

Clothier, B.E., and I. White. 1981. Measurement of sorptivity and soil water diffusivity in the field. Soil Sci. Soc. Am. J. 45:241-245.

Engberg, R.A., and R.F. Spalding. 1978. Groundwater quality atlas of Nebraska. Resource Atlas No. 3, Conservation and Survey Div., Univ. of Nebraska.

Exner, M.E., and R.F. Spalding. 1990. Occurrence of pesticides and nitrates in Nebraska's groundwater. WC1. Univ. of Nebraska Water Center.

Ferguson, R.B., A.P. Christiansen, D. Krull, D. Gosselin, M. Kuzila, E. Barnes, and T. Murphy. 1992. Root zone and intermediate vadose zone nitrate accumulation as influenced by nitrogen and irrigation management. Agron. Abs. 1992, pp. 227.

Green, R.E., L.R. Ahuja, and S.K. Chong. 1986. Hydraulic conductivity and sorptivity of unsaturated soils: field methods. *In* Methods of soil analysis Part 1, Physical and Mineralogical Methods. A. Klute (ed.), Am. Soc. Agron., Madison, WI.

Spalding, R.F. and L.A. Kitchen. 1988. Nitrate in the intermediate vadose zone beneath irrigated cropland. Ground Water Monitoring Rev. 8:89–95.

Watts, D., A. Christiansen, K. Frank, and E. Penas. 1991. The impact of nitrogen and vadose zone conditions on ground water contamination by nitrate-nitrogen. EC91–735. Univ. of Nebraska.

31 Nitrogen Specific Management by Soil Condition: Managing Fertilizer Nitrogen in Corn

J. A. Vetsch
G. L. Malzer
P. C. Robert
D. R. Huggins

Soil Science Department
University of Minnesota
St. Paul, Minnesota

Best management practices for nitrogen (N) fertilization of corn (*Zea mays* L.) are being improved continuously. Environmental and economic concerns are considered when evaluating a fertilizer practice or program, but a sound agronomic practice must be open to new ideas. Site-specific fertilization is seen as a link between sound agronomic practices and new technology. Its appeal comes in the form of perceived economic and environmental benefits. The benefits appear justified, since, in theory, fertilizer recommendations can be calibrated at the site-specific scale as well as the field scale. By fine tuning the recommendations to inherent soil variability, a site-specific fertilizer program should produce equal or greater yields while reducing the risk for N losses to the environment due to over application.

There are several questions involving research in site-specific N fertilization. This paper will address four: Is there significant variability in the N supply of production fields? Does this variability affect yields? Does this variability have any spatial character (can it be mapped) or is it random variation? If this variability is significant enough to warrant site-specific application, what soil–crop–other parameters can be used to make site-specific N recommendations? Field research experiments were designed in 1990 to address these questions. The objectives of these experiments were: to quantify yield and nutrient variability in production fields, evaluate yield response of applied fertilizer N and differential N loss as influenced by soil conditions, and to determine what soil parameters can be used to make site-specific N fertilizer recommendations for corn.

MATERIALS AND METHODS

The research protocol consisted of several parts; only selected parts will be discussed in this paper. The selection of sites was based on size, inherent soil variability, and cropping systems (rotations) common to the area. Soil variability, both nutrient and taxonomic, was measured extensively. Treatments of different rates of N fertilizer were applied in replicated strips. Grain yield was measured by combine harvest and yield response to fertilizer N was calculated using response curve analysis. Finally, soil data were correlated with grain yield response.

Sites

Four research sites (two different locations in 1991 and 1992) were located in Southwestern (Southwest Experiment Station, Lamberton) and Central (near the Sandplain Experiment Station, Becker) Minnesota. The sites ranged in size from 8 to 19 ha. The soils at the Lamberton sites (L91 and L92) were predominately Mollisols that developed under prairie vegetation on Des Moines Lobe glacial till. The soils at the Becker sites (B91 and B92) were Mollisols that developed in sandy glacial outwash. These sandy outwash soils produce high yields when managed at high levels of fertilizer and water (irrigation) inputs.

Treatments

Five constant N rates (four plus a check, five plus a check at L92 site) were applied as replicated strips across the entire field in a split-block design. A variable rate treatment (two at Lamberton) was included to evaluate current site-specific recommendations. Nitrogen rates and timing of treatment application varied over site yr (See Table 31–1). Anhydrous ammonia was the N source for all site yr.

Table 31–1. Nitrogen rates and timing of anhydrous ammonia application by site.

Sites	Timing	Rates
		kg N ha^{-1}
Lamberton 91	VE (May 22 and 23)	0, 67, 101, 134, 168
Lamberton 92	VE (May 15)	0, 67, 101, 134, 168, 202
Becker 91	V6 (June 6)	0, 134, 168, 202, 235
Becker 92	V4 (June 12)	0, 101, 134, 168, 202

Measurements

Soil measurements included grid soil sampling in the fall preplant, spring preplant and fall post harvest. Sampling grids ranged in size from 15 by 18 m for the spring sampling of the 1992 sites to 61 by 61 m for the spring sampling of the 1991 sites. No fall preplant samples were taken at Becker. Fall preplant grid spacing at Lamberton was 30 by 30 m. Soil samples were analyzed for NO_3-N, NH_4-N, pH, total Kjeldahl N, total carbon and mineralizable N. Only spring preplant soil data will be discussed in this report.

Grain yield and moisture were measured from segmented harvest strips ≈1.5 m wide and 12 m long. Alternate samples were taken for measurement of N concentration in the grain. Yields presented herein were calculated at 15.5% moisture.

Site characteristics were also taken, including a 1:5000 scale soil survey, digital elevation model (DEM) and aerial photographs of bare soil and crop. The storage and analysis of all data was assisted by a geographic information system.

Statistical Analysis

Yield data were fitted to a linear-plateau model (Fig. 31–1), calculated with the NLIN procedure in SAS®, 1991. The data from the segmented harvest areas were partitioned into blocks one replication of treatments wide (55 m and 75 m at Lamberton and Becker, respectively) by one-half a replication long (Fig. 31–2). The model was used to calculate the response to fertilizer N at each of these small blocks in the field. Model predicted parameters (intercept=check yield, slope=yield response to N fertilizer increment, y axis value at plateau of model=maximum yield and x axis value at intersection of the plateau and linear sections of the model=economic optimum N rate, EONR) were graphed to evaluate variability, spatial structure, and correlation in parameters. Soil data, taken from the same areas in space as the response parameters, were then regressed on the model parameters to determine if any of the measured parameters were correlated with the response to fertilizer N.

RESULTS AND DISCUSSION

The results and discussion are broken down into three parts. Variability in nutrient supply, response to fertilizer N over space, and the relationship between fertilizer response and soil data will be discussed herein.

Nutrient Supply

Yields ranged from 2.3 to 6.8 Mg ha^{-1} in a check plot yield transect at the L92 site and from 4.7 to 7.2 Mg ha^{-1} at B92 (Fig. 31–3). High variability in check yield transects was evident at all sites. Variability in check plot yields was attributed to variability in the soil N supplying capacity.

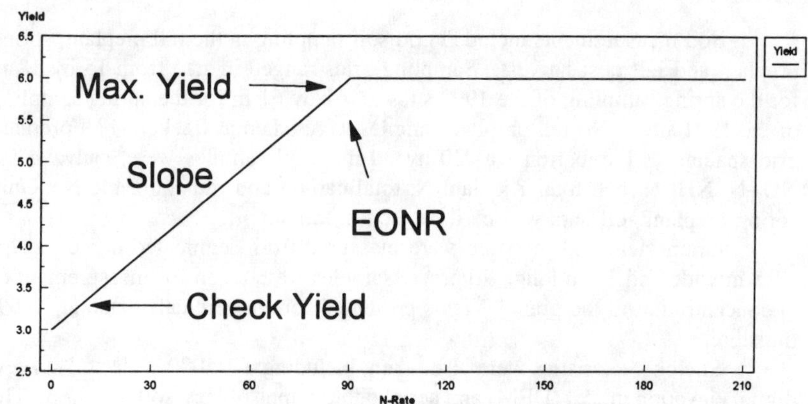

Fig. 31–1. Definitions of model parameters of interest from a linear plateau model.

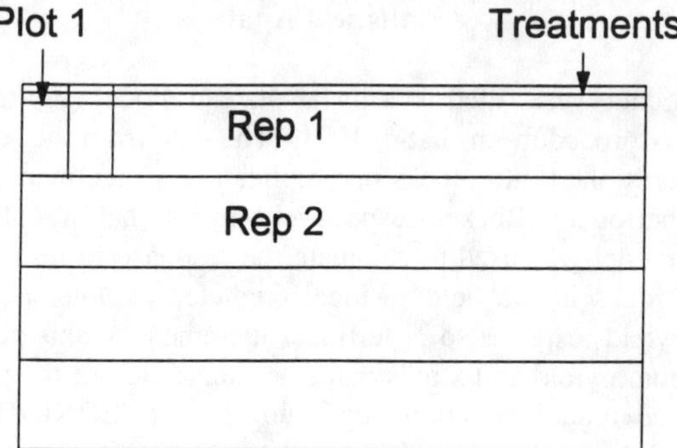

Fig. 31–2. Schematic diagram of subplot locations for response curve analysis.

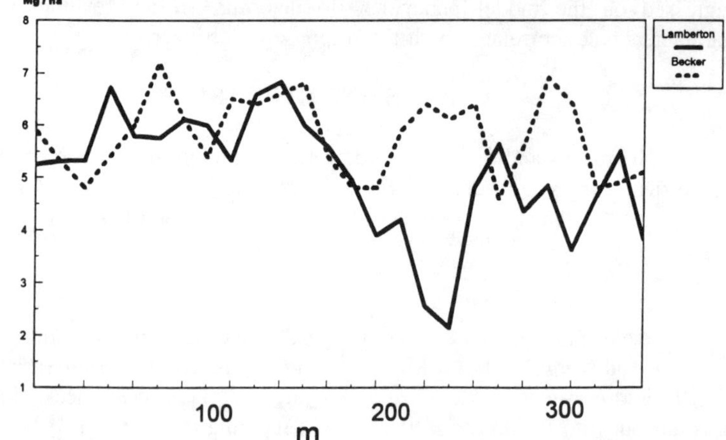

Fig. 31–3. Check plot (zero N) yield variability in transects from Lamberton and Becker, 1992.

Fertilizer N Response

Response parameters from the fitted linear plateau models were graphed for spatial similarities (Fig. 31–4 & 31–5). Several important observations can be made from these graphs. Generally, these observations apply to all four sites, only the B92 and L92 sites are presented here. Considerable variability was evident in check plot yield, response slope and EONR (Fig. 31–4 & 31–5). Little variability was found in maximum model predicted yield (Fig. 31–4 & 31–5). Spatial character between adjoining subplots was evident in both sites for all parameters (Fig. 31–4 & 31–5). More random variation was found in response parameters from the L91 site (data not presented here). Variability in check plot yields, as discussed earlier, was attributed to variability in the soil N supplying capacity. Variability in response slope suggests that soil conditions affect fertilizer usage and fertilizer efficiency. Variability in EONR helps support the need for site-specific nitrogen application. The lack of variability in maximum predicted yield suggests that for these data yield monitoring would be of little value for predicting site-specific N rates. Spatial structure in the response parameters, specifically EONR, shows that variation is not purely random, thus site-specific N fertilization is feasible.

The average EONR calculated from the response analysis of the subplot areas ranged from 61 to 91 kg N ha^{-1} less than the conventional (single rate) application over all sites (Table 31–2). Also, the maximum value in some portion of each field was higher than the conventional rate in two of the four sites (Table 31–2). These data suggest that site-specific N fertilization has tremendous environmental and economic potential. The wide range (maximum–minimum) in values shows that considerable variability in N requirement was recognized at all sites.

Table 31–2. Summary of the response parameter economic optimum N rate for Lamberton.

Site	Economic Optimum N Rate (EONR)			
	Minimum	Maximum	Mean	Conv.†
	kg N ha^{-1}			
Lamberton 1991	0	168	77	168
Lamberton 1992	28	202	107	168
Becker 1991	75	235	133	202
Becker 1992	24	202	124	202

†Conventional, single rate application from University of Minnesota recommendation.

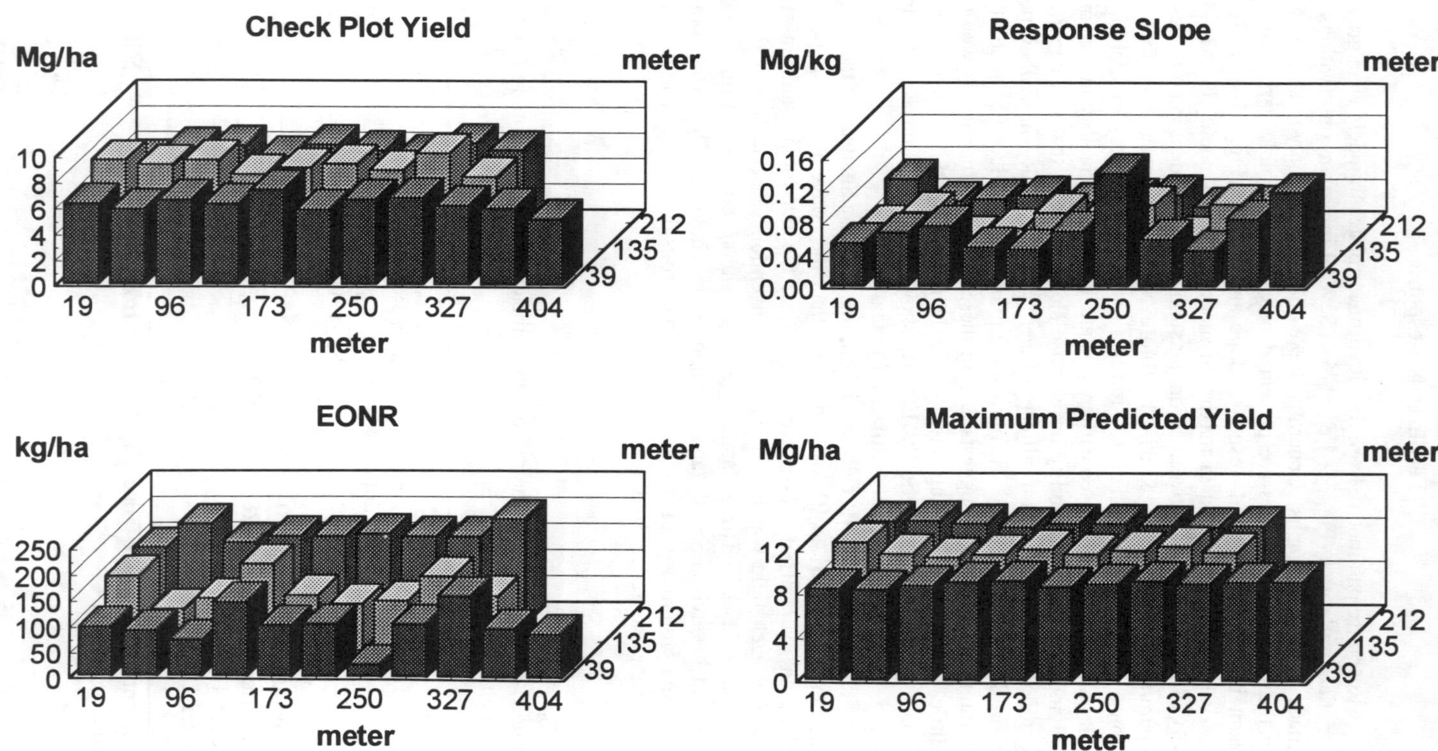

Fig. 31–4. Spatial variability of response parameters from linear plateau model for Becker, 1992.

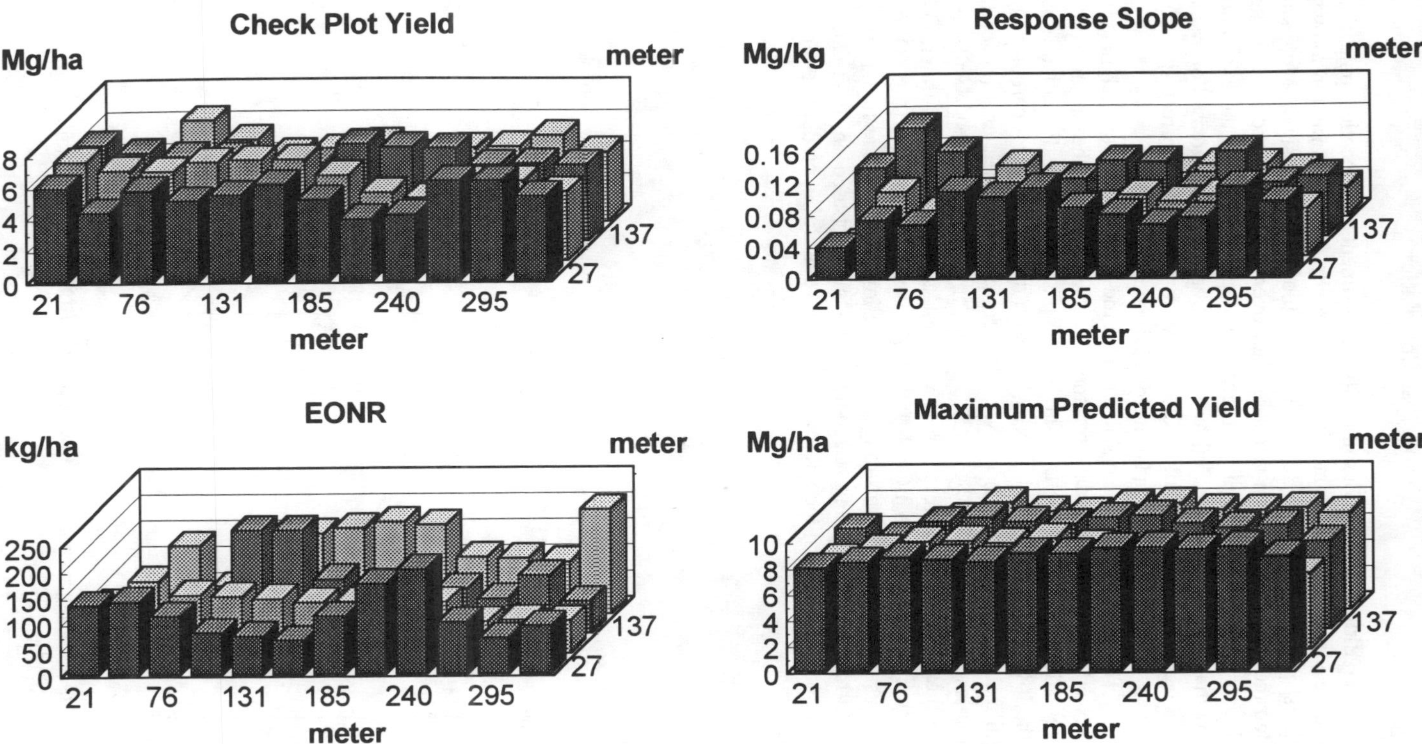

Fig. 31–5. Spatial variability of response parameters from linear plateau model for Lamberton, 1992.

Soil Test Correlation

Traditional small plot soil fertility experiments assume yield potential and response slope to be constant within a given site. This assumption allows the soil test parameter to be regressed on check plot yield over sites. More simply stated the differences in soil test parameter over site yr are used to calibrate the soil test. Because of variability in response slope within sites, these data do not allow that type of analysis. Soil test information must be regressed on *response curve parameters*, specifically EONR, because it is the correct site-specific nitrogen fertilizer recommendation for each site.

The response parameters, taken from the subplot areas of each field, were regressed on all the soil data for the given site. Only the spring soil data regressed on EONR for the L92 site will be presented here. Elevation (relative to the lowest part of the field), NH_4-N, total nitrogen and total carbon were all significantly correlated with EONR. The correlation coefficients (r) for these variables ranged from 0.23 to 0.32. Elevation had the highest r value (0.32) from the whole field regressions so we evaluated this variable further. When elevation is regressed on EONR within a given soil map unit (Fig. 31–6) the relationship becomes much stronger. Correlation coefficients as high as 0.99 for the Normania soil map unit (L92 site) were found. Based on these data early indications suggest that topographic data may be useful in predicting variable N rates.

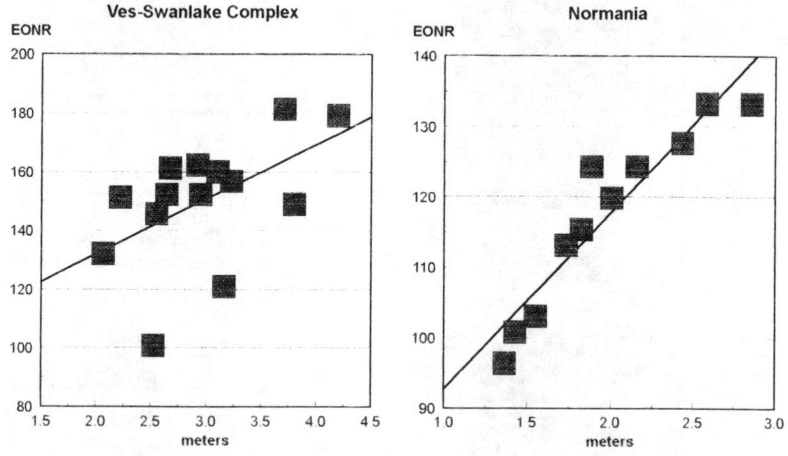

Fig. 31–6. Regression of elevation on EONR within soil map units at Lamberton 1992.

CONCLUSIONS

Considerable variability in check plot yield (N supplying power) and response slope (fertilizer efficiency) suggests that site-specific application of N may be warranted. Spatial structure in response parameters, specifically EONR, support the feasibility of site-specific N fertilization. The average model calculated EONR for each site was lower than the conventional application and in two of the four sites the maximum EONR value was higher than the conventional application. Thus, site-specific N fertilization has the potential for both environmental and economic benefits. The lack of variability in maximum predicted yield suggests that for these data yield monitoring would be of little value in predicting variable N rates. Early indications suggest that topographic data, like elevation, may be effective in predicting site-specific N fertilizer recommendation. These parameters may be improved if evaluated within soil map units.

The variability in response slope within sites in these experiments raises questions about the effectiveness of current soil N tests obtained from small plot research. Rennie and Clayton (1960) concluded that care should be used in interpretation and application of fertility data obtained from small plot research experiments as the data are applicable to only one particular soil type, in most instances the soil member on which the soil test was established. These data would support Rennie and Clayton's conclusion. Thus, our current recommendations may not be appropriate for making site-specific N fertilizer recommendations.

REFERENCES

Rennie, D.A., and J.S. Clayton. 1960. The significance of local soil types to soil fertility studies. Can. J. Soil Sci. 40:146–156.

SAS Institute Inc., SAS® STAT, Version 6.02, Second Edition, Cary, NC: SAS Institute Inc., 1991.

32 Use of Row Fertilizer With Corn for the Management of the Effects of Field P and K Variability

R. P. Wolkowski
N. C. Wollenhaupt

Dep. of Soil Science
University of Wisconsin-Madison
Madison, Wisconsin

Farmers and agricultural professionals recognize that soils vary in the landscape due to changes in topography, parent material, drainage, and vegetation. Superimpose the activity of man (e.g., cropping practices, manure and fertilizer use, and soil erosion) and an appreciation can be gained that plant nutrient levels will vary within fields. Traditional soil sampling methods only estimate the field-average soil test level and may not detect the true nutrient need of the field. Therefore, portions of fields that have soil test nutrient levels in the responsive range will have significant profit loss if these areas are under-fertilized. Fertilizer applied to non-responsive areas decreases profit, while unnecessarily maintaining high soil test levels.

Recent intensive sampling studies in Wisconsin have documented within-field variability of available plant nutrients (Wollenhaupt & Wolkowski, 1993). Soil test P and K variation in some fields existed within a range that would be expected to respond well to fertilizer, and in others soil tests were variable but in the very high or excessively high range, where the probability of response to fertilizer additions was low. An examination of the data from the responsive fields demonstrated that 30 to 40% of the spatial variability in soil test P and K could be mis-classified unless the soil sampling was conducted on a 30 m or finer grid (Wollenhaupt et al., 1993). This work also showed that incorrect mapping of responsive areas, especially those mapped less responsive than they actually were, would lead to large profit losses. Because it would be difficult and expensive to correctly delineate all soil test variability, the use of some starter fertilizer could compensate for yield losses in improperly categorized field areas.

University of Wisconsin fertilizer recommendations encourage the use of row-placed or starter fertilizer for corn, applied with the planter, at a minimum rate of 11 kg N ha^{-1}, and 22 kg each of P_2O_5 and K_2O ha^{-1} (Kelling et al., 1991). Starter application is recommended even at high soil test levels because most corn is planted into cool soils where conditions are less than favorable for nutrient uptake, especially P (Fixen & Lohry, 1993). These conditions are exacerbated as residue levels or compaction increase, in which case K has been found to be important (Wolkowski et al., 1987). The standard placement is in a band 5 cm to the side and 5 cm below the seed. Starter fertilizer applications provide greater nutrient use efficiency than broadcast applications because of reduced fertilizer contact with the soil and proximity to the root system. Therefore, starter fertilizer applications have been shown to produce similar yields when compared to higher rates of broadcast fertilizer (Welch et al., 1966). Ghodrati (1983) demonstrated that corn yields measured at low soil test with starter fertilizer were often equivalent to those found at much higher soil test levels, even when the high soil test treatments received starter.

This paper evaluates the potential of using starter fertilizer application as a means of compensating for soil test variability in fields. While starter application would not be expected to significantly adjust within-field soil test variability, it could reduce the large yield differences found between areas of low and high fertility. Obviously a starter fertilizer system could be fine-tuned if planters were equipped with variable starter application attachments that permitted fertilizer application according to properly documented soil test variation.

WISCONSIN P AND K RECOMMENDATIONS

The University of Wisconsin Soil Test Recommendation Program is designed to utilize plant nutrients from both the soil and fertilizer or creditable sources. The interpretive descriptions of soil test P and K range from "very low to excessively high" with the "optimum" category being the most desirable from both an economic and environmental standpoint. Corn grown at soil test P and K levels in the optimum range should receive annual phosphate and potash applications equal to removal in the grain. For example, at a yield goal of 9.4 Mg ha^{-1} the phosphate and potash recommendation for most southern Wisconsin soils testing in the optimum soil test category would be 61 and 39 kg ha^{-1}, respectively (Kelling et al., 1991). Where soil tests are below the optimum level, additional nutrients are recommended over a period of years, while at levels above optimum the recommended application rates are reduced. It would be feasible for many Wisconsin farmers to supply all phosphate and potash needs with a starter fertilizer application, thus meeting crop need and increasing nutrient use efficiency. A description of the soil test P and K categories according to Kelling et al. (1991) for the locations reported in this paper are shown in Table 32–1.

Table 32–1. Categories of soil test P and K for field corn production at the site-specific management studies (Kelling et al., 1991).

Site	Nutrient	Soil Test Category				
		VL[†]	L	O	H	EH
		mg kg^{-1}				
Trinrud	P	< 5	5-10	11-15	16-25	> 25
	K	<60	60-80	81-100	101-140	>140
Kohel	P	< 8	8-12	13-18	19-28	> 28
	K	<70	70-100	101-130	131-160	>160
Metcalf Sommers Waller	P	<10	10-15	16-20	21-30	> 30
	K	<70	70-90	91-110	111-150	>150

[†]VL, L, O, H, EH = very low, low, optimum, high, and excessively high.

WISCONSIN SITE-SPECIFIC MANAGEMENT STUDIES

Studies to evaluate management of within-field soil test variation were conducted in Wisconsin at two central locations in 1992 (Trinrud & Kohel) and at three southern locations in 1993 (Metcalf, Sommers, & Waller). The areas studied were a portion of larger fields. A description of the predominant soil type and the fertilizer program for each site is given in Table 32–2. The plow-layer of each field was point-sampled (eight cores in a 3–m radius) at the intersection of 30-m grid spacings and mapped for soil test P and K categories using a contouring procedure. Summary statistics and the distribution of soil test P and K for each field over the soil test categories are shown Table 32–3. These data show that the Trinrud and Kohel fields had significant areas of soil test P and K in responsive soil test categories, while the Metcalf, Sommers, and Waller fields each tested in nonresponsive P and K ranges. The Trinrud field had the widest distribution of soil tests in the responsive range. This is demonstrated in Fig. 32–1 which shows the soil test category maps for P and K for the Trinrud field. The higher levels of P and K in the lower portion of each map are adjacent to a barnyard area. It is speculated that manures were routinely applied to this portion of the field because of its proximity. This field permitted the best opportunity to evaluate yield response to fertilizer. The Kohel field was in fact very responsive to fertilizer additions, but because of its high requirement for K throughout the field it was not possible to assess fertilizer response across soil test.

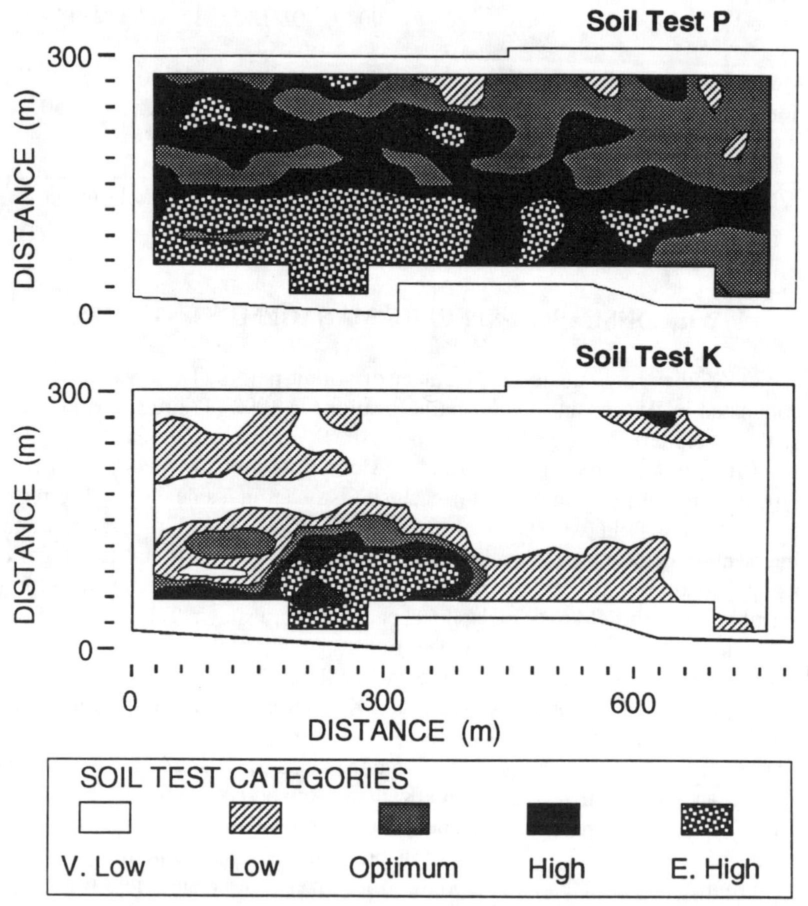

Fig. 32–1. Trinrud field soil test P and K category maps based on the 106 ft grid-point data, contoured with the Delaunay triangulation method.

Table 32–2. Field characteristics and fertilizer treatments in the site-specific management studies.

Site	Predominant soil type		Field size	N applied	Starter N	Starter P_2O_5	Starter K_2O
			ha	kg ha^{-1}			
Trinrud	Zurich	silt loam	22.7	165	11	28	62
Kohel	Rosholt	loamy sand	17.0	165	10	25	72
Metcalf	Plano	silt loam	17.6	179	6	22	90
Sommers	Elburn	silt loam	13.3	99	7	21	7
Waller	Plano	silt loam	23.0	149	12	35	12

Fields were planted to corn by farmer-cooperators with starter fertilizer at the rates shown in Table 32–2, except for several strips where no starter or a one-half rate of starter was applied (1992 only). All P and K applied to thc field in the study years was applied by the planter in a 5x5 cm placement with the exception of the Metcalf field, where the P and K was surface banded on the row positions prior to planting.

Yield check plots were established in adjacent areas of the strips planted with and without starter fertilizer. Each check plot consisted of two rows, 9 m long. Distance between check plots, across strips varied by site; however, the distance along strips was always 90 m. This resulted in 24 yield comparisons at the Trinrud and Kohel sites, 36 yield comparison at the Sommers and Waller site, and 56 yield comparisons at the Metcalf site. Grain yield was determined by hand harvesting all the ears in the two 9–m rows and shelling it in a portable sheller. Grain yield is reported at 15.5% moisture.

Table 32–3. Summary statistics and P and K distribution by soil test category in the site-specific management studies.

Field	n	Nutrient	Min	Max	Mean	Median	Soil test category VL	L	O	H	EH	2(EH)†	3(EH)†
			mg kg^{-1}				Number of observations						
Trinrud	199	P	7	140	24	17	0	17	67	61	36	11	7
		K	28	300	71	55	107	48	14	15	13	2	0
Kohel	149	P	9	54	18	15	0	10	90	36	13	0	0
		K	25	97	48	45	134	15	0	0	0	0	0
Metcalf	189	P	22	180	46	38	0	0	0	30	128	18	13
		K	105	570	181	150	0	0	1	27	49	98	14
Sommers	143	P	47	105	77	78	0	0	0	0	6	114	23
		K	195	350	271	270	0	0	0	0	0	110	33
Waller	247	P	36	175	64	60	0	0	0	0	123	103	21
		K	115	450	205	195	0	0	0	4	1	231	11

† 2(EH) and 3(EH) are arbitrary categories equivalent to multiples of the low point of the EH range.

YIELD RESULTS

Figure 32–2 shows the yield response to starter averaged over all measurements for each field. The response to starter was notable at the Trinrud and Kohel fields and was undetectable at the Metcalf, Sommers, and Waller fields. The lack of response in the latter group of fields may be explained by their high soil test levels and the fact that their location is approximately 250 km south of the other sites. Much of the response at the Trinrud and Kohel sites was achieved at the half-starter rate. The full rate at these locations matched the recommendation to meet removal in the grain. These results should not be interpreted as a suggestion to reduce fertilizer rates to a small amount of P and K fertilizer in the starter, but do demonstrate the improvement in yield observed when starter applications are used at responsive locations.

The relationship between soil test P and K and the response to starter for the Trinrud field is shown in Table 32–4. Quadratic expressions of these data did not significantly improve the R^2 value of the equations and are therefore not shown. A significant relationship between both soil test P and K and the response to starter was found at both rates of application. The equations appear independent of starter rate. Because the individual effects of P and K could not be determined in this study (P and K simultaneously applied in a complete fertilizer), an approach of establishing an index was used. Depending on soil test category, a discrete value was given for each comparison. A very low (VL) category was given a value of 5, a low (L) was given a value of 4, and so on. A P index, K index, and a PK index (sum of the P and K indices) were related to response. This provides a better relationship by reducing the variation within soil test categories. The indices also show the relative importance of K in the starter. These indices show that for each increase of one index unit a response of 0.45 to 0.75 Mg ha^{-1} can be expected.

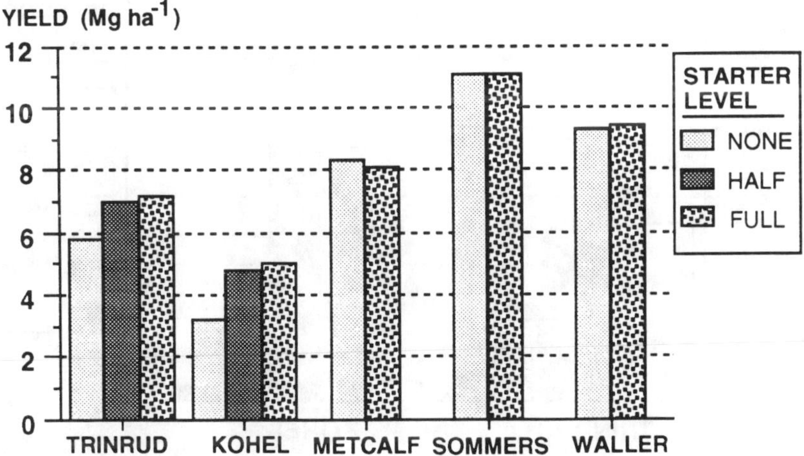

Fig. 32–2. Corn grain yield response to starter fertilizer in the Wisconsin site-specific studies.

Table 32–4. Relationship between P and K soil test and corn grain yield response to starter fertilizer at the Trinrud site.

Factor	Half starter Intercept	$b_†$	$r^‡$	Full starter Intercept	$b_†$	$r^‡$
Soil test P (mg kg^{-1})	3.78	-0.11	0.32**	3.33	-0.09	0.19*
Soil test K (mg kg^{-1})	2.40	-0.02	0.16**	2.57	-0.02	0.17*
P index^2	-0.18	0.75	0.17*	0.31	0.56	0.10
K index	-1.69	0.70	0.33**	-1.52	0.69	0.33**
P index + K index	-1.74	0.49	0.35**	-1.34	0.45	0.29**

†Equation has the form yield response (mg kg^{-1}) = intercept/b_1(factor).
‡P or K index: Soil test category VL, L, O, H, EH; then index = 5,4,3,2,1, respectively.

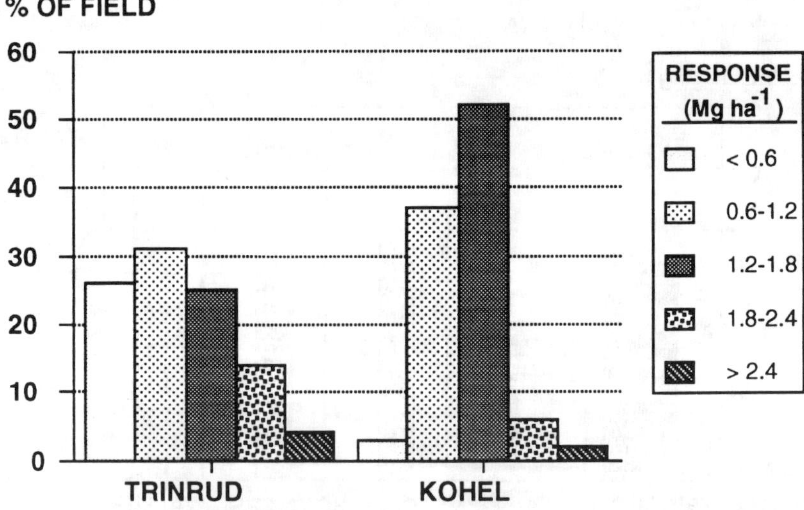

Fig. 32–3. Percent of field area at Trinrud and Kohel locations responding to the full rate of starter fertilizer.

An economic response to starter was observed at both the Trinrud and Kohel sites. Assuming a corn price of \$35.87 Mg^{-1} and a fertilizer cost of \$181 t^{-1}, the yield response needed to pay for the fertilizer would be 0.36 and 0.38 Mg ha^{-1} at the Trinrud and Kohel sites, respectively. Figure 32–3 shows the starter response in 0.6 Mg ha^{-1} increments for both fields. This figure shows that 26% of the Trinrud field would respond in the 0 to 0.6 Mg ha^{-1} range. Using the midpoint of this range, which is quite close to the break-even yield, it could be assumed that about 15% of the field was not economically responsive to the fertilizer. At the Kohel site only 3% of the field fell into the 0 to 0.6 Mg ha^{-1} response category, suggesting that only about 2% of that field was not responsive to fertilizer.

Environmental considerations may someday determine how and where fertilizers are applied. Applying nutrients only where they are needed makes both economic and environmental sense. Site-specific soil test maps developed by an appropriate sampling and mapping scheme can determine the responsive and nonresponsive portions of fields. The placement of fertilizer in a manner that maximizes efficiency by reducing loss or unavailability and increasing plant uptake should be encouraged. While this may slow the planting process because of the time needed to handle the fertilizer, it would seem to be the best P and K placement option for corn growers in the northern corn belt.

ACKNOWLEDGMENTS

The authors wish to thank the Wisconsin Fertilizer Research Council and Growmark for their financial contribution to this research. The cooperation of the individual growers is gratefully appreciated.

REFERENCES

Fixen, P.E., and R.D. Lohry. 1993. The state of the art of starters. Presented at the Fluid Forum Proc. Fluid Fertilizer Foundation, St. Louis, MO.

Ghodrati, M. 1983. Effect of broadcast nitrogen and soil test phosphorus and potassium on the response of corn to row N, P, and K fertilizers. M.S. thesis, Univ. of Wisconsin-Madison.

Kelling, K.A., E.E. Schulte, L.G. Bundy, S.M. Combs, and J.B. Peters. 1991. Soil test recommendations for field, vegetable and fruit crops. Univ. of Wisconsin Ext. Publ. A2809. 40 p.

Welch, L.F., P.E. Johnson, G.E. McKibben, L.V. Boone, and J.W. Pendleton. 1966. Relative efficiency of broadcast versus banded potassium for corn. Agron. J. 58:618–621.

Wollenhaupt, N.C., and R.P. Wolkowski. 1993. Soil sampling for developing variable-rate fertilizer programs. Proc. Fert., Aglime, and Pest Mgmt. Conf. 32:1–15. Madison, WI. 19–21 Jan. 1993.

Wollenhaupt, N.C., R.P. Wolkowski, and H.F. Reetz. 1993. Variable-rate fertilizer application: Update and economics. *In* Proc. 23rd North Central Extension-Industry Workshop. Bridgeton, MO. 27-28 Oct. 1993. Potash and Phosphate Inst., Manhattan, KS. 9:139–151.

Wolkowski, R.P., L.G. Bundy, and B. Lowery. 1987. Compaction-soil fertility interactions. Proc. Fert., Aglime, and Pest Mgmt. Conf. 26:142–150. Madison, WI. 20–22 Jan. 1987.

33 Variable Rate Water Application Through Sprinkler Irrigation

B. A. King

Dep. of Agricultural Engineering
University of Idaho
Aberdeen, Idaho

R. A. Brady

Agricultural Consultant
Santa Cruz, California

I. R. McCann

Dep. of Agricultural Engineering
University of Idaho
Moscow, Idaho

J. C. Stark

Dep. of Plant, Soil and Entomological Sciences
University of Idaho
Aberdeen, Idaho

The potential to optimize water and chemical application by combining site-specific crop management techniques with continuous-move irrigation systems exists. All of the techniques that can be used to define crop management zones for variable rate chemical application by conventional ground based equipment can be used with continuous-move irrigation systems. Irrigation management zones can be initially defined by in-field sampling of soil texture, soil depth and slope which remain essentially unchanged. Remote sensing using satellite images or low-altitude aerial videography can be used to aid in determining soil texture distribution and establishing crop management zones characterized by similar responses to water or fertilizer application. Remote sensing along with real time field measurements can be used to provide in-season modification to a base site-specific crop management plan. Yield monitoring and crop yield histories can be used to further define crop management zones. The same mapping systems and databases developed for site-specific crop management using conventional variable rate chemical application equipment can be utilized.

The application of agricultural chemicals through an irrigation system, known generically as chemigation, is commonly practiced as a means to economically apply appropriately labeled chemicals. Sprinkler irrigation systems are used to apply a wide variety of chemicals including fertilizers, insecticides, herbicides, nematicides and fungicides. Chemigation using sprinkler irrigation systems offer several distinct advantages when compared to ground application equipment. These include:

1. In-season application of crop nutrients after lay-by allowing nutrients to be applied as needed to optimize utilization.
2. Potential reduction of chemical requirements and environmental effects due to improved timing and sometimes higher efficiency.
3. Reduction of field equipment traffic resulting in less compaction and crop damage.
4. Reduction of operator hazards.
5. Reduction of application costs.

The use of continuous-move irrigation systems for variable rate application of chemicals offers some additional advantages over conventional ground based variable rate application equipment. These are:

1. Multiple use capability - water and chemicals
2. Built-in locating system in the case of center-pivots (using the angle of the first tower) avoiding the need for a global positioning system.

As with most any system, there are disadvantages associated with using continuous-move irrigation systems to apply chemicals either uniformly or in a variable rate. A list of these disadvantages include:

1. Chemical application uniformity is controlled by water application uniformity.
2. Wind affects the distribution of applied chemicals and application can not normally be limited to low wind conditions in arid and semi-arid areas during high water use periods.
3. Chemicals must be suitable or labeled for application through continuous-move irrigation systems.
4. Generally, only one chemical can be injected and distributed in a spatially variable manner unless the relative amounts of the chemicals to be applied remain constant across the field.
5. Distributing chemicals in a spatially variable manner using variable amounts of applied water will likely require additional operational hours on the equipment or increased system capacity.

The growing popularity of chemigation will likely overcome the limited availability of chemicals with time.

The latest commercially available control systems for center-pivots enable the machine to be stopped or reversed, on–off control of chemigation pump(s), and the speed of the machine to be programmed as a function of position (angle of first tower). Hence, it is possible to apply different amounts of water and to control on–off chemical application to different "pie slices" (sectors) of the field. In order to vary water and chemical application to an area of any shape within the field, the water application rate along the length of the lateral must also be variable. This in turn requires that an electronically controlled variable chemigation pump be used to maintain a constant chemical concentration in the applied water as system flow rate varies. Additionally, a variable speed pump system is needed if energy is to be conserved and system pressure is to be controlled. A graphical comparison of commercially available center-pivot systems with that needed for site-specific crop management is shown in Fig. 33–1.

The control system hardware to perform all the functions needed to develop a center-pivot irrigation system capable of site-specific crop management is currently available with the exception of one important item, an economical and reliable electronically controlled variable rate sprinkler. The lack of a commercially available variable rate sprinkler represents the single biggest obstacle to commercial development of a center-pivot irrigation system capable of site-specific crop management. The second biggest obstacle to commercial development is the lack of a service infrastructure to generate and deliver the maps needed to control and manage the irrigation system throughout the season. Development, evaluation and demonstration of a center-pivot irrigation system capable of site-specific crop management is the focus of research being conducted at the University of Idaho.

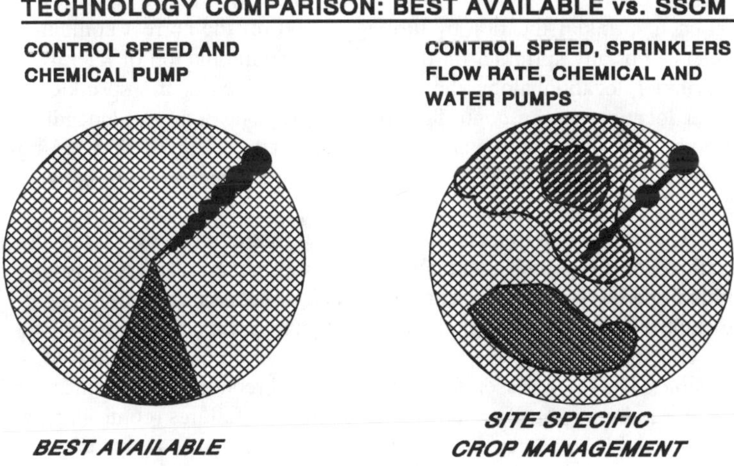

Fig. 33–1. Comparison of current center-pivot technology to that needed for site-specific crop management.

In the absence of a commercially available variable rate sprinkler, the immediate solution has been to replace each sprinkler along the lateral with multiple sprinklers. Replacing a single sprinkler with two sprinklers, each with a solenoid valve, and sizing the sprinkler nozzles to provide 1/3 and 2/3 of the original single sprinkler flow rate effectively provides a step-wise variable application rate of 0, 1/3, 2/3 and full application rate through on–off control of each solenoid valve. This concept can be taken a step further using three sprinklers at each location. Replacing the single sprinkler with three sprinklers, each with a solenoid valve, and sizing the sprinkler nozzles to provide 1/7, 2/7 and 4/7 the original single sprinkler flow rate effectively provides a step-wise variable application rate of 0, 1/7, 2/7, 3/7, 4/7, 5/7, 6/7, and full application rate. This approach is acceptable for research purposes but is likely too costly for commercial development. Assuming ≈100 sprinklers on a typical 402 m (1320 ft) long low pressure center-pivot, each additional sprinkler assembly costs ≈\$3000 to \$4000 in addition to the \$1500 to \$2500 for solenoid valves for the original sprinkler. Thus, ≈\$10 000 in hydraulic components is needed to provide an 8–level step-wise variable sprinkler arrangement. This cost doesn't include the electronic hardware needed to control the hydraulic hardware.

A prototype control system enabling site-specific crop management using continuous-move irrigation systems has been developed to facilitate needed research, evaluation and demonstration of the technology. The prototype control system primarily consists of a microprocessor based controller, a location sensor, and solenoid valves under the control of individually addressable relays which switch solenoid current on and off. The controller includes digital input and output lines, digital to analog and analog to digital conversion, two RS232C serial communication ports, a 4x5 keypad and a 2–line 40 character LCD. The controller monitors the location of the system and maintains the appropriate flow rate at each sprinkler location by turning on and off the correct combination of sprinklers. For any sprinkler on the system, the combination of lateral position and sprinkler location along the lateral spatially locates the sprinkler. Each sprinkler location can be controlled individually or as a group depending upon the number of individually addressable relays used, up to 255. Graphical representation of the prototype control system for a center-pivot is shown in Fig. 33–2.

A second generation control system is currently under development which will allow bi-directional communication between a master controller and individually addressable slave controllers placed anywhere on the system. The slave controllers provide digital input and output capability for actuation of relays and input from any type of transducer through analog to digital conversion. The main reason for developing bi-directional communication is to have a closed loop control system so that equipment failures (controllers, wiring, valves, solenoids, etc) can be detected by software and an operator warning issued. An immediate advantage of bi-directional control is the ability to locate fixed points in the field by use of proximity sensors to signal when the location has been reached by the system. This will allow differences between the angle of the first tower and the actual location of the last tower of the center-pivot system to be taken into account for more precise system locating. This second

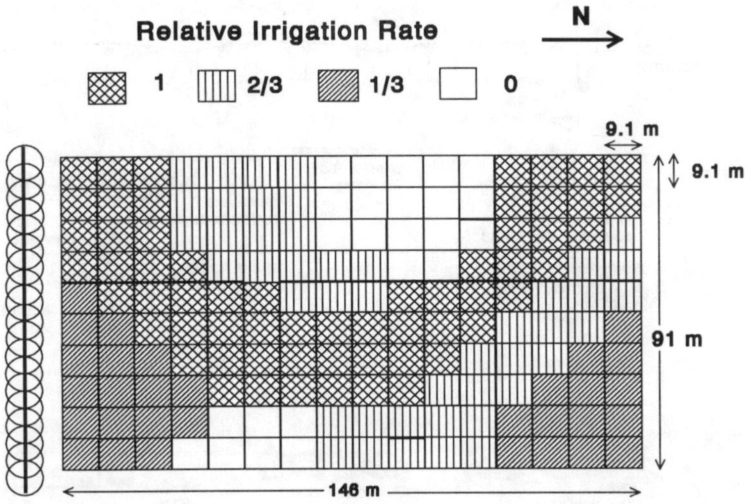

Fig. 33–2. Arbitrary irrigation map used to measure performance of linear-move system.

generation control system will be field tested throughout the 1994 irrigation season on a center-pivot irrigation system.

TESTING AND RESULTS

To evaluate the operational feasibility of variable rate irrigation and chemigation using the original prototype control system, a 100 m linear-move irrigation system was equipped with low pressure (20 psi) spray drops and used to irrigate a 91 m by 146 m (300 by 480 ft) field. The field was divided into 9.1 m square cells, and an arbitrary variable irrigation depth map was imposed on the field by assigning one of four target irrigation depths to each cell, as shown in Fig. 33–2. These target irrigation depths corresponded to 0, 1/3, 2/3 and full irrigation, and were achieved using two sprinklers on each drop, each equipped with an individually addressable relay and solenoid valve. The nozzles of the two sprinklers were sized to apply 1/3 and 2/3 of the original sprinkler flow rate, and operated in one of the four possible on–off combinations to achieve the four application depths. The full irrigation amount was used to determine the speed of the system (percentage timer) which remained unchanged throughout an irrigation. Eleven irrigations were applied to the field using the arbitrary map of Fig. 33–2, and the performance measured using catch cans placed at the center of each 9.1 m square cell. Table 33–1 shows the results in terms of the calculated coefficient of uniformity (CU) (ASAE, 1987) for each of the four target irrigation depths normalized to a depth of 12.7 mm (0.5 in). The resulting CUs are less than those attainable by the linear-move system operating conventionally. This is due to the wetted diameter of the sprinklers being slightly larger than the cell size. This resulted in over- or under-application in

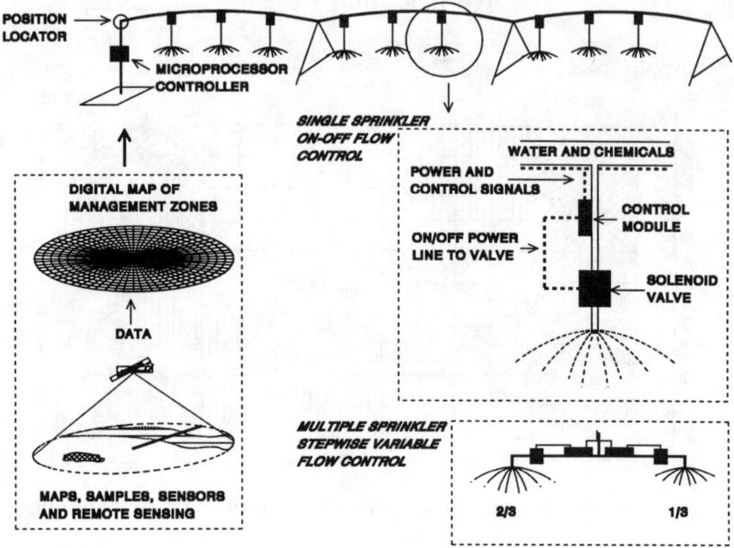

Fig. 33–3. Prototype microprocessor control system.

cells along the boundary between differing target application amounts. This is evident by the average measured depth of 2.5 mm (0.1 in) in cells having a zero target amount, which can be attributed to overspray from sprinklers located over adjacent cells. The large percentage of boundary cells in the arbitrary map resulted in the low CU values for all target depths. The results are encouraging despite the low CU values because the areas receiving applications of 1/3 and 2/3 of the full amount have nearly equal CUs demonstrating that the system was able to differentially apply water with the same precision as a conventional system under the given conditions.

Near the end of the irrigation season, the size of the cells was increased to 18.2 m (60 ft) square by consolidating four adjacent 9.1 m square cells and changing the target application amount in some of the original 9.1 m square cells. The performance of the system was measured using a single catch can at the center of the 18.2 m square cell. Composite results over four irrigations are shown in Table 33–2. The calculated CU values increased as a result of the reduction in overspray from sprinklers over adjacent cells. The lower CU value for the 1/3 target amount and the small average depth measured for the zero target amount are likely due to drift of the fine spray and windy conditions during some irrigation events. The 18.2 m square cell size (corresponding to 0.06% of the area under a typical center-pivot) is likely the minimum size appropriate for this system and sprinkler type. The minimum cell size will likely be greater on full size center-pivots, and will partially depend on sprinkler characteristics.

Table 33–1. Composite irrigation results for a 9.1 m by 9.1 m cell size.

Relative Application	Number Samples	Mean Depth	Coefficient of Uniformity on
		mm	%
1	534	14.2	74
2/3	404	10.9	75
1/3	163	5.6	75
0	262	2.5	--

The ability of the automated control system to apply variable rates of a chemical was also evaluated. Using the 18.2 m square cell size, a chemigation test was performed using bromide as the tracer. Chemical application amounts depend upon both the irrigation amount and the chemical concentration in the irrigation water. The control system continuously adjusted the chemical injection rate to maintain a constant concentration in the irrigation water, based on the computed flow rate using the nominal flow of each operating sprinkler. The actual concentration of bromide in the applied water was measured using four catch cans placed in a square pattern near the center of each cell, and the bromide concentration of the captured water measured using flow injection analysis (Lachat Instruments, Milwaukee, WI, Method No. 10–135–21–2–A). The target concentration of bromide in the irrigation water was 20 mg/l. The actual mean concentration was 19.0 mg/l with a standard deviation of 2.5 mg/l (n=140). The uniformity of water application values for the chemigation trial are included in the composite results presented in Table 33–2. The relative chemical application target amounts follow the same proportion as the relative irrigation targets, namely 0, 1/2, 2/3 and 1. In both variable rate and conventional systems, the uniformity of chemical application depends on the application uniformity of the water. The results showed that different relative target chemical applications were achieved with ≈ the same accuracy as a conventional system would be able to apply a uniform target amount.

Following the end of the growing season, the control system was installed on a 7–tower, 210 m (690 ft) long center-pivot system. The center-pivot was equipped with low pressure rotator type sprinklers, three per span with 9.5 m (31 ft) spacing. The first and last sprinklers on each span were mounted on 4.6 m (15 ft) offsets. An initial test of the system using American Society of Agricultural Engineers standards (ASAE, 1987), resulted in a CU of 83 % excluding the first span and outer overhang, and 76 % when these were included. The relatively large wetted diameter (≈18.2 m (60 ft)) of the sprinklers, combined with the offsets, produced a sprinkler pattern length of 22.9 m (75 ft) in the direction of travel.

Table 33–2. Composite irrigation results for a 18.2 m by 18.2 m cell size.

Relative Application	Number Samples	Mean Depth	Coefficient of Uniformity
1	32	17.0	90
2/3	54	10.7	87
1/3	58	6.1	77
0	16	1.8	--

With a center-pivot, a polar coordinate system is more appropriate than the rectangular coordinate system used on the linear-move system. Cells corresponding to 5.6 degree angular component (circle subdivided into 64 radial sectors) and a 28.7 m (94 ft) linear component (measured along the lateral) which corresponded to the length of each span where used for the control map and to measure system performance. An arbitrary application map, consisting simply of either full or zero irrigation for each cell was imposed on the field. Due to time constraints, the tests were conducted only on a quarter circle, or 16 sectors. In the first two sectors, all the cells were targeted to receive irrigation, while in the remaining 14 sectors, various combinations of cells were targeted to receive zero irrigation. The first sector represented a conventional system, the second sector served as a buffer between conventional and variable rate control. Catch cans were aligned radially in each sector at a 9.1 m (30 ft) linear spacing to measure system performance. Figure 33–4 shows typical results for a single sector under variable rate control. The solid strip in the foreground is the target,

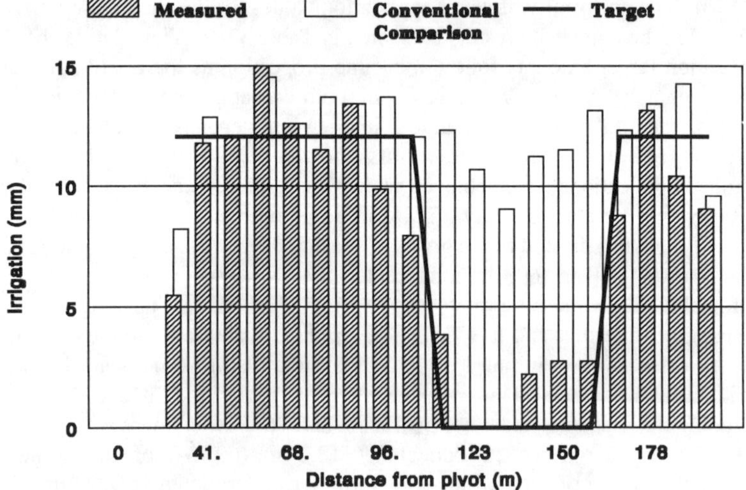

Fig. 33–4. Typical field test results on center-pivot system.

the shaded bars in the middle are the measured application amounts under variable rate control, and the empty bars at the back are the measured application under conventional control. The control system was generally successful in limiting irrigation in the cells targeted for zero application. There was some overspray from adjacent irrigated cells, but the total water applied to the zero application cells was low. This is in marked contrast to the application that would have occurred with a conventional system.

CONCLUSIONS

A prototype control system enabling site-specific application of water and chemicals using continuous-move irrigation system has been developed and evaluated. Preliminary results show that water and chemical application uniformity equal to that of conventional uniform application system can be achieved. Commercial development of such a system is impeded by the lack of a variable rate sprinkler and an existing service infrastructure to produce the maps needed to operate the system throughout the growing season. Research is needed to determine the information required to manage and operate the system throughout the season and evaluate potential benefits and system performance.

REFERENCES

American Society of Agricultural Engineering. 1987. ASAE standard: S436, Test procedure for determining the uniformity of water distribution from center pivot, corner pivot, and moving lateral irrigation machines equipped with spray or sprinkler nozzles. pp 546–547. ASAE, St. Joseph, MI.

34 Mapping Wild Oats Infestations Using Digital Imagery for Site–Specific Management

L. D. Hanson

Field Control Systems
White Bear Lake, Minnesota

P. C. Robert

Department of Soil Science
University of Minnesota
St. Paul, Minnesota

Marvin Bauer

Department of Forest Resources
University of Minnesota
St. Paul, Minnesota

Remote sensing has provided agricultural crop and soil condition information for over 50 yr. In most cases however, farm managers have not able to fully use the detailed photographic or other imagery data because crop operations were carried out on a full-field scale basis and the time requirement for processing–analysis of the imagery. Now with the availability of powerful site-specific crop management tools (Robert et al., 1993), it is appropriate to take another look at remote sensing, and particularly digital imagery, as a source of in-field variability information.

Wild oats (*Avena fatua* L.) infestations cause substantial yield reductions in small grains and other crops throughout the northern Great Plains (Fig. 34–1). North Dakota researchers reported wild oats reduced the state wheat (*Triticum aestivum* L.) yield 600 million L (17 million bu) in 1978 and 194 million L (5.5 million bu) in 1979 (Jenny, R. 1992). Barley (*Hordeum vulgare* L.) yield loss was 194 million L (5.5 million bu) and 98.7 million L (2.8 million bu), respectively, in the two yr (Dexter et al., 1979).

Crop–weed interactions can be considered in two broad categories: a crop–centered and a weed–centered approach (Van Groenendael, 1988, Thorton et al. 1990). In the first category, most common approach of a crop–field model, uniform weed control practices are carried out for the whole field. With a weed–

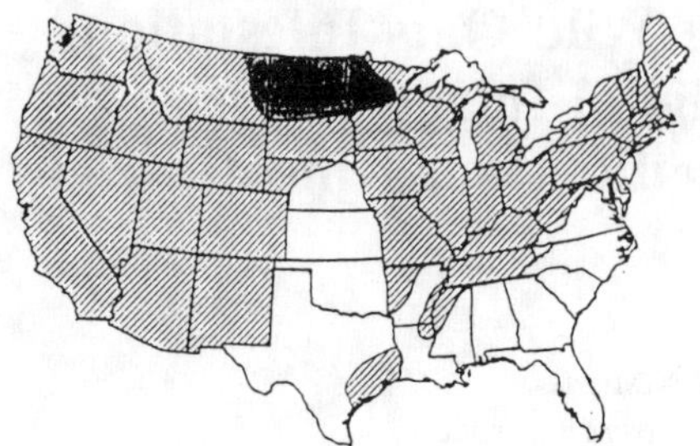

Fig. 34–1. Location of wild oat infestations. Close pattern denotes area where wild oats are a problem in small grains (USDA-ARS, 1970).

centered approach, weed distribution and site-specific controls are evaluated and applied. In their review paper, "Weed distribution in agricultural fields" for the 1992 Soil Specific Crop Management Workshop, Mortensen et al. (1993) traced research related to the heterogeneous nature of weed infestations. They also reported a high degree of weed aggregation as indicated by low chi–square K values for several weed species in a soybean (*Glycine max* L.) research field.

Conventional color film aerial imagery, color infrared film, videography and digital imagery are media for a synoptic perspective of a crop field. These data can be acquired from airborne and satellite platforms. Everitt has pointed out that video has many attractive attributes for remote sensing (Everitt et al., 1993). These include real–time monitoring, immediate availability of the electronic signal for digitization, and low cost of imaging operation compared to film. In the last 20 mo new digital cameras have become available. The Kodak DCS 200 offers about six times higher resolution than a Charge Coupled Device (CCD) video camera with frame grabber digitizing software (Kodak, 1992). Cost–performance progress has been made in film related technology involving computers, computer peripherals and software (Taylor, 1994). The combination of new, high quality color scanners, low–cost RAM, and high capacity computer disk drives can provide digital data from film images.

Aerial imagery is particularly suited to providing weed infestation locations within a crop field because:

1. Film and digital images can show many important weeds from non–weed areas at the critical time for weed management.
2. Cost of aerial imagery is low compared to scouting and can be provided in a timely manner.
3. New technologies - rapid film development, digital cameras, high quality color scanning devices and CD ROM drives - provide opportunities for automation of data processing and interpretation.
4. Aerial image information can be rectified to an accurate map representation of spatial patterns with computer software.
5. Increased use of post–emergence herbicides to reduce prophylactic, broadcast soil applications at higher use rates, is a trend in weed management.

OBJECTIVES

The research objectives were to: i) determine the feasibility of aerial digital mapping of wild oat infestations in wheat for same–yr weed management, and ii) develop a site-specific herbicide application controller to utilize the weed map data.

MATERIALS AND METHODS

The study was organized into two phases:

1. Field operations - aerial data acquisition, weed scouting and site-specific herbicide application.
2. Interpretation and evaluation of digital and scanned slide imagery.

Methodology-Field Operations

The study was carried out by the authors in the May to November, 1993 period. Two fields were surveyed for wild oat infestation and aerial digital and film imagery were collected and interpreted during the course of this study. The two fields used were part of the W 1/2 of NE 1/4 of NE 1/4 of Section 29 (referred to as Field 1) and parts of the N 1/2 of SE 1/4 of Section 31 (referred to as Field 2). Both parcels are in Norway Township, Traill County, North Dakota. Fields were selected for the study by the farmer cooperator, Gary Kaldor, because of a history of patchy wild oat infestations. Field 1 was 13.6 ha (34 acres) in size and irregular in shape. A 7.2 ha (18 acre) rectangular area within the field was measured and marked for study. Field 2 was 34.4 ha (86 acres) in size, rectangular in shape, and the whole field was a study area. The date of wheat seeding was 26 April for Field 1 and 8 May for Field 2. Tram lines, 18.3 m (60 ft) apart, were established at the time of seeding.

Base maps were drafted and ground location reference flags were installed for both fields on 17 and 18 May. Field corner "targets" were also installed. These targets were air–media visible, 1 X 2 m (3.3 X 6.6 ft) white plastic film, staked into place to prevent movement from wind.

An experienced crop consultant from AGVISE Laboratories, Northwood, North Dakota, surveyed the two fields and compiled weed maps, primarily wild oat infestations, on 26 May. The Field 2 weed map is Fig. 34–2.

Imagery was acquired on 25 May and 2 June using a Kodak DCS 200 digital camera and a Nikon F–1 35–mm camera with Kodak Ektachrome slide film and carried on a Cessna 172 equipped with a 15 cm (6 in) belly hole. Both cameras were equipped with 28–mm lenses. Imagery was acquired at altitudes of 335 m (1100 ft), 670 (2200 ft) and 936 m (3079) ft) as a means of testing the coverage of the digital camera, and the effect of resolution on the ability to interpret image features.

Kaldor's spray–rig equipment consisted of a 8.3 m (60 ft) boom, pump, and water supply tank mounted on a pickup truck. He has a chemical injection system, a combination of Micro-Trak Systems Trak-Net spray controller and AGSCO

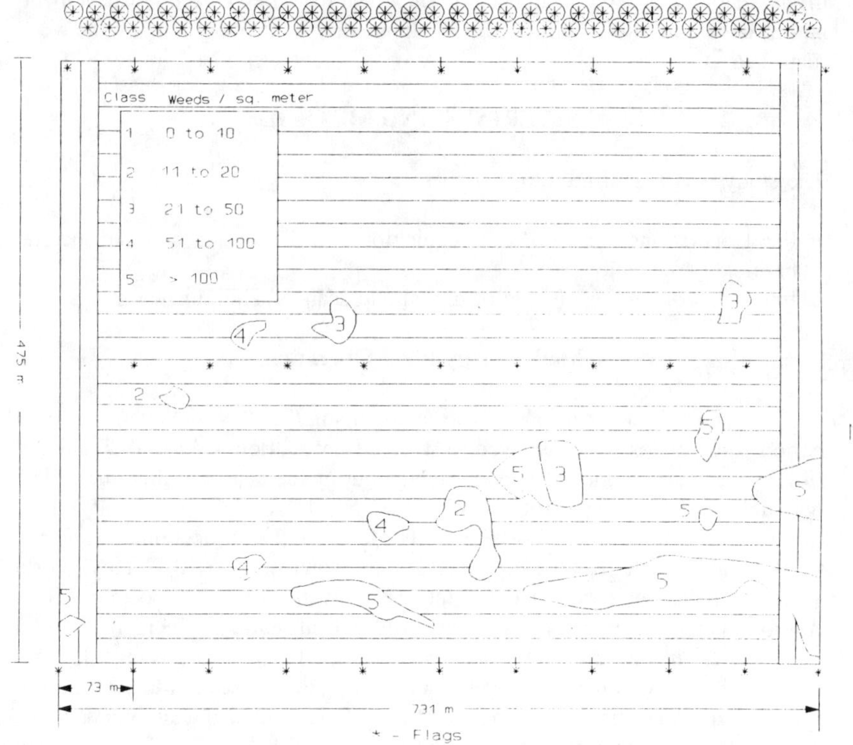

Fig. 34–2. Wild oat infestations, Field 2. Five classes of weed density per meter^2 are indicated.

(an agricultural chemical supply company in Grand Forks, N. D.) stainless steel pressurized tanks containing full strength herbicides. His system allows him to independently manually apply Bronate for broadleaf and Tiller for wild oat control. Bronate is a mixture of Bromoxynil and MCPA. Kaldor selectively sprayed Tiller on wild oat areas rated Class 3 (Table 34–1) and above using manual control and the AGVISE weed map reference on Field 1 on 31 May. The cost of wild oat herbicide treatment in the Red River Valley was about \$35.50/ha (\$15/acre). Weed map data for Field 2 was converted to 8.3 m (60 ft) cell format with the rule that a cell with 25% Class 3 wild oats or more be considered a weed cell to be sprayed. In future research, we will strive to detect and control Class 2 infestations.

Table 34–1. Classes used in the interpretations.

Field Survey (weeds/m^2)			Digital Photo-Interpretation	
	0 - 10	(1)	Not infested	(1)
Infestation	11 - 20	(2)	Not infested	(1)
Classes	21 - 50	(3)	Infested	(2)
	51 - 100	(4)	Infested	(2)
	101 - 200	(5)	Infested	(2)

A Toshiba laptop computer running Field Control Systems (FCS) software, Weed Tracker 3, was interfaced with Kaldor's injection control system and the weed cell map was entered as a new field file in Weed Tracker 3. Weed Tracker 3 is a FCS program that can control or simulate site-specific field operations. This allowed automatic spraying of the designated weed cells. Field position was by dead reckoning and tram line reference. FCS has been working with site-specific herbicide application in cooperation with several North Dakota and Minnesota farmers since 1987. A result of this work was the Repeatable Pattern Field Spraying Control patent issued to FCS in 1991. The automatic site-specific Tiller herbicide application in the experiment was an application of this patent.

The AGVISE crop consultant again surveyed the two fields for wild oat infestations on 3 August 1993.

Methodology–Interpretation and Evaluation of Imagery

This part of the research was carried out at the University of Minnesota using facilities of the Soil and Landscape Analysis Laboratory, Department of Soil Science. Images captured from the Kodak DCS 200 digital camera and digitized 35–mm color slides were transferred to a Kodak writable Photo CD, imported into Adobe PhotoShop on a Macintosh computer, converted to TIFF format, then transferred to a Sun workstation and converted to the Khoros TIFF file format as an intermediary. ERDAS 7.5 was used for analysis, importing the Khoros TIFF format file and displaying it on a 48.3 cm (19 in) 24–bit color monitor. No spectral enhancement of the image was performed before interpretation.

Screen measurements were taken of the four ground targets visible in the extreme corners of the field. The scale of the digital image was calculated from the image coordinates of the field corners and the measured length of the field along the long axis, provided by the ground reference team. A picture element (pixel) in the interpreted digital image was roughly 61 cm (2 ft) across. A grid of 18.3 m (60 ft.) cells (the effective resolution of the herbicide delivery system) was generated automatically for the field and rotated into alignment with the digital image using custom software written by the interpreter.

The treatment grid was overlaid visually on the displayed 24–bit graphic image. Treatment cells having a visible population of higher-density vegetation were selected using the computer mouse, which recorded the coordinates of a point within cell for later reference. Interpretation was done with the image displayed at twice normal resolution, and the interpreter sat about one m (3 ft) back from the image display, as this was found to improve the interpretability of the image compared to normal proximity to the screen.

Imagery was captured at several altitudes, as a means of testing the coverage of the digital camera, and the effect of spatial resolution on interpretability. Imagery used in this analysis generally corresponded to a film scale (measured on the film plane of the camera, digital or film) of about 1:16000. In the case of Kodak Photo CD imagery for Field 2, a higher altitude (about 1:32000 scale) image had to be used, as full field coverage was not available in the higher resolution imagery. All other digital and film imagery was of comparable scales and resolutions.

RESULTS - FIELD OPERATIONS

Wild oats infestations of >20 weeds/m^2 could clearly be delineated with both Ektachrome film and digital images. Crop and weed development in a 15–d period from 20 May to 5 June was appropriate for ground scouting and aerial mapping. In Field 1, wild oats were in the 1.5 to 2.5 leaf stage 30 d after crop seeding and wheat was in a 4 leaf stage, 5.1 on the Haun scale, (Jenny, 1992) on the day of field scouting. In Field 2, wheat was in the 3 to 4 leaf stage, 3.5 Haun; wild oats were in 1.5 to 2.5 leaf stage. In Field 1, 9% and in Field 2, 11% of the spray boom cells required application of Tiller herbicide. The criteria for classification as a herbicide treatment cell was 25% or more of the cell have wild oat infestation of >20 plants/m^2. Figure 34–3 is a Field 2 map showing the spray boom cells designated for control.

The site-specific controller and Weed Tracker 3 software performed as designed and anticipated. The dead reckoning field position information was adequate for the operation. Data on lagtime for herbicide delivery to an application cell were not collected. The operator anticipated this factor in the manual application on Field 1. This factor needs to be considered in automatic site-specific applications. The spray rig traveled at 19.6 km/hour (12 mph).

Weed scouting of the fields on 3 August indicated that < one ha (2.5 acres) were rated with > a Class 3 level of wild oats in Field 2 and no areas in Field 1.

Accuracy assessment was undertaken by manually transferring the cells identified as weed infested to a paper map of the same scale as the field–surveyed reference map. Correspondence between the reference and interpreted maps was manually determined, with a membership threshold of about 25% as the agreement criterion. A correspondence matrix was created using the ERDAS program MATRIX, and classification accuracy statistics, including the coefficient of agreement (Kappa), were calculated.

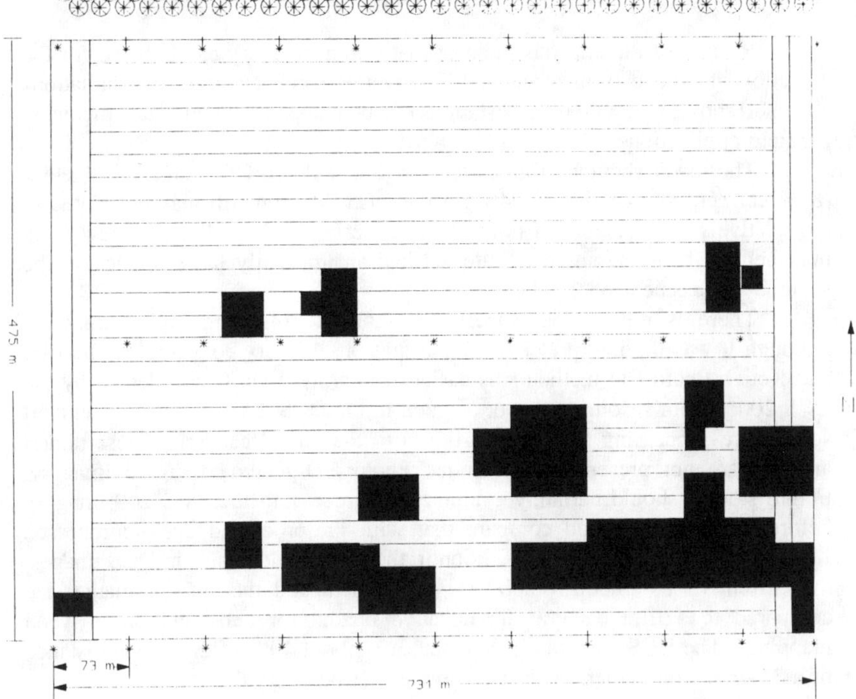

Fig. 34–3. Site-specific spray-boom cells, Field 2.

Table 34–2. Summary of classification accuracy statistics.

Media	Overall % Correct
DCS 200 Field 1	94
Kodak Photo CD Field 1	80
DCS 200 Field 2	87
Kodak Photo CD Field 2	87

DISCUSSION

Results of the analysis indicate that digital imagery can be used for the detection of wild oats in wheat fields. Overall percent correct interpretations above 80 shows that wild oat infestations can be mapped with adequate precision for field applications.

The lower accuracy figures for the Kodak Photo CD imagery in Field 1 (80% correct, compared with 94% for the DCS 200 camera) may be related to the relatively small extent of infestation in the field, about 10%. A few extra management cells classified as infested had an unusually large impact on the accuracy assessment of that field.

There is not a clear superior method of collecting digital imagery, although it was the interpreter's opinion that the imagery collected by the film and Kodak Photo CD method was easier to interpret than that collected by the DCS 200. Color resolution is not as good in the DCS 200, at least not without special pre–processing to remove artifacts of the digital camera. These digital artifacts are not present in the Kodak Photo CD-generated digital imagery. Future studies should certainly examine pre–processing the DCS 200 imagery before interpretation, and compare that with pre-processed and unprocessed Kodak Photo CD imagery. In addition to the color deficiencies in DCS images, the current cost of about $8,000 for the camera and the smaller image size compared to a film camera, are negative factors for consideration in weed mapping. The DCS camera image distance is about 40% of the image distance of a 35–mm film camera with the same lens (Kodak, 1992).

There are two major advantages to developing mapping methods with digital imagery: (i) compatibility with GIS, GPS and other digital data base systems, and (ii) potential for image enhancement and automated classification. Digital imagery of the fields can be accomplished by scanning and digitizing color or color infrared film, or by direct acquisition of digital imagery using a digital camera such as the Kodak DCS 200.

By scanning 35-mm slides most of the higher resolution of film is available while gaining the advantages of digital imagery. Low cost, but high resolution slide scanners are now readily available. For example, the Nikon CoolScan has a spatial resolution of 2700 dpi and 24–bit color resolution, and can be purchased for ≈ $2500. At this resolution there is little difference between the original film and the digital image produced from it. Either color or color infrared film may be scanned.

Use of the digital camera eliminates film from the system, saving the costs of film and processing and the time needed for processing film. It should be possible to generate interpreted data more quickly than when film has to be processed and then scanned. The digital camera option currently has less spatial resolution, although our results indicate that this is not a limitation for mapping wild oat infestations. Assuming improvements in CCD technology continue, the resolution should increase and the cost decrease in the future. There is also potential to modify a camera to detect near infrared radiation. By applying image enhancement techniques such as contrast stretch or histogram equalization, subtle differences in the spectral-radiometric responses of weeds vs. crop can be accentuated, making it easier to distinguish and map the weed infestations.

At this time the option of digitized film images is the preferred method of processing remotely sensed wild oat infestation weed data. A combination of aerial imagery and limited field scouting for ground–truth reference, is a feasible method of obtaining accurate, timely maps for site-specific weed management.

ACKNOWLEDGEMENT

The authors appreciate the financial support of the Small Business Innovative Research (SBIR) program, CSRS, USDA, for the research; it was a Phase I project. They also appreciate the controller equipment made available by Micro-Trak Systems, Inc., Mankato, Minnesota.

REFERENCES

Dexter, A.G., J.D. Nalewaja, D.D. Rasmusson, and J. Bukli. 1979. Survey of wild oats in North Dakota, 1978 and 1979. Rep. 79, N.D. Agric. Exp. Stn. and Ext. Serv.

Everitt, J., D.E. Escobar, R. Villarreal, M.A. Alanz, and M.R. Davis. 1993. Canopy light reflectance and remote sensing of Shin Oak (*Quercus havardii*) and associated vegetation. Weed Science 41:291–297.

Jenny, R. 1992. A closer look at the spring wheat cropping system for more efficient yield (MEY) and sustainability. Extension Bull. 58. North Dakota State Univ., Fargo, ND.

Kodak. 1992. Professional DCS 200 digital camera. Eastman Kodak Company. Rochester, NY.

Mortensen, D.A., G.A. Johnson, and L.J. Young. 1993. Weed distribution in agricultural fields. Proc. of the first workshop soil specific crop management, A workshop on research and development issues. 14–16 April 1992. p. 113–124. Minneapolis, MN. ASA–CSSA–SSSA, Madison, WI.

Robert, P.C., R.H. Rust, and W.E. Larson. 1993. Proceedings of the first workshop soil specific crop management. A workshop on research and development issues. 14–16 April 1992. Minneapolis, MN. ASA, CSSA, SSSA, Madison, WI.

Taylor, W. 1994. Camera meets computer. PC Computing. Feb. p. 198-204.

Thornton, P.K., R.H. Fawcett, J.B. Dent, and T.J. Perkins. 1990. Spatial weed distribution and economic thresholds for weed control. Crop Prot. 9:337–342.

USDA–Agricultural Research Service. 1970. Selected Weeds of the U.S. Agric. Handb. 366. p. 38. USDA-ARS. Washington, DC.

Van Groenendael, J.M. 1988. Patchy distribution of weeds and some implications for modelling population dynamics: A short literature review. Weed Res. 28:437–441.

35 Remote Sensing to Detect Nitrogen Deficiency in Corn

T. M. Blackmer

Department of Agronomy
University of Nebraska
Lincoln, Nebraska

J. S. Schepers

USDA-Agricultural Research Service
University of Nebraska
Lincoln, Nebraska

G. E. Meyer

Department of Biological Systems Engineering
University of Nebraska
Lincoln, Nebraska

Nitrogen fertilizer mismanagement has detrimental economical and environmental consequences. A study was conducted to determine how well remote sensing techniques can identify N deficient areas in fields, which could lead to improved N management practices. Fertilizer N response trials involving multiple corn (*Zea mays*, L.) hybrids and soil types were investigated in central Nebraska during 1992 and 1993. Light reflectance measurements were made on clear d using a spectroradiometer, photometric cell, and aerial photography. Spectroradiometer results showed areas of the visible spectrum that were sensitive to crop N stress. An optical bandpass filter was used on the camera to isolate the green region of the spectrum for black and white photographs. The photometric cell provided the capability of detecting reflected light in a portion of the spectrum where N deficiencies could be detected. Negatives from standard color and black photographs were digitized and images were analyzed to identify areas of N stress. All light reflectance measurements taken from the corn canopy were capable of identifying N deficient plots, as verified with a chlorophyll meter. Both aerial photography and photometric sensors show promise as techniques for identifying N deficient areas. These methods could be valuable in identifying areas of a field that require special management practices.

INTRODUCTION

Spatial variability in N availability complicates N fertilizer management in corn production. The first step towards improving management is to identify areas of a field that are significantly different in N status. Much attention has been given to soil sampling strategies to identify areas within fields that have different patterns of available N (NH_4^+ and NO_3^-). Because soil nitrate concentrations can vary considerably over relatively short distances, collecting a representative soil sample can be difficult and expensive.

LEAF SAMPLING TECHNIQUES

Several analytical techniques are available to characterize crop N status and nutrient availability to the plant. Leaf N concentration has been evaluated for corn with special emphasis on the ear leaf (or one leaf below) near silking time (Tyner & Webb, 1946; Viets et al., 1954; Cerrato & Blackmer, 1991). This method is complicated by luxury consumption, frequently provides information too late for correction of deficiencies, and may not represent the N status of the entire field. In addition, cost and time restraints required for analyzing plant tissue in the laboratory may deter wide scale sampling of an entire field required to identify small, but significant N deficient portions of a field.

Chlorophyll meters represent a new technology to detect N deficiencies in corn instantaneously (Piekielek & Fox, 1992; Schepers et al., 1992; Wood et al., 1992). The instantaneous results from the meter permit extensive sampling without laboratory expense, but are plagued by not knowing how well the data represent the entire field. Also, differences in hybrids and other conditions make calibration of the meter to a reference area desirable. Because chlorophyll meters are relatively insensitive to luxury consumption, reference areas need only be non-limiting in N. Reference areas and other areas of a field can be compared to detect N deficiency. Using a reference must be considered in contrast to needing to identify a critical level of N for optimum yield. Considerable labor and time as well as reference areas are still required to accurately sample an entire field.

Leaf reflectance measurements are also sensitive to chlorophyll concentrations (Thomas & Gausman, 1977; Maas & Dunlap, 1989; McMurtrey et al., 1994) and can be related to differences in N status (Al-Abbas et al., 1974; Thomas & Oerther, 1972). Reflectance from two corn leaves differing in N status were compared by expressing reflectance at each wavelength as the ratio of an N deficient leaf to a sufficient leaf (Fig. 35–1). Reflectance around 550 nm and 710 nm provide two portions of the spectrum sensitive to N stress. Thus, differences in the reflectance ratio can be used to detect N deficiencies.

CANOPY REFLECTANCE MEASUREMENTS

Reflectance data from optical sensors specially designed to measure portions of the spectrum sensitive to N stress could be of value by measuring a section of a canopy in contrast to an entire field. Canopy reflectance measurements offer the advantage of integrating a larger sampling area than an individual leaf and do not require that an instrument be clamped onto an individual leaf. Canopy reflectance measured with a spectroradiometer from an N response trial (Fig. 35–2) showed a similar pattern to that observed for leaf reflectance (Fig. 35–1). Canopy level reflectance measurements have been shown to detect N stress in corn (Walburg et al., 1982), wheat (*Triticum aestivum*, L.) (Stanhill et al., 1972; Hinzman et al., 1986), and rice (*Oryza sativa*, L.) (Takebe et al., 1990).

As shown (Fig. 35–1 and 35–2), 550 nm is a portion of the spectrum where reflectance measurements are sensitive to N stress. The photometric cell is a relatively inexpensive sensor that has peak sensitivity to light near 550 nm. The use of such a sensor on a clear d can provide a good prediction of N stress that can reduce grain yield (Fig. 35–3). One major problem with absolute reflectance measurements that should be noted is that they are dependent upon the sun as the light source, which is seldom constant over time. Several strategies can be used to deal with these problems, but they are beyond the scope of this report.

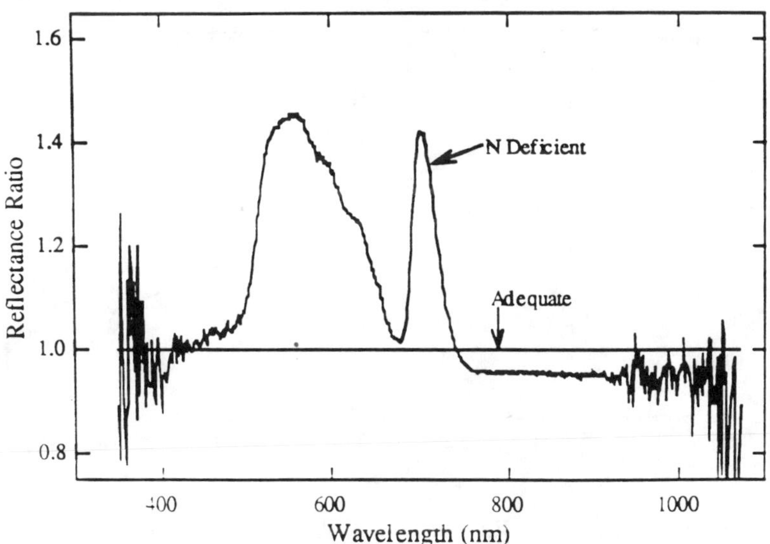

Fig. 35–1. Light reflectance of an N stressed corn leaf expressed as the ratio to an N sufficient leaf.

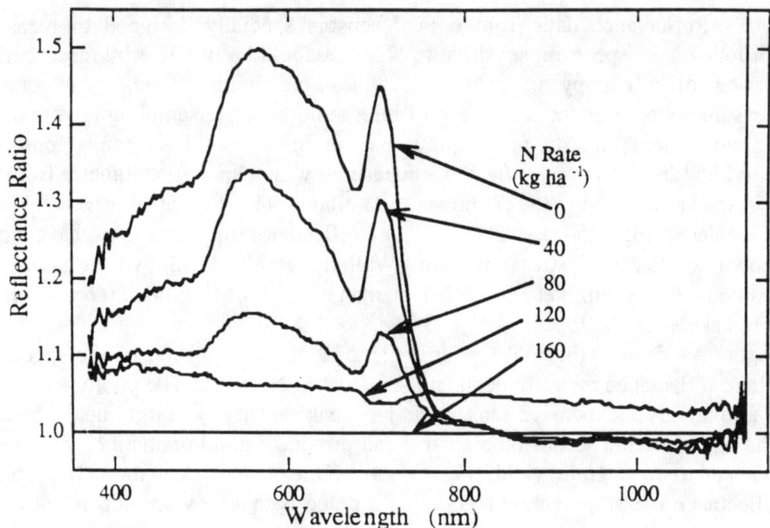

Fig. 35–2. Canopy reflectance for one corn hybrid grown at five different N rates expressed as a ratio of reflectance from the highest N rate.

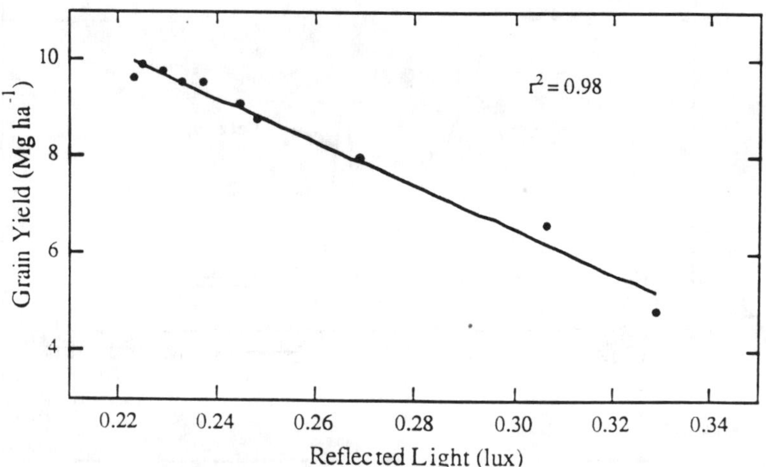

Fig. 35–3. Relationships between canopy reflectance measurements obtained using a photometric cell and grain yield for site with one hybrid and 10 N rates.

AERIAL PHOTOGRAPHS

Aerial photographs have the potential to measure differences in light reflectance from a canopy for an entire field at one time. Because the benefit of measuring light reflectance centered at 550 nm has been documented, a technique was developed in which black and white photographs were collected using a filter that restricted light entering the camera to only that portion of the spectrum.

By using black and white film, any difference in light reflectance would result in a difference in exposure which could be categorized using a gray scale with a range of 0-255. Digitization of negatives was done using a Nikon LS[1] 3500 scanner with the Photostyler Software. The gray scale assessments were made using the Image Pro Plus software package. A similar technique has been used for plant canopy cover evaluation (Thomas et al., 1988). Although the described equipment was used for this study, other equipment could be used to improve data collection and interpretation. Results using this technique illustrate the feasibility of detecting N stress in a corn canopy (Fig. 35–4).

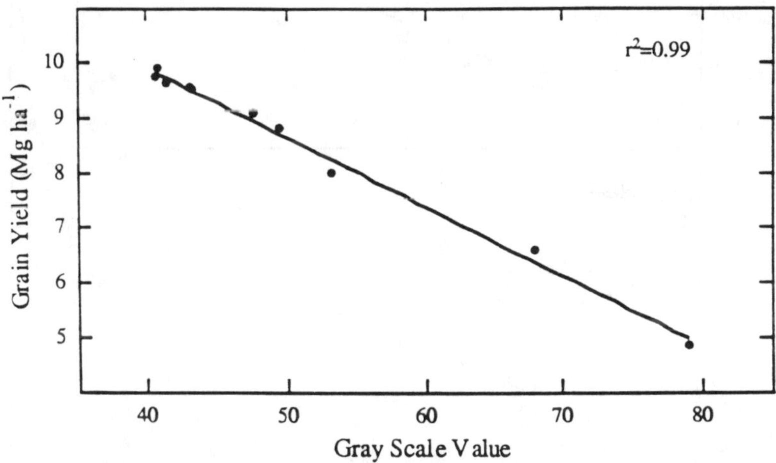

Fig. 35–4. Relationship between canopy reflectance measurements obtained from a black and white photograph obtained while using a filter and grain yield for a site with one hybrid and 10 N rates.

[1]Mention of trade names or proprietary products does not indicate endorsement by USDA, and does not imply its approval to the exclusion of other products that may also be suitable.

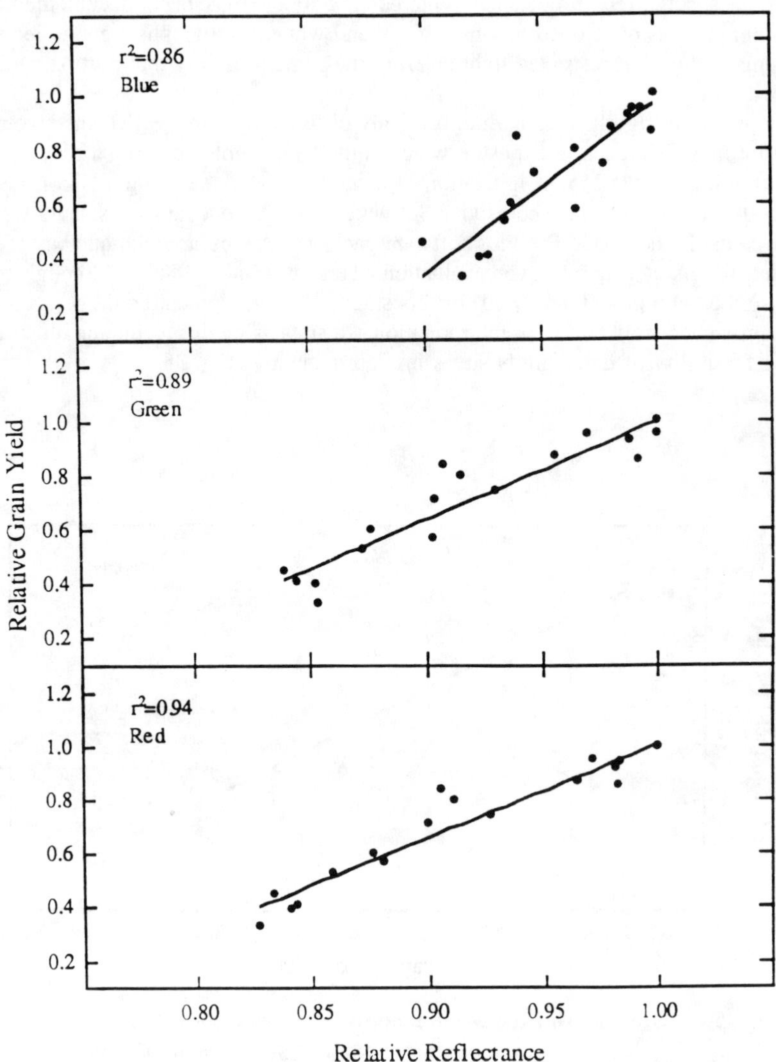

Fig. 35–5. Relationship between canopy reflectance measurements obtained from a color photograph by measuring the blue, green, and red gray scale that has been normalized to the highest of the five N rates for each of the four hybrids and grain yield.

Reflectance measurements from aerial photographs makes identification of N deficient areas in a field possible without collecting ground samples. These atypical portions of the field may have been missed by ground sampling techniques. Once identified, site-specific ground sampling can be used to evaluate the need for unique management practices.

Depending on the application, different types of photographs may be considered. Color-infrared photography has been historically used to identify differences in a field, but this technique is sensitive to differences in soil moisture status, which is not the focus of this report. Standard color film captures an image representing the entire visible spectrum. Earlier it was demonstrated that wavelengths differed in sensitivity to N stress, so a single measurement of the entire visible spectrum would be expected to reduce sensitivity. Narrowing the visible spectrum can occur by measuring a gray scale that resulted from digitization of an image using a light source that provided three separate primary colors (red, blue, and green). Because differences in procedures (processing, lighting, etc.) can result in different gray scale values, detection of N stress is most sensitive for comparisons within the same image.

Within the same image, hybrid differences can affect gray scale values, despite similar N status (Fig. 35–5). A simple conversion to a relative gray scale, done by the same method as is commonly used with grain yields, provided a satisfactory normalization for the data across four hybrids and five N rates. The relationship for all three primary color gray scales show a difference in sensitivity. The red color gray scale had the highest coefficient of determination.

APPLICATIONS

For canopy reflectance measurements, a photometric sensor could be mounted on a high clearance vehicle, and measurements of all portions of a field could be collected by driving through the field. Aerial photographs collected late in the growing season can detect portions of a field that are N deficient. This information can be used to generate a map that guides specific ground sampling to further investigate a problem. In addition, the image created might be incorporated into a geographical information system database that can be part of an expert system to help manage N in the future.

REFERENCES

Al-Abbas, A.H., R. Barr, J.D. Hall, F.L. Crane, and M.F. Baumagardner. 1974. Spectra of normal and nutrient-deficient maize leaves. Agron. J. 66:16–20.

Cerrato, M.E., and A.M. Blackmer. 1991. Relationships between leaf nitrogen concentrations and the nitrogen status of corn. J. Prod. Agric. 4:525–531.

Hinzman, L.D., M.E. Bauer, and C.S.T. Daughtry. 1986. Effects of nitrogen fertilization on growth and reflectance characteristics of winter wheat. Remote Sens. Environ. 19:47–61.

Maas, S.J., and J.R., Dunlap. 1989. Reflectance, transmittance, and absorptance of light by normal, etiolated, and albino corn leaves. Agron. J.

81:105–110.

McMurtrey, J.E., III, E.W. Chappelle, M.S. Kim, J.J. Meisinger, and L.A. Corp. 1994. Distinguishing nitrogen fertilization levels in field corn (Zea *mays*, L.) with actively induced fluorescence and passive reflectance measurements. Remote Sens. Environ. 47:36–44.

Piekielek, W.P., and R.H. Fox. 1992. Use of a chlorophyll meter to predict sidedress nitrogen requirements for maize. Agron. J. 84:59–65.

Schepers, J.S., D.D. Francis, M.F. Vigil, and F.E. Below. 1992. Comparisons of corn leaf nitrogen concentration and chlorophyll meter readings. Comm. Soil Plant Anal. 23:2173–2187.

Stanhill, G., V. Kalkofi, M. Fuchs, and Y. Kagan. 1972. The effects of fertilizer applications on solar reflectance from a wheat crop. Israel J. Agric. Res. 22:109–118.

Takebe, M., T. Yoneyama, K. Inada, and T. Murakam. 1990. Spectral reflectance of rice canopy for estimating crop nitrogen status. Plant and Soil. 122:295–297.

Thomas, D.L., F.J.K. daSilva, and W.A. Cromer. 1988. Image processing technique for plant canopy cover evaluation. Trans. ASAE. 31:428–434.

Thomas, J.R., and H.W. Gausman. 1977. Leaf reflectance vs. leaf chlorophyll and carotenoid concentration for eight crops. Agron. J. 69:799–802.

Thomas, J.R., and G.F. Oerther. 1972. Estimating nitrogen content of sweet pepper leaves by reflectance measurements. Agron. J. 64:11–13.

Tyner, E.H., and J.W. Webb. 1946. The relation of corn yields to nutrient balance as revealed by leaf analysis. J. Am. Soc. Agron. 38:173–185.

Viets, F.G., C.E. Nelson, and C.L. Crawford. 1954. The relationship among corn yields, leaf composition and fertilizers applied. Soil Sci. Soc. Amer. Proc. 18:297–301.

Walburg, G., M.E. Bauer, C.S.T. Daughtry, and T.L. Housley. 1982. Effects of nitrogen nutrition on the growth, yield, and reflectance characteristics of corn canopies. Agron. J. 74:677–683.

Wood, C.W., D.W. Reeves, R.R. Duffield, and K.L. Edmisten. 1992. Field chlorophyll measurements for evaluation of corn nitrogen status. J. Plant Nutr. 15:487–500.

36 Some Applications of Temporal Climate Probabilities to Site-Specific Management of Agricultural Systems

Mark Seeley

Department of Soil Science
University of Minnesota
St. Paul, Minnesota

Applications of climatic data bases or weather forecast and outlook guidance to site-specific management invoke the use of both temporal and spatial probabilities. Much of the current research and development of technology is focused on understanding and applying principles related to the spatial scales and variability inherent to agricultural systems. Nevertheless, most researchers and agricultural practitioners recognize that temporal climatic scales and variability have tremendous impact on site-specific management as well, and in many instances offset or amplify any adjustments made for spatial variability.

Concurrent development of on-line climatological data bases at various regional climate centers and better weather forecast and outlook guidance from the National Weather Service provoke the question as to whether or not better management decisions might result from greater utilization of information concerning temporal probabilities (primarily weather and climate).

Many hypotheses related to this question are yet to be tested. Therefore, there are little comparative data to show, much less to evaluate on specific decisions. The modernization of the National Weather Service (currently in progress) will produce more accurate and timely forecast and outlook information which will be more accessible and widely distributed than ever before (Seeley, 1994). There are already a significant number of agricultural producers receiving weather information in near-real time via satellite communications through such systems as the Data Transmission Network (Richter, 1994). In addition, there is a growing community of users who are accessing the regional climate centers throughout the country. Their data bases provide users with the means to calculate probability distributions for climatic thresholds or events by location and for any d or period of the yr. Thus, temporal distributions of frosts, growing degree d, precipitation or other climatic variables can be determined for either specific or groups of climatic stations. Spatial extrapolation of these probability distributions is possible using kriging techniques or digital terrain analysis (U.K. Meteorological Office, 1989). These developments suggest a conceptual framework of topics which might be studied and field tested in the near future using new technologies to run decision-aid models from climatic data bases and weather forecast products. Currently, university researchers, extension specialists, crop consultants and agricultural industries are jointly testing and demonstrating a wide range of precision farming technologies and strategies, including the use of temporal probabilities (Robert, 1994).

Tillage, planting, fertilizing, cultivating, spraying, irrigating and harvesting are all field operations which are dependent on weather and soil conditions. The timeliness, efficiency and efficacy of these operations are highly regulated by the environmental conditions. Ignoring microclimatic differences, the variability of environmental conditions under which these operations are performed is primarily temporal in nature. One h may be distinctly different from another, a windy d followed by a calm d, a wet wk followed by a dry wk.

HISTORICAL DISTRIBUTION AND PROBABILITY

As mentioned above, many of the regional climate data centers in the United States (now numbering 6) have comprehensive data bases (especially concerning precipitation and temperature) available for users via dial-up or Internet connections. Even at the state level, similar data bases exist, sometimes encompassing the entire period of record for all observations, many of which go back over 100 yr, such as the Minnesota Centennial Climatic Data Base (Seeley & Zandlo, 1992).

Using on-line data bases of daily climatic observations, various production decisions or questions concerning the temporal distribution of environmental conditions at any particular location can be addressed. Several topics which are important to agricultural producers can be examined in the context of historically-based probability distributions, including the probability distribution of field working d throughout a growing season, the probability of receiving sufficient precipitation to activate a surface applied or shallow incorporated herbicide in the spring, the probability of heat stress for confined livestock during July and August, the probability of consecutive dry d for making hay, or the probability of a freeze occurring prior to projected harvesting dates for any particular crop.

The use of historically-based probability distributions is one of the traditional applications in climatology (Changnon, 1991). Such probability distributions provide a bounded framework from which several management contingencies can be examined. The topics of field working d distributions in the spring, moisture probabilities for the activation of soil applied herbicides, and temperature conditions for the volatilization of postemergence (foliar applied) herbicides can be used as examples to illustrate this traditional approach.

Field Working Days

A diary of field working d has been kept by the staff of the University of Minnesota Southwest Experiment Station since 1974. Field working d were noted in the diary when staff were actually in the field working for either the entire d or portion of the d, as well as when conditions were favorable for field work even though staff had no particular reason to be working in the field. This record of field working d for the mo of April, May and June is shown in Fig. 36–1 as a probability distribution based on the records from 1974 to 1991. Both 3-d and 10-d smoothing functions have been used to show the temporal pattern. Though there is quite a bit of variability evident in the daily probability values of these mo, the smoothed lines show a clear temporal trend of increasing probability throughout the mo of April and lasting into the first wk of May when it starts to level out. Negative deviations from this trend (reduced probability) are noted around April 13, 28, and May 14. Peak probabilities can be seen around the dates of May 6 and May 11. These dates are very close and on either side of the historical mean corn planting date for southwestern Minnesota of May 9 (Baker et al., 1984).

Previous research by Rosenberg (1982), Acharya et al. (1983) and others has suggested that field working d should be closely associated with soil moisture conditions. Staff of the Southwest Experiment Station have also kept a measurement record of soil moisture covering the same period as the field working d diary. This record includes biweekly sampling of soil layers, corresponding to depth increments of 15, 30, 45, 60, 90, 120, and 150 cm, respectively. Gravimetric moisture content was determined for each layer from among five samples and then averaged by layer. The soil sampled was a Webster clay loam (Typic Haplaquoll) with a field capacity of 32.5 mm in the top 30 cm. Baker et al. (1979) have described in more detail the soil moisture measurements at the Southwest Experiment Station.

Seventy-three measured soil moisture values taken during the mo of April, May and June (1974-1991) were studied in conjunction with the field working d diary. These data are shown in Fig. 36–2 for the top 30 cm of soil. Forty-four of the 73 dates were denoted as field working d (FWD=1), while 29 dates corresponded to non-field working d (FWD=0). Examination of Fig. 36–2 shows that 33 cases, or 70% of the field working d occurred on dates when the measured soil moisture was below field capacity. Eighteen of the non-field working d, or 62% occurred on dates when the measured soil moisture was above field capacity. Though these distribution characteristics substantiate an association between soil moisture and field working activity, the relationship is not as strong as expected. There are a number of factors that might explain this. For example, the concept of a "forced field working day" is denoted on Fig. 36–2 (FWD=F). This is based on indications in the field working diary that certain field operations had to be performed regardless of the weather in order to meet criteria associated with experimental design. Like many agricultural experiment stations, the Southwest Experiment Station conducts dozens of field studies on several thousand hectares of land. Many experiments require precisely timed field operations which must be carried out in spite of unfavorable

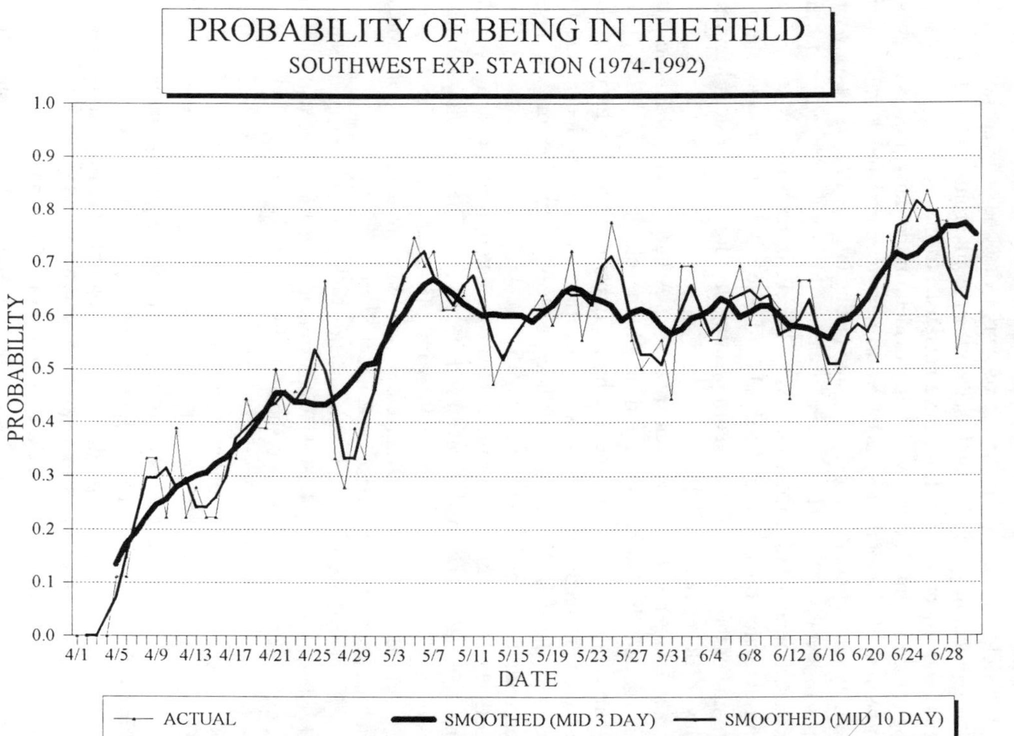

Fig. 36–1. Actual and smoothed field working d probability - April to June - Southwest Exp. Stn. in Lamberton, MN.

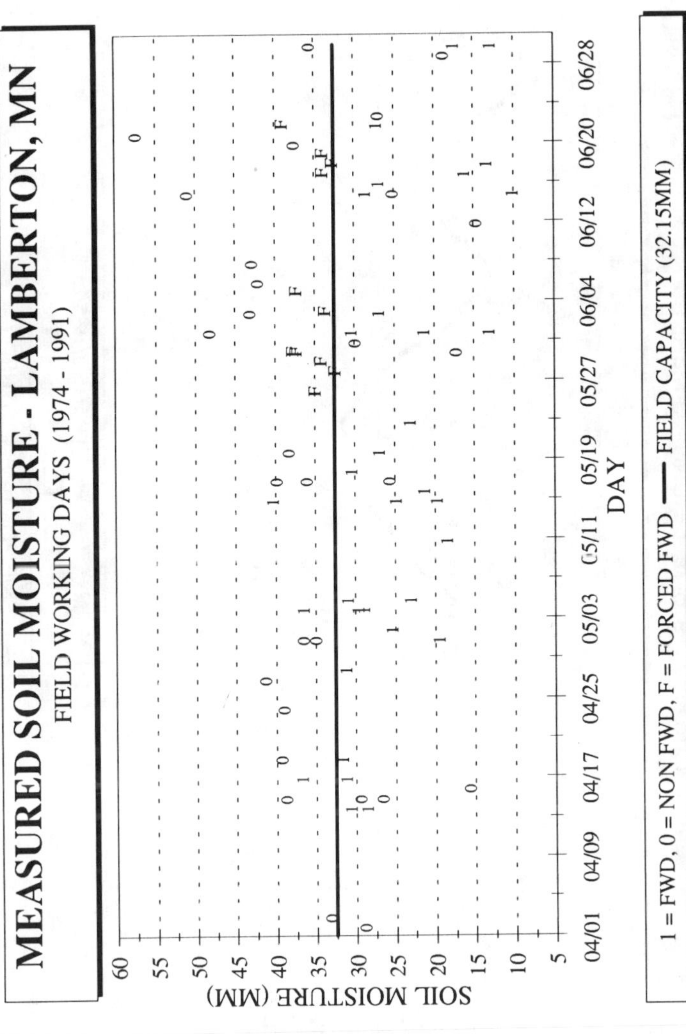

Fig. 36–2. Relationship between Field Working D (FWD) and soil moisture values at Southwest Exp. Stn., Lamberton, MN - 1 April to 30 June (measurement from top 30 cm of soil profile, equally divided into zone 1 and zone 2).

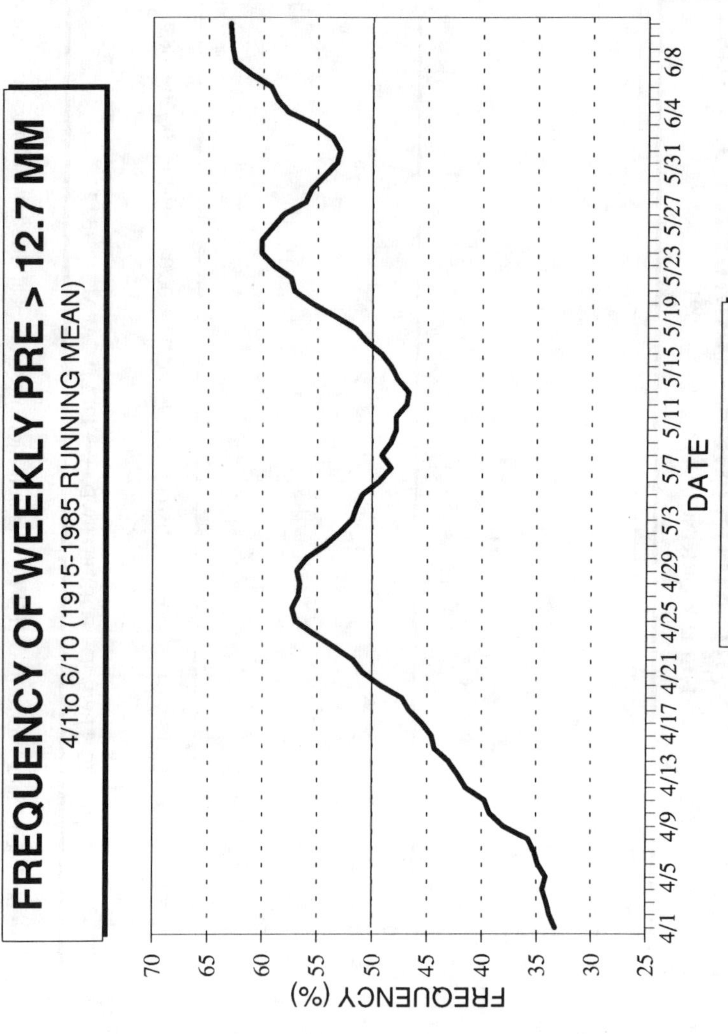

Fig. 36–3. Smoothed weekly frequency of precipitation for Fairmont, MN (1 April - 30 June).

climatic conditions. In addition, errors associated with soil moisture measurement in conjunction with the spatial variability in soil moisture can produce unrepresentative values with respect to the experiment station at large. Further studies will be conducted using soil moisture models to make estimates of daily soil moisture content, which in turn can be related to field working d. The correspondence between the temporal pattern of daily soil moisture estimates in the spring will be checked against the temporal pattern of field working d.

Activation of Pre-emergence Herbicides

Another application of historical climatic distribution and probability relates to the use of pre-emergence herbicides for early season weed control. A number of soil applied herbicides require moisture for activation. In irrigated crop systems this is controlled, but in rainfed agricultural production, the effectiveness of a soil applied herbicide is highly dependent on the timing of rainfall events, a climatic characteristic that can be examined temporally using historical data bases. In Minnesota, rainfall varies in frequency and amount from west to east and south to north, with the lesser values found in the west and north.

In recent yrs, Minnesota weed scientists have used weekly rainfall frequencies as a sorting criteria to delineate geographic boundaries for the suitability of using soil applied herbicides for early season weed control. Examples are shown in Fig. 36–3 and 36–4 using a weekly rainfall amount which is > or equal to 12.7 mm, an amount determined necessary to activate a number of pre-emergence herbicides. Seventy yrs of daily climatic records were used to construct the temporal pattern for this rainfall threshold in the mo of April and May when most of the soil applied herbicides are used. Figure 36–3 shows the pattern for Fairmont, MN located in the south-central portion of the state, while Fig. 36–4 shows the pattern for Milan, MN located in the west-central portion of the state. By 20 April, the frequency of rainfall exceeding 12.7 mm is >50% at Fairmont, then somewhat surprisingly, falls below 50% during the period from 6–18 May, before increasing again. This pattern suggests that in this region of the state soil applied herbicides would likely receive sufficient moisture for activation during late April and early May, then again over the latter half of May. Quite a different temporal pattern is evident at Milan, where the frequency of weekly rainfall exceeding 12.7 mm does not go past the 50% mark until 26 May, but then increases sharply in June. This period of time, however, is generally too late to use soil applied herbicides because most weed species are far too developed to be effectively controlled by these compounds. The distinct difference between the temporal pattern of rainfall at these two sites suggests that soil applied herbicides are a viable weed control strategy in southern, but not western Minnesota. This is substantiated by an analysis of several other southern and western locations which have long-term climatic records.

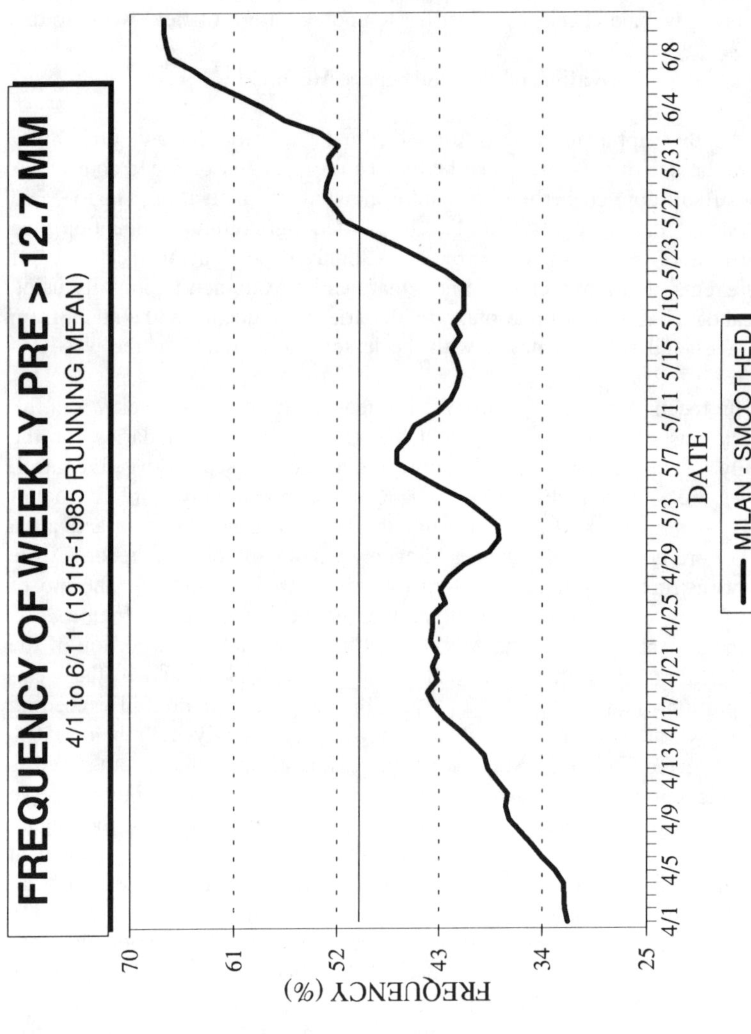

Fig. 36–4. Smoothed weekly frequency of precipitation for Milan, MN (1 April–30 June).

Postemergence Herbicides

Other climatic sorting criteria could be used and applied to specific compounds as well. For example, some postemergence herbicides which are applied by drop nozzles to the foliage of actively growing weeds are subject to volatilization and drift if applied when temperature conditions are too high. One such compound is dicamba, which is commonly used on corn in Minnesota for postemergence weed control. This compound can be volatilized at temperatures of 29 degrees Celsius or greater, however, and the resulting vapors may drift and cause injury to nearby soybeans. Utilizing a climatic data base to construct the temporal probability for maximum temperatures of 29 degree C or greater during the period of time when dicamba is most frequently used can show the relative risk of having volatilization occur. Figure 36–5 illustrates the temporal pattern for this temperature threshold at Worthington, MN, covering the period from 25 May, which corresponds to early use of postemergence herbicides, to 25 June which corresponds to late use of such herbicides. Note that the probability of volatilizing temperature conditions to occur, though relatively low (<50%) is, nevertheless, 4 times greater in late June than it is in late May, based on the historical distribution of temperature at this particular location (1894–1987). Such information can be valuable to producers who are variably adverse to risk and want to evaluate weed control options among compounds or perhaps among field operations (rotary hoe or field cultivator).

Granted the above historical probability distributions are site-specific and their degree of spatial extrapolation needs to be tested. This is limited by the network density inherent to the climatic data base, but increasingly models can be used to test the coherence of spatially derived variables. Baker and Skaggs (1984) and Hubbard (1994) have discussed such techniques in detail. Additionally, in soil moisture modeling, specific soil properties must be accounted for, as is the case with the study of field working d at Southwest Experiment Station at Lamberton, MN. Lastly, there may also be dominant landscape features which greatly influence local environmental conditions (e.g. nearby lakes, valleys, sharp elevation changes) and which are not picked up in the available climatic data base.

CONDITIONAL DISTRIBUTION AND PROBABILITY

Because of the great variability of climate, certainly in the mid latitude regions, purely historical temporal probability can be improved upon by using conditional probability. That is to say, initial climatic conditions such as those of today, influence the probability distribution of what conditions will be tomorrow, next wk or even next mo. This is referred to as the principle of *persistence* by meteorologists who often adjust forecasts and outlooks based on current conditions. This is especially true in considering soil moisture values. An example can be drawn from the regional assessment of over-winter soil moisture conditions and their potential to cause a higher risk of spring flooding along major drainage basins in the north-central region or to delay or disrupt the spring planting season.

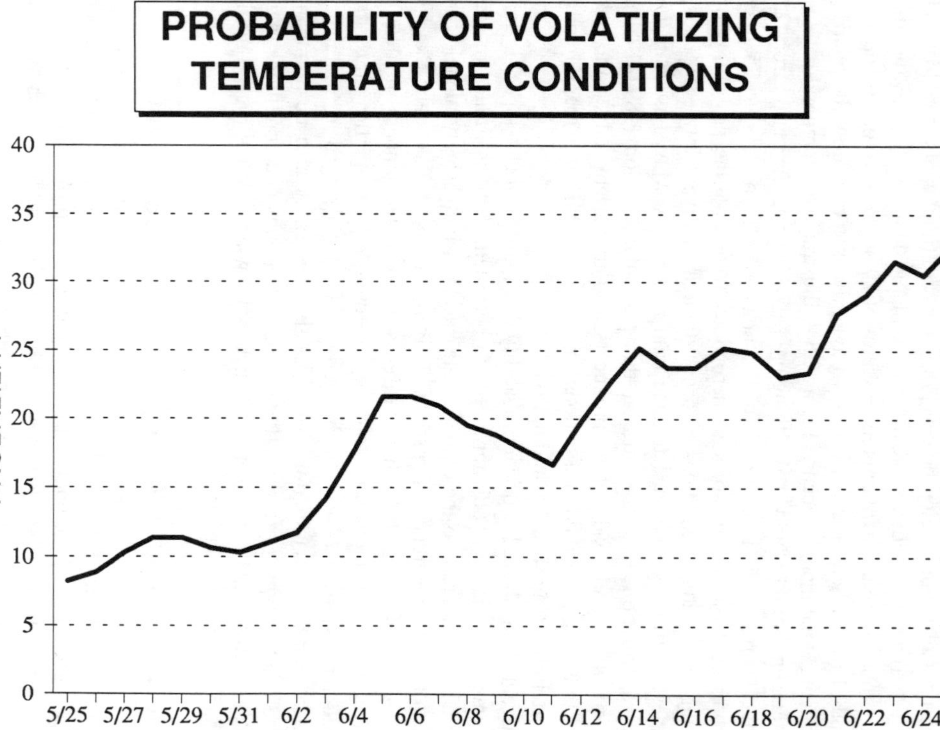

Fig. 36–5. Temporal probability (using a 3–d running mean) of temperatures equal to or exceeding 29 degrees C at Worthington, MN (1894-1987), 25 May to 25 June, inclusive.

Regional assessments of soil moisture conditions are periodically described in news releases by the Midwest Climate Center (MCC) in Champaign, IL. Kunkel (1990) has described the MCC soil moisture model and its' use in making regional assessments of field conditions for the agricultural season. In February of 1994, a regional assessment of soil moisture was released by MCC which employed the use of conditional probability to address the question, "what is the probability that soil moisture values will still be above historical normals by May 1st?" At the time this assessment was released, both measured and estimated values of soil moisture across the region were 3 to 10 cm above normal in the top 180 cm of soil.

These values were used as initial conditions to run the MCC soil moisture model through a series of forward looking analyses to estimate May 1st values. Using the most recent 44 yrs of daily climatic data (1950–1993) from a network of regional stations, the model estimated May 1st soil moisture values for each yr, based on initial conditions for February 1, 1994 and subsequent climatic data for each February through April in the time series. The frequency distribution of these 44 May 1st values was then examined with respect to the historical normal soil moisture for this date. A probability distribution was determined relative to the normal. The model analysis is depicted by crop reporting district for the north-central region in Fig. 36–6. Note that there was a 95% probability in western Minnesota that soil moisture values would still be above normal by 1 May of 1994 based on the moisture conditions of 1 February. Only two yrs (scenarios) of the 44 used in the model produced conditions which would have diminished soil moisture values to normal levels by 1 May.

This conditional probability assessment by MCC was used by the North-Central River Forecast Center to project the spring flood outlook on major rivers of the region and thereby prepare communities to deal with high flows. It was also used by agronomists to anticipate frequent disruptions in the field planting season during the spring and to advise crop producers to plan for an early start (if possible) and long working-d in order to achieve appropriate planting dates for selected maturities and to compress planting activity into fewer d. This advice was indeed followed, as for example Minnesota Agricultural Statistics reported 20% of the field corn acreage in the state was planted by 24 April, with some southern counties nearly completely planted due to utilizing long field working d.

The MCC soil moisture model assessments are but one example of utilizing conditional probability to provide strategic information about distant planning horizons of a few wks to a few mo in length. There are perhaps many other applications waiting to be tested with other models.

Fig. 36–6. Probability (%) that soil moisture in the top 180 cm will be above the long-term average on May 1, 1994. These are climatological probabilities based on 1 February conditions, without assumptions about future weather conditions up to May 1, 1994 (with permission of Dr. Ken Kunkel, Midwest Climate Center "Special Soil Moisture Update and Outlook for Winter and Spring 1994 Conditions", released Feb. 1, 1994, Champaign, IL).

FORECASTED DISTRIBUTION AND PROBABILITY

Irrigation design capacities and scheduling are in part based on soil properties and statistical climatology (Bergsrud et al., 1982). Evapotranspiration (ET) estimates based on historical climatology provide some guidance with respect to the seasonal pattern of ET, but for tracking any given season they may be seriously in error (Stegman et al., 1977). Figure 36–7 illustrates an example comparing daily estimates of potential ET as determined from three different methods for Morris, MN during the 1991 crop season. These methods are described by Wright (1986) and Jensen (1973). The bold line shows the standard checkbook method or Minnesota table for estimating daily values of ET. This method is based on the historical climatology of daily temperature conditions and a coefficient governed by the number of wks after emergence. The modified Jensen-Haise method utilizes both real time daily temperature and solar radiation along with a crop coefficient (based on growth stage) to estimate potential ET. The Penman method employs the most comprehensive set of real-time climatic variables, including temperature, vapor pressure, wind speed and solar radiation, along with a crop coefficient (based on growth stage). Both the daily Jensen-Haise and modified Penman methods show substantial deviations from the climatologically based estimates from the Minnesota ET table. This is not surprising, in that the ET table is fundamentally based only on historical climatology of temperature (which on a daily basis has a standard deviation ranging from 4.5 to 7.5 degrees C), while the Jensen-Haise and Penman methods rely more on the combined contribution or effect of the real-time climatic variables. The climatic variables used in these three methods then may have different distribution characteristics. Regardless of which method of estimating real-time ET values is used, however, the irrigator still requires some knowledge of what future weather conditions might be.

Forecasting for Irrigation Management

It is increasingly common (at least in this country) to find that irrigators utilize daily ET estimates to provide guidance for scheduling water applications, based on some criteria to maintain a given soil water content or to minimize stress to the crop. This is one of the chief uses of real time climatic data from automated station networks in many states (Meyer & Hubbard, 1992).

Perhaps a somewhat overlooked area to improve precision irrigation management with respect to temporal variability is greater utilization of forecast and outlook products of the National Weather Service. The Weather Service provides medium range guidance (out to 196 h) and outlooks which estimate important climatic variables. Utilizing the example of irrigation scheduling, the application of forecast and outlook probabilities can be illustrated. Figure 36–8 shows the seasonal pattern of rainfall and irrigation applied during the 1992 growing season by an irrigator in west central Minnesota. It is evident that in mid August this irrigator applied over 38 mm of water, which contributed to the cumulative excess of precipitation and irrigation water that particular season. This irrigation was applied just prior to a wet period which was predicted in both the medium range forecast and the 6-10 outlook by the National Weather Service. In fact, August 22, 23, 24 and 26 all had >60% chance of showers and thundershowers in the forecast guidance. This raises the question of should the irrigator have believed this information and either limited irrigation or not used it at all. The temporal probabilities expressed in guidance products of the National Weather Service need to be evaluated for their value in such decisions. In addition, agricultural meteorologists and climatologists should be advocates for the availability of such guidance products. One hundred-ninety six h medium range forecasts derived from model output statistics (ranging from 80 km to 100 km grid spacings) formerly used only as in-house guidance by the National Weather Service, but now publicly available, are issued at 6:00 a.m. each d. Six-to ten-d outlooks are released at 3:00 p.m. EST every Monday, Wednesday and Friday. These products and others are available over various Internet resources and could be utilized in near-real time to run models for irrigating, spraying, crop drying, etc.

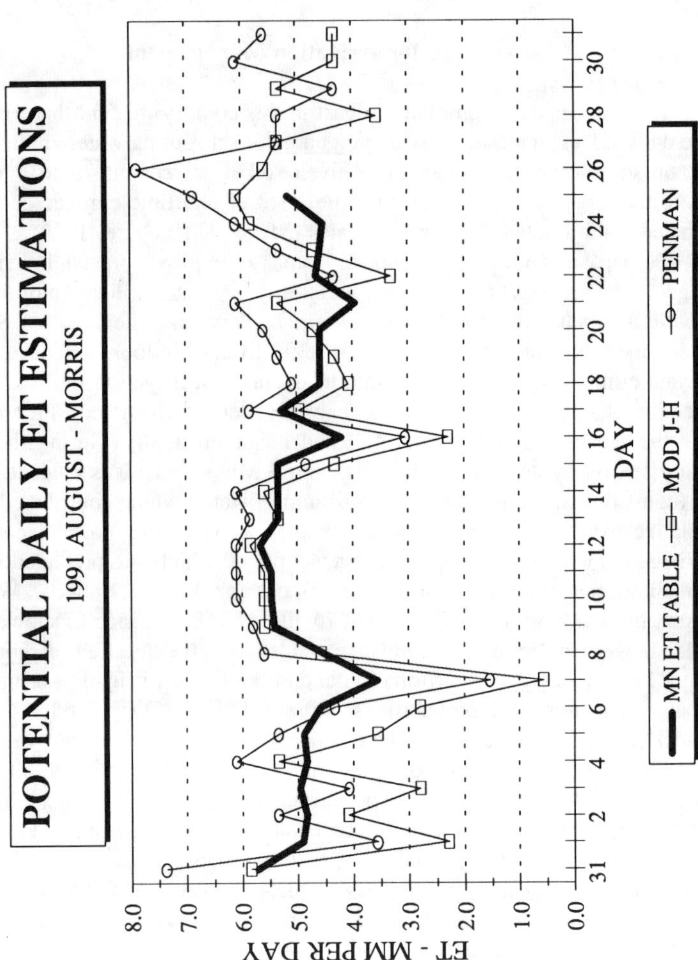

Fig. 36–7. Comparison of daily evapotranspiration (ET) estimates by three methods.

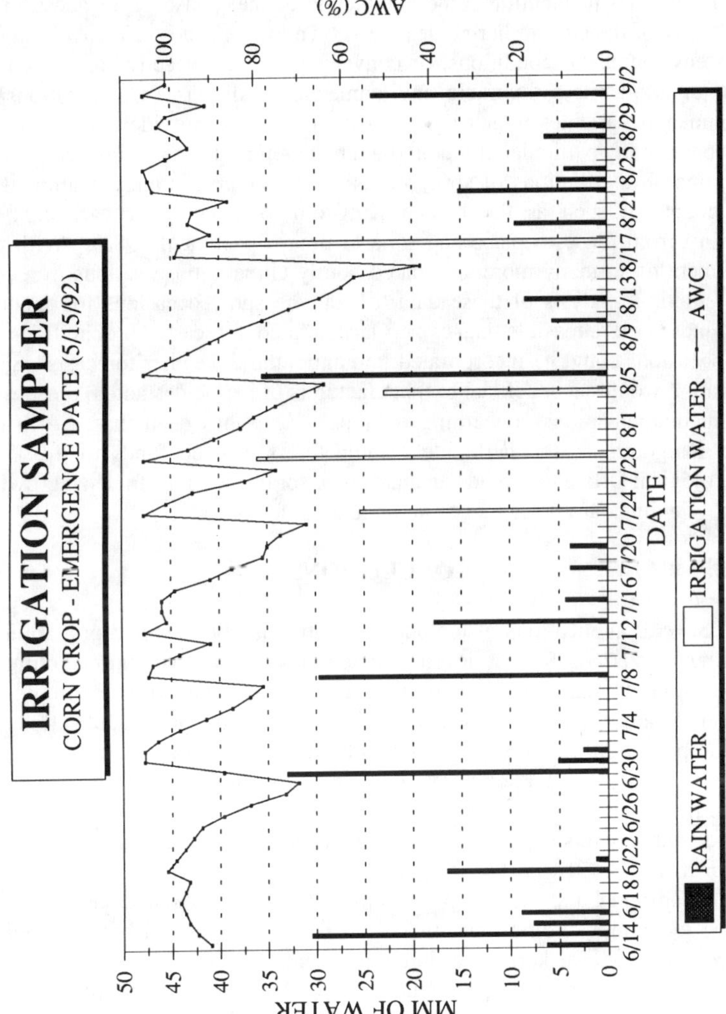

Fig. 36–8. Monitored rainfall and irrigation for a corn field in west-central MN 1992 (center pivot irrigation). AWC refers to available water holding capacity of soil root zone.

Forecasting for Herbicide Application

Regarding crop spraying, one of the true tests for a meteorologist is to be able to provide a producer or custom applicator with a precise forecast for wind and temperature conditions by h (Thompson, 1983). This requires a high degree of competence in the interpretation of upper air observations and wind profiler data, not to mention some knowledge of the landscape in question as well. Many of the current herbicides are known to work most effectively under certain environmental conditions, or conversely to be ineffective and to cause crop injury under less favorable environmental conditions. Combinations of environmental variables which meet labeled criteria or provide near optimum conditions for any particular compound might be referred to as 'spray occasions.' The temporal distributions of spray occasions are more routinely examined in some countries, such as the United Kingdom to assess their variability or suitability from place to place and time to time (Spackman, 1983). With the deployment of so many more automated hourly climatic observations in recent yrs a temporal analysis of these data to examine spray occasions for specific compounds is more feasible than ever. Once defined, forecasted probabilities for spray occasions could be incorporated into agricultural weather forecasts. Such information would be beneficial to manufacturers of herbicides and to those who make recommendations concerning their use. This has been the case in the United Kingdom, as the herbicide manufacturers provide funds to the U.K. Meteorological Office to provide an analysis of spray occasions for each growing season. Findings are discussed at an annual conference.

CONCLUSIONS

Several applications of temporal climatic distributions and probabilities have been discussed. Recent developments in comprehensive regional climatic data bases, communications technologies (access), and availability of improved National Weather Service forecasts and outlooks permit more applications of temporal probability to be tested in agricultural production systems.

Historically-based temporal climatic probabilities can be tailored for specific field operations if required environmental conditions are well defined or can be adequately modeled. There are existing climatic data bases which can be used for this.

Conditional and forecast probabilities can be implemented and tested for strategic planning of some field operations if their output can be delivered to producers (decision makers) in a timely manner.

Some caution is advised, however. Areal limitations of temporal probabilities need to be tested and defined as well. How far can maximum or minimum temperature be extrapolated? How about precipitation, solar radiation or wind? What are the limits of areal extrapolation and how are they affected by landscape? In addition, the value of forecast probabilities for specific decisions can only be tested when producers (decision makers) believe the forecasts and weigh the probabilities against the consequences of their actions (or inactions). Despite seasonal bias and other drawbacks, National Weather

Service products have varying degrees of skill. The skill level of each product needs to be evaluated for specific regions and operations.

ACKNOWLEDGMENTS

Much of the material presented and discussed was funded in part by the Northern Cornbelt Sand Plains Management Systems Evaluation Area (MSEA), a program of the Midwest Initiative for Water Quality through the USDA Special Water Quality Grant. This is a multiagency project involving the USDA-CSRS, USDA-ARS, USGS, EPA, USDA-ES and USDA-SCS. The able assistance of Anne Barsch, civil engineering student at the University of Minnesota and Dennis Fuchs, soil scientist at the Southwest Experiment Station, Lamberton, MN is gratefully acknowledged for the field working d analyses. Jerry Wright, irrigation engineer with the Minnesota Extension Service provided examples of irrigation system monitoring. Dr. Jeff Gunsolus, weed scientist with the University of Minnesota Department of Agronomy and Plant Genetics supplied valuable information about weed control measures and alternatives.

REFERENCES

Acharya, B.P., J.C. Hayes, and L.C. Brown. 1983. Available working days in the mid-south. Proc. ASAE meeting, Paper 83–4529, Chicago, IL. 26pp.

Baker, D.G., W.W. Nelson, and E.L. Kuehnast. 1979. Climate of Minnesota. The hydrologic cycle and soil water. Tech. Bull. 322. Agri. Exp. Stn., Univ. of Minnesota.

Baker, D.G., and R.H. Skaggs. 1984. The distance factor in the relationship between solar radiation and sunshine. J. Climatol. 4:123–132.

Baker, D.G., M.W. Seeley, E.L. Kuehnast, G.J. Spoden, J.A. Zandlo, and D.L. Ruschy. 1984. Crop calendars and agroclimatic data for Minnesota crop reporting districts: southwest district. MN Univ. Ext. Serv. AG-BU-2240.

Bergsrud, F., J. Wright, H. Werner, and G. Spoden. 1982. Irrigation systems design capacities for west-central Minnesota as related to the available water holding capacity and irrigation management. Proc. ASAE Paper 82–101, 14pp. St. Joseph, MI.

Changnon, S.A. 1991. Applied climatology: Atmospheric sciences biggest success story faces major new challenges. *In* Applied Climatology Research: Misc. papers, SWS Misc. Publ. 134:60–62. Illinois State Water Survey, Champaign, IL.

Hubbard, K.G. 1994. Spatial variability of daily weather variables in the high plains of the USA. Agric. and Forest Meteor. 68(1):29–42.

Jensen, M.E. 1973. Consumptive use of water and irrigation water requirements. Amer. Soc. Civil Eng., 215pp.

Kunkel, K.E. 1990. Operational soil moisture estimation for the midwestern United States. J. Appl. Meteor. 29(11):1158–1166.

Kunkel, K.E. 1994. Special soil moisture update and outlook for winter and spring 1994 conditions. News release, Feb. 1, 1994, from Midwest Climate Center, Champaign, IL, 6pp.

Meyer, S.J., and K.G. Hubbard. 1992. Nonfederal automated weather stations networks in the United States and Canada: A preliminary survey. Bull. Amer. Meteor. Soc. 73:449–457.

Richter, S. 1994. Rain or shine? Cooperative Partners 6(2):2–5, Cenex/Land O'Lakes, Inc.

Robert, P. 1994. The Minnesota precision farming initiative. Proc. Second International Conference Site-Specific Management of Agricultural Systems, Mar. 27–30, 1994, Bloomington, MN. ASA, CSSA, SSSA, and ASAE.

Rosenberg, S.E. 1982. Prediction of suitable days for field work. Proc. ASAE Meeting. Paper 82–1032, 17pp. Madison, WI.

Seeley, M.W. 1994. The future of serving agriculture with weather/climate information and forecasting: some indications and observations. Agric. and Forest Meteor. 68(5):322–335.

Seeley, M.W. and J. Zandlo. 1992. Some multidisciplinary applications of Minnesota's centennial climatic data base. *In* Abst. Minnesota Water '92, University of Minnesota Water Resources Research Center, Minnesota Environmental Quality Board and USGS.

Spackman, E.A. 1983. Weather and the variation in numbers of spraying-occasions. Aspects of App. Biology 4:317–327.

Stegman, E.C., A. Bauer, J.C. Zubriski, and J. Bauder. 1977. Crop curves for water balance irrigation scheduling. North Dakota State University Agric. Exp. Sta. Tech. Bull. 66, 11pp.

Thompson, N. 1983. Meteorology and crop spraying. Meteorology Magazine 112:249–259. Royal Meteorological Society.

United Kingdom Meteorological Office, 1989. Climatological data for Agricultural Land Classification. Crown Publishing, 119pp.

Wright, J. 1986. Irrigation scheduling checkbook method. Minnesota Extension Service, Univ. of Minnesota. AG-FO-1322. 12pp.

37 Grid Soil Sampling in Ohio

Walter Schmidt
Greg La Barge

Ohio State University Extension
Findlay, Ohio

A grid soil sampling project was initiated in the eight western counties of northern Ohio to assess variability in soil fertility levels during 1992 and 1993. Farmers and local agri-business were interested in site specific management but unsure of the actual variability in soil fertility and whether this practice is economically or environmentally justified in Ohio.

The soils of northwest Ohio resulted from glacial till and deposits in glacial lakes. Soil texture classes range from loamy sand to silty clay.

MATERIALS AND METHODS

Point samples were taken on a 360 foot by 360 foot grid resulting in 3 acres represented per sample. Five random 8 inch cores were taken at each sample point and mixed. No-till fields in the study had a second 2 inch sample collected to identify surface pH levels. Samples were processed at the Research, Extension and Analytical Laboratory in Wooster for soil pH, buffer pH, phosphorous, potassium, magnesium and calcium. Nutrient maps were then generated. Low-technology methods consisting of measuring wheels, ground driven units or radar units were used to establish grids. GPS and other advanced positioning technologies were not used in these initial investigative stages.

Maps were generated using the following parameters:

pH

High	> 7.0
Medium	5.9 to 7.0
Low	< 5.9

Phosphorous

High	> 40 pounds/acre P_2O_5
Medium	21–40 pounds/acre P_2O_5
Low	< 21 pounds/acre P_2O_5

Potassium

High	> 350 pounds/acre K_2O
Medium	301–350 pounds/acre K_2O
Low	250–300 pounds/acre K_2O
Very Low	< 250 pounds/acre K_2O

RESULTS

Over 3000 acres representing 64 fields have been tested in the eight county area. Results are shown in Fig. 37–1 through Fig. 37–3 for the zero to eight inch cores for 34 fields that have been summarized to date. Current findings do not show a correlation between soil type and soil test value. Both pH and potassium values have been highly variable. Overall field phosphorous levels have been above the yield response range of 40 pounds per acre recommended for corn/soybean production in the state of Ohio.

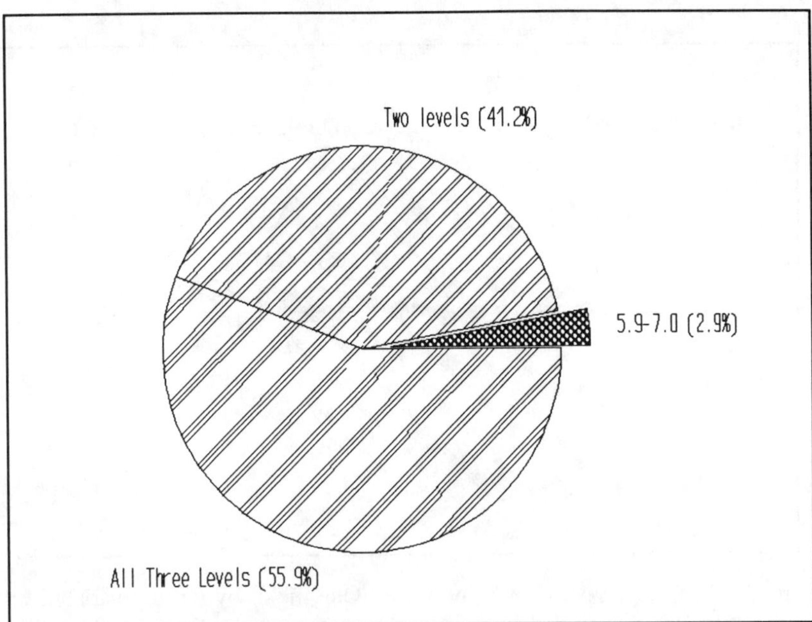

Fig 37–1. Categorization of 34 northwest Ohio fields by fertility variability at grid points. Fields were placed into categories with all grid points falling within one pH range, or with grid points falling into two or three different pH level ranges. Sample depth zero to eight inches.

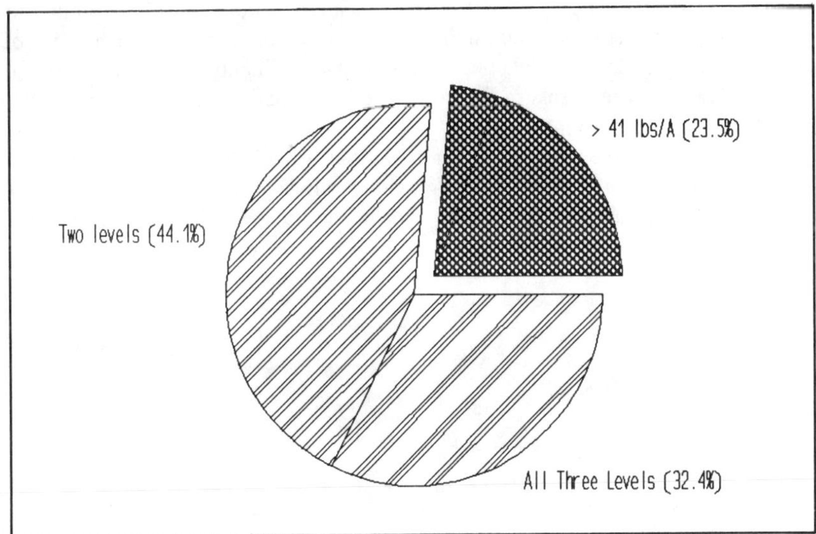

Fig. 37–2. Categorization of 34 northwest Ohio fields by fertility variability at grid points. Fields were placed into categories with all grid points falling within one P range, or with grid points falling into two or three different P level ranges. Sample depth zero to eight inches.

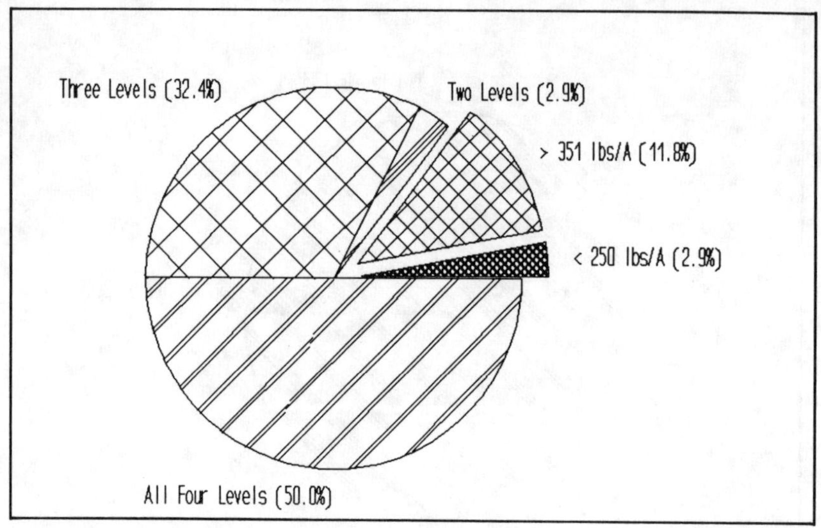

Fig. 37–3. Categorization of 34 northwest Ohio fields by fertility variability at grid points. Fields were placed into categories with all grid points falling within one K range, or with grid points falling into two, three or four different K level ranges. Sample depth zero to eight inches.

Fertilizer dealers and farmers are using results to explain production problems in isolated areas and herbicide carryover damage associated with pH sensitive chemicals. The data has also being used to identify and correct low pH areas. The major nutrients of P and K have been applied site specifically where needed based on nutrient maps.

Generally the practice has not resulted in large fertilizer input savings. Fertilizer usage is similar or < previous whole field treatments however applications are targeted to locations where the greatest crop benefit would be expected.

38 Sampling For Site-Specific Application

David W. Franzen
Ted R. Peck

Agronomy Department
University of Illinois
Urbana, Illinois

The number of soil samples needed to represent the variability of a field has been a matter for discussion since at least the 1920's. The nearly simultaneous development of the high speed computer, global positioning technology and variable rate fertilization equipment has spurred interest in variable rate fertilization. Fertilizer rate change boundaries need to reflect much of the variability in the field. Some commercial variable rate fertilization programs use sampling densities of between 11 and 16 samples per 16.2 ha. to help define these boundaries. Most programs, however, have relied on assumptions rather than knowledge of background sample support to derive their sampling density. The objective of this study was to examine the background variability in a 256 sample per 16.2 ha. grid in two Illinois fields and use this information to compare less dense grid fertility features with the original features to determine what sampling density would adequately describe the variability in these fields. The original 25 meter grid was compared to a 100 meter grid and a 66 meter grid. The 100 meter grid did not identify many important features of pH, P1 and available K maps as described by the 25 meter grid. The 66 meter grid, however, best identified features and approximate fertility boundaries of the 25 meter grid. The 66 meter grid could be adopted without a great cost in economics, while improving the probability for returns by better identifying areas which would benefit from variable rate application.

When the professors within the Land-Grant colleges first met to discuss their agricultural schools' objectives in 1877, the very first experimental subject proposed was the variability in crop yields due to soil heterogeneity (Hatch, 1967). Farmers have acknowledged that different parts of fields yield differently. Some of this variability can be attributed to soil type (Beckett & Webster, 1971), but some is the result of unseen differences in soil fertility levels (Reed & Rigney, 1947; Petersen & Calvin, 1986).

Researchers recognized long ago that often extensive sampling of a field could form the basis for spot fertilizer application, resulting in a savings of money and labor (Linsley & Bauer, 1929). As fertilizers became less expensive money and labor (Linsley & Bauer, 1929). As fertilizers became less expensive and easier to apply, it was simpler to treat entire fields than complicate the process with spot application. Soil sampling patterns and sampling point density became based again on finding a central tendency (Peck, 1990).

Economics and environmental forces are again making localized fertilizer applications attractive. Application technology is now in place which can vary the rate of fertilizer automatically as the applicator moves across a field, based on a digitized map derived from a soil sampling pattern (Luellen, 1985). Site specific farming techniques will soon allow certain organic matter related variables such as seeding and herbicide rates to change based on a sensor. No sensors, however, are currently available to meaningfully change the rate of lime, phosphate or potassium in the field. Determining these soil properties rely mostly on a soil test.

Soil sampling patterns currently used to guide variable rate machines are based on many assumptions, but little scientific basis. Some variable rate sampling uses random sampling within soil types as its basis. Others use a systematic grid consisting of one sample for each 1 to 2 hectare area (Reichenberger, 1992; Buchholtz, 1992).

Combine monitors are now being used to map yield variability across a field (Berling, 1992; McMullin 1992). If fertilizer recommendation mapping does not closely approximate the yield variability, this potentially valuable tool will lose its effectiveness. Every effort should therefore be made to develop a sampling pattern which will allow fertilizer recommendation maps to show a true picture of the variability of plant nutrients (Peck & Melsted, 1965).

The objectives of this study were to:

1. Characterize the spatial variability of soil pH, phosphorus, and potassium in two 16.2 hectare fields.
2. Determine how many samples and what patterns are necessary to identify important field soil test features, which will aid in meanfully directing a variable rate fertilizer application.

MATERIALS AND METHODS

Two fields were examined in this study. The first is a square 16.2 ha, 40 acre field located southwest of Mansfield, Il. The field location is the SE 1/4 of the SE 1/4 of section 9, Blue Ridge township, Town 20N, Range 6E, Piatt County, Illinois (Fig. 38–1). The field was selected by Dr. J. C. Laverty from the University of Illinois to begin evaluating a systematic sampling plan in 1961 (Peck & Melsted, 1965). Soil samples were taken in this field in a 25 meter grid. The pattern of sampling was a 16 by 16 grid for a total of 256 samples. The sampling was repeated in 1982, and each year from 1986 through 1992. The history of the sampling support at Mansfield is therefore well established. Each soil sample consisted of five individual soil cores taken with a commercial 2.5 cm diameter soil tube, 17.5 cm deep. The five individual cores were taken in a loose grid pattern with one core centered in the middle of the plot and the

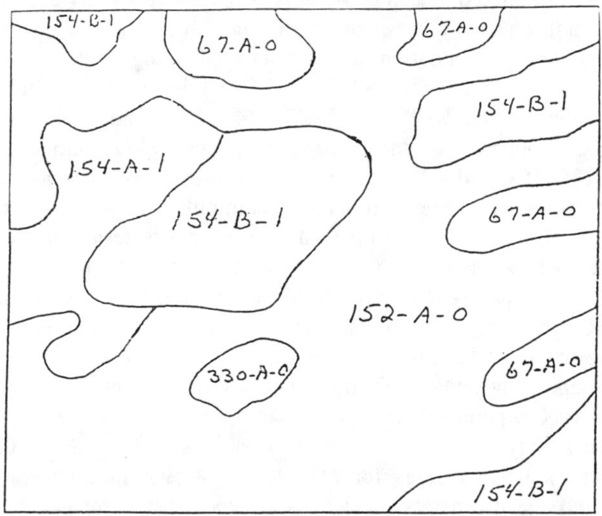

Fig. 38–1. Soil type map of Mansfield tract.

other four cores taken roughly from the corners of a 5 meter square surrounding the central core. The five cores were placed in a common bag, dried at 36° C, pulverized (Eick & Gelderman, 1988) and analyzed for soil pH (1:1 soil paste) (Eckert, 1988), Bray P1 and available potassium. Bray P1 level was determined as outlined by Knudsen and Beegle (1988). Available potassium was determined using 1N ammonium acetate, pH 7.0, as the extracting solution with a 5 minute shaking time. The potassium extract was then analyzed using a flame photometer.

The second field in the study was a 16.2 ha., 40 acre tract northwest of Thomasboro, Illinois, in Champaign County. The field location is the SE 1/4 of the SE 1/4 of section 20 in Rantoul township, Town 21N, Range 9E. The soil type map developed as a result of our sampling is shown in Fig. 38–2. The field history before the 1940's is vague, however, from WWII until 1982 the field received no agricultural limestone or fertilizer up to 1982. In 1982, the field was turned over to the University of Illinois, and the first soil sampling was performed in the same 16X16 sampling grid as at Mansfield. The 25 meter grid was also used from 1983 through 1992 to direct sampling.

In both fields, from the original sample point values it was possible to determine which of the points also would represent a 4X4 grid with approximately 1 sample per hectare. Using these points, grid patterns were mapped and compared to the original map of the 16X16 grid using both conventional statistics and geostatistics.

Both fields were also sampled an extra time on the same day as the primary 16X16 sampling using a 6X6 sampling. The new grid placed samples 66 meters apart, or one sample per 0.45 ha. The 4X4 grid and the 6X6 sampling grid were evaluated by comparing with the 16X16 grid using linear regression.

The relevance of the regression was examined by dividing the soil test levels into ranges which correspond to different lime, phosphate and potassium fertilizer rate recommendation levels. The pH ranges were >7.0, 6.5–7.0, 6.2–6.4, 5.9–6.1, 5.6–5.8, 5.3–5.5 and <5.3. P1 levels were grouped into the following ranges: >90, 56–90, 45–55, 34–44, 23–33 and <23 kg ha^{-1}. The K levels were grouped into the following ranges: >672, 336–671, 302–335, 269–301, 235–268, and <235 kg ha^{-1}. These ranges are intervals under which different fertilizer recommendations are commonly made in Illinois. The 4X4 map was superimposed on the original 16X16 grid pattern and each of the 16 points which surrounded the 4X4 sampled point would assume the value of the point. When comparing the 6X6 grid to the original 16X16 grid, each point in the 16X16 grid would also assume the value of the 6X6 grid square superimposed on it. The values predicted by the less dense samplings were evaluated depending on their ability to fall within the same soil test level range as the real 16X16 point value.

The soil type maps were also superimposed over the 16X16 sampled points. The soil test values for pH, P1 and K within each soil type were averaged and the estimates based on soil type were compared to the values represented by the 16X16 grid.

Geostatistical analysis and mapping of each sampling was performed using the Surfer grid program in combination with a GS+ program. Classical statistics were obtained with Lotus 1-2-3 and Systat. The purpose of the geostatistical analysis in this part of the study was to help identify areas in each field with common characteristics with some statistical certainty and view the characteristics on a map of the fields. The features identified by the less dense mapping were compared to the complexity of features in the dense 16X16 grid maps.

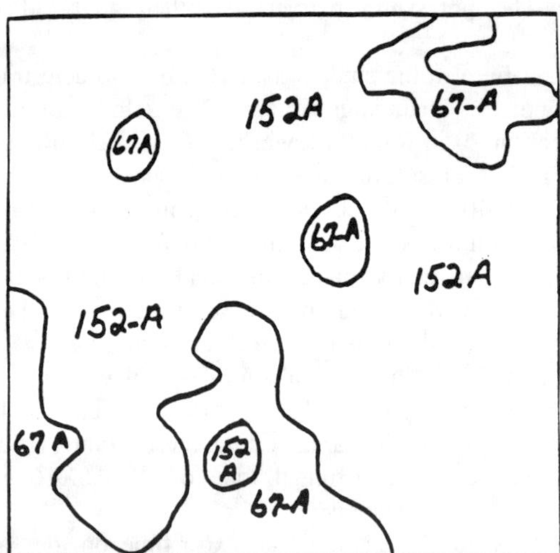

Fig. 38–2. Soil type map of Thomesboro tract.

Table 38–1. 1992 r^2 values for Mansfield and Thomasboro 4X4 and 6X6 sampling grids compared to the 16X16 grid.

Sampling Grid	Mansfield	Thomasboro
pH 4X4	0.292	0.208
pH 6X6	0.484	0.261
P1 4X4	0.025	0.222
P1 6X6	0.174	0.283
K 4X4	0.029	0.166
K 6X6	0.379	0.345

LSD at the 5% levels between grid patterns within location is 0.156.

RESULTS AND DISCUSSION

In 1992, both fields were sampled not only in the 16X16 pattern, but with a separate sampling in a 6X6 grid as previously described. Table 38–1 shows the correlation between the 256 estimates from the 4X4 and 6X6 grids compared to the 16X16 sampling.

The 6X6 pattern was superior to the 4X4 pattern in 1992 at Mansfield when comparing pH and K levels. At Thomasboro, the K levels were superior using the 6X6 grid. The 4X4 pattern was not superior to the 6X6 pattern at either location or with any soil test.

Because the exact correlation of each sample estimate made using a less dense grid with each actual sample value is not necessarily important because of the fertilization of ranges of values, an evaluation was made to determine if predicted values fell into the same range as the actual values. Table 38–2 shows the results of the evaluation which includes figures from soil type sampling estimates.

There were no significant differences between sampling grids either within the same range or within the same range or within one range of values.

The final method for determining the value of sampling grids was to compare fertility level features identified with the 4X4 and 6X6 grid maps to those shown with the 16X16 grid maps. The 16X16 Mansfield 1992 pH map in Fig. 38–3 contains three areas of pH higher than 7.0. The Mansfield 1992 pH 4X4 map in Fig. 38–5 shows the area in the southeast and the high pH area in the northeast, but neglects the high pH area in the northeast. The 6X6 Mansfield 1992 pH map in Fig. 38–4 shows all three high pH areas, closely approximating the boundaries shown by the 16X16 grid. The 6X6 mapping reveals more pH level features than the 4X4 grid.

Table 38–2. Evaluation of soil pH, P1 and K predictions from 4X4, 6X6 and soil type mapping compared to the 16X16 grid fertilizer recommendation ranges.

1992 Pattern	% Within Same Range	% Within ± 1 Range
Mansfield pH 4X4	54.9	95.7
Mansfield pH 6X6	55.5	96.1
Mansfield P1 4X4	43.8	89.5
Mansfield P1 6X6	50.0	87.1
Mansfield K 4X4	34.4	88.7
Mansfield K 6X6	41.4	80.5
Thomasboro pH 4X4	30.8	60.9
Thomasboro pH 6X6	26.7	61.3
Thomasboro P1 4X4	38.7	67.2
Thomasboro P1 6X6	36.0	77.5
Thomasboro K 4X4	45.9	81.8
Thomasboro K 6X6	60.5	88.5
Thomasboro soil type P1	29.6	72.3
Thomasboro soil type K	54.9	83.4

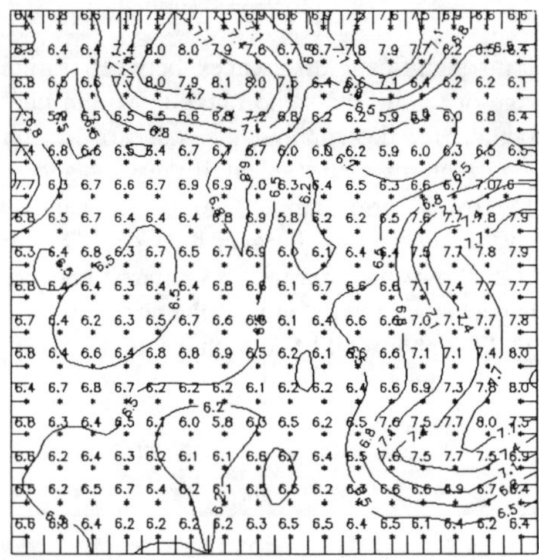

Fig. 38–3. Mansfield pH 1992. Kriging based on a linear variogram model with r^2=0.777. Contour lines 0.3 pH units apart.

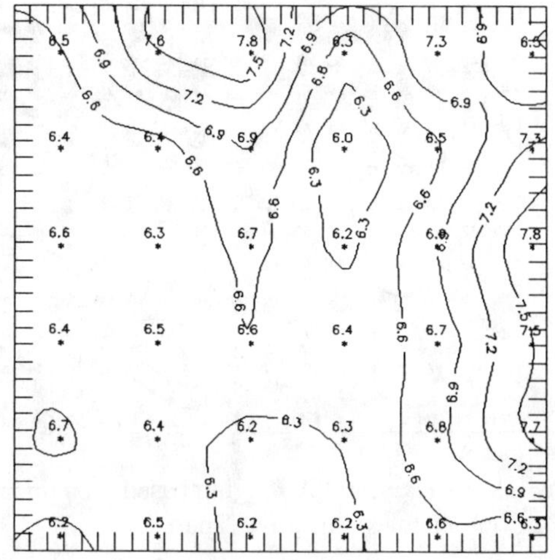

Fig. 38–4. Mansfield pH 1992 6X6, 66 meter grid.

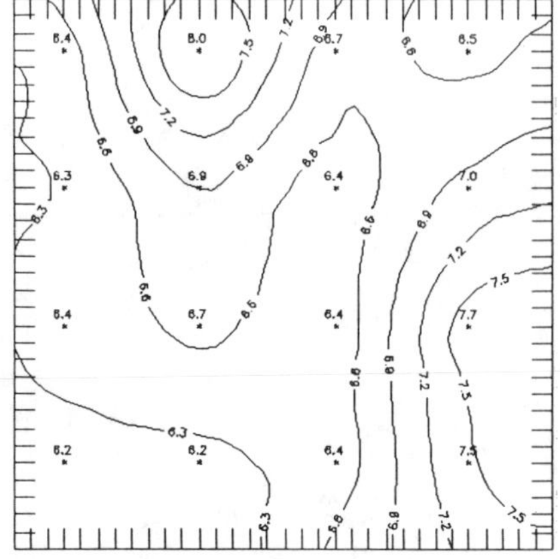

Fig. 38–5. Mansfield pH 1992 4X4, 100 meter grid.

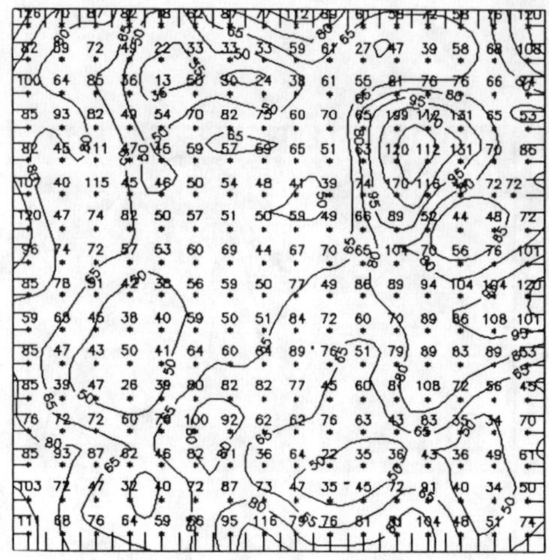

Fig. 38–6. Mansfield P1 levels 1992. Kriging based on a linear variogram model with r^2=0.633. Contour lines 15 kg ha^{-1} apart.

The 1992 Mansfield P1 levels in the 16X16 grid are shown in Fig. 38–6. There are three main features in this map. There is an area of high P in the northeast, and relative low P areas in the northwest and southeast. The Mansfield 4X4 P1 level map in Fig. 38–8 does not show the area of high P in the northeast, but does show the other two features. The 6X6 map in Fig. 38–7 shows all three main features.

The 1992 Mansfield K level map in Fig. 38–9 shows the 16X16 grid. The main feature of the 16X16 map in the high K levels in the old building lot area.

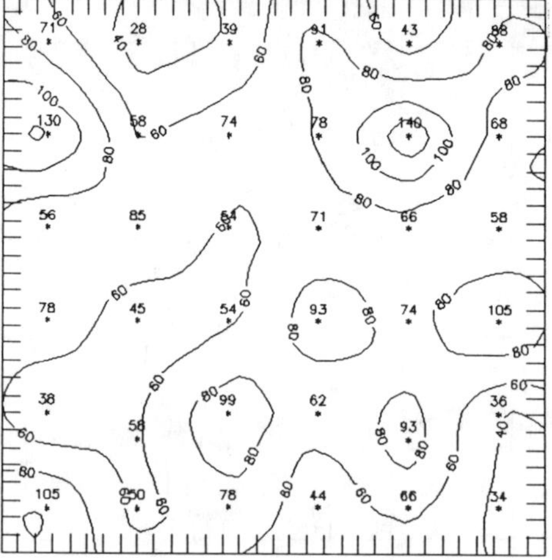

Fig. 38–7. Mansfield P1 level 6X6, 66 meter grid.

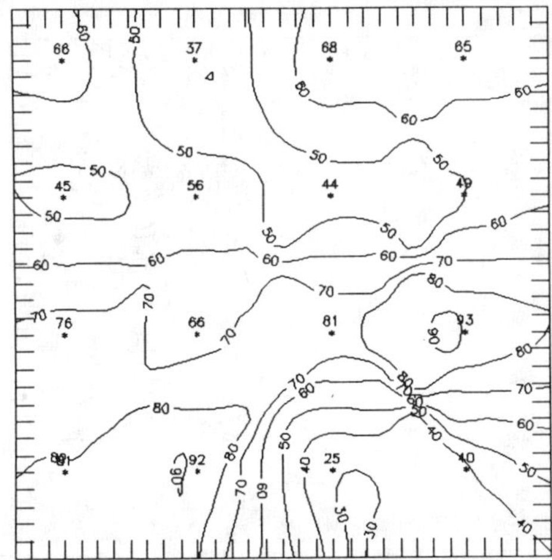

Fig. 38–8. Mansfield 1992 P1 level 4X4, 100 meter grid.

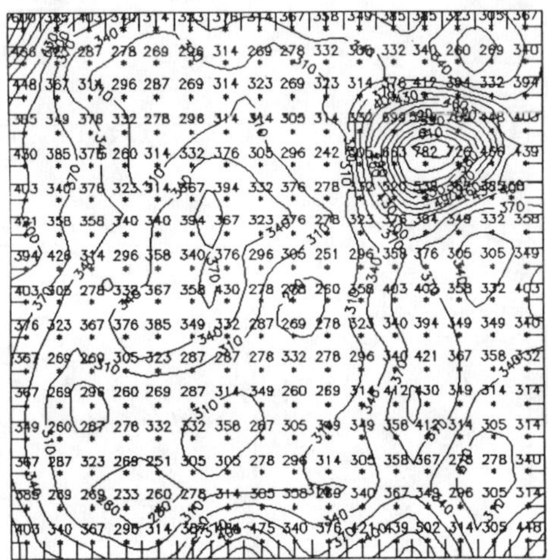

Fig. 38–9. Mansfield K levels 1992. Kriging based on a linear variogram with r^2=0.584. Contour lines 30 kg ha^{-1} apart.

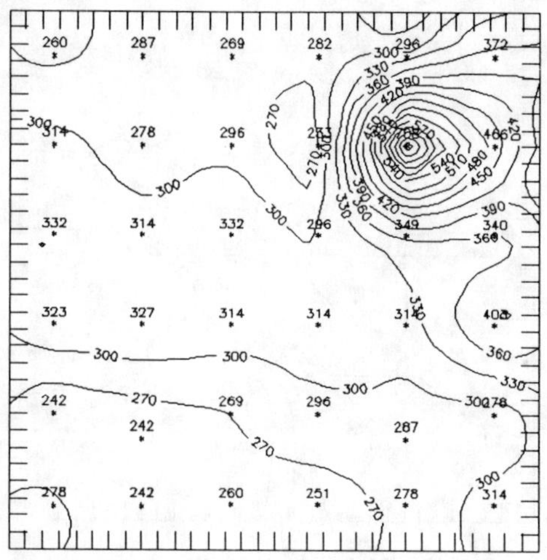

Fig. 38–10. Mansfield 1992 K levels 6X6, 66 meter grid.

The 1992 Mansfield 4X4 map in Fig. 38–11 does not show this area. The 6X6 K level map in Fig. 38–10 shows the high K region clearly and approximates the boundaries as well. The 6X6 map again represents soil test levels better than the 4X4 grid.

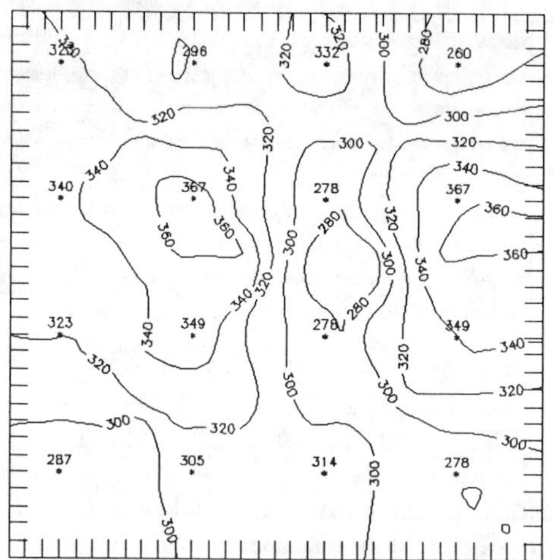

Fig. 38–11. Mansfield K levels 4X4, 100 meter grid.

The Thomasboro 1992 16X16 pH map in Fig. 38–12 shows a complex pattern of high and low pH areas. There are high pH regions in the south, central, northeast and southwest, and low pH areas in the east and northwest. The 4X4, 1992 Thomasboro pH mapping in Fig. 38–14 identifies only the high pH areas in the south and northeast, but shows both low pH areas. The 6X6 grid in Fig. 38–presents all six main features. The 6X6 Thomasboro pH level map represents the pH features in the field better than the 4X4 map.

Figure 38–15 shows the 1992 Thomasboro P1 16X16 grid map. This map shows three low P features and three high P areas. The 4X4 Thomasboro P1 grid in Fig. 38–17 shows an area of low P in the south and central regions, but misses the low P area in the northwest. The 4X4 map also overestimates the area of high P in the west. It closely approximates the high P areas in the northeast and southeast. Figure 38–16 shows the 6X6 Thomasboro P1 map. This map identifies all six features of the 16 X16 map. The 6X6 grid better represents the Thomasboro P1 features than the 4X4 grid.

The 1992 Thomasboro 16X16 K levels are shown in Fig. 38–18. The mapping shows two high K areas in the southeast and northeast. Figure 38–20 shows the high K levels in the southeast and northeast, but this map also depicts a large relatively low area of K not present in the 16X16 map. The 6X6 Thomasboro K grid mapped in Fig. 38–19 shows the high K levels in the southeast and northeast, while approximating the boundaries of the areas better than the 4X4 mapping.

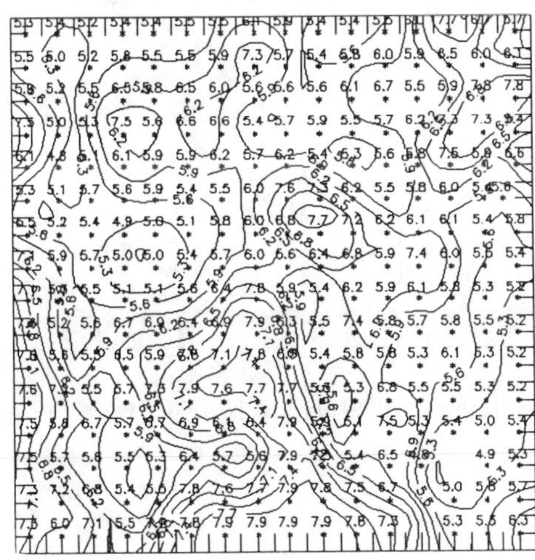

Fig. 38–12. Thomasboro pH levels 1992. Kriging based on a linear variogram with r^2=0.739. Contour lines 0.3 pH units apart.

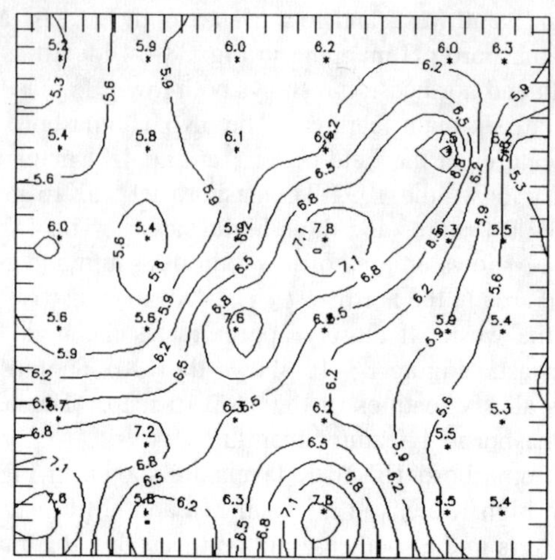

Fig. 38–13. Thomasboro 1992 pH levels 6X6, 66 meter grid.

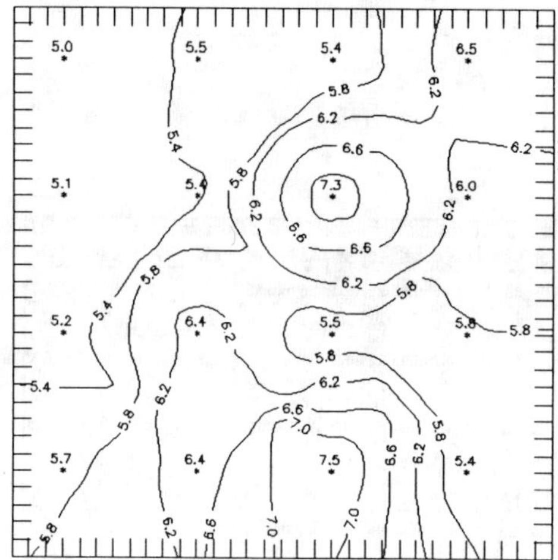

Fig. 38–14. Thomasboro 1992 pH levels 4X4, 100 meter grid.

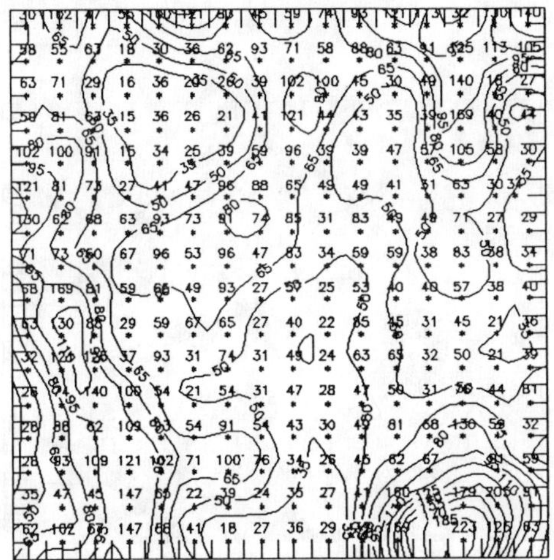

Fig. 38–15. Thomasboro P1 levels 1992. Kriging based on a linear variogram with r^2=0.824. Contour lines 15 kg ha^{-1} apart.

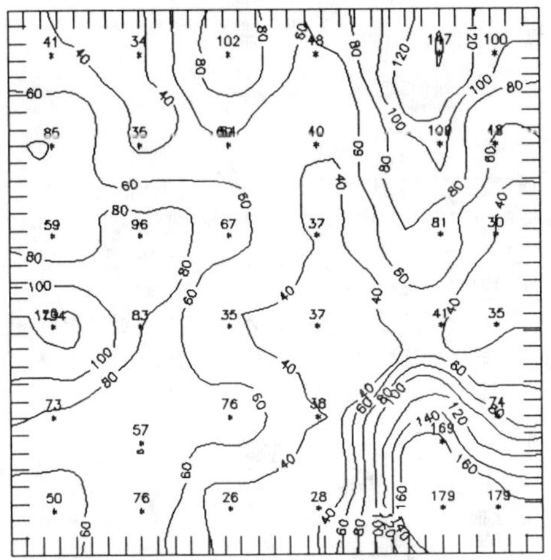

Fig. 38–16. Thomasboro P1 levels 6X6, 66 meter grid.

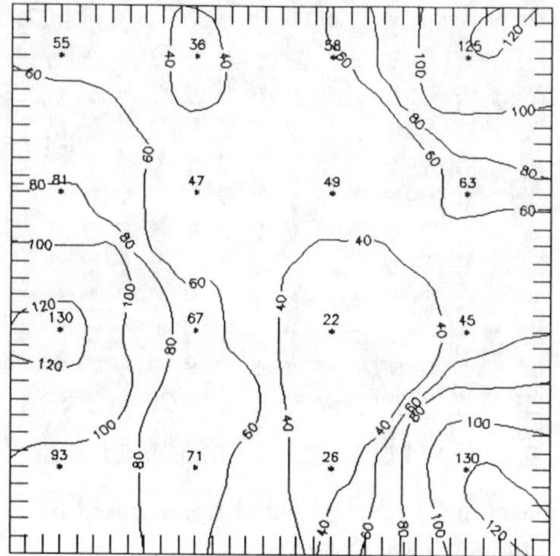

Fig. 38–17. Thomasboro P1 levels 4X4, 100 meter grid.

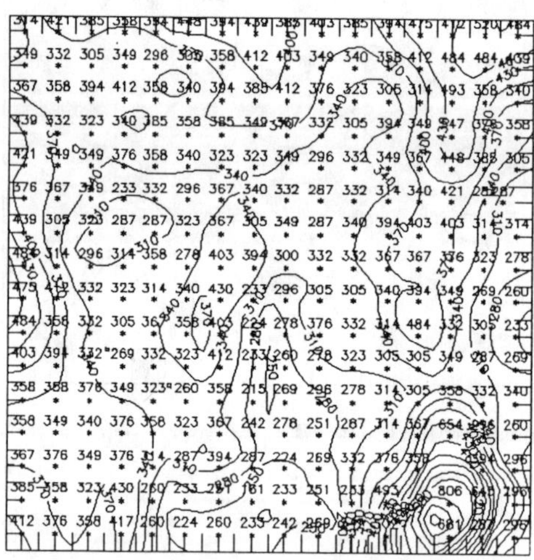

Fig. 38–18. Thomasboro K levels 1992. Kriging based on a linear variogram with r^2=0.905. Contour lines 30 kg ha^{-1} apart.

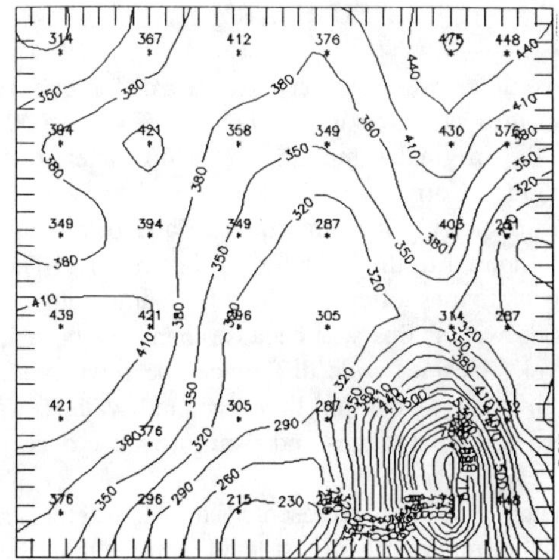

Fig. 38–19. Thomasboro K levels 6X6, 66 meter grid.

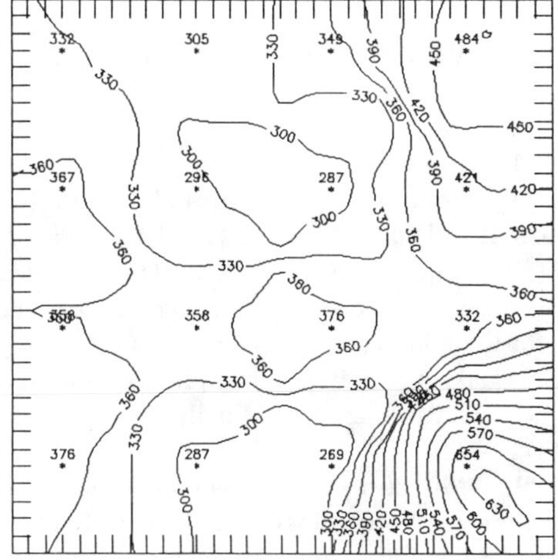

Fig. 38–20. Thomasboro 1992 K levels 4X4, 100 meter grid.

Overall, the 6X6 mapping represented the original 16X16 grid best by identifying important soil test level features and approximating the boundaries of those features. The 4X4 mapping often showed little resemblance to the 16 X16 grid map.

SUMMARY

Two Illinois fields were sampled by two different sampling schemes. The dense 25 meter grid sampling was compared to both a 100 meter sampling grid derived from the original sampling, and a 66 meter grid from a second independent sampling. Correlations between levels predicted from mapping prepared from the 66 meter grid were equal to and often superior to the 100 meter grid when compared to the 25 meter grid sampling. The 100 meter grid and 66 meter grid estimates often fell into the same range of fertilizer rate determined ranges, however, this was because many of the pH, P1 and K tests were at a high level. Despite wide differences between predicted and actual levels, especially in the 4X4 grid, the predicted values were still in the maintenance or zero fertilizer ranges and therefore were classified in the same fertilizable range.

The 66 meter grid best represented the 25 meter grid mapping by displaying more of the important soil test level features than the 100 meter grid in the two farms. The boundaries shown by the 66 meter grid also approximated those of the 25 meter sampling. The 66 meter grid offers improvements in the soil test level prediction over the 100 meter or greater grids now being used to direct variable rate fertilization. The 66 meter grid could be adopted without a great sacrifice in economics, while improving the probability for returns by better identifying and defining areas in fields which would benefit from variable rate application.

REFERENCES

Beckett, P.H.T., and R. Webster. 1971. Soil variability: A review. Soils and Fertilizers. 34:1–11.

Berling, J. 1992. New products make precision farming simple. Farm Industry News. 25(8):4–6.

Biggar, J.W., and D.R. Nielsen. 1976. Spatial variability of the leaching characteristics of a field soil. Water Res. Res. 12(1): 78–84.

Brown, J.R., and D. Warncke. 1988. Recommended cation tests and measures of cation exchange capacity. *In* Recommended chemical soil test procedures for the north central region. No. Central Reg. Publ. No. 221 (Revised). N. Dak. Agric. Exp. Stn. Bull. 499 (Revised).

Brunoehler, R. 1992. Grid fertilizing rig finally hits pay dirt. Farm Industry News. 25(4):22–23.

Buchholz, D.D. 1992. Missouri grid soil sampling project. *In* 1992 Illinois Fertilizer Conference Proceedings. R. G. Hoeft (ed.) p.1–7.

Burrough, P.A. 1991. Sampling designs for quantifying map unit composition. *In* Spatial Variabilities of Soils and Landforms. SSSA Special Publ. No.

28. SSSA, Madison, WI.

Eckert, D.J. 1988. Recommended pH and lime requirement tests. *In* Recommended chemical soil test procedures for the North Central Region. No. Central Reg. Publ. No. 221 (Revised). N. Dak. Agric. Exp. Stn. Bull. No. 499 (Revised).

Eick, K., and R.H. Gelderman. 1988. Soil sample preparation. *In* Recommended chemical soil test procedures for the North Central Region. No. Central Reg. Publ. No. 221 (Revised). N. Dak. Agric. Exp. Stn. Bull. No. 499 (Revised).

Hatch, R.A. 1967. An early view of the Land-Grant Colleges. Convention of Friends of Agricultural Education in 1871. University of Illinois Press, Urbana, Il.

Knudsen and Beegle. 1988. Recommended phosphorus tests. *In* Recommended chemical soil test procedures for the North Central Region. No. Central Reg. Publ. 221 (Revised). N. Dak. Agric. Ext. Stn. Bull. 499 (Revised).

Leo, M.W.M. 1963. Heterogeneity of soil of agricultural land in relation to soil sampling. Agric. Food Chem. 11(5): 432–435.

Linsley, C.M. and F.C. Bauer. 1929. Test your soil for acidity. Univ. of IL, Col. of Agric. and Agric. Exp. Stn. Circ. 346.

Luellen, W.R. 1985. Fined-tuned fertility: tomorrow's technology here today. Crops and Soils. 38(2):18–22.

McMullin, E. 1992. Now you can know exact yield on the go. Farm Industry News. 25(7):32–34.

Peck, T.R. 1990. Soil testing: past, present, and future. Commun. in Soil Sci. Plant Anal. 21(13–16), 1165–1186.

Peck, T.R. 1991. Variability of soil pH, phosphorus, and potassium levels. p. 107–112. *In* 1991 Illinois Fertilizer Conference Proc. R. G. Hoeft (ed.)

Peck, T.R., and S.W. Melsted. 1965. Field sampling for soil testing. p. 25–35. *In* Soil Testing and Plant Anal. 1st ed. Soil Sci. Soc. Amer.

Petersen, R.G., and L.D. Calvin. 1986. Sampling. p. 33–50. *In* Methods of soil analysis, Part I. Physical and Mineralogical Methods. Am. Soc. Agron.-Soil Sci. Soc. Am.

Reed, J.F., and J.A. Rigney. 1947. Soil sampling from fields of uniform and nonuniform appearance and soil types. J. Amer. Soc. Agron. p. 26–39.

Reichenberger, L. 1992b. Picture your soil's fertility. Farm Journal. Mid-February, p.12–14.

Webster, R. 1985. Quantitative spatial analysis of soil in the field. Adv. Soil 3:2–66.

39 Effects Of Fertilizer On Yield At Different Soil Landscape Positions[1]

S. C. Nolan
T. W. Goddard
D. J. Heaney
D. C. Penney
R. C. McKenzie

Alberta Agriculture
Food and Rural Development
Edmonton, Alberta, Canada

This preliminary field study is part of a larger study to assess the utility of Differential Global Positioning Systems (DGPS) and Geographic Information Systems (GIS) for individually managing soil landscapes. To characterize a test site, two parallel strips of i) uniform application of 61 kg ha^{-1} (54 lbs ac^{-1}) N and 30 kg ha^{-1} (27 lbs ac^{-1}) P and ii) no fertilizer were laid out on strongly rolling topography on a clay loam soil near Drumheller, Alberta, Canada. Yields of spring wheat were measured from segments within the fertilized and unfertilized strips. These were located using DGPS and classified into summit/shoulder (S), back (B) or foot (F) slope landscape position. The accuracy of the DGPS was within 70 cm. All yields ranked F > B > S. For the fertilized treatments, there were no significant differences (P=0.10) between yields at any slope position. For the unfertilized treatments, F significantly outyielded S by 34%. No differences were found between unfertilized yields on B or F slopes, though F yielded 23% more. The effects of fertilizer on productivity were estimated using the ratio of unfertilized to fertilized yield at corresponding landscape positions. Yield ratios were 31% greater from F than from S. There were no differences between yield ratios from B and S, though B was 11% higher. These results suggest that savings in fertilizer costs (and reduced offsite environmental impacts) could be made by reducing rates or changing to more appropriate blends on summit/shoulder and back slopes where marginal yield was limited.

---

[1]Canada/Alberta Environmentally Sustainable Agriculture Agreement (CAESA), Lethbridge, Alberta, Canada.

INTRODUCTION

Yield has been shown to vary over landscapes in the western Canadian Prairies. Verity and Anderson (1990), Malhi et al. (1993), Larney et al. (1991) and others document decreased yields in eroded landscapes. McKenzie et al. (1983) and others reported decreased yields in areas of high salinity. Elliot and De Jong (1992) found higher yields on foot slope and level complexes than on shoulder complexes. Pennock and Anderson (1992) found that the effect of landscape changes on yield were overwhelmed by climatic conditions present at the time of the study.

In the United States, uneven patterns in crop growth have been related to spatial variation in soil properties and decreased efficiency of uniformly applied fertilizer (Miller et al. 1988, Larson & Robert 1991). Ciha (1984) found that slope position and cultivar significantly influenced grain yield. Soil type had a significant influence on production of barley (Bennett et al. 1980), even when uniform production practices were used (Lee & Spillane, 1970).

Kachanoski et al. (1985) recommend that resources should be concentrated on the lower slope positions where yield potential is greatest. Most farm fields have several soil landscape units that could benefit from individual management decisions to achieve optimum productivity and sustainability. This preliminary field study is part of a larger study to assess the utility of Differential Global Positioning Systems (DGPS) and Geographic Information Systems (GIS) for individually managing soil landscapes. Our objectives were to characterize a test site at different slope positions in terms of i) yield variation, and ii) fertilizer effects, and iii) to conduct a preliminary assessment of the accuracy of DGPS for locating sample areas within large fields.

MATERIALS AND METHODS

The study site is located on strongly rolling topography near Drumheller, Alberta, Canada (Fig. 39–1). The soil is a Dark Brown Chernozemic clay loam developed from calcareous till. Average annual precipitation in the area is 323 mm, with an average of 178 mm of rainfall occurring during the growing season. The growing season of 1993 was unusually moist with 307 mm of rainfall. Spring wheat was direct seeded in May of 1993 using a one-pass Concord 3000 air delivery system on a 42 foot New Nobel 9000 Seedovator (shank type opener). The previous crop had been spring wheat.

Granular fertilizer (34–17–0) was applied at seeding into 2.4 m wide parallel transects of i) uniform application of 61 kg ha^{-1} (54 lbs ac^{-1}) N and 30 kg ha^{-1} (27 lbs ac^{-1}) P_2O_5 and ii) no fertilizer. A Wintersteiger Nursery Master Elite 2000 plot combine was used to measure yield from both the fertilized and unfertilized strips at different landscape positions along three transects. A total of 24 segments per treatment were measured. Segments were an average of 28 m in length (ranging from 15 to 47 m). The ends of each segment were positioned using DGPS. Each segment was classified into water and sediment shedding summit/shoulder slope (S); water and sediment transporting back slope (B); or water and sediment receiving foot slope (F) positions (Fig. 39–2), based

Fig. 39–1. Location of study site at approximately 52°N and 113°W near Drumheller, Alberta, Canada.

on an interpretation of the three general soil-landform regimes in the Prairies as described by Pennock et al. (1987).

An old fence line with soil accumulation from wind erosion resulted in unusually high yields. These 2 (of 24) samples were omitted from further analysis as they were not typical of the management of the rest of the field. The remaining data were analyzed by slope position, using the Fisher's Least Significant Difference (LSD) separation of means test for the normally distributed data and the Kruskal-Wallis test for the non-normally distributed data.

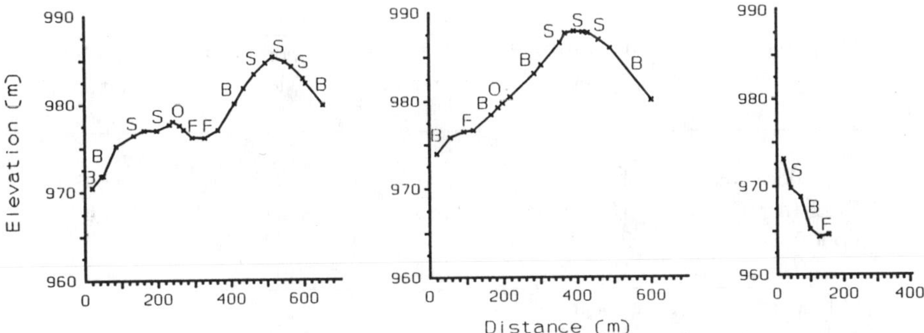

Fig. 39–2. Elevation and classified landscape position of summit/shoulder slope (S), back slope (B), foot slope (F), and old fenceline, omitted, (O) segments along three transects.

RESULTS AND DISCUSSION

The unfertilized yields were normally distributed, while the fertilized yields were not. Both fertilized and unfertilized yields ranked F > B > S (Table 39–1). For the fertilized treatments, F yielded 8% >S, however, there were no significant differences (P = 0.10) between yields at any slope position. The lack of significance could be due to the unusually wet growing conditions of 1993. In drier years greater differences may be apparent. For the unfertilized treatments, F significantly outyielded S by 34%. No differences were found between unfertilized yields on B or F slopes, though F yielded 23% more.

The effects of fertilizer on productivity were estimated using the ratio of unfertilized to fertilized yield at corresponding slope positions (Fig. 39–3). Increases in yield due to fertilizer additions were significantly (P=0.11) greater from F slopes; 22% greater than B and 31% greater than S. Though ratios from B were 11% higher, there were no differences between yield ratios from B and S. As 1993 was an unusually moist growing season, in more common drier growing conditions, even greater differences in response to fertilizer at varying landscape positions would be expected. These preliminary results suggest that savings in fertilizer costs could be realized by reducing rates or changing to more appropriate blends on back and summit/shoulder slopes where marginal yield was limited. Resources should instead be concentrated on the lower foot slope positions where yield potential was greatest.

Acceptable 3-D positioning was performed by DGPS to locate the transects and superimpose those positions on a digital elevation model. The accuracy of the DGPS was verified independently through a crossover point analysis and a comparison with an ambiguity resolution on-the-fly (OTF) solution (Gehue et al. 1994). Locational accuracy was within 0.15 m in the east-west direction, within 0.30 m in the north-south direction, and 0.70 m in elevation. DGPS shows promise as a valuable tool for landscape based research.

Table 39–1. Yield variation at different slope positions.

Slope Position	Unfertilized Yield (kg ha^{-1})		Fertilized Yield (kg ha^{-1})	
	Mean	Standard Error†	Mean	Standard Error
Shoulder (n=9)	1379.7 b‡	259.8	2963.1 a§	155.4
Back (n=9)	1602.3 ab	180.0	3077.7 a	155.6
Foot (n=4)	2086.0 a	237.0	3222.7 a	298.0
Total (n=22)	1599.2	141.4	3057.2	101.2

†Standard error of the mean.

‡Means within this column followed by the same letter are not significantly different at the 0.10 probability level using the Fishers Least Significant Difference test.

§Means within this column followed by the same letter are not significantly different at the 0.10 probability level using the Kruskal-Wallis test.

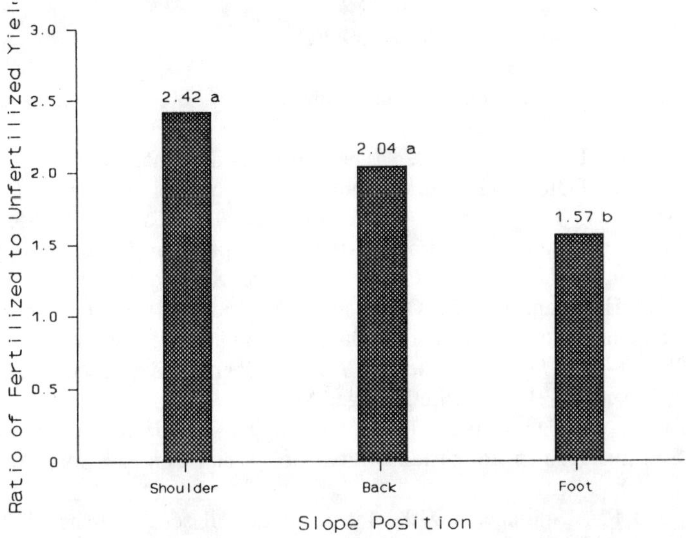

Fig. 39–3. Ratio of unfertilized to fertilized yield at corresponding slope positions.

SUMMARY

Increases in yield due to fertilizer additions ranked summit–shoulder > back > foot slope position. Yield responses were significantly greater on foot slopes than on back (22%) or summit/shoulder (31%) slope positions. As 1993 was an unusually moist growing season, in more common drier conditions, even greater differences in response to fertilizer at varying landscape positions would be expected. The DGPS was able to locate sample areas with sub-meter accuracy and shows promise as a valuable tool for landscape research. These preliminary results suggest that savings in fertilizer costs (and reduced offsite environmental impacts) could be realized by reducing rates or changing to more appropriate blends on back and summit/shoulder slopes where marginal yield was limited. Resources should instead be concentrated on the foot slope positions where yield potential was greatest.

ACKNOWLEDGEMENTS

Financial support from the Canada-Alberta Environmentally Sustainable Agriculture Agreement (CAESA) is gratefully acknowledged. The authors thank the other members of our multidisciplinary research team: Hazen Gehue, Research Assistant, University of Calgary; Gerald Lachapelle, Geomatics Engineering Professor, University of Calgary; Tim Martin, Computer Applications Specialist, University of Alberta; Germar Lohstraeter, Technologist, Alberta Agriculture, Food and Rural Development (AAFRD); and Karen Skarberg, Agronomist, AARFD.

REFERENCES

Ciha, A.J. 1984. Slope position and grain yield of soft white winter wheat. Agron. J. Vol. 76. 193–196.

Elliot, J A., and E. De Jong. 1992. Landscape-based variable rate fertilization. *In* Great Plains Soil Fertility Conference Proc. Vol. 4. Denver, CO.

Kachanoski R.G., R.P. Veroney, E. De Jong, and D.A. Rennie. 1985. The effect of variable and uniform N-fertilizer application rate on grain yield. *In* Proc. of the Soils and Crops Workshop. Saskatoon, Sask.

Larney, F.J., H.H. Janzen, B.M. Olson, and C.W. Lindwall. 1991. The impact of simulated erosion on soil productivity and methods for its amendment. P. 277–285. *In* Proc. of the 28th Annual Alberta Soil Science Workshop; February 19-21, 1991. Lethbridge, Alberta.

Larson, W.E., and P.C. Robert. 1991. Farming by soil. *In* R. Lal, et al. (ed.) Soil management for sustainability. Soil Water Conserv. Soc., Ankeny, IA.

Lee, J., and P.A. Spillane. 1970. Influence of soil series on the yield and quality of spring wheats. Irrig. J. Agric. Res. 9:239–250.

Malhi, S.S., R.C. Izaurralde, M. Nybourg, and E.D. Solberg. 1994. Influence of topsoil removal on soil fertility and barley growth. J. Soil and Water Cons. 49(1): 96–101.

McKenzie, R.C., C.H. Sprout, and N.F. Clark. 1983. The relationship of the yield of irrigated barley to soil salinity as measured by several methods. Can. J. Soil Sci. 63: 519–528.

Miller, M.P., M.J. Singer, and D.R. Nielsen. 1988. Spatial variability of wheat yield and soil properties on complex hills. Soil Sci. Soc. Am. J. 52:1133–1141.

Pennock, D.J., B.J. Zebarth, and E. De Jong. 1987. Landform classification and soil distribution in hummocky terrain, Saskatchewan, Canada. Geoderma 40: 297–315.

Pennock, D.J., and D.W. Anderson. 1992. Landscape-scale effects of cultivation on soil quality and crop yields in the black soil zone. *In* Proc. of Soils and Crops Workshop '92. February 20–21. Saskatoon.

Verity, G.E., and D.W. Anderson. 1990. Soil erosion effects on soil quality and yield. Can. J. Soil Science. 70:471–484.

40 Assessment Of Beef Cattle Manure Management In The Southern High Plains

J. A. Daniel
S. J. Smith
A. N. Sharpley

USDA-Agricultural Research Service
Durant, Oklahoma

B. A. Stewart

Dryland Agriculture Institute
West Texas A & M University
Canyon, Texas

Established confined beef operations in the Southern Plains have stored manure in or near playas; natural, circular, shallow depressions with relatively impermeable bottoms. Because of growing concern over water quality as impacted by manure storage in playas, recent regulations require that new confined animal operations must compost or land apply manure. This study assesses past use of playas and examines current Southern Plains manure management policies that impact water quality. Soil cores from a feedlot and playa that receives runoff from a feedlot were analyzed for nitrogen and phosphorus forms. Elevated nutrient concentrations were found at the surface of both sites; however, a decrease of 80 to 90% below 1.5 m depth indicates N and P movement below the soil surface is limited. This implies beef feedlot management and short-term use of playas for manure storage are adequate. Manure stored in a playa, however, should remain near the impermeable center because the playa rim is more permeable and nutrient movement is possible at the rim.

Potential leaching of N to groundwater was discussed, while P movement from land-applied manure in surface runoff was estimated by phosphorus enrichment and kinetic desorption equations. Predicted total P content in runoff from experimental watersheds receiving 22 Mg ha^{-1} beef manure for 8 years was 600% greater than from untreated watersheds. Also 89% of this total P was algal-available compared to 46% in runoff from fields without manure application. In areas where runoff directly contributes to surface waters, manure application rates should be guided by soil P levels and crop P requirements. In isolated or hydrologically closed areas, higher rate of application may be an alternative management practice.

INTRODUCTION

Texas has seen an increase in confined animal operations, primarily beef feeder cattle over the last decade. In the Texas panhandle near Amarillo, five million beef animals are fattened annually, a total of 25% of the nation's beef (Southwestern Public Service, 1992). A problem associated with confined animal operations, though, is the disposal of generated manure. There are 6.5 million ha of cultivated fields and 5.7 million ha of rangeland available for manure disposal by land application in the Southern High Plains (Jones et al., 1985). The cost of transport of manure from the site of production, however, can become prohibitive (Bosch & Napit, 1992). During the development of concentrated beef animal operations in the Southern High Plains in the 70's, feedlots were located adjacent to playas that served as containment units for feedlot runoff.

Playas are natural, circular, shallow depressions which range in areal extent from 1 to 100 acres. Approximately 19,000 of them dot the landscape of the Southern High Plains (Lehman, 1972). Because playas are the lowest topographic points in the region, surface runoff from rainfall events flow into playas and form shallow lakes. Playas catch an estimated 2500 m^3 to 3700 m^3 (2 to 3 acre-ft) of runoff annually, but most runoff is lost to evaporation (Lehman, 1972). Hydrologic studies, initiated when playas were first used for manure storage, determined that movement of manure-derived nutrients would be minimal because of the relatively impermeable clay bottom of a playa, the high evaporation rate of the region and the deep groundwater table (Lehman et al., 1970; Clark, 1975). Follow-up studies of playas receiving long-term beef feeder manure applications indicate that playas do an effective job in restricting nutrient transport (Smith et al., 1993).

Recently, the U.S. Environmental Protection Agency (EPA) amended confined animal waste discharge permits to include playa lakes along with rivers, streams, lakes and wetlands as "waters of the U.S." (Federal Register, 1992). According to this classification, discharge of wastes from concentrated animal production operations into a playa is prohibited, and removal of solid and liquid manure is to be managed using recognized agricultural practices, such as composting or land application. Rates of land-applied manure can not exceed the nutrient plant uptake requirements unless land application sites are isolated from surface waters. If sites are isolated from surface water, then land applications rates can be greater than nutrient plant uptake. The intent of the new regulations is to prevent significant pollutants from reaching "waters of the U.S." (Federal Register, 1992).

The objective of this study is to assess the impact of long-term manure management in an active beef feedlot and adjacent playa prior to the implementation of the EPA National Pollutant Elimination Systems permit (March 1993) and to evaluate the potential impact of manure-derived N and P from land application as an alternative to manure storage in playa.

METHODS

Study sites were selected to examine the past effectiveness of the beef feedlot and storage playa for manure management. Sampling sites were chosen on the following criteria:

1. each site must be currently active
2. each site has been receiving beef manure for 20 years or longer
3. each site has a relatively large-scale beef operation (>20,000 head).

Beef Feedlot

Active for over twenty years, a confined beef operation in Wheeler County, Texas was selected to represent the beef feedlot site (Fig. 40–1). With a lot capacity of 20,000 to 25,000 head of cattle, the feedlot generates between 14,000 Mg to 17,500 Mg of dry weight manure annually (Gilbertson et al., 1979). Normal manure handling procedure involves the temporary piling of accumulated manure in the center of the pens and the periodic removal of these piles to storage areas located on the playa watershed.

The feed lot is situated on Devol fine-sandy loam (coarse-loamy, mixed, thermic, Udic Haplustalfs). Because of the coarse-textured soil, this site represents a worst case scenario and considerable nutrient movement beneath the feedlot is expected. Other beef operations situated on fine-textured soils would be expected to have less potential for nutrient movement beneath the feedlot. Soil cores were collected using a truck-mounted hydraulic push probe. Twelve feedlot pens were cored to a depth of 3 m. Locations for core sampling are shown in Fig. 40–1.

Playa

A playa receiving runoff from feedlot for over 20 years, located in Deaf Smith County, TX, was chosen to represent the playa site (Fig. 40–1). The playa is associated with a feedlot with a lot capacity of 40,000 head and generates 28,000 Mg of dry weight manure annually (Gilbertson et al., 1979). An adjacent feedlot was not chosen to represent the beef feedlot site because the lot is situated on a finer textured soil and less nutrient movement is expected than in the coarser textured Devol soil. Typical manure handling procedure involves storing feedlot runoff in an untilled playa.

The floor of the playa is Randall Clay (fine, montmorillonitic, thermic, Udic Pellusterts). Samples of the playa soil were obtained by drilling four holes with a truck-mounted percussion-type drill rig. Holes were drilled to a depth of 15.2 m (50 ft) and sample cores were taken - two near the center of the playa, one near the rim of the playa, and one on the playa ridge. Core sampling locations are seen in Fig. 40–1.

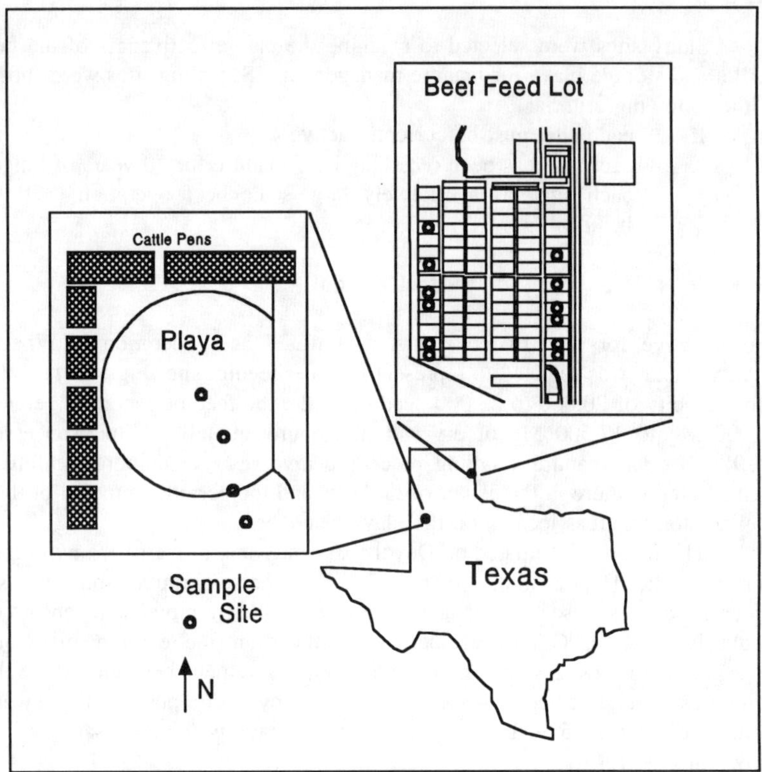

Fig. 40–1. Location of sample sites for project.

Chemical Analyses

Cores were divided into 30 cm increments and analyzed for total Kjeldahl nitrogen (TKN), nitrate-nitrogen (NO_3-N), ammonium-nitrogen (NH_4-N), total phosphorus (TP), and algal-available P (P available for uptake by plants and aquatic biota). Analyses of inorganic N forms (NO_3-N and exchangeable NH_4-N) involved KCl extraction using procedures described by Keeney and Nelson (1982). TKN and TP were determined by a semi-micro Kjeldahl procedure (Bremner and Mulvaney, 1982). Algal-available P was determined using iron oxide-impregnated paper strips (Sharpley, 1991).

Chemical results of the soil samples are averaged by depth and used to construct a composite distribution of nutrients beneath the ground surface. All twelve core sites for the beef feedlot are used to show the vertical distribution of nutrients beneath the ground surface, while for the playa site, the chemical results of three cores were used to construct a composite diagram. A core taken from the playa ridge above the feedlot is used as a control at this site.

RESULTS AND DISCUSSION

Beef Feedlot

Results of the coring and chemical analyses show that elevated concentrations of all nutrients are found at the surface. Since the Devol soil is a moderately permeable fine sandy loam, extensive downward nutrient movement was expected; however, nutrient movement is shown to be limited.

Vertical distribution of N forms is shown in Fig. 40–2. An elevated TKN concentration of 500 mg kg^{-1} is found in the top 30 cm, but levels decrease to < 100 mg kg^{-1} at a depth of 2 m - a decrease of 80%. Nitrate-N and NH_4-N have respective average values of 17.8 mg kg^{-1} and 175.5 mg kg^{-1}. Nitrate-N decreases to 1.7 mg kg^{-1} while NH_4-N drops to under 10 mg kg^{-1} at a depth of 1.5 m, a 90% decrease from surface values (Fig. 40–2). Ammonium-N values are high because the feedlot is still active. If the feedlot becomes inoperative and is not cleaned, potential NO_3-N leaching problems may occur.

Vertical movement of P is restricted with elevated values confined in the top 0.5 m (Fig. 40–3). Values for TP decrease from 138 mg kg^{-1} to 58 mg kg^{-1} and algal-available P drops from 24 mg kg^{-1} to 2.2 mg kg^{-1}.

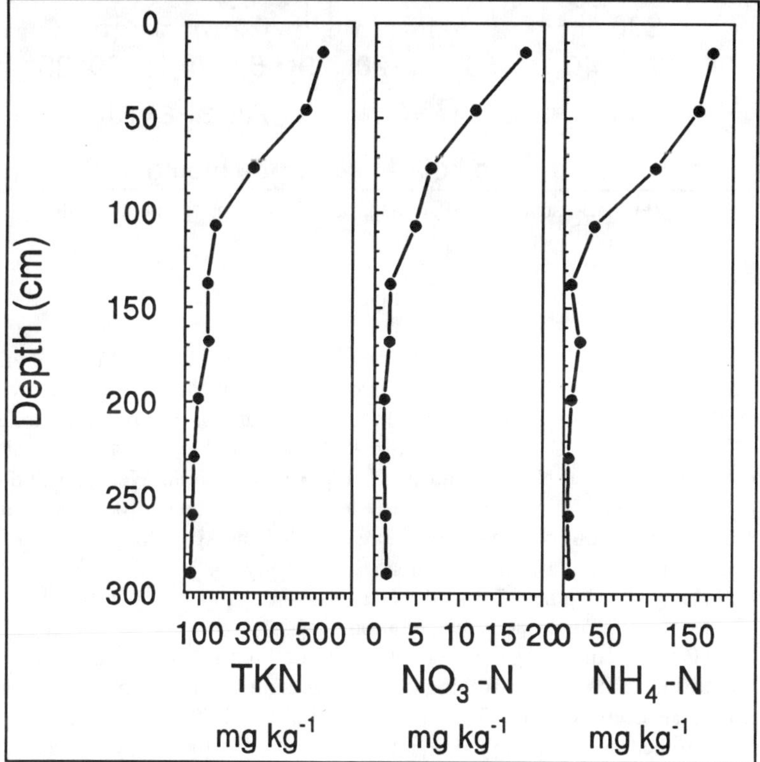

Fig. 40–2. Vertical distribution of nitrogen forms for the beef feedlot.

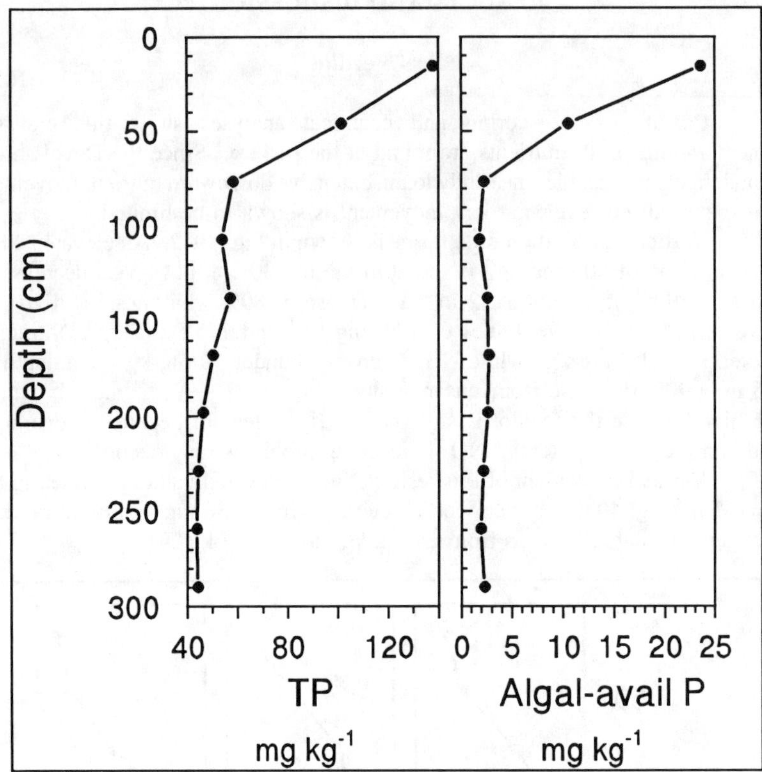

Fig. 40–3. Vertical distribution of phosphorus forms for the beef feedlot.

Playa

At the playa center all the nutrients analyzed show elevated levels at the surface that decrease to background levels with depth (Fig. 40–4). Surface levels of TKN show an average value of 2450 mg kg^{-1} for the top 30 cm that decrease to 250 mg kg^{-1} at 1.3 m and 162 mg kg^{-1} at 3 m (Fig. 40–4). Elevated concentrations of NO_3-N are found in the top 30 cm with an average value of 20 mg kg^{-1}. Under 30 cm depth, NO_3-N levels drop to under 10 mg kg^{-1} and stabilize. Ammonium-N levels show increases (141 mg kg^{-1} to 375 mg kg^{-1}) for the top 1 m of the playa floor compared to 30 mg kg^{-1} for the control site. Nevertheless, at a depth of 2.3 m the NH_4-N values drop to under 10 mg kg^{-1}, a 96% decrease from surface concentrations.

TP shows an average concentration of 1050 mg kg^{-1} in the top 30 cm of the playa floor compared to 508 mg kg^{-1} for the control (Fig. 40–5). At a depth of 0.5 m, TP decreases to 453 mg kg^{-1}. Algal-available P values, like TP, are elevated in the top 30 cm of the playa floor compared to the control site (78 mg kg^{-1} and 22 mg kg^{-1}, respectively). Vertical movement of P is restricted to the top 30 cm.

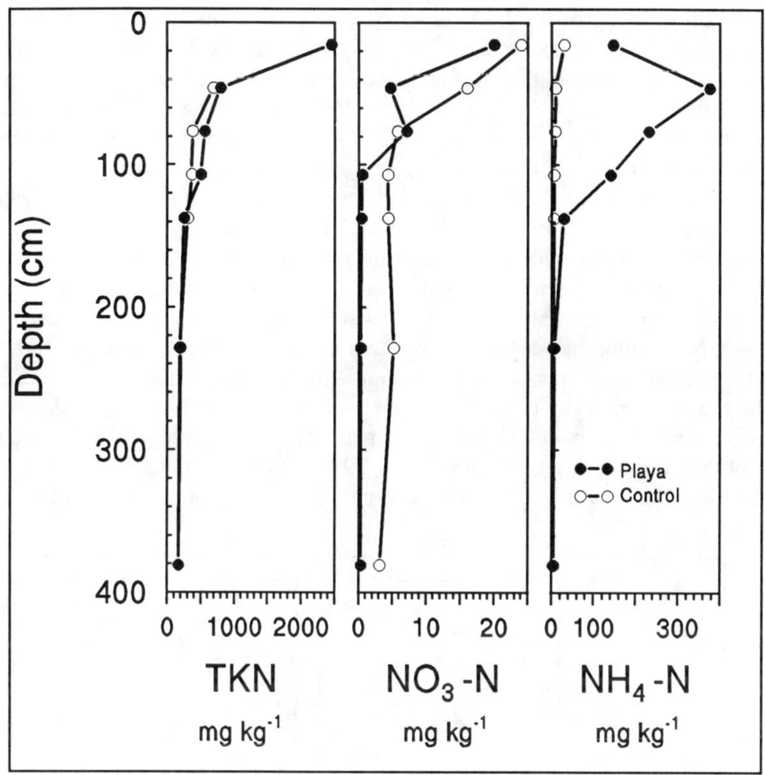

Fig. 40–4. Vertical distribution of nitrogen forms for the storage playa.

Soil samples taken from the playa rim indicate that the soil texture is coarser (Smith et al., 1993) and thus, probably more permeable than the center. Care must be exercised during feedlot management to keep runoff levels low so nutrients are contained in the center of the playa, thus prevent nutrient leaching at the playa rim.

Potential N and P Movement from Land Application

With the addition of nutrients to a soil by land application of manure, both N and P concentrations increase in the soil surface and both can affect water quality. To evaluate the impact of land application of manure on water quality, estimations of the downward transport of N in the vadose zone and lateral movement of P in surface runoff are made.

Potential Vertical Movement of N

Addition of manure to a soil is a source of nitrogen, an important chemical for plant growth. If there is too much NO_3-N, however, plants can not use the excess and the potential for NO_3-N leaching increases. Generally NO_3-N losses are not substantial in surface runoff (Kissel et al., 1976; Smith et al., 1983; Alberts et al., 1978; Dunigan et al., 1976), but excess NO_3-N may leach to groundwater. The period that NO_3-N is most susceptible to leaching is between harvest and the next growing season (Smith et al., 1980).

Nitrate-N leaching affects groundwater quality only if it reaches the water table. The principal ground aquifer in the High Plains region is the Ogallala with a depth to water table at approximately 30 m (100 feet) or deeper (Carr & Bergman, 1976). Potential evapotranspiration in the Southern High Plains exceeds the mean monthly precipitation which indicates that climatic leaching of NO_4-N is improbable (Smith & Cassel, 1991). The chance of NO_4-N leaching, however, increases with irrigation. Playas are the only locations where surface water can accumulate, consequently, leaching potential in the playas is greater. But with the slowly permeable playa bottom, leaching is still unlikely except near the playa edge (Daniel et al., 1993). Continued manure application on the watershed will increase the amount of NO_3-N available for downward leaching.

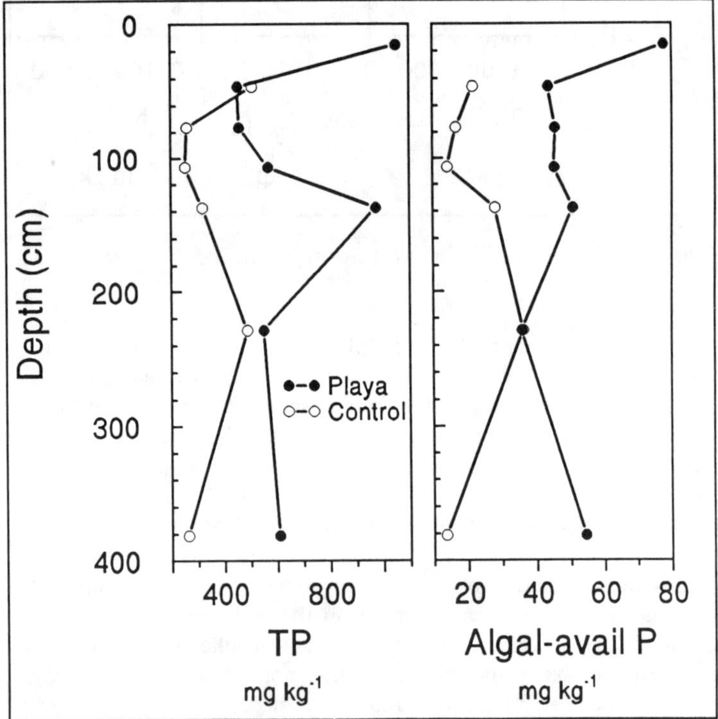

Fig 40–5. Vertical distribution of phosphorus forms for the storage playa.

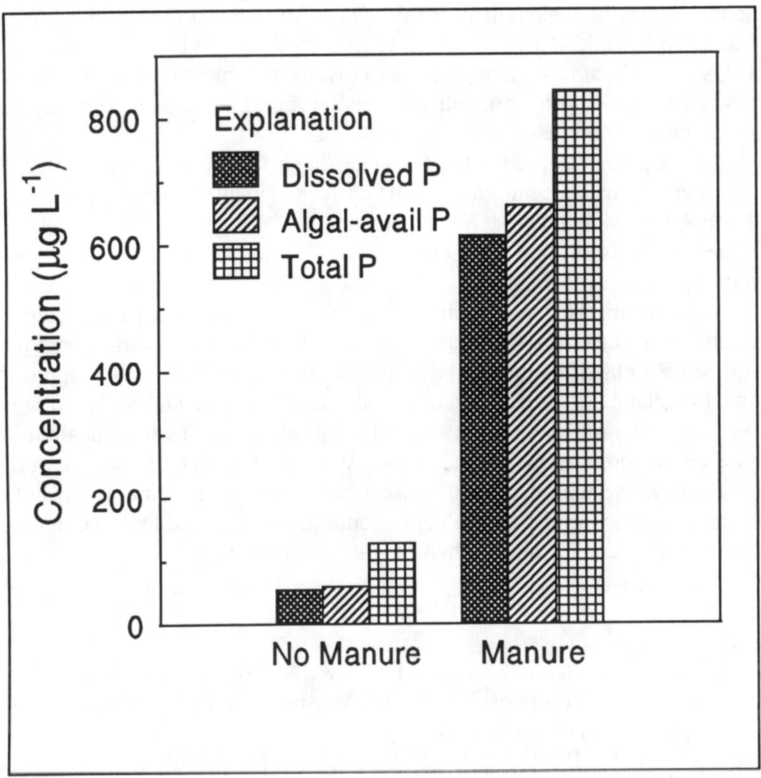

Fig. 40–6. Predicted P concentrations in runoff from control fields and fields receiving beef manure.

CONCLUSIONS

Results of soil sampling of a beef feedlot and storage playa indicate that currently used feedlot management strategies have been effective in minimizing manure-derived N and P movement, even though the feedlot sampled is situated on moderately permeable soil. This strongly suggests that management techniques used prior to the implementation of the EPA regulations have minor impact on the movement of N and P in the soil and subsoil. Nevertheless, groundwater samples were not available, so a more definite assessment of long-term manure application on groundwater quality can not be made.

Vertical movement of nutrients under the sampled feedlot is limited. N movement for the feedlot is limited to the top 1 to 2 m while P was contained in the top 0.5 to 1m. We believe that the periodic cleaning of pens helps to reduce nutrient movement.

Nutrient movement in the playa is also restricted. The high clay content playa bottom effectively seals the leaching of nutrients from downward migration. Nitrogen movement for the playa was limited to the top 1 to 1.3 m surface depth while P was confined to the top 0.5 m. While the center of the playa is effective at minimizing nutrient movement, a more permeable soil at the playa rim exists. Improper handling of feedlot runoff at the playa rim may result in unwanted nutrient leaching.

Simulations to predict the P content in surface runoff indicate that all P forms increase in concentration for fields receiving beef manure compared to fields not receiving manure. The percentage of TP that is algal available in surface runoff is higher for fields receiving manure (78%) than those not receiving manure (46%).

Playas are the lowest points on the Southern High Plains landscape and they collect surface runoff, eroded sediments and nutrients from nearby fields. With the EPA classification of playas as "waters of the U.S.", manure application rates to cropland could be dictated by soil P and crop P requirements to prevent or minimize the movement of P, even though playas are hydrologically-closed landscape features. This would intensify local area land limitations already facing many livestock farmers. Consequently, long-term manure and nutrient management plans in the High Plains should account for the fact that negligible P losses will occur from these hydrologically-closed playas.

REFERENCES

Alberts, E.E., G.E. Schuman, and R.E. Burwell. 1978. Seasonal runoff losses of nitrogen and phosphorus from Missouri Valley loess watersheds. J. Environ. Qual. 7:203–208.

Bosch, D.J., and K.B. Napit. 1992. Economics of transporting poultry litter to achieve more effective use as a fertilizer. J. Soil Water Conserv. 47:342–346.

Bremner, J.M., and C.S. Mulvaney. 1982. Nitrogen - total. *In* A. L. Page et al. (ed.) Methods of soil analysis, Part 2. 2nd ed. Agronomy 9:595–624.

Carr, J.E., and D.L. Bergman. 1976. Reconnaissance of the water resources of the Clinton Quadrangle west-central Oklahoma. Oklahoma Geological Survey Hydrologic Atlas 5, 4 sheets.

Clark, R.N. 1975. Seepage beneath feedyard runoff catchments. *In* Managing livestock wastes. Proceedings Third International Symposium on Livestock Wastes. ASAE, St. Joseph, MI, p. 289–295.

Daniel, J.A., A.N. Sharpley, S.J. Smith, and B.A. Stewart. 1993. Animal manure management and potential impacts of N and P on water quality. p. 915–928. Proc. FOCUS Conf. on Eastern Regional Ground Water Issues, National Ground Water Association.

Dunigan, E.P., R.A. Phelan, and C.L. Modart Jr. 1976. Surface runoff losses of fertilizer elements. J. Environ. Qual. 5:339–342.

Federal Register. 1992. 57(141): 32476–32499.

Gilbertson, C.B., F.A. Norstadt, A.C. Mathers, R.F. Holt, A.P. Barnett, T.M. McCall, C.A. Onstad, and R.A. Young. 1979. Animal waste utilization on cropland and pastureland. USDA Util. Res. Rep. No. 6. 135 p.

Jones, O.R., H.V. Eck, S.J. Smith, G.A. Coleman, and V.L. Hauser. 1985. Runoff, soil, and nutrient losses from rangeland and dry-farmed cropland in the Southern High Plains J. Soil and Water Conser. 40:161–164.

Keeney, D.R., and D.W. Nelson. 1982. Nitrogen-inorganic forms. *In* A. L. Page et al. (ed.) Methods of soil analysis, Part 2. 2nd ed. Agronomy 9:643–698.

Kissel, D.E., C.W. Richardson, and E. Burnett. 1976. Losses in nitrogen in surface runoff in the Blackland Prairie of Texas. J. Environ. Qual. 5:288–293.

Lehman, O.R. 1972. Playa water quality for groundwater recharge and use of playas for impoundment of feedyard runoff. p. 25–30. *In* C. C. Reeves, Jr. (ed.). Playa Lake Symposium, ICASALS Pub. No. 4. Lubbock, TX.

Lehman, O.R., B.A. Stewart, and A.C. Mathers. 1970. Seepage of feedyard runoff water impounded in playas. Texas Agric. Exp. Stn. MP-944.

Sawyer, C.N. 1947. Fertilization of lakes by agricultural and urban drainage. New England Water Works Assoc. 61:109–127.

Sharpley, A.N. 1991. Soil phosphorus extracted by iron-aluminum-oxide-impregnated filter paper. Soil Sci. Soc. Am. J. 55:1038–1041

Sharpley, A.N., and R.G. Menzel. 1987. The impact of soil and fertilizer phosphorus on the environment. Adv. Agron. 41:297–324.

Sharpley, A.N. and S.J. Smith. 1989. Prediction of soluble phosphorus transport agricultural runoff. J. Environ. Qual. 18:313–316.

Sharpley, A.N., and S.J. Smith. 1992. Prediction of bioavailable phosphorus loss in agricultural runoff. J. Environ. Qual. 21:32–37.

Smith, S.J., and D.K. Cassel. 1991. Estimating nitrate leaching in soil materials. p. 165–188. *In* R.F. Rollett, et al. (ed.) Managing nitrogen for groundwater quality and farm profitability. SSSA, Madison, WI.

Smith, S.J., D.E. Kissel, and J.R. Williams. 1980. Chapter 13: Nitrate production, uptake, and leaching. p. 493–508. *In* W.G. Knisel (ed.) CREAMS: A Field-scale model for chemicals, runoff, and erosion from Agricultural Management Systems. U.S. Department of Agriculture, Conserv. Res. Rep. No. 26.

Smith, S.J., R.G. Menzel, J.R. Rhoades, J.R. Williams, and H.V. Eck. 1983. Nutrient and sediment discharge from Southern Plains grasslands. J. Range Mgt. 36:435–439.

Smith, S.J., B.A. Stewart, A.N. Sharpley, J.W. Naney, T. McDonald, M.G. Hickey, and J.M. Sweeten. 1993. Nitrate and other nutrients associated with playa storage of feedlot wastes. Texas Agric. Ext. Serv. Bull. B-5081.

Southwestern Public Service. 1992. Cattle feeding capitol of the world. 1992 Feed Cattle Survey, Amarillo, TX.

Vollenweider, R.A., and J. Kerekes. 1980. The loading concept as a basis for controlling eutrophication: Philosophy and preliminary results of the ACCEDE program on eutrophication. Progr. Water Technol. 12:5–38.

41 Site-Specific Management By Soil Condition: Managing Fertilizer Nitrogen With Nitrapyrin[1]

G. L. Malzer
P. C. Robert
T. J. Graff
S. L. Tomer
T. W. Bruulsema

Soil Science Department
University of Minnesota
St. Paul, Minnesota

J. A. Vetsch

Southern Experiment Station
University of Minnesota
Waseca, Minnesota

D. R. Huggins

Southwest Experiment Station
University of Minnesota
Lamberton, Minnesota

Efficacy of the nitrification inhibitor nitrapyrin (N-Serve®) is influenced by a number of soil factors. Soil specific management has potential for identifying these factors to predict where nitrapyrin may be useful in minimizing N losses. This study was conducted to relate corn yield response to N fertilizer with and without nitrapyrin to soil factors varying spatially within a field. Both positive and negative yield responses to nitrapyrin occurred, and appeared to be spatially clustered. Nitrapyrin response was weakly negatively correlated to relative elevation within each field in two separate years.

---

[1]The assistance and financial support of the Minnesota Agricultural Experiment Station, USDA-CSRS, Dow Elanco, Soil Teq, Inc., and DMI is gratefully acknowledged.

INTRODUCTION

Soil factors including texture, pH, organic matter and temperature have been shown to influence the effectiveness of the nitrification inhibitor nitrapyrin or N-Serve® (Schmidt, 1982). These factors vary within fields owing to differences in soil physical characteristics and landscape position. Soil and landscape parameters may be mapped using soil specific management technologies to predict where nitrapyrin may be useful in minimizing losses of N and maximizing crop N uptake.

The objectives of this study were to evaluate within-field variability of yield response to N fertilizer with and without nitrapyrin, and to relate such variability to soil and landscape factors within the field.

METHODS AND MATERIALS

The experiments were conducted in 1991 and 1992 in two different irrigated fields near Becker, in Sherburne county, MN (45° 23'N; 93° 53'W). These were the same experiments reported in Vetsch et al. (1994). In each year, treatments were applied in 8 m wide strips along the length of the field. The treatments were replicated within three separate blocks; each block extending the length of the field. In each field, the previous crop was corn in a corn-corn-soybean rotation.

Nitrogen rates were chosen to bracket the rate previously used by the farmer. After the first season, it was discovered that this rate was well above the optimal level, partly due to the presence of substantial concentrations of nitrate (30 mg NO_3-N L^{-1}) in the irrigation water.

The 1991 field was 18 ha in size and included soil textures ranging from sandy loam to sand (Table 41–1). The experimental plot measured 410 m by 220 m. Corn (cv. 'Pioneer 3569') was planted 29 April. Soil was sampled on a grid of 30 m by 30 m. Ten nitrogen treatments were applied on 5 and 6 June as anhydrous ammonia. The treatments comprised 4 rates of nitrogen (134, 168,

Table 41–1. Soil map unit data for 1991 experimental plot in Becker, Sherburne County, MN.

Soil Series	Area	Textural	Slope	Carbon†	USDA Subgroup
	ha	class	%	%	
Hubbard A	1.7	LS	0–2	1.1	Udorthentic Haploboroll
Hubbard B	4.0	LS	2–6	1.4	Udorthentic Haploboroll
Hubbard C	0.5	LS	6–12	2.4	Udorthentic Haploboroll
Mosford A	2.7	SL	0–2	1.5	Udic Haploboroll
Mosford B	2.3	SL	2–6	1.2	Udic Haploboroll
Sandberg	0.3	LCOS	6–12	1.4	Udorthentic Haploboroll
Sand/Peat	1.5	S/P	0–2	2.3	

† soil carbon measured in 0-30 cm depth increment

Table 41–2. Soil map unit data for 1992 experimental plot in Becker, Sherburne County, MN.

Soil Series	Area	Textural	Slope	Carbon†	USDA Subgroup
	ha	class	%	%	
Hubbard A	3.4	LS	0-2	1.0	Udorthentic Haploboroll
Hubbard B	2.5	LS	2-6	0.7	Udorthentic Haploboroll
Hubbard C	0.4	LS	6-12	0.8	Udorthentic Haploboroll
Mosford A	5.5	SL	0-2	1.0	Udic Haploboroll

† soil carbon measured in 0-30 cm depth increment

202 and 235 kg N ha^{-1}) with and without nitrapyrin, one variable rate, and one check without any nitrogen fertilizer other than that in the starter fertilizer (32 kg N ha^{-1}). Grain was harvested 14-18 October in 12 m long segments.

The 1992 field was 12 ha in size and included soil textures ranging from sandy loam to loamy sand (Table 41–2). The experimental plot measured 560 m by 200 m. Corn (cv. 'Pioneer 3751') was planted 1 May. Soil was sampled on a grid of 43 m by 15 m. Eleven nitrogen treatments were applied on 12 June as sidedressed anhydrous ammonia. The treatments comprised 4 rates of nitrogen (101, 134, 168, and 202 kg N ha^{-1}) and one variable rate, each with and without nitrapyrin, and one check as in the 1991 field. Corn grain was harvested in 15 m long segments.

In both years, soils were sampled at grid points to a depth of 90 cm in 30 cm depth increments. Each depth increment was analyzed for NO_3-N, NH_4-N, total nitrogen, carbon, and mineralizable nitrogen (phosphate-borate extractable N and hot KCl extractable N; Gianello & Bremner, 1986).

Statistical Analysis

Both linear-plateau and quadratic yield response models were fitted. In general, the linear plateau fit more closely and gave lower estimates of the economically optimum N rate than did the quadratic. For this reason, all response parameters reported are from the linear-plateau models as shown in Vetsch et al., (1994). The intersection of the linear and plateau portions of the response was taken as an approximation of the economically optimum N rate (ONR).

Response models were fitted to yield data from square blocks of yield segments within transects as in Vetsch et al., (1994). The non-linear regression model used at least two data points for each level of N fertilizer. Response parameters for each block were then correlated to soil parameters measured at the nearest grid point.

Table 41–3. Correlation coefficients of difference in maximum corn yield (with minus without nitrapyrin) with soil and landscape factors.

	1991 (n=122)		1992 (n=147)	
	r	p	*r*	p
Soil Color (aerial photo)	-0.21	*	-0.05	NS
Relative Elevation	-0.24	**	-0.23	**
Soil Nitrate§	0.29	**	-0.07	NS
Phosphate-Borate N§	0.17	+	0.10	NS
Hot KCl N§	0.01	NS	0.12	NS
Soil Total Nitrogen§	0.21	*	0.16	+
Sil Carbon§	0.28	**	0.06	NS

+,*,**, and *** refer to values of the correlation coefficient *r* significant at the 0.1, 0.05, 0.01, and 0.001 levels, respectively.

§ soil parameters were measured in 0-30 cm depth increment.

RESULTS AND DISCUSSION

Grain yield on check plots receiving no N fertilizer was variable both among and within soil map units. Averaged over the entire field, there was no significant response to nitrapyrin in terms of the response curve parameters-maximum yield, ONR and response slope. Within the field, however, yield responses to nitrapyrin in partitioned harvest areas ranged from a benefit of 1.0 Mg ha^{-1} to a detriment of 0.9 Mg ha^{-1} . The ONR in these harvest areas ranged from 56 kg N ha^{-1} less to 52 kg N ha^{-1} more with nitrapyrin.

In both years there was a negative correlation between relative elevation and nitrapyrin effect on maximum yield (Table 41–3). The lower elevation parts of the field appeared to benefit more from nitrapyrin than those higher in the landscape. The correlation, although significant statistically, explained < 6% of the total variation in nitrapyrin response. It is unclear why the correlation was significant for the effect of nitrapyrin on the maximum yield rather than on fertilizer efficiency-related parameters such as ONR or the slope of the linear portion of the response (Table 41–4 and Table 41–5). There were no consistent correlations between effects of nitrapyrin on ONR or response slope and any of the other soil and landscape properties measured. Although only simple correlations are reported here, it is possible that further analysis using multiple regression with interactions among soil parameters will yield stronger relationships. Spatial clustering of nitrapyrin responses was apparent but objective techniques to distinguish spatial clustering patterns from random distributions are lacking.

Table 41–4. Correlation coefficients of difference in optimum N rate (without minus with nitrapyrin) with soil and landscape factors.

	1991 (n=122)		1992 (n=147)	
	r	p	*r*	p
Soil Color (aerial photo)	-0.03	NS	0.18	*
Relative Elevation	-0.03	NS	-0.09	NS
Soil Nitrate§	0.09	NS	0.07	NS
Phosphate-Borate N§	-0.02	NS	0.08	NS
Hot KCl N§	0.01	NS	-0.11	NS
Soil Total Nitrogen§	0.06	NS	0.01	NS
Soil Carbon§	0.07	NS	-0.01	NS

+,*,**, and *** refer to values of the correlation coefficient *r* significant at the 0.1, 0.05, 0.01, and 0.001 levels, respectively.
§ soil parameters were measured in 0-30 cm depth increment.

Table 41–5. Correlation coefficients of difference in response slope (with minus without nitrapyrin) with soil and landscape factors.

	1991 (n=122)		1992 (n=147)	
	r	p	*r*	p
Soil Color (aerial photo)	-0.05	NS	0.05	NS
Relative Elevation	-0.13	NS	0.03	NS
Soil Nitrate§	0.23	*	0.06	NS
Phosphate-Borate N§	-0.05	NS	0.13	NS
Hot KCl N§	0.04	NS	-0.03	NS
Soil Total Nitrogen§	0.16	+	-0.01	NS
Soil Carbon§	0.18	*	0.11	NS

+,*,**, and *** refer to values of the correlation coefficient *r* significant at the 0.1, 0.05, 0.01, and 0.001 levels, respectively.
§ soil parameters were measured in 0-30 cm depth increment.

The correlation with relative elevation implies some relationship between the landscape position of an area and its responsiveness to nitrapyrin. Elevation is likely to be correlated with many other soil factors including soil map unit and soil organic matter. Other topographic parameters including slope curvature and landscape position may also be important in conjunction with relative elevation.

CONCLUSIONS

Both positive and negative yield responses to nitrapyrin occurred, and appeared to be spatially clustered. Response of maximum yield to nitrapyrin was weakly negatively correlated to relative elevation within each field in two separate years. Among soil and landscape parameters measured, no single parameter gave a good prediction of within-field areas responsive to nitrapyrin. Further analyses using multiple regression with interactions among soil parameters and utilizing more measures of landscape properties including slope curvature may yield stronger relationships.

REFERENCES

Gianello, C., and J.M. Bremner. 1986. Comparison of chemical methods of assessing potentially available organic nitrogen in soil. Comm Soil Sci and Plant Anal 17:215–236.

Schmidt, E.L. 1982. Nitrification in soil. p.277. *In* Nitrogen in Agricultural Soils. F.J. Stevenson, (ed). American Society of Agronomy, Madison, WI.

Vetsch, J.A., G.L. Malzer, P.C. Robert, and D.R. Huggins. 1995. Nitrogen specific management by soil condition: managing fertilizer N in corn. Proc. of the 2nd International Conference on Site Specific Management for Agricultural Systems, 27–30 March 1994, Bloomington, MN. (in press).

42 An Overview Of Site-Specific Management Research Activities In Nebraska

T. A. Peterson
J. S. Schepers

USDA-Agricultural Research Service
University of Nebraska
Lincoln, Nebraska

R. B. Ferguson

University of Nebraska
Clay Center, Nebraska

G. W. Hergert

University of Nebraska
North Platte, Nebraska

D. A. Mortensen

Dep. of Agronomy
University of Nebraska
Lincoln, Nebraska

Scientists at the University of Nebraska have initiated a number of studies that relate to the rapidly developing discipline of site-specific agricultural management. Integration of technologies that provide location in the field along with automated control of agricultural inputs may become common in the near future. Much research effort in Nebraska is devoted to learning the potential environmental and economic benefits of variable rate application of fertilizer and pesticides. Perhaps more important to the science of agriculture, the automation of input control also generates a record of input application at each point in the field. These input records along with yield maps make possible the collection of much more crop response data and measures of input efficiency than has previously been available. Our ability to collect vast quantities of production data suddenly surpassed our ability to interpret those data and use the information to improve crop management decisions. An overall objective of our combined research efforts is to develop techniques to analyze layers of geo-referenced data to improve the way we evaluate crop management decisions. The following is a list of major research projects and associated activities that relate to site-specific agricultural management in Nebraska.

Variable Rate Application Technology-Demonstration Project

Anhydrous ammonia has been applied using variable rate equipment (Soilection by SoilTeq) at several cooperators sites as a part of the Mid-Nebraska Water Quality Demonstration Project. Uniform application of the UNL recommended N rate, recommended +50 lbs N/acre, and the recommended rate -50 lbs N/acre were compared to variable rate applications on replicated field length strip plots. Corn (*Zea mays* L.) grain yields in 1992 and 1993 were determined using a combine equipped with on-the-go yield monitoring and GPS equipment. Initial results are reported by Ferguson (1993) in these proceedings. The variable rate anhydrous applicator was the focus of several popular field days, tours, and was displayed at Husker Harvest Days. Investigators: Drs. Richard Ferguson, and Joel Cahoon, University of Nebraska and Mark Schrock, Kansas State University .

Factors Influencing Spatial Yield and N Use Efficiency of Furrow-Irrigated Corn.

Concern about nitrate contamination of Nebraska's ground water continues to increase. Areas experiencing elevated nitrate levels in ground water are characterized by continuous furrow-irrigated corn overlying shallow water tables. Non-uniform water application common in furrow-irrigated systems complicates nitrogen management. Evaluation of spatial variability of grain yield and factors that affect N use efficiency is especially interesting in these systems. Strategies to utilize variable rate N application technology in response to those factors will also be evaluated for their potential to reduce nitrate leaching from furrow-irrigated corn. Investigators: Drs. Richard Ferguson, Gary Hergert, Joel Cahoon, Todd Peterson, and Carol Gotway.

Evaluating the Use of Variable Rate Application Technology to Minimize Nitrate Pollution of Water

Five different N management strategies utilizing variable rate N application equipment will be evaluated and compared to currently recommended BMP's utilizing uniform N applications. Large replicated plots will be established near Shelton, NE. on a field of continuous sprinkler-irrigated corn. Treatments will include evaluation of several different ways to utilize spatially dependent data to establish the rate and method of N application using the variable rate approach. One treatment will vary N rates based on a map of soil nitrate and organic matter levels. Other treatments will build on this approach by incorporating previous yield maps and other factors to vary N rate across the field. Unique to this study will be utilizing in-season N status evaluation to react to potential N deficiencies using variable rate simulated fertigation. Investigators: Drs. James Schepers, Derrel Martin, Todd Peterson, Gary Hergert, Leonard Bashford, Richard Ferguson, Darrell Watts, Jerome Pier, Albert Sims, Wayne Woldt, James Merchant, and Carol Gotway.

Spatial Distribution of Weed Species

A multi-yr weed species distribution survey is being conducted in farmers' fields just prior to post-emergence herbicide application. Weed distributions are mapped and subjected to geostatistical analysis. The stability of weed distributions over years may suggest site specific weed management strategies or utilization of optical weed sensing technologies for intermittent herbicide application. Investigators: Dr. David Mortensen.

Spatial Distribution of Soil Nitrate

A research project was initiated in 1987 to help determine how many soil samples are required to accurately represent the amount of nitrate in a production field. Approximately 40 farmer fields across the major corn-growing areas in Nebraska have been grid sampled with each increment of every sample analyzed separately (Hergert, 1990). Average soil nitrate in the 0–1.2 m profile ranged from 18 to 134 kg ha^{-1} with an overall coefficient of variability of 57%. When five to eight soil cores are collected to represent a field (which is the common practice), the estimate of soil nitrate may be no better than +/- 25% of the mean with an alpha level of 0.3. Geostatistical analysis of these data is proceeding to attempt to learn common characteristics of fields that exhibit similar soil nitrate distributions. Investigators: Drs. G.W. Hergert and R.B. Ferguson.

Biometrics and Geostatistics Research

The University of Nebraska Department of Biometry is known nationally and internationally for their staff expertise in the areas of Geostatistical and Fractal analysis techniques. Dr. David Marx, Department Head, and other faculty present numerous workshops and training courses for scientists and professional engineers on the use of geostatistics in their respective fields. The Biometry department cooperates in many of the studies and provides invaluable service and expertise to the site-specific management investigations underway.

University of Nebraska Conservation and Survey Division

This unit within the Institute of Agriculture and Natural Resources includes a team of faculty utilizing Geographic Information Systems and remote sensing data acquisition for a variety of applications. Their expertise and willingness to cooperate with other researchers has been beneficial to site specific management activities at the University of Nebraska.

Evaluating Crop Nitrogen Status Using Remote Sensing Techniques

Use of the chlorophyll meter has demonstrated the potential value of in-season N status and the ability to correct N deficiencies before they reduce crop yield (Peterson et al., 1993). Recommendations developed at Nebraska for using the chlorophyll meter to improve N management are being utilized across the Corn Belt and in Europe. Extension of these principles has led to cooperation with several military remote sensing laboratories in evaluating specific approaches for scanning reflectance from irrigated corn fields followed by simulated fertigations to correct developing N deficiencies. N status information may be obtained by scanning standard color photographs as well as black-and-white images using selective wavelength filters. Investigators: Drs. James Schepers, Wallace Wilhelm, Gary Varvel and predoctoral Fellow, Tracy Blackmer.

REFERENCES

Hergert, G.W., 1990. Spatial variability of nitrate in farmer's fields. Soil Science News Vol. 12, No.11. Univ. of Nebraska Cooperative Extension, Lincoln, NE.

Ferguson, R.B., 1993. Variable Rate Nitrogen Application - A Progress Report. Soil Science News Vol. 15, No. 6. Univ. of Nebraska Cooperative Extension, Lincoln, NE.

Peterson, T.A., T.M. Blackmer, D.D. Francis, and J.S. Schepers. 1993. Using a Chlorophyll Meter to Improve N Management. NebGuide G93-1171-A, Univ. of Nebraska, Lincoln.

43 Stochastic Modeling Of Soil Condition During Tillage

E. Roytburg
J. Chaplin

Department of Agricultural Engineering
University of Minnesota
St Paul; Minnesota

A stochastic model is proposed to describe the changes in soil physical condition during tillage using soil dynamic properties. It is anticipated that this model will predict the soil condition resulting from tillage and generate a number of ways to reach the desired soil condition with the different probabilities. The soil resistance force is chosen as a measured parameter, reflecting the changes in the soil condition. The value of the soil resistance force will be used as an indicator of the soil condition state and will serve as an input to the stochastic model. A tillage-control system, which utilizes the random walk model and the tillage transducers to control the application of tillage by continuously sensing the current soil conditions and adjusting the tillage control parameters, is proposed.

INTRODUCTION

Soil tillage is probably the world's largest materials handling operation. Yet most current tillage systems till the total surface area of a field irrespective of the variation in soil conditions. It is doubtful whether this "blanket treatment" is efficient in terms of energy, capital, and other important resources.

The purpose of tillage is to manipulate soil from an initial soil condition (ISC) into a desired soil condition (DSC) by the application of a tillage implement. Soil conditions should be altered with the smallest expenditure of energy required to operate the tool, and the final soil condition (FSC) must be acceptable. This goal can be reached if we have the quantitative descriptions of changes in the soil condition and soil condition is controlled. Assessments of soil physical condition are presently qualitative rather than quantitative, and soil reaction cannot be predicted, let alone controlled.

Quantitative descriptions of soil condition are difficult because no method has been developed for describing soil conditions adequately. Tool forces can be readily measured and reported so that the force aspect of changes in soil condition is relatively easy to describe. Quantitative descriptions of soil conditions, on the other hand, are not well developed. In fact, specific conditions usually cannot be quantitatively described with regard to the intended use of the soil. We can see how the inability to characterize both actual and desired soil conditions has hampered the performance of tillage tools.

Lack of quantitative descriptions of soil conditions precludes establishing an accurate performance goal for a particular tool. As a result, in seedbed preparation that portion of the required soil manipulation not accomplished by moldboard plow is often completed by subsequent harrowing, leveling, or packing. Without a specific performance goal, the inadequacies of the moldboard plow go unnoticed and are compensated by additional tillage operations, which may have a negative effect on the soil physical conditions. The development of methods to describe changes in soil condition due to the tillage and to evaluate the final soil condition in quantitative terms will improve the design and use of tillage tools.

Control of soil compaction is a continual requirement in modern agriculture. A major task of soil management is to minimize soil compaction to the furthest possible extent. An approach to the prevention of soil compaction is the avoidance of all but essential pressure-inducing operations. This calls for reducing the number of operations involved in primary and secondary tillage and achieving desired soil condition in a single operation rather than by a repeated sequence of passes.

For tillage to satisfy an efficiency requirement, a tillage system should use the most efficient implement at the most appropriate time so as to effect the desirable soil condition in a single pass. Optimal performance could be attained by sensing the local soil physical condition and automatically adjusting the tillage machine. Therefore, methods which would sense the soil condition "on the go" and mathematical models which would predict changes in soil condition are desirable.

The new method for predicting final soil condition after tillage, proposed in this paper, may lead to a better understanding of the soil response to tillage, a better prediction of forces produced by the operating implement, and therefore, a forecast of soil condition produced by the implement passage.

The specific objectives of this paper are:

1) To summarize previous work on the soil tillage modelling.
2) To describe a proposed stochastic model.

Literature Review

Tillage and cropping practices have developed by trial and error at specific sites. Field research focuses on problems that will improve a system in place, but attempts to transfer the improved system to another site have often failed. Usually the soil, climate, and management factors have not been taken into consideration fully, so that quantitative transfer of the results to another site was not possible.

Modeling tilled soil is a relatively new topic (Gupta et al., 1991). Most of the models predicting tillage effects on soil physical properties and processes have been developed in the 1970s and 1980s. Presently, there are few comprehensive reports dealing with the modeling of processes in tilled soil. Unger et al., (1982) reviewed works describing experiments conducted to evaluate the effects of tillage on soil physical properties rather than provide models to predict these parameters. Most of the models deal with soil physical processes and assume that either the models for predicting tillage effects already exist or the input of the tilled soil physical properties can be measured experimentally. Much of the research on the modeling of tillage effects on soil physical properties has been on the prediction of soil compaction due to the passage of a tractor tire (Sohne, 1953; Perumpral et al., 1971; Gupta & Larson, 1982; Raper & Erbach, 1990; Gupta & Raper, 1994). Even in this area, the prediction has been limited to load effects on volumetric strain or bulk density and not to other soil physical properties related to flow processes or soil strength.

Tillage affects several soil properties, and these, in turn, influence root growth, plant growth, and ultimately crop yield. Therefore, a crop growth model intended to simulate tillage effects on yield must operate at the physical process level and must contain appropriate linkages and feedback mechanisms (Whisler et al., 1982). Several such complex simulation models have been developed including NTRM (Shaffer & Larson, 1987), GOSSYM (Whisler et al., 1982), SOYMOD (Meyer et al., 1979), and MUTillS (Porter & McMahon, 1990).

Conceptual ISC → FSC models.

In agriculture, we apply active force systems-tillage to prepare seedbeds and rootbeds, incorporate residue and amendments, control weeds, control pests, enhance infiltration and control erosion. The state of the soil is changed from its initial condition to some final condition as the result of applied forces and of the resulting soil movement (Schafer & Johnson, 1982). Cooper and Gill (1966) illustrated that idea with the conceptual relation:

$$Sf = h\ (Si, F) \quad [1]$$

where

Sf = final soil condition
Si = initial soil condition
F = mechanical forces applied to the soil.
h = functional relation between Sf, Si, F.

Gill and Vanden Berg (1967) and Vanden Berg and Reaves (1966) expressed two additional generalized relations which reflect aspects of the tillage machine system.

$$F = f\ (Ts, Tm, Si) \quad [2]$$

and

$$Sf = g\ (Ts, Tm, Si) \quad [3]$$

where

Ts is the tool shape factor
Tm is the manner of tool movement factor
f is functional relation between F, Ts, Tm, Si
g is functional relation between Sf, Ts, Tm, Si

Equation [2] is referred to as the force-tillage equation and equation [3] as the soil-condition equation.

Equations [2] and [3] represent the most general situation because the functional relations f and g are completely arbitrary, and may or may not be different. If F and Sf are functionally related, equations [2] and [3] can be combined and soil resistance forces can be served as a measure of final soil condition.

Past research on soil-machine relations and soil dynamics related to equation [2] have been aimed at relating a differential change in the force as influenced by a change in one or more soil static physical properties (e.g., moisture content, bulk density). This is analogous to characterizing the current flow through a piece of wire by measuring its length, cross-sectional area, chemical composition and temperature (Young et al., 1984). The description of the dynamic behavior (current flow) is complicated unless the dynamic property of resistance, as defined by Ohm's law is applied. Once dynamic properties are known, it is possible to characterize dynamic properties without regard to the physical properties. That is, the wire may be schematically represented as a black box. It is not necessary to have a knowledge of the physical properties of the wire (information inside the box) in order to describe the current flow if the resistance, a dynamic property, is known.

Similarly in the case of soil, the dynamic properties are precisely determined by the physical properties. The soil may be represented as a black box, and according to equation [3], the factors tool shape and manner of tool movement act on a soil to produce the final soil condition. A change from initial soil condition to final soil condition is caused by soil movement. This change involves strain and yielding of the soil. Therefore, force-movement relations of soil are important when soil condition changes. Such research has addressed soil stress and strength characteristics, and rigid body movement (Schafer & Johnson, 1990).

Soil Resistance-Force (SR-F) models

Based on the above discussion about using dynamic soil properties when modeling changes in soil physical condition during tillage, it is appropriate now to present models developed using a dynamic parameter, namely, soil resistance force. This is a force which opposes the forward movement of a tillage tool, and is cyclic in nature because of the development of major shear failures in the soil.

Soil-stress characteristics during failure can be used to solve equations of stresses and strains developed when an implement is operated in a given soil body. Most of these models were developed for wide vertical bodies or two-dimensional models, such as earth-moving scrapers (Reece, 1965;

Hettiaratchi & Reece, 1967, 1974; Yong & Chen, 1970) or for three-dimensional bodies such as narrow tines (Hettiaratchi & Reece, 1967; Godwin & Spoor, 1977; McKyes & Ali, 1977; Perumpral et al., 1983). McKyes (1985) made a comparative prediction of draft forces using field data for sand (Luth & Wismer, 1971) and a clay soil (Payne, 1956).

Generally, the predicted values followed the same trends as measured ones, and the agreement was better for the clay soil. Experiments, however, conducted with various soil conditions showed that real values of draft force measured by means of dynamometers and transducers differ significantly from those obtained from theoretical models. This is because all formulae, proposed for calculating draft forces, are based on static models which assume the homogeneous conditions of tillage operation and all soil forces reacting to an inclined tillage tool. Modifications must be made in these analytical models if they are to account for dynamic changes, i.e. changes in forward velocity of the implement, rate of stress loading as it changes shear strength, shear or compressive strains, inertial effects and spatial changes in soil strength. The calculated draft force range was based on average values without regard to the distribution of the values about mean. The random nature of soil strength - the variation in dynamic soil properties - makes it impossible to predict the FSC in a deterministic manner. Thus, the application of stochastic processes may be more appropriate.

It is well documented (Gill & Vanden Berg, 1967) that the mean draft varies with changes in soil condition, however, the mean draft by itself probably does not characterize all of the dynamic properties of the soil. In the past, most studies have been concerned with the mean draft. Several researchers (Osman, 1964), however, have noted the cyclic variation in draft. Although most of these researchers reasoned that the instantaneous draft forces were related to dynamic soil properties, they did not attempt to establish a relationship. Other researchers (Belykh, 1972; Lur'e, 1972; Iofinov & Savel'ev, 1972; and Summers et al., 1985) have attempted to characterize soil condition using spectral density analyses of the draft of a tool. The spectral density function shows how the variance of a sequential process is distributed over frequency (Bendat & Piersol, 1989). Iofinov and Savel'ev proposed using the spectral density of the draft of a standard tool as an index of soil physical condition. Results of their research indicated that the spectral density varied with changes in soil physical condition.

Young et al. (1984) found that the spectral density of the draft of a narrow vertical chisel varied with changes in depth of operation and soil compaction. This type of analysis, as well as other frequency response methods are useful; however, they require the user to make many subjective decisions during the analysis procedure in order to arrive at a model of the system being studied. For instance, since the spectral density function yields a nonparametric description of the variation in draft, it is difficult to make quantitative comparisons of different soil conditions. Such comparisons are essential for tillage control applications, therefore, a quantitative description of SPC is needed.

Young et al. (1987) developed autoregressive models of tillage tool draft and the results of experiments indicated that the mean draft, residual draft, and autoregressive coefficients can be used to characterize soil physical condition.

Some major deficiencies, however, remain in this model. The first is the knowledge necessary to specify the DSC based on agronomic considerations, and the second is an understanding of how to adjust the tool shape and manner of tool movement to achieve desired results. The model dealt with forces and no connections to FSC have been made. Unless FSC is measured in terms that are directly related to an intended use, the model has a limited value.

Modeling final soil condition (FSC)

All models mentioned above deal with forces acting on the machine and the resultant energy or draft required, yet no model attempts to predict the structural condition of the soil once the implement has passed through soil elements, i.e., predict FSC. Changes in soil conditions on response to implement and tillage have been measured by many researchers. Some of these measurements are: random roughness of the surface (Allmaras et al., 1977); relative loosening, change in soil bulk density (Allmaras et al., 1977; Ojeniyi & Dexter, 1979a, b; McKyes & Desir, 1984; Cassel & Nelson, 1985); changes in hydraulic properties and infiltration (Ehlers, 1976; Edwards, 1982; Klute, 1982; Mielke et al., 1986); soil temperature and heat balance (Allmaras et al., 1977; Cruse et al., 1982; Gupta et al., 1984); changes in mechanical impedance (Bradford, 1980; Cassel, 1982; Cassel & Nelson, 1985); and aggregate-size distributions (Hadas et al., 1978; Hadas & Wolf, 1983). Many of these measured properties have characteristic ranges associated with particular implement types.

Dexter (1976) proposed a prediction of tillage-induced soil porosity, a prediction which uses a stochastic procedure to predict voids and solid. This soil-tilth prediction accounts for implement characteristics, management practice, soil water content at the time of tillage, consecutive implement passes, depth of operation, and compaction. Applicability has not been tested adequately because many measurements are needed to provide a new calibration for each crop management site and parameters involved.

Rawls et al. (1983) integrated tillage-induced porosity with other soil properties (contents of organic matter and mechanical parameters) to predict water flow-related properties in a soil subjected to tillage operations such as moldboard plowing, chiseling and wheel traffic after moldboard plowing. Brakensiek and Rawls (1983) then used typical values of random roughness to compute soil-crust effects in infiltration.

The NTRM model (Schaffer & Larson, 1987) developed at the University of Minnesota is the most comprehensive simulation model of soil-crop system. It has not yet been applied in a sufficiently wide range of situations to evaluate its effectiveness adequately. It does not include a component that focuses on the machine as it passes through the soil and on the soil physical condition that resulted after this pass. Continuing improvements in measured soil structure responses to tillage and models such as NTRM are mutually supporting.

The problems caused by natural and artificially-induced spatial variability of soil were not considered in models for predicting soil condition due to the tillage. Yet, inclusion of these variations using a probabilistic approach (with respect to time and space) will improve the incorporation of parameters related to machine and soil.

The interface between machine and the soil properties produced by the machine is the interface most needed for improving understanding for managing the machine-soil-plant system. It is an interface in which there are many processes and a deficiency of theory to obtain predictions. Research on this interface has much promise for the development of tillage systems to meet the needs of production, conservation and energy savings.

1. Modeling state changes in soil condition.

The soil physical system is continually being subjected to external forces, and therefore, soil physical condition is dynamic. These external forces may be natural (climate, plants, animals and micro-organism) or mechanical (forces applied by man using some type of machine). Soil in many ways is a much more complex medium than air or water, but any structure, be it a tillage tool, an airplane, or a ship is subjected to a spectrum of fluctuating forces on its surface as it passes through a medium.

Therefore, the tillage tool operates in conditions of random external influence. Among the factors concerning the external medium, the important ones are random fluctuations in the *soil resistance forces*. Thus, it is clear that the only satisfactory type of prediction about the soil condition produced after the tillage tool has passed through the soil must be on a probability basis. That is, we should want to be able to specify the probability distribution of the soil condition at any time. We might also want to go further and make probability statements about the actual times of occurrence of the various soil conditions, including desired soil condition. A model which specified the complete joint *probability distribution* of the possible soil conditions at each point of time would be a *stochastic model*, and the whole process, conceived as a continuous development in time, would be called a *stochastic process*.

1.1. Tillage - control process

For the proposed stochastic model, the following description of the tillage-control process will be considered (Fig. 43–1):

1. Initial soil condition (ISC) is measured, and the tillage-implement system is chosen to create desired soil condition (DSC).

2. When the first pass is completed, the final soil condition (FSC) is measured and compared with the DSC. If the DSC is reached, the tillage process is stopped.

3. In most cases, however, due to a soil variability and, sometimes, an inappropriate use of tillage implements during poor soil conditions, the DSC is not achieved and more passes are needed.

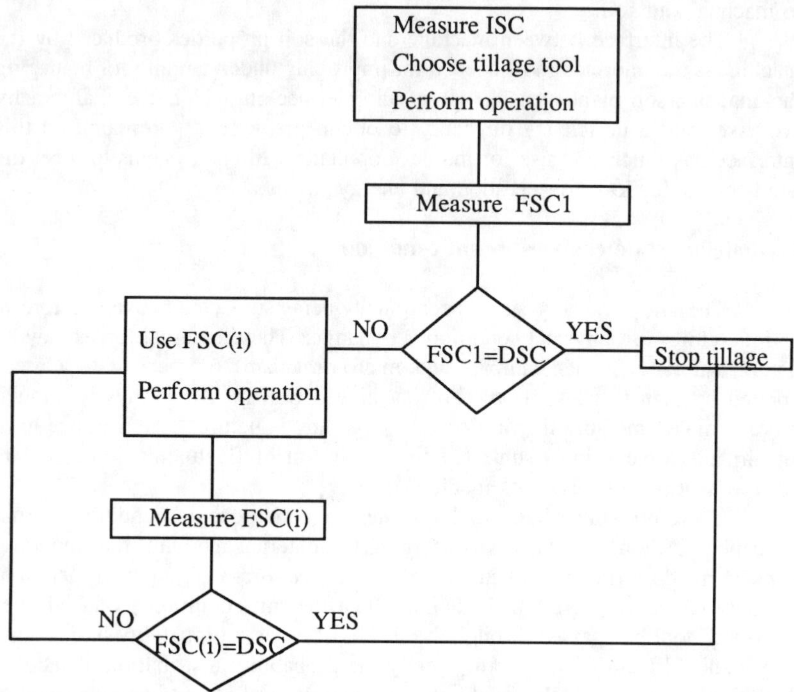

Fig. 43–1. Tillage process.

4. Tillage-implement parameters are adjusted and the second pass is performed.

5. The tillage process is repeated until the DSC is reached or until no significant differences in soil condition before and after the tillage are detected.

1.2. Random Walk Model

1. Soil condition states resulting from the tillage will be defined as $\mathbf{X_0, X_1, X_2, ..., X_n}$ where $\mathbf{X_0}$ is ISC, $\mathbf{X_i}$ is the soil condition state occurring after the tillage is applied to $\mathbf{X_{i-1}}$, and $\mathbf{X_n}$ is DSC (Fig. 43–2).

2. All possible transitions among soil condition states will be defined as $\mathbf{dX_0^1}$, $\mathbf{dX_0^2}$, $\mathbf{dX_i^{i+1}}$ where the subscript denotes soil condition states before the tillage and the superscript denotes soil condition states after the tillage. $\mathbf{dX_0^1}$, for example, means a transition between $\mathbf{X_0}$ and $\mathbf{X_1}$.

3. We will consider the transition process $\{\mathbf{dX_i}\}$ associated with $\{\mathbf{X_i}\}$ as an independent differently distributed (i.d.d) sequence with unknown transition probabilities.

4. The soil condition state process will be given by

$$X_0 + \sum_{n-1}^{i=0} dX_i^{i+1} = X_n$$

which represents a *random walk model.*
The name "random walk" is motivated by the idea of regarding the random variable $\mathbf{X_i}$ as being the position at time i of the soil condition changing on a straight line in such a manner that at each step the soil condition either remains where it is or jumps to the right or left according to the distribution functions of $\mathbf{dX_i^{i+1}}$ and $\mathbf{dX_{i-1}^{i}}$.

5. Evaluation of distribution functions of $\mathbf{dX_i^{i+1}}$, $\mathbf{dX_0^{i,..,n}}$, and $\sum_{n-1}^{i=0} dX_i^{i+1}$ will be the major points of our interest.

6. Some possible outcomes of the model:
   a) Prediction of FSC.
   b) Probability of reaching DSC in n steps.
   c) Minimum number of steps to ensure achieving DSC.
   d) Probability of overtilling the soil.
   e) Estimation of different ways to achieve DSC and choosing the optimum one.

1.3. Soil resistance-force (SR-F) as a parameter used in model

In light of discussing the use of dynamic soil properties when describing a dynamic behavior of the soil (see Literature Review), the *soil resistance force (SRF)* will be chosen as a parameter to reflect the changes in the soil condition during tillage.

1. The value of the SR-F will be used as an indicator of the soil condition state and will serve as an input to the random walk model.

2. The SR-F will be measured using an extended octagonal ring transducer (Cook & Rabinowisch, 1963; Godwin, 1975; Gebresenbet, 1989)

3. To receive information about two consecutive soil states $\mathbf{X}_i$ and $\mathbf{X}_{i+1}$, two SR-F sensors will be mounted, ahead and behind the tillage tool and SR-F data will be continuously collected (Fig. 43–2). The transition $\mathbf{dX}_i^{i+1}$ will be estimated as a difference $\mathbf{X}_{i+1}$-$\mathbf{X}_i$ for each measured point.

4. The transducers will provide continuous time signals. For data acquisition, the signals will be sampled periodically, the sample frequency being as the reciprocal of the time interval between two successive samples.

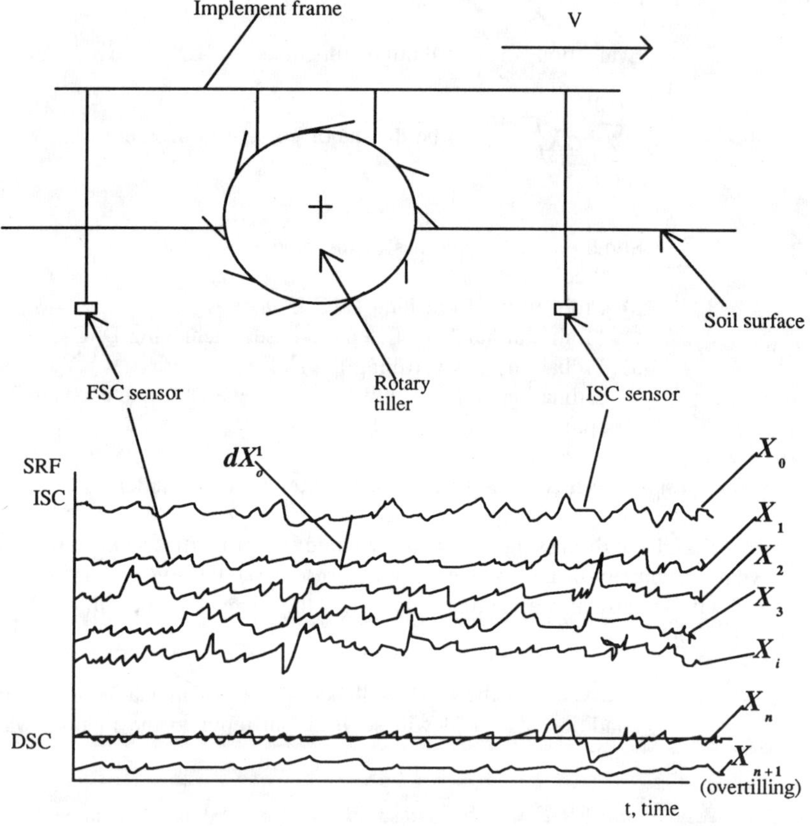

Fig. 43–2. Soil condition states.

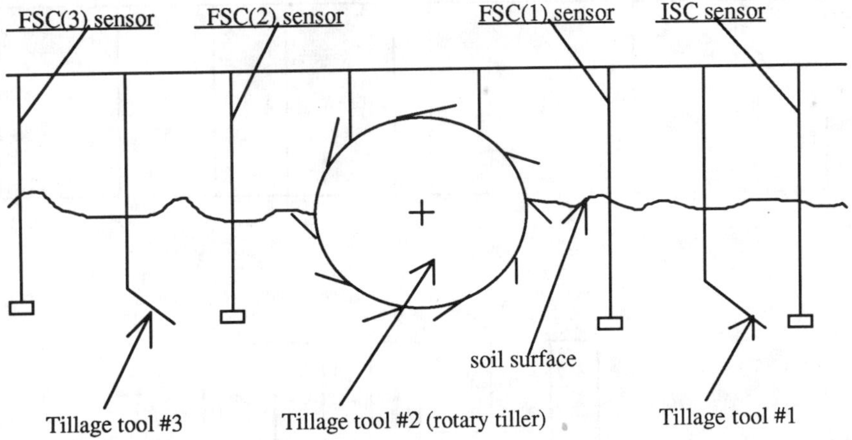

Fig. 43–3. Multipass-tillage tool.

5. As far as the sample frequency is concerned, the following considerations will be taken into account. Shannon's rule (Priestley, 1981) states that, the sample frequency should be twice the highest frequency occurring in the signal. On the other hand the autocorrelation of the signals should also be taken into consideration.

6. The SR-F data will be analyzed by using spectral analysis (Priestley, 1981) to account for a great variability of soil conditions.

1.4. Creating soil condition states for a model.

1. A rotary tiller with two operating parameters (ω, angular velocity, rad/s and υ, forward velocity, km/h) and a multipass-tillage tool with the ability to change the number of tillage tools engaging simultaneously in tillage (Fig. 43–3), will be used as tillage tools.

2. The range of angular, forward velocities, and the number of tillage tools of the multipass-tillage tool will be used to create different soil condition states (including DSC) so that the relations between control parameters and the probability of certain soil condition states can be obtained.

3. Two extreme soil types, for example, sandy loam and silty clay, will be used to account for the different reactions to the movement of a tillage tool.

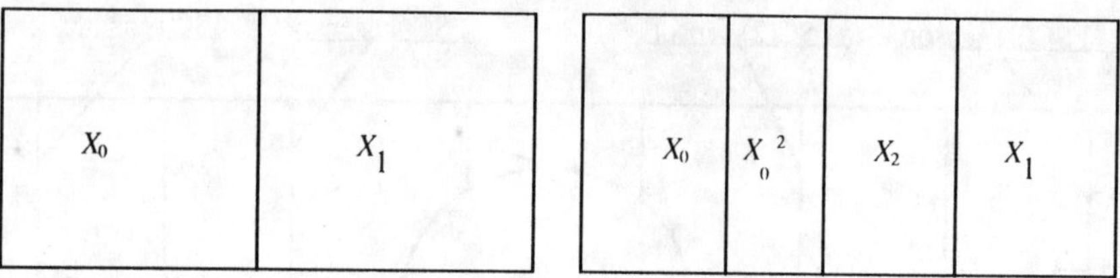

Fig. 43–4a.

Fig. 43–4b.

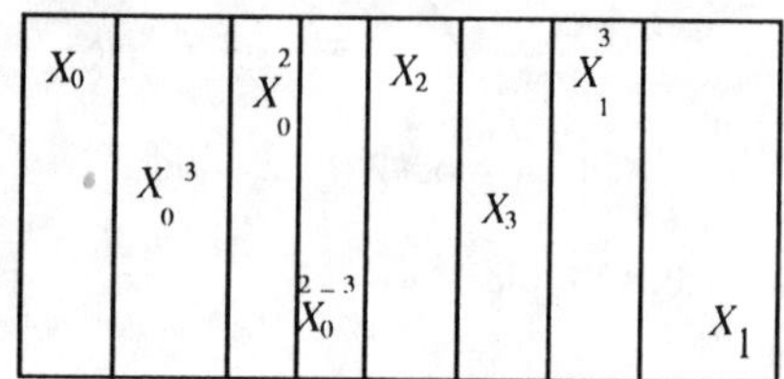

Fig. 43–4c.

Fig. 43–4. Soil condition states preparation.

4. Soil physical properties (particle size distribution, bulk density, soil moisture content, and cone index) will be measured to account for the difference in the initial soil conditions.

5. Initial soil conditions will be maintained the same throughout all experiments.

6. Tillage plots will be prepared as follows:
   a) The entire plot will be divided into two equal parts.

   b) One part of the plot will be tilled, creating $\mathbf{X_1}$ and second part will remain undisturbed, representing $\mathbf{X_0}$ (ISC) (Fig. 43–4a). The transition $\mathbf{dX_0^1}$ can be evaluated.

   c) With adjusted operating parameters ω and υ, the second pass will be performed on part of $\mathbf{X_0}$ and $\mathbf{X_1}$, creating $\mathbf{X_2}$ and $\mathbf{X_0^2}$(the soil condition state resulting directly from $\mathbf{X_0}$) (Fig. 43–4b). The transitions $\mathbf{dX_1^2}$ and $\mathbf{dX_0^2}$ can be evaluated.

d) With the different operating parameters, the third pass will be performed on the parts of already existed soil condition state plots, creating $\mathbf{X_3}$, $\mathbf{X_0^3}$, $\mathbf{X_1^3}$ and $\mathbf{X_0^{2\text{-}3}}$ (Fig. 43–4c). The transitions $\mathbf{dX_2^3}$, $\mathbf{dX_0^3}$, $\mathbf{dX_1^3}$, $\mathbf{dX_0^{2\text{-}3}}$ can be obtained.

e) The tillage will be stopped when the DSC is reached or when soil overtilling is occurring (this will happen when the sensor is showing the SR-F data below DSC range (Fig. 43–2)). If *n* is a number of passes needed to complete experiment, then $\mathbf{2^n}$ is a number of plots needed to be tilled to obtain all transition probabilities.

2. Model validation experiments.

1. Using ISC (measured in the field) and DSC, the proposed model will generate a number of paths to reach the DSC with the different probabilities. Each path corresponds to a certain set of tillage control parameters.

2. Several sets of parameters will be chosen to verify the possibility of reaching DSC.

3. The set of parameters, which gives the maximum probability of achieving DSC according to model, will be used to till the soil and resulting soil conditions will be compared with the DSC.

3. Proposed tillage-control system.

1. A tillage-control system will be based on a proposed model with possibilities to change the soil condition "on the go". The on-board computer will recommend a optimum path of reaching the DSC based on continuous measurements of ISC and FSC resulting after the tillage.

2. A small part of the field will serve as a test area (Fig. 43–5), where the tillage-control system will adapt to the local soil conditions, and the recommendations from computer about the tillage control parameters will be made (Fig. 43–6).

3. As soon as the FSC coincides with the DSC, the soil-implement system will move to the main part of field with continuous monitoring of initial and resulting soil conditions.

4. If the DSC has not been attained, the model will adjust to a new situation and will provide a tillage-control system with a different set of tillage parameters, which has a high probability of reaching DSC (Fig. 43–7).

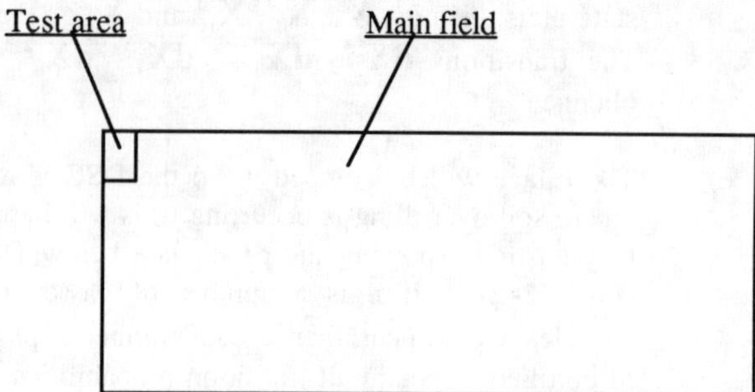

Fig. 43–5. The location of test area.

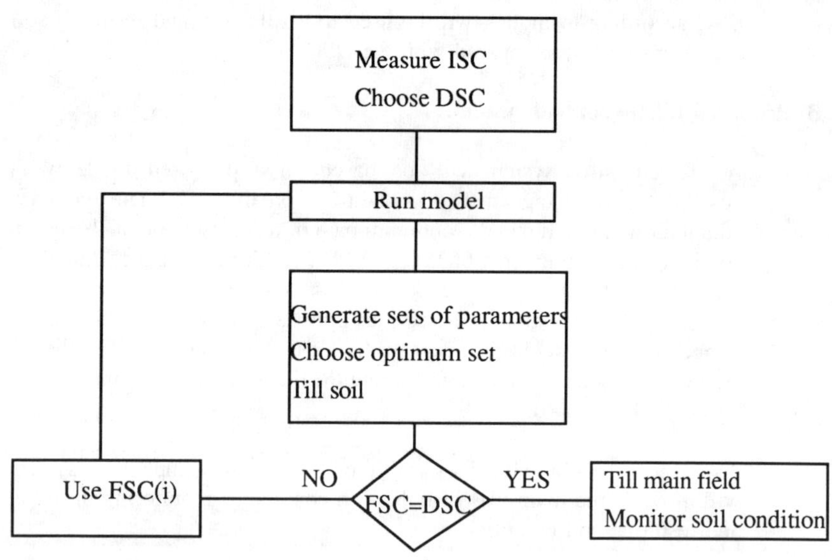

Fig. 43–6. Tillage-control system on test area.

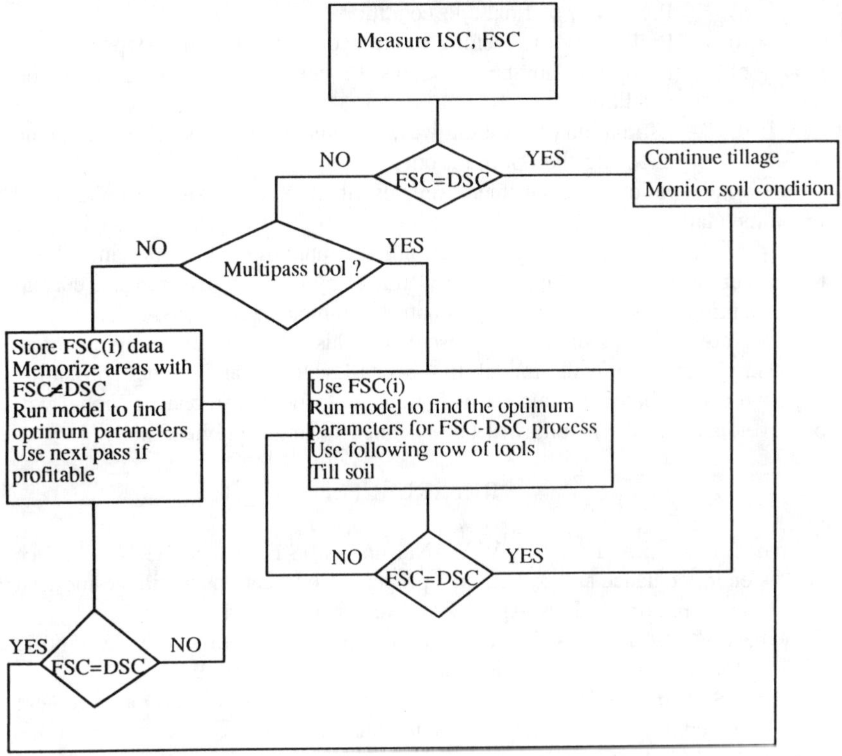

Fig. 43–7. Tillage-control system on a main field.

5. If the soil is tilled using the multipass-tillage tool (two or more rows of tillage tools), the inadequacies of the first pass (not reaching DSC) will be compensated by additional rows of tillage tools with adjustable control parameters.

6. If a soil is tilled with a single implement, inadequacies of the first pass will be compensated by making additional passes if it is worthwhile, that is, if the areas with soil conditions different from DSC make up a significant part of the entire field.

SUMMARY

The purpose of the research described in this paper is to model changes in soil physical condition using dynamic properties as parameters defining soil condition in the model. Specifically, changes in soil physical conditions as they relate to soil tillage are investigated.

A random walk model is proposed to describe the random transitions of soil conditions resulting from tillage. Some possible outcomes of the model:

a) Prediction of final soil condition.
b) Probability of reaching desired soil condition in n steps.
c) Minimum number of steps to ensure achieving desired soil condition.
d) Estimation of different ways to achieve desired soil condition and choosing the optimum one.

Model validation experiments are described to verify random walk model recommendations.

A tillage-control system, which utilizes random walk model and tillage transducers to control the application of tillage by continuously sensing current soil conditions and adjusting tillage control parameters, is discussed.

Note: The research plan, presented in this paper, is being implemented. The simple random walk model has been developed and model validation experiments are being conducted. The more sophisticated stochastic model will be developed and the results will be presented in future publications.

REFERENCES

Allmaras, R.R., E.A. Hallauer, W.W. Nelson, and S.E. Evans. 1977. Surface energy balance and soil thermal property modifications by tillage-induced soil structure. Minn. Agric. Exp. Stn. Bull. 306. 44 pp.

Belykh, V.V. 1972. Statistical evaluation of loads on a plow body. Scientific works of the Kazan Agricultural Institute 15(1): 16-19.

Bendat, J.S., and A.G. Piersol. 1989. Random data: Analysis and measurement procedures. John Wiley and Sons, Inc. New York.

Bradford, J.M. 1980. The penetration resistance in a soil with well-defined structural units. Soil Sci. Soc. Am. J. 44: 601-606.

Brakensiek, D.L., and W.J. Rawls. 1983. Agricultural management effects on soil water processes. Part 2: Green and Ampt parameters for crusting soil. Trans. A.S.A.E. 26: 1753-1757.

Cassel, D.K. 1982. Tillage effects on soil bulk density and mechanical impedance. pp. 45-67. *In* P.W. Unger, et al, (eds.) 1982. Predicting tillage effects on soil physical properties and processes. Am. Soc. Agron. Madison, WI. Spec. Publ. 44.

Cassel, D.K., and L.A. Nelson. 1985. Spatial and temporal variability of soil physical properties of Norfolk loamy sand as affected by tillage. Soil Tillage Res. 5: 5-17.

Cook, N.H., and E. Rabinowisch. 1963. Physical measurement and analysis. Addison-Wesley Publishing Co. Reading, MA. pp. 160-164.

Cooper, A.W., and W.R. Gill. 1966. Characterization of soil related to compaction. Grundforbattring AGR 19: 77-80 NRI. Uppsala, Sweden.

Cruse, R.M., K.N. Potter, and R.R. Allmaras. 1982. Modeling tillage effects on soil temperature. pp. 133-150. *In* P.W. Unger, et al, (eds.) 1982. Predicting tillage effects on soil physical properties and processes. Am. Soc. Agron. Madison, WI. Spec. Publ. 44.

Dexter, A.R. 1976. Internal structure of tilled soil. J. Soil Sci. 27: 267-278.

Dexter, A.R. 1988. Advances in characterization of soil structure. Soil and Tillage Research. 11: 199-238.

Edwards, W.M. 1982. Predicting tillage effects in infiltration. pp. 105-155. *In* P.W. Unger, D.M. Van Doren Jr., F.D. Whisler, and E.L. Skidmore (eds.) 1982. Predicting Tillage Effects on Soil Physical Properties and Processes. Am. Soc. Agron., Madison, WI, Spec. Publ. 44.

Ehlers, W. 1976. Rapid determination of undisturbed hydraulic conductivity in tilled and untilled loess soils. Soil Sci. Soc. Am. J. 40: 837-840.

Gill, W.R., and G.E. Vanden Berg. 1967. Soil dynamics in tillage and traction. USDA Agric. Handb. 316. U.S. Gov. Print. Office. Washington, DC.

Gebresenbet, G. 1989. Multicomponent dynamometer for measurement of forces on plough bodies. J. Agric. Eng. Res. 42: 85-96.

Godwin, R.J. 1975. An extended octagonal ring transducer for use in tillage studies. J. Agric. Eng. Res. 20(4): 345-352.

Godwin, R.J., and G. Spoor. 1977. Soil failure with narrow tines. J. Agric. Eng. Res. 22: 213-228.

Gupta, S.C., B. Lowery, J.F. Moncrief, and W.E. Larson. 1991. Modeling tillage effects on soil physical properties. Soil Tillage Res. 20: 293-318.

Gupta, S.C., and W.E. Larson. 1982. Modeling soil mechanical behavior during tillage. pp. 151-178. *In* P.W. Unger, D.M. Van Doren Jr., F.D. Whisler, and E.L. Skidmore (eds.) 1982. Predicting tillage effects on soil physical properties and processes. Am. Soc. Agron. Madison, WI. Spec. Publ. 44.

Gupta, S.C., W.E. Larson, and R.R. Allmaras. 1984. Predicting soil temperature and soil heat flux under different tillage-surface residue conditions. Soil Sci. Soc. Am. J. 48: 223-232.

Gupta, S.C., and R.P. Raper. 1994. Prediction of compaction under vehicles. *In* B.D. Soane, and C. Van Ouwerkerk (ed.) Soil Compaction in Crop Production. Elsevier, Amsterdam.

Hadas, A., D. Wolf, and I. Meirson. 1978. Tillage implements, soil structure relationships, and their effects on crop stands. Soil Sci. Soc. Am. J., 42: 633-637.

Hadas, A., and D. Wolf. 1983. Energy efficiency in tilling dry clod-forming soils. Soil Tillage Res. 3: 47-59.

Hadas, A., W.E. Larson, and R.R. Allmaras. 1988. Advances in modeling machine-soil-plant interactions. Soil Tillage Res. 11: 349-372.

Hettiaratchi, D.R.P., and A.R. Reece. 1967. Symmetrical three dimensional soil failure. J. Terramech. 4: 45-67.

Hettiaratchi, D.R.P., and A.R. Reece. 1974. The calculation of passive soil resistance. Geothechnique 24: 289-310.

Iofinov, A.P., and A.V. Savel'ev. 1972. The statistical evaluation of quality of work of soil working machines. Reports of the Chelyabinsk Institute of Mechanization and Electrification of Agriculture 57: 35-42.

Klute, A. 1982. Tillage effects on the hydraulic properties of soil: A review. pp. 29-42. *In* P.W. Unger, et al. (eds.) 1982. Predicting tillage effects on soil physical properties and processes. Am. Soc. Agron. Madison, WI. Spec. Publ. 44.

Lur'e, A.B. 1972. Statistical evaluation of the draft resistance of soil working machines. Mechanization and Electrification of Socialist Agriculture 10: 50-52.

Luth, H.J., and R.D. Wismer. 1971. Performance of plane soil cutting blades. Trans. A.S.A.E. 14: 255-259.

McKyes, E. 1985. Soil Cutting and Tillage. Developments in Agricultural Engineering. Vol 7. Elsevier, New York. 217 pp.

McKyes, E., and O.S. Ali. 1977. The cutting of soil by narrow blades. L. Terramech. 14: 43-58.

McKyes, E., and F.L. Desir. 1984. Prediction and field measurements of tillage tool draft forces and efficiency in cohesive soils. Soil Tillage Res. 4: 459-470.

Meyer, G.E., R.B. Curry, J.G. Streeter, and H.J. Mederski. 1979. SOYMOD/OARDC; A dynamic simulator of soybean growth, development and seed yield: 1. Theory, structure and validation. Res. Bull. 113. Ohio Agric. Res. and Dev. Cent. Wooster, OH. 36 pp.

Mielke, L.N., J.W. Doran, and K.A. Richards. 1986. Physical environment near the surface of plowed and non-tilled soils. Soil Tillage Res. 7: 355-366.

Ojeniyi, S.O., and A.R. Dexter. 1979a. Soil factors affecting the macro-structures produced by tillage. Trans. ASAE. 22: 339-343.

Ojeniyi, S.O., and A.R. Dexter. 1979b. Soil structure change during multiple pass tillage. Trans. ASAE. 22: 1068-1072.

Osman, M.S. 1964. The mechanics of soil cutting blades. J. Agric. Eng. Res. 9:4 p. 313.

Payne, P.C. J. 1956. The relationship between the mechanical properties of soil and the performance of simple cultivation implements. J. Agric. Eng. Res. 1: 23-50.

Perumpral, J.V., J.B. Liljedahl, and W.H. Perloff. 1971. The finite element method for predicting stress distribution and soil deformation under tractive device. Trans. ASAE. 14: 1184-1188.

Perumpral, J.V., R.D. Grisso, and C.G. Desai. 1983. A soil-tool model based on limit equilibrium analysis. Trans. ASAE. 26: 991-995.

Porter, M.A., and T.A. McMahon. 1990. MUTillS - The Melbourne University Tilled Soil Model. Trans. ASAE. 33(2): 419-431.

Priestley, M.B. 1981. Spectral analysis and time series. Vols. 1 and 2. Academic press. London.

Raper, R.L., and D.C. Erbach. 1990. Prediction of soil stresses using finite element analysis. Trans. ASAE 33: 725-730.

Rawls, W.J., D.L. Brakensiek, and B. Soni. 1983. Agricultural management effects on soil water processes. Part 1: Soil water retention and Green and Ampt infiltration parameters. Trans. ASAE. 26: 1747-1752.

Reece, A.R. 1965. The fundamental equation of earth moving mechanics. Symposium on earth moving machinery. Institute of Mechanical Engineering, 179, Part 3F. London, Gt. Britain. 179 pp.

Schafer, R.L., and C.E. Johnson. 1982. Changing Soil Condition - The Soil Dynamics of Tillage. pp. 13-28. *In* P.W. Unger, et al, (eds.) 1982. Predicting tillage effects on soil physical properties and processes. Am.

Soc. Agron. Madison, WI. Spec. Publ. 44.

Schafer, R.L., C.E. Johnson, C.B. Elkins, and J.C. Hendrick. 1985. Prescription tillage: The concept and examples. J. Agric. Eng. Res. 32: 123-129.

Schafer, R.L., and C.E. Johnson. 1990. Soil dynamics and cropping systems. Soil Tillage Res. 16 (1990) 143-152.

Shaffer, M.J., and W.E. Larson (ed.) 1987. NTRM. Soil-crop simulation model for nitrogen, tillage, and crop-residue management. Conservation Research Report 34-1. USDA-ARS. Washington, DC. 103 pp.

Sohne, W.H. 1953. Pressure distribution in the soil and soil deformation under tractor tires. Grundlagen, Landtech. 5: 49-84.

Summers, J.D., M.F. Kocher, and J.B. Solie. 1985. Frequency analysis of tillage tool forces. Proc. International Conference on Soil Dynamics. Auburn, Al.

Vanden Berg, G.E., and C.A. Reaves. 1966. Characterization of soil properties for tillage tool performances. Grundforbattring AGR 19: 49-58 NRI. Uppsala, Sweden.

Unger, P.W., D.M. Van Doren, Jr., F.D. Whisler, and E.L. Skidmore. (ed.) 1982. Predicting tillage effects on soil physical properties and processes. Am. Soc. Agron. Madison, WI. Spec. Publ. 44, 198 pp.

Whisler, F.D., J.R. Lambert, and J.A. Landivar. 1982. Predicting tillage effects on cotton growth and yield. pp. 179-198. *In* P.W. Unger, et al. (ed.) 1982. Predicting tillage effects on soil physical properties and processes. Am. Soc. Agron. Madison, WI. Spec. Publ. 44.

Yong, R.N., and Chen. 1970. Analytical and experimental studies of soil cutting. ASAE. St. Joseph, MI. ASAE Paper No.70-676.

Young, S.C., C.E. Johnson, and R.L. Schafer. 1984. Real-time quantification of soil physical condition. ASAE Paper No. 84-1037. ASAE. St. Joseph, MI.

Young, S.C., C.E. Johnson, and R.L. Schafer. 1987. Quantifying soil physical condition for tillage control applications. ASAE Paper No. 86-1620. ASAE. St. Joseph, MI.

SESSION III

ENGINEERING TECHNOLOGY

44 Engineering Systems for Site–Specific Management: Opportunities and Limitations

Stephen W. Searcy

Department of Agricultural Engineering
Texas A&M University
College Station, Texas

Site specific crop management (SSCM) is a technology that depends heavily on engineering systems. This new means of managing agronomic production is based on the availability of detailed data describing growing conditions on a plant level. The use of extensive data in the management process requires engineering systems for the data collection, data analysis, decision support, and in-field operations. An additional requirement for successful site specific management is the ability to locate machines and people as they travel across fields. These technologies rely heavily on electronics and the use of long range positioning systems.

The concept of site specific crop management springs from the recognition that spatial variability occurs in the many parameters that affect crop production (soil type, fertility, slope, microclimate, etc.). Equipment is available for varying production inputs (chemicals and fertilizers) as a function of location in a field. Techniques for monitoring the yield during harvest allow us to create yield maps that document the variability and can be used in future management decisions. With SSCM, the producer can vary the amounts and placement of inputs, so that their use is optimized.

While broadly discussed and understood, SSCM does not yet have a universally accepted definition. For the purposes of this talk, I will use the following definition.

> *Site specific crop management is the use of local soil and crop parameters to make precise applications of production inputs to small areas with similar characteristics.*

Like all good definitions, this one is broad enough to accommodate several situations and practices. SSCM can encompass different practices as required by regional conditions. There are some commonalities, however, regardless of crop or region. The phases of greatest importance, and which will be the focus of my talk, are "prescription application" and "small areas with similar characteristics."

Agriculture's Challenge

In recent yr, agriculture's image has become tarnished. The food and fiber producers whose productivity have allowed increasingly high standards of living, are accused a variety of sins. The American public widely holds the opinion that too much emphasis has been placed on increasing productivity, while too little emphasis has been placed on environmental conservation. Agriculture has been documented as serious polluter of surface and groundwaters (CAST, 1992). Much of this pollution is in the form of sediment that occurs during runoff events. There are significant amounts of chemicals and nutrients, however, which move into water sources from agricultural fields.

The use of agricultural chemicals has created a food safety concern with many consumers. The result of these various concerns has been a gradual loss in confidence in agriculture by the consuming public in the United States. K. Rickeldorfer (1991), in a speech given before the American Society of Agricultural Engineers, addressed agriculture's contract with society, and specifically the question, "What does society want?" She presented a number of statistics, some of which may be familiar to you. In a 1985 poll, 60% of Americans agreed with the statement "Farmers use too many pesticides." Even closer to home, in a poll conducted in Iowa, three-quarters of the Iowa farmers surveyed, agreed that modern farming relies too heavily on insecticides, herbicides, and chemical fertilizers. Perhaps an observation by D. Cloud (1990) best sums up where we in agriculture find ourselves today.

> *"Environmentalists have spent a quarter of a century repudiating the notion that tillers of the soil are the best stewards of the land, water, and food supply. They finally have the public nodding its head and farmers shaking theirs, bewildered by their sudden unpopularity."*

While we in agriculture may wish to debate many of these points, and to disagree with the "bad guy" image, we must deal with the reality of these perceptions. In this task, a little historical perspective may be useful.

In the 1970s and 1980s, U.S. manufacturing firms also came under hard times. Many manufacturers found that they were losing their market share to foreign competition that was producing higher quality products and selling them at a lower price. As a result, profit margins were decreasing, consumers were losing confidence in the "Made in America" label, and many firms had a questionable long-term future. Manufacturing firms turned to several possible solutions to attempt to solve this dilemma. One of them was a concept developed in America, but first adopted by our foreign competitors.

"Just In Time" (JIT) manufacturing is a system of eliminating waste in all parts of the product design, construction, and marketing system. Waste is eliminated by balancing the material availability with the need for product as determined by the marketing operation. This results in less inventory, both in-process and completed. In addition, an emphasis on quality as determined by the customers is central to the JIT concept. A focus on the needs and perceptions of the customer is critical to designing an effective and profitable manufacturing business. Matching the product quality with the customer expectations minimizes the expense associated with returns and warranty operations.

Figure 44–1 shows the basic steps involved in the "Just In Time" philosophy. The philosophy starts with an emphasis on quality as determined by the consumers. A simplified, synchronous production schedule helps to eliminate the cost of excessive inventory. The production rate is synchronized with the marketing rates. If product demand declines, that information is immediately transferred to the production line. Close attention to the marketing demand results in less inventory and fewer storage losses.

A process oriented flow organizes the manufacturing plant around the product requirements, rather than manufacturing functions. An example is a plant producing five different products, each of which requires some welding assembly. Through a process-oriented flow, five sections of the plant would be organized for the individual products, each with welding capability. This is in contrast with the old style of plant design where each of the five products would be brought to a common welding area. This design often resulted in partially completed products sitting in a queue for long time periods.

Advanced procurement technology describes the means of interacting with material suppliers. "Just In Time" management requires that the manufacturing plant be in close contact with the suppliers, and that those suppliers know what exactly is needed and when it will be needed. This requires good communication and a cooperative relationship. Employee involvement is a critical aspect throughout all facets of the "Just In Time" management philosophy.

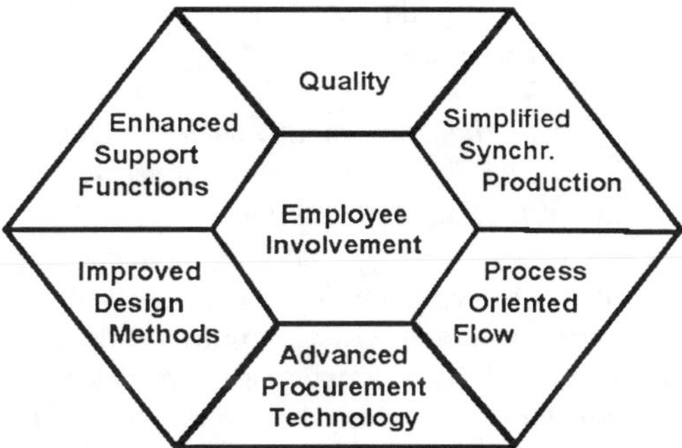

Fig. 44–1. Functions of the Just In Time manufacturing philosophy.

All of the above functions and activities, taken as a whole, implement the "Just In Time" philosophy. Experts on JIT management indicate that a strong commitment to the entire concept is necessary for JIT to be successful. It is not difficult to see similarities between JIT and SSCM. There is a need to recognize and respond to the quality expectations of the customer. Production inputs such as seed, fertilizer, chemicals and machine operations are comparable to sheetmetal, plastic, bolts and welders. Through well-informed and knowledgeable management, the cost of these inputs can be minimized and waste eliminated. Decision support tools and expanded databases allow the manager to be more precise in decision-making. The relationships with service and material vendors need to be more coordinated in order to implement that precise management.

SSCM PHILOSOPHY

Site specific crop management calls for a change in thinking similar to that necessary when manufacturing plants convert to the "Just In Time" philosophy. Producers adopting SSCM need to view their fields as a collection of small management units, each of which can be optimized. Conventional agriculture techniques are based on prophylactic applications of chemicals and fertilizers. We aren't sure how the field varies, where the weeds will come up or the amount of nutrients needed, so we try to apply enough material to ensure profitable production. This approach has a number of limitations. Without detailed knowledge of the variability in the field, a farmer either has to meet the needs of the most productive soils (resulting in over application in other areas) or apply to the average conditions (resulting in both over- and under-application).

Site specific crop management techniques allow the manager to have prescription application of materials for his crop. Through the use of controllers, database software, and positioning systems, its possible to apply a precise amount of chemical or nutrients based on the local conditions of the soil and crop. Just as the JIT manufacturing approach requires a new way of thinking and planning, SSCM requires the producer to adjust his thinking as well. SSCM is a collection of technologies that must come together to create a production system that responds to crop needs and addresses the necessity for sustainable systems. SSCM should be viewed not as just a set of tools, but as a philosophy and a system to implement that philosophy. That philosophy is one based on trying to eliminate waste and increase efficiency.

SSCM Philosophy: Production inputs, such as seed, fertilizers and chemicals, should be applied only as needed and when needed to maintain profitable production.

Figure 44–2 illustrates the various knowledge areas, tools and techniques that, taken together, form the site specific crop management system. Each of these items will be discussed in turn, looking at issues that present themselves, the current status of development, and needs for the future. The emphasis will be on the engineering needs and opportunities.

Quality in SSCM

Quality is a buzzword currently being embraced in many businesses. While there may be some overuse of the term and its various different schemes for implementation, it behooves us to look at quality in terms of agronomic production. A basic step in analyzing the quality needed for a product, is to ask the questions "Who are my customers?" and "How do they define quality?" Agricultural producers can be grouped into two categories. The first would be the traditional customers that we are all familiar with - grain elevators, millers, textile plants, food processors. These are the people that we've been dealing with for yr and we think we understand their desires. We are hearing more and more from non-traditional customers, such as government agencies, neighbors, consumer and environmental groups. We know reasonably well what the traditional customers want. They need a high quality commodity, but without chemical residues that may cause difficulty for them in marketing the products. Our non-traditional customers are no longer concentrating on cheap, plentiful food and fiber, but in sustainability of the agricultural production systems, minimal environmental impact of that production, and assurances of a safe food.

Agriculture has yet to broadly embrace the customer's view of quality production. Until those concerns are acknowledged and adequate responses developed, agriculture will continue to have a negative image. SSCM is a technology that has the potential to address the quality concerns of both our traditional and non-traditional customers. At the same time, it can benefit financially, the producer who adopts site specific crop management. The practice of applying only those materials necessary to meet the needs in a given location, is well received within the non-agricultural community.

The environmental advantages of prescription applications seem to be implicit. We currently lack sufficient hard data, however, that demonstrates the impact of SSCM on the environment. This is particularly true for the wide range of crops and regions in which the management practice may be adopted. The question remains to be answered, "Does SSCM address the concerns of society over agricultural production?"

Fig. 44–2. Functions in a site specific crop management system.

Management Information

SSCM is dependent entirely on the availability of accurate information for managing agronomic production. Systems, devices and techniques are needed which can collect the needed data and analyze it into more useful forms. Information is needed for the soil (type, water holding capacity, moisture content, nutrient concentrations, organic matter, slope), crop (nutrient or water stress, maturity, yield) and pests (weeds, insects). To date, most data collection has relied on the use of soil sampling in large grids that overlay a field. Grid spacings of 60–120 meters (200–400 feet) have been suggested as a balance between accuracy and cost. This approach has a high cost for both gathering and analyzing the large numbers of samples.

On-the-go sensors have the potential for greatly reducing costs, but have had limited success in practice. While nitrate sensors have been investigated for control of fertilizer application (Adsett & Zoerb, 1991), no device is currently accepted as reliable. An infrared reflectance, soil organic matter sensor has seen commercial use for fertilizers and herbicides (Tyler, 1994). While research to recognize weeds in standing crops has been underway for several yr (Thompson, et al., 1990; Guyer, et al., 1993), commercial developments are still in the future.

Grain yield sensors have been developed by several manufacturers, and are beginning to have broader acceptance by producers. Currently, most of these sensors are installed as monitors rather than mapping devices. There are many other crops which need similar yield sensing developments. Certainly in the South, cotton is a major crop with potential for SSCM. For all of the sensed parameters, data quality is a great concern. If management decisions are to be made on the basis of site specific data, it is imperative that data be as accurate as possible.

The collection of many field parameters will result in millions of bytes of information. Since this is far more data than an individual can process manually, data analysis tools are necessary. Site specific data analysis tools are becoming more available, and they seem to fall into two basic types, generic and ag-specific. The first is a generic geographical database, or geographic information system (GIS). These systems are intended for broad geographic management applications, typically have great capabilities to display and analyze the available data and are the preferred tools of researchers. The expanded capabilities can be harder to learn and not as attuned to agricultural needs. Agricultural mapping software packages have been oriented toward display of fixed features (waterways, fences, roads, buildings, etc.). These packages are also being enhanced with geographical database analysis tools that are needed for SSCM. Both types of software packages are trending toward a common set of capabilities.

Geographical data processing takes basically two forms. For site specific management, a field is broken into management units that have similar properties. Each of these units must have parameter values assigned. Sparse datasets, such as soil samples, require interpolation of the data to fill in the unsampled areas. Dense datasets, such as yield maps and aerial photographs, require data reduction in order to assign values to the management units. Standardized and validated techniques are needed for both of these functions.

Techniques for combining multiple data layers (both current and historical) to generate additional layers should be developed. An example would be the combination of soil type, organic matter and historical weed data layers to generate a herbicide application rate layer.

Agronomic Sciences

Agronomic sciences, including soil science, provide a basis for understanding the significance of soil and crop parameters. That understanding provides tools and relationships that can be used to manage the inherent variability of the soil and crop. Not all variability is understood. Several questions impact directly on the central focus of site specific management.

What are the sources of the yield variability, and how can they be recognized?
How do we handle uncontrollable variability, such as weather and soil type?
What is the significance of each source of variability in determining the profitability of a crop?
What variability can we manage and what can't we manage?

Previous speakers in this conference have questioned the appropriateness of using general relationships, developed over a wide range of soil types, when dealing with specific soil types. SSCM as an agronomic practice must also be compatible with other management practices. For example, how compatible are integrated pest management and site specific crop management? I think there's potential for a synergistic linkage, but I am unaware of any work in this area. Today we have a very large body of general knowledge, but yet the sources and effects of yield variability are not fully understood. Before SSCM will be widely accepted, agronomists must distinguish deterministic sources of yield variability from those that are stochastic. SSCM is likely to only address that variability that can be predicted and controlled. While most SSCM discussion has concentrated on spatially variable crop needs, temporal variation should also be considered. The practice of using the soil to hold excess nutrient inventory until the plant requires it may not be the best approach when considering producer risk and the potential for environmental impact.

Decision Support Systems

A large body of knowledge and extensive datasets make decision support systems (DSS) essential for SSCM. When a field that was formerly considered to be a single unit is managed as a collection of distinct areas, the number of management decisions is greatly increased. This increased workload will deter adoption of SSCM if adequate decision support systems are not available. In many cases, the producer may rely on consultants to assist implementing SSCM. Since consultants handle multiple fields for multiple customers, the need for DSS tools will likely be greatest for agronomic consultants.

Plant growth models have been developed for many crops, but few are directly applicable to the simultaneous management of the multiple growing conditions expected in site specific management. Most models are one-dimensional, predicting the performance of a single plant under a given set of soil and moisture conditions. Hydrologic models are available for determining the movement of water and soil borne nutrients, both above and below ground. While these transport models consider watershed areas, they are less concerned about the accuracy of the crop yield predictions. Multiple simulations in a field with varying conditions have been attempted (McCauley, et al., 1994), but this work has been limited and required a level of computing power that is unrealistic in commercial practice. Running parallel simulations is inefficient because weather (a highly significant model input) would normally be considered constant over a field. Improved models are need to simulate plant growth and hydrologic phenomenon, while considering those parameters that vary geographically.

Few decision support systems are available for SSCM today. Most capabilities are limited to manual combination of data layers in a GIS. While new linkages between GIS's and models are being developed, more effort is needed to target SSCM. The human input in the decision support systems should be limited to a supervisory role, rather than the more mundane tasks of data entry and manual layer combination. This will be especially true as yield mapping becomes more widely available. When producers have maps of historical crop yield, they will ask the question, "What does it mean?" No methods have been developed for evaluating the results of site specific management. A great potential exists for rule-based and expert systems to assist the producer or consultant in making management decisions that are optimized for each unique area of a field. Without good decision support, the wide spread adoption of SSCM will be limited by the ability of managers to handle and use the spatial data.

Supplies and Services

An important part of a SSCM system is the availability of supplies and services that will support a producers management plan. While some of the site specific activities will be performed by the producer, many others will be on a contract basis. Much of the collection of spatial data will likely be done on a fee-for-service basis. Field scouting, remote sensing and grided soil sampling are all data collection services that are available today. Site specific fertilizer and chemical applications are primarily performed on a contract basis. While the demand for application services will increase with wider adoption of SSCM, inexpensive Global Positioning System (GPS) receivers and implement controllers will allow farmers to apply more of the materials themselves.

The reliance on contracted services raises several questions regarding the relationship between the producer and service agency. When the data is gathered by an independent company, issues of ownership and transfer arise. Service contracts need to clearly state who has ownership and right to the data. Historical data for a piece of ground will have a monetary value, and potential conflicts could result between the company that collected and maintained the data, and the producer who paid for it. A related issue will be the custody of field data. For producers who operate their own site specific database, data interchange standards will be important. While a variety of GIS and mapping software is available, standard geographic data formats should be used to ensure easy data transfer. The issue of who owns the data may be less clear when a contractor collects, maintains and analyzes the data for the producer. In this case, conflict may result if contracts do not specifically include the data ownership rights.

Variable Rate Technology

The ability to vary application rates while traveling through a field is critical to the SSCM concept. Recognition of in-field variability is useless if no means exists to respond to those variations. Variable rate technology (VRT) equipment is probably the best developed part of the SSCM system. Two forms of VRT applicators are possible, map-based and sensor-based. Map-based applicators for dry granular materials have been in use for several yr. The use of sensors in closed loop application control has been limited to organic matter sensors, and, more recently, weed detecting sensors. While these machines can be designed for farm level or custom applicators, the commercial developments have emphasized the custom application market. A significant potential market exists for lower cost, ground applicators as well as aerial and irrigation delivery systems. Impediments to rapid development have included the lack of a reasonably priced, accurate positioning system and an uncertainty over the breadth of patent claims in the VRT area. With differential corrections for the Global Positioning System becoming more widely available and less costly, positioning has become less of a problem.

As more manufacturers produce VRT machines for the farm market, an open architecture will be needed. Communication between the tractor and various control modules will be needed, regardless of the paint color of the tractor or implement. The developing Construction–Agricultural standards for use of the SAE J1939 standard for in-vehicle communication addresses both hardware and data protocol compatibility. Use of this standard by manufacturers should be encouraged. Participants in the SSCM technologies should encourage innovation in the development of new devices and techniques. The availability of a diverse set of VRT machines will more rapidly expand the entire SSCM market.

Manager Commitment

As in most changes made by humans, a strong commitment to the SSCM philosophy of just-in-time and just-enough management is needed for SSCM to be successful. This includes a commitment to quality products and production through the use of prescription farming techniques. Unfortunately, not all of the necessary tools and techniques are in place. The site specific concept has not been tested in many regions and with many crops. While significant numbers of producers have embraced site specific management, these are still the early technology adopters, and are primarily located in the north central corn belt.

The engineers and scientists who support the SSCM technology have a responsibility to address the needs for better understanding of variability sources, improved data collection methods and better decision support systems.

SUMMARY

In this talk, I have given you my view of the current status and needs of site specific crop management. I have tried to emphasize the opportunities that exist for the development or expansion of engineering inputs to SSCM. Others may disagree with some of what I have said. Regardless of viewpoint or biases, I think all attending this conference will agree that SSCM holds a great potential for improving agronomic production. Whether or not that potential is realized will depend on the efforts of scientists and engineers. In time, the complete system of knowledge, tools and techniques will be available.

Site specific crop management is a technique that can improve both economic and environmental sustainability. It is a win–win technology.

REFERENCES

Adsett, J.F., and G.C. Zoerb. 1991. Automated field monitoring of soil nitrate levels. pp 326–335. *In* Proc. of the Symposium on Automated Agriculture for the 21st Century. ASAE, St. Joseph, MI.

Council for Agricultural Science and Technology. 1992. Water Quality: Agriculture's Role. Report 120. Council for Agric. Sci. and Tech., Ames, IA.

Cloud, D.S. 1990. Farmers reap a crop of scorn from anti-chemical forces. Congressional Quarterly 48:166–170.

Guyer, D. E., G. E. Miles, L. D. Gaultney, and M. M. Schreiber. 1993. Application of machine vision to shape analysis in leaf and plant identification. Trans. ASAE 36(1):163–171.

McCauley, J.D., A.D. Whittaker, and S.W. Searcy. 1994. Sampling resolutions for prescription farming and their effects on cotton yield. Trans. ASAE (in press).

Reichelderfer, K. 1991. What does society want? pp 11-21. *In* Agriculture's Contract with Society. ASAE, St. Joseph, MI.

Thompson, J.F., J.V. Stafford, and B. Ambler. 1990. Weed detection in cereal crops. ASAE Paper No. 9097516. ASAE, St. Joseph, MI.

Tyler Limited Partnership. 1994. S.M.A.R.T. system soil probe. Tyler Limited Partnership, Benson, MN.

45 Positioning Aspects of Site-Specific Applications

R. J. Palmer

University of Regina
Regina, Saskatchewa
Canada

Positioning is an important part of any site-specific management scheme for agriculture. Many different schemes and systems have been tried with limited success. This paper discusses the different schemes and suggests specifications for what is required from such a system. GPS, differential GPS, inertial, dead-reckoning and a new system called Accutrak are reviewed in regard to their appropriateness to site specific farming.

It is clear that different levels of precision are required for the various components of site specific application. A system that is accurate to 5 meters is probably more than adequate to determine a specific block of land and the rate of application can be determined accordingly. This increase in efficiency, however, is only reasonable if the application is being done with no overlap or missing in which the rate could be double or zero respectively. Such accurate driving would require a much more accurate system in the order of 15 cm.

Taking site specific to the limit requires a site in the vicinity of each plant. The nutrient placement and individual care of each plant is what really counts. To get to this level a system that could achieve accuracies of 1 or 2 cm would be needed. To achieve this a vision system could be used in which the plants themselves become the reference points.

INTRODUCTION

Variable Rate Application Technology (VRAT) or Site-Specific Farming requires real time position information for the application vehicle. There are different aspects to VRAT, that require different degrees of accuracy. Farmers and custom applicators have quickly come to recognize the merits of VRAT; however for those that are contemplating adopting VRAT in their operations there remains the practical questions as to what positioning system should be used? --With what accuracy? --For what cost?

Level I Accuracy

To just vary the rate of an application to match the conditions in the field, only requires a position accuracy of several meters since the gradient of the soil parameters is usually small.

Level II Accuracy

To drive the applicator accurately requires a higher degree of accuracy in order to guide the operator in making passes that neither overlap or miss. An overlap is essentially double the rate and a miss would be no rate at all. Without formal substantiation, it is suggested that positioning systems used for driving, guidance systems, should have accuracies of better than 15 cm.

Level III Accuracy

The ultimate degree of site-specific farming, requiring the highest level of accuracy could be called **plant specific**. Each plant would be regarded as a site and be treated on an individual bases. Inputs would be placed in an optimum position to give the greatest advantage to the plant and the least advantage to the weeds. At this level, positioning accuracies of 1 cm are required.

BACKGROUND

VRAT has been used by farmers in a crude fashion for years. If the farmer were applying a herbicide and noticed a heavy infestation of weeds, he would slow the applicator to increase the effective application rate. In the mid 1980's, this was refined somewhat by having a variable rate applicator in which the operator controlled the application rate by referring to a rate map and estimating his position on the map (Bens & Girwin, 1987). This was not deemed satisfactory by the operator. It was a frustrating and tedious task to manually adjust the rate. It was clear that an automatic method of adjusting the rate from the rate map was required. This then required a positioning system that could locate the application vehicle in order to get the correct rate from the rate map.

Dead Reckoning was used as a crude positioning system by measuring the distance from the headland from the odometer or by integrating the speed from a radar sensors, and by having the operator increment a lap counter to indicate how many laps the vehicle was into the field. This method was tried by many (Bulani, 1993) but it was generally not deemed to be a practical solution for several reasons. It relied on the operator to increment the lap count on the end of each pass, and this became a hopeless situation when a field had several obstacles in it. It was suggested that dead reckoning could be enhanced by adding an electronic compass and perhaps even a crude inertial system. The extra cost and complexity dissuaded researchers in doing further work in this area.

It was clear that positioning information was needed for VRAT; but, there was little to choose from. Some suggested Loran C, but with a questionable accuracy of hundreds of meters, it was never a serious candidate. Inertial systems were being used in the military and some survey applications with a fairly high degree of accuracy, but such systems cost hundreds of thousands of dollars. Rotating lasers provided a precise position; but the short range of only a few hundred meters made the number of reference lasers on the edge of the field prohibitively high. A 160–acre field could easily need a dozen or more reference lasers. There were also the problems of dust, hills and obstacles.

The Global Positioning System (GPS) was viewed by many as the system that would be appropriate for VRAT. It would just be a matter of waiting for the full complement of satellites to be installed and waiting for the price to drop.

Global Positioning System (GPS)

Much has been written on GPS (Lachapelle, 1992), and in terms of providing the positioning information for VRAT, one could consider a GPSD system. The D standing for the differential receiver that is used to determine the anomalies of the RF signal which are then used to correct the signal of the mobile. With accuracies of 2 to 5 meters, this would be considered a LEVEL I system. To achieve LEVEL II accuracy for guidance of the application vehicle, an even more sophisticated technique is used. Carrier cycle counting is used (Lachapelle et al., 1993) to achieve sub-meter accuracy. There are many pieces to this system and an accordingly higher price of about $40 000.

Latency is another issue that aggravates this solution. GPSD commonly has a latency of a second or more. This means that for a moving vehicle, the position reported is not the current position, but the position the vehicle was in a s or more ago. For example, for a vehicle moving 16 km/hr. (10 MPH.), a one s latency results in a position report that is 4.5 m behind the vehicle.

Accutrak System

The Accutrak System (Morris and Palmer, 1994) is a RF (Radio Frequency) system that has accuracies in the order of 15 cm. It uses land based reference beacons and is a completely independent land based system. It provides up to 10 position reports a second and has a latency of about a tenth of a s. The system currently has a range of 3 km. A higher power system will provide ranges to 20 km. Low range systems were tested last season and it is anticipated that high range systems will be under test this season. The Accutrak System was designed to address the needs of the LEVEL II VRAT, namely the guidance and accurate placement of the application vehicle.

Vision-Based/Image Processing Positioning

A high contrast between plant and non-plant material can be achieved by contrasting two specific bands of infrared. An inexpensive CCD camera with the proper filters can provide the video images that contrast plants against soil and trash. A simple algorithm (Cai and Palmer, 1994) has been developed that can locate the plants and specific inputs could be applied exactly where they are most beneficial. If the crop was seeded in precise separation patterns, the crop itself could provide the local navigational information for plant specific applications and guide the application vehicle through the crop pattern.

CONCLUSIONS

GPSD or Accutrak could be used for Level I VRAT. Accutrak is probably the best candidate for Level II; and Accutrak and a vision system could be used for Level III VRAT.

Level II Enhancements

The primary objective in using a Level II system would be to minimize the lateral overlap and missing. Overlap was measured (Palmer, 1984) with many different farmers using a number of implements of varying widths. Although there was a large variation from 5% to 15%, the overlap typically 9-10%. It is possible that a good operator using a foam marker could achieve a 4% to 5% average overlap, but with a positioning system this could be lowered to 1% or 2%. This lowered overlap would result in proportionately less inputs and less application time.

A Level II system would have a display that would guide the operator in staying on line. An auto-pilot or auto-steering actuator could be added to alleviate the operator in steering on the straight passes. This has been done on an experimental basis (Biberdorf, 1990) but is not yet available commercially.

With many farm operations, there is a very small time window in which the work is to be done. An auto-pilot system would allow an operator to "drive by instrument" at night. This would extend the operating hours of an applicator significantly.

It is important for custom applicators to demonstrate exactly what they have done for a customer. To achieve this the positions of the applicator could be logged to a file and later shown to the customer to demonstrate the real path followed. The exact number of acres could be calculated from this path (Matheson, 1991) and used for more accurate billing as well as more accurate rate input information. Returning to do missed stripes would be eliminated.

The operator is now responsible for turning the applicator on and off at the headlands. If the system was logging the position information, it would "know" what has been covered and could turn the applicator on and off at precisely the right time and minimize the headland overlap.

If the outline of the field and the obstacles in it were previously mapped. An optimum field path could be computed before the application (Liu, 1988) and followed either manually or automatically to cover the field with the shortest possible path. These field generated paths can be 10% to 15% shorter than the ad hoc courses.

To reduce the lateral overlap to zero is difficult in that a rather large piece of equipment moving relatively quickly must be kept on course to within centimeters. The positioning system may "know" that the applicator is off course but it can do little other that to suggest corrective steering corrections. By having on/off control of the outermost nozzles or flutes, it is possible to create a "virtual sliding boom" that essentially can move laterally to compensate for vehicle misalignment error. This has been tried experimentally (Palmer, 1989), but is not available commercially.

The quantifiable merits of VRAT are not yet known. To measure these results, it will be necessary to accurately measure the "instantaneous yield". This requires a log of the positioning and yield data. This data would be used to determine the actual profit of each piece of the field, and this, in turn, would be useful information in making subsequent rate maps for the area.

If a digital data link were used between the mobile and a control center, the position of the applicator and other pertinent information could be used to organize and control an efficient fleet of applicators.

With a computer generated field course and auto-steering, it is not difficult to imagine an automatic vehicle that has no operator. Such applicators could be made smaller and more numerous, since one is now not relying on a qualified operator. At this point, a Level III positioning system would be considered.

REFERENCES

Bens, A; and I. Girwin. 1987. Private correspondence. Novametric, Inc., Saskatoon, Saskatchewan, Canada.

Biberdorf, T. 1990. Design and development of an automated steering system for agricultural spray truck. EN 053 Report (unpubl.). ES Eng., Univ. of Regina, Saskatchewan, Canada.

Bulani, D. 1993. Dead reckoning trials. Bulani Agro, Biggar, Saskatchewan, Canada.

Cai, T, and R.J. Palmer. 1994. An image processing algorithm for row crop following. CSAE Annual Meeting. Regina, Saskatchewan, Canada.

Lachapelle, G. 1992. Journal of Institute of Navigation, Alexandria, VA.

Lachapelle, G., et al. 1993. GPS System Integration and Field Approaches for precision farming. Eng. Dep. of Geomatics, Univ. of Calgary, Alberta, Canada.

Liu, G. 1988. Computer generation of efficient farm field courses. M.SC. thesis, ES Eng., Univ. of Regina, Saskatchewan, Canada.

Matheson, S. 1991. Computer calculation of the area of an irregularly shaped field with obstacles. EN417 Final Report. Univ. of Regina, ES Engineering. (Unpubl. report).

Morris, R, and R.J. Palmer. 1994. A short range, high precision navigation system for automated farming. Accutrak Systems, Ltd., Regina,

Saskatchewan, Canada.

Palmer, R.J. 1984. Optimization of farm field operations, Energex Proceedings. Regina, Saskatchewan, Canada. Pergamon Press, NY.

Palmer, R.J. 1989. Applying farm chemicals with a truck outfitted with electronic navigation. SAE Tech. Paper Series 891840.

46 A Site-Specific Expert System With Supporting Equipment for Crop Management

D. E. McGrath

Tyler Limited Partnership
Benson, Minnesota

A. V. Skotnikov

Farm Machinery Department
Byelorussian Agri-Technical University
Minsk, Republic of Belarus

V. A. Bobrov

Laboratory of Optimization of Agrocenous Productivity
Institute of Experimental Botany
Belarus Academy of Science
Minsk, Republic of Belarus

Many of the contemporary approaches in solving problems in Site Specific Crop Management (SSCM) are based on existing maps of fields based on soil characteristics, e.g. soil types or fertility test results. It is known, however, that the chemical and physical attributes (factors) of soil can change inside the mapping units and may also overlap between units.

A direct method of economically determining certain key soil attributes will contribute to the development of SSCM. This concept has drawn significant attention. As a response to this interest an automated system of equipment, hardware and software has been designed and constructed. This system consists of:

- Automated soil sampler with packaging systems.
- Automated laboratory work station for the analyses of the physical and chemical attributes of the sample including pH, organic matter, particle size, sesquioxides, carbonates and nutrient elements.
- An expert system to consider the soil test results combined with data bases of crop models, fertilizers, chemicals, weather, field history, seed characteristics, yield economics and other inputs. The

system then uses dynamic simulation to forecast requirements of seeds, nutrients and chemicals.

- A system to implement the program.
- An automated applicator to carry out the program.

The expert system is designed to be self instructing. As data on results are accumulated algorithms are modified to improve forecasting capability.

INTRODUCTION

A major problem in site specific agriculture (also known as precision farming, site specific management, etc.) has been the economical collection, correlation and processing of the data necessary in order to automate the decision making and application of the appropriate nutrients. Spatially variable maps such as the organic matter map (Gaultney et al, 1988, McGrath et al., 1990, Alsip & Ellingson, 1991) may be used as an indicator of the approximate rate of fertilizer to apply. Applicators have used USDA soil maps, grid sampling or other data as a basis for estimating fertilizer requirements.

It is known, however, that the chemical and physical attributes of soil can change inside these zones and may overlap between zones.

A direct method of economically determining these attributes will contribute to the development of SSCM. This concept has drawn significant attention.

Work done at the Institute of Experimental Botany of the Academy of Science of Belarus has suggested a possible method of data collection, evaluation and decision making which can be combined into a whole package consisting of an expert system and support equipment. In this work a complex study of the key factors and mechanism of the production process of plants (taking into account the mechanical composition and fertility of soil) in accordance with the methodology of the agro system simulation has been done. Based on this work, hardware, software and support equipment has been designed and constructed. The research objective is to develop an economically feasible system that incorporates the theoretical work done on optimizing production of field crops.

METHODS AND MATERIALS

The site specific expert system with supporting equipment designed to optimize product capacity of a field for a given crop (See Fig. 46–1) is a new approach. It is a complete package designed by the TYLER firm and includes automated units consisting of:

- desktop computer (PC)
- packages of software programs
- automated soil sampler (AS)
- automated workstations for sample analysis (AWAS)
- yield monitoring system (YMS)
- automated system for fertilizer and pesticides application (ASFA)
- automated system for seed (ASR) variable rate planting

DISCUSSION

Conception of the Automated Work Station For Analysis of Nutrients and Other Characteristics of Soil

One of the objectives of this project is to develop an automated work station for analysis of specific nutrients and certain physical properties. The workstation design includes algorithms for processing the results of the analysis and an algorithm to determine the nutrient requirements to apply on a site specific basis to the field analyzed that will optimize the economical growth of a specified crop. An algorithm for the correction of the constants and coefficients of the model on the basis of the result of yield monitoring is also included.

Since the direct dependence between the total content of elements in the soil and their concentration in the plants is not observed, different soil extracts are used to approximate the plant available nutrients. The most suitable extractants have been studied very widely. Research in different areas of the world have found the best extractants for use in those areas. Thus, the recommendations for the application of fertilizers have essentially local (regional) essence dependent on both the selected method of obtaining the soil extract and the physical properties of the soil samples.

Generally, all soils may be subdivided into mineral, organic or a mixture of the two according to the content of the organic material. Depending on the content of the particles with a size < 10 μm, the mineral soils are subdivided into the four groups (Rinkis et al., 1989): sandy (up to 5%), sandy loam (up to 20%), loamy (up to 50%) and clay (up to 90%). But there are also other components such as iron and aluminum hydroxides (sesquioxides), calcium and magnesium carbonates, powder sand (particles size 10...100 μm) as well as the soil acidity (pH), each of which has a different binding selectivity relative to the macro- and micronutrients.

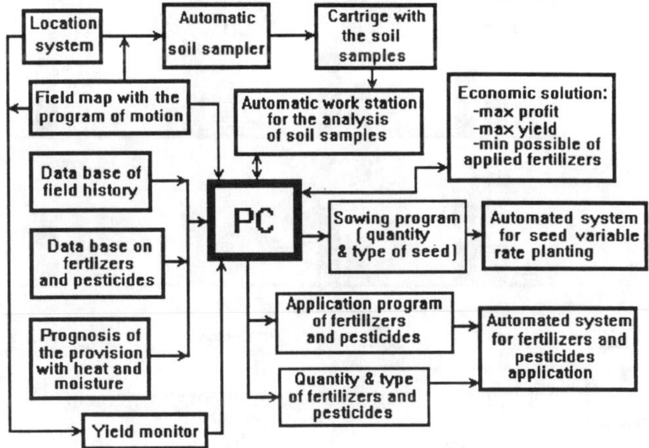

Fig. 46–1. Diagram of the expert system with supporting equipment

Figure 46–2, (Rinkis et al., 1989) represents the data on the relationship of the quantities of some macro- and micronutrients which are available to the plants and bound in a soddy, calcareous clay soil. Nitrogen and potassium are the more mobile macroelements; the soil absorbs only 44–45% of their optimum content (100%) and 87-88% of the less mobile phosphorous and calcium. Of the microelements, boron is absorbed to smaller degree (88%) while copper and manganese are absorbed to greater degree (93-95%). Another view of the influence of the soil factors on the plant nutrition is given in Fig. 46–3 (Rinkis et al.,1989).

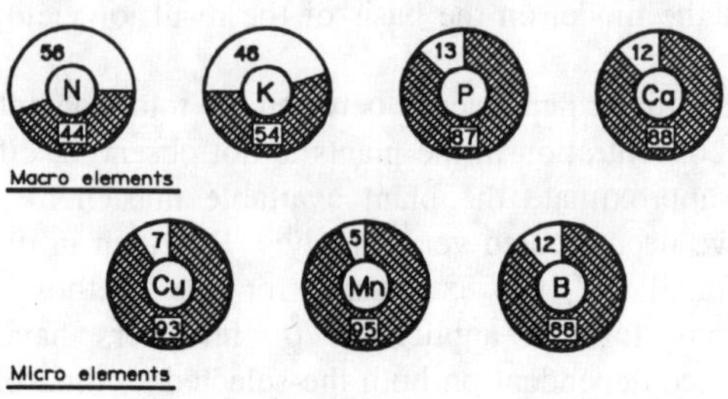

Fig. 46–2. Elements available to the plants ▿ and bound with soil ▾. quantities in the soddy-calcareous soil.

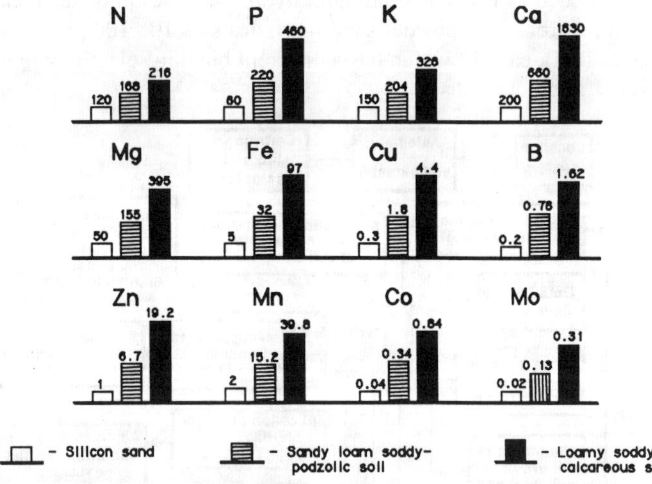

Fig. 46–3. The optimum for plant concentrations of the elements in the different soils (mg/l).

Table 46–1. Quantity of elements bound by certain soil parameters.

1991-1994
THE SYSTEM OF BALANCED MINERAL NUTRITION OF PLANTS
(C) LABORATORY OF OPTIMIZATION OF THE AGROCENOUS PRODUCTIVITY
THE INSTITUTE EXPERIMENTAL BOTANY. THE BELARUS ACADEMY OF SCIENCE

IDEALIZED EXAMPLE

ELEMENT	BIOLOGICAL OPTIMUM FOR CROP (mg/kg)	QUANTITY OF ELEMENTS THAT ARE BOUND UP BY CERTAIN SOIL COMPONENTS (mg/kg)							CALCULATION CONCENTRATION REQUIRED (mg/kg)
		(a) PARTICLE SIZE 10 > 100 μ (19%)	(b) PARTICLE SIZE > 10 μ (21%)	(c) GENERAL HUMUS (2.5%)	(d) DISSOL. HUMUS (0.7%)	(e) SESQUI-OXIDES (2.5%)	(f) CARBONATES (1.2%)	(g) pH 6.2	
N	60.00	5.70	12.60	0.00	0.00	0.00	3.00	0.00	81.30
P	40.00	17.10	37.80	12.50	7.00	50.00	14.40	28.80	207.60
K	100.00	5.70	12.60	12.50	7.00	0.00	7.20	18.00	163.00
Ca	300.00	114.00	252.00	500.00	280.00	150.00	0.00	0.00	1596.00
Mg	120.00	15.20	31.50	62.50	35.00	30.00	24.00	12.00	330.20
S	50.00	Not Determined							
Fe	50.00	11.40	21.00	25.00	14.00	0.00	12.00	24.00	157.40
Cu	0.50	0.17	0.38	0.50	0.28	0.45	0.07	0.22	2.57
Zn	1.00	0.57	1.05	0.38	0.21	0.88	0.60	1.20	5.88
Mn	20.00	1.90	2.10	1.00	0.56	10.00	4.80	12.00	52.36
Co	0.10	0.04	0.08	0.06	0.03	0.07	0.05	0.05	0.48
Mo	0.02	0.01	0.01	0.02	0.01	0.07	0.00	0.01	0.13
B	0.30	0.06	0.13	0.10	0.06	0.25	0.02	0.05	0.96

Concentrations of the 13 nutrient elements in the inert substrate being optimum for the crop being grown at the normal provision with moisture, heat and light.	+	Calculation of the quantity of elements that will be absorbed based on: a) particle size > 10 μ b) particle size 10-100 μ c) humus d) soluble fraction of humus; e) sesquioxides f) carbonates g) acidity	=	Optimum concentration of the nutrients for the crop being grown	-

-	Contents of nutrients in the given soil (1 M HCl)	=	Quantity of nutrients to make up the deficiency

Fig. 46–4. Binding selectivity of the macro- and micronutrients by selected soil factors (Calculated by Program ARMagro).

Thus, the optimum concentrations of the nutrient elements in the soddy-podzolic loamy soil are 1.5–8.5 times and in the clay calcareous soil are up to 21 times higher than in the silicon sand.

The data presented in Table 46–1 are an additional argument in favor of the necessity of accounting for both nutrient composition of the soil extract and the ion-exchange properties of the soil.

It is seen from the example given in Fig. 46–4 that the measurable soil factors; particle size 10 > μ (21)%, particle size 10–100 μ (19%), general humus (2.5%), alkali-soluble portion of humus (0.7%), sesquioxides (2.5%), carbonates (1.2%) and soil acidity (pH 6.2) have different binding selectivity relative to the selective nutrients. It is seen from the data represented in Fig. 46–4 that when taking into account the adsorbing capability of humus, it is necessary to estimate quantitatively the alkali-soluble humus fraction which has a relative absorbing capability several times higher than that of the general humus fraction. This relative availability of the nutrients has been used as the basis for the design of an automated workstation. The workstation analyzes for these specific parameters:

1) Determination of the concentration of the macronutrients (NO_3,NH_4,K,P,Ca,Mg) and, in the future, micronutrients in a 1 M HCl soil extract;

2) Determination of particle size 10 > μ, particle size 10 - 100 μ, overall humus, alkali-soluble fraction of humus, sesquioxides, carbonates and acidity.

On the basis of the obtained analytical data, a high-resolution program for the basic nutrient requirements is computed. The following algorithm will be used for the computation of the quantity and type of the nutrients to be applied (Rinkis et al., 1989).

An Expert System

The simulation mathematical model of soil fertility (FERT).

Soil fertility in this model (Kan, 1989; Bobrov at al., 1994) is associated with main forms of nitrogen, phosphorus and potassium in the cultivated soil layer independent of soil type, where plant root activity is significant. The model includes five submodels (see list of symbols for models in Fig. 46–5.):

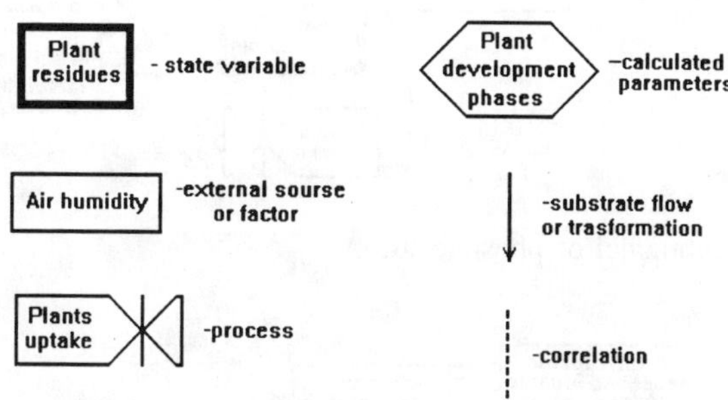

Fig. 46–5. List of symbols.

Submodel of nitrogen (Fig. 46–6) transformations describes daily balance and resultant seasonal dynamics of nitrate and ammonium for plant available ions and applied solid inorganic fertilizers.

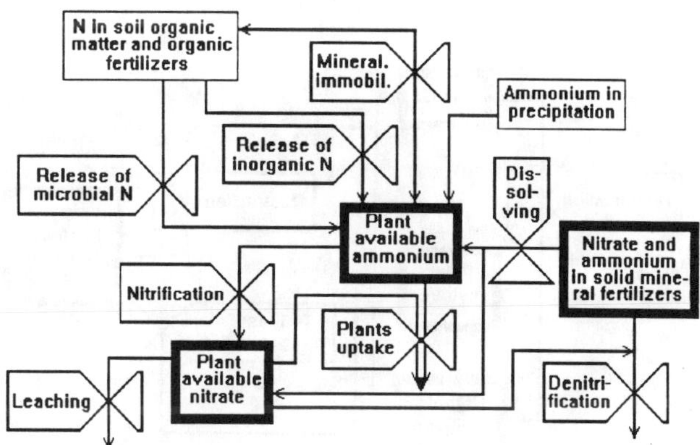

Fig. 46–6. Submodel of nitrogen.

Phosphorus (Fig. 46–7) and potassium (Fig. 46–8) submodels also simulate balance and dynamics of these nutrition elements in plant available form and applied inorganic fertilizers.

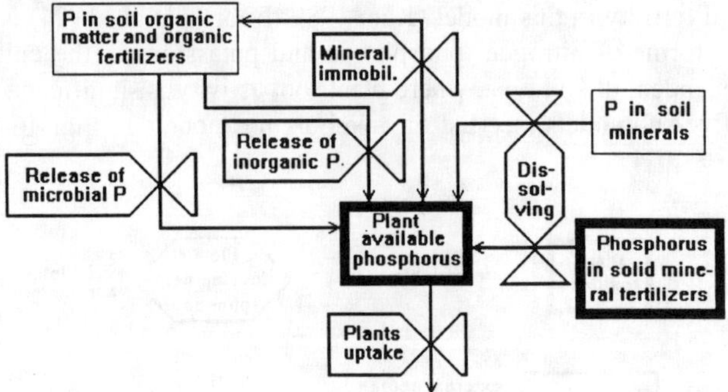

Fig. 46–7. Submodel of phosphorus.

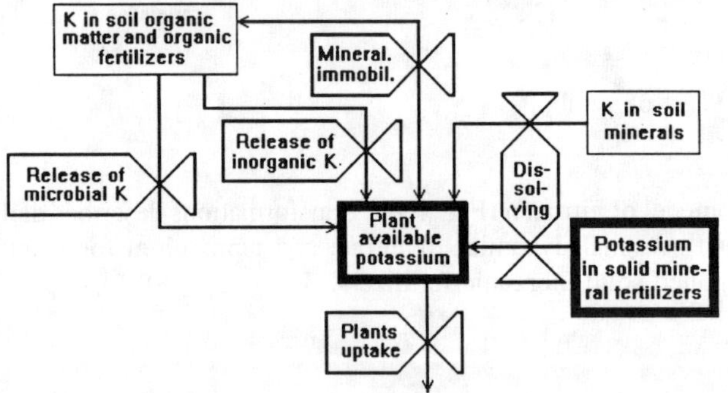

Fig. 46–8. Submodel of potassium.

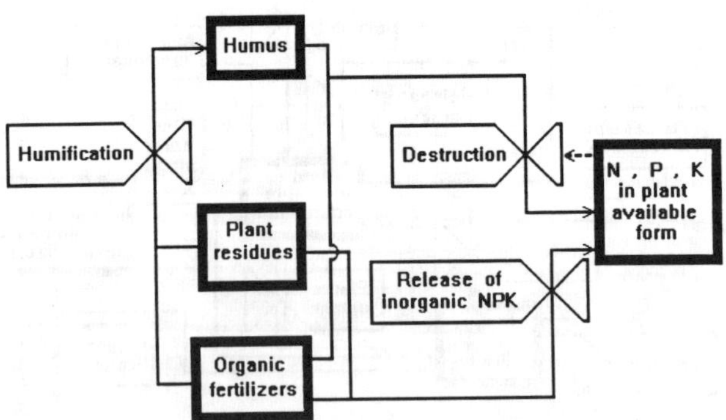

Fig. 46–9. Organic matter submodel.

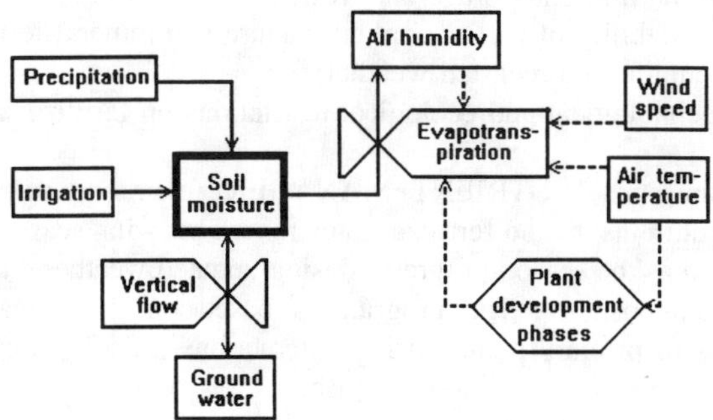

Fig. 46–10. Soil moisture model.

Organic matter submodel (Fig. 46–9) deals with transformation of humus, plant residues and organic fertilizers.

Intensity of N, P, K and organic matter transformations depends on soil moisture and temperature. Simple regression equations are developed for temperature dynamics in upper and lower parts of soil layer. Daily balance and dynamics of soil moisture are provided by particular model shown (Fig. 46–10).

Special software supports the data base of the model. One can set numerous physical and agrochemical parameters, manure, and crops. The model can calculate dynamics of primary variables under various meteorological conditions.

There are 11 dynamic primary variables: amount of soil ammonium, nitrate, phosphorus and potassium in form of plant available ions and solid mineral fertilizers, humus, plant residues and manure.

Daily meteorological data are:

- average daily temperature at 2 m height,
- average daily air humidity at 2 m height
- daily maximum wind speed at 12 m height
- daily total precipitation

Constant parameters are:

- determined soil factors, plant uptake, plant residues, organic and mineral fertilizers,
- relative velocities of transformation processes dependent on soil temperature and moisture.

"NUTRIENT MANAGER" - Computer Program for Fertilizer Management in Crop Production

The program takes into account numerous circumstances of decision making in real fertilizer management:

- plant requirements of nutrient supply;
- characteristics of previous crop and planned manure;
- initial nutrient storage and availability
- availability of earlier applied manure and mineral fertilizers;
- actual and forecasted weather;
- technological and ecological restrictions on fertilizer management.

Using the "NUTRIENT MANAGER" plan one can predict optimal nutrient requirements and fertilizer rates for the growing season. This plan may be corrected as the season progresses using updated weather data and the results of plant diagnostics. The program is based on theoretical and empirical estimations of primary plant nutrient interactions (FERT program) and the use of information from a special data base.

Inorganic Fertilizers and Manure Requirement

The program calculates yearly balance of humus considered as humification of plant residues and manure minus humus decomposition. User may set wanted yearly increase of humus percent in soil (may be positive, negative and zero) and thus get appropriate recommendations on applying manure.

Annual values of plant residues and organic fertilizer decomposition are also calculated. Relative rates of decomposition and humification processes dependent on weather are obtained from computer experiments with simulation model of soil fertility program "FERT".

Mineral fertilizers requirement

Calculated annual decomposition of organic fertilizer and plant residue allows an estimate of corresponding values of net N, P and K mineralization from these substrates. Other positive components of plant available NPK balance are:

- release of inorganic N and K from manure;
- residual P and K fertilizers applied in previous yr.

The requirements for nutrients are calculated by subtracting the anticipated uptake of the plants from the available nutrients.

Before using the determined requirement to calculate the crop demand, the mineral fertilizers program makes some corrections to take into account:

- initial NPK storage,
- total losses at the expense of absorption, leaching and denitrification,
- influence of acid-forming PK fertilizers on pH,
- possible utilization of N from humus.

Technological and Ecological Restrictions on Mineral Fertilizer Rates

First, the program calculates total required rates of NPK. Then entire rate must be distributed between several applications.

These considerations are used here:

- nitrogen fertilizers with high solubility must be applied close to beginning of plant growth to avoid losses;
- a part of P and K may be applied in the autumn for spring planting due to low solubility and lesser losses;
- upper limit for application rate is defined to avoid high concentrations of nutrients in plants beyond what is necessary for optimum yield and significant leaching to groundwater;
- there exists a lower limit for each application rate under which the cost of yield increase is less than expense of application.

Special data base is developed for needed field information. The minimum data requires only four characteristics:

- name of crop to be grown,
- name of previous crop,
- determined soil factors,
- type of manure planned for application.

In this case the program finds > 20 other field characteristics needed for calculation in special reference files, which must be created during adaptation procedure.

BIOTOP - Topographic Model to Analyze the Chart of the Field Yield Capacity.

The reliable multidimensional dependencies of the production factors on the rate and relationship of the biological factors in the nutritional substrate (topographic characteristics of the production process) may be obtained with the help of the multi-factor field experiment.

The original semi-empirical method - the Method of Systematic Variants, which allows the drastic reduction of the volume of the multi-factor experiment has been described in the literature (Homes & Van Schoor,1982). Thus, in order to solve the above task on the determination of the optimum relationship of the six macronutrients, it is sufficient to perform an experiment with 8 variants. The effect of reducing the volume of the multi-factor experiment is achieved here owing to introduction into the model of mathematical concepts using a dome like dependence of the yield on the relationship of the nutrients. The correctness of the method has been confirmed by the investigations in the course of > 30 yr, but its widespread use was hampered by the difficulties in the theoretical interpretation and routine methods of the experimental data processing. Moreover, the use of the method is restricted by the variety of relationships among the nutritional elements in the soil.

In order to process the data of the multi-factor experiment, the Nelder-Mid simplex method of non-linear approximation has been used (Bobrov et al., 1994). The following analytical expression was selected as a basic function approximating the dependence of the morphometric and biochemical factors (Y) on the relationship of the macroelements in the nutritional medium:

$$Y = P(0) * \overset{n}{\underset{k=1}{\text{Å}}} [X(k)^{P(k)}]$$

where n is the number of optimized elements,

P(0) and P(k) are the parameters of approximation,

$\overset{n}{\underset{k=1}{\text{Å}}}$ designates the multiplication of the contents in the brackets n times,

X(k) is the portion of the k-th element in the nutritional substrate to be calculated from the formula:

$$X(k) = C(i) / \sum_{i=1}^{n} C(i)$$

here C(i) is available macroelements for the plant concentration of the element being optimized in the substrate.

The proposed mathematical expression simulates satisfactorily the dependence of some morphological and biochemical factors on the relationship of the macronutrients. Especially the experiment is valid for those factors which correlate with the increase of the biomass (vegetative matter, plant height at the initial stage of growth, grain yield, etc.).

The content of the approximation process consists in the minimization of the contradiction between the theoretical and experimental values of the characteristic factors at the concentration values specified by the scheme of experiment or determined from the analysis of the heterogeneous soil areas. The approximation method developed is notable for its great flexibility and reliability.

The calculated approximation parameters P(i) allow the user to ascertain immediately the optimum relationship of elements in the nutrition substrate for the studied factor. In this case the optimum portion of the k-th element ($X(k)_{opt}$) is calculated from the formula:

$$X(k)_{opt} = P(k) / \sum_{i=1}^{n} P(i)$$

The convenient form of mapping the joint influence of the three elements on the parameters of the production process is a triangular matrix (Fig. 46–11).

We have developed the program projecting the topographic characteristics (optimum relationship of the three selected elements, contours of zones with the specified range of values of the production parameters) on the matrix field. The method can handle up to eight variables but three are used to help in visualizing the process. The serviceability of the "BIOTOP" topographic model has been checked with the help of the multi-factor experiments of three types (Yermaloyev at al., 1991):

- the hydroponic laboratory experiment at the fixed illumination, temperature and humidity values;
- the vegetative period in the phytotron with the fixation of the illumination, temperature and humidity values;
- field experiment.

The obtained results indicate that the "BIOTOP" model is suitable for the detail study of the influence of the relationship of the mineral nutrition elements on the production process of plants. The model allows us to obtain the information on the nutrition regimes favorable for the development of the whole plant and its separate organs. The model may be used to estimate the potential possibilities of maximum crop yield and to develop the energy-saving ecologically clean agricultural technologies.

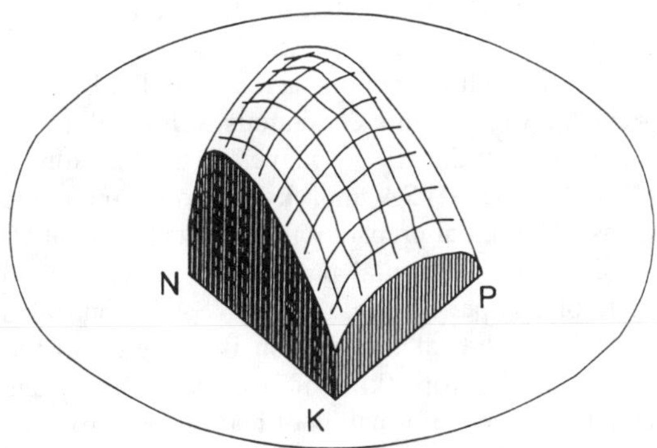

Fig. 46–11. Biotop model.

The obtained data indicate also that the developed computer system will reveal, on the one hand, the optimum nutrient regime for the achievement of the maximum biological productivity and, on the other hand, the optimum regimes for the assurance of the high quality of production according to a number of factors.

The expansion of the possibilities of the computer system is connected with the investigations in two directions:

1. A more detailed study of the ion-exchange mechanism in the soil and modernization on this basis of the empirical submodel of the influence of the soil components on the availability of the elements to the plant.

2. A more precise definition of the quantity of the available biogenic soil nutrients providing the optimum conditions of the growth, development and productivity of the plants (expansion of the data base on the biological optimum of crops).

The biological optimum of crops is determined by means of the multi-factor laboratory-and-field experiment in the course of a number of yrs. The drastic reduction of the volume of works when determining the biological optimum of crops is possible if the "BIOTOP" system is used for the computations on the basis of the high-resolution field productivity (yield map) and the agrochemical factors obtained as a result of the realization of the conception of the project.

Supporting Equipment

In use the expert system requires detailed sampling and analysis of the field to provide a beginning set of data. This data is provided by supporting equipment described below.

The automated soil sampler (AS) is moved in a direction based on existing field map. The width of the sampling coverage is selected based on the swath of the fertilizer application machine. The data on sampling grid spacing and sampling depth are entered into the AS memory.

The AS samples while moving across the field, with predetermined intervals and packs the samples in a continuous web of film. The containers of packaged soil samples from the AS are moved to the location of the automated workstation (AWSA). The AWSA unpacks the soil sample and prepares each sample for analysis. The initial preparation period takes about 90 minutes in the prototype machine. After the initial preparation period the AWSA produces analysis at the rate of one per minute. The AWSA is controlled by a Personal Computer (PC). The analytical information is moved into the PC. The PC memory also stores data on crops (kind of crop, seed variety and price, fertility requirements, estimated price information of harvested crop), on fertilizers (type, quantity of the active substance per mass unit, application cost); herbicides (type, unit price, application cost), forecast concerning heat and moisture expected.

Based on the analysis and the stored data the seeding rate, number of applications, fertilizer type and volume per each application are calculated for each sample representing the given portion of the field. Using the expert programs three options for applications of seed, fertilizer and chemicals to the field are worked out. The three options give maximum profit, maximum yield or minimum fertilizer and herbicide application for given profit level. If the total result obtained is satisfactory, a total program for seeding and chemical and fertilizer application or separate programs for seeding or chemical or fertilizer application are processed.

The results obtained for each field are translated to match the specific application equipment. Then final results on seed type and quantity, and on the fertilizer and chemicals (type and quantity for the given field) are totaled and reported. The application programs are outputted and stored on PCMCIA cards. Strip application and seeding for diverse crops may be programmed. For example, soy–corn combinations or diverse grasses may be planted and fertilized in strips. These programs may be then used in automatic systems for seed sowing and fertilizer application. The bin dimensions and quantity on the planter or applicator are matched to the adopted program.

After crop harvesting, the yield monitoring data may be entered into the PC. The data on yield is compared with the planned data on yield used to apply fertilizer. The comparison is used to check and correct the regression equations' constants and factors and to determine new required rates depending on the determined analytical information, content of nutrients, pH, etc. The yield monitoring data plus information on residue removal can be used to calculate the next season's application with little or no further soil sampling. This calculation is done by subtracting the elements used by the crop and residue as part of an input–output calculation.

The AS is an attachable machine designed for automatic on-the-go sampling from a field to be cultivated. The AS is pulled through the field by a self-propelled sprayer or equivalent with foam marker on the boom having a hydraulic system and electric power supply system. The AS consists of two wheel trailer equipped with a sampling mechanism, sample packer, packed sample collector, trace marker, control and alert system. The AS samples the soil from the places spaced > 15 m apart while traveling across the field and packs samples into the plastic bags to form the band consisting of the bags and a connecting web. The band length is determined by the dimensions of the container (about 30 m or 300 samples). The beginning of each run is marked by sealing 3 empty bags into the band. The motion along the first run is implemented according to the route planned on the field map. During each run of the AS a trace is made by a track marker placed along the longitudinal axis of the AS. This trace is used to orientate the fertilizer application machine during the subsequent runs. The foam marker is set up to the width equal or a multiple of the coverage width of the fertilizer application machine. A location system may also be used.

The samples packaged in the container web are fed into the AWAS. The above mentioned analytical information is determined in the workstation.

On the basis of obtained analytical data, the net nutrient requirement plan is computed. This plan is made into a fertilizer application program. The program obtained in such a way may be used in the Automatic System for Fertilizer Application (ASFA). The ASFA has an on-board computer having a PCMCIA card. The outputs of the material availability sensors and distance sensors are input to the on-board computer. The on-board computer outputs drive the respective actuators.

The ASFA operates as described. Prior to the operation, an operator makes all necessary input and output connections of the on-board computer and respective sensors and actuators of the bins containing respective products.

Then the program of fertilizer application is entered into the on-board computer from the PCMCIA card. The applicator with ASFA then begins its motion to the predetermined program (traces left by the AS). The ASFA is turned ON or OFF synchronously with turning ON or OFF of the operation process of the applicator. During the operation, depending on the applicator position within the field, the proper amount of each fertilizer, micro element or herbicides is applied according to the program.

The program of applicator motion and accordingly, the fertilizer application can be visualized as if it were a long tape, which, to prevent error accumulation, the marks of the "beginning of the run" and "end of the run" are made. These imaginary tape portions represent a record of necessary digital instructions for each product being applied. Error is limited to the operator starting each run accurately and the small possible error in the radar speed and distance reporting.

A GPS, radio ranging system or radar vehicle tracking system may be used to determine current position of the ASFA. The ASFA has feed sensors, sensors of material availability and actuators along with a radio receiver and transmitter that respond to instructions based on location. Radio telemetry can be used to replace the on-board computer. A centrally located computer could send instructions to several fertilizer applicator vehicles.

This expert system is self-correcting through the use of yield monitor information. It opens a wide range of management possibilities for growers and dealers exploring site specific agriculture. The models of the system are designed and have been tested for functionality. Full scale prototypes have been designed. They are scheduled to be tested at the University of Minnesota during the spring and summer of 1994.

ACKNOWLEDGEMENT

This study was funded in part by Funds of Fundamental Research of Belarus. (Director Vasiljev, E.I.)

Many suggestions and comments from Dr. Nikolaj Kan (Research and Technological Center "Biomodel", Novocherkask, RUS) and Dr. Tomara Janchevskaja (Laboratory of Optimization of the Agrocenous Productivity of the Institute of Experimental Botany of the Belarus Academy of Science, Minsk, Republic of Belarus) have been incorporated into this chapter and the authors acknowledge this contribution.

A grant from the "Special American Business Internship Training (SABIT) program supported the preparation and presentation of this paper.

SABIT is a program of the U.S. government private sector partnership, providing technical assistance to the Independent States of the former Soviet Union. Managed by the U.S. Department of Commerce in cooperation with the Agency for International Development. Tyler Limited Partnership of Benson, Minnesota provided conceptual design and funding for equipment development.

REFERENCES

Alsip, C., and J. Ellingson. 1991. Computer correlation of soil color sensing with positioning for application of fertilizer and chemicals. Automated Agriculture for the 21st Century Proc. ASAE, St. Joseph, MI.

Bobrov, V.A., V.P. Samsonov, L.M. Panifedova, Yu.S. Yermolayev, T.G. Yanchevskaya, N.A. Kan, and Ye Kan. 1994. Optimization of the productivity of cereals on the base of the dynamic model. *In* News of the Academy of Science of Republic of Belarus, Biological Series, Minsk, Belarus.

Gaultney, L.D., J.L. Shonk, and Z. Yu. 1988. Automatic soil organic matter mapping. ASAE Paper No. 881607, ASAE, St. Joseph, MI.

Homes, M.U., and G.H. Van Schoor. 1982. Alimentation et Fumure Minerales des Vegetaux. Bruxelles.

Kan, N.A. 1989. Methodological problems of imitational simulation of agrocenosae. (Preprint/Centr. Sci. Biol. Inst. Acad. of Sci. Pushino, USSR.

McGrath, D.M., J. Ellingson, and A.O. Leedahl. 1990. Variable application rate based on soil organic matter. ASAE Paper No. 901598, ASAE, St. Joseph, MI.

Rinkis, G.YA., H.K. Ramane, G.V. Paegle, and T.A. Kunitskaya. 1989. The optimization system and diagnostical methods of mineral plant nutrition. Riga.

Yermaloyev, Uy., S. Alshanikava, N.A. Vyarbitskaya, and N.G. Dabrakhotava. 1991. Computer mathematical models of the influences of doses of mineral nutrient elements on growth, development and productivity of plants. *In* News of the Academy of Science of Republic of Belarus, Biological Science Series, 4:92-95. Minsk, Belarus.

REFERENCES

47 Sensing Grain Yield With a Triangular Elevator[1]

M. D. Schrock
D. K. Kuhlman
R. T. Hinnen
D. L. Oard
J. L. Pringle

Kansas State University
Department of Agricultural Engineering
Manhattan, Kansas

K. D. Howard

USDA-Agricultural Research Service
App. Tech. Res. Unit
Stoneville, MS

Knowledge of grain yield on a spatial basis is a key ingredient in the implementation of site specific crop management. Spatial yield variations are important both as input data for management decisions as well as for feedback and evaluation of decision processes.

A number of methods have been developed to determine grain yields in real time as the crop is harvested. Potential grain flow measurement concepts include volumetric paddle-wheel sensors (Searcy et al., 1989), impact sensors (De Baerdemaeker et al., 1985) optical measurement of elevator content (Pfeiffer et al., 1993), piezo-film impact (Pang & Zoerb, 1990), radiometric techniques (Auernhammer et al., 1993), and a pivoted auger (Wagner and Schrock, 1989). Additional grain flow measurement concepts were reviewed by Borgelt (1992).

This paper reviews the development of a triangular elevator concept developed at the Department of Biological and Agricultural Engineering at Kansas State University. The concept is a mass-based system that requires a moderate level of modification to a basic grain combine.

---

[1]This research was sponsored by J. I. Case and the Kansas Agricultural Experiment Station.

MEASUREMENT CONCEPT

A number of sensor concepts were initially considered, including impact sensors, the pivoted auger, and optical techniques. Each sensor concept appeared to offer a compromise between expected accuracy, complexity, and susceptibility to changes in crop properties.

Functional requirements outlined at the start of the projects included:

1. A continuous grain flow signal was desired.

2. The sensor should not restrict combine capacity, and should function in corn (*Zea mays* L.), wheat (*Triticum aestivum* L.), soybeans (*Glycine max* (L.) Merr.), and grain sorghum (*Sorghum vulgare* Pers.).

3. The accuracy goal was established at +/- 5%, as evaluated by comparing bin load weights to the integrated sensor reading.

4. The sensor should not require extensive modifications to the standard combine.

5. The influence of grain properties on the accuracy of the sensor should be minimal.

Ultimately, the decision to begin development of a triangular elevator concept was based on a combination of analysis and subjective judgement.

The clean grain elevator used on most combines consists of an enclosed roller chain having attached rubber paddles to convey grain from the bottom of the cleaning shoe to the grain bin. The chain and paddles normally run between two sprockets, the driver typically being located at the top of the elevator, where the grain is unloaded into a bin-filling auger.

The triangular elevator was created by adding a third sprocket to the standard elevator, the sprocket being located to create a third horizontal leg in the elevator (Fig. 47–1 and 47–2). The housing containing the horizontal leg was constructed with an "active section" that allowed the grain to be weighed as it passed through. The grain entry end of the active section was supported by a simple pivot, while the grain discharge end was supported by a standard strain-gage load cell.

It is desirable that the load cell reading be unaffected by chain tension. Accordingly, the pivot point for the active section was located in the plane of the chain, and "dead sections" were added leading into and out of the active section. The dead sections had length sufficient to ensure that at least one rubber paddle was borne on each of the dead sections at all times. The length of the active section was 102 cm., although shorter active sections were evaluated in the laboratory.

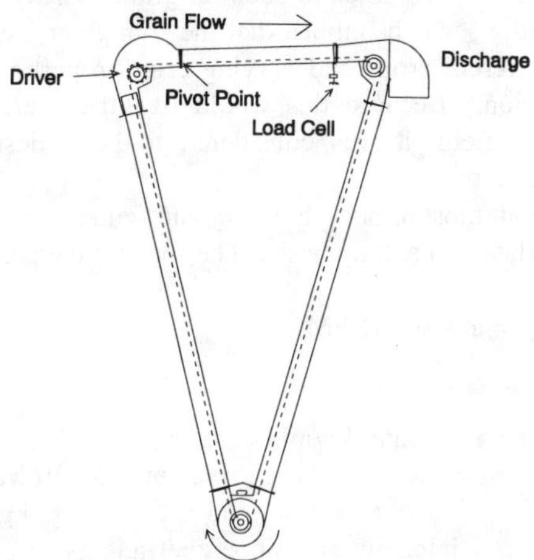

Fig. 47–1. Schematic Triangular Elevator Measurement Concept.

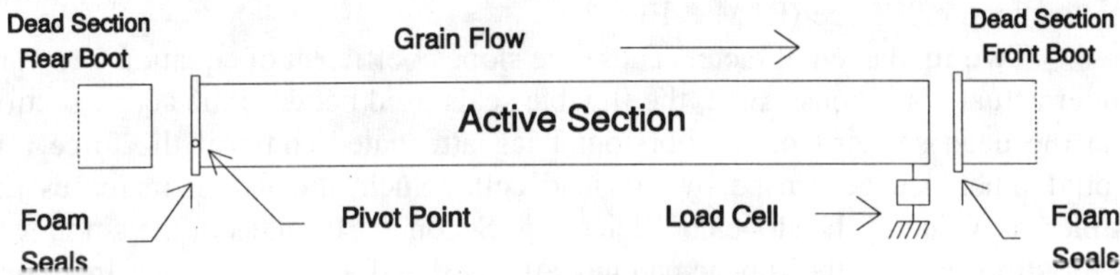

Fig. 47–2. Orientation of active section and dead sections.

Assuming that no grain slips past the rubber paddles, the theoretical flow in a paddle elevator can be expressed as:

$$Q = W * V \tag{1}$$

where:

Q = flow rate, kg/min
W = mass of grain per meter of conveyor length, kg/m
V = paddle velocity, m/min

When specific dimensions, chain pitch, and sprocket size of the active section are considered, equation 1 can be revised to calculate the theoretical grain flow as a function of load cell reading and elevator sprocket rpm. For the unit used in this study, the result is:

$$Q = 0.695 * (RPM * F) \tag{2}$$

where:

Q = flow rate, kg/min
RPM = elevator chain drive sprocket RPM
F = net active section load cell reading, kg_f

Note that crop parameters such as grain density do not enter into the theoretical equation, which implies that the triangular elevator may be capable of measuring different crops and varying crop properties while using a single prediction equation. Because this would free the user from measuring crop properties in the field, it was considered to be a desirable feature of this measurement concept.

Throughout most of our laboratory and field testing, grain flow prediction models were updated on a daily basis. The form of the prediction equation was:

$$Q = a + b * (RPM * F) \quad (3)$$

where:

Q = flow rate, kg/min
RPM = elevator chain drive sprocket RPM
F = net active section load cell reading, kg_f
a, b = intercept and slope coefficients

The intercept and slope coefficients were determined by regressing known gravimetric grain flows (Q) against the product of elevator speed and net active section load cell force (RPM * F).

Two fundamental factors cause the slope coefficient of equation 3 to vary under actual conditions. First, the flexible seals used between the active section and the dead sections of the horizontal leg attenuated some of the force that would otherwise be sensed by the load cell. Such attenuation increases the numerical value of the slope coefficient, b. Secondly, changes in the amount of grain slippage past the rubber paddles affects the slope coefficient. Increased slippage tends to increase the load cell reading at a given flow, thereby decreasing the slope coefficient.

Signal Conditioning

Because the active section is weighed while operating on a mobile machine having many rotating components, vibration requires that substantial signal conditioning be performed on the load cell output.

In order to determine appropriate filtering parameters, the load cell output was read at 300 hz during a stationary run-up of the combine. A Fourier transform revealed several sources of vibration (Fig. 47–3), with a very large, distinct spike at 60 hz, corresponding to the pitch frequency of the elevator chain.

The excessive noise in the load cell signal was removed in two phases. First, a signal conditioning card (Calex 160-MK) was set to perform analog low-pass filtering with a 10 hz cutoff. Next, the output from the card was oversampled at 70 hz by our data acquisition system, and the raw 70 hz data was later digitally filtered to a .5 hz cutoff. The .5 hz cutoff rejected most noise while retaining acceptable response to changes in grain flow rate.

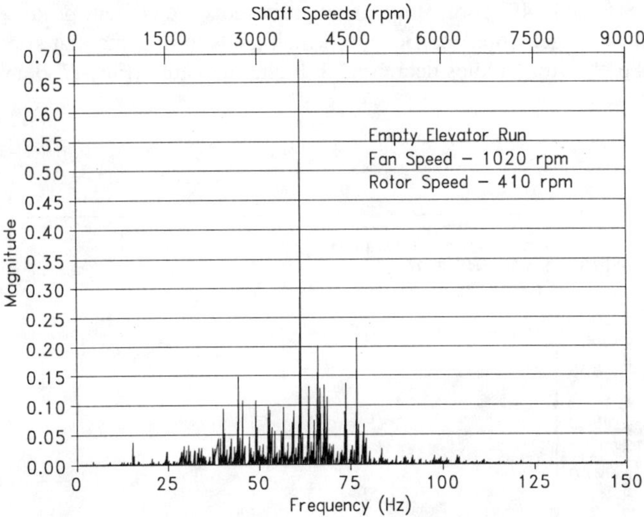

Fig. 47–3. Fourier analysis of the active section load cell signal.

LABORATORY TESTS

A triangular elevator was constructed and used to validate the overall concept in the laboratory prior to field testing on the combine. During laboratory tests, grain flow rate was determined by weighing the bin that supplied grain to the elevator as well as the catch bin at the elevator outlet.

Laboratory flow models verified that the "RPM*F" factor could be used to predict flow, with R^2 (coefficients of determination) values generally over 0.98. The slope (b) of the flow models, however, were generally above the theoretical value of 0.695. The increase in the flow model slope was caused by the seal attenuation mentioned previously.

The laboratory unit was used to evaluate the effects of different active section lengths, side slopes, fore-aft slopes, and grain moisture content (Howard, 1992).

1. Active sections 25 and 51 cm long were compared in the laboratory to the standard 102 cm active section. R^2 values for the 51 cm. and 102 cm. flow models were consistently above 0.99, while the 25 cm. active section flow models produced R^2 values of 0.95 to 0.97.

2. Side slopes of 12% had minimal influence on either the flow model coefficients or the linearity and fit of the model (Fig. 47–4 and 47–5). Up-and-downhill inclinations of 12% were also found to be negligible.

3. Lab tests were conducted with 10, 14, and 18% wheat and with 12%, 18%, and 24% corn. Increasing grain moisture content produced higher flow model slope coefficients, which tends to indicate that slippage past the elevator paddles decreased at higher moisture (Fig. 47–6 and 47–7).

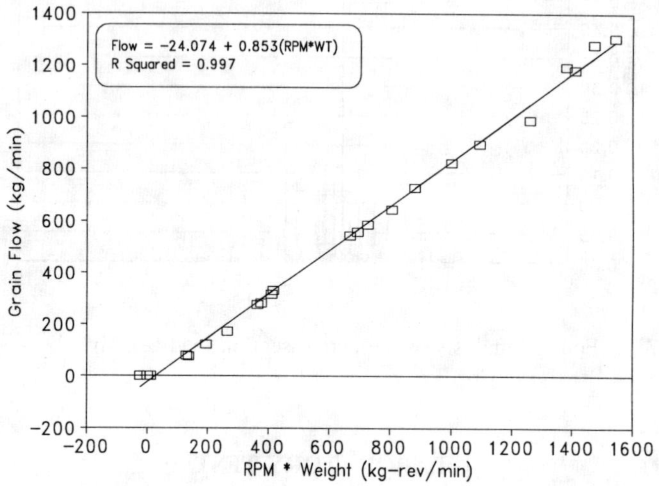

Fig. 47–4. Linear prediction model for the laboratory slope test of 0% slope with wheat.

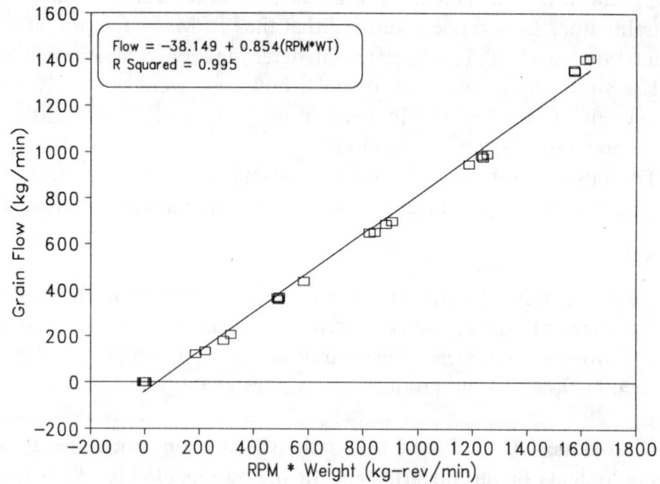

Fig. 47–5. Linear prediction model for the laboratory slope test of 12% side slope with wheat.

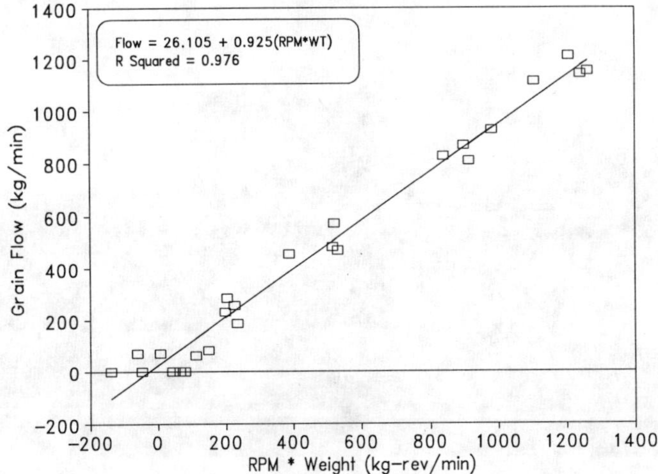

Fig. 47–6. Linear prediction model of the laboratory grain moisture test of 12% corn.

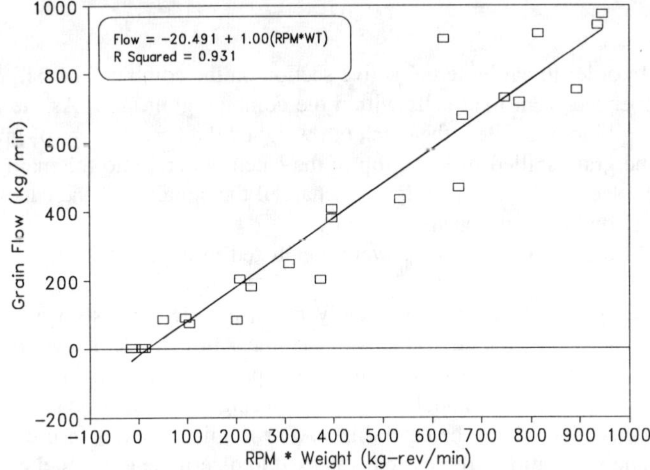

Fig. 47–7. Linear prediction model for the laboratory grain moisture test of 24% corn.

FIELD TESTS

For field tests, the combine (Fig. 47–8) was equipped with a triangular elevator assembly that was functionally identical to the laboratory unit. The combine was equipped with a 102 cm. active section. Initially, the triangular elevator was driven at the top front sprocket, but the drive was later moved to the top rear boot to reduce chain tension in the active section.

Fig. 47–8. Case IH model 1680 grain combine equipped with a triangular paddle elevator flow sensor.

In order to calibrate the active section on the combine, a small catch bin was suspended from load cells within the combine grain bin. As the main bin of the combine was filled, the catch bin weighed the first 0.53 m^3 of grain, after which the grain spilled over the top of the catch bin and into the main combine bin. An electrically-actuated door discharged the contents of the catch bin into the main combine bin during unloading.

Two basic types of tests were conducted in the field.

1. Flow Models. On roughly a daily basis, a series of tests was conducted to determine the coefficients of the linear flow model shown in equation 3. Each test point was obtained by operating the combine at a constant ground speed for about 30 seconds, during which the true grain flow was determined from the catch bin. After running a series of ground speeds and replications, regression was used to determine the model coefficients for equation 3.

2. Bin Load Accuracy. To verify the overall accuracy of the system, individual loads of the main bin of the combine were weighed. The predicted grain flow obtained from the triangular elevator was then integrated for that bin and compared to the actual grain weight.

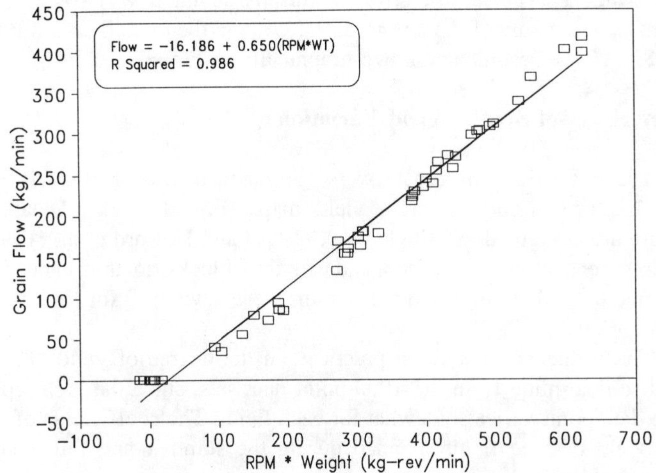

Fig. 47–9. Example field test linear prediction model, wheat.

Flow Models

Field tests conducted in summer 1991 concentrated on the development of grain flow models for wheat. Twelve d of testing revealed flow model slope coefficients ranging from 0.58 to 0.71. All but one of the flow models illustrated coefficients of determination of 0.97 or higher. Figure 47–9 illustrates typical results from one day's testing (Howard, 1992).

Flow models were developed for soybeans, grain sorghum, and sunflowers (*Helianthus annus*) during fall 1991. The slope of the flow coefficients for grain sorghum coincided very well with the 0.69 theoretical value. The slope coefficients for soybeans were about 0.55, indicating large amounts of slippage past the paddles. Large amounts of soil were ingested into the machine, due to the fact that no rain had fallen between the last cultivation and harvest of the beans. Also, the beans had matured unevenly and contained large amounts of foreign material in addition to the soil. The slope coefficients for sunflowers were about 0.83, indicating less slippage than with the other two crops. All of the crop flow models had coefficients of determination values of 0.95 or higher.

Bin Load Accuracy

An example of the accuracy obtained from the triangular elevator is shown in Table 47–1 (Pringle et al., 1993). Each line of the table corresponds to a single combine swath through a rectangular field of irrigated corn in Hamilton Co., NE. The second column shows the grain weight as determined from the integrated triangular elevator readings, based on a flow model of the equation 3 form. The third column shows the grain weights as determined by a weigh wagon. The largest error on this particular d was -6.72%, and the average absolute value of the error was 2.45%, and the overall error for the field was -0.85%. The results are shown graphically in Fig. 47–10.

Characteristics of Spatial Yield Variation

The raw flow sensor data were correlated to differential GPS position signals and processed into grain yield maps (Fig. 47–11). Details of the procedure are contained in Pringle et al. (1993) and Howard et al. (1993). The grain flow sensor data were consolidated into blocks so that each field was represented by 1000 points, and variograms were developed for each field based on those points.

The influence of cultural practices on the degree of yield variability is difficult to determine from the 1000-point data sets, since the area represented by each data point varies somewhat for each field. The coefficients of variation for the yield of nine fields mapped during the summer and fall of 1992 are shown in Table 47–2.

The dryland wheat yields showed the highest coefficients of variation, ranging from 19.98 to 32.57%. The highest coefficient of variation was for a field that had experienced winter-kill that varied across the field.

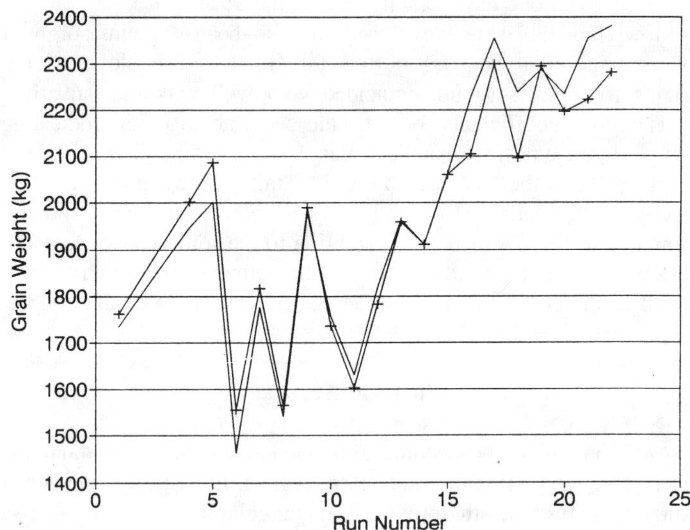

Fig. 47–10. Example "Bin Load" comparison, corn.

Table 47–1. Flow sensor predicted weight and actual weight comparison, corn, Hamilton, Co., NE.

Run #	Flow Sensor Weight (kg)	Weigh Wagon Weight (kg)	Percent Error
1	1735	1762	1.52
4	1948	2003	2.78
5	2002	2086	4.02
6	1463	1555	5.92
7	1775	1817	2.30
8	1541	1565	1.51
9	1980	1990	0.48
10	1756	1736	-1.12
11	1631	1604	-1.67
12	1819	1783	-2.04
13	1964	1960	-0.21
14	1911	1912	0.05
15	2062	2061	-0.09
16	2227	2106	-5.75
17	2354	2299	-2.39
18	2238	2097	-6.72
19	2288	2295	0.32
20	2235	2197	-1.71
21	2355	2223	-5.94
22	2380	2281	-4.37
Total	39674	39340	-0.85
Largest Error			6.72
Ave. Error			2.45

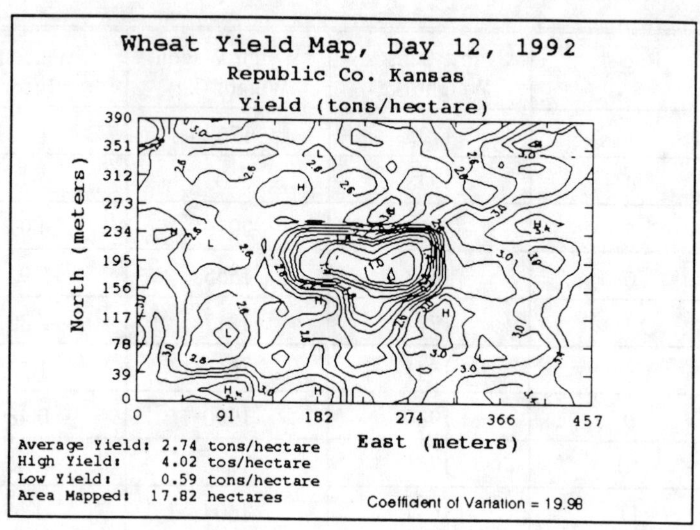

Fig. 47–11. Day 12 wheat yield map, Republic County, Kansas.

Table 47–2. Coefficients of variation for 1992 harvest yield data.

Crop	Day	Water Management	Coefficient of Variation
Wheat	12	Dryland	19.98
Wheat	13	Dryland	26.40
Wheat	13	Dryland	32.57
Wheat	15	Dryland	20.32
Corn	1, 2, 3	Irrig. Moderate	15.42
Corn	4, 5, 6	Irrig. Limited	14.19
Corn	6	Irrig. Moderate	8.20
Corn	7	Irrig. Full	10.26
Corn	8	Irrig. (Replant)	22.58

All of the corn fields shown in Table 47–1 and 47–2 were irrigated, but the method and management of the irrigation varied from field to field. In general, the irrigated corn had lower coefficients of variation than the dryland wheat, and the variability appears to have been reduced as the level of irrigation was increased. The highest variation was evident in a field that experienced flooding and replanting.

FLOW MODEL MODIFICATIONS

Although equation 2 indicates that the triangular elevator should be insensitive to crop conditions, the fact that the flow model slope coefficients vary by crop indicates the need for additional terms in the flow prediction model. Howard (1992) investigated a number of additional terms, including grain density, grain moisture, and the percentage of light material in the grain. Statistical analysis led to a flow equation that included crop density as well as squares and cross products of the sprocket RPM and net load cell reading. The equation had the form:

$$Q = a + b(RPM) + c(F) + d(DEN) + e(F*RPM) + f(RPM^2) + g(F^2) + h(RPM*F^2) \quad (4)$$

where:

Q = flow rate, kg/min
RPM = elevator chain drive sprocket RPM
F = net active section load cell reading, kg_f
DEN = crop density, kg/m^3
a...h = regression coefficients

After the coefficients of the model were determined by regression, the model was validated by applying a single set of model coefficients to a variety of crops and test conditions. A total of 24 bin load tests were conducted in wheat, soybeans, and grain sorghum. Each test compared the integrated flow sensor reading to scale tickets for each combine binload. Errors ranged from -8.67% to 10.38%, and average absolute value of the errors was 4.49%.

CONCLUSIONS

1. The triangular elevator measurement concept has achieved errors with an average absolute value of under 3% during "bin load" tests, when the flow prediction coefficients were determined in the field being harvested.

2. When a single flow prediction model was used to predict flow in wheat, soybeans, and grain sorghum at varying grain moisture contents, the average absolute value of the error in "bin load" tests was under 5%.

3. The seals at the ends of the active section are a major contributor to inaccuracy of the system. Future development should focus on the redesign or elimination of the seals.

4. No mechanical problems or combine capacity restrictions have become apparent in over 800 ha. of testing with the triangular elevator in wheat, corn, grain sorghum, soybeans, and sunflowers.

REFERENCES

Auernhammer, H., M. Demmel, T. Muhr, J. Rottmeier, K. Wild. 1993. Yield measurements on combine harvesters. ASAE 931506. ASAE, St. Joseph, MI.

Borgelt, S.C. 1992. Sensor technologies and control strategies for managing variability. Proc. Soil Specific Crop Management. University of Minnesota Extension Service.

De Baerdemaeker, J., R. Delcroix, and P. Lindemans. 1985. Monitoring the grain flow on combines. Paper presented at Agri-Mation Conference and Exposition, Feb. 1985, Chicago. ASAE Publ. 01-85. St. Joseph, MI.

Howard, K.D. 1992. The development of an empirical model for predicting factorial influences on the weight section of a triangular paddle grain elevator. Unpubl. Ph.D. Dissertation, Kansas State Univ.

Howard, K.D., J.L. Pringle, M.D. Schrock, D.K. Kuhlman, and D.L. Oard. 1993. An elevator based combine grain flow sensor. ASAE Paper No. 931504. ASAE, St. Joseph, MI.

Pang, S.N., and G.C. Zoerb. 1990. A grain flow sensor for yield mapping. ASAE Paper No. 901633. ASAE, St. Joseph, MI.

Pfeiffer, D.W., J.W. Hummel, and N.R. Miller. 1993. Real-time corn yield sensor. ASAE Paper No. 931013. St. Joseph, MI.

Pringle, J.L., M.D. Schrock, R.T. Hinnen, K.D. Howard, and D.L. Oard. 1993. Yield variation in grain crops. ASAE Paper No. 931505. St. Joseph, MI.

Searcy, S.W., J.K. Schueller, Y.H. Bae, S.C. Borgelt, and B.A. Stout. 1989. Mapping of spatially variable yield during grain combining. ASAE Trans. 32(3):826–829.

Wagner, L.E. and M.D. Schrock. 1989. Yield determination using a pivoted auger flow sensor. ASAE Trans. 32(2):409–413.

48 Measuring the Spatial Performance of Chemical Applicators

Jonathan Chaplin
Eugene Roytburg

Dep. of Agricultural Engineering
University of Minnesota
St. Paul, Minnesota

Joseph Kaplan

Agricultural Machinery Research Institute
Minsk, Byelorussia

A review of the ASAE standard used in measuring the distribution of dry material during field application. Presentation of a methodology and field testing of a new procedure that quantifies the spatial distribution of material during field operation of chemical applicators. A grid sampling technique is compared to a random sampling procedure to provide information on longitudinal and lateral variability of thc application rate. This paper provides insight into how the quality of material distribution might be evaluated for spreaders used in site specific application of materials.

INTRODUCTION

The ASAE Standard S341.2 "Procedure for measuring distribution uniformity and calibrating granular broadcast spreaders" (1993) presents field test procedures to determine the effective swath width of the machine and application rate. Tests are to be conducted under field conditions that represent normal use with wind velocities <8 kmph.

Collection trays are placed across the swath to collect material that is deposited by the applicator. The size of the collection trays is specified having a minimum length of 30 cm and a width not exceeding 10% of the effective swath width. At least 10 collecting trays must be placed within the effective swath. Spacing between trays must be uniform. Additional trays are placed on either side of the effective swath up to 50% of the effective swath width.

The tests are conducted to determine the application rate and the uniformity of distribution across the swath. To determine the uniformity of distribution, a simulated field application of multiple adjacent swaths is used to compute the coefficient of variation across the swath. The test does not account for variations that are attributable to driving precision, the effect of metering efficiency for varying application rates, or the nature of the field surface.

As a result of interest in site specific farming and application of agricultural chemicals according to position in the field, a revised procedure is needed that accounts for the variability of driving precision and varying application rates. It is appropriate to look at this operation as a total system that includes the driver, machine and field surface. The ASAE standard is sufficient to determine some general characteristics for a particular machine, however, variations due to driver–guidance precision and other field errors are not determined.

This paper focuses on variability attributable to driving precision and presents the theory, methodology and practical testing for a single disk mounted fertilizer spreader. The theory is applicable to sprayer and dry chemical applicators.

LITERATURE REVIEW

Spread pattern evaluation and collection methods have been investigated by Parish (1986). Twelve methods were used to collect samples for spreader pattern tests under controlled conditions using one spreader and two materials. The results indicated that the method of collection can influence the observed application rate, effective swath width, amount of skewing, and the coefficient of variation across overlapped swaths. In this test, traditional baffled collection trays (as specified by ASAE S341.2) yielded coefficients of variation of 8 to 22% compared to coefficients of variation of 27 to 57% for the same spreader and materials using long, narrow pans and 21 to 41% for floor collection. The differences were attributed to the different amounts of material bouncing in or out of the various collection devices.

Parish and de Visser (1989) conducted a theoretical study of the relationship between the collection tray width and spread pattern. A prediction equation was used to normalize spread pattern data obtained using different pan widths. The relationship between collection pan width and coefficient of variation had little effect on the results obtained using ASAE S341.2.

Contrary to the ASAE Standard, spreaders are often tested on smooth surfaces that are not representative of field conditions. Parish (1991) showed that operating a rotary spreader on a rough surface had a detrimental effect on spread pattern uniformity. The coefficient of variation, as determined by ASAE Standard S341.2, increased from 10% for a spreader operating on a smooth surface to 30% for the same machine on a bumpy surface.

Coates (1992) investigated the effect of swath width, feed gate opening, and hopper level on the spread pattern of a pendulum spreader (oscillating spout). In two thirds of the tests spreading performance was unacceptable since the coefficient of variation for the distribution exceeded 20%. A coefficient of variation in excess of 20% was considered unacceptable for field application. Coefficients of variation increased as the hopper emptied and decreased as swath width decreased. When operating at the effective swath width the range of application rates indicated poor spreader performance as they frequently were >120% or <80% of the mean rate.

Roth et al. (1985) showed that spread patterns are frequently asymmetrical and rough. If machine adjustments did not improve the spread pattern then a pattern analysis was conducted to determine the best application method and maximum effective swath width to achieve an acceptable level of distribution uniformity. Three characteristic non-symmetrical distribution patterns (pattern-edge disturbance, mid-pattern disturbance and offset machine centerline) were analyzed. The relationships between the coefficient of variation and swath width for each pattern were presented. A variable swath width was proposed for machines with an asymmetric spread pattern.

Many investigations have been conducted to determine how the method of assessment, and the size of the collection trays affect the measured distribution. This analysis incorporates lateral and axial distribution together with driving precision in the estimation of field variability of material applied. Factors which have a direct effect on material distribution:

1. Driving precision.
2. Field surface
3. Metering efficiency and variation with hopper level.
4. Systematic errors associated with machine calibration.
5. Wind speed and direction.

OBJECTIVES

To develop a statistically sound model that explains the spatial variability of materials applied with a field applicator. This paper emphasizes variability due to driving precision. The following is a list of sources of variation that may account for spatial variability in an overall model.

THEORY

Differential Model

The following model was developed as an extension of that used in the standard.

$$X = X_s \pm \Delta X_w + \epsilon \qquad [1]$$

Where:

X = actual application rate (kg ha^{-1})
X_s = desired application rate
Δx_w = field deviation from x_x
ϵ = error

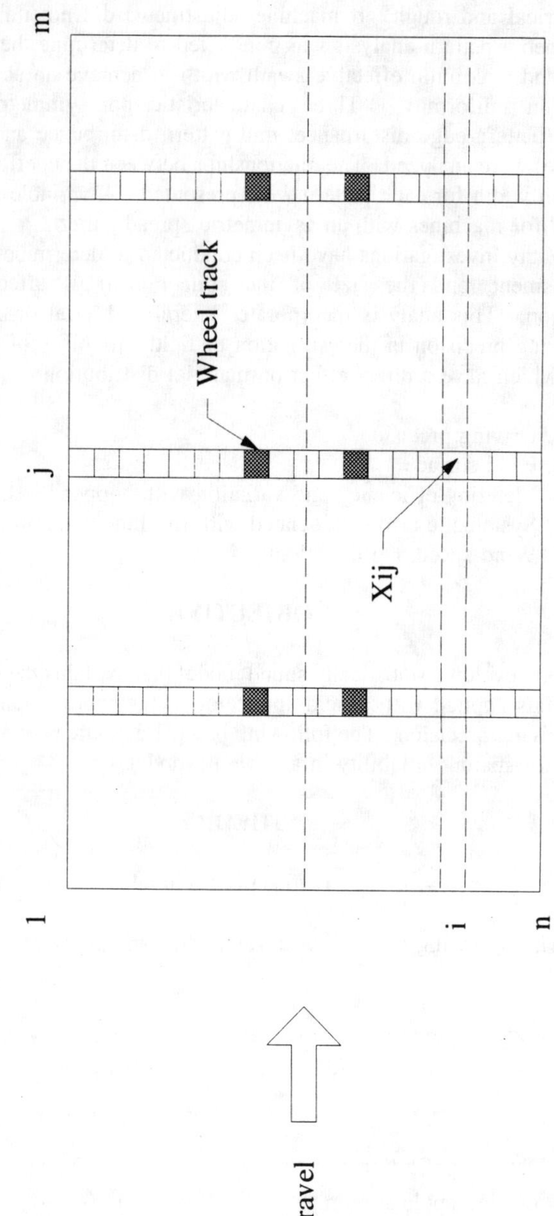

Fig. 48–1. Schematic of collection tray layout.

A schematic of a collection tray layout is shown in Fig. 48–1. The trays are placed perpendicular to the direction of travel according to ASAE Standard S341.2. There are "n" trays across the swath and "m" rows of trays along the swath. Let X_{ij} be the weight of material collected in tray, where "i" is the lateral index and "j" is the axial index.

The overall mean material mass is:

$$\overline{X} = \frac{1}{mn} \sum_{1,1}^{n,m} X_{ij} \qquad [2]$$

An estimation of the variance of material distributed along and across the swath is given by:

$$D_w = \frac{1}{nm-1} \sum_{1,1}^{n,m} (X_{ij} - \overline{X})^2 \approx \frac{1}{nm} \sum_{1,1}^{n,m} X_{ij}^2 - \overline{X^2} \qquad [3]$$

(assuming $nm > 100$)

X_{ij} can be expressed as follows:

$$X_{ij} = \overline{X}_i + \Delta_{ij} \qquad [4]$$

Where:

$\overline{X}_i = \frac{1}{m} \sum_{j=1}^{m} X_{ij}$ is the mean of the ith row of trays.

Δ_{ij} = deviation of the ijth observation from the swath mean.

The first component in Equation [3] can be rearranged as follows:

$$\frac{1}{nm} \sum_{1,1}^{n,m} X_{ij}^2 = \frac{1}{mn} \sum_{i=1}^{n} \sum_{j=1}^{m} (\overline{X}_i + \Delta_{ij})^2$$

$$= \frac{1}{mn} \sum_{i-1}^{n} \sum_{j=1}^{m} (\overline{X_i^2} + \Delta_{ij}^2)$$

$$= \frac{1}{n} \sum_{1}^{n} \overline{X} \qquad [5]$$

Where:

n_i = number of trays in the effective swath.

n = number of trays across overall swath.

$\sigma_i^2 = \frac{1}{m}\sum_{j=1}^{m} \Delta_{ij}^2$ is the standard deviation of the weight in the ith row of trays

Let $\overline{X}i = \overline{X} + \Delta_i$ [6]

Where:

$\overline{X}$ is the overall mean as defined in Equation [2].
Δ_i is the deviation of the ith observation from the overall mean.

Then Equation [5] can subsequently be rearranged:

$$\frac{1}{nm}\sum_{1,1}^{ij} X_{ij}^2 = \overline{X^2} + \frac{1}{n}\sum_{1}^{n} \Delta_i^2 + \frac{1}{n}\sum_{1}^{n} \sigma_i^2 \quad [7]$$

Substituting Equation [7] into [3] yields:

$$D_w = \frac{1}{n}\sum_{1}^{n} \Delta_i^2 + \frac{1}{n}\sum_{1}^{n} \sigma_1^2 \quad [8]$$

The field variance can now be separated into two parts, D_a the axial variation and D_t being the lateral or cross variation. The summation index is changed to reflect the number of trays in the effective swath width.

$$D_a = \frac{1}{n_1}\sum_{i=1}^{n} \sigma_i^2$$

$$D_t = \frac{1}{n_1}\sum_{i=1}^{n_1} \Delta_i^2$$

The variance can be expressed in terms of kg^2m^{-4} by using the following transformation from mass to application rate:

$$D_a = \frac{1}{B_w l^2 b} \sum_{i=1}^{B/b} \sigma_i^2 \quad [9]$$

$$D_t = \frac{1}{B_w l^2 b} \sum_{i=1}^{B_w/b} \Delta_i^2 \quad [10]$$

Where

B = overall swath width (m)
B_w = effective swath width (m)
b = collecting tray width (m)
l = collecting tray length (m)

$$n_1 = \frac{B_w}{b} \text{ and } n = \frac{B}{b}$$

The overall mean application rate can be expressed in kg ha^{-1} as:

$$\overline{Q} = \frac{1}{B_w l} \sum_{i-1}^{B_w/b} \overline{X_i} \quad [11]$$

The coefficient of variation of the application rate can now be determined:

$$CV_a = \sqrt{\left(\frac{B_w}{bq^2} \sum_{i=1}^{B/b} \sigma_i^2\right)} \quad [12]$$

$$CV_t = \sqrt{\left(\frac{B_w}{bq^2} \sum_{i=1}^{B_w/b} \Delta_i^2\right)} \quad [13]$$

$$q = \sum_{i=1}^{B/b} \overline{X_i} \quad [14]$$

Equation [12] can be used to determine the coefficient of variation of application along the swath. Equation [13] yields the variation across the swath. Equations [12] and [13] do not account the variations in swath width that result from driving errors.

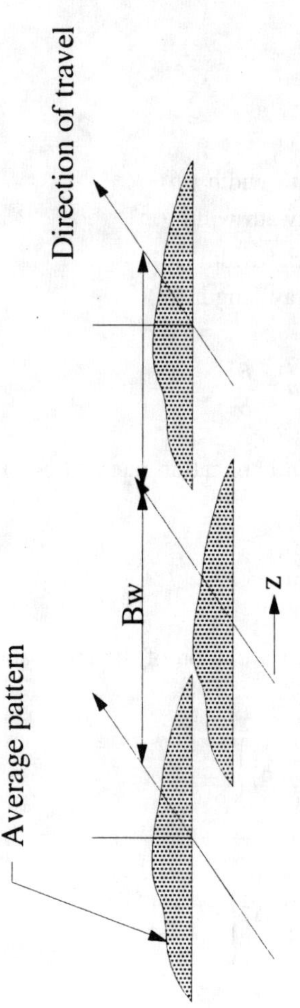

Fig. 48–2. Field pattern schematic.

The coefficient of variation across the swath can also be determined by lateral displacement of a central pattern as shown in Fig. 48–2. The coefficient of variation can be approximated by a hyperbola for uni-directional application with overlapping swaths. This general relationship has been developed by Kaplan & Rumyantcev (1986) following a study of many spread patterns for single and multi-disk type spreaders.

$$CV_i^2 = CV_o^2 + 5K_b \frac{Z^2}{B_w^2} \qquad [15]$$

Where:

CV_i = actual coefficient of variation

CV_o = coefficient of variation with zero displacement

Z = lateral displacement of the center swath

K_b = B_w/B

The average coefficient of variation for the whole field can be determined as the sum of all incremental lateral variations multiplied by the probability of the distribution occurring:

$$\overline{CV} = \sqrt{\left(\sum_1^n P_i CV_i^2\right)} \qquad [16]$$

Assuming a normal distribution, P, for the population of lateral displacements of the center pattern, and using Equation [15] then the average coefficient of variation can be determined:

$$\overline{CV}^2 = \overline{\left(CV_c^2 - CV_o^2\right)}$$

$$= 2\int_o^\infty \left(\frac{5K_b^2 Z^2}{B_w \sigma_z \sqrt{2\pi}} \; e^{-\frac{z^2}{2\sigma_z^2}} \right) dz$$

$$= 5K_b^2 \frac{\sigma_z^2}{B_w^2}$$

$$\overline{CV}_c^2 = CV_o^2 + 5K_b^2 \omega^2 \qquad [17]$$

Where:

$\omega = \dfrac{\sigma_2}{B_w}$ is the coefficient of variation of the swath width.

Equations [12], [13], [14] and [17] can be used to determine the coefficient of variation for the field. This is an incremental model since the variation is explained by parts. The axial variation has a random origin and can be evaluated using Equation [12]. The lateral variation is a summation of variation due to driving error and the minimum variability expected when operating at the effective swath width. This may be determined using Equations [13] and [14].

Integral Model

For an integral estimation of the variation over the field the individual measurements of the spread patterns sampled are arranged on center lines representing the average swath width. The even patterns are displaced laterally by the equivalent shift determined from the probability function. The coefficient of variation is now determined from the cumulative distribution of the entire set as depicted in Fig. 48–3.

The driving precision can be characterized by examining the variation of the effective swath width. Deviation from the effective swath width can be measured from field trials. This is used in turn to determine the expected coefficient of variation for the equivalent center swath lateral shift. The average being determined using Equation [16].

TEST PROCEDURE

Field tests were conducted at the Rosemount Station using a Herd F-160 160 kg, single disk spreader. The spreader was mounted on a small 4WD tractor and operated on a level cultivated field. The same tractor driver conducted all tests at a forward speed of 8 kmph. Tests were conducted using a uni-directional pattern with the tractor heading into the wind. Wind speed varied from 2 to 5 m/s. Wheat was used in all tests (1000 grain wt = 18.3 g, size distribution - 3.4% 1-2 mm, 96.6% 2-4 mm).

Collection trays were made from egg trays measuring 25 x 28 cm. Although this collection tray size does not meet specific ASAE Standards it does offer the advantage of built in baffles.

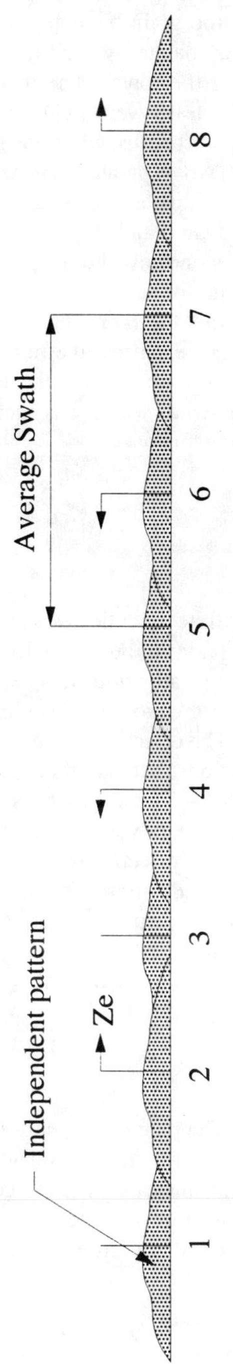

Fig. 48–3. Integral method schematic.

A sticky backed tape 5 cm high was attached around the edge of each tray to stop grain bouncing out onto the ground. The inside surface of the tape was spray painted so that the wheat would not stick and cause difficulties when emptying the trays. The trays were arranged in rows as shown in Fig. 48–1. Since the trays were light, weights were placed in the trays to keep them from being disturbed by wind gusts.

Two separate tests were conducted. The first consisted of multiple distribution pattern tests according to the ASAE Standard. The path for each run was measured and flagged so that the driver had a visual guide across the plot. In the second test the trays were placed at random throughout the plot and the driver did not have any flags to determine the swath width. Trays that were run over in this test were discarded.

For each test the number of grains were counted in each tray and then converted to mass using the 1000 grain weight. In the unflagged portion of the test the swath width was measured (wheel track to wheel track) to determine driving precision.

RESULTS

Effective Swath Width

Sixteen replicates were used to measure the spread pattern. Two adjacent spread patterns are used to determine the effective swath width. Here the effective swath width was defined as the swath width minimizing the coefficient of variation across the swath.

The effective swath width was determined both by overlapping individual distribution patterns and by using an overall average distribution as shown in Fig. 48–4. There was no statistical difference (α=0.05) between the average effective swath widths determined by either method.

The effective swath width was determined to be 6.9 m (690±9.5 cm). The method minimizes the coefficient of variation across the swath for incremental displacements of the center pattern as shown in Fig. 48–5. In this case each increment was 25 cm since this was the width of each collection tray. The minimum coefficient of variation for this spreader when working at the effective swath width is ≈15%.

Distribution Function

The relationship between the coefficient of variation and the displacement of the center pattern was determined. Three adjacent swaths were positioned at 6.9 m. In this case the average distribution pattern was used. The center pattern was moved in increments of 25 cm to the left and right. For each position the coefficient of variation was calculated for the entire swath width (see Fig. 48–6).

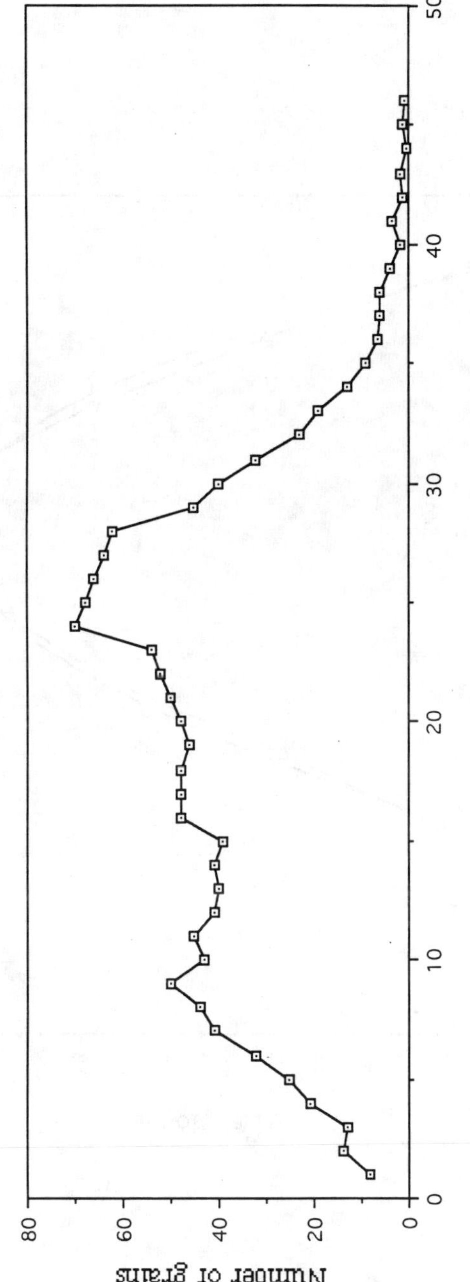

Fig. 48–4. Average spread pattern (n=16).

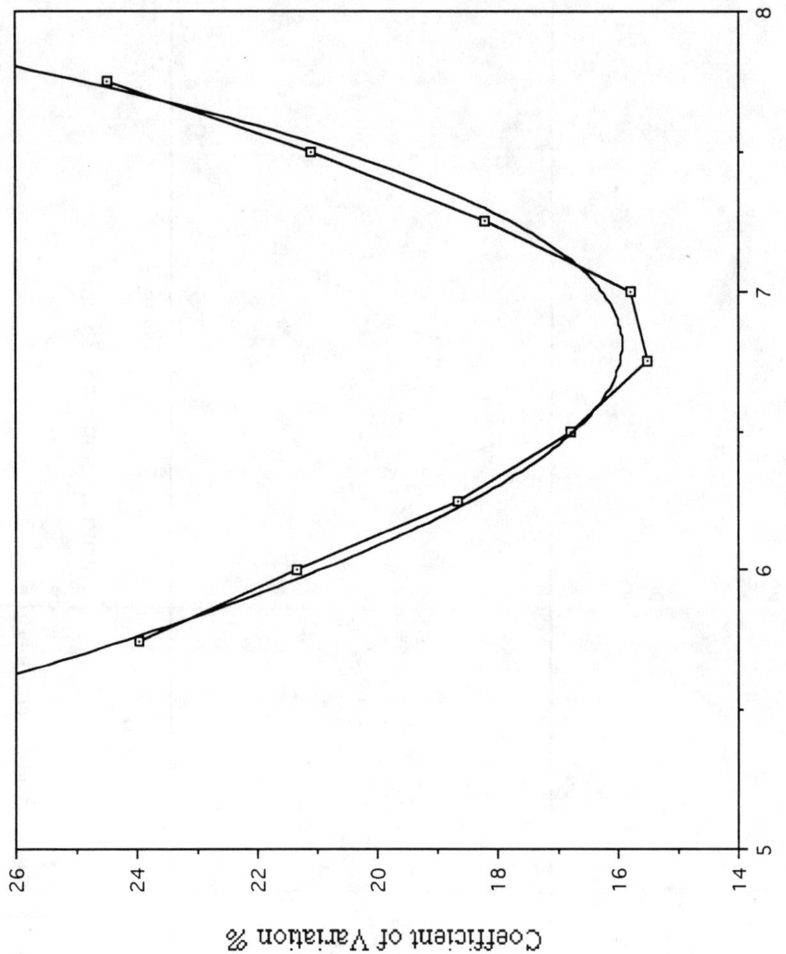

Fig. 48–5. Effective swath width determination.

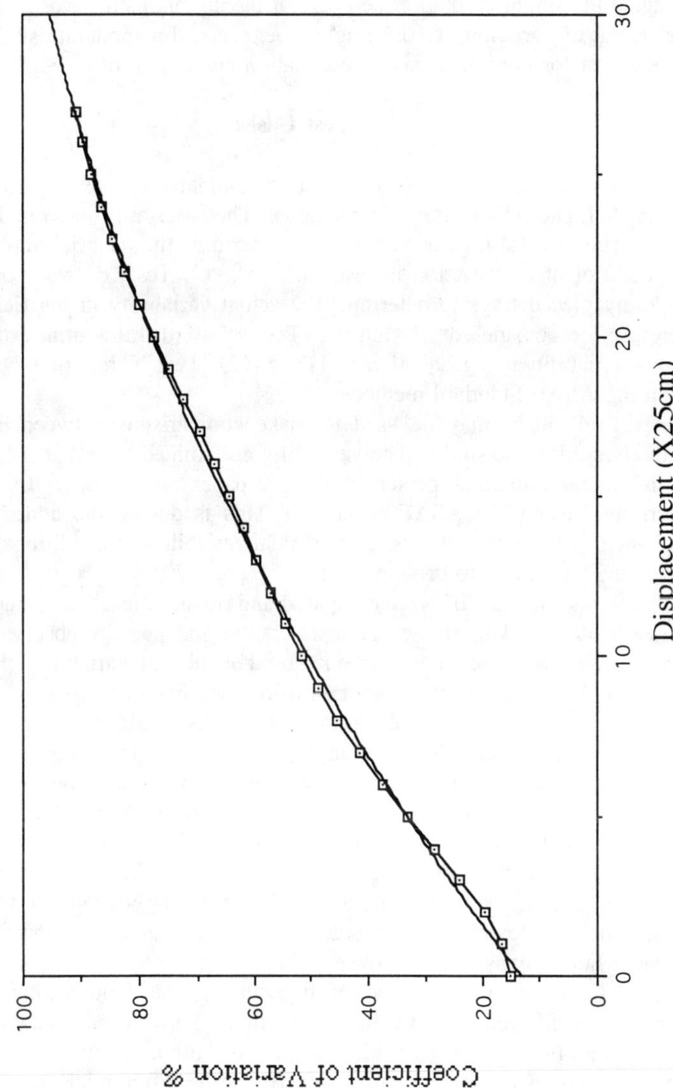

Fig. 48–6. Coefficient of variation vs center swath displacement.

Driving Precision

Measurements taken in the unflagged portion of the field resulted in information on driving precision. The average swath width was 6.96±0.05 m. The coefficient of variation for the driving test was 5.3%. This is lower that expected but might be due to the operator having high skill levels. Normally a coefficient of variation of 10% might be expected for moderate skill levels. A Rankits test for normality gave a correlation coefficient of 0.988.

Test Cases

Six test cases were conducted to compare the results from integral, differential, and ASAE Standard methods. The cases are shown in Table 48–1.

The integral method was used to determine the material application rate and coefficient of variation in tests "a", "b", "c". Test "d" was conducted on randomly placed trays to determine the actual variability in the field and can, therefore, be considered a "control". Test "e" used differential estimation for spatial variability using Equations [11] and [12]. Test "f" reported the variability using the ASAE Standard methods.

Table 48–1 may be used to make comparisons between each of the methods used in the study. The variability determined by either the differential or the integral method presented in this paper are ≈50% > the variability determined using the ASAE Standard. This is due to the added variability explained by the new methods, namely axial variability and additional variability across the swath due to driving error.

Tests "a" and "b" were compared and no significant difference could be shown (α=0.05). This shows that a set of eight independent observations of the spread pattern is sufficient to determine field or sub-plot variability. Figure 48–7 shows the lateral material distribution using data from test "b".

Test "c" showed that reasonable results could be obtained without estimating the amount of material in the wheelings. Attempts are often made to ramp the spreader over the collection trays during a distribution test. This, however, may well upset the equilibrium of the metering mechanism.

The random placement of trays in test "d" attempted to measure over-field variation of the distribution. No significant difference in variability was found between this and the other tests. The mean application rate was found to be significantly lower than the other tests. This was due to the fact that the spreader hopper was run low towards the end of the test.

The differential method of determining the field variation was not significantly different from the integral method. The error attributable to driving did not significantly influence the lateral distribution in this test. The integral method is not as computationally intensive as the differential method.

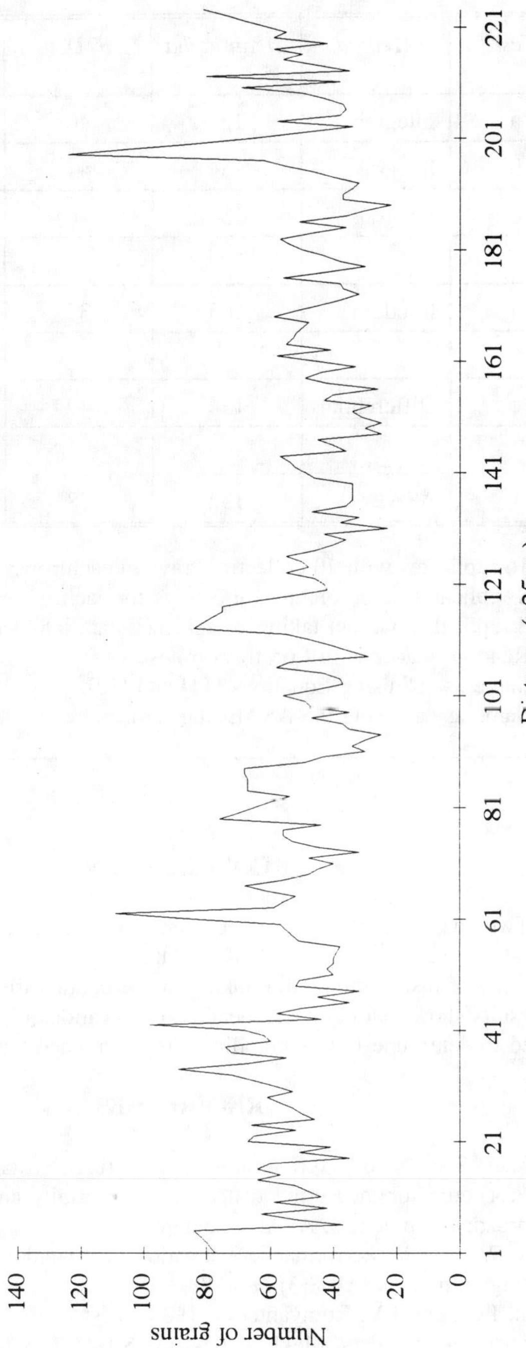

Fig. 48–7. Lateral distribution using the integral method (n=8).

Table 48–1. Field test results.

Test	*Method*	*Mean kg/ha*	*STDev*	*CoeffVAR %*
a	Integral	132	40	30
b	Integral	134	44	33
c	Integral	134	46	34
d	Random	105	35	33
e	Differential	134	49	33
f	ASAE	132	28	21

a. 16 replicates with 40 collection trays in each row.
b. 8 replicates, interpolating for 3 trays for each wheel.
c. 8 replicates without taking wheel track data into account.
d. Random placement of 65 trays in test area.
e. Same as "b" using Equations [11] and [12].
f. Same as "a" using the ASAE Standard.

CONCLUSIONS

Two analysis methods were developed to determine the field variability of material applicators. These methods of data analysis make it possible to estimate the quality of distribution while taking into account varied working conditions. The results show that eight independent observations of the spread pattern can be used to determine field variability when combined with driving precision.

REFERENCES

American Society of Agricultural Engineering. Standard S341.2. 1993. Procedure for measuring distribution uniformity and calibrating granular broadcast spreaders. ASAE Standards.

Coates, W. 1992. Performance evaluation of a pendulum spreader. Applied Eng. Agric. ASAE 8(3):285-288.

Kaplan, I.G., and I.V. Rumyantcev. 1986. Distribution quality estimation for fertilizer spreaders. Mekh. i Elektriph S.H. N7. p 3-6.

Parish, R.L. 1986. Comparison of spreader pattern evaluation methods. Applied Eng. Agric. ASAE 2(2):89-93.

Parish, R.L. 1991. Effect of rough operating surface on rotary spreader distribution pattern. Applied Eng. Agric. ASAE 7(1):61-63.

Parish, R.L., and Piet de Visser. 1989. Experimental verification of the effect of collection pan width on apparent spread pattern. Applied Eng. Agric. ASAE 5(2):163-164.

Roth, L.O., R.W. Whitney, and D.K. Kuhlman. 1985. Application uniformity and some non-symmetrical distribution patterns of agricultural chemicals. Trans. ASAE 28(1) 47-50.

49 Electromagnetic Induction Sensing as an Indicator of Productivity on Claypan soils

Kenneth A. Sudduth

USDA-Agricultural Research Service
University of Missouri
Columbia, Missouri

Newell R. Kitchen
David F. Hughes

Department of Soils and Atmospheric Sciences
University of Missouri
Columbia, Missouri

Scott T. Drummond

Department of Agricultural Engineering
University of Missouri
Columbia, Missouri

Claypan soils have a dense, high-clay content layer and cover large areas of the Midwest. Erosion and deposition processes on claypan soil landscapes often determine the depth of the claypan horizon in the soil profile. The claypan may be at or near the surface on eroded sideslopes, yet covered by a significant depth of overwash sediments at lower landscape positions. Productivity is often highly correlated to the thickness of the topsoil (or depth to claypan, DTC) and the physical characteristics of the claypan horizon itself. Variability in these claypan properties must be considered in development of site-specific nutrient management strategies for claypan soils.

Standard practice for quantifying claypan depth and characteristics involves using a manual or powered soil probe at each of numerous sampling points, a slow and laborious process. Consequently, mapping of the claypan has not been practical for use in production agriculture decision-making. An automated, preferably noninvasive, measurement technique would be required to quantify claypan variations over a large area. This paper reports on the implementation and evaluation of a commercially available electromagnetic induction (EM) ground conductivity sensor to map claypan soil variability for site-specific nutrient management.

LITERATURE REVIEW

There are about 4 million tilled hectares on claypan soils in the midwestern United States. Claypan soils developed in loess parent material and, other than the topsoil above the claypan, are usually very low in natural fertility. Clays in the claypan are montmorillonitic and characteristically swell and shrink with wetting and drying. The wet bulk density of the claypan is usually low (1.2 to 1.3 g/cm^3), but because of the high clay content and fine porosity, absorption and oxygen diffusion is slow. Plant available water from the claypan is low because a large portion of the stored water is retained with the clay at the wilting point (Jamison & Kroth, 1958). Many small and some large cracks develop during drier periods of the growing season (Baer, 1993). With subsequent rains, these cracks are hypothesized to be primary pathway for recharge into the underlying glacial till aquifer (Alberts el al., 1993). The claypan also restricts the growth of crop roots (Gantzer & McCarty, 1987). With these characteristics, the claypan can have a great impact on soil productivity.

One approach to considering the effects of soil properties (such as those associated with the claypan) on production has been a soil productivity index (PI). A PI developed for Missouri soils included factors, such as plant available water and aeration which are negatively affected by a claypan, thus lowering the soil PI. Gantzer and McCarty (1987) found that 82% of the variation in yields on claypan soils could be predicted by PI and total crop water use. Although not a parameter of PI, they also observed that 62% of the variation in PI could be related to depth to the claypan.

Electromagnetic induction (EM) sensing, originally developed for geophysical surveying applications, uses electromagnetic energy to measure the apparent conductivity of earthen materials. This method has been used extensively to measure the apparent conductivity of saline (Corwin & Rhoades, 1990; Rhoades & Corwin, 1981) and sodic (Ammons et al., 1989) soils. In addition, this technology has been used to map thickness of clays (Palacky, 1987), measure soil water content (Kachanoski et al., 1988), and for groundwater investigations (Greenhouse & Slaine, 1983). Jaynes et al. (1993) recently reported on the correlation of field-scale EM maps to variations in soil type and crop yield in central Iowa.

Variations in electromagnetic response are produced by changes in the ionic concentration of earthen materials. Factors influencing the ionic concentration and conductivity of earthen materials include: (i) the volumetric water content, (ii) the amount and type of ions in soil water, and (iii) the amount and type of clays in the soil matrix. In non-arid areas, soil texture and moisture content are often the principal factors determining apparent conductivity (Kachanoski et al., 1988).

OBJECTIVE

The objective of this study was to evaluate electromagnetic induction sensing as an indicator of crop productivity variations on claypan soils. It was hypothesized that (i) crop productivity on claypan soils would be non-uniform and mediated by a non-uniform DTC, (ii) EM response could accurately estimate DTC, and (iii) EM response could accurately estimate crop production.

MATERIALS AND METHODS

Electromagnetic Induction Sensor

The instrument used in this study was the EM38 (Fig. 49–1) manufactured by Geonics Limited, Mississauga, Ontario, Canada (Mention of trade names or specific products is made only to provide information to the reader and does not constitute an endorsement by the USDA Agricultural Research Service or the University of Missouri). The EM38 uses the principle of electromagnetic induction to quantify soil conductivity. Unlike other methods of soil conductivity measurement, which require insertion of electrodes into the soil mass and may consume significant amounts of energy in the process, electromagnetic induction is a non-invasive, low power measurement technique (McNeill, 1980).

The instrument can be operated either upright (vertical dipole mode, Fig. 49–1) or on its side (horizontal dipole mode). In the vertical mode, the effective measurement depth of the unit is approximately 1.5 m, compared to 0.75 m in the horizontal mode. The instrument response as a function of depth is quite different for the horizontal and vertical modes. Sensitivity in the horizontal mode is highest at the surface, while vertical mode sensitivity peaks at ≈0.4 m. The apparent conductivity as measured by the instrument is determined by the soil conductivity with depth, as weighted by the instrument response functions (McNeill, 1980). The standard technique for measuring soil conductivity calls for placing the instrument on the ground surface in either the horizontal or vertical orientation, or both. More advanced techniques also include readings taken with the instrument held at varying heights above the ground (McNeill, 1980).

Fig. 49–1. Geonics EM38 electromagnetic induction ground conductivity sensor.

To further improve ease and speed of use of the EM38, an Omnidata Model 516C Polycorder was connected to the manufacturer-supplied analog output port. This portable data logger was used to store site ID information along with the conductivity data obtained from the EM38. To further automate the data collection process, a cart has been fabricated to tow the EM38 behind an all-terrain vehicle for on-the-go surveys of large areas, in a manner similar to that reported by Jaynes et al. (1993). Due to timing and weather difficulties, this automated system has not yet been used to collect data from the field areas where production information was also available. The data used in this report were all obtained at discrete points on the soil surface, with the meter carried and positioned by hand (Fig. 49–1).

Study Area

The study was located on a set of 30 large plots at the Missouri Management Systems Evaluation Area (MSEA) near Centralia, Missouri. These plots, each 18 m by 190 m (Fig. 49–2) were situated on a west-facing backslope of a broad, low, upland divide with relatively low relief (< 2.5 m). Soils of the area are primarily of the Mexico-Putnam association (fine, montmorillonitic, mesic Udollic Ochraqualfs). The study site included both eroded and overwash phases of Mexico soils on 1 to 3 % slopes.

Mexico-Putnam soils formed in moderately-fine textured loess over a fine textured pedisediment. Surface textures range from a silt loam to a silty clay loam. The subsoil claypan horizon(s) are silty clay loam, silty clay or clay, and may commonly contain as much as 50 to 60% montmorillonitic clay. Because of extensive weathering, the claypan soil is usually low in natural fertility and pH. Contrasts in electrical conductivity were assumed to exist between the surface horizon and claypan horizon, due to the differences in clay content. The response of the EM meter to these contrasting layers was expected to be a function of both the thickness and conductivity of surface and subsurface claypan layers. Within the study site, DTC ranged from < 10 cm to > 100 cm.

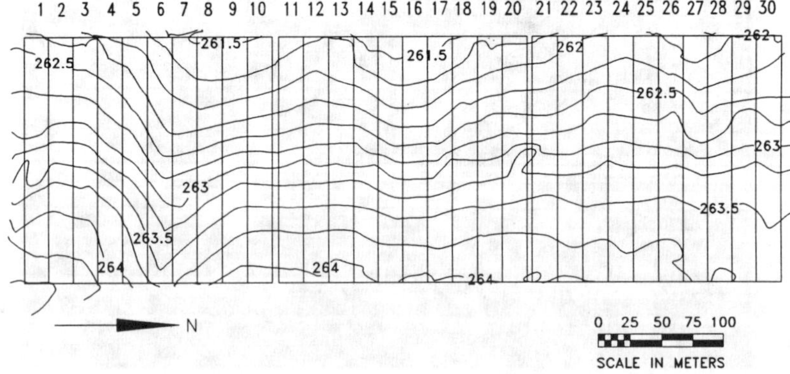

Fig. 49–2. Layout and topography of the 30 Missouri MSEA research plots.

Five farming systems have been assigned to these plots in a randomized complete block design with three replications (Alberts et al., 1993). Systems present included a minimum-till corn-soybean rotation, minimum-till corn-soybean-wheat rotation, ridge-till corn-soybean rotation, no-till corn-soybean rotation, and continuous grass–legume. All crops in each rotation were present each yr of the study. Numerous parameters have been measured on this site to spatially describe landscape variability, including soil nutrients and organic matter content, DTC, thickness of claypan, characteristics of soil cracking, soil water content, permeability, and slope. Harvesting of the plots has been accomplished using a combine-mounted continuous grain flow sensor (Birrell et al., 1993), allowing mapping of the yield variability within each plot.

Data Collection

A preliminary EM evaluation was carried out in August, 1992 along a transect established on the south border of Plot 7 (Doolittle et al., 1994). Previous observation and soil property information had indicated that DTC varied greatly along this transect, making it an ideal location for evaluation of the EM technique. A 190 m transect line with 31 observation points at 6.1 m intervals was established and soil probing was used to determine the DTC at each observation point. With the EM38 meter placed on the ground surface, both horizontal and vertical dipole orientation measurements were taken at each of these same observation points.

EM mapping of the entire 11 ha MSEA plot area was accomplished during a relatively dry period in the summer of 1993. Two transects, again with 31 observation points spaced 6.1 m apart were established in each of the 30 plots, 4.5 m from the north and south edges of the plot. DTC measurements were obtained by soil probe along the south transect in each plot, and EM38 conductivity measurements were taken along all transects. Grain yields for the 1991 and 1992 crop years were determined from a lengthwise harvest transect within each plot.

RESULTS AND DISCUSSION

Preliminary EM Evaluation

Along the Plot 7 transect line, the DTC averaged 39.6 cm and ranged from 7 cm to 105 cm (Fig. 49–3). The claypan typically appeared between depths of 10 cm and 20 cm, with noticeable troughs extending to depths of about 100 cm in several areas where the transect crossed a buried channel meandering across the plot (Sudduth & Kitchen, 1993). Corn grain yields along this transect gave an indication that DTC was a significant factor in productivity, with major increases in yield occurring at the locations of greatest DTC (Fig. 49–3).

At a given location, values of apparent conductivity obtained with the horizontal and vertical coil orientations were highly correlated. Since a stronger relationship existed between the vertical dipole measurements and DTC, the vertical dipole data were used to develop regression equations to predict DTC

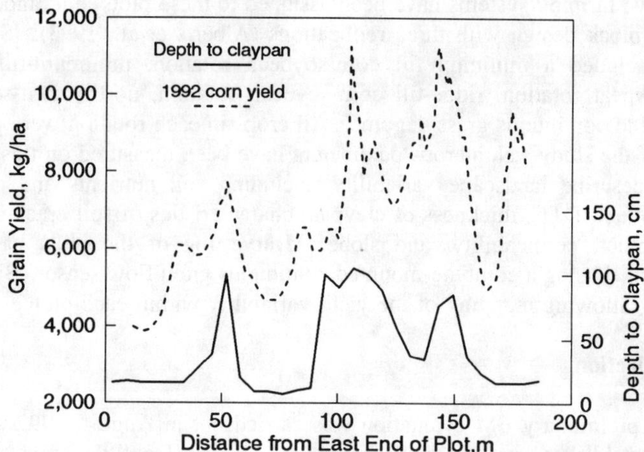

Fig. 49–3. Correspondence of depth to claypan and grain yield along the south edge of Plot 7.

from values of apparent conductivity (Doolittle et al., 1994). Based on thirty-one observation points, an exponential regression on apparent conductivity was found to be the best estimator of DTC ($r^2 = 0.81$):

$$DTC = 1662 * 10^{-0.033\ EM} \quad (1)$$

where: DTC = depth to claypan, cm
EM = vertical dipole EM reading, mS/m.

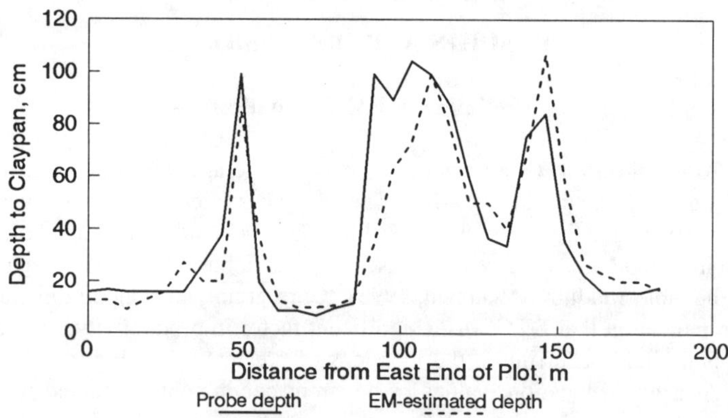

Fig. 49–4. Correspondence of measured and EM-estimated depth to claypan along transect line at the south edge of Plot 7.

There was excellent agreement between the soil probe depth and EM-estimated depth along most of the transect (Fig. 49–4). The major differences were along the edge of one buried channel, and may have been due to the fact that the EM38 measurement was spatially integrated over approximately 1 m, as compared to the point measurement obtained by soil probe.

Relationship of Production to Claypan Features and EM Data

An isoline representation of soil probe DTC over the entire MSEA plot area showed the expected trends (Fig. 49–5). The claypan was nearest the surface on the sideslope position, extending from north to south in the center of the area. Somewhat deeper areas were located at the summit (east edge of the area) and toeslope (west edge of the area) positions. A buried channel was apparent in plot 7, as were several areas of greater DTC along the west edge of the plots. Comparison of these data with the surface relief of the area (Fig. 49–2) showed that these were depositional areas located along paths of concentrated water flow.

DTC and production data were correlated based on previously determined landscape positions. Each of the 30 plots was divided into summit, upper sideslope, lower sideslope, and toeslope positions (n=120). DTC was calculated as an arithmetic mean of all probe data points falling within each plot position, and yield was determined from the appropriate portion of the harvest transect. Correlations of yield to DTC were positive and significant (at the 0.05 level) for 1991, but were not significant for 1992 (Fig. 49–6). Deficient plant-available water was a significant crop stress in 1991 at this site, while timely rains during the 1992 growing season made the water-supplying capacity of the soil (and therefore DTC) a less important factor. Corn yields were more highly correlated with DTC than were soybean yields (Fig. 49–6). Wheat yields, obtained only in 1992, were not significantly correlated with DTC.

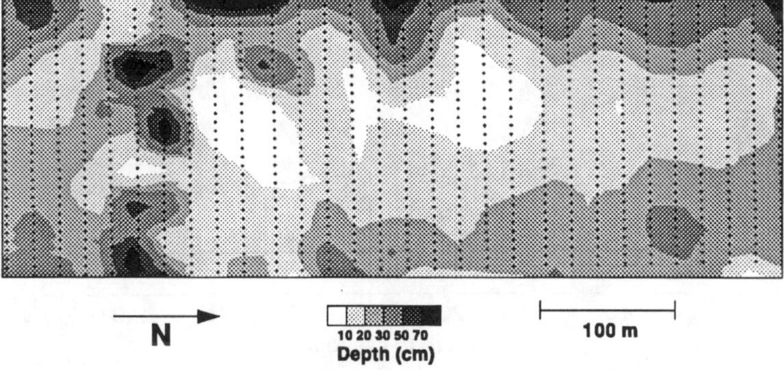

Fig. 49–5. Variation in depth to claypan obtained by soil probe over the MSEA research plots. Dots indicate measurement locations.

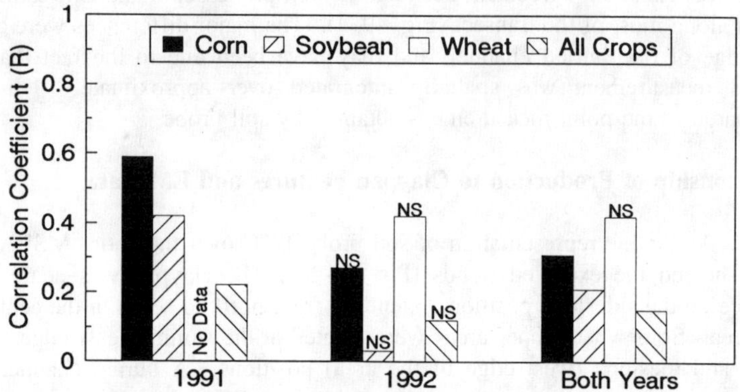

Fig. 49–6. Correlations between yield and depth to claypan. Correlations labeled "NS" were not significant at the 0.05 level.

Regressions of 1991 corn yield on DTC were developed for each individual farming system. In this analysis, each plot was divided into 10 equal areas along its length, and mean yield and DTC data were calculated. Where yield was not limited by other factors, yield and DTC were strongly related, as in the minimum-till corn-soybean rotation (Fig. 49–7). Yield was relatively insensitive to variations in DTC for the no-till corn-soybean rotation since nitrogen fertilizer application problems caused nitrogen stress to be the primary yield limitation on this farming system (Fig. 49–7).

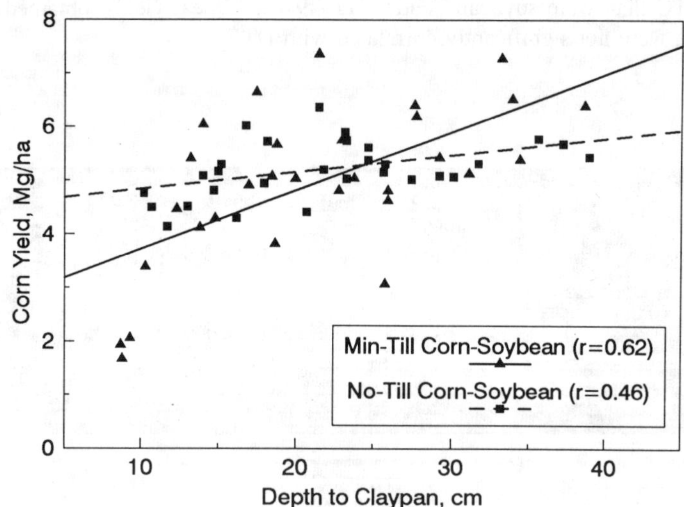

Fig. 49–7. Corn yield (1991) as a function of depth to claypan for two farming systems.

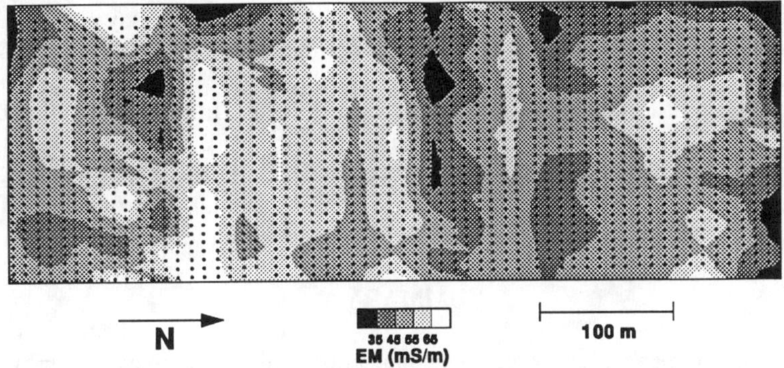

Fig. 49–8. Variation in EM vertical dipole readings over the MSEA research plots. Dots indicate measurement locations.

Raw EM vertical dipole data (Fig. 49–8) exhibited little resemblance to the probe DTC data (Fig. 49–5). The patterns in the EM data tended to run from west to east across the study area, in contrast to the north-south patterns shown by the probe data. Further examination showed that the EM data were strongly dependent on the crop and farming system under which data were obtained (Sudduth & Kitchen, 1993). This dependency was the cause of the east-west patterns in the EM data, corresponding to the plot orientation (Fig. 49–2). The mean EM reading was also statistically dependent on the individual plot in which the data were taken. EM-based DTC maps could only be made to resemble the probe DTC maps when using a separate regression to estimate DTC within each individual plot (Sudduth & Kitchen, 1993).

Correlations were developed between EM and DTC data averaged for the 4 landscape positions per plot. For all 30 plots, the correlation coefficient (R) obtained was -0.50. For correlations developed within a single plot, R-values ranged from -0.99 to -0.51. Since EM data did not completely describe DTC variations, the relationship between yield and EM was not well defined. None of the correlations developed between yield and EM data were significant at the 0.05 level.

The apparent need to develop EM vs. DTC relationships on a plot-by-plot basis indicated some unmodeled variability between the three replications of the data. This bias did not appear to be related to the data collection sequence or technique. There was an apparent trend based on crop. The wheat plots, which had been harvested several weeks prior to the time of data collection, had the least rep-to-rep variation, followed by soybean and grass–legume plots. The highest variation was observed in the corn plots. This may indicate that the EM38 was also responding to changes in the conductivity of the plant biomass, either above or below ground, or both. As these data were collected in late summer, a full canopy was present on both the corn and soybean plots, and nothing in the design of the EM instrument would make it insensitive to changes in above-ground conductivity. Variations in soil water content under the different crops and tillage systems may also have been a contributing factor.

The intended goal for EM measurement of DTC is to provide an estimation of yield potential on claypan soils. As such, it may not be as necessary to optimize DTC estimation as to optimize the regression of yield on EM data. Several of the other factors influencing EM readings, such as clay density and profile water content, may also affect yields. Initial investigation of the relationship between EM response and yield on the MSEA research plots has shown weaker correlations than those between DTC and yield. An EM dataset obtained under drier soil conditions and with better spatial correspondence to yield measurement points will be collected and used for additional investigation.

SUMMARY

A commercial electromagnetic induction soil conductivity meter was investigated for sensing within-field variability in DTC. It was hypothesized that these data could be used to develop site-specific nutrient management strategies for claypan soils, where productivity (expected yield or yield goal) is often strongly correlated to DTC.

EM readings were compared with soil probe DTC measurements obtained at the Missouri MSEA research site. EM data collected along a transect line were highly predictive ($r^2 = 0.81$) of DTC. EM and probe data were obtained in a grid pattern across the entire 11 ha MSEA plot area. EM data were correlated not only with probe DTC but also with the particular crop and farming system present on each plot. Successful EM-based estimations of DTC were not obtained with a single regression applied over the entire plot area, perhaps due to collection of the EM data within a standing crop and when the soil profile was too moist for best DTC resolution.

Correlations of yield with DTC were variable. DTC seemed to influence yields more in a relatively drier yr, and had a greater affect on corn yields than on soybean yields. The relationship between yield and DTC was also dependent on farming system. Estimation of yield based on EM data was not successful with this dataset. The lack of significant correlations between yield and EM data was likely due to the relatively low correlation between probe DTC and EM data. Additional EM data will be collected under drier soil conditions and without the presence of a crop canopy in an attempt to refine these relationships.

REFERENCES

Alberts, E.E., T. Prato, N.R. Kitchen, and D.W. Blevins. 1993. Research and education to improve surface and ground water quality of a claypan soil. p. 21–33. *In* Agricultural Research to Protect Water Quality Conference Proceedings - February 21–24, 1993. Soil and Water Conservation Society, Ankeny, IA.

Ammons, J.T., M.E. Timpson, and D.L. Newton. 1989. Application of an above ground electromagnetic conductivity meter to separate Natraqualfs and Ochraqualfs in Gibson County, Tennessee. Soil Surv. Horiz. 30(3):66-70.

Baer, J.V. 1993. Landscape effects on dessication cracking behavior of a Missouri claypan soil. p. 200. *In* 1993 Agronomy Abstracts. Am. Soc. Agron., Madison, WI.

Birrell, S.J., K.A. Sudduth, and S.C. Borgelt. 1993. Crop yield and soil nutrient mapping. ASAE paper 931556, Am. Soc. of Agric. Engrs., St. Joseph, MI.

Corwin, D.L., and J.D. Rhoades. 1990. Establishing soil electrical conductivity - depth relations from electromagnetic induction measurements. Commun. Soil Sci. Plant Anal. 21 (11&12): 861-901.

Doolittle, J.A., K.A. Sudduth, N. R. Kitchen, and S. J. Indorante. 1995. Estimating depths to claypans using electromagnetic inductive methods. J. Soil Water Conserv. (accepted for publication).

Gantzer, C.J., and T.R. McCarty. 1987. Predicting corn yields on a claypan soil using a soil productivity index. Trans. ASAE. 30(5): 1347–1352.

Greenhouse, J.P., and D.D. Slaine. 1983. The use of reconnaissance electromagnetic methods to map contaminant migration. Ground Water Monitoring Rev. 3(2): 47-59.

Jamison, V.C. and E.M. Kroth. 1958. Available moisture storage capacity in relation to textural composition and organic matter content of several Missouri soils. Soil Sci. Soc. Amer. Proc. 22: 189–192.

Jaynes, D.B., T.S. Colvin, and J. Ambuel. 1993. Soil type and crop yield determinations from ground conductivity surveys. ASAE paper 933552, Am. Soc. of Agric. Engrs. St. Joseph, MI.

Kachanoski, R.G., E.G. Gregorich, and I.J. Van Wesenbeeck. 1988. Estimating spatial variations of soil water content using noncontacting electromagnetic inductive methods. Can. J. Soil Sci. 68: 715-722.

McNeill, J.D. 1980. Electromagnetic terrain conductivity measurement at low induction numbers. Technical note TN-6, Geonics Limited, Mississauga, Ontario.

Palacky, G.J. 1987. Clay mapping using electromagnetic methods. First Break 5(8): 295-306.

Rhoades, J.D., and D.L. Corwin. 1981. Determining soil electrical conductivity-depth relations using an inductive electromagnetic soil conductivity meter. Soil Sci. Soc. Am. J. 45: 255-260.

Sudduth, K.A., and N.R. Kitchen. 1993. Electromagnetic induction sensing of claypan depth. ASAE paper 931550 Am. Soc. of Agric. Engrs., St. Joseph, MI.

50 Variable Rate System For Side-Dressing Liquid N Fertilizer

Michael D. Cahn

Agricultural Engineering Department
University of Illinois
Urbana, Illinois

John W. Hummel

Crop Protection Research Unit
USDA-Agricultural Research Service
Urbana, Illinois

Economical applicators that accurately and rapidly modulate fertilizer rates on-the-go are needed for site-specific crop management. Application error reduces the benefits of site-specific fertilization. Presently, most conventional applicators must be manually reset, and do not record actual application rates. Site-specific fertilizer rates may need to change every 10 to 20 m for nutrients, such as N, that vary across short distances (Cahn *et al.,* 1994). At such short intervals the response time of the system will limit the accuracy of the application. For example, to vary fertilizer rates at 20-m intervals while traveling at 8 km h^{-1}, a system with a 2-s response time would apply the correct rate to only 78% of the interval distance. Costs of variable-rate application systems for site-specific crop management might be minimized by using components that are already available in the marketplace.

Variable rate applicators vary in complexity, ranging from rigs that control flow rate, change chemicals, and vary application pattern to those that only modulate flow rate (Schueller, 1992). Liquid applicators that vary flow rate in response to nozzle pressure and ground speed are perhaps the easiest to adapt for rate-only variable application (Schueller, 1989). The essential requirements for this type of applicator are rapidly responding pumps or valves to change fertilizer flow rates and minimize delivery time to the nozzles (Schueller, 1989; Schueller & Kulkarni, 1988). This paper describes a rate-only variable applicator that uses the DICKEYjohn[1] control system for site-specific side-

[1] Trade names are used in this paper solely for the purpose of providing specific information. Mention of a trade name, proprietary product, or specific equipment does not constitute a guarantee or warranty by the U. S. Department of Agriculture or the University of Illinois.

dressing of liquid N fertilizer in maize and evaluates its accuracy and response time.

MATERIALS AND METHODS

Equipment and Instrumentation

We modified an 8-row (76-cm spacing) anhydrous nitrogen applicator for site-specific side-dressing of liquid N fertilizer in maize (Fig. 50–1). Liquid N fertilizer was carried in a 133 L stainless steel tank, mounted on the applicator. Pressurized air, regulated to maintain 207 kPa in the liquid N tank, was used to drive the flow of fertilizer. A DICKEYjohn CCS100 sprayer control system was used for real-time control of fertilizer rates. The DICKEYjohn control system included a rotary quarter-turn ball valve, a magnetic flow sensor, a ground speed radar sensor, and a control console (Fig. 50–2). The rotary valve was modified with a rotary potentiometer to record valve position. The control system and potentiometer were interfaced through a Metrabyte data acquisition board to a lap-top computer. Software written in Microsoft Quickbasic language allowed a user to change fertilizer rates on-the-go, and record flow rates, valve position, and response time of the system. Ball check valves above the knives kept the manifold and delivery lines filled when the control valve was closed.

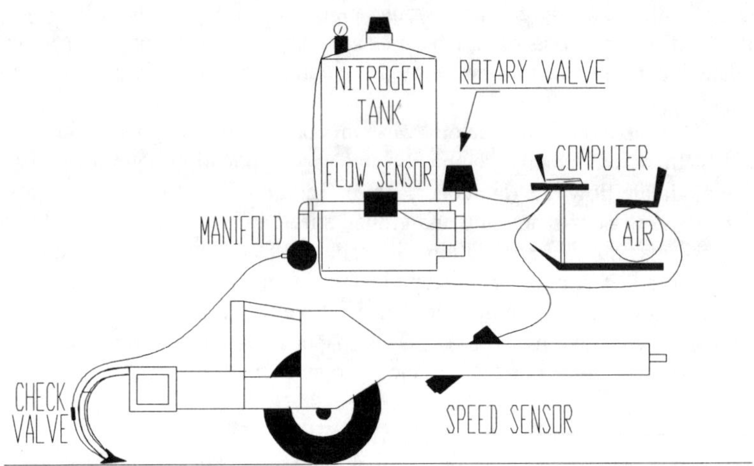

Fig. 50–1. Schematic diagram of the variable rate applicator used to side-dress liquid N fertilizer.

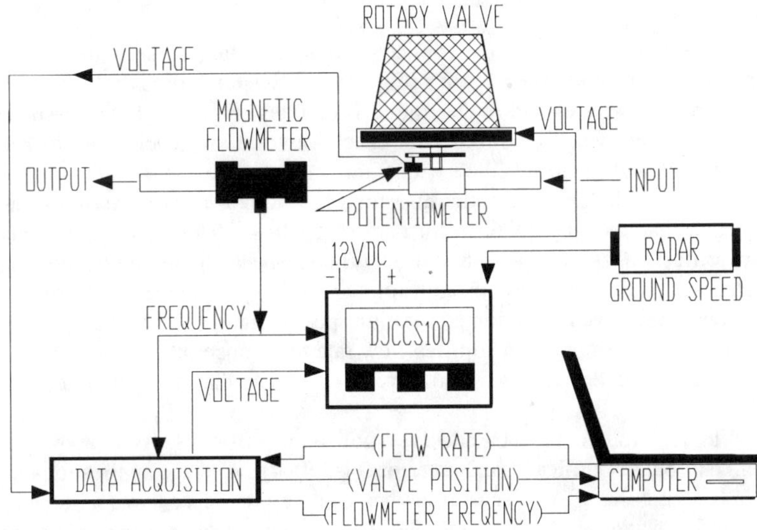

Fig. 50–2. Schematic diagram of the variable rate applicator control system.

Calibration and Laboratory Testing

The valve and flow meter were calibrated for H_2O and liquid N flow rates, ranging from 5.7 to 37.1 L min^{-1}, by measuring outflow for 30-s intervals. The accuracy and response time of the system were evaluated while changing H_2O flow rates every 8 to 10 s to simulate modulating fertilizer rates at 20-m intervals in the field. More specifically, initial flow rates of 0, 5.7, and 9.5 L min^{-1} were varied to attain target flow rates ranging from 7.6 to 20.8 L min^{-1}. Response time of the valve to attain target flow rates with $< a \pm 10\%$ error and average target flow rate error during the 8- to 10-s interval were determined 4 times for each flow rate.

Field Testing

The variable rate applicator was used to side-dress liquid 28 % N ammonium nitrate-urea fertilizer in maize grown for a field experiment investigating the agronomic benefits of variable-rate N application at the University of Illinois, Agricultural Engineering Farm. Maize was planted in 76-cm spaced rows 12 May 1993. Liquid N fertilizer was side-dressed at rates varying between 0 and 225 kg N ha^{-1} 6–8 weeks after planting. The applicator traveled at 8 km h^{-1} (2.22 m s^{-1}) during application, and fertilizer rates were changed at 14-m intervals. Flow meter data were recorded at 1 s intervals. Mean application rates were calculated for each 14 m interval using flow meter data and an assumed ground speed of 8 km h^{-1}.

RESULTS AND DISCUSSION

The response time of the applicator to attain a target rate with < a 10% error usually ranged between 1 to 2 s (Fig. 50–3). Response time decreased with higher target flow rates and higher initial flow rates (Fig. 50–3). The response lag in the system was minimized with check valves above each knife to keep delivery lines and manifold filled with fluid.

Under field conditions, the measured N fertilizer rates were on average 5% greater than the target N fertilizer rates (Fig. 50–4). This positive error may be an artifact of assuming an 8 km h^{-1} ground speed in the application rate calculations. Operator error in adjusting ground speed, due to adverse field conditions, may have resulted in application speeds slightly higher than the 8 km h^{-1}. Since the control system adjusted flow rate to compensate for ground speed effects, measured flow rates would have been higher than the calibrated flow rates.

In addition to the net positive application error discussed above, the system exhibited significant random error (Fig. 50–4). Average application error was ± 6.4 kg N ha^{-1}. As discussed above, since ground speed was assumed to be 8 km h^{-1}, variation in speed may have introduced error into the application rate calculation. Also, a significant portion of the error may be attributable to shortcomings in the control system. The control dynamics differed depending on the initial flow rate. Target flow rate errors became increasingly negative as initial flow rates increased from 0 to 9.5 L min^{-1} (Fig. 50–5). Furthermore, application errors were dependent on the direction of the movement of the valve.

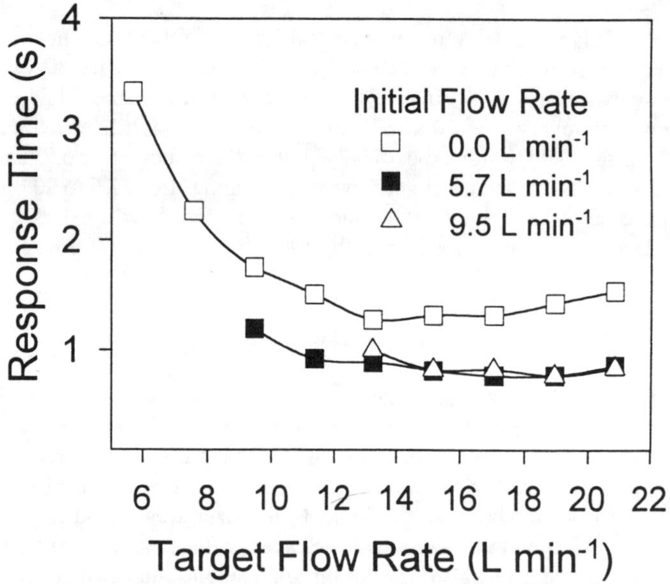

Fig. 50–3. Response time for the control valve to attain target flow rates with < a 10% error as a function of initial flow rate. Initial flow rates varied from 0 to 9.5 L min^{-1}.

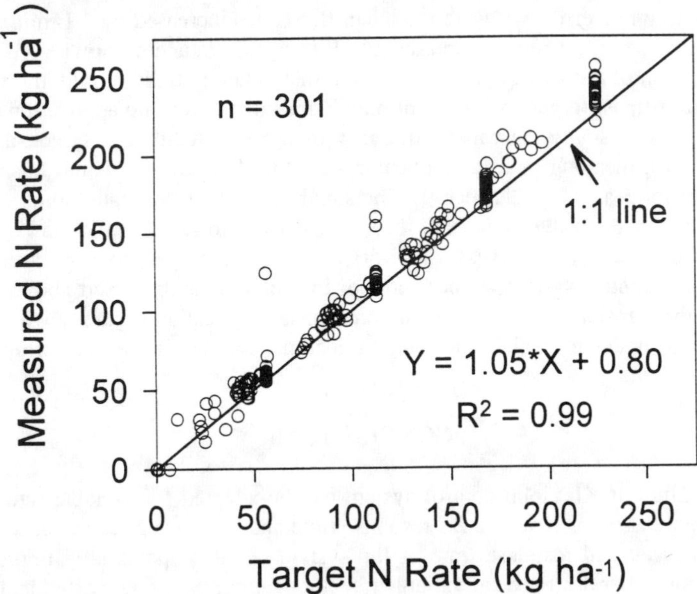

Fig. 50–4. Accuracy of liquid N fertilizer application using the variable rate applicator under field conditions. Measured N rates were obtained from flow meter data and were not adjusted for variations in ground speed.

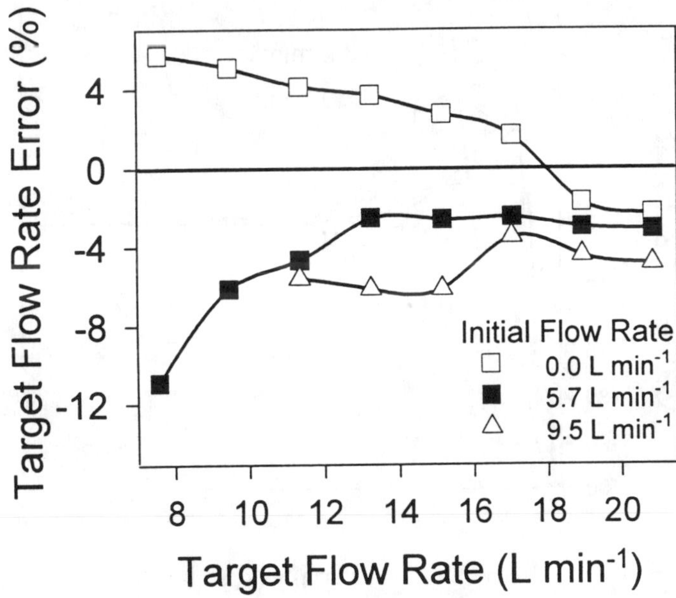

Fig. 50–5. Target flow rate error as a function of initial flow rate. Error was computed from data collected for 8 to 10 s after changing from an initial to a target flow rate.

Target flow rate error was negative when flow rates increased by 1 L min^{-1} and positive when flow rates decreased by 1 L min^{-1} (data not shown). Closer examination of the valve dynamics showed that, when initially closed, the valve tended to surpass target positions, introducing positive error into application rates (Fig. 50–6). Usually the valve required > than 5 s to stabilize at target rates (Fig. 50–6), meaning that the applicator, traveling at 8 km h^{-1}, would misapply fertilizer for > an 11-m distance. We presume that at low flow rates some of the instability in the valve may have been caused by noise from the flow meter signal. The family of curves in Fig. 50–6, however, is indicative of an under-dampened control system. Since most of the limitations discussed above exist during the transient phase of the valve response, application accuracy would be acceptable under normal single-rate applications for which the control system was designed.

CONCLUSIONS

The DICKEYjohn control system can be adapted for variable rate side dress application of N with a 2 s response time and a 95% application accuracy. The accuracy and response time of the system could be potentially improved with a controller designed for variable rate application. Some suggested features for a variable rate controller include:

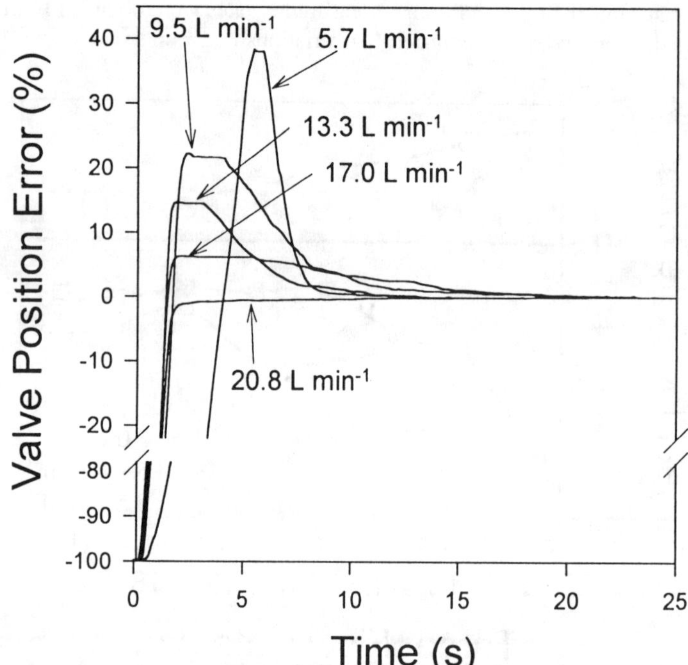

Fig. 50–6. Dynamics of the control valve: valve position error as a function of time and target flow rate. The valve was initially closed and target flow rates ranged between 5.7 and 20.8 L min^{-1}.

1. Capacity to rapidly regulate flow rate.
2. Accuracy and response time of the system should be independent of the flow rate and the direction of change in the flow rate (increasing or decreasing flow rate).
3. Feedback inputs, such as from a flow meter, should have high signal/noise ratios and rapid response.

ACKNOWLEDGEMENTS

We would like to thank DICKEYjohn for their assistance and for providing the control system for the variable rate applicator. We also want to thank Dennis King for his assistance in developing and testing the variable rate applicator.

REFERENCES

Cahn, M.D., J.W. Hummel, and B.H. Brouer. 1994. Spatial analysis of soil fertility for site-specific crop management. Soil Sci. Soc. Am. J. 58:1240-1248.

Schueller, J.K. 1992. A review and integrating analysis of spatially-variable control of crop production. Fert. Res. 33:1–34.

Schueller, J.K. 1989. Spatially-variable fluid fertilizer applicator design concepts. J. Fert. Issues. 6:100–102.

Schueller, J.K., and R.S. Kulkarni. 1988. Spatially-variable fluid fertilizer mobile applicator design concepts. SAE Paper 88–1290.

51 Mapping Corn And Soybean Yields Using A Yield Monitor And GPS

L. L. Bashford
S. Al-Hamed
M. Schroeder
M. Ismail

Department of Biological Systems Engineering
University of Nebraska
Lincoln, Nebraska

Yield maps for corn and soybeans were developed using data obtained from a simple yield monitor and by using the GPS system for site location. A yield monitor was installed in the exit portion of the clean grain elevator of a combine. The principle used for the yield monitor was a series of light emitters and receivers such that the greater the grain flow rate, the smaller the amount of light seen by the receivers. The flow rate and voltage changes from the monitor appeared to be linearly related in a laboratory study. The monitor, however, needed to be calibrated with the dynamics that occur under actual field conditions.

A differential GPS system was used so that corrected location data were obtained in real time. The speed of the combine was obtained using the speed output device of the combine. The voltage from the yield monitor and the frequency transducer for speed were digitized using a Daytronic signal conditioner. A program was written for a laptop computer that queried the GPS receiver and the signal conditioning unit to obtain the data. The data were stored in a spreadsheet format with each row of data containing a reading for GPS time, latitude, longitude, GPS speed, combine speed, and yield monitor mV output.

Calibration was accomplished by weighing the grain collected over a period of time to obtain a scale factor that converted the mV output of the yield monitor to a flow rate. Knowing combine speed and cutting width, instantaneous yield could be calculated.

Trade names are used in this paper solely for the purpose of providing specific information. Mention of a trade name, proprietary product, or specific equipment does not constitute an endorsement, guarantee or warranty by the University of Nebraska, to the exclusion of other products that may be suitable. Yield maps of corn and soybeans were prepared using SAS graph techniques. The combine path could be overlaid to the yield maps to visualize the relationship between the yield and data collection. Where data were missing, linear interpolation techniques were used.

INTRODUCTION

Farming by the foot, prescription farming, site-specific farming and other terminology have surfaced in recent years in popular press articles and scientific meetings crossing many disciplines. This terminology is associated with the development of methods and techniques to sense, control and measure both inputs and outputs at a particular site so that the maximum return can be achieved depending on some criteria or constraint such as available water or environmental concerns or some other factor.

Included in this protocol is application technology where product, be it seed, fertilizer, water, pesticide, etc., can be applied to a field based on requirements of subunits of a field instead of an average over the whole field. This includes the selection and utilization of devices in the metering, distribution, placement and covering of product, be it liquid, pellet, granular, or some combination.

A major output to be measured and site located is yield. This also becomes an input parameter in the following production cycle. In order to determine the correct amount of product to apply at any specific site, a decision support system must be developed that recognizes the relative importance of all the inputs at a particular site in order to achieve the desired output. Therefore, there must be techniques developed that permit the measurement, presentation and location of those outputs.

Pfeiffer et al. (1993) provides a good review of yield monitors being developed or on the market. They also briefly describe some of the problems they see with some of the present monitors. They discuss the yield monitor that they have developed, which is based on the measurement of the height of the grain on the individual flights of the clean grain elevator. They used light to sense the grain height on the individual flights.

The purpose of this paper is to discuss the testing of a prototype yield monitor that uses light as a sensing device and the development of the yield maps resulting from the use of the yield monitor.

Yield Monitor

The concept of a yield monitor is very simple. In order to obtain instantaneous yield, the grain flow rate and velocity and cutting width of the combine are required. The yield can then be determined by: Y = K*M/(W*V)

where
Y = instantaneous yield, kg/m^2
M = grain flow rate, kg/s
W = cutting width, m
V = velocity, km/h
K = units coefficient

Illustrated in Fig. 51–1 is a graphical representation of these inputs used to calculate yield. The cutting width of crops planted in rows such as corn and soybeans is constant. Measuring cutting width for crops such as standing wheat, however, would pose another problem of determining cutting width-on-the-go. The velocity of the combine should be easy to obtain using the combine electronics or some other device such as a radar unit. In addition, there are dynamics involved in the system as there are delays from the time material enters the combine until a yield monitor senses the flow rate. The actual delay will depend on the location of the yield monitor and other combine dynamics which are not constant, Vansichen and DeBaerdemaeker (1992). Therefore, location sensing and yield sensing are out of phase. Also, there is a time lapse at the end of a row as the combine is clearing material during a turn. Therefore, a yield monitor will detect pseudo yield values at these locations until the combine material flow rate becomes relatively static.

A simple monitor was constructed to fit in the discharge side of a clean grain elevator. The monitor consists of light emitters and receivers that are mounted in a rectangular frame. As grain passes through the monitor, the grain blocks light proportional to the flow rate and there is a voltage change. The hypothesis was that the flow rate was directly proportional to the voltage output.

Simple laboratory tests indicated that voltage output from the monitor was directly related to flow rate for corn and soybeans. Field tests, however, would be required to calibrate the monitor.

Equipment and Instrumentation

The yield monitor was installed in a CaseIH 1660 combine. The voltage output from the yield monitor was digitized by a voltage conditioner card installed in a Daytronic signal conditioner system. The combine forward speed was to be measured by using the frequency transducer that the combine uses for speed display. There were problems, however, and this was not accomplished in the field. Speed was then recorded during the trials using the combine speed display and the travel speed from the GPS system.

Location was determined by the use of a GPS system, operating in the differential mode, so that location was determined in real time. A Navstar base station and a Trimble roving receiver were used with send and receive radios. The base station was located within 2.5 km of the research sites.

The Daytronic unit and the GPS roving receiver were connected to a laptop computer through the serial ports on the computer. A program was written so the computer queried the receiver for the GPS information and then the Daytronic unit for speed and yield. The data were stored in spreadsheet format with each row of data containing a reading from the GPS receiver of GPS time, latitude, longitude, GPS speed, and from the Daytronic unit of combine speed and the yield monitor reading in mV. This was done in sequence. Once the combined entered the field, the computer continuously recorded the information. In many instances, there were > one sequence of data with the same GPS coordinates.

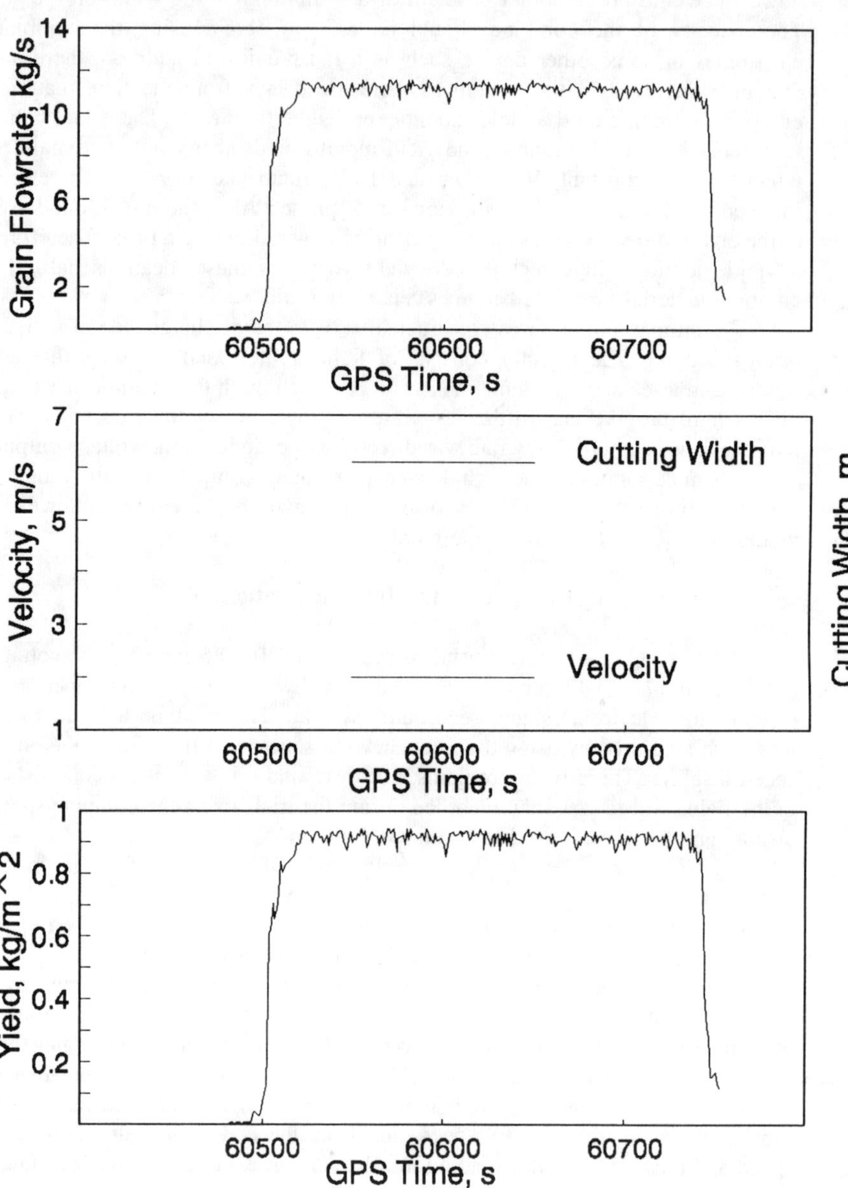

Fig. 51–1. Instantaneous variables needed to calculate yield.

A commercially available harvest yield monitor, Micro-Trak, was purchased and installed in the combine. This monitor has two flat fingers that set in the grain flow at the discharge end of the clean grain elevator. The fingers measure an impact force which is calibrated to grain flow rate. There was no opportunity to use the monitor for instantaneous yield, but a number of field trials were made to evaluate the totalizer component of the monitor.

Field Procedure

For corn, the combine used an 8-row corn head with 0.76 m (30–in) row spacing for a swath width of 6 m (20 ft). A number of tests were devised to allow calibration of the monitor and to evaluate the response of the monitor to various throughput. A weigh wagon was used to determine the total yield for each plot. These data combined with the mV reading from the yield monitor were used to determine the scale factor to convert the mV signal into units of kg/s.

Three fields were harvested using the yield monitors with the combine traveling at various speeds and alternating swaths being harvested. This permitted some evaluation of the tracking of the combine paths in the field and some variation in the yield map preparation.

In field 14B, every other eight rows across the field were harvested with both monitors installed. The combine traveled at different speeds, 1.12, 1.34, 1.56, 1.79, 2.01, 2.24, and 2.46 m/s (2.5, 3, 3.5, 4.0, 4.5, 5, and 5.5 mph) to evaluate different flow rates through the monitor. The remaining alternate 8 rows were harvested at normal combine operating speeds with the Micro-Trak monitor removed.

In field 14D, harvesting with both yield monitors installed was accomplished by harvesting 16 adjacent rows then skipping 16 rows. The Micro-Trak was then removed and the remaining rows harvested.

In field 9S, harvesting across the field was accomplished by harvesting 16 adjacent rows then skipping 16 rows. This was accomplished with both monitors installed. The remainder of the field was harvested without yield monitors.

One soybean field was harvested with a 6.2 m (20 ft) header on the combine. The prototype yield monitor was the only one evaluated in soybeans.

RESULTS AND DISCUSSION

The total bushels harvested as indicated by the Micro-Trak and compared to bushels determined by a weigh wagon were observed on 125 different tests with amounts ranging from approximately 50 bushels to one over 300 bushels. The results of the comparisons are illustrated in Fig. 51–2. The Micro-Trak yield monitor in most instances, underestimated the total amount of corn harvested. The monitor did seem to respond to changes in yield and probably was a good indicator of yield variation, but did not provide an accurate measurement of total yield.

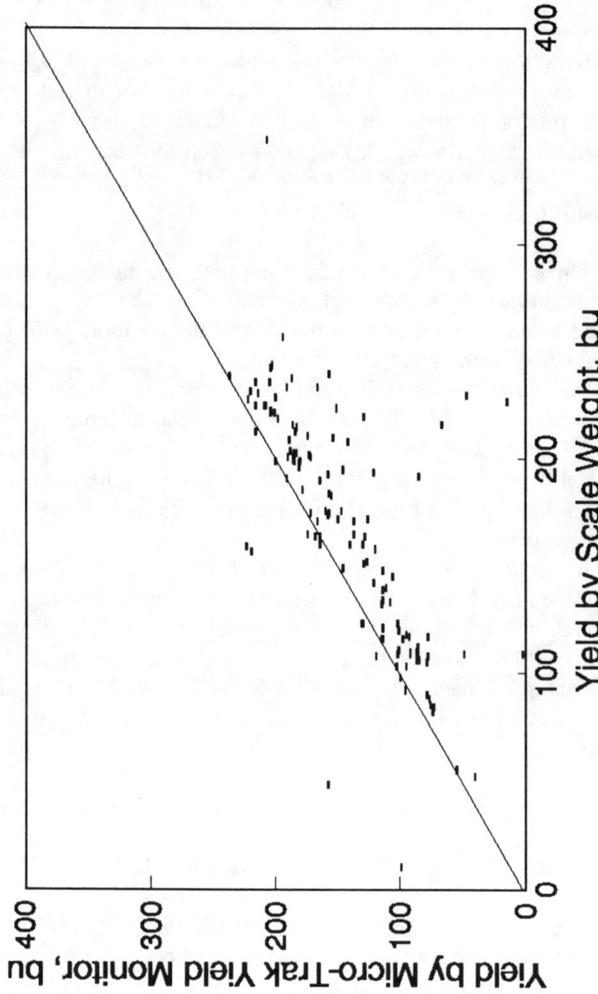

Fig. 51–2. Output of Micro-Trak yield monitor compared to weigh wagon.

The output for the light sensing yield monitor for corn and soybeans can be seen from Fig. 51–1 and 51–3, respectively. Figure 51–1, for corn, has been converted to kg/s by multiplying by a scale calibration factor, whereas the raw yield monitor output from soybeans is illustrated in Fig. 51–3. The yield monitor output varies more for soybeans than from corn. The monitor measures a volume of material thrown from the paddle elevator. In corn, with higher unit yield, these paddles are delivering much more grain than with soybeans or other small grains.

Calibration of the yield monitor was done by use of a curve as in Fig. 51–3 and knowing the measured grain weight represented by the curve. With the proper scale factor, the area under the curve of Fig. 51–3 equals the weight of grain harvested. The yield monitor mV reading can then be converted to a flow rate.

A scale factor was calculated for corn with the combine traveling at seven different speeds. Illustrated in Fig. 51–4 are the scale factors as a function of combine velocity. It was hoped that the scale factor would be independent of combine velocity. This was not the case, however. Therefore, the scale factor had to be adjusted for the combine velocity. Also, the one test run at 2.46 m/s (5.5 mph) was very non-linear with the rest of the observations. It is believed that the yield monitor may have been saturated in that all light was blocked because of the high volume of corn being moved. Therefore, the monitor would provide a constant reading after saturation or be very non-responsive. A regression of the scale factor on the velocity up to 2.24 m/s (5 mph) was accomplished.

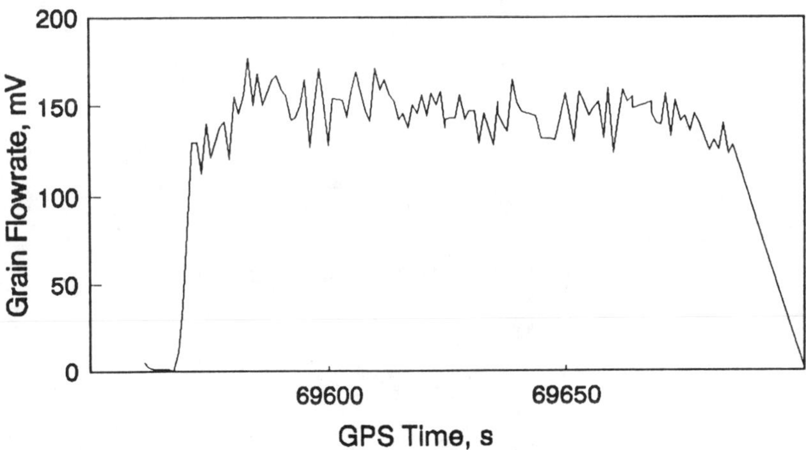

Fig. 51–3. Yield monitor raw data for soybeans.

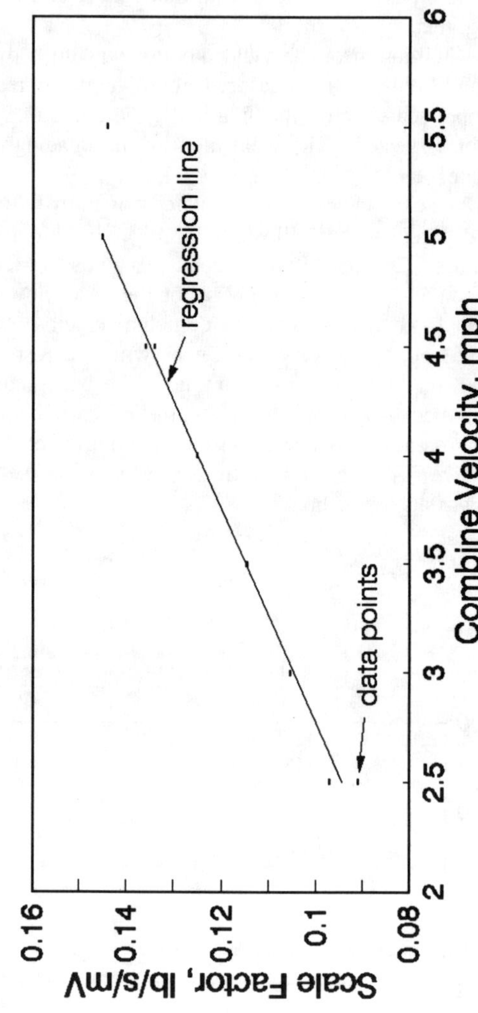

Fig. 51–4. Scale factor as a function of speed for corn.

$$SF = 0.04361 + 0.020317 * V$$
$$r^2 = 0.992$$

where

SF = scale factor, lb/s/mV
V = combine velocity, mph

The regression equation was used to determine the scale factor during the rest of the harvest, using GPS speed as the combine speed indicator. The GPS speed is not satisfactory for speed as required in these calculations, for both the scale factor and the instantaneous yield calculations. It is necessary to measure speed to a finer precision than GPS provides , which was the nearest whole mph.

There is considerable manipulation of the data required before the yield maps can be developed, Birrell et al. (1993). Data were sometimes continuously recorded when the combine was stopped and transferring grain to the weigh wagon. Also, because the data were continuously streamed there were a lot of double observations with > one yield observation occurring at the same GPS coordinates. Therefore, a Quickbasic program was written to scan the data for the doubles and average any yield observations with the same coordinates. Also, all data obtained when the combine was stopped were removed.

Illustrated in Fig. 51–5 are the combine travel map and two yield maps based on those paths for field 14B. The yield data and path maps for the two trips across the field were combined to form one map for travel and two for yield. The spacing for the cells for the middle yield map is geared to the combine path, about 6.2 m (20 ft) by 10 m in the direction of travel. The map coordinates are in meters. The second yield map has cell sizes 12.5 m by 10 m. The yield increments in the maps are in bushels/acre ranging from 100 to 160. The SAS graphic software was used to draw the maps. All observed data points that fall within a given cell were averaged to arrive at the yield value of the cell. No correction was attempted for the delay in position measurement and yield measurement. The cell size relationships to combine swath width and the spatial yield data are very important. Each cell is an average of from 4 to 6 observations depending on the speed of the combine. Half of this field was combined as part of the calibration trials. It is obvious that cell size is very important in defining the yield ranges for a field.

From Fig. 51–5, one can observe the two distinct paths across the field. There seems to be considerable deviation in the path. Other paths from these trials were better defined. Both yield maps illustrate that in the middle of the field there appears to be a low yielding trough. This is a pseudo low yield and is influenced by the path data. The path map indicates a gap at this location and this lack of data probably influenced the graphics.

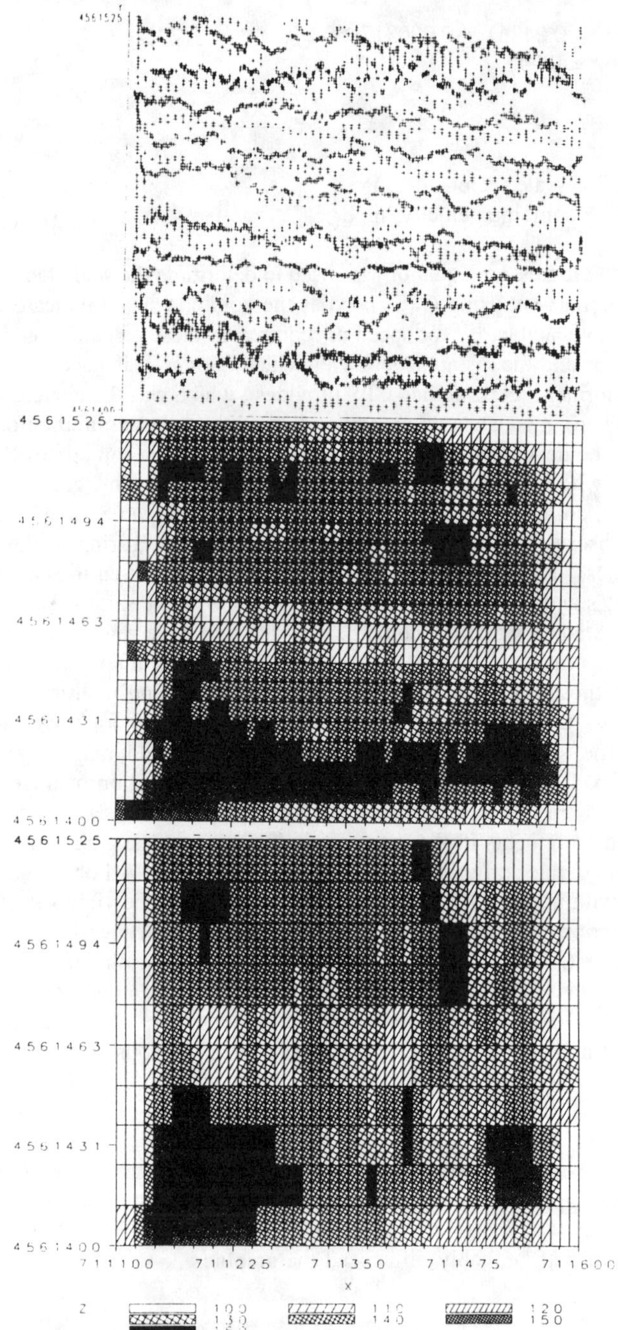

Fig. 51–5. Combine path and yield maps for field 14B. The map coordinates are in meters and the yield ranges in bu/ac.

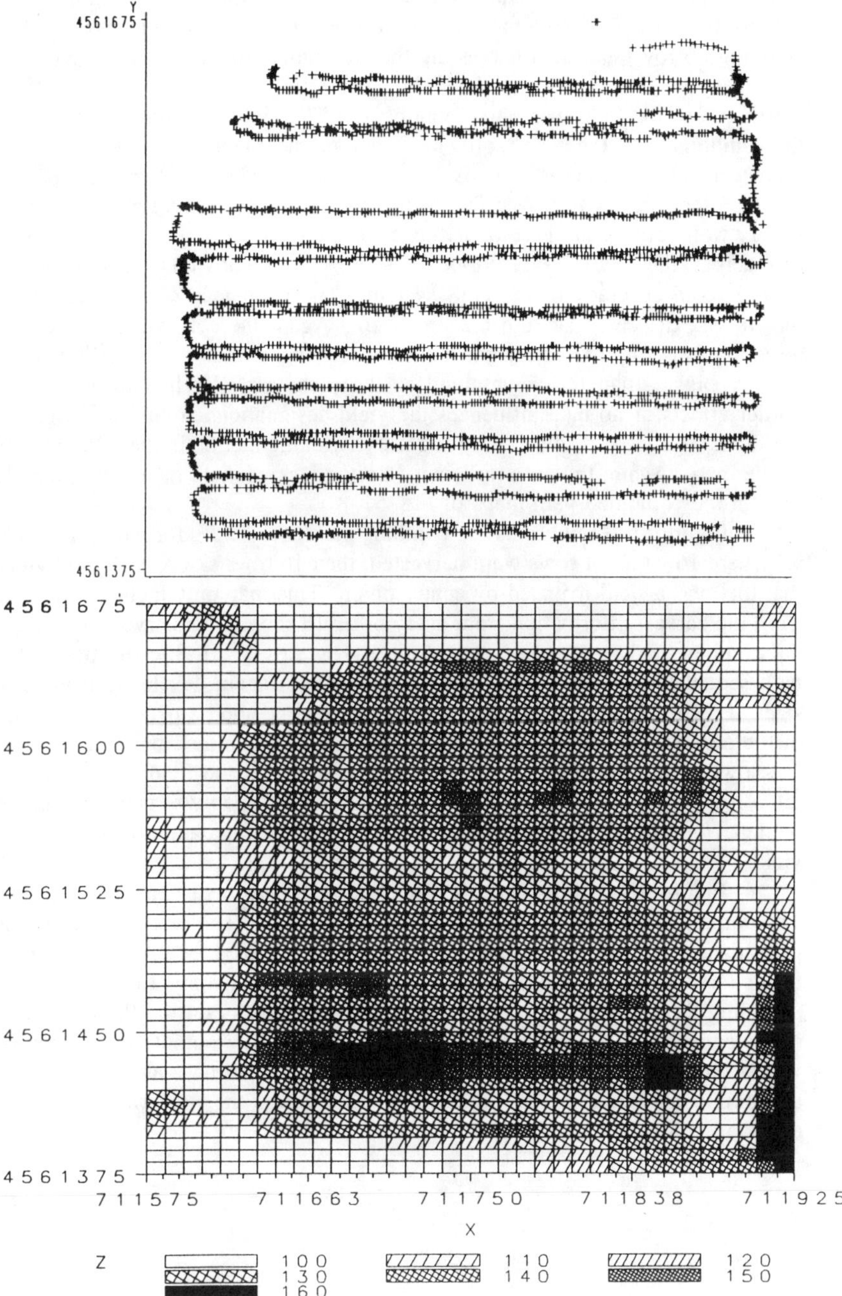

Fig. 51–6. Combine path and yield map for the first pass across field 14D. The map coordinates are in meters and the yield ranges in bu/ac.

Figures 51–6, 51–7 and 51–8 represents the combine paths and the yield data from field 14D. These paths represent harvesting 16 adjacent rows and then skipping 16 rows and then harvesting the alternating 16 rows. Figure 51–6 illustrates the first trip across the field. The wide space between rows at the North end of the field was a tree windbreak. The yield map did not recognize the windbreak as the SAS software filled in the blank area. Figure 51–7 represents the second pass across the field and Figure 51–8 represents the combination of the two trips. The path for one trip across the field, Fig. 51–6, is very well defined and is certainly better than the path described for the second trip across the field, Fig. 51–7. There is no obvious reason for the differencesin accuracy of the GPS system. The yield map coordinates are in meters and the cell sizes are 6.25 by 7 m. Again the cell width in the N-S direction was approximately the combine cutting width. The yield map is slightly bigger than the actual field as the path map clearly indicates. The border affects of no data influences the yield designations around the edges of the field. The yield map certainly suggests a yield reduction in the NE quadrant of the map. Again, this was influenced by the lack of yield data that is clearly seen in the combined path maps of Fig. 51–8.

Figure 51–9 represents the combine path and yield information for field 9S where 16 adjacent rows were harvested, then 16 rows skipped. It is obvious that this was a field irrigated by center pivot. This map only includes the data from the harvest information from every other 16 rows being harvested. Again, the combine path was not defined as clearly as in Fig. 51–8, even though the path sequence was the same. The yield map was again developed using cell sizes of 6.25 x 10 m, with the 6.25 m representing the approximate cutting width of the combine. It is obvious that the combine was turning around in the field as shown in the low yield area in the NW quadrant of the field as there was a ditch running in the NE by SW direction. The yield map was generated based on data from harvesting alternating 16 rows. The missing data influenced actual yield.

Figure 51–10 illustrates the combine path and yield map for soybeans. The combine followed adjacent rows across the field. The blank spaces in the path data occurred where position data were garbled. Speed data were unavailable from the data files, so the map illustrates only trends in yield, assuming the combine speed was approximately constant during the harvesting of the entire field. In this instance the lower the monitor output, the greater the yield. For these tests, the combine path was well defined, as the paths are approximately 6.2 m (20 ft) apart. The yield across the field was apparently constant.

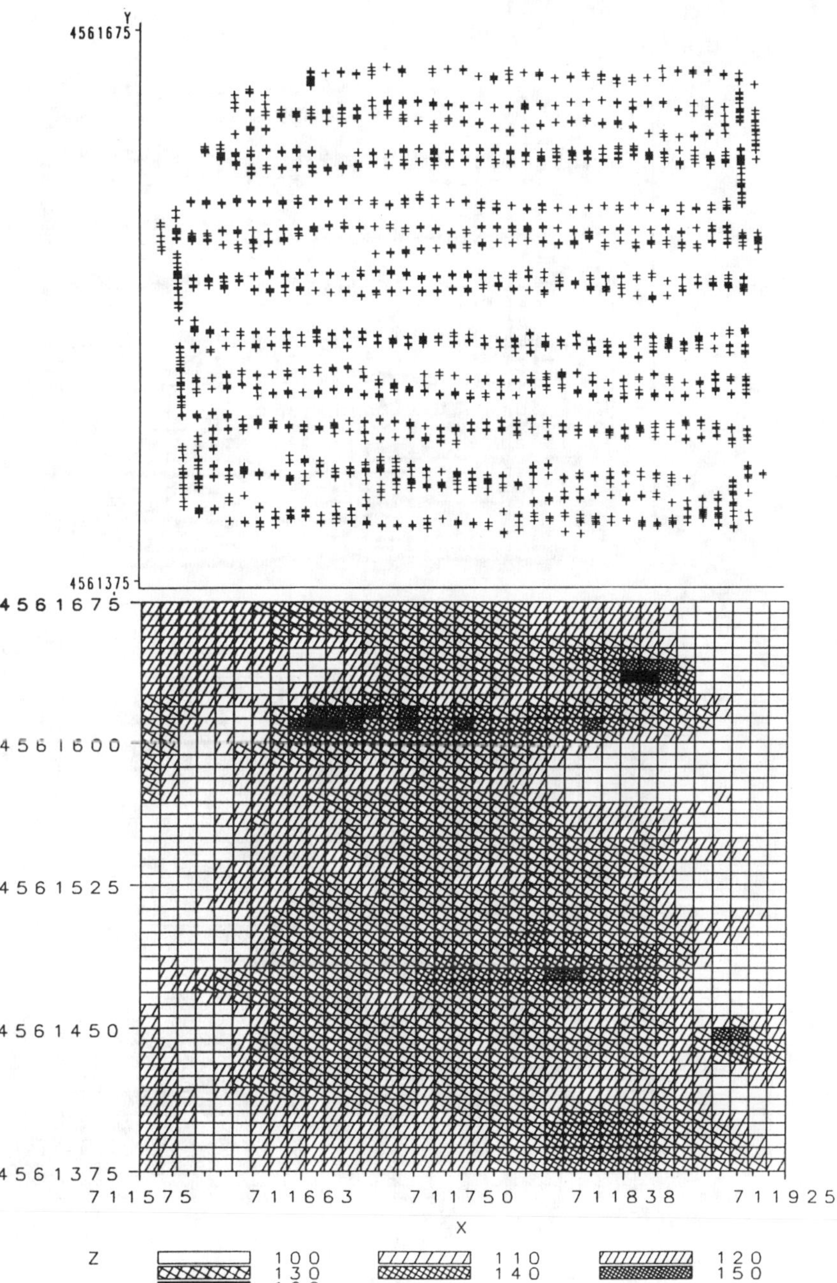

Fig. 51–7. Combine path and yield map for the second pass across field 14D. The map coordinates are in meters and the yield ranges in bu/ac.

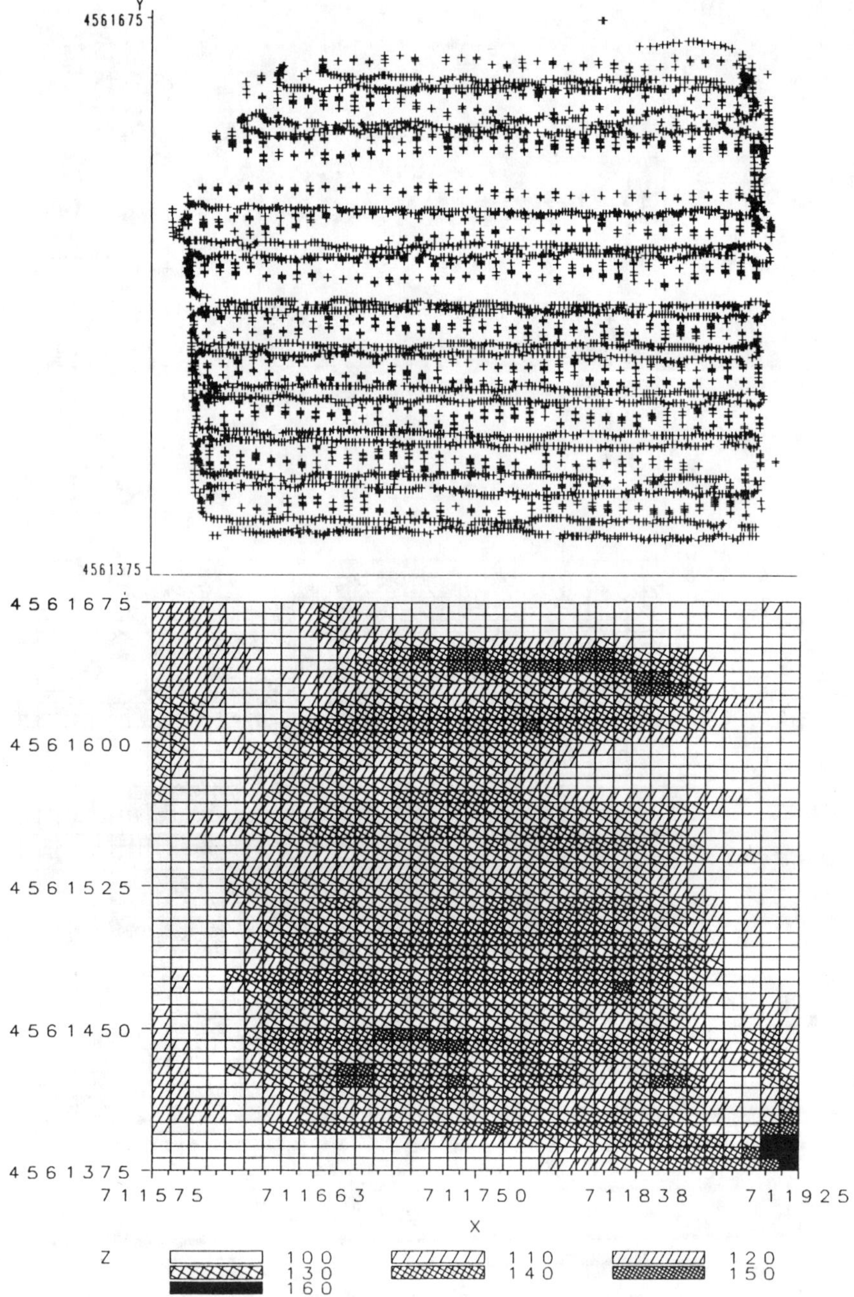

Fig. 51–8. Combined path and yield map for both passes across field 14D.

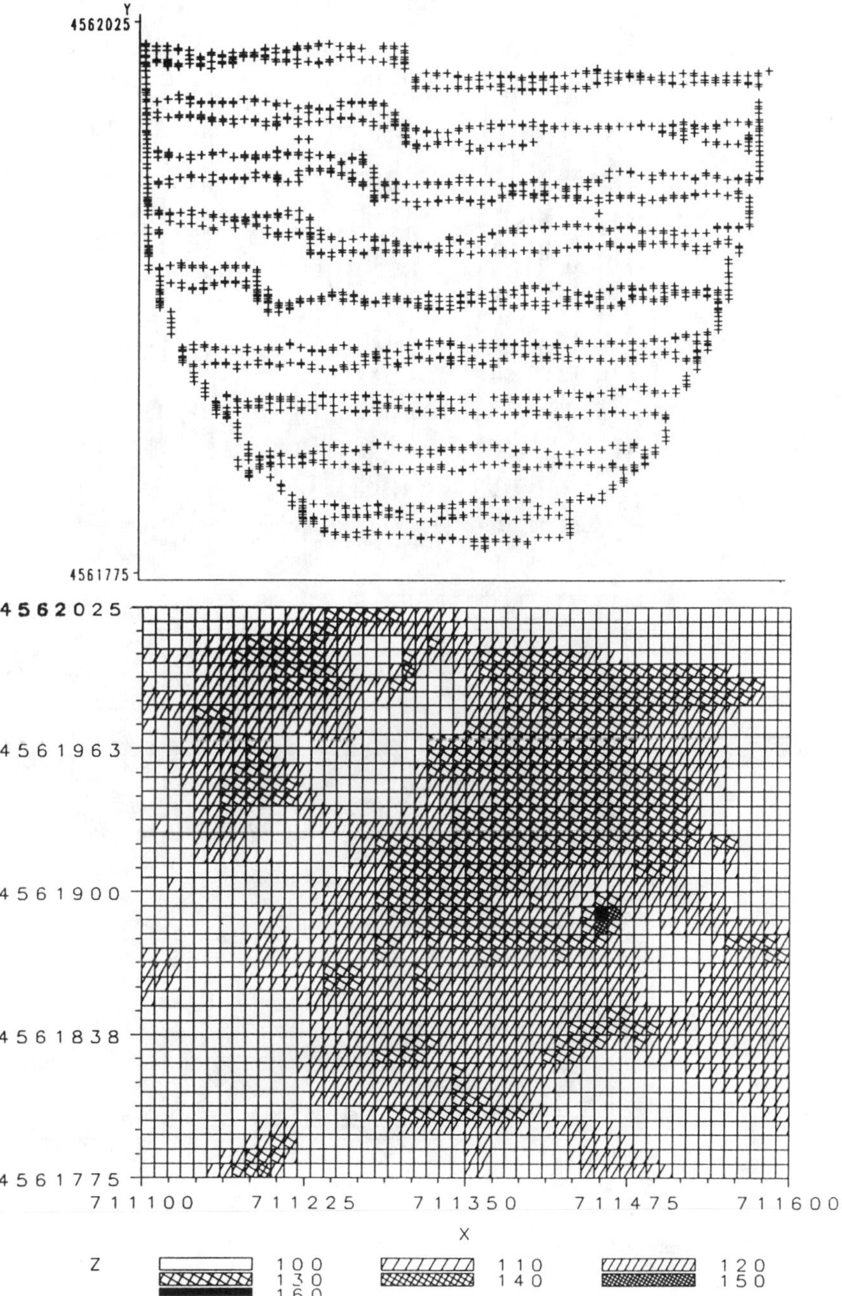

Fig. 51–9. Combine path and yield map for field 9S. Only every other 16 rows were harvested.

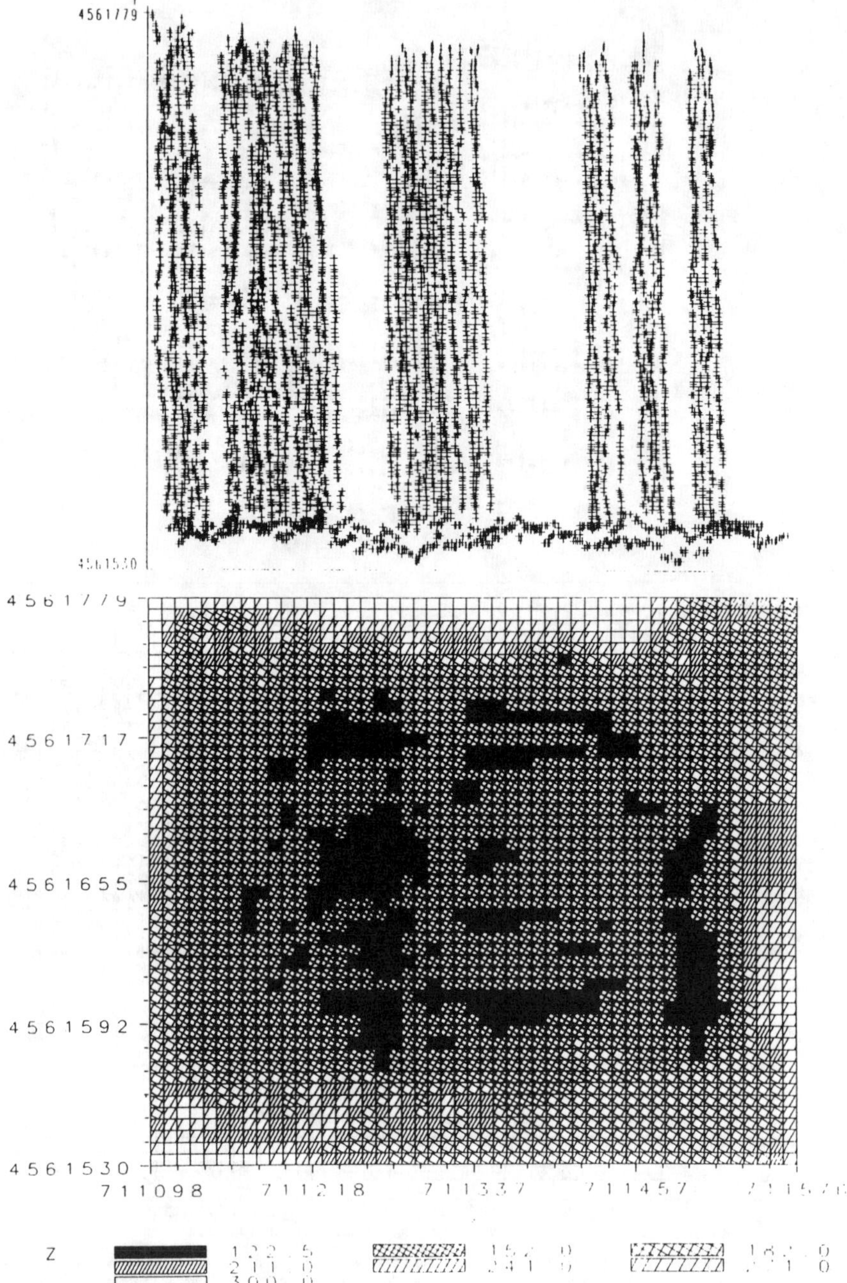

Fig. 51–10. Combine path and yield map for soybean field. The map coordinates are in meters and the yield ranges in yield monitor output. The lower the value the higher the yield.

Figure 51–10 illustrates the combine path and yield map for soybeans. The combine followed adjacent rows across the field. The blank spaces in the path data occurred where position data were garbled. Speed data were unavailable from the data files, so the map illustrates only trends in yield, assuming the combine speed was approximately constant during the harvesting of the entire field. In this instance the lower the monitor output, the greater the yield. For these tests, the combine path was well defined, as the paths are approximately 6.2 m (20 ft) apart. The yield across the field was apparently constant.

CONCLUSION

The combine path maps location data play a major part in the generation of the yield maps. Several of the combine paths generated were quite good, but some were not good, with large deviations from the actual path. The contribution to the determination of the yield via position becomes very important on a micro-scale. There needs to be some standard method or procedure for the development of yield maps. Certainly, methods need to be established that take into account border effects and time delays within the combine because of the system dynamics. Cells sizes and the range of the major divisions in the yield map also need to be defined.

The calibration procedure used during the corn harvest needs to be improved to be sure that all grain is allowed to clear the combine during the trials. An external marker for the data to indicate when material starts through the combine is also needed to assist in the calibration procedure.

The yield monitor needs to be insensitive to combine travel speed for any crop harvested. This will need further evaluation. The yield monitor will be evaluated in small grains this production season and again in corn and soybeans.

The combine speed as measured by the GPS system is not adequate. A method, such as radar or other link to the combine speed sensor is necessary in order to calculate instantaneous yield data.

REFERENCES

Birrell, S.J., K.A. Sudduth, and S.C. Borgelt. 1993. Crop yield mapping using GPS. ASAE Paper No. MC93–104. St. Joseph, MI.

Pfeiffer, D.W., J.W. Hummel, and N.R. Miller. 1993. Real-time corn yield sensor. ASAE Paper No. 93–1013. St. Joseph, MI.

Vansichen, R., and J. DeBaerdemaeker. 1992. Signal processing and system dynamics for continuous yield measurement on a combine. Paper No. 920601 presented at the AG ENG 1992 Agricultural Engineering International Conference, Uppsala, Sweden, June 1–4.

52 The Accuracy Of Site-Specific Fertilizer Application By Means Of A Spinning-Disc Fertilizer Spreader

R. Olieslagers
H. Ramon
H. Delcourt

Department of Agricultural Engineering
Katholieke Universiteit
Leuven, Belgium

L. Bashford

Department of Biological Systems Engineering
University of Nebraska
Lincoln, Nebraska

Nutrient maps were developed for test fields in several regions in Belgium. These maps illustrated the enormous variation in nutrient demands within a field. It is clear that there is a need for a control system to regulate the application of fertilizer on a site basis. By means of a simulation program, that includes the entire model of a spinning disc spreader, simulation of the distribution profile for the spinning disc applicator can be executed. Up to now, the width and the profile of the distribution pattern change drastically as the application rate changes, which results in a non-accurate application. The simulation program provides the opportunity to modify the spreader so that the spreader pattern remains insensitive to changes in mass flow rate.

INTRODUCTION

Usually, a uniform fertilizer application rate is applied across an entire field. Nutrient maps, together with yield maps, developed for several test fields in Belgium, show that there is a need for site-specific application of different nutrients (N,P,K,...). When fertilizer is applied accordingly to site-specific requirements, leaching to groundwater and surface water is minimized. Therefore, an accurate control system for site-specific application of fertilizer provides an economic advantage and reduces the environment risk.

For site-specific application, fertilizer demand maps are used as an input parameter. The location of the spreader on the field has to be determined using a system such as GPS. An available control parameter for site-specific application with a spinning-disc spreader is the mass flow rate (kg/s). This parameter can be controlled by adapting the orifice dimensions of the spreader bin. The desired fertilizer mass flow rate can be calculated from the site-specific requirement (kg/ha), by multiplying the site-specific flow with the travel speed (m/s) and the spreading width (m).

OBJECTIVES

The objective of this paper is to present a simulation model to evaluate a prototype spreader for which the spreader pattern remains constant (constant pattern shape and spreading width) when applying fertilizer to site-specific requirements. The spreading width and the shape of the distribution pattern have to remain constant for an efficient control system for site-specific application. In reality, this is not the case. The prototype spreader will be the basis for an optimized control system for site-specific fertilizer application.

Review of Literature

The simulation program is partially based upon formulas, developed during years of research by several investigators. Patterson and Reece (1962) described the motion of a particle on the disc of the spreader, considering a near-center feed. In this analysis, bouncing of the particle against the vanes was neglected because the particle was fed on the disc and vanes without impact. A distinction was made between rolling and sliding of the particles. Inns and Reece (1962) developed a model for the motion on the disc, considering off-center feed. In this situation, the particles are impacted by the vanes. Therefore, bouncing is considered in the analysis. Mennel and Reece (1963) modelled the particle trajectory through the air. Cunningham and Chao (1967) generalized the basic equations for the particle trajectory on the disc for curved-shaped vanes.

Starting from the basic formulas for particle trajectories through the air, approximating equations were derived by Pitt et al. (1982). They compared the approximate and exact solution of the particle trajectory calculations.

Griffis et al. (1983) simulated the trajectory of the particle on the disc, based upon the equation of Patterson and Reece (1962).

The equations were developed, considering a non-conical disc shape and radial vanes. Most of the spinning-disc fertilizer spreaders, however, have cone-shaped discs and non-radial, pitched vanes. Therefore, there is a need for a more general equation, taking the pitch of the vanes and the cone shape of the disc into account.

Hofstee and Huisman (1990) examined the behavior of particle motion on the disc and in the air for changing physical properties of fertilizer (particle size, particle size distribution, coefficient of friction between fertilizer and structural surface, coefficient of restitution, aerodynamic resistance, particle strength).

Based upon earlier research, together with additional analysis, general equations which describe the trajectory of the particle on the disc and in the air, were included in a simulation model, used to predict the distribution pattern for site-specific application.

Simulation Model

In this section, the different stages in the simulation model will be discussed. For the simulation, 'Matlab'-software was used.

- Input of particle and spreader characteristics

The simulation program requires particle- and spreader characteristics. Particle characteristics are density, size distribution and the coefficient of friction. Spreader characteristics are the disc dimensions (cone-shape, radius), orifice dimensions and position, height of the disc above the ground, vane and disc shape and disc speed.

- Positioning of the valve for desired mass flow

The relation between the orifice opening (segment-shape) and the mass flow rate is known from calibration tests and is linear. The mass flow rate is calculated from the aimed site-specific requirement by multiplying the site-specific requirement with the travel speed and the desired spreading width. The opening size of the orifice is calculated from the desired mass flow rate by means of the known linear relation.

- Particle trajectory on the disc

The equation which describes the particle trajectory on the disc is derived from the forces, acting upon the particle. These forces are the centrifugal force, the gravity force, the coriolis force and the friction force, which is a function of the 3 other forces. The most general differential equation is:

$$\frac{d^2r}{d^2t}\cos\beta + 2\,\mu_p\,\omega\,\frac{dr}{dt}\cos\beta$$

$$-\omega^2\,r\,(\cos\alpha\,\cos\beta - \mu_p\,\sin\beta + \mu_f\,\sin\alpha) \qquad [1]$$

$$+g\,(\sin\alpha\,\cos\beta + \mu_f\,\cos\alpha + \mu_p\,\sin\alpha\,\sin\beta) = 0$$

where:

r = distance between particle and disc centre, [m]
ω = angular velocity of disc, [rad/s]
r_0 = pitch of vanes, [m]
β = $\arcsin(r_0/r)$, [°]

μ_p = friction coefficient for particle-vane interaction
μ_f = friction coefficient for particle-disc interaction
α = cone angle of disc, [°]
g = gravitational acceleration, [m/s^2]
m = particle mass, [kg]

Equation [1] can only be solved by numerical techniques (e.g. Runge-Kutta).

When α and β are 0 (radial vanes, non-conical disc), the equation becomes (cfr. Patterson and Reece (1962):

$$\frac{d^2r}{dt^2}+2\ \omega\ \frac{dr}{dt}\ \mu_p-\omega^2\ r=-\mu_f\ g \qquad [2]$$

Equation [2] can also be solved analytically. The numerical simulation approximates very closely the exact analytical simulation (0.15 % difference). Figure 52–1 shows the simulation of the particle trajectory on the disc.

• Particle trajectory through the air

After leaving the disc, the particle has an initial velocity v_0, induced by forces acting on the particle during its trajectory on the disc. This velocity has a radial and a tangential component. The angle β_u between the radial velocity component and v_0 must be calculated to determine the position of the particle on the field.

The forces acting upon the particle in the air are the air resistance force and the gravitional force. This force balance results in differential equations [3] and [4].

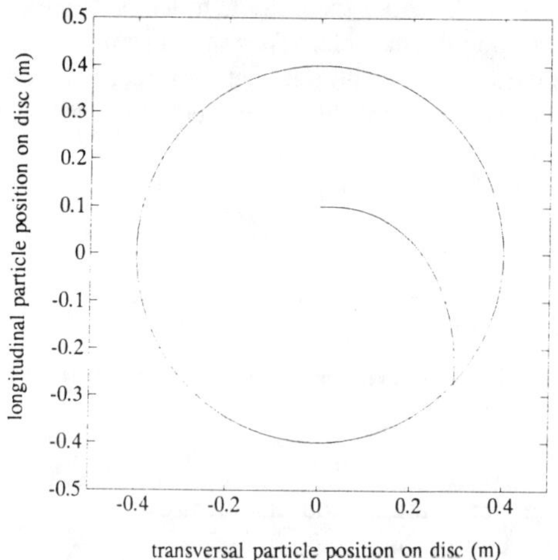

Fig. 52–1. Particle trajectory on the disc (μ_f=μ_p=0.2; ω=1000 t/min).

$$\frac{d^2x}{dt^2} = -k \frac{dx}{dt} \sqrt{\frac{dx^2}{dt} + \frac{dy^2}{dt}} \quad [3]$$

$$\frac{d^2y}{dt^2} = -g - k \frac{dy}{dt} \sqrt{\frac{dx^2}{dt} + \frac{dy^2}{dt}} \quad [4]$$

$$k = \frac{3}{4} \frac{C_w \rho_l}{D \rho_d}$$

where:

ρ_l = density of air, [kg/m_3]
ρ_d = particle density, [kg/m^3]
C_w = air resistance coefficient, f(Re)
Re = Reynolds number = ρ_a*v*d/ν
d = particle diameter, [m]
v = particle velocity, [m/s]
ν = dynamic viscosity of air, [kg/m.s]
A = resistance surface of particle, [m^2]
x = the horizontal position of the particle, [m]
y = the vertical position of the particle, [m]

These non-linear differential equations can only be solved numerically (Runge Kutta) which requires a lot of calculation time (Matlab). Therefore, an approximating equation, developed by Timoshenko and Young (1948) is used:

$$y = g*\left(\frac{4*\rho_d*d}{3*\rho_l*C_w*v_0}\right)^2*\left[e^{\frac{3*\rho_l*C_w}{4*\rho_d*d}*x} - \frac{3*\rho_l*C_w}{4*\rho_d*d} - 1\right] \quad [5]$$

The simulation results of equation [5] correspond very well with the numerical simulations (< 0.2 % difference). Figure 52–2 shows the simulation of a particle through the air. An important input parameter here is the height H of the disc above the ground.

- Determination of the particle location on the field

When the starting position is known and both the angle β_u and the particle trajectory through the air are calculated, the particle location on the field, referred to the disc center can be calculated.

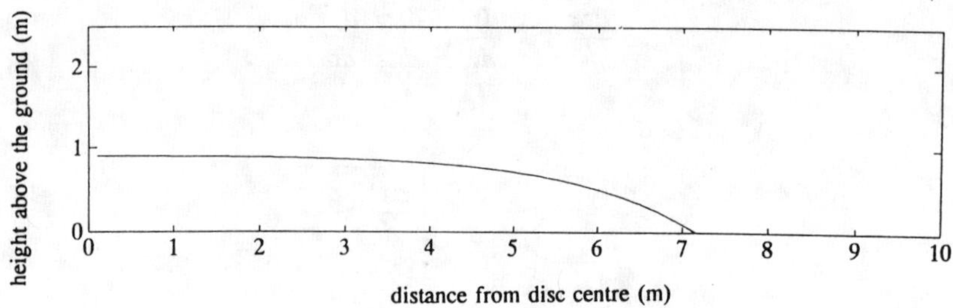

Fig. 52–2. Particle trajectory through the air (H=0.9 m).

• Simulation of the spreader pattern

The calculation of the particle trajectory can be done for a whole range of particles, corresponding with the orifice dimensions, which are equivalent to the mass flow. Figure 52–3 shows the resulting spreader pattern. The program provides the opportunity to simulate spreader patterns for single- or two-disc spreaders.

• Calculation of the distribution pattern

Figure 52–4 shows the static distribution pattern, considering a constant travel speed.

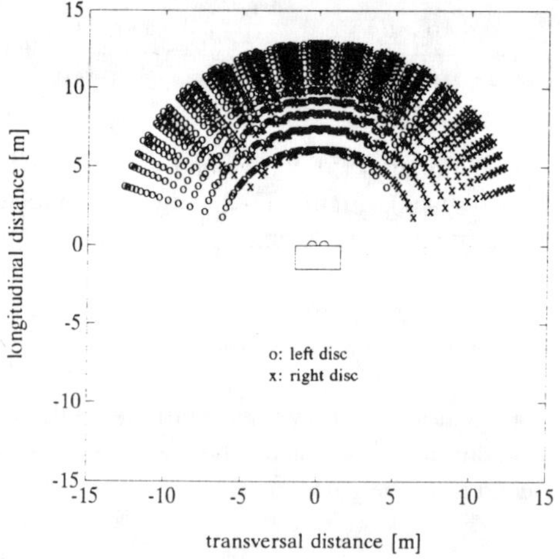

Fig. 52–3. Spreader pattern (static) for a spreader with two discs (ϕ_0=0°).

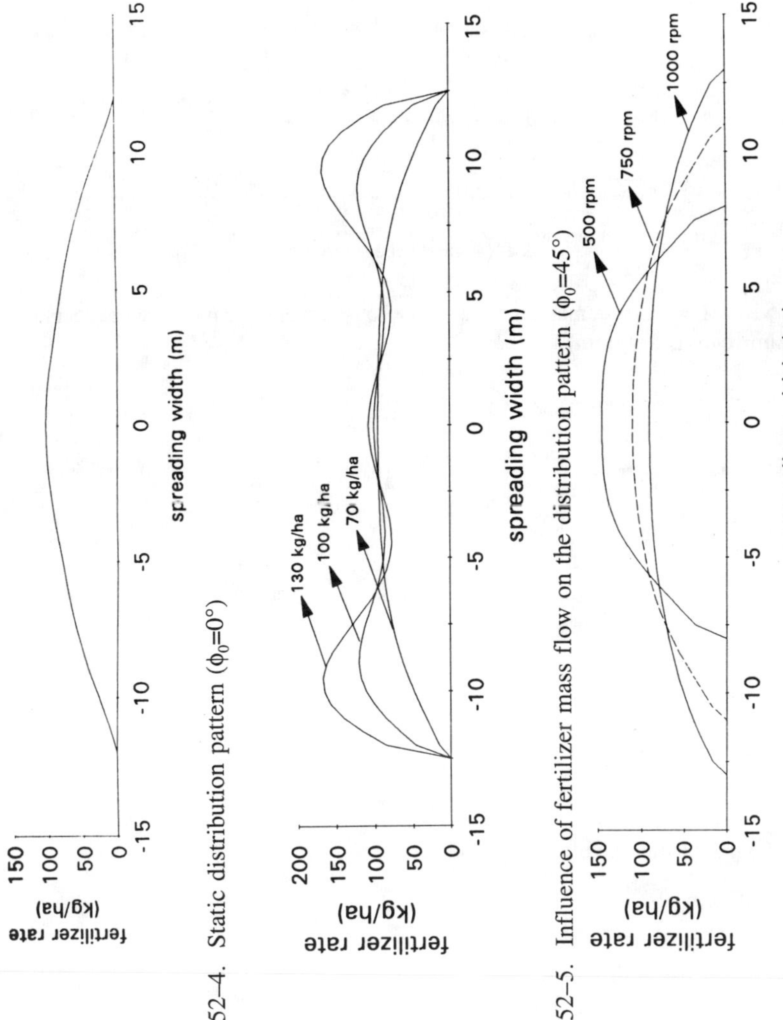

Fig. 52–4. Static distribution pattern (ϕ_0=0°)

Fig. 52–5. Influence of fertilizer mass flow on the distribution pattern (ϕ_0=45°)

Fig. 52–6. Influence of a changing disc speed (rpm) on the distribution pattern (ϕ_0=0°)

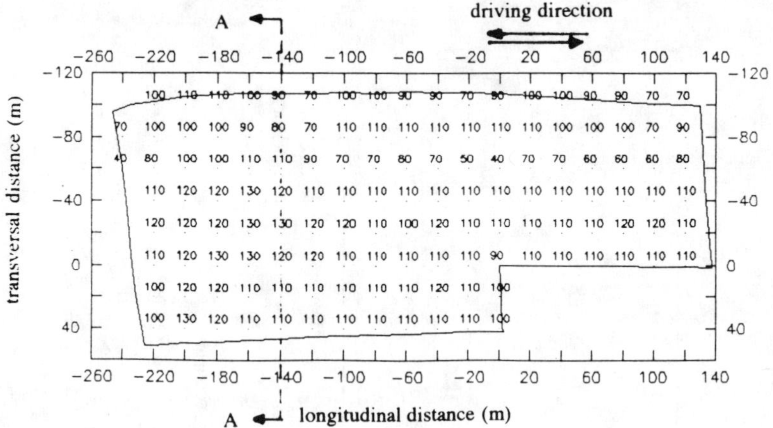

Fig. 52–7. Fertilizer demand map; phosphorous demand for winter wheat(Sandy-loam region, Belgium)

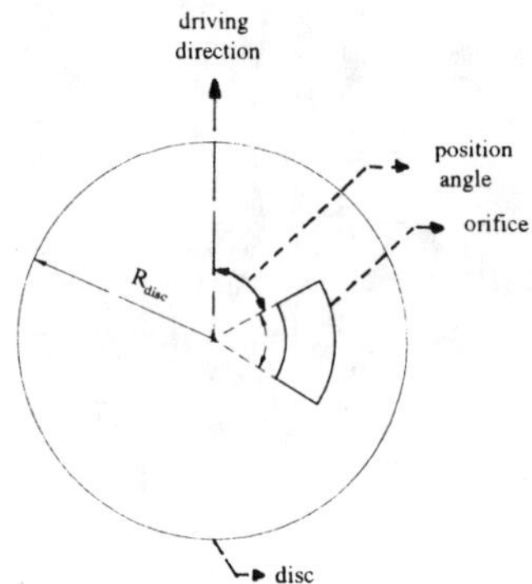

Fig. 52–8. Orifice position angle ϕ_0

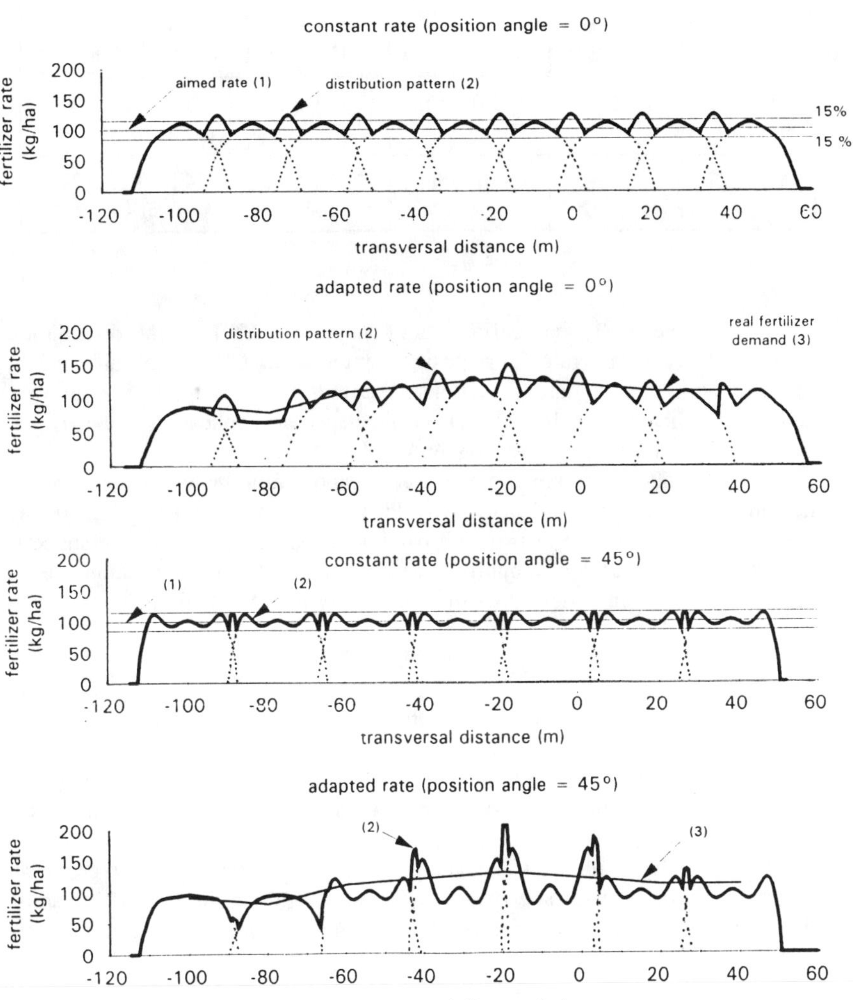

Fig. 52–9. Total distribution pattern (section A-A of Fig. 52–6); constant flow, compared with adapted flow for 2 position angles ϕ_0 (0° and 45°) of the orifice

Table 52–1. Comparison of the standard deviation (S.D.) of the distribution pattern (2) to the aimed rate (1) for uniform application and (3) for site-specific application (coefficient of variation C.V. of distribution pattern for constant rate is also mentioned).

	rate:	constant	adapted	S.D. worsening at site-specific
	C.V. (2)	S.D. (1)–(2)	S.D. (2)–(3)	application
$\phi_0 = 0°$	7.84 %	10.90 kg/ha	16.51 kg/ha	5.61 kg/ha (51 % ⇓)
$\phi_0 = 45°$	6.80 %	6.99 kg/ha	31.34 kg/ha	24.35 kg/ha (448 % ⇓)

In Table 52–1, the coefficients of variation (C.V.) for the distribution patterns at a constant rate for a position angle ϕ_0 of 0° and 45° are shown, together with the standard deviations (S.D.) of the aimed rate for uniform application (line (1) of 100 kg/ha) and site-specific application(line (3), real fertilizer demand, based on section A-A in Fig. 52–7).

Table 52–1 shows that a site-specific application increases the standard deviation, which of course has to be avoided. The increase of S.D. is caused by the drastic change of the distribution pattern shape. The change of the shape is relatively small when ϕ_0=0°, but has a dramatic effect on the distribution pattern at 45°. The spreading width doesn't change very much in both cases.

CONCLUSIONS

The shape of the distribution pattern and the spreading width are sensitive to changes in mass flow. This sensitivity is a function of the spreader characteristics. The negative influence on the spreader pattern by changing only one parameter, should be avoided by adapting other controllable parameters simultaneously.

A simulation model was developed to evaluate spreaders and to construct an adapted prototype for site-specific application with an insensitive spreading width and a constant shape of the distribution pattern for a changing mass flow.

REFERENCES

Cunningham, F.M., and E.Y.S. Chao. 1967. Design relationships for centrifugal fertilizer distributors. Trans. ASAE 10(1): 91–95.

Griffis, C.L., D.W. Ritter, and E.J. Matthews. 1983. Simulation of rotary spreader distribution patterns. Trans. ASAE 26(1): 33–37.

Hofstee, J.W., and W. Huisman. 1990. Handling and spreading of fertilizers part 1: physical properties of fertilizer in relation to particle motion. J. Agric. Eng. Res. 47:213–234.

Inns, F.M., and A.R. Reece. 1962. The theory of the centrifugal distributor II: Motion on the disc, off-centre feed. J. Agric. Eng. Res. 7(4):345–353.

Mennel, R.M., and A.R. Reece. 1963. The theory of the centrifugal distributor III: Particle trajectories. J. Agric. Eng. Res. 8(1):78–84.

Patterson,D.E., and A.R. Reece. 1962. The theory of the centrifugal distributor I: Motion on the disc, near-centre feed. J. Agric. Eng. Res. 7(3):232–240.

Pitt, R.E., G.S. Farmer, and L.P. Walker. 1982. Approximating equations for rotary distributor spread patterns. Trans. ASAE 25(6):1544–1552.

Timoshenko, S., and D.H. Young. 1948. Advanced dynamics. McGraw-Hill Book Co., Inc., New York, NY.

SESSION IV

PROFITABILITY

53 Incorporating Economic Analysis Into On-Farm GIS

Duane Griffith

Extension Agricultural Economics
Montana State University
Bozeman, Montana

Agriculture has enjoyed a respite after coming through the farm crisis of the mid 1980s. The farm crisis was brought about by low per unit profit margins and relatively high nominal interest rates. Many farm operators were forced out of business during this financial crisis and many of the younger generation decided agriculture was not a good career choice. Economic conditions in the agricultural sector, however, continue to be tight. Low profit margins, increased regulations, and world wide competition for agricultural products require producers to manage all aspects of their businesses to maintain economic viability and remain financially sustainable. Producers assimilate both physical and economic information (rotation, soil moisture, soil fertility, soil type, farm programs, market prices, etc.) from many sources into daily production decisions without the aid of comprehensive analysis tools.

Geographic Information System (GIS) provides the framework for a comprehensive analysis tool for physical and economic factors affecting agricultural production and hence its profitability and financial sustainability. The evolution of this tool will be slow, progressing from simple mapping packages that are spatially aware, to full-featured GIS programs for farm operators. The spatial management of data within GIS allows analysis of changes in profitability as production practices change due to increased understanding of how physical attributes affect the production process and impact profitability.

Site-specific management, by its name, requires that an individual producer be able to analyze the production and economic implications of moving away from the "average for all acres" approach. Although there is considerable interest in site-specific management, producers want to know: "Is it economically feasible?"

A key element in being able to satisfy producers' questions about economic feasibility, is understanding how producers make and implement production decisions. Obviously research plays an important role in the process. Producers consider the available research information and make production decisions. These production decisions, however, are based on probable outcomes. Producers can only estimate probable outcomes before implementing production practices. The same will be true of site-specific farming practices. Producers will only require an indication of the probable outcome of implementing site-specific management. Part of that indication will come from research, but as mentioned earlier, the term site-specific demands an indication of the probable outcome on a particular site.

Site-specific farming has taken on the goal of being able to incorporate all relevant factors in the production process. Previous emphasis of agricultural research has been to control all factors of production except one variable of interest. The result is a response to the variable of interest while controlling other variables to the extent possible. This type of research does not lend itself well to the nature of site-specific farming decisions. Site-specific farming requires the integration of all limiting factors affecting each producer. In most cases these factors are both physical (soil type, slope, moisture) and economic (limited funds for operating inputs and capital purchases). Currently, producers assimilate this information and implement their decisions based on the probable outcome.

GIS systems can incorporate both static and dynamic data layers. Static layers include soils, slope, aspect, elevation, and others. Dynamic layers are associated with data that can change over time - crop cover, seeding or fertilizer rates, annual precipitation, etc. The goal of site-specific farming is to combine relevant data from all relevant disciplines. GIS provides the tool to combine physical and economic information sets to answer questions about production practices and their effects on profitability.

Economists typically use enterprise budgets to analyze economic implications of production practices. A typical enterprise budget is shown in Table 53–1. The level of inputs, i.e. seed, fertilizer, chemicals; the type of inputs, such as a specific seed variety, the type of machinery and equipment used, and whether the machinery is owned, leased or rented are all important considerations affecting the profitability of each crop (enterprise). Enterprise budgets incorporate all of these variables, analyzing the profitability of all aspects of production.

The process of developing enterprise budgets is relatively simple. Costs associated with all inputs are collected, or calculated from other inputs, on a per hectare basis. In the past only one budget was needed for each enterprise because all hectares were treated alike. In some instances per hectare input costs are easily collected for items such as seed, fertilizer, and chemicals. Some costs, such as repairs and fuel for machinery, are not easily determined for a given enterprise on a per hectare basis. There are two items in Table 53–1 that are difficult for most producers to generate. These are the operating and ownership costs of machinery per hectare. Production inputs are divided into two categories to help complete enterprise budgets - operating and ownership. Operating inputs are those that vary with the level of production (seed, fertilizer, chemicals, fuel, etc.). As more hectares are planted, total operating costs increase.

Table 53–1. Typical Enterprise Budget for Dryland Winter Wheat.

Dryland Winter Wheat	Per Hectare
Expected Yield	64.2 bu.
Expected Price	$3.30
Expected Government Payments	$69.38
Total Income	$281.24
Operating Costs	
Seed	$14.83
Chemicals	$9.88
Chemical Application	
Fertilizer	$29.65
Fertilizer Application	
Crop Insurance	$12.35
Hired Labor	
Energy Costs	
Water Costs	
Machinery Operating Costs	$42.45
Interest on Operating Costs	$8.72
Total Operating Costs	$117.88
Ownership Costs	
Taxes	$2.47
Insurance	$.79
Depreciation	$.96
Opportunity Costs of Real Estate	$25.01
Machinery Ownership Costs	$55.89
Total Ownership Costs	$85.12
Total Costs (Ownr. + Oper.)	$203.00
Net Income Per Hectare	$78.24

Operating costs per hectare, however, remain relatively constant. Similar amounts of seed, fertilizer, and chemical are used on each additional hectare planted. Total ownership costs do not change as the level of production changes. Taxes, insurance, interest, and considered ownership costs. For an existing piece of machinery, the taxes must be paid even if the machinery is not used during a production cycle. Per hectare ownership costs vary inversely with the total hectares in production. As a piece of machinery is used on more hectares, taxes per hectare decline. This simple approach of accumulating costs per hectare for each enterprise makes GIS an attractive tool for site-specific profitability analysis. Since GIS systems are spatially aware, site-specific information when collecting operating and ownership input costs used in the production process provides the GIS system with the information necessary to generate a site-specific budget. A spatially aware system will produce site-specific profitability analysis for each enterprise, a subset of an enterprise, or a given site, rather than just an entire enterprise. Profitability analysis may include something simple, such as changing the rate of one particular input, e.g. fertilizer, within fields, or a more complex adjustment, such as changing an entire production system, including machinery, equipment, and cultural practices.

In order for a site-specific analysis system to be effective, it must include all resources used in the production process: machinery and equipment, seed, fertilizer, labor, chemicals, land, etc.. This type of system cannot be so complex that it eliminates all but a few producers from using GIS for economic analysis. Another problem with on-farm use of such a system is the current knowledge base to drive a farm level GIS. While the dynamic data layers for an individual farm can be supplied by the user, static data layers of soil type, topography, aspect, etc., are usually not readily available. Most static data layers will not be available at the individual field level for many yrs, nor can individual producers readily supply these data layers. An on-farm GIS, however, would allow the producer to enter necessary data for the producer's farm. Another problem closely related to the lack of necessary data layers for individual producers is the lack of functional relationships that tell the producer how a particular crop will perform under site-specific conditions. Even if the data layers were available to drive an on-farm GIS system, algorithms describing the functional relationships of all data layers are not available. With these limitations, producers are faced with a challenge to utilize GIS for site-specific management. Spatial awareness and data management capability will lead producers to adopt GIS for site-specific management even without existing data to drive the system. Remember, producers are currently assimilating hundreds of variables in the daily decision making process, most with uncertain outcomes.

There are literally hundreds of enterprise budget generators available, most producing similar results. These programs could be modified to allow for site-specific analyses, but this ignores the potential benefits of using a visual interface provided by GIS. The visual interface for both input and output, has been and is a key element in the quick adoption of GIS. The visual interface allows producers to specify different input levels, tillage practices, and entire machinery complements for specific areas. During the process of selecting areas to be treated differently, the producer specifies the spatial attributes necessary to

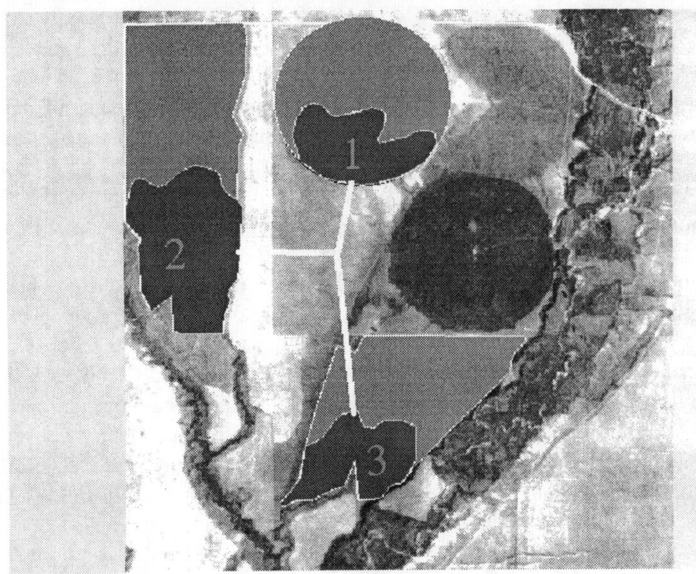

Fig. 53–1. Depiction of three fields, each with two separate treatment areas.

calculate costs and revenues for a site, or for the entire enterprise. Delineation of site-specific treatment areas is the first step in site-specific economic analysis. Although this sounds straight forward, it is not quite as easy as it sounds.

The visual interface, which allows users to select specific sites for treatment, also requires users to enter the data necessary to generate site-specific budgets, Fig. 53–1. Data requirements for site-specific analysis may or may not be difficult to supply. If the terrain is rolling and contains dramatic differences in soil types, slope, aspect, etc., the producer may be able to easily delineate areas to treat differently. In some cases, yield differences on this type of terrain are obvious and different input levels may also be easy to specify. It is more often the case that a relatively featureless terrain provides little or no indication of the variability present in input attributes that are manifested in yields. The producer must then draw on information, other than visual clues, to specify how a site will be treated and how yields will respond to the treatment. Current site-specific farming practices deal mainly with variable fertilizer rates. The information source most commonly used for fertilizer management is grid sampling. Grid sampling provides the user with detailed information about areas within a field. Fertilizer rates, given yield goals by grid attribute, are then specified for each grid sampled. Yield goals can also be specified for each grid. Variable fertilizer application has been our tentative first step towards site-specific farming. Site-specific farming, however, incorporates much more than variable fertilizer application. All inputs in the production process can be varied if there is an indication it will benefit the individual, environment, or society. GIS can provide this indication.

GIS systems come in varying degrees of abilities and complexity. A complex system is not required for economic analysis. Currently, the producer must provide most, if not all, of the data for a site-specific economic analysis. A good mapping package, in lieu of a GIS system, which allows entry of data for all operating inputs and services obtained from quasi-fixed or fixed factors, is sufficient to track the information necessary to produce site-specific enterprise budgets. A producer can specify fields and sites, Fig. 53–1, and specify all operating inputs by site in the production process. The remaining task is to provide an algorithm that will produce the enterprise budgets for the site or enterprise in question. The capability of the software used will determine how this algorithm must be supplied. Spatially aware programs will aid producers in the task of allocating machinery and equipment costs by enterprise or site. Allocation of machinery and equipment costs is the most difficult task in preparing enterprise budgets. While it is relatively easy to allocate the estimated costs of owning and operating machinery based on how the machinery is used, producers do not keep historical records showing machine use by crop, enterprise, or site. A visual interface would allow the specification of fields and sites within fields, such as that shown in Fig. 53–1. This visual interface, if attached to an appropriate mapping package or GIS, would also allow producers to specify all inputs in the production process used on each area specified. The appropriate algorithm could then produce an enterprise budget showing the economic implications of the proposed production practices by site, field, or entire enterprise. Essentially, a visual budget generator. This analysis would answer questions about the profitability of site-specific farming for any site.

Mapping packages and GIS systems are not necessarily the same. Typically, GIS systems have far more capability than mapping packages. Mapping packages, however, contain many attributes of GIS systems. The development of site-specific economic analysis for agricultural operations should always keep a point far into the future as its focus. The evolution of this type of analysis within spatially aware systems will likely progress from simple two dimensional maps that are largely user generated and supported, to sophisticated three dimensional maps driven by data supplied by someone other than the user. These later models will include all aspects of GIS and have algorithms that specify complete functional relationships for all data layers within the GIS. Currently, work is progressing on individual elements of these future models. NLEAP, CREAMS, AGNPS, EPIC, WEPS, SPUR, are just a few of the models being developed that aim to define the behavior of a small portion of the total production process. At some point, all of these models will be integrated into one comprehensive analysis tool. Should producers wait for a complete data driven tool rather than venture off on their own, utilizing the tools currently available? If they wait, they should be young. It will be a long wait. Producers, however, currently have tools necessary to implement site-specific management practices. They must provide their own data, but this is not as big an obstacle as it seems. Producers can implement grid sampling, purchase yield monitors, and continue to implement the technology as it develops. There will be continuous feedback to the analysis system. Yield monitors are a key element to this feedback and to fine tune site-specific profitability analysis. All variables at work in the production process are manifested in yields. With the ability to monitor crop yields spatially, the revenue side of the costs and returns equation is provided and producers can analyze site-specific farming practices. As new research provides indications on variable seeding rates, seeding depth, chemical application, variable tillage systems, and other input variations, functional relationships will be specified for all of the data layers and GIS will provide a powerful analysis tool. Producers can take the first step towards answering their own question about the profitability of site-specific farming by using today's simple mapping programs.

54 Site-Specific Soil Test Interpretation Incorporating Soil and Farmer Characteristics

P. E. Fixen
H. F. Reetz, Jr.

Potash and Phosphate Institute
Brookings, South Dakota

Site-specific nutrient management has been evolving for several decades and was the initial objective of soil testing. More recently, the concept of site specific nutrient management has evolved to include assessment of the nutrient supplying ability of areas within fields by use of grid or soil type sampling. A critical component of modern site-specific nutrient management that is often neglected is the interpretation of the soil test results in a site-specific manner. Even where the latest positioning, sampling, yield measuring and application technology is employed, general or average recommendations are often used to determine the nutrient rate to apply. It is very likely that the weakest link in such systems is how the soil tests are interpreted. The focus of this paper is on an alternative approach in which the interpretation of soil tests considers the grower and soil characteristics of the specific site.

THE TRADITIONAL SOIL TEST INTERPRETATION PARADIGM

Today's soil test interpretation and recommendation paradigm has changed very little since its development in the 1950's. Interpretation and recommendation writing are now done by computer rather than by ballpoint pen but the approach is still the same. A singular soil test level - fertilizer recommendation relationship is assumed to hold for all individuals and sites with adjustments sometimes made for yield level and soil association. A soil test level goes into a *black box* and a recommended rate comes out. The contents of the *black box* are not known or understood by most of the users of soil testing making adjustments in the rate based on site-specific factors very difficult. An alternative paradigm is needed for site-specific management.

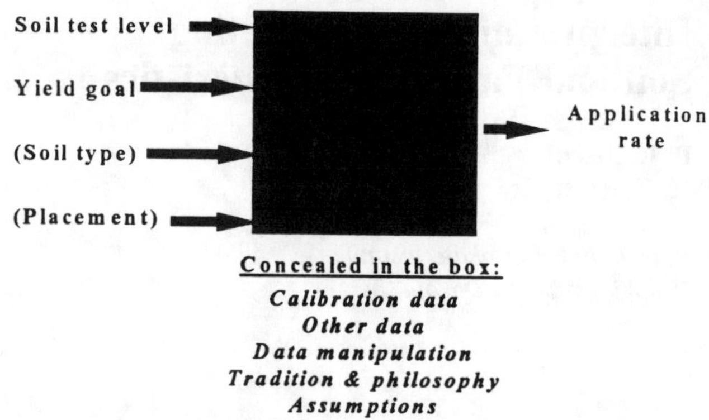

CAPITALIZING ON THE STRENGTHS OF SOIL TESTING

Soil tests, especially for P and K, are indices reflecting the average relative yield or probability of response. They frequently do not accurately predict the rate necessary to give a certain yield in any given season. Therefore, they are most effective when a long-term approach is used to P and K management and to soil test interpretation. Also, only a fraction of the P or K applied in any one yr is used by the crop in that yr. In most cases, the majority of applied P or K remains in the soil in forms that are available for future uptake. The substantial residual effect of applied P and K and the short term uncertainty of soil test predictions argue for a long-term approach to economic evaluation and management of P and K .

SITE-SPECIFIC SOIL TEST INTERPRETATION USING PKMAN

PKMAN (Potash & Phosphate Institute, 1994) is a computer program intended to facilitate and document the personalization of P and K management decisions by substituting site and grower information for as many assumptions and generalizations as possible. The program's focus is on the optimization of long-term profitability. The most important program output is the optimum soil test level which becomes the target for P and K management.

Although the intent of PKMAN is to minimize assumptions, some are made in order to simplify the program and input demands. The major assumptions are:

1. Soil test P and K are capital investments and build-up costs should be amortized.
2. Maximum economic yield (MEY) for the grower's farming system occurs at the soil test level where marginal return equals the minimum acceptable return on investment.
3. Long term yield is a function of soil test level.
4. The optimum soil test level can be maintained over time if nutrients removed by crops are replaced.

All of the potential factors that influence P and K management decisions cannot be described by mathematical equations in a computer program. Winter kill and quality factors can be major concerns for some crops. Uncertainty always exists concerning the actual P or K soil test relationships in a specific field under a given set of management practices. How an individual reacts to that uncertainty is a factor. Tax management concerns, temporary opportunities caused by fertilizer price fluctuations, an individual's personal land ethic, attitudes concerning risk, over-all management style, and other factors all impact actual decisions made. Therefore, program output should be viewed only as a guide.

Factors Considered by PKMAN in Site-specific Soil Test Interpretation

Soil Test Calibration.

Soil test calibration defines the relationship between soil test level and relative yield. The calibration used needs to be appropriate for the soils, climate, and management practices of the grower.

Yield potential.

Yield potential is the average yield over the land tenure period if the nutrient in question is not limiting yield. The yield potential is used to determine the economic value of a percentage change in relative yield. This is not the same as the yield goal used in most soil testing programs. Yield potential of individual soils would be used in the case of soil-specific management.

Net crop price.

The average market value of the crop over the land tenure period minus the cost of maintenance per yield unit. Normally the value of the P or K contained in the harvested portion of the crop is the cost of soil test level maintenance.

Total uptake per unit.

The total amount of P or K taken up by the crop per unit of yield (Table 54–2). This number and the yield potential determine the amount of nutrient that must be supplied to the crop from either the soil or the fertilizer.

Removal per unit.

The amount of P or K that is actually removed from the field in the harvested portion of the crop per unit of yield. Usually this is the amount of nutrient required to maintain soil test levels. If good data are available suggesting that applying less or more than removal is required to maintain soil test levels, the adjusted values should be used rather than actual removals.

First yr fertilizer nutrient recovery.

The rate of fertilizer needed at suboptimal soil test levels is influenced by the recovery percentage of applied nutrients. This is usually estimated from research data by subtracting the uptake of the check (no fertilizer applied) from the uptake of the fertilizer treatment and dividing by the rate applied with the result expressed as a percent recovery. Several factors influence first-yr recovery including soil test level, rate applied, placement, nutrient, crop type, growing conditions, and soil adsorption/fixation properties. Normal ranges for first yr recoveries at very low soil test levels and near optimum rates are 10-35% for P and 20-60% for K.

Level to stop fertilizer application.

The soil test level at which the fertilizer rate to apply drops to zero. If you intend for a small amount of starter to be applied even when the recommended rate is zero, this level could be lower than if you assume starter nutrients are included in the rate.

Minimum ROI (acceptable return on investment).

The minimum acceptable return per dollar invested will vary with the attitude, investment opportunities, and financial condition of the individual. A value of 1.00 indicates that profit from P or K was maximized and the last dollar spent should increase crop value by one dollar. A value of 1.50 would return $1.50 on the last dollar spent and be equivalent to a 50% net return on investment. The goal of a manager with limited capital is to maximize return on the last dollar spent considering all alternative investments and their associated risk.

Interest rate.

The actual interest rate applicable if money is borrowed to purchase P or K.

Tenure.

Land tenure refers to the period of time the grower will be farming the field. Since in most soils, residual P or K should not be depleted if removed nutrients are replaced, expected time of ownership or operation in most cases substitutes for the life expectancy of the capital investment in the amortization process.

P or K/ppm.

Quantity of fertilizer required to change the soil test (buffer potential). Soils differ in the amount of fertilizer required to change soil test levels. Soil test P or K levels are typically easier to change on coarse textured sandy soils than on medium or fine textured soils. Some low pH and some high pH soils fix applied P readily and increasing soil test P levels is more costly, decreasing the optimum level. This assumes such soils have the same P yield response relationships as normal soils. Typically 3 to 4 kg P (14 to 22 lb P_2O_5) are required to increase the Bray P1 test by 1 mg kg^{-1} to a depth of 17 cm (6 2/3 in). A reasonable estimate of buffer potentials is 3.6 kg P per mg kg^{-1} (18 lb P_2O_5/ppm) for P and 3 kg K per mg kg^{-1} (8 lb K_2O/ppm) ammonium acetate extractable K. If a good history of soil tests, crop removal, and fertilizer/manure application is available, this figure can be estimated for the field in question. Soil test procedures are also available to estimate these values. When soil specific management is being used and soils of diverse texture, chemistry, or erosion potential are found within the field, buffer potential should be estimated for each soil within the field.

P or K cost.

The average cost of the fertilizer to be purchased.

AN EXAMPLE OF SITE-SPECIFIC SOIL TEST INTERPRETATION

Data from a USDA study located near Boone, Iowa will be used to illustrate site-specific soil test interpretation. The study area was grid sampled and corn and soybean yields have been measured by combining areas 12 to 15 m (40 to 50 ft) in length. The yield and soil test data summarized by soil mapping unit were supplied by Dr. D. Karlan of the USDA Tilth Lab, Ames, IA.

Although corn yields varied by as much as 2.5 Mg ha^{-1} (40 bu/A) among soils in certain yrs, the 4-yr average yields for the soils varied by only 0.44 Mg ha^{-1} (7 bu/A) (Table 54–1). Soybean yields when averaged across yrs were also quite uniform with soil averages varying only 0.14 Mg ha^{-1} (2 bu/A) (Table 54–2). The lack of yield variability among soils when averaged across yrs does not at first give an optimistic outlook for soil-specific management.

Table 54–1. Corn yield by soil mapping unit in the Clarion-Nicollet-Webster Association.†

	Yr				
Soil	1989	1990	1991	1992	AVG.
			Mg ha^{-1}		
Clarion	9.4	7.9	8.7	11.4	9.3
Nicollet	9.6	7.8	9.5	11.0	9.5
Canisteo	9.6	7.4	9.9	11.8	9.7
Webster	8.8	6.9	11.2	10.1	9.2
Field avg.	9.2	7.6	9.7	11.3	9.5

† Data supplied by Dr. D. Karlen, USDA Tilth Lab, Ames, IA.

Table 54–2. Soybean yield by soil mapping unit in the Clarion-Nicollet-Webster Association.†

Soil	Yr			
	1990	1991	1992	Average
	Mg ha^{-1}			
Clarion	3.6	2.9	3.0	3.2
Nicollet	3.3	3.0	3.2	3.2
Canisteo	2.8	3.0	3.2	3.0
Webster	3.4	3.0	3.0	3.1
Field average	4.5	3.0	3.1	3.0

† Data supplied by Dr. D. Karlen, USDA Tilth Lab, Ames, IA.

Evaluation of soil test levels revealed considerable differences among soils with pH ranging from 5.2 to 7.4, P from 16 to 37 mg kg^{-1}, and K from 91 to 168 mg kg^{-1} (Table 54–3). Measured soil test levels can be associated with average relative yield through the process of soil test calibration. The calibration data used to determine the relative yield percents in Table 54–3 for P were those summarized for the Webster/Canisteo soil association by Fixen (1992) and for K those reported by Melsted and Peck (1977). The relative yield percents for pH were estimated by the authors.

The soil test levels measured when viewed in light of soil test calibration, indicate that multiple limiting factors were reducing yields over the 4 yrs of measurement for two of the soils in the study. Multiplying the relative yield percents for pH, P, and K, as suggested by Wallace (1993), gives the products reported in the last column of Table 54–3. These calculations indicated that the Nicollet soil is producing at only 73% of its potential yield. It is important to note that if field-average soil tests were used to estimate the current relative yield for this field, one would incorrectly conclude that it was producing at 96% of its yield potential rather than 88%.

Table 54–3. Estimated relative corn yields.

Soil	Current Levels pH	P	K	Current relative yields pH	P	K	Total
	mg kg^{-1}			%			
Clarion	5.5	16	95	95	96	85	78
Nicollet	5.2	22	91	90	99	82	73
Canisteo	7.4	17	168	100	97	100	97
Webster	6.3	37	134	100	100	98	98
Field	6.5	20	134	97	97	93	88
Based on field averages:				100	98	98	96

Table 54–4. Estimated corn yield potentials after soil test correction.

Soil	Past yield	Relative yield	Yield Potential
	Mg ha^{-1}	%	Mg ha^{-1}
Clarion	9.1	78	11.8
Nicollet	9.5	73	13.1
Canisteo	9.7	97	10.0
Webster	9.2	98	9.5
Field	9.5	88	10.9
Based on field average:		96	9.9

Table 54–5. PKMAN inputs used for field average (conventional).

Yield, Mg ha^{-1}	[0][†]	[3.6 kg P per mg kg^{-1}]
Corn, $/Mg	84.56	3 kg K per mg kg^{-1}
Tenure, yrs	10	$0.85 per kg P
ROI, $/$	1.50	$0.29 per kg K
Interest, %	10	3.5 kg P per Mg
Recovery, %	P=15 K=40	4.3 kg K per Mg Zero at 30 mg kg^{-1} P Zero at 200 mg kg^{-1} K

Average optimum STL P = 23 mg kg^{-1}; K = 156 mg kg^{-1}

[†][] = values to be varied by soil type within field.

Dividing the measured average past yield for each soil by the estimated relative yield gives the soil yield potentials reported in Table 54–4. These estimates suggest that managing the entire field as a single unit has resulted in the depletion of soil fertility in the potentially most productive soils causing them to yield no more than the lesser productive soils in the field. Using these estimates, the correction of soil fertility problems in this field should increase field average yield by ≈1.3 Mg ha^{-1} (20 bu/A). Other inputs such as N would need to be managed to allow these altered yield potentials to be attained.

With this background information, PKMAN can be used to establish the first approximation of a long term nutrient management plan for this field. Table 54–5 contains the input values assumed for this example. If this field was managed as one unit (conventionally), the optimum soil test levels for the field for P and K would be 23 and 156 mg kg^{-1}, respectively. Thus, the long-term plan would be to either build to or draw down to these levels by varying P and K rates.

In a site-specific management scenario, yield potential and soil test P or K buffer potential would be varied by soil type, if appropriate, in order to estimate the optimum soil test level for each soil area (Table 54–6). In this specific case, the Canisteo soil was assumed to have a higher P buffer potential due to its high pH and due to experience with similar soils. The higher buffer potential of this soil makes building soil test P a more expensive process, reducing the optimum soil test level. The resulting optimum P and K soil test levels become the targets in P and K management. Some areas are currently above optimum and will be drawn down by applying less than maintenance fertilizer while other areas need to be built up.

Table 54–6. Soil-specific optimum soils test levels for P and K.

Soil	Yield potential	P buffer potential	Soil test levels: Current P	Current K	Optimum P	Optimum K
	Mg ha^{-1}	kg P/mg kg^{-1}	mg kg^{-1}			
Clarion	11.8	3.6	16	95	25	158
Nicollet	13.1	3.6	22	91	26	160
Canisteo	10.0	4.8	17	168	20	156
Webster	9.5	3.6	37	134	23	155

Table 54–7. Soil-specific P and K fertilizer rates.

Soil	Yield Potential	Fertilizer Rates: 1st yr P	1st yr K	Maintenance P	Maintenance K
	Mg ha^{-1}	kg ha^{-1}			
Clarion	11.8	74	196	41	50
Nicollet	13.1	57	228	45	56
Canisteo	10.0	44	33	35	43
Webster	9.5	0	86	33	41

First yr fertilizer rates are extremely variable but eventually application rates will approach maintenance levels and be determined by the yield potential of the soil (Table 54–7). Fertilizer to be applied in yr 2 and subsequent yrs until the field is resampled can be determined by estimating the soil test level using harvested yield, removal, and soil test buffer potentials for the soils.

SITE SPECIFIC INFORMATION FEEDBACK

The process discussed above is full of estimates and assumptions, some of which may be wrong. As the plan is followed or as the circumstances of the farmer change, however, new information becomes available and is used to make corrections in the plan. The estimated corrected yield potentials can be adjusted as the yields resulting from modified management reveal their actual impact. The nutrient rates needed for soil test maintenance or to change the soil test by one unit can be calculated using a nutrient budget approach. Even soil test calibration can be accomplished on-site.

Ideally, soil tests should be calibrated for the specific soils, crop rotation, tillage system, and crop varieties of the farmer. Until now, that has never been possible. With today's technology, however, site-specific soil test calibration **is** possible. Strategically placed perennial calibration strips for immobile nutrients like P, K, and Zn or for lime can be used to generate calibration data for a certain region or set of farming practices. For example, a pair of nutrients like P and K can be selected and three strips laid out. In the center strip P and K would be built up to a soil test level where 100% relative yield is assured. The P is left out on one side of the center strip and K left out on the other side. The site-specific soil test level is obtained from the soil test map produced from grid sampling the field or from a special sampling of the strip transect. Yield is measured along each strip using a combine yield monitor.

The result is a database of paired "with" vs "without" yields for P and K across a range of soil test levels existing in the field. These data can then be used to generate a soil test level vs relative yield curve or table that is specific for the local circumstances. Over several yrs, the resulting data can be examined to determine what local factors influence response other than soil test level. For example, it may be necessary to have a separate K calibration for no-till or ridge-till fields. In other cases, certain soils within an association may differ in response. The resulting calibration is input to programs like PKMAN as the basis of future P and K management. The longer such a program operates, the more accurate predictions will become.

SUMMARY

The full yield and profit potential of site specific technology cannot be realized without site-specific soil test interpretation.

- Site-specific interpretation should consider both soil and farmer characteristics.

- Site-specific technology may <u>increase</u> yield variability by expressing the full yield potential of the most productive members of a soil association.

- Information feedback, including site-specific soil test calibration, will cause site-specific management to continuously improve over time.

REFERENCES

Fixen, P.E. 1992. Personalizing soil test interpretation. p. 20-30. *In* Proc. soils, fertilizer, and agricultural pesticides short course and exposition. Univ. of Minnesota Educational Development System, St. Paul, MN.

Melsted, S.W., and T.R. Peck. 1977. The Mitscherlich-Bray growth function. p. 1-18. *In* T. R. Peck (ed.) Soil testing: Correlating and interpreting the analytical results. ASA Special Publ. No. 29. ASA, Madison, WI.

Potash & Phosphate Institute. 1994. PKMAN: A tool for personalizing P and K management, Version 1.0. Potash & Phosphate Inst., Norcross, GA.

Wallace, A. 1993. The law of the maximum. Better Crops 77(2):20-22.

55 Economic Analysis of Site-Specific Nutrient Management Systems

Harold F. Reetz, Jr.
Paul E. Fixen

Potash & Phosphate Institute
RR2 Box 13
Monticello, Illinois

Nutrient management systems continue to become more sophisticated in light of economic pressures and increasing public concern about agriculture's impact on the environment. Farmers' skills and technical education have improved markedly in recent years. New technology is now available that enables them to monitor production systems and to more precisely control application rates and timing of production inputs.

To meet the challenges of this rapidly changing industry and to better utilize the new technology available to them, farmers are learning to emphasize the specific characteristics of the soil and climate resources on their individual fields---and even to capitalize on the variability within fields to improve their management systems. The days of statewide recommendations for crop management systems are gone. The only way for today's farmer to remain competitive and meet the demands placed upon his operation is to focus on site-specific management systems that best utilize his unique resources and experience.

National, or even regional or state-level, nutrient management planning cannot effectively address the problems associated with nutrient management, nor the opportunities for solving those problems through better management and application of technology. A key element of any national, regional, or state plan must be the support of flexibility at the farm level to employ the best available technology on a site-specific basis designed to complement the resource base of the individual farmer.

Farm-level plans, in turn, must be based upon detailed records and monitoring systems that help to best utilize the resources of individual fields and individual managers. Farmers are increasingly turning to outside assistance from consultants that can complement their own skills and experience to form a management team needed to develop the best plan for the farm. Depending upon the farmer's own skills and experience, he may seek outside help in general agronomics, pest management, nutrient management, marketing, etc., working with local dealers, public agencies, and contracted consultants to complete the team. Each member of the team brings more expertise and technology to the crop production system. On larger farms, some of these team players may actually be part of the farm staff, but there will always be a need for outside support as well. All of these team members must focus on the site-specific management for the individual fields---a level of management that has not been common on farms in the past, but which will be essential to their viability in the future.

The agronomic aspects of site-specific nutrient management must be carefully studied and documented to be sure the technology is scientifically sound and can be practically applied at the farm level. We have made significant progress in meeting that need in the past few years. Soil sampling on a planned grid basis, coupled with geographic information systems (GIS) and global positioning satellite (GPS) technology puts a whole new perspective on soil testing and nutrient management recommendations. Variable-rate application of production inputs (seed, fertilizer, and chemicals) using GPS and GIS systems, and on-the-go yield measurements cataloged with GPS and GIS, make it possible to manage and monitor a wide range of production decisions that have been impossible to handle on a sub-field basis in the past.

The technology has been greatly improved and the details of statistically-valid sampling for mapping and evaluation are being worked out to help ensure the technology is properly implemented on the farm. Adding new technology comes with many benefits, but also many questions, not the least of which is the economic impact of the respective changes. The cost–benefit relationship must be carefully evaluated to determine the viability of any new technology or management system.

ECONOMIC ANALYSIS

The Need for Economic Analysis of New Technology

The key to acceptance of site-specific management by farmers, farm managers, and landowners will be the economic analysis of the profitability of the technology. As in any business, new technology in crop production must have a positive effect on the bottom line of the balance sheet if it is to be widely accepted.

- The economic analysis should provide sufficient detail to evaluate the individual component practices of the technology package and show how each affects the costs and returns per acre as well as per unit costs of production. As much as possible, the analysis should address not only the costs and returns on the field-by-field basis, but also the economic implications from an environmental impact basis.

- The economic analysis should be easily updated for price changes of inputs and of the crop produced. In addition to normal financial records and analytical tools, the farmer needs some simple economic analysis tools for individual enterprises and individual fields to be able to quickly evaluate the impact of price changes on the profitability of site-specific practices. Often weather and market changes force farmers to consider alternative strategies to what may have been planned for an individual field. Here again, an easy-to-use analytical tool is important to the decision-making process.

- The economic analysis package should be able to help the farmer evaluate various options through ***"What if...?"*** scenarios. By having an economic analysis set up for each field in advance, the economic effects of changes in the prices or in individual practices can be easily evaluated.

The *MEY Analysis* Software.

The *MEY Analysis* software package was developed by the Potash & Phosphate Institute in cooperation with staff members of the University of Illinois Departments of Agronomy and Agricultural Economics, to provide farmers and their advisers a tool to be used in economic analysis of agronomic management options on an individual field basis. Many crop budget analysis packages group practices into categories, such as "fertilizers", "pesticides", "machinery", etc. Even enterprise analysis packages usually group all fields of the same crop together.

The name ***MEY Analysis*** was chosen to serve as a reminder that the ultimate goal of a management system should be to improve the economic viability of the crop production enterprise for the farmer, with the objective of approaching the management system combination that produced the most profit --- the ***maximum economic yield*** (MEY) for that farm. There are no built-in economic optimization features or linear programming models for least-cost analysis. It is simply a straightforward, detailed crop system budget analysis tool that allows the user to easily study the economic impact of changes in an agronomic management plan. While MEY may not ever be achieved, since it is a moving target that changes as technology and economic factors change, working *toward* MEY management is the best way to ensure the long-term economic viability of the farm.

The *MEY Analysis* package allows the user to evaluate the economics of each individual input on each individual field. Up to four fields may be analyzed at a time with a base plan and an alternative plan for each field. Comparisons may be made on an individual field basis and summarized over the four fields.

The *MEY Analysis* may also be used to study the economic implications of a set of research results. If a farmer wants to evaluate how the results of a particular research project may fit his farm, for example, he could set up the budget for his current system (Base Plan), then set up the alternative (MEY Plan) incorporating the improved practice. Based on the research results, he can estimate the change in yield to be expected on his farm. Then the *MEY Analysis* will help him determine not only the yield impact, but also the economic impact of the new practice relative to the costs of implementation. Of course, he has to be able to provide a realistic estimate of the yield response for his farm. Once he has implemented the practice and has real yield response data for his farm, he can re-run the *MEY Analysis* with the real data for a more specific evaluation of whether he should apply the practice on a broader scale in the future.

Availability of detailed field-by-field records of all inputs and all crop yields is essential to make the most use of the *MEY Analysis* and other new technology tools becoming available to farmers and their advisers. Farmers who do not have detailed records on individual fields will find it difficult to evaluate site-specific technology for their farms. Records of production inputs and yields may become a legal requirement for some farmers, but even if they are not, they are a good management practice and should be a part of every farmer's way of doing business.

CASE STUDIES

Example Farms for Intensive Management Systems

Several example case histories are being developed for studying the details of intensive mangement systems from the standpoint of agronomic responses, economic viability, and environmental impact. As more years of experience with these systems are added to the data base, better interpretations can be made. All farmers should be thinking ahead to information needs they will have to make sound decisions for their management systems and should upgrade their record-keeping systems to provide that information. Decisions can only be as sound as the information used to make them. The longer the time period covered by the data base, the more growing seasons that are covered and the better the ability to predict the response to be expected.

Three farms[1] where such data bases are being developed are used in the following examples.

1. Walker Farm--Central Illinois

   The example used to illustrate the economic comparison of two different agronomic management systems for a farm in central Illinois. The practices, production inputs, and yields are from actual records of a real farm.

---

[1]The names and exact locations of these farms have been changed to protect the farmers' identities and the prices used have been updated to reflect 1994 prices for inputs and grain produced.

2. Larson Farm, North Central Iowa
   The Larson Farm example is a combination of management data from two neighboring farms in Iowa where detailed records were available---one with grid sampling and site-specific management and the other with conventional field-level management.

3. Bevan Farm, Central Illinois
   The Bevan Farm example is a real farm situation in east central Illinois, where grid sampling, site-specific fertilizer application, and on-the-go yield measurement have been implemented in the early 1990s. It is another example of a highly productive cash grain operation that is striving to employ intensive management to help maintain economic viability and protect the natural resource base upon which the business is dependent.

Economic comparison of two management systems for a farm.

The impact of increasing yield through better management is clearly reflected in the improved profitability of the new system. As shown in Table 55–1, management changes included increasing plant population, reducing tillage, changing method of herbicide application, and increasing nitrogen rate.

Table 55–2 summarizes the economic effects of these management changes as computed with the *MEY Analysis*. The changes resulted in an increase in production costs of $28 per acre, but increasing yields by 30 bushels per acre in the process resulted in a net reduction of 10 cents per bushel in cost of production.

Agronomic responses are weather dependent, but tend to be relatively stable over time compared to price fluctuations. The *MEY Analysis* makes it easy to attach a new set of prices to an existing management system budget and evaluate the economics under a new price scenario. Some people prefer to use a 5–yr average set of agronomic data to run an economic analysis. This approach is also supported by the *MEY Analysis* package.

Table 55–1. Comparison of inputs for previous and improved management systems for the Walker Farm, Central Illinois.

Practice	Previous Management	Improved Management
Plants/ha (Plants/acre)	59,280 (24,000)	64,220 (26,000)
Tillage	Chisel & Field Cultivate	One-Pass
Herbicide Use	Pre-Plant Incorporated	Post-Emerge when needed
Nitrogen kg/ha (lb/acre)	136 (152)	188 (210)

Table 55–2. Comparison of production and economic returns two management systems for the Walker Farm, Central Illinois.

Practice	Previous Management	Improved Management
Yield Mg/ha (bu/acre)	9.4 Mg/ha (150 bu/acre)	11.3 Mg/ha (180 bu/acre)
Production on 324 ha (800 acres)	7,524 Mg (120,000)	9,029 Mg (144,000)
Production Costs per ha (acre)	$556 ($225)	$625 ($253)
Production Costs per kg (bu)	$.09 ($1.50)	$.084 ($1.40)

Economic comparison of site-specific management with a field-scale system.

Applying a similar analysis to the Larson Farm example, the economic impact of site-specific, variable-rate fertilizer application can be evaluated. Table 55–3 summarizes this analysis.

Table 55–3. Agronomic and economic comparison of field-average and variable-rate nutrient management systems for a Central Iowa farm.

Larson Farm Example		Variable-Rate Plan			
	BASE PLAN	A	B	C	D
Hectares (Acres)	395 (160)	190 (77)	86 (35)	72 (29)	47 (19)
Yield Mg/ha (bu/acre)	8.2 (131)	10.0 (159)	11.7 (187)	13.0 (208)	9.5 (151)
N kg (lb)	73.1 (161)	75.8 (167)	95.3 (210)	99.9 (220)	70.8 (156)
P_2O_5 kg (lb)	31.3 (69)	31.8 (70)	37.2 (82)	41.8 (92)	30.0 (66)
K_2O kg (lb)	20.0 (44)	21.0 (46)	24.5 (54)	27.2 (60)	20.0 (44)
Net Income/ha (acre)	$247 (100)	$235 (95)	$339 (137)	$435 (176)	$200 (81)
Cost/kg (bu)	$.087 (2.19)	$.088 (2.22)	$.079 (2.00)	$.073 (1.84)	$.092(2.32)

The site-specific, variable-rate system resulted in an increase in the total amount of fertilizer needed. This is commonly the case where field-average soil tests have been used to determine fertilizer needs in the past. When rates are based on field-average soil tests, influence of high testing areas tends to negate the effect of low-testing areas of the field. As a result, less fertilizer is applied than is required to maintain soil tests at optimum levels.

Implementing a grid-sampling pattern and basing nutrient applications on this grid, low testing areas are given larger amounts of fertilizer and high-testing areas are given lesser amounts. But the net result is usually that higher total nutrient applications are needed.

Table 55–4 provides a summary of the economic impact of changing from a field-average to a site-specific, variable-rate nutrient management system on the Larson Farm. The more intensive nutrient management plan resulted in a 17.5% increase in net income, for a total profit increase of $2,798 for 160 acres.

Table 55–4. Economic comparison of field-average (Base Plan) and variable-rate (MEY Plan) nutrient management for a Central Iowa farm.

Larson Farm Supply	**Base Plan**	**MEY Plan**	**% Change**
Receipts/ha (acre)	$1097.29 (444.25)	$1186.54 (480.38)	8.1
Fertilizer Cost/ha (acre)	$ 84.15 (34.07)	$ 95.31 (38.59)	13.3
Total Variable Costs/ha (acre)	$ 323.37 (130.92)	$ 369.41 (149.56)	14.2
Net Income per hectare (acre)	$ 246.83 (99.93)	$ 290.03 (117.42)	17.5
Total for 64.8 Ha (160 Acres)	**$15,988**	**$18,787**	**$2,798**

Table 55–5. Soil test P and K variability within a field on a Central Illinois cash grain farm.

Bevan Farm	**A**	**B**	**C**	**D**	**E**
P_1 Test kg/ha (lb/acre)	127 (113)	96 (86)	87 (78)	75 (67)	57 (51)
Hectare (Acres)	10.9 (27)	4.3 (10.7)	4.9 (12.2)	4.7 (11.7)	4.6 (11.5)
K Test kg/ha (lb/acre)	808 (721)	495 (442)	427 (381)	401 (358)	363 (324)
Hectare (Acres)	2.2 (5.4)	5.5 (13.6)	6.9 (17.1)	7.6 (18.7)	7.4 (18.2)

Application of grid-sampling and variable-rate nutrient management

On the Bevan Farm in central Illinois, grid-sampled soil tests were used to develop a site-specific, variable-rate nutrient application plan for P and K. Table 55–5 shows the soil test variability in one of the fields.

Based on these tests, fertilizer was applied in five different zones in the field, each getting a different P and K combination. Table 55–6 shows the rates applied and the area covered with each. There was no area treated with a uniform rate for comparison, but these data illustrate the kind of variability that is found in a relatively uniform field.

The Bevan Farm was one of about 15 Midwest farms where on-the-go-yield measurements were successfully made and stored in computer files for spatial analysis. Yield checks were made at one-s intervals and recorded with grain moisture and GPS[2] coordinates. These data were used to develop field maps showing the spatial distribution of the yield level corrected for moisture. Spatial analysis techniques were applied to group areas of the field with similar yields into yield zones.

These data can be used to help determine nutrient removal rates and to help in planning nutrient application rates for future years. On-the-go yield measurement provides farmers and their advisers with the first accurate means of determining the spatial distribution of yields within a field. This will allow the correlation of yield with other variables such as soil type, soil test levels, nutrients applied, and other factors. Table 55–7 shows the mean yields for five different areas of the Bevan Field.

Table 55–6. Variable-rate fertilizer application scheme for a grid-sampled field on a Central Illinois cash grain farm.

Zone	Hectares (Acres)	N kg (lb)	DAP kg (lb)	KCL kg (lb)
1	8.5 (21.1)	85.4 (188)	0 (0)	49 (108)
2	9.8 (24.2)	85.4 (188)	47.2 (104)	81 (178)
3	7.3 (18.1)	85.4 (188)	104 (230)	81 (178)
4	3.9 (9.8)	85.4 (188)	104 (230)	81 (178)
5	3.0 (7.6)	85.4 (188)	176 (387)	108 (238)

[2]GPS = Global Positioning Satellite

Table 55–7. Yield variability within a field on a Central Illinois cash grain farm as measured by a GPS-based on-the-go yield measuring system.

Yield Zone	Hectares (Acres)	Mean Yield Mg/ha (bu/acre)
A	5.7 (14.0)	11.1 (177)
B	5.9 (14.6)	10.1 (161)
C	5.9 (14.6)	9.0 (143)
D	5.9 (14.6)	7.5 (119)
E	5.9 (14.6)	5.2 (83)

The challenge with such data bases now becoming available is to be able to relate the yields to the various production factors, and then use those relationships to help in planning management for future years. This challenge creates a need for more research into such relationships, but the new technology also presents the opportunity for new approaches to research which might utilize field scale studies and multi-variable analysis in ways that have been impossible in the past.

Such research is going to be essential if site-specific nutrient management plans are to be developed for individual fields. Realistic plans must be adapted to local soil-crop-environment systems, and to the specific skills, goals, and resources of the individual farmer. The plans must also address local agronomic and environmental problems and specific opportunities unique to the local area. National, regional and state plans cannot meet such needs.

Detailed analysis is essential for such plans, to determine management systems that are agronomically sound, economically viable, and environmentally responsible. All potential nutrient sources should be included, with appropriate credits given for applied livestock manure, industrial by–products, and carryover from previous crops.

Long-term production records are very helpful, but few farmers have records with enough detail to be of real value in site-specific management planning. Likewise, detailed soil testing, preferably on a 2.5-acre (or smaller) grid basis, is needed to prepare accurate plans. Few farmers have such soil test information. But the sooner such records and soil tests are started, the sooner the detailed planning can begin.

CONCLUSIONS

- Site-Specific Management must include a detailed economic analysis to make it acceptable to farmers, landowners, and financial advisers.

- Cost–Return analysis on specific practices is important so that their individual contribution to profitability may be determined accurately.

- Yield response must be tied directly to variables, again to help determine accurate cause-effect relationships impacting on yield and profitability. Detailed yield records and on-the-go yield measurement are important tools for such determinations.

- Agronomically sound, economically viable, environmentally responsible site-specific crop management is possible through careful implementation of new technologies now available to farmers and their advisers.

56 Precision Farming Concepts: An Industry's Perspective and Experience Since 1986

Dean Fairchild

White Bear Lake, Minnesota

Involvement with variable rate concepts since 1986 has resulted in various successes and failures associated with development of precision farming programs. This presentation will summarize these experiences from the various phases of variable rate concepts including: map development, soil testing, multiple yield goals and recommendation, application equipment, marketing and profitability. Information will be summarized from various fertilizer dealers, crop consultants, equipment manufacturers and farmers.

Each phase of variable rate farming (map development, soil testing, etc.) will be discussed from different methods and approaches used by various locations. The advantages, disadvantages and results of each program will be presented.

To provide conference participants economic benchmarks, profitability and costs comparison will be discussed for variable rate services. Profitability will be approached from a farmer's, dealer's, and crop consultant's perspective.

Based on previous results in marketing variable rate technology, ideas and suggestions will be offered for future needs of precision farming program development and growth.

57 How a Crop Consulting Firm is Using Precision Agriculture Technology

Ron Olson

Top Soil Testing Service Co.
Frankfurt, Illinois

Top-Soil Testing Service has been taking soil samples using 2.5 acre grids since 1975 and has sampled over 1.4 million acres of farmland in Illinois, Indiana, Wisconsin and Iowa. In 1992, Top-Soil Testing Service began a relationship with Applications Mapping, Inc. to demonstrate precision agriculture as a feasible process with added economic benefit, agronomic benefit and environmental benefit to farming operations.

Since 1975, the author has worked with hundreds of farmers that understand the importance of crop management and who have been finding through a crop management relationship that site specific crop management and variable rate applications or materials are beneficial to their operation. This paper relates the authors experience of providing soil sampling and testing services, crop management consulting services and support of those services. Top-Soil Testing Service also works with fertilizer dealers that have purchased the fully automated variable rate application equipment. This paper will discuss how a consultant interfaces with a dealer who is working to implement precision agriculture processes into his operation.

Beginning in 1992, Top-Soil Testing Service installed GPS equipment and industrial computers on six-wheel Polaris vehicles in order to begin taking soil samples by GPS technology, sampling approximately 10 000 acres. In 1993, over 100 000 acres were sampled with GPS technology. This paper provides examples of issues to be addressed when GPS technology is integrated into a consulting program and services offered to farmers as well as fertilizer dealers.

SESSION V

ENVIRONMENT

58 Land Targeting and Agricultural Policy

C. Ford Runge

Department of Agricultural and Applied Economics
University of Minnesota
St. Paul, Minnesota

Agricultural competitiveness and environmental quality are increasingly consensus objectives in American agriculture. · Yet the institutional interests undergirding agricultural policy are often at odds with those promoting improved environmental quality. This paper examines ways in which institutional reforms in land targeting can improve both agricultural competitiveness and the environment. Before focusing on the Conservation Reserve Program in particular, some general remarks are in order.

Farmers, like other business managers, make decisions based on information received from markets and government sources. These include signals from commodity markets, federal agricultural policies, federal and state environmental regulations, and private sector and university extension recommendations. The management problems of farmers, already difficult due to the vagaries of nature, are made worse when these signals are contradictory and change unexpectedly over time.

Farm policy is a particular source of confusion. Agricultural policy signals have been overriding in recent yrs as evidenced by high rates of program participation: roughly two-thirds of all cropland in the U.S.A. has been enrolled in agricultural programs during the past five yrs. From 1990 to 1994, an average of 83% of wheat base acres were enrolled in the government price support program; in corn the five-yr average was 78%; in rice it was 95%; in sorghum 76%; in barley 74%; in oats 30%; and in upland cotton 88%.

Because of the overriding importance of maintaining commodity specific *base acres* as an entitlement to federal subsidies, farmers are encouraged to grow the above crops on a continuous basis even if other cropping alternatives, or rotations, would make more sense economically or agronomically. At the same time, farmers participating in the programs have been compelled to set aside to *conserving crops* a percentage of base under the Acreage Reduction Programs (ARPs) and some 36.5 million acres have been enrolled in 10-yr land retirement under the Conservation Reserve Program (CRP).

Agricultural policy signals thus include production subsidies which have increased the number of acres under cultivation at the same time that ARPs and the Conservation Reserve Program have lowered them; markets for certain *non-program* crops have sent signals to increase production at the same time that farm programs have discouraged any plantings that would decrease program base. Agricultural price support policies have also encouraged the conversion of wetlands, while other programs penalize farmers who drain and plow these wetlands. When public policies are so confused, it seems unreasonable to lay the problem of environmental quality in agriculture solely at the feet of farmers. While it is naive to suppose that perfect consistency can be found, a variety of substantial changes in these institutional signals can be made which would make agricultural and environmental goals more complementary.

Apart from the *quid pro quo* of price supports in return for acreage reduction, one of the most significant policy signals sent to farmers has been the Conservation Reserve Program (CRP), which has paid landowners to idle over 36 million acres since 1985.

The Conservation Reserve Program (CRP)

In the face of major crop surpluses in the early 1980s and as a result of new demands from environmental groups, the CRP became part of the Food Security Act of 1985 (PL 99-198, 1985). It was reauthorized in the 1990 farm bill. The CRP, like its 1956–62 precursor the Soil Bank, pays volunteering farmers to retire land from field crop production. The original Soil Bank paid farmers to retire cropland for 3–10 yrs (10–15 yrs for trees). In return farmers received an annual rental payment and 80% *cost-sharing* to plant cover crops or trees. No limits were placed on individual acreage enrollment and *whole farm* retirement was rewarded with a 10% rental bonus. Where trees were planted (2.1 million acres) especially in the South, nearly 90% remained planted to trees in 1976 (Alig, et al., 1980). Much of the rest of the Soil Bank, however, especially in the Midwest, was returned to field crops in the 1970s and 1980s.

As CRP contracts begin to expire in 1996, a question arises: will the CRP, like the Soil Bank, simply end up as a temporary measure to remove acres from production? Or can more sustainable methods be found to protect vulnerable lands? The answer to these questions requires disentangling the two primary objects of the CRP: surplus crop reduction and environment protection.

From the outset, the CRP has attempted to do two things at once: reduce surpluses and protect highly erodible lands. Enrollment through 1994 is roughly 36.5 million acres. All acres enrolled are in 10–yr contracts, during the life of which landowners must agree to plant and maintain approved cover crops or trees, and must receive approval from the county offices of the Agricultural Stabilization and Conservation Service (ASCS) and Soil Conservation Service (SCS) for a plan which restricts harvesting, grazing, or other commercial use of these crops unless specifically authorized by the Secretary of Agriculture.

The CRP is part of a package of conservation provisions contained in the 1985 and 1990 farm bills, generally referred to as "conservation compliance," *sodbuster* and *swampbuster* provisions. *Conservation compliance* restricts eligibility for future farm program payments if existing highly erodible cropland is farmed without an approved conservation plan. This plan must be fully implemented by January 1, 1995. *Sodbuster* provisions restrict farmers from plowing up erodible cropland unless conservation practices are installed consistent with the aforementioned plan, while *swampbuster* language similarly restricts draining or cultivating designated wetlands. Violating either *sodbuster* or *swampbuster* provisions theoretically results in loss of eligibility for USDA program payments.

In practice, these provisions have been only sporadically enforced, and relatively few farmers have actually been denied eligibility for subsidy payments. In a 1989 survey under the U.S. Freedom of Information Act, only 25 farmers in the United States had been subjected to denial of program benefits due to the conservation measures of the 1985 farm bill, and many of these cases were subsequently overruled on appeal.

The essential problem is that these environmental requirements are in conflict with the farm subsidy programs to which they are tied. As the Economic Research Service (ERS) of USDA itself noted in a 1990 report:

The effectiveness of the conservation provisions depends upon the attractiveness of Federal price and income support programs. If Federal commodity support programs become less attractive due to such factors as higher market prices or increased set-aside requirements, the conservation provisions will become less effective (Young & Osborn, 1990, p. 31).

The CRP reflects many of the same inconsistencies. This lack of fit arises largely from conflicts between farmers' incentives to grow more when prices rise, and the CRP's efforts to control supplies. First, the opportunity cost of the 10–yr contract is set by the market price of the commodities which could be grown on CRP acres, and thus is related to deficiency payments, which fall with rising market prices. When market prices were weak and deficiency payments high (as in the 1985-86 period when the CRP began), the program looked relatively *cheap* to USDA, since it obviated the high deficiency payments on the acres enrolled. In order to attract farmers into the program, however, rental payments had to be competitive with these same deficiency payments. As a result, USDA had to pay rental rates well-above market rents in most areas of the country in order to induce enrollment (in some cases 200-300% higher) and even offered bonuses for corn base.

These bonuses reflected a second major problem: because the CRP was understood as a mechanism for supply control the lands targeted for retirement explicitly included base acres. Specifically, farm base was reduced when land was enrolled in the CRP according to the ratio between acreage put in the CRP and total acreage for program crops on the farm. For example, if a farmer had a 200–acre farm (all of which are *crop acres*) and a 100–acre corn base, and put 50 acres into the CRP, the ratio of CRP acreage to total cropland is 50/200, or 1:4, and corn base is reduced by 25%, or from 100 to 75.

The result of this *base bite* was to further increase the *reservation rent* which would induce farmers into the program. The CRP will cost about 19.2 billion dollars between fiscal yrs 1987 and 2003. A 1989 GAO report found that it could have been much less costly and more effective, and that USDA was focusing mainly on getting acres into the CRP, rather than on fulfilling its environmental objectives (GAO, 1989). Despite some broadening of program design, a 1993 GAO report concluded:

CRP is an expensive way to reduce the environmental problems linked to agricultural production. The program will require budget outlays of about $19 billion to take 36.5 million acres out of production; however, not much is known about the dollar value of the environmental benefits purchased or about the extent to which removing the land from production will alleviate environmental problems associated with agriculture.

Furthermore, CRP postpones rather than resolves these problems, and additional costs may be incurred to maintain the program's objectives when the contracts expire. Other USDA conservation initiatives use less costly strategies to achieve lasting environmental objectives; however, unlike CRP, these other programs are not intended to curb excess production, and they only indirectly support farmers' incomes (GAO, 1993, p. 8).

While lands eligible for the CRP were intended to be highly subject to water and wind erosion, the evidence suggests that the preponderance of CRP-enrolled acres are not those considered least suitable for cropping. In Minnesota, for example, ≈90% of the 1.9 million acres enrolled fall into capability classes 2, 3 and 4. In general, a capability class below 4 (classes 1-3) can be cropped without significant problems as long as suitable conservation practices are employed. Only about 25% of Minnesota's CRP contracts fall in capability classes 4 and above.

Finally, as surpluses have dwindled and market prices have risen, both farmers and the government find the CRP less attractive as a supply control measure, and the desire to be done with it grows. Its impact on total acres in production is significant (Table 58–1). Yet simply eliminating it would do nothing more than repeat the Soil Bank experience, at considerable cost. Better alternatives exist, which allow agricultural and environmental objectives to be made more complementary, at lower overall cost.

What To Do

The essential problem of the CRP revolves around the question of land targeting. Specifically:

- Which lands now under CRP contract should be returned to active cropping (although still subject to the 1995 conservation plans under "conservation compliance")?
- Which lands now under CRP contract should remain under restrictive contract, and what form should this contract take?
- Which lands not now under CRP contract should come under some form of environmental restrictions?

Table 58–1. Commodity Market Changes Under the CRP.

Item	1988	1990	1992	1994
		Percent Change		
Wheat:				
Acres	–8.3	–13.3	–12.7	–10.8
Production	–7.0	–13.5	–12.6	–10.6
Stocks	–3.0	–14.8	–29.7	–26.8
Prices	5.8	11.8	15.7	22.6
Corn:				
Acres	–4.7	–8.6	–7.2	–5.7
Production	–4.0	–7.7	–6.4	–4.9
Stocks	–5.7	–23.4	–35.0	–36.3
Prices	.1	2.3	11.8	18.4
Sorghum:				
Acres	–12.3	–21.6	–19.5	–18.4
Production	–9.3	–19.0	–16.3	–14.2
Stocks	–10.3	–82.2	–176.4	–188.6
Prices	2.2	7.1	18.7	24.3
Barley:				
Acres	–13.2	–24.7	–22.3	–23.8
Production	–11.1	–22.4	–19.8	–20.6
Stocks	–13.3	–72.5	–140.3	–226.3
Prices	7.6	12.0	20.7	32.9
Oats:				
Acres	–4.7	–8.1	–8.2	–8.2
Production	–7.8	–12.3	–11.2	–10.5
Stocks	–15.2	–41.5	–45.7	–45.1
Prices	4.4	12.0	19.8	23.3
Cotton:				
Acres	–9.3	–14.4	–14.2	–17.2
Production	–4.9	–8.6	–7.9	–12.1
Stocks	–9.2	–18.6	–22.1	–30.3
Prices	6.0	11.0	15.0	17.6
Rice:				
Acres	.1	–.6	–.9	–1.1
Production	.1	–.6	–.9	–1.0
Stocks	.1	–1.0	–1.5	–1.9
Prices	–.2	4.3	6.5	9.1
Soybeans:				
Acres	–5.6	–8.4	–8.2	–8.2
Production	–4.8	–7.2	–6.8	–6.6
Stocks	–.9	–1.2	–1.9	–2.5
Prices	4.5	7.2	10.8	13.0

Source: C. Edwin Young and C. Tim Osborn, *The Conservation Reserve Program: An Economic Assessment*, USDA/ERS Agricultural Economic Report Number 626, Washington, DC, February 1990, p. 31.

In order to answer these questions, a targeting distinction needs to be made between land that is *marginal* from a supply control perspective (because it is unproductive) and land that is *marginal* from an environmental perspective (because it is vulnerable to erosion damage or otherwise in need of retirement for conservation reasons). In Fig. 58–1, these two dimensions, identified as *productivity potential* and *environmental vulnerability* are described as continuous variables, but are divided into *high* and *low* categories for purposes of discussion.

Productivity Potential†		Environmental Vulnerability‡: High	Environmental Vulnerability‡: Low
	High	Category 2	Category 3
	Low	Category 1	Category 4

Fig. 58–1. Categories of land based on productivity potential and environmental vulnerability.

†Measurable either as yields per acre on the basis of historical data or in terms of productivity indices calculated for various soils.

‡Environmental vulnerability can be expressed in term of erosion potential, or more broadly to reflect land parcels subject to groundwater contamination, strips along protected streams, wetlands, or areas of special value as wildlife habitat, etc.

Category 1: Low productivity–High vulnerability land

Land in this category has limited potential for supply control, but is highly vulnerable to environmental damage. It is thus land which, if currently enrolled, should remain in the CRP or under some form of restrictive easement, preferably on a permanent basis. Land not currently in the CRP, but falling in this category, should also be targeted for permanent retirement from cropping through land-use restrictions. Because its productivity is low, the opportunity cost of its removal from cropping is also low. Hence payments to farmers to retire it or not to crop it should also be relatively low.

Category 2: High productivity–High vulnerability land

Land in this category has high productivity potential, but is also highly vulnerable to environmental damages. If it is currently in the CRP, it should remain so, protected by a restrictive easement which will require a higher price paid to participating landowners than Category 1 lands. There may be reason not to seek permanent easements, simply because on a short-term basis, its productivity might be needed. If it is not currently in the CRP, then efforts should be made to designate it through swampbuster, sodbuster and conservation compliance criteria, restricting cropping in most cases. Some additional CRP contracts might be offered where needed.

Category 3: High productivity–Low vulnerability land

As noted above, a surprising amount of the land in the CRP already falls in this category, at least if it corresponds to capability classes 1, 2 and 3. In these cases, current CRP contracts should be allowed to expire, and only limited land-use restrictions should be imposed in the future, consistent with conservation compliance requirements. Land in this category not currently in the CRP should be granted special status as *sustainable cropland*, and cropping practices that maintain its high productivity should be encouraged.

Category 4: Low productivity–Low vulnerability land

Lands in this category are of relatively limited value for either agricultural production or environmental conservation. If they are now in the CRP, such contracts should simply be allowed to expire. This land may be especially well-suited, however, for non-agricultural industrial or residential uses, and land-use restrictions, zoning ordinances, and land-use planning could all reflect these considerations.

Implementing CRP Targeting Criteria

Categorizing land into such divisions is not especially demanding from an analytical or official point of view. The approach is adaptable to local conditions, so that relative rather than absolute standards could be set for given geographic areas. Such divisions could go a long way toward tailoring policies which would

- maximize the environmental benefits of land use restrictions, even where these benefits are difficult to quantify;
- reduce budget expenditures for land retirement contracts or easements;
- release highly productive cropland from the CRP to produce competitively in a world market.

Apart from the CRP, these categories could also be used to calibrate penalties for non-compliance with conservation requirements, including sod- and swampbusting. These penalties could be divorced from the commodity programs, and would be paid like traffic tickets, so that such punishments would *fit the crime*. Specifically, penalties would be highest on Category 2 lands, followed by Categories 1, 3 and 4.

An Offer Not to be Refused

In terms of the post-CRP regime, those lands on which CRP contracts were not simply allowed to expire (mainly Categories 3 and 4) would be subjected to strict enforcement of conservation compliance, with high penalties for violations. A single offer would be made for either a permanent easement (mainly in Category 1) or a limited-term contract (mainly Category 2) which would be set at roughly one-third the current value for Category 1 lands and one-half the current value for Category 2 lands, reflecting productivity differentials.

The farmer would thus face the choice of either accepting the offer, together with the conservation compliance restrictions, or not accepting it, but still facing the conservation compliance restrictions. Choosing whether to accept the offer under these circumstances should not require much thought.

Principles of Policy for the CRP

While this proposal is detailed, it follows some general principles which are likely to meet with support from commodity groups and sustainable agriculture advocates alike. These are that:

- Land which is highly productive but not vulnerable to environmental damages should be allowed to produce on a sustainable basis largely free of restrictions on cropping practices. If such land is currently in the CRP, the contract should end at expiration.
- Land which is highly vulnerable to environmental damages, whether productive or not, should be subject to strict conservation compliance, and should be retired on a permanent basis if low in productivity, and on a time-limited contract if higher in productivity.
- Penalties for violations of conservation requirements should be adjusted to "fit the crime," and should be decoupled from eligibility for commodity program benefits.
- Payments for land retirement on either a permanent or time-limited basis should reflect the productivity of the land, with lower payments for lower productivity lands.

REFERENCES

Alig, R.J., et al. 1980. Most soil bank plantings in the South have been retained; Some need follow-up treatment." Southern J. Applied For. 4:1: 60–64.

GAO. 1989. Farm programs: Conservation reserve program could be less costly and more effective GAO/RCED 90–13.

GAO. 1993. Conservation reserve program: Cost-effectiveness is uncertain. GAO/RCED–93-132. p. 8.

Public Law 99-198. Subtitle D, Title XII of the Food Security Act of 1985. PL 99-198.

Young, C.E., and C.T. Osborn. 1990. The conservation reserve program: An economic assessment, USDA/ERS Agric. Econ. Rep. No. 626. February 1990. Washington, DC. p. 31.

59 Roles for Site-Specific Research in U.S. Pollution Prevention Initiatives

C. Ogg[1]

U.S. Environmental Protection Agency
Washington, District of Columbia

After decades of heavy investment in urban and industrial point source pollution control, we now recognize that agricultural chemicals represent a pervasive environmental problem (USEPA & USDA, 1990). Herbicide levels in spring runoff cause drinking water concerns in the major rivers draining Midwestern watersheds (Goolesby & Thurman, 1990). Nitrate levels exceed health standards in 17–28% of wells in major agricultural states like Iowa (Kross, et. al., 1990), Nebraska (Exner & Spalding, 1990), and Kansas (Stiechen, et. al., 1988), for which data is available (Spalding & Exner, 1993). Groundwater interflow results in nitrate pollution of surface water, and agricultural nutrient sources cause severe ecological damage to Eastern estuaries, accounting for over 80% of nutrient loadings, for example, to the upper Chesapeake Bay (USEPA, 1983).

Over the past two decades, some scientists took advantage of the time available to develop Crop Management (CM) technologies which provide very low cost, but highly effective responses to farmers' and environmentalists' needs for more efficient use of chemicals (Bouldin, et. al., 1971; Fox, et. al., 1989; Olson, et. al., 1967). As policy makers focus more attention on agricultural pollution, however, pressure mounts to deal with this remaining major source of pollution.

CM technologies are the focus of major environmental initiatives in actual Federal and State programs, as well as in farm and environmental legislative proposals. A strategic, problem solving approach to research, with a larger applied component will likely make substantial contributions to water quality, as will basic, site specific CM research. This paper will describe the major State and Federal environmental initiatives which depend for their success on CM research, describe strategic research initiatives, and describe some critical CM research needs.

---

[1]The opinions expressed in this paper are those of the author only and not of any governmental agency.

Environmental Programs Which Depend on CM Research for Their Success

Research has shown that addressing agricultural pollution caused by soil erosion would bring more stream segments up to certain postulated water quality "standards" than would urban and industrial point source controls. Conservation Compliance and other soil conservation programs provide low cost conservation measures, such as reduced tillage, in contrast to the billions of dollars devoted to urban and industrial sources. Crop management programs for chemicals address problems caused by leaching and run-off of soluble chemicals, which are almost as pervasive as soil erosion (USEPA & USDA, 1990), and often cost even less than soil conservation; in fact, crop management systems for managing nutrients may save U.S. farmers hundreds of millions of dollars by crediting nutrients from livestock waste, crop residues, and carry-over of previously applied fertilizer, which farmers generally do not now credit (Tractenberg & Ogg, 1994).

State Initiatives and Findings

The net effect of increased nutrient use efficiency is often a substantial reduction in nutrients applied. Improved crop management, such as that provided by the late spring soil test (Magdoff et al., 1984) produced nearly 50% savings in nitrogen use for the 36% of Pennsylvania farmers who have adopted the technology (Shortle et. al., 1993). Potential nitrogen fertilizer savings from this soil testing approach to reduce farmers' uncertainty about nitrogen availability in Iowa range from 28-38% (Babcock & Blackmier, 1992). Clearly, development and dissemination of currently available nitrogen technologies could greatly increase efficiency of chemical use (Hallberg, et. al., 1991; Fox & Piekielek, 1983; Ferguson & Christiansen, 1991; Olson, et. al., 1967).

Similarly, wider use of available crop management technologies may greatly reduce phosphorous run-off. Unlike nitrogen, phosphorus remains in the soil for many years (Kamprath, 1967; Pothuluri et al., 1991). Phosphorus build-up to the "high" range has been documented on most cropland in many states such that crops would generally not respond to phosphorus applications for many years. These include the following: Iowa (Killorn, 1990), North Carolina (Novais & Kamprath, 1978), Connecticut, Maine, Delaware, Maryland, Virginia, and West Virginia (Sims, 1992), and many others.

Elevated levels of nitrogen in the soil result in leaching (Ruhl, 1987; Keeney, 1982, 1986; Exner & Spalding, 1979), while elevated levels of phosphorus lead to increased phosphorus in surface runoff (Korentajer et al., 1991). Reducing the amount of nitrogen applied by some proportion may produce much larger reductions in the nitrogen available for leaching (National Academy of Science, 1993). For example, a study using the GLEAMS model found that a 12% reduction in nitrogen use can result in a 27% reduction in nitrate leaching (Davis & Heatwole, 1990).

The above analysis indicates a considerable environmental role for crop management research, including site specific research successes described in this volume:

- o Agricultural pollution accounts for most impairments to water uses, and these impairments are largely nutrient related.
- o Farmers rapidly adopt the more efficient chemical use practices when they become available.
- o Efficient chemical use, at least in the case of nitrogen, dramatically reduces leaching.

National Environmental Initiatives

It should come as no surprise, therefore, that CM technologies are the focus of major environmental initiatives. Nutrient management measures, which consist of following a nutrient recommendation and timing nutrient applications, represent an important part of the measures proposed by the U.S. Environmental Protection Agency (EPA) for use by the States in carrying out programs called for in the Coastal Zone Management Act.

In addition, Clean Water Act proposals, including President Clinton's Clean Water Initiative, draw on the strength of these same management measures, which depend for their success on development of efficient and profitable CM technologies, and their dissemination. 1995 Farm Bill discussions and various internal, strategic initiatives for research and education focus on advancing CM technologies. Advancing CM to meet specific environmental goals and commitments, as exemplified by the Chesapeake Bay Program, is gaining attention, as well.

In the past, the U.S. Department of Agriculture (USDA) offered research and education as a major vehicle for addressing agricultural pollution (U.S. Department of Agriculture, 1990). In response to perceived successes of recent CM research, EPA has begun to share in these initiatives, working jointly with several USDA agencies on highly focused research and education initiatives and enthusiastically supporting them. Examples of this new cooperation include a jointly developed request for research proposals in 1992, led by the Cooperative State Research Service (CSRS), and supported by the Agricultural Research Service (ARS) as a Special Project, and subsequent funding of 25 research projects which focus on providing farmers improved recommendations for nitrogen fertilizer use.

President Clinton's Climate Change Initiative also contains a research initiative, led by ARS, to support reduction in global gas emissions by increasing efficiency of nitrogen fertilizer use. The CSRS led project supported site specific CM and other CM projects; the ARS project is still in the planning stages.

Roles for CM Research within U.S. Environmental Research Agendas

Some CM technologies take 3–5 years for a state to adapt to its soils and climate. The CSRS Special Project for nitrogen, therefore, included a mix of basic research and applied research, with the latter aimed at supporting states' efforts to adapt available technologies to their soils and climate. This emphasis on adaptive, applied research appears particularly critical to make the benefits of CM discoveries available to farmers, as it was, decades ago, in developing the use of fertilizer and plant species that led to our high yielding, chemical intensive agriculture.

Although CM research is expected to play a large role in all major environmental initiatives, dollars committed to the above research supporting improved fertilizer recommendations are very small when compared, for example to sustainable agriculture research and fate and transport research. This relatively low cost may reflect, in part, the strategic nature of the research. For example, sustainable research and fate and transport research is expected to support environmental programs by, respectively, addressing farming systems for the whole farm and for the whole watershed.

Relation to Sustainable Agriculture Research

Sustainable–organic agriculture research may include CM systems, but the emphasis is on the more ambitious goal of employing integrated cropping systems that substitute natural pest control and nutrient sources on the farm for chemicals. Although sustainability is intended to provide for economic sustainability of the farm and for environmental sustainability, analysis has shown that widespread reliance only organic nutrient and energy sources would reduce the availability of food and other agricultural products and make some environmental problems worse (Taylor, et. al., 1991). Under a total shift to organic agriculture, substitution of land for chemicals results in more soil erosion and sedimentation, while reliance on nutrients from legumes does not address nutrient leaching.

The sustainable agriculture research program, therefore, may not offer the strategic role claimed by CM research in rapidly advancing the water quality initiatives, described above. Due to its heavy emphasis on cropping systems that largely eliminate pesticide use, as opposed to simply gaining greater chemical use efficiency, sustainable–organic agriculture may soon offer consumers a choice regarding pesticides and fresh produce. Such a choice is clearly the intent of the Organic Certification initiative authorized by the 1990 Farm Bill.

Research suggests that consumers would be willing to pay to attain greater choice regarding pesticides and food. One study found that about a fourth of fresh apple consumers have deeper concerns about pesticides than scientific evidence would warrant; another forth of the consumers had much lower concerns, more in line with available expert opinion--but would still pay a substantial premium for foods certified by the government to be low in pesticides (van Ravenswaay & Hoehn, 1991). Organic and sustainable agriculture research into whole farm systems potentially offers much to consumers looking for a much wider choice regarding pesticides and food. CM research, on the other hand, is more strategically adept at addressing water quality needs, because of the relative ease for farmers to ultimately adopt CM technologies.

Relation to Fate and Transport Research

CM research attempts to provide chemical use efficiency, and thus, pollution prevention, over large areas. However, not all pollutants lend themselves to low cost, CM remedies. Many large livestock producers, with insufficient land to apply their livestock nutrients, face considerable costs in properly managing manure. Fate and transport modeling serves a number of planning and policy analysis needs, generally related to targeting scarce resources and incentives where they will do the most good.

Critical CM Research Needs

CM research expenditures have been relatively small but are nonetheless expected to play a pivotal role in virtually all the major water quality initiatives under consideration. Since nutrient pollution is widespread, CM technologies for addressing nutrient use efficiency are particularly needed (National Academy of Science, 1993).

Similarly, the widespread occurrence of herbicides in surface water suggests a need to focus more research emphasis on gaining greater efficiency in herbicide use. Highly focused, problem solving research, directed at the most promising technologies for gaining efficiency in herbicide use, could use the above, national CM research initiatives for nitrogen as a model. The recent report by the National Academy of Science (1993) suggests that increased chemical use efficiency, combined with riparian sinks which intercept chemicals before they reach streams, might greatly improve the quality of the nation's water resources.

Comparing the degree of sophistication of some of the site specific CM research with the actual techniques in use in many states for providing nutrient recommendations suggests that the greatest need is for applied research. Studies show, for example, that most livestock operations do not credit nutrients from animal waste and legumes (Hallberg, et. al., 1991). As a result, hundreds of millions of dollars are spent annually on redundant fertilizer use. The gap between the highly sophisticated CM technologies identified in this volume, compared to those practiced by the majority of U.S.A. farmers, remains very wide. This gap, and the several years of research required by one state to adopt a new technology developed in another state, highlights the need for adaptive or applied research.

This is not to suggest a lack of response to emerging environmental CM opportunities. On the contrary, the highly focused research initiatives, described above, have been very responsive and timely, given that many of the CM technologies are new. Evidence that these technologies could greatly reduce pollution and gain farmer acceptance became available in the literature even more recently. But more applied research is clearly needed, and greater regional coverage should be a goal in evaluating future research proposals.

Rarely has such a modest investment offered so much to environmental and agricultural programs. Water quality benefits eminating from inexpensive CM research investments may rank with the benefits obtained through some of the very large past investments addressing pollution in the U.S.

REFERENCES

Babcock, B.A., and A.M. Blackmer. 1992. The value of reducing temporal input nonuniformities. J. Agric. Res. Econ. 17(2):335–347.

Bouldin, D., W. Reid, and D. Lathwell. 1971. Fertilizer practices which minimize nutrient loss. Agricultural wastes: Principles and guidelines for practical solutions. Proc. Cornell Univ. Conference on Agricultural Waste Management at Syracuse, NY. pp. 192–206.

Davis, P.E., and C.D. Heatwole. 1990. Modeling impacts of alternative agricultural systems on groundwater in the Virginia Coastal Plain. Presented at the International Winter Meetings. ASAE, December 18–21, Paper No. 902592, Chicago, IL.

Exner, M.E., and R.F. Spalding. 1979. Evolution of contaminated ground water in Holt County, NE. Water Resour. Res. 15:139–147.

Exner, M.E., and R.F. Spalding. 1990. Occurrence of pesticides and nitrate in Nebraska's ground water. Water Center Publ. 1. Inst. of Agric. and Nat. Resour., Univ. of Nebraska, Lincoln.

Fox, R.H., and W.P. Piekielek. 1983. Response of corn to nitrogen fertilizer and the prediction of soil nitrogen availability with chemical tests in Pennsylvania. Penn. Agric. Exp. Stn. Bull. 843. Pennsylvania State Univ., University Park, PA.

Fox, R.H., G.W. Roth, K.V. Iversen, and W.P. Piekielek. 1989. Soil and tissue nitrate tests compared for predicting soil nitrogen availability to corn. Agron. J. 81:971–974.

Ferguson, R.B., and A. Christianson. 1991. Producer adoption of water quality management practices in South-Central Nebraska--1991. Soil Sc. News 13, No. 12.

Goolesby, D.A., and E.M. Thurman. 1990. Herbicides and pesticides in rivers and streams of the Upper Midwestern United States. Proc. 46th Annual Meeting of the Upper Mississippi River Conservation Committee. Washington, DC.: U.S. Geological Survey.

Hallberg, et. al. 1991. A progressive review of Iowa's agriculture - energy - environmental initiatives: Nitrogen management in Iowa. Iowa Dep. of Natural Resources, Geological Survey Bureau, Tech. Info. Series 22. Ames, IA.

Iowa State University Extension. 1990. Summary of soil test results, 1986–1989. Ames, IA.

Kamprath, E.J. 1967. Residual effect of large applications of phosphorus on high phosphorus fixing soils. Agron. J. 59:25–27.

Keeney, D.R. 1982. Nitrogen management for maximum efficiency and minimum pollution. pp.605–649. *In* Nitrogen in agricultural soils. Agron. Monograph No. 22. Agron. Soc. of Am. Madison.

Keeney, D.R. 1986. Sources of nitrate to ground water. CRC Crit. Rev. Environ. Control 16:257–304.

Killorn, R. 1990. Trends in soil test P and K in Iowa. Presented at the 20th North Central Extension-Industry Soil Fertility Workshop, November 14–15, Bridgeton, MO.

Korentajer, L., P.C. Henry, and B.E. Eisenberg. 1991. A simple effects analysis of the effects of phosphogypsum application on phosphorus transport to runoff water and eroded soil sediments. J. Environ. Qual. 20:596–603.

Kross, B.C., G.R. Hallberg, C.R. Bruner, R.C. Libra, and others. 1990. The Iowa state-wide rural well-water survey. Water quality data: Initial analysis. Iowa Dep. of Nat. Resour. Tech. Inf. Ser. 19. USEPA Rep. 440/5–85–001. Office of Water Reg. and Standards. Washington, DC.

Magdoff, F.R., D. Ross, and J. Amadon. 1984. A soil test for nitrogen availability to corn. Soil Sci. Soc. of Am. J. 48:1301–1304.

Novais, R., and E.J. Kamprath. 1978. Phosphorus supplying capacities of previously heavily fertilized soils. Soil Sci. Soc. Am. J. 42:931–935.

Olson, R.A., A.F. Dreier, C. Thompson, K. Frank, and P.H. Grabouski. 1967. Using fertilizer nitrogen effectively on grain crops. Nebraska Agric. Exp. Stn.

Pothuluri, J.V., A.A. Whitney, and D.E. Kissel. 1991. Residual value of fertilizer phosphorus in selected Kansas soils. Soil Sci. Soc. Am. J. 55:399–404.

Ruhl, J.F. 1978. Hydrologic and water quality characteristics of glacial drift aquifers in Minnesota. Water-Resour. Invest. U.S. Geol. Surv. 87–4224.

Shortle, J.S., W.N. Musser, W.C. Huang, B. Roach, K. Kreahling, D. Beegle, and R.M. Fox. 1993. Economic and Environmental Potential of the Pre-sidedressing Soil Nitrate Test. Report Prepared for EPA.

Sims, J.T. 1992. Environmental management of phosphorus in agricultural and municipal wastes. *In* Sikora, F.J. (ed.) Future Directions for Agricultural Pollution Research. TVA Bull. Y-224. Muscle Shoals, AL.

Spalding, R.F., and M.E. Exner. 1993. Occurrence of nitrate in groundwater--A review. J. Environ. Qual. 22:393–402.

Steichen J., J. Kowlliker, D. Grosh, A. Heiman, R. Yearout, and V. Robbins. 1988. Contamination of farmstead wells by pesticides, volatile organics, and inorganic chemicals in Kansas. Ground Water Monit. Rev. 8:143–160.

Taylor, C.R., J.B. Penson, Jr., E.G. Smith, and R.B. Knutson. 1991. Impacts of chemical use reduction in the South. S. J. Ag. Econ. 23: 15–23.

U.S. Department of Agriculture. 1990. Water quality and nitrate: Agricultural sources of nitrate and approaches to reduce nitrate contamination of waters. Prepared by the Working Group on Water Quality.

U.S. Environmental Protection Agency (USEPA). 1983. Chesapeake Bay: a framework for action. Chesapeake Bay Program, Region 3.

________ and USDA. 1990. National water quality inventory: Report to

Congress. Washington, DC.
Van Ravenswaay, E., and J.P. Hoehn. 1991. Consumer willingness to pay for reducing pesticide residues in food: Results of a nationwide wurvey. Staff Paper No. 91–18. Dep. of Agri. Econ., Mich. State Univ.

60 The Role of Precision Farming in Sustainable Agriculture: A European Perspective

B. S. Blackmore
P. N. Wheeler
J. Morris

Silsoe College
Cranfield University
Silsoe, Bedfordshire, United Kingdom

R. M. Morris

The Open University
Walton Hall
Milton Keynes, United Kingdom

R. J. A. Jones

Soil Survey and Land Research Centre
Cranfield University
Silsoe, Bedfordshire, United Kingdom

Agricultural managers have always looked for better ways to grow crops. The dependency of some European countries on food supplies from North America during and after the second world war encouraged national governments to foster agricultural systems that could grow as much food as possible. The success of scientific researchers in providing a basis for achieving this aim by maximizing yields in Europe is well known but the pressure is now to seek financially sustainable cropping systems that minimize impact on the environment.

This paper reviews the current European research effort in the field of precision farming. The authors describe the potential benefits of yield mapping and the need for Management Information Systems (MIS). In the latter context, the spatial data sets needed are best handled by a Geographical Information System (GIS).

The authors conclude that there is considerable scope for using the techniques of precision farming to develop agricultural systems, particularly in the European Union countries, which are much less wasteful and less damaging to the environment than existing practices.

BACKGROUND

Sustainability is a concept which has been developed for a number of yrs and is the centre of the Government's environmental policy. The concept is that the environment should be so regarded and maintained that it does not erode or degrade and is handed on to future generations in the same condition or possible enhanced or developed. Therefore, no operation should be allowed to take place that would damage the environment without restoring or replenishing the damage. (Ian Lang, Secretary of State for Scotland: House of Commons, 11 February 1991).

This paper is the result of a project undertaken by Cranfield University in collaboration with The Open University for Scottish Natural Heritage who are developing a set of programs designed to evaluate approaches to sustainable development of the rural environment. The TIBRE project (Targeted Inputs for a Better Rural Environment), an important component of this set, concentrated on intensive agricultural systems. So far most attempts to resolve such problems have involved *turning the clock back* to less intensive systems of agricultural production, with financial support from government on a scale that is unlikely to be economically sustainable in the long term. The TIBRE project will explore opportunities for *turning the clock forward faster* - it asks how can we use new technology to reduce the environmental impact of intensive agriculture while retaining, or improving upon, current levels of productivity. (SNH, 1994)

In this context a report was written addressing the issues of Information Technology (IT) and arable management, which included the concept of precision farming. To do this a study of the European situation was undertaken in terms of both commercial and research developments, with particular reference to Germany and Denmark. This paper has been drawn from that report (Blackmore et al., 1994).

INTRODUCTION

Precision Farming is the term used to describe the goal of increased efficiency in the management of agriculture. It is a developing technology that modifies existing techniques and incorporates new ones to produce a new set of tools for the manager to use. Inevitably it integrates a significant amount of computing and electronics but higher levels of control require a more sophisticated system approach.

Therefore, Precision Farming is not simply the ability to apply treatments that are varied at local level, but must be considered as the ability to precisely monitor and assess the agricultural enterprise at a local and farm level and to have sufficient understanding of the processes involved to be able to apply the inputs in such a way as to be able to achieve a particular goal. This is not necessarily maximum yield but may be to maximize financial advantage while operating within environmental constraints.

Furthermore, the techniques of precision Farming should be regarded as indivisible from the concept of sustainable land management. In sustainable development, the requirement is for non-negative changes in the stock of natural

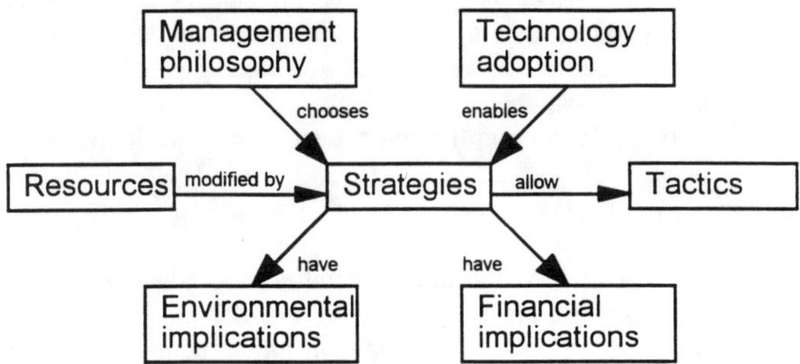

Fig. 60–1. Strategy formation and impact.

resources and in waste capacity of the receiving environment (Pearce et al., 1988). This definition is helpful in Precision Farming as it contains the concept of stewardship of the land for future generations, preserving potential for multi-use, and the importance of off-site as well as on-site effects. Sustainable land management implies more than maintaining crop yields.

To be able to identify the implications of new technologies and practices, we must identify the strategies being adopted by the managers. These, in turn, are influenced by the management philosophy and the level of technology adoption as shown in Fig. 60–1.

This paper, therefore, considers precision farming by defining three possible technology adoption levels and outlines the associated possible farming management strategies. Each will be dealt with separately.

Technology Level One represents conventional practice with no IT and is taken as a reference. *Technology Level Two* has some IT investment and provides the farmer with an increased understanding of the enterprise, but does not include the ability to vary application rates automatically. Farmers can, however, achieve patch application variation by manually influencing machinery settings. This technology level is seen as an interim to *Technology Level Three* which has fully supported variable application rate capability. The financial and environmental implications associated with each technology level and farming strategy are outlined.

TECHNOLOGY LEVEL ONE TRADITIONAL PRACTICE WITH NO IT

Description of Technology Level

Conventional crop management practices have evolved as a product of many yrs personal experience by growers observing the consequences of management decisions, and the adoption over time of the conclusions of scientific research as interpreted by the individual grower, usually via some intermediate advice or extension agency. This formula has been successful over several decades in that it has generated large increases in yield and quality, although not always in financial performance. Indeed, the current economic situation deems increases in crop performance essential to avoid decline in financial performance.

Potential Environmental and Operational Implications

Farming strategy

Traditional practice is the management strategy that has been built up over generations based on craft experience. Particular knowledge of the local environs is built up, but no sudden changes in infrastructure or conditions can be tolerated. Often decision making will tend to be made collectively; i.e. watch to see when other people are cutting hay, etc. Strategy options are limited to minimizing risk of crop losses and there is a tendency to over treat land with agrochemicals.

Environmental implications

Traditional practice has been associated with environmental problems in four areas, and these would be expected to remain.

Deleterious effects on the farmland ecosystem itself. Examples include the build-up of pesticide resistance among target organisms, loss of predators that may be exerting a stabilizing effect on pest numbers, loss of desirable visible species, and possible damage to soil structure through physical and biological changes.

Pollution of watercourses and underground aquifers represents a waste of applied inputs, and causes problems for other organizations, such as water suppliers.

Direct effects on human health particularly from pesticides, are controversial, but their potential harm should not be overlooked.

Reduction or changes in desirable non-farmed habitats such as woodlands and heaths within rural areas. These may be valued aesthetically, and also have value as reservoirs of desirable species.

Social effects include the costs of non-renewable-resources needed as inputs to traditional agricultural practice, and social structure.

Financial Implications

This scenario reflects the current state of affairs and as such has no investment hence no benefits. It could be considered as losing money through wastage, particularly of agrochemicals (i.e. nitrates, pesticides and phosphates). We shall take this scenario as the datum.

TECHNOLOGY LEVEL TWO

MANAGEMENT INFORMATION SYSTEMS, 'WHAT IF' MODELS AND LIMITED INSTRUMENTATION

Description of Technology Level

A farm operating at this technology level would use an on farm computer and software capable of supporting stock keeping, historical records and models to help predict different scenarios, for instance (a) the effect if prices went up by 3%, or (b) the break-even period relating to a particular investment in machinery. Most of these types of models tend to be spreadsheet based as these give the manager the greatest flexibility, certainly more than the more highly structured packages can offer. On the other hand the more structured the program the more it forces the manager into formalizing the record keeping which may be an advantage in itself. Any form of computerization always forces the data entry into certain organized formats which in itself helps the manager get organized beforehand. This level of adoption also encompasses the use of limited agricultural instrumentation and control. There is a large retrofit market for electronics to existing equipment. The most predominant sector has tended to be chemical sprayers.

Strategy options tend to be more directed than in traditional farming with smaller risk margins.

High value/risk cropping is practiced occasionally as heightened control, financial models and break-even budgets help to identify profitable areas. Patch application of inputs is possible and a number of farmers are now considering this option. It seems likely that this is an interim stage before moving to a fully spatially variable system. This is the current level of technology used on commercial farms in Europe today.

IT components and descriptions

Computer

Personal computer of a 486 standard.

Arable farm software (Main module)

An arable farm management software system allowing the farmer to build up an easily accessible information base through the yr. Provides detailed information and access to historical field records and total stock control. The base package allows for inclusion of additional software packages (for planning, budgeting, fertilizer prediction and invoicing).

Planning software (Sub module)

Creates plans defining fertilizer and chemical application requirements on particular crops or fields, produces operator instructions, calculates commodity requirements, records actions taken with dates, times, and weather conditions.

Fertilizer requirement prediction software (Sub module)

The program calculates nutrient requirements field by field based on proposed cropping, anticipated yield and soil details recorded in the Main Module. By using product prices and nutrient value the most suitable fertilizer is selected.

Budgeting software (Sub module)

The farmer builds a record of the farm into the system that includes field areas, crops, varieties and treatments. The program accesses data already within the Main Module and produces budgets that are updated as information is gathered. Shows cash balances, margins, commodity requirements and field applications.

Tractor with monitoring facility.

A tractor with in cab display that includes forward speed, PTO speed, distance travelled, fuel per hectare, and work rate. The machine can also control implement depth to maintain the wheel slip at a value selected by the operator.

Sprayer with control and data logging facility.

Sprayer controller and monitoring system with data logging facilities.

Market Availability and Costs of Technology

Individual IT components with assessment of availability and costs

The costs outlined in this section apply only to the IT components associated with this technology level. This means that the costs shown for the tractor and sprayer refer to the additional cost over and above those of the basic machines (Table 60–1).

Table 60–1. Availability and estimated costs of product components for Technology Level Two.

Product	Availability	Cost£
Computer	Now	1200
Arable Farm Management Software	Now	850
Planning Software	Now	400
Fertilizer Requirement Prediction Software	Now	400
Budgeting Software	Now	400
Tractor with Monitoring facility	Now	1100
Sprayer with Control and Data logging facility	Now	1050

†At present exchange (1994) one U.S. Dollar ($1.00) is equivalent to ≈1.5 English pound (£).

Potential Environmental and Operational Implications

Operational benefits to the farmer

The farmer using this technology level would have a better understanding of the farming enterprise both in terms of day-to-day management needs and in on-going financial performance. The use of management information and control systems that acquire data and use information to reduce risk are a significant improvement. Also the efficiency of decision making, and assistance with the implementation and evaluation of improved business management strategies are an added bonus. Improved stock control, operational timeliness and timing of materials procurement should be seen. Accounting methods should be streamlined and may result in some cost savings. The heightened understanding this gives the manager may give him/her the confidence to treat on a patch basis.

Scenario 1

Farming strategy

The same as conventional practice (as outlined above) but with improved understanding.

Environmental implications

This scenario would involve little change in the environmental effects compared to traditional practice. More effective management of inputs could decrease agrochemical losses to watercourses and aquifers, but management of pesticide inputs could result in either an increase or decrease in pesticide use, depending on the farmers' attitude to risk. The technology *per se* would not affect the management of non-farmed areas, although any gains in financial efficiency might allow some money to be spent on environmental improvements.

Scenario 2

Farming strategy

Two options have become apparent in recent yrs. These are to increase the yield using the same inputs or to maintain the yield while using reduced inputs. In practice this involves small reduction of inputs (by manually influencing machine settings) in areas of low yield to reduce waste. The saved input may or may not be applied elsewhere in areas likely to give a return.

Environmental implications

The two options possible under this strategy would have contrasting environmental effects. A yield increasing strategy would be likely to increase deleterious effects since it is likely that wasted inputs would also increase. The technology would also allow the farmer to bring more areas of currently non-farmed land into productive use (subject to any area or conservation controls currently in force).

A waste-reducing strategy would produce some environmental benefits, through reductions in pollution by biologically active inputs. More accurate assessment of risk, particularly the pollution of drinking water, might amplify the reductions in inputs; higher equilibrium levels of pest species might be expected, with reductions in the development of pesticide resistance and increases in stability of pest numbers from the presence of predators and other species. Any increases in farm income which are achieved could be used in part for active conservation of non-farmed areas, with potential benefits, and the reduction of inputs reducing the off-farm resource costs. These changes, however, are likely still to be small, and in the nature of *fine-tuning*.

Financial Implications

Obviously there are different financial implications depending on the management approach taken. If the farming strategy is the same as for conventional practice there will be no significant savings in inputs, the benefits will relate mainly to gains in farmer confidence / understanding of the enterprise and better timeliness and accounting methods. If the farming strategy involves a small reduction of inputs there will obviously be some savings although these are also difficult to quantify.

The implications associated with this investment level (£5400) in IT can be summarized as follows. Taking 200 ha. as a representative area for cereal cropping on a medium size arable farm, additional average costs are in the order of £11/ha., this represents about 1.7% of the value of the wheat output per hectare. In order to recover the additional cost of IT at *Technology Level Two*, farmers would need to achieve savings in total variable costs of seeds, fertilizers and chemicals of ≈5%.

Obstacles to Implementation Including Public Attitude

Public attitude has changed towards farmers in recent yrs. Farmers were seen as the providers of food for the nation but, as long as food supplies are adequate, they are seen more as custodians of the countryside. As the populace has become more aware of the rural environment it is demanding that intensive agriculture should have a less damaging effect on the environment. Any changes in agricultural practice that are likely to benefit the environment are likely to be encouraged rather than discouraged by the public. The attitude towards increased use of IT as defined in *Technology Level Two* will depend on whether it is perceived as having a positive or negative effect on farming practice by reference to environmental influence.

Farmers will adopt IT if it offers some relative advantage and can be introduced into their system of farming without too much difficulty or expense, and if it is reliable and easy to use. As the components associated with this technology level are already partially established and accepted as being of benefit it is likely that uptake will increase significantly.

Technologies are often taken up quickly if they meet a perceived need or if there are incentives to encourage adoption. The likelihood of increased need for reporting for legislative control and monitoring may well provide a catalyst for increased uptake.

Use of information technology involves new skills, although the familiarity with computers has increased rapidly among many sectors of the public [Open University, 1993]. The level of complication involved in the components of *Technology Level Two* should not create a major barrier to adoption.

TECHNOLOGY LEVEL THREE
FULL SPATIAL UNDERSTANDING AND TREATMENT

Automated Data Collection and Integrated MIS/DSS

Description of technology level

The ultimate goal is a full understanding of the whole process and rational decision making following the management philosophy. This will never happen as there are too many unknowns in a very complex process, the weather being the most obvious. Nevertheless, IT can still automate many of the repetitive tasks and help to provide accurate information at the right time. Levels of automation are bound to rise, as are the use of computers and controllers. The trend is towards a better understanding and a smaller unit of treatment. Information sources will be on-farm as well as off-farm. Examples of on-farm sources are; yield maps, soil maps, weather station data, quantities held in storage bins, weigh bridge, etc. Examples of off farm sources are; market prices, pest prediction models, expert advice, etc. Both of these sources could be integrated into an information system that allows the manager easy and timely access to the data, precisely when the decision must be made.

IT components and descriptions

Many of the IT components in this section are similar to those used in *Technology Level Two* but will have enhanced capability.

Computer

Personal computer of a 586 standard.

Arable farm software (Main module)

Similar to that outlined in *Technology Level Two.*

Planning software (Sub module)

Similar to that outlined in *Technology Level Two.*

Fertilizer requirement prediction software (Sub module)

Similar to that outlined in *Technology Level Two.*

Budgeting software (Sub module)

Similar to that outlined in *Technology Level Two.*

Mapping software(Sub module)

Allows downloading of Ordnance Survey topographic maps from disk or by digitizing the paper map. Display facility to show, for example, soil type, drainage layouts, weed patches, pH hot spots etc. Yield mapping data from combine can be imported and displayed graphically. All these spatial data sets require the manipulation tools normally found in a GIS. Treatment maps can be constructed.

Modelling and decision support software

Allows detailed 'what if ?' type analysis that predicts financial and also environmental effects. By accessing a weather records database probability ratings could be given to potential effects of changes in treatments/farming approach. The package could also warn the farmer if there is likelihood of legislation infringement resulting from decisions taken.

Tractor with monitoring and recording facility.

Similar to that outlined in *Technology Level Two* but with agricultural data bus system. [de Haan, 1990]

Combine with yield mapping ability

A combine that is equipped with a full GPS based instrumentation system to provide yield mapping data acquisition capability.

Variable rate sprayer

Similar to that outlined in *Technology Level Two* but with ability to vary local application rate in response to treatment maps.

Variable rate fertilizer applicator

Machine with control and monitoring of solid fertilizer application rate that can be linked to treatment mapping requirements.

Market Availability and Costs of Technology

Individual IT components with assessment of availability and costs

As in the description of *Technology Level Two* the costs outlined in this section apply only to the IT components i.e. the additional cost over and above those of basic machines (Table 60–2).

Table 60–2. Availability and estimated costs for product components for Technology Level Three.

Product	Availability	Cost (£)
Computer	Now	2500 RRP
Arable Farm Software Shell	Now	850 RRP
Planning Software	Now	400 RRP
Fertilizer Requirement Prediction Software	Now	400 RRP
Budgeting Software	Now	400 RRP
Mapping Software (Yield and Treatment)	Within 1 Yr	350 EST
Modelling and Decision Support Software	Yrs +	1500† EST
Tractor with Monitoring and Recording facility	Within 1 yr	1100 EST
Combine with Yield Mapping Ability (includes GPS)	Now	13500 EST
Variable Rate Sprayer	Yrs +	2000 EST
Variable Rate Fertilizer Applicator	Yr	2000 EST
	TOTAL	£25,000

†If this Software contains expert knowledge this price may be increased to what the market will stand.

Potential Environmental and Operational Implications

Operational benefits to the farmer

All the advantages associated with *Technology Level Two* (improved stock control, operational timeliness, timing of materials procurement and streamlined accounting methods etc.) apply to *Technology Level Three* but with additional features, mainly linked to decision support systems and the ability to automatically vary treatment rates at local levels.

As environmental constraints may be imposed in terms of the use of inputs and practices such as nitrogen use, or on limits on environmental impacts such as leachate in ground or surface water farmers will have to record and report that they are meeting the set criteria. *Technology Level Three* would greatly assist with automated reporting and confirming compliance. For example, a water company or authority might insist on strict controls on intensive farming in areas where the soils are susceptible to high leaching losses.

If environmentally sensitive low input–low output systems are adopted they are unlikely to sustain a viable farm sector without significant levels of income support. IT at this level, however, could help to formulate and implement locally relevant farming strategies that, through better targeted input use, achieve financially viable farming systems that simultaneously comply with environmental quality criteria. Spatially variable field operations allow localized optimization and flexibility thus helping to reducing the business risk inherent in environmentally sensitive farming systems.

POSSIBLE SCENARIOS

Three possible scenarios are identified, depending on the management approach. The major characteristics of the three scenarios are given below.

Scenario 1

Farming strategy; Yield protected, high physical input.

The farmer is assumed to wish to enhance current yield and to protect this yield to the maximum level. *Environmental* criteria are deliberately not considered in the decision making process and high physical inputs are used. Spatially variable inputs are used for fertilizer and pesticide application, the rates of which are determined to be economically optimal and sufficient to maintain low/zero threshold levels of pest species.

Environmental implications

The high level of control implied by this strategy could result in increased damage to the farmland ecosystem by putting additional stress on areas capable of higher productivity. It is possible that better soil physical management practices could mitigate some of these effects, and increased control of pesticide inputs could reduce losses to water bodies. Similarly, the increased level of control would be expected to improve the handling of hazardous substances, with reduced risks of accidental exposure of operatives or others and reductions in other accidental pollution effects. Resource costs would increase, and the prevailing changes in landscape would continue. These would be welcome to some farmers, but in general are unwelcome to the wider public.

Scenario 2

Farming strategy; Reduced input, optimal return, low environmental concern

In Scenario 2, a higher degree of risk of current yield loss may be accepted, and input use is restricted to economically optimal levels consonant with the degree of risk accepted. No explicit consideration is given to environmental criteria but pesticide application rates are such as to maintain moderate threshold levels of pest species. Fertilizer is still applied at economically optimal rates.

Environmental implications

This strategy would exhibit small, but positive effects on most of the current concerns for the environment. Accepting higher pest thresholds, and restricting inputs to more rigid economic optima would benefit all the fauna and flora of farmed areas, and the refined management should maintain or improve soil status. Benefits to non-farmed habitats would be largely incidental, since in the absence of incentives, there would be no specific efforts to enhance these, except where (as in the case of beetle banks–conservation headlands) they can be demonstrated to have some economic advantages to the farm. Effects on the landscape would be relatively neutral, but the improved control of inputs might reduce accidental damage to humans and to the wider environment.

Scenario 3

Farming strategy; Reduced input, high environmental concern

The third Scenario assumes that the farmer gives a high rating to *environmental* aspects in decision making, either as a result of particular values–beliefs or in response to external income support. For example, a water company or authority might insist on strict controls on intensive farming in areas where the soils are susceptible to high leaching losses.

Again, pesticide application rates are such as to maintain moderate threshold levels of pest species. Fertilizer may be applied at rates lower than economically optimal but these may still be too high to prevent significant losses. Potential nitrate leaching losses have been estimated for the arable farming areas in England and Wales by Jones and Thomasson [Jones & Thomasson, 1990].

Environmental implications

This strategy has potentially the most environmental benefits, since these are expressly included in the decision making process of the farmer. Non-target organisms, and non-farmed habitats are likely to benefit greatly from being explicitly considered as part of the management of the farm, and all the other benefits of more precise control of the farming operation would also be achieved. The most positive aspect of this strategy is the opportunity, given suitable incentives, for the creation of desired biological communities or entire landscapes. For example, natural chalk grasslands or lowland heaths could be better preserved within a general arable environment with more precise management.

Such benefits, however, would only be achieved given appropriate incentives, since economic survival of the farm unit would still have to be a priority.

Financial Implications

As in *Technology Level Two* there are different financial implications depending on the farming strategy. If the aim is to enhance current yield and to protect this yield to the maximum level using variably applied high inputs there is potential to offset costs with increased revenues. Also, by localized treatment the average input level may be reduced giving savings. At the other extreme if a high rating is given to *environmental* aspects in decision making there will be a risk of financial penalties without regional or state intervention.

The implications associated with this level of investment in IT (£25000) can be summarized as follows: Taking 200 ha. as a representative area for cereal cropping on a medium size arable farm, the additional average costs are not inconsiderable (in the order of £50/ha.). This represents about 7% of the value of the wheat output per hectare. In order to recover the additional cost of IT at *Technology Level Three*, farmers would need to achieve savings in total variable costs of seeds, fertilizers and chemicals of ≈22%.

It should be noted that some of the costs of IT components have been estimated as they are not commercially available at this time. Also, the costs of the components that are in the marketplace have been entered at full recommended retail price. The most expensive component is the GPS (£10,000) that is incorporated in the combine. It is possible that in future economies of scale and pressure from competitors will reduce this and other costs.

Obstacles to Implementation Including Public Attitude

As in *Technology Level Two* the public attitude towards increased use of IT will depend on whether it is perceived as having a positive or negative effect on farming practice and consequently on the quality of the environment.

The level of capital investment is significant at this Technology level and at this time the potential environmental and financial benefits have not been fully quantified. Until the subject has been more fully researched and developed, both in terms of benefits and in terms of gaining confidence in component reliability, farmers will not be prepared to make the necessary investment. IT will only be adopted if it offers some relative advantage and can be introduced into the system of farming without too much difficulty, and if it is reliable and easy to use.

Farmers are usually impressed if new technologies can be shown to perform well locally, particularly on neighboring farms. More experience is needed in running farms using fully spatially variable technology.
Technologies are often taken up quickly if they meet a perceived need or if there are incentives to encourage adoption. In some cases, financial incentives may be appropriate to encourage early adoption, particularly amongst those who are likely to influence others.

The level of complication of the individual components in this technology grouping could be discouraging and extension services could be of benefit.

Managing the spatial variability of agricultural operations both require and with suitable technology, can provide an enormous amount of data. The acquisition and interpretation of these data raise questions of data protection (Data Protection Act) and of the opportunities for detailed control of farming operations that conflict with the individualistic nature of farmers. The increasing use and demand for quality standards such as BS5750, BS7750 [BSI, 1987 1992], ISO9000, however, is engendering a climate in which the need for control information is more accepted.

SUMMARY OF ENVIRONMENTAL EFFECTS

Environmental effects of the various scenarios are summarized in Table 60-3. Effects are rated according to whether they are negative (-), where the effect of the adopted scenario is likely to be a perceived reduction in environmental quality, neutral (0) where no detectable effect is expected, or positive where effects that are perceived to be beneficial (+) or highly beneficial (++) might be obtained.

This constitutes the elements of an environmental risk analysis system for field, farm and regional level providing a rational basis for the evaluation of environmental risks from different farming systems and land use practices.

CONCLUSIONS

Precision Farming has the potential to make a major contribution towards improving agricultural practice in order to reduce the impact on the environment from agrochemical wastage. Adoption of technology of sufficient capability to provide fully supported precision farming is likely to be an evolutionary process, with farms gradually improving their IT capability.

Most of the component parts needed for fully supported precision farming are sufficiently developed to be used in the near future by farmers themselves. The area that is likely to need continuing development for the foreseeable future is the software to support the farmer when making decisions concerning appropriate treatments and levels of treatment to be spatially applied.

Table 60–3. Estimate of environmental effects for different scenarios at Technology Levels Two and Three.

Technology Level	2	2	3	3	3
Scenario	1	2	1	2	3
Farmland ecosystem					
Target organisms	0	+/-	-	+	+
Non-target organisms	0	+/-	-	+	++
Soils	0	0	0/+	+	+
Non-farmed habitats					
Terrestrial	0	-/0	-	0	++
Water	0	+	0/+	+	++
Direct effects on humans	+	+	+	+	+
Social costs					
Resource costs	0	0/+	0/-	+	+
Pollution	+	0/+	+	+	+
Rural economy	0	0/+	0/+	0/+	0/+
Landscape	0	0/-	0/-	0	++

The costs of IT components are likely to fall as more companies enter the marketplace.

It is possible that increased revenues may result from precision farming due to more appropriate input levels with less wastage.

If however, precision farming is used within a farming strategy that is driven by environmental considerations there will be a need for grants–subsidies to offset financial penalties.

As environmental issues are given higher priority when considering the levels of inputs the financial risks to the enterprise could increase.

ACKNOWLEDGEMENT

This work was partially funded by Scottish natural heritage; TIBRE (Targeted Inputs for a Better Rural Environment) project.

REFERENCES

Blackmore, B.S., P.N. Wheeler, R.M. Morris, J. Morris, and J. Graham 1994. Information technology in arable farming. Report for Scottish Natural Heritage; TIBRE Project.

British Standards Institution. 1987. BS5750: British Standard Quality Systems BSI.

British Standards Institution. 1992. BS7750: Specification for Environmental Management Systems. BSI

de Haan, P. 1990. Connection between on-board computer and management computer. van Hall Institute, Groningen, The Netherlands. Unpubl.

Jones, R.J.A., and A.J. Thomasson. 1990. Mapping potential nitrate leaching losses from agricultural land in the U.K. *In* Nitrates and Pollution (P.J.C. Hamer, and P. Leeds-Harrison (ed.) UK Irrigation Association Technical Monograph 3, 35–46, Cranfield.

Pearce, D., E. Barbier, and A. Markanda. 1988. Sustainable development and cost-benefit analysis. Paper 88–01. London Environmental Economics Centre, UK.

Scottish Natural Heritage. 1994. TIBRE Project. Cranfield University, Scotland.

61 Methodology for Predicting Agrochemical Leaching on a Watershed Basis

R. de Jong
W. D. Reynolds

Centre for Land and Biological Resources Research Centre, Research Branch
Agriculture Canada, Ottawa, Ontario
Canada

Most pesticide contamination of ground water in Southern Ontario is below Canadian drinking water guidelines, but there are growing public concerns over potential health hazards related to long-term exposure to low levels of pesticides (Agriculture Canada, 1990). Pesticide residues, especially atrazine, have been detected in surface, ground and tile drainage waters of many agricultural watersheds in the Great Lakes basin, particularly where there is some combination of high pesticide usage, intensive agriculture, high precipitation, coarse and other highly permeable soils, high water tables, and sloping topography (Millette & Torreiter, 1992).

Pesticide contamination of ground water has traditionally been considered to be due primarily to spills, and to improper storage, disposal and application practices. There is increasing evidence, however, that normal agricultural practices can also result in low-level, non-point source contamination of ground water via the downward migration of pesticides through the soil profile (Agriculture Canada, 1990). Consequently, there is a need to determine how important and widespread this type of contamination might be, what the controlling soil, land use and weather factors are, and what agricultural practices are required to maintain this type of pollution at acceptable levels. Essential steps in obtaining this information include identification of the primary mechanisms controlling pesticide movement through the soil profile, and development of the capability to characterize and predict the pesticide movement in space and time with acceptable accuracy. To achieve these steps, a methodology was developed which incorporates a solute transport simulation model, pedotransfer functions, geostatistics and a geographic information system (GIS).

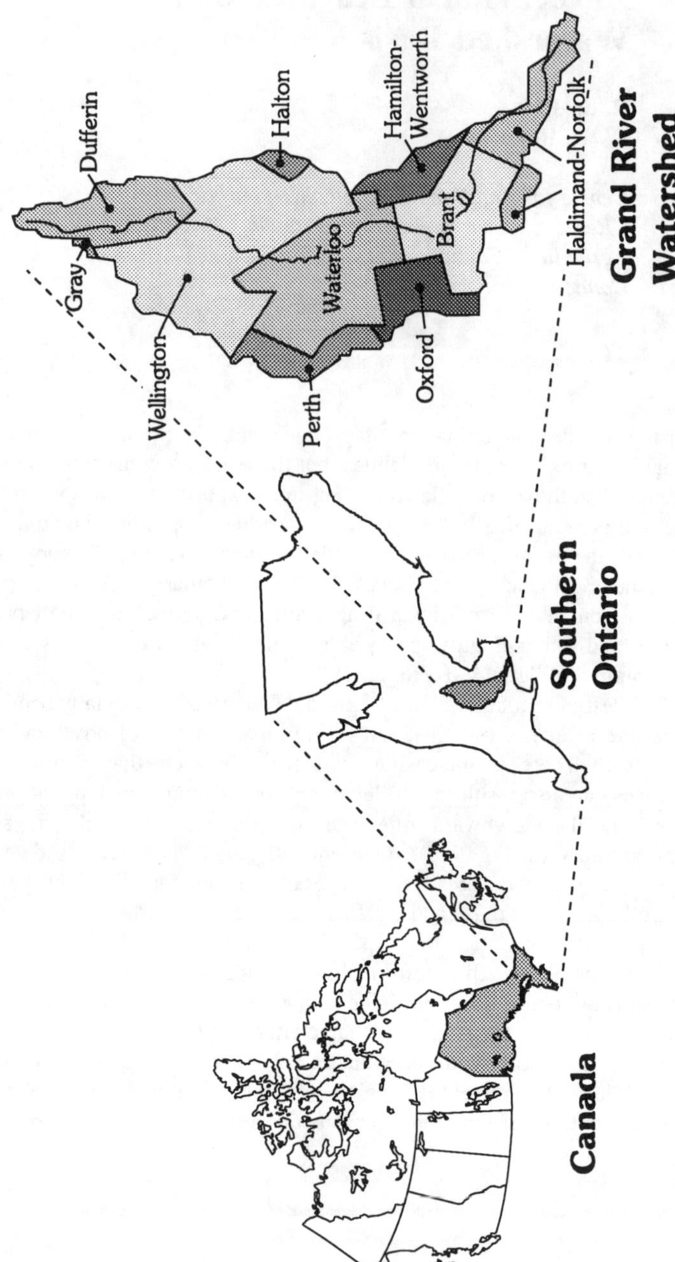

Fig. 61–1. Location map for the Grand River watershed.

Although the methodology can be applied to virtually any landscape unit (e.g. plot, field, watershed, region, etc.), the focus in this study is on the watershed. This scale was chosen because it is a manageable natural landscape unit in Southern Ontario, it is compatible with the scale and density of the most readily available and complete soil and weather databases, and it provides a convenient basis for estimating agrochemical loadings to the Great Lakes. Predictions of atrazine leaching through the soil profiles of the Grand River watershed (Fig. 61–1) are presented as a preliminary test and demonstration of the methodology.

METHODOLOGY

The solute transport model is an in-house modification of the well established and tested modelling package LEACHM (Leaching Estimation And CHemistry Model, Hutson & Wagenet, 1989). The model integrates the major processes that occur in the soil-plant-atmosphere system, including soil horizonation; saturated, unsaturated, steady and transient water flow; crop growth and transpiration; solute sorption, degradation, advection and dispersion; precipitation and evaporation; soil heat flow; and changes in water table elevation. LEACHM is a mechanistic, one dimensional solute transport model, which uses Richards' equation to predict water flow and the convection-dispersion equation to predict solute fluxes. Its submodels include LEACHW which describes soil water flow only; LEACHP, which describes sorption, migration and degradation of pesticides; LEACHN, which describes nitrogen transport and transformations; and LEACHC, which describes the movement of inorganic salts. The submodels consider a variety of processes that occur in the plant root zone, including transient fluxes of water, solutes and heat; alternating periods of rainfall and evapotranspiration; and variable soil conditions with depth. All submodels utilize similar numerical solution schemes, based on procedures developed from several earlier models (Bresler, 1973; Nimah & Hanks, 1973; Tillotson et al., 1980). Only the LEACHW and LEACHP submodels are used in this study. The reader is referred to Hutson and Wagenet (1989) for a more detailed discussion of the entire LEACHM package.

Modifications made to LEACHW and LEACHP

Soil Hydraulic Properties

The soil hydraulic property functions used in LEACHW poorly reflect the often observed rapid decrease in water content and hydraulic conductivity at pore water pressure heads between 0 and -0.2 kPa (see e.g. Topp et al., 1980). The functional relationships between pressure head (h), soil water content (θ) and hydraulic conductivity (K) proposed by Van Genuchten (1980), on the other hand, usually give a more accurate representation of the observed behavior. Consequently the Van Genuchten functions, defined by Eq. 1 and 2 were incorporated into LEACHW as an optional alternative:

$$\theta = \theta_r + \frac{\theta_s - \theta_r}{[1 + |\alpha h|^n]^{[1-(1/n)]}} ; \qquad \theta_r \leq \theta \leq \theta_s \tag{1}$$

where α [L^{-1}] and n [dimensionless] are empirical parameters, and θ_s and θ_r refer to the saturated and residual volumetric water contents, respectively [L^3L^{-3}]. The K(h) relationship has the form:

$$K(h) = K_s \frac{[(1 + |\alpha h|^n)^{[1-(1/n)]} - |\alpha h|^{n-1}]^2}{[1 + |\alpha h|^n]^{[1-(1/n)](L+2)}} \tag{2}$$

where L = 0.5 is usually assumed.

Water Extraction by Plants

Preliminary runs with LEACHW indicated that in wet soil profiles, (i.e. pore water pressure heads greater than -10 kPa), the Nimah and Hanks (1973) water extraction function, U(z,t), would predict unrealistic (oscillating) water uptake patterns. An alternative function based on the work by Feddes et al. (1978) was therefore added as an option. This function has the form:

$$U(z,t) = R_{df}(z,t)\ \beta(h)\ S_{max} \tag{3}$$

where R_{df} is a root distribution function [dimensionless] which is already calculated within a submodel called GROWTH, $\beta(h)$ is a dimensionless sink term variable, ranging from zero to 1 as a prescribed function of soil water pressure head and S_{max} represents the maximum possible rate at which plant roots can extract water from the soil, and is given by:

$$S_{max} = PT / Z_r \tag{4}$$

where PT is the potential transpiration rate [LT^{-1}] and Z_r [L] is the rooting depth. Root water uptake is adjusted to reflect non-optimal conditions using the β function and specified pressure head limits, h_1, h_2, h_3 and h_4 (see Feddes et al., 1978). Values of $h_1 \rightarrow h_4$ in this function were adapted from Dierckx et al. (1988) and Veenhof (1993): h_1 = -1 kPa, h_2 = -2 kPa, h_{3min} = -60 kPa, h_{3max} = -40 kPa, h_4 = -1500 kPa. The R_{df} function used for the Grand River watershed was developed by Tillotson et al. (1980), with the provision that Z_r increased linearly from 5 cm depth at planting to 90 cm depth at root maturity.

Dissipation Rates

The dissipation rate constant, k, of most pesticides (including atrazine) is known to exhibit a substantial dependence on soil water content and temperature. Empirical water content and temperature corrections for k were therefore added to LEACHP. The water content correction has the form (Walker, 1978):

$$k_w = \ln 2 / A\ (100\ \theta)^{-B} \quad (5)$$

where k_w is the water content corrected dissipation rate constant [T^{-1}], and A and B are dimensionless constants. The temperature correction, T_{cf} [dimensionless], is given by:

$$T_{cf} = Q_{10}^{\ 0.1(t-tbase)} \ ; \qquad 0\ ^0C < t < 35\ ^0C \quad (6)$$

where Q_{10} [dimensionless] is the factor by which k changes over a 10 ^{0}C temperature interval, t [^{0}C] is soil temperature, and tbase [^{0}C] is the base temperature from which the temperature correction was determined. The water content - temperature corrected dissipation rate constant, k_{wt}, is then calculated as:

$$k_{wt} = k_w T_{cf} \quad (7)$$

For the Grand River watershed application, practical constraints dictated that Eqs. (5) and (6) could be calibrated only for the 3 predominant soil types in the watershed (Table 61–1). Consequently, every soil in the watershed was assigned one of the 3 calibrated k_{wt} functions on the basis of which predominant soil type they were the most similar to.

Table 61–1. Dissipation rate constants, k^*, Q_{10} values and constants A and B (Eqs. 5 and 6) for atrazine. (Data provided by E. Topp and W.N. Smith, CLBRR, Agriculture Canada, Ottawa, Ont., Canada).

Variable	Soil Type		
	Clay loam	Loam	Sandy loam
k^* (d^{-1})	0.0468	0.0217	0.0215
Q_{10}	4.896	3.715	3.687
A (d)	686.9	122.9	198.4
B	1.061	0.369	0.514

k^* is a "reference level" dissipation rate constant (determined at tbase = 25 ^{0}C and $\theta = 0.70\theta_s$) which is required for the calculation of Q_{10}, A and B.

Bypass Flow

It is well established that preferential, or bypass, flow in macropores (i.e. cracks, worm holes, root channels, etc.) can cause rapid movement of water and solutes through the soil profile and into the ground water (e.g. Thomas & Phillips, 1979; Hendrickx & Dekker, 1991). As a result, many attempts have been made to incorporate bypass flow into the traditional mechanistic water and solute transport models, which neglect this phenomenon. All of these representations, however, have serious limitations (Beven, 1991), and consequently a simplistic, but measurement based, approach to bypass flow was adopted for this work.

Chloride tracer breakout curves measured in situ in the soil profile often show a $0.5C_0$ concentration (C_0 = chloride concentration in input spike of tracer) occurring before one pore volume (PV) of soil water is eluted (Fig. 61–2). This is indicative that some of the water in the soil profile is bypassed by the chloride solute. The proportion of soil water bypassed by the solute was therefore estimated as:

$$F = 1\text{-}(\theta_T/\theta) = 1\text{-}PV_{0.5}\ ; \qquad \theta_T \leq \theta;\ \ 0 < F \leq 1 \tag{8}$$

where F is a "bypass flow" factor [dimensionless], θ_T is the amount of soil water that participated in solute transport [L^3L^{-3}], and $PV_{0.5}$ is the number of pore volumes at which the $0.5C_0$ tracer concentration occurred. This factor was incorporated into the advection and mechanical mixing terms of the convection-dispersion equation, i.e.:

$$qC \rightarrow qC/(1\text{-}F) \tag{9}$$

$$D_m(q) = \lambda q/\theta \rightarrow \lambda q/\theta(1\text{-}F) \tag{10}$$

where q [LT^{-1}] is water flux density, C [ML^{-1}] is solute solution concentration, D_m [L^2T^{-1}] is the mechanical mixing component of the hydrodynamic dispersion coefficient and λ [L] is the soil dispersivity. The above implies that all of the soil water participates in solute sorption, dissipation and diffusion, but only a fraction of the water (specified by F) participates in advection and mechanical mixing. The effect of F on the convection-dispersion equation is to increase the average velocity with which solute migrates through the soil profile relative to the average linear pore water velocity (q/θ). This in turn causes the predicted $0.5C_0$ concentration of a nonreactive solute (e.g. chloride) to occur at < one PV (i.e. $PV_{0.5} \leq 1$).

Bypass flow factors (F) have been determined from field measured chloride breakout curves collected in Southern Ontario for several soil types, water contents and depths in the soil profile. The F values were highly variable (CV = 168 %) and did not show any consistent patterns with soil texture, water content, pore water flux, or depth in the soil profile. Consequently, the mean value of F = 0.20 (n = 56) was used in the simulations to represent bypass flow in the Grand River watershed application.

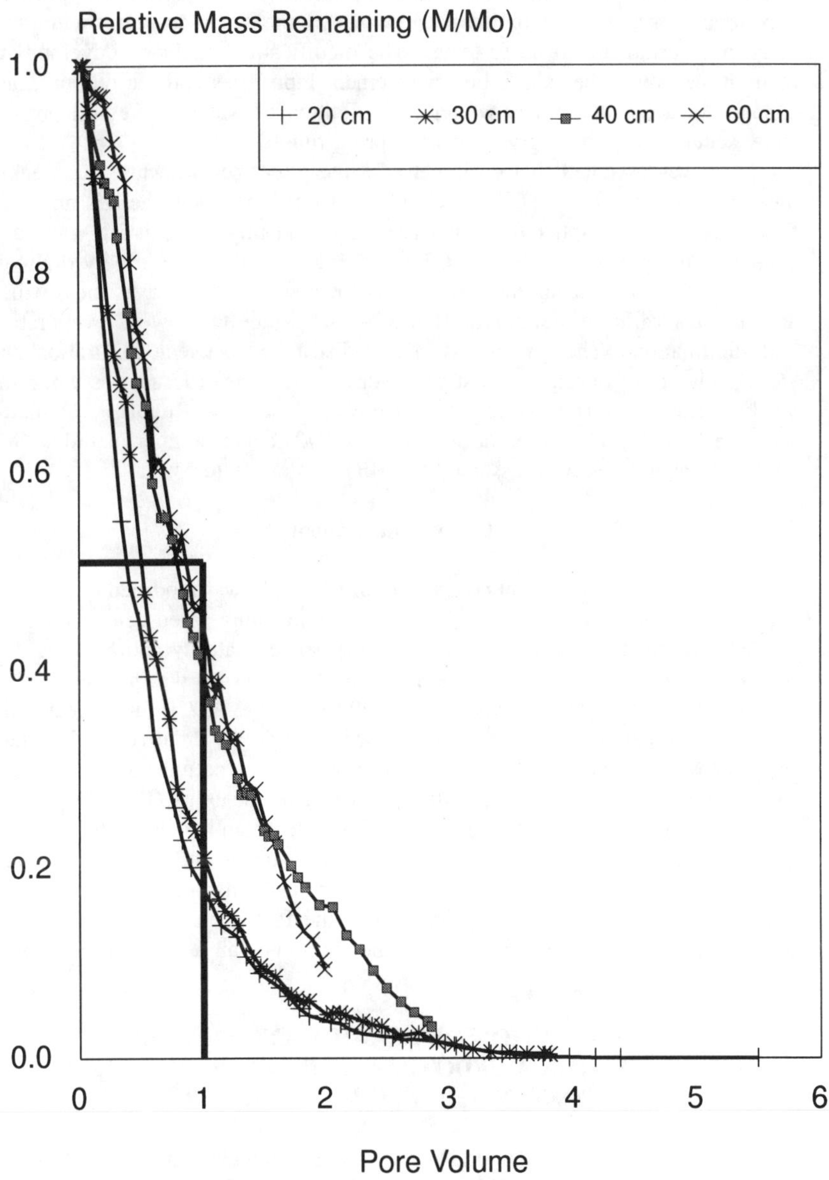

Fig. 61–2. Field averaged (n=30) chloride breakout curves for a Brookston clay loam soil.

Multiple Yr Simulations

LEACHW and LEACHP were modified to run over a number of consecutive yrs to determine if the annual pesticide mass loading at a specified depth becomes constant with time, continuously increases or decreases with time, or changes erratically from yr to yr. The modifications included development of multiple yr weather and crop management input files and various program changes to account for over-winter redistribution of water in the soil profile, snow accumulation during winter, and spring runoff.

It was assumed that during the "winter" (defined as when the weekly mean air temperature is < 0°C), the soil surface is frozen and there is no water flux across the atmosphere/soil interface. Consequently, the existing water and solutes in the profile slowly redistribute during the winter via gravity drainage only. Precipitation during the winter period is assumed to occur as snow which accumulates with the assumption that 30% is lost due to blowoff, evaporation and sublimation. The remaining 70% is distributed between infiltration and snow melt runoff during the first seven days of "spring" (defined as when the weekly mean soil surface temperature is greater than 0°C). Runoff is calculated with the USDA Soil Conservation Service (1972) curve number method. Soil surface temperatures were assumed to equal air temperatures.

Crop Management

The crop management component of LEACHP was modified to take into account crop type, soil permeability and air temperature when determining the dates for pesticide application, planting, emergence, maturity and harvest. For the Grand River watershed application, corn (*Zea mays* L.) was planted in soils with intermediate surface permeability (100 mm/d < K_s < 250 mm/d) when the mean air temperature reached 12.8°C (Brown, 1976). Relative to this date, planting was advanced 7 d on soils with high surface permeability (K_s > 250 mm/d), and delayed 7 d on soils with low surface permeability (K_s < 100 mm/d). Emergence was assumed to take place 7 d after planting, and atrazine was applied two weeks after emergence. Crop maturity (full crop cover and maximum root depth) occurred when 1250 corn heat units had been accumulated since planting. The crop was harvested in the fall when either the mean air temperature dropped below 12°C, or the minimum air temperature dropped below -2°C.

APPLICATION OF LEACHW AND LEACHP WITHIN A PEDOTRANSFER FUNCTION - GEOSTATISTICS - GIS FRAMEWORK

The LEACHW and LEACHP models simulate water and solute movement in the vertical direction only, i.e. they are one-dimensional. Extension of the models to an areal basis requires running them at a number of georeferenced locations distributed throughout the area, and then applying interpolation procedures that account for the inherent spatial variability within the area. This was accomplished using archived soil survey and weather databases, pedotransfer functions, geostatistical analyses, and a GIS.

A large (≈ 3 million ha), arbitrarily defined, "map window" encompassing the Grand River watershed was actually used for the application, rather than just the watershed alone (Fig. 61–3). This was done to increase the amount of data available, and thereby, precision of the geostatistical calculations; and to eliminate inaccuracies in the geostatistical and GIS calculations along the watershed boundaries due to border effects.

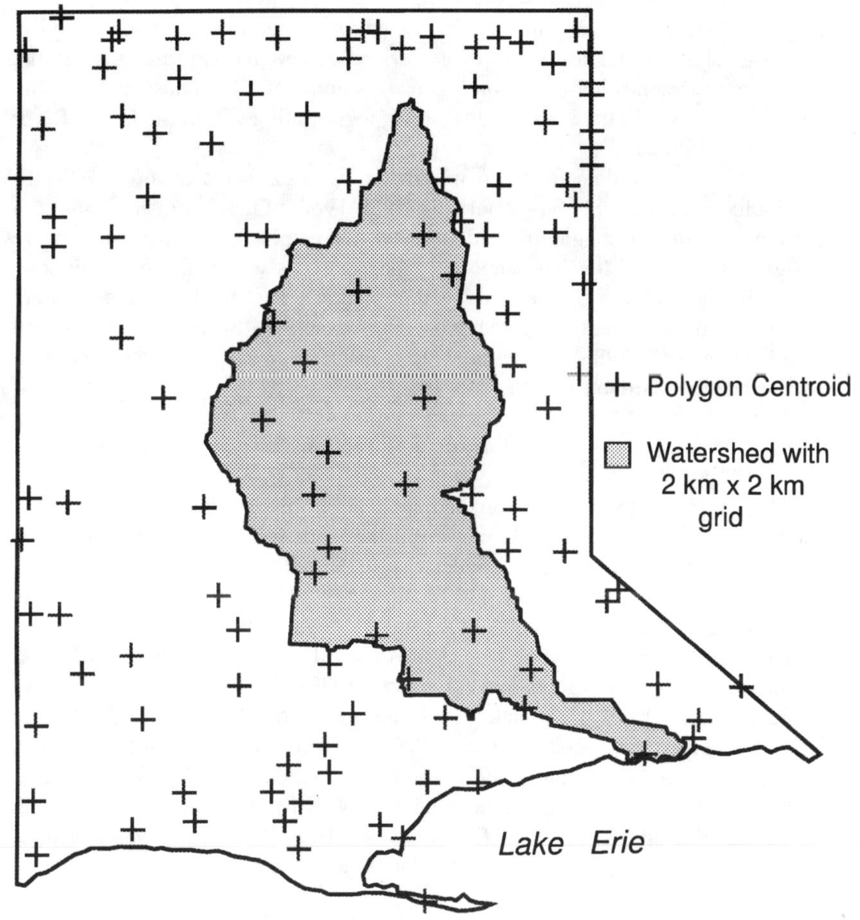

Fig. 61–3. Map window encompassing the Grand River watershed with locations of soil landscape polygon centroids.

Soil Survey and Weather Data

The input requirements of the LEACHW model include (among other things) soil physical and hydraulic properties, and weather data. When the model is being applied to large areas, such as watersheds, the main sources of these data in Canada are the National Soil Data Base (NSDB) and the Archived Weather Data Base (AWDB).

The soil input data required for the Grand River watershed application was identified and extracted from the NSDB on the basis of the dominant soil type in the 1:1 million scale soil landscape polygons (Soil Landscapes of Canada, Shields et al., 1991) that fell within the map window. The data for each polygon were extracted for 3 soil layers (approximately A, B and C horizons), and included the upper and lower depths of the layer; sand, silt and clay contents; bulk density; organic carbon content; saturated hydraulic conductivity; and 2-4 points on the soil water characteristic. The data were assigned to the georeferenced centroids of the polygons. A total of 119 landscape polygon centroids fell within the map window, 18 of these falling within the Grand River watershed (Fig. 61–3).

The climate data for the map window were extracted from the AWDB, and included the monthly 30–yr normals (1950-1980) of maximum and minimum air temperature, precipitation, and days with precipitation. These data were assigned to each of the landscape polygon centroids within the map window, using the values from the nearest weather station. The monthly normals were converted to daily data using the procedures described by Brooks (1943) and Van Diepen et al. (1988). Potential evapotranspiration was calculated according to Baier and Robertson (1965).

Pedotransfer Functions

For the Grand River watershed application missing data from the NSDB included: bulk density (61% missing); the soil water characteristic (61% missing); saturated hydraulic conductivity (91% missing); and the unsaturated hydraulic conductivity function (100% missing). These were estimated using pedotransfer functions based on soil texture and organic carbon (OC) content, for which there were no missing values in the NSDB. Bulk density (BD) was estimated using the Gupta and Larson (1979) model. The soil water characteristic, $\theta(h)$, was estimated using the model of McBride and Macintosh (1984), and then least squares fitted to the Van Genuchten (1980) function (Eqn. 1). Saturated hydraulic conductivity, K_s, was estimated using the model of Jabro (1992). The unsaturated hydraulic conductivity, $K(h)$, was estimated from Eq.2 using the $\theta(h)$ and K_s results. The estimated data, as with the available data, were all assigned to the georeferenced landscape polygon centroids.

Simulation and Output

The LEACHW - LEACHP simulations were conducted for all 119 soil landscape polygon centroids in the map window, using the appropriate soil and weather input data at each centroid. The simulations were run for 10 consecutive "simulation" yrs, assuming an initially atrazine - free soil profile and repeating the 30–yr normal weather each yr. Corn was grown every yr over the entire map window using the crop management scheme described above. Atrazine was applied each yr at the recommended rate of 150 mg/m^2 (Ontario Ministry of Agriculture and Food, 1993), 3 weeks after planting. A representative mean soil dispersivity (λ = 15.5 cm, unpublished field data) and atrazine partition coefficient (K_{oc} = 160 ml/g, Jury et al., 1984) was assumed for all soil types and depths in the soil profile. Atrazine dissipation rate constants were determined using Table 61–1 and the procedures described above. A constant water table depth of 120 cm was assumed because of inadequate water table data in the NSDB. The predicted annual mass loading of atrazine at the 90 cm depth at the end of the 10–yr simulation was used as an estimate of ground water contamination.

Geostatistical Analyses

A geostatistical technique known as kriging was used to account for spatial variability when extending the soil and weather input data and model predictions from a point basis to an areal basis (e.g. watershed). Kriging is essentially a weighted moving-average technique for interpolating between known data values at georeferenced locations, with the weighting factors determined from a semivariogram. The latter characterizes the spatial variability of the known values by defining both the maximum distance over which the values are related to each other (i.e. the range of the semivariogram), and the functional nature of this relationship (i.e. the shape of the semivariogram).

Kriging was used within the modelling-geostatistics-GIS framework to convert the relatively small number of irregularly spaced and highly variable point values of soil properties, weather attributes and predicted agrochemical loadings, into a large number of interpolated values that extend throughout the area of interest on a regular, fine-mesh grid. The interpolations also provided the required extension from a point basis to an areal basis, and also retained the spatial variability characteristics of the original data sets. The interpolations provided the spatial detail necessary for effective use of a GIS.

Semivariograms of soil properties (texture, BD, OC, θ_s, θ_r, K_s, α, n), precipitation (spring, summer, fall, winter) and predicted atrazine loadings (mg atrazine/m^2/yr) were calculated using the 119 locations in the map window. The data were then kriged (interpolated) on a 2 km x 2 km grid to produce a total of 7381 georeferenced grid points (1657 within the watershed) containing estimates of soil hydraulic properties, precipitation and predicted atrazine loadings. These kriged data formed the input to the GIS.

Geographic Information System (GIS)

The main function of the GIS was to produce, quantify and overlay maps of the kriged soil, weather and agrochemical loading data. This allowed estimation of the importance and distributions of agrochemical contamination of

Table 61–2. Basic statistics for the Grand River watershed application, based on the 18 landscape polygon centroids within the watershed. Thickness = layer thickness; remaining parameters defined in text.

Parameter	Unit	Mean	CV	Min. Value	Max. Value	Skewness	Kurtosis
LAYER 1							
Thickness	cm	14.4	33.3	5.0	25.0	0.204	2.60
Sand	g kg^{-1}	32.9	68.9	11.0	75.0	0.901	2.13
Silt	g kg^{-1}	43.8	32.9	17.0	64.0	-0.682	2.13
Clay	g kg^{-1}	23.2	54.1	8.0	45.0	0.345	1.75
BD	g cm^{-3}	1.34	11.8	1.0	1.57	-0.433	2.11
OC	g kg^{-1}	1.83	36.0	0.5	3.10	-0.218	2.18
θ_s	g kg^{-1}	49.5	13.9	40.0	63.0	0.227	1.80
K_s	cm/s	5.8E-4	156.7	2.7E-5	3.8E-3	2.537	9.22
LAYER 2							
Thickness	cm	16.7	48.3	5.0	30.0	0.525	1.79
Sand	g kg^{-1}	34.5	72.5	4.0	80.0	0.693	2.07
Silt	g kg^{-1}	38.8	34.7	15.0	62.0	-0.361	2.16
Clay	g kg^{-1}	26.7	65.0	5.0	61.0	0.422	1.98
BD	g cm^{-3}	1.45	9.0	1.25	1.70	0.311	1.98
OC	g kg^{-1}	1.21	47.4	0.17	1.91	-0.306	1.63
θ_s	g kg^{-1}	45.7	11.2	35.9	56.2	-0.116	2.50
K_s	cm/s	1.2E-4	153.4	1.2E-7	8.2E-4	3.217	12.76
LAYER 3							
Thickness	cm	88.3	11.2	70.0	100.0	-0.213	1.67
Sand	g kg^{-1}	33.9	85.1	3.0	87.0	0.766	2.00
Silt	g kg^{-1}	32.1	43.9	9.0	64.0	0.206	2.61
Clay	g kg^{-1}	34.1	65.1	4.0	64.0	0.015	1.33
BD	g cm^{-3}	1.50	8.1	1.30	1.71	-0.143	2.31
OC	g kg^{-1}	0.62	64.0	0.10	1.72	0.977	3.90
θ_s	g kg^{-1}	42.1	11.1	35.5	53.0	0.779	2.92
K_s	cm/s	6.5E-5	79.7	1.2E-7	1.6E-4	0.418	1.95
Atrazine Loading	$mg/m^2/yr$	0.50	136.9	0.0	1.88	0.956	2.30

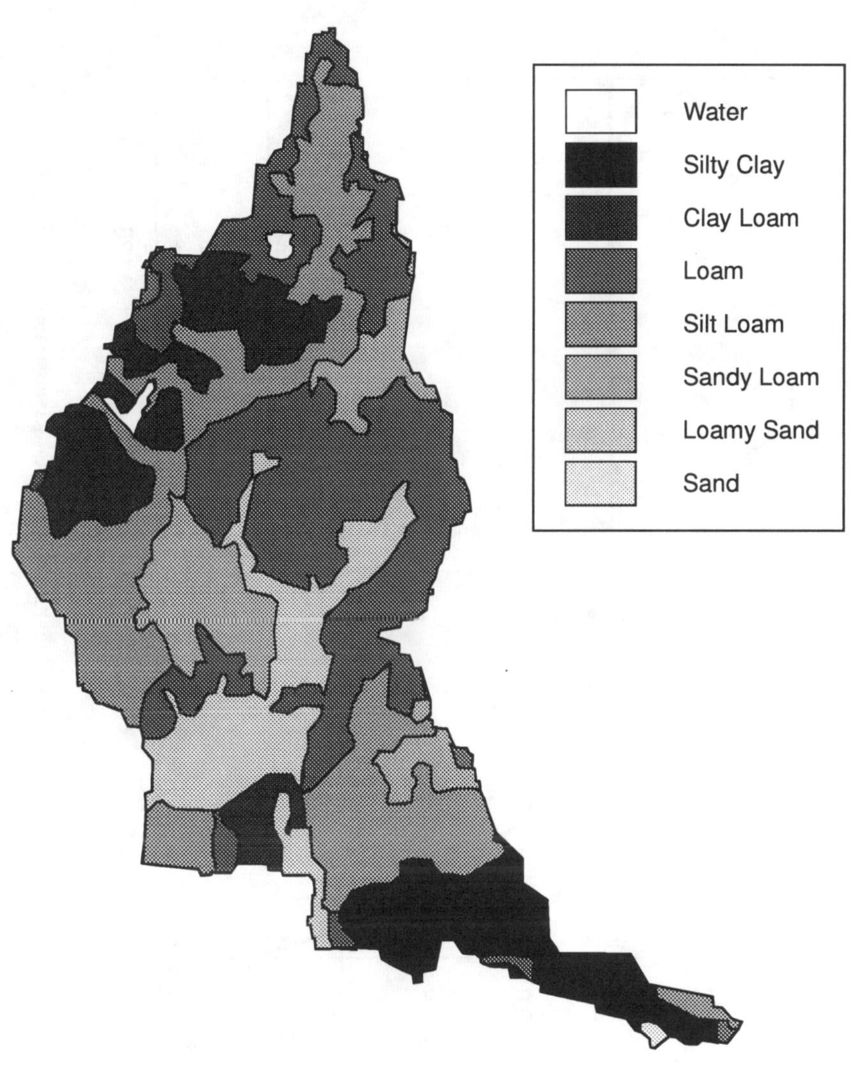

Fig. 61–4. Surface texture map of the Grand River watershed.

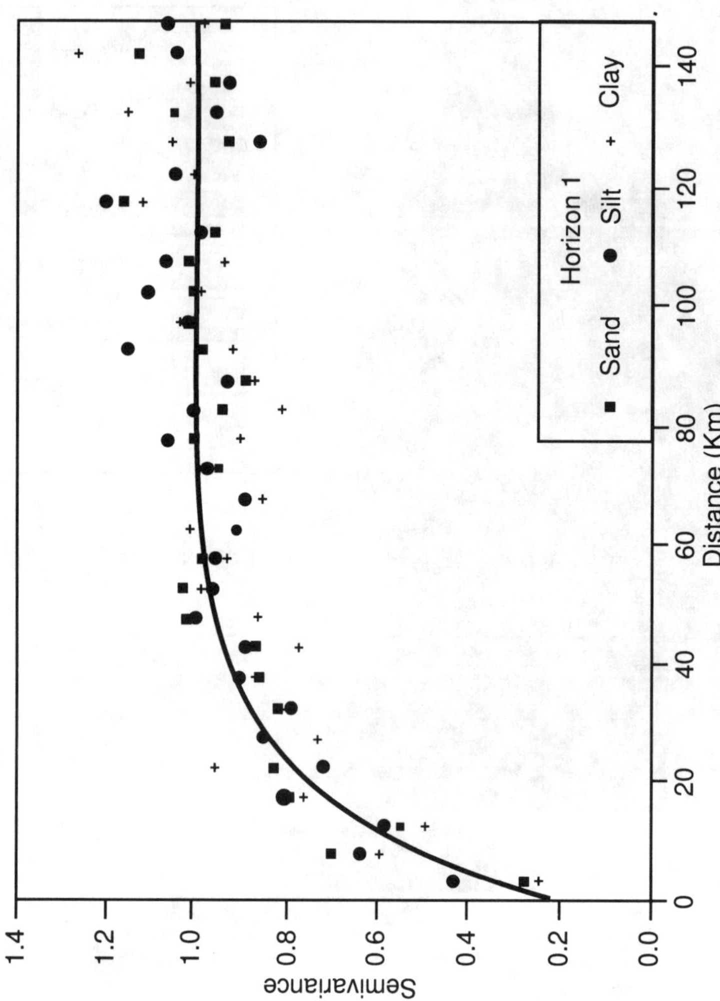

Fig. 61–5. Semivariogram for the A horizon soil texture in the Grand River watershed, scaled according to variance.

ground water, as well as determination of the soil, weather and land management factors that control the contamination. The GIS used in this study was the Integrated Land and Water Information System (ILWIS, version 1.3), developed at the International Institute for Aerospace Survey and Earth Sciences (ITC), Enschede, The Netherlands.

RESULTS AND DISCUSSION OF THE GRAND RIVER WATERSHED APPLICATION

Soil texture (sand, silt, clay content), bulk density and θ_s exhibited moderate to high variability across the watershed (CV = 7.1 to 85.1 %), but only modest changes with depth (Table 61–2). The high lateral variability in soil texture is also reflected in the surface texture map for the watershed (Fig. 61–4), which shows a wide range in soil types (sand to silty clay), as well as very complex spatial distributions. The soil texture semivariograms for the 3 soil horizons (A horizon shown in Fig. 61–5) are similar, with a small nugget (20–40% of the variance) and a correlation distance (range) of about 60 km. The similarity between the semivariograms, coupled with the extreme and intricate lateral variability in soil texture, probably reflects the complex glacial origin of most soils in this watershed.

The OC and K_s values are moderately to extremely variable across the watershed (36.0% $\leq$ CV $\leq$ 64.0% for OC; 79.7% $\leq$ CV $\leq$ 156.7% for K_s) at any particular depth, and also decrease substantially in mean value with increasing depth (Table 61–2). The K_s distributions for soil layers 1 and 2 also have very large positive skewness and kurtosis values (2.537 $\leq$ skewness $\leq$ 3.217; 9.22 $\leq$ kurtosis $\leq$ 12.76), indicating that many low K_s values exist close to the mean value, many large K_s values exist far above the mean value, and relatively few K_s values fall in between. The decrease in mean OC and K_s with increasing depth is not surprising because of the usual decrease in biological activity and soil structure with depth. The decrease in K_s with increasing depth may favor reduced atrazine contamination of ground water by decreasing the pore water velocity (q/θ) and thus allowing more time for degradation, adsorption, etc. This may be largely offset, however, by an accompanying decrease in atrazine retardation with depth due to the rapid decrease in OC and hence atrazine sorption.

The simulated atrazine loadings across the watershed are highly variable (CV = 136.9 %) and form a statistical distribution that is positively skewed (skewness = +0.956) and flat (kurtosis = 2.30), which indicates that many high loading values exist far above the mean value. The loading distribution also appears to be multimodal, as several peaks occur in the loading histogram and many of the histogram classes contain no values. In contrast to this extreme and complex spatial variability, is the temporal (yr to yr) variability which declines to zero as the predicted annual atrazine loadings become constant at any particular location after about 5–8 simulation yrs (example given in Fig. 61–6). Evidently, the spatially and/or temporally distributed weather and soil attributes interact in such a way as to enhance the spatial variability, but eliminate the annual variability, of atrazine loading at the 90 cm depth.

It should also be noted from the example soils in Fig. 61–6 that the rate and path by which atrazine loading stabilizes, as well as the final loading value, appear to be determined by complex interactions among weather, soil properties and solute transport mechanisms. The predicted atrazine loadings all start at zero, reflecting the fact that initially atrazine-free soil profiles were assumed. For a few soils (e.g. Haldimand clay), the loadings stay at zero for the entire 10 yr simulation, which implies that the atrazine applied to these soils is either degraded entirely or sufficiently retarded in its movement that it does not reach the 90 cm depth after 10 simulation yrs. The majority of soils, however (e.g. Fox sand, Guelph loam, Huron clay loam), contribute increasing quantities of atrazine with time until a plateau is reached after about 5-8 simulation yrs. These constant final loadings are somewhat soil dependent, increasing with coarser textures, but exceptions are frequent. For example, Fig. 61–6 shows that the annual atrazine loading is initially greater in the Fox sand (yr 2 $\rightarrow$ 4) than in the Guelph loam, which is consistent with the much higher sand content of the Fox soil ($\approx$ 80% sand in Fox sand; $\approx$ 35% sand in Guelph loam). After 5 yrs, however, the trend reverses, and the Guelph loam contributes a greater final annual atrazine loading than the Fox sand. This can be explained in terms of interacting water and solute transport properties. The average atrazine migration velocity at field capacity ($h = -10$ kPa), which depends on q, K(h), θ(h), bulk density, F, atrazine sorption, etc., is about 3 times greater in the Fox sand than in the Guelph loam. Atrazine therefore reaches the 90 cm depth sooner in the Fox sand. The concentration of atrazine in solution, however, is greater in the Guelph loam than the Fox sand, due to an $\approx$22% lower average atrazine sorption for Guelph loam (22% lower OC content). The final annual loading is consequently greater (by about 14%) for the Guelph loam soil.

An ILWIS - generated map of kriged atrazine loading throughout the Grand River watershed is given in Fig. 61–7. It confirms, both the high variability and the complex spatial distribution of loadings indicated in Table 61–2. Visual comparison of this map with the surface texture map (Fig. 61–4) and the summer (June, July, August) precipitation map (Fig. 61–8) shows that the lowest atrazine loadings (0 - 0.1 mg/m^2/yr) tend to correlate with clayey soils and low summer precipitation; and the intermediate to high loadings (0.8 - 2.5 mg/m^2/yr) tend to correlate with sandy to loamy soils and moderate to high summer precipitation. Correlation analysis shows further that atrazine loading is significantly correlated with many soil and weather attributes, but the magnitudes of these correlations are low. This supports the indication in Fig. 61–6 that atrazine loading tends to be determined by complex interactions among several soil, weather, crop management and solute transport factors, rather than by one or two dominant factors.

The predicted atrazine loading to the ground water in the Grand River watershed (based on the ILWIS compilation of the 1657 kriged loading values) ranges from 0 to 2.5 mg/m^2/yr (Fig. 61–7) with a mean value of 0.67 mg/m^2/yr. The maximum and mean loadings given here are somewhat higher than those in Table 61–2, because the kriging interpolations take into account the LEACHP-simulated loadings at all 119 polygon centroids in the map window (Fig. 61–3), several of which are considerably higher than the loadings for the 18 centroids

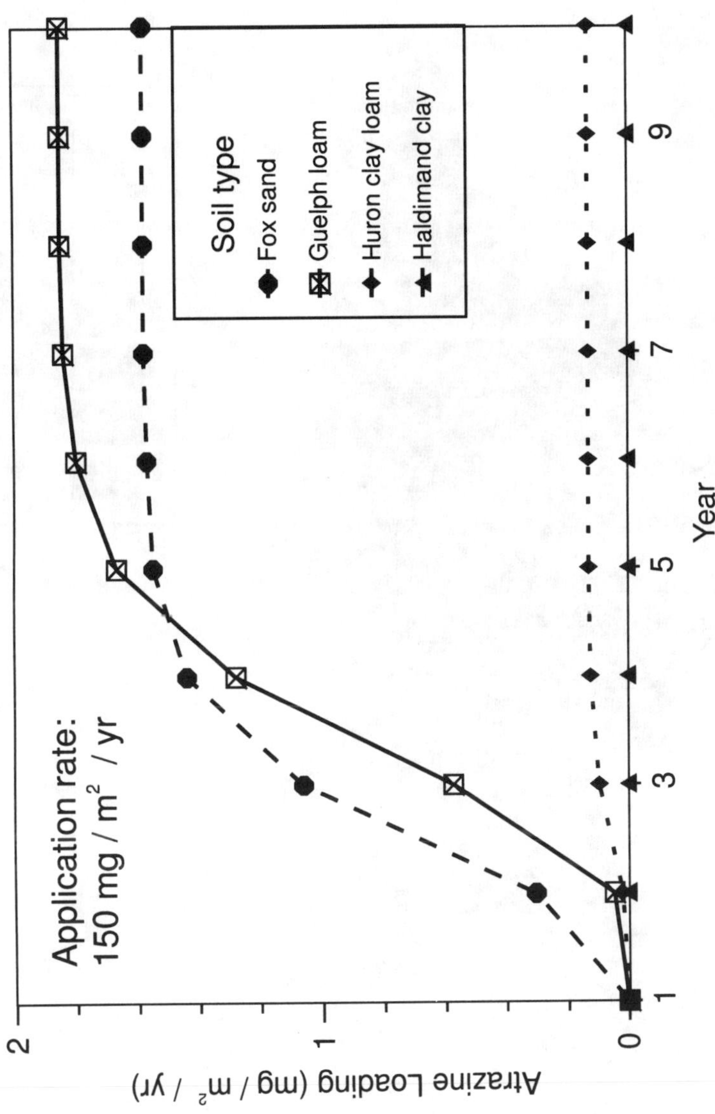

Fig. 61–6. Predicted atrazine loading (mg/m^2) versus time (yr) for Fox sand, Guelph loam, Huron clay loam and Haldimand clay.

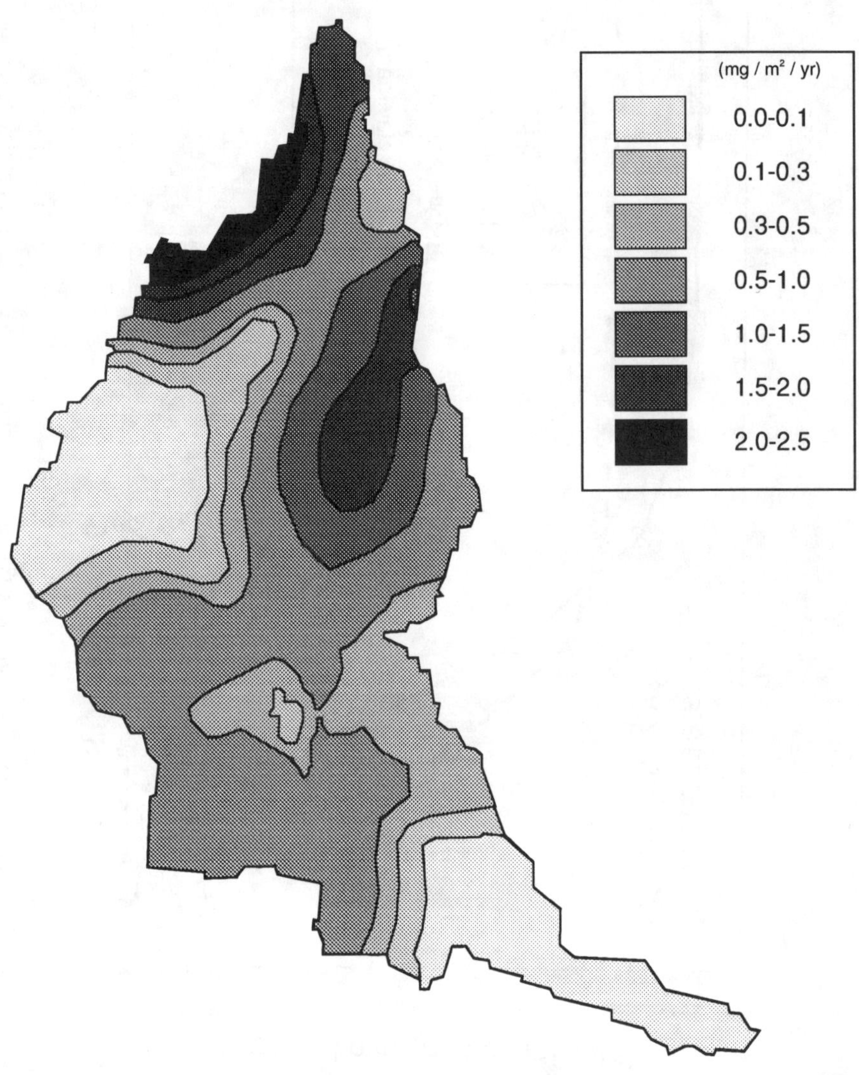

Fig. 61–7. Predicted annual atrazine loading ($mg/m^2/yr$) at the 90 cm depth for the Grand River watershed.

within the watershed. (The maximum LEACHP-simulated loading in the map window was 8.29 mg/m^2/yr.) Both the mean and maximum predicted atrazine loadings for the watershed (i.e. 0.67 mg/m^2/yr and 2.50 mg/m^2/yr, respectively) are quite low relative to the application rate of 150 mg/m^2/yr, suggesting that atrazine sorption and dissipation are extensive within the soil profile. The total predicted atrazine loading to the ground water (90 cm depth) for the watershed is estimated (via ILWIS) to be 4500 kg/yr, which is only 0.44% of the total specified surface application of 1.02 million kg/yr.

The concentration of atrazine in the soil water at the 90 cm depth are also predicted to be generally low throughout the watershed. The former 60 ppb Canadian drinking water guideline for atrazine (Canadian Water Quality Guidelines, 1989) was never exceeded at the 90 cm depth during the 10–yr simulation. The 3 ppb USEPA standard (USEPA, 1987) was exceeded, however, on or before the 10th simulation yr in about 27% of the watershed area (Fig. 61–9). The areas where this occurs also have predicted annual atrazine loadings that fall within the top half of the range (0.5 - 2.5 mg/m^2/yr, Fig. 61–7), which suggests that atrazine concentration and loading rates are related, but this relationship is by no means direct or simple. It also suggests that the areas where the 3 ppb concentration is exceeded (Fig. 61–9) represent regions of potentially significant low-level non-point source contamination of ground water by the downward percolation of atrazine through the soil profile. Figure 61–9 thus demarks regions in the watershed where more detailed investigation may be warranted.

The Grand River watershed predictions are consistent with the results of a recent survey of ground water quality in Southern Ontario (Agriculture Canada, 1992), which included over 900 farm water supply wells (~100 fell within the Grand River watershed) and 144 monitoring wells (≈11 fell within the watershed). The survey concluded (as did the predictions) that non-point atrazine contamination of ground water was infrequent, highly variable spatially, and low-level; that the former Canadian drinking water guideline for atrazine (60 ppb) was never exceeded; that the USEPA drinking water standard (3 ppb) was exceeded occasionally; and that the contamination was not strongly related to soil type or land use, implying control by many interacting soil, weather and land use factors rather than one or two dominant factors. This lends credibility to the LEACHM-Kriging-ILWIS methodology, notwithstanding that more extensive comparisons with field data are required before definite conclusions can be drawn.

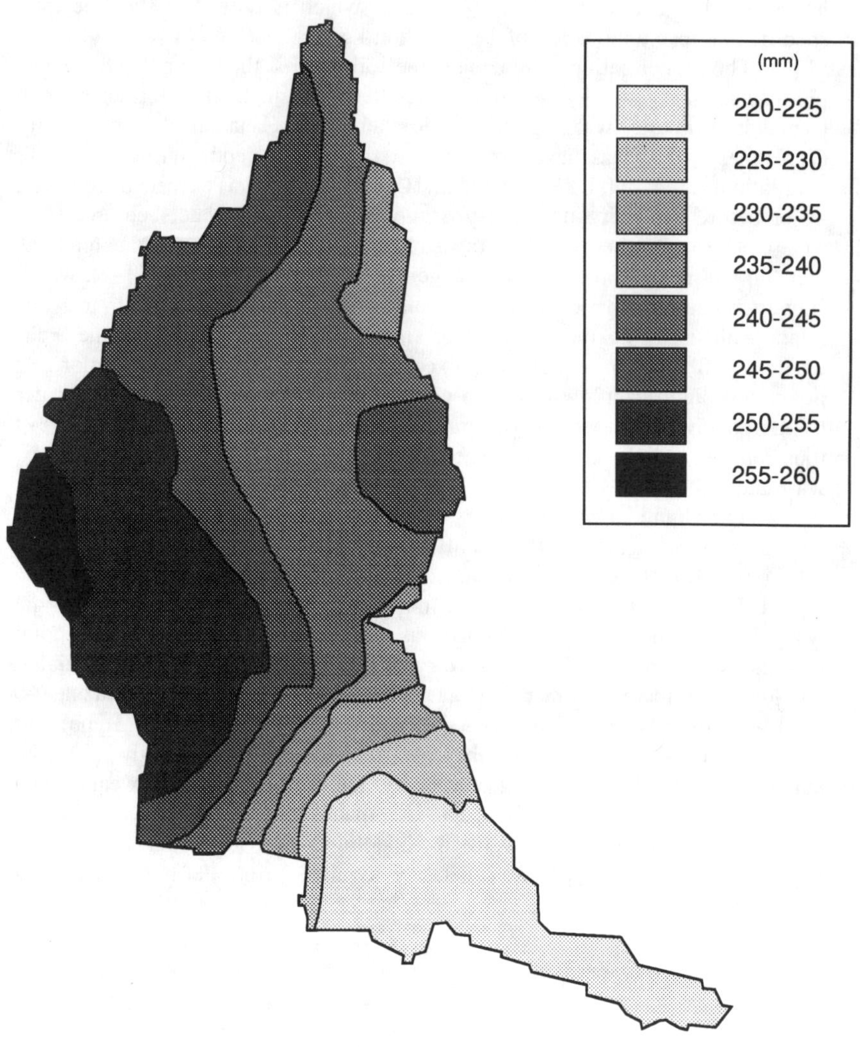

Fig. 61–8. Thirty–yr normal (1950-1980) summer precipitation (June, July, August) for the Grand River watershed.

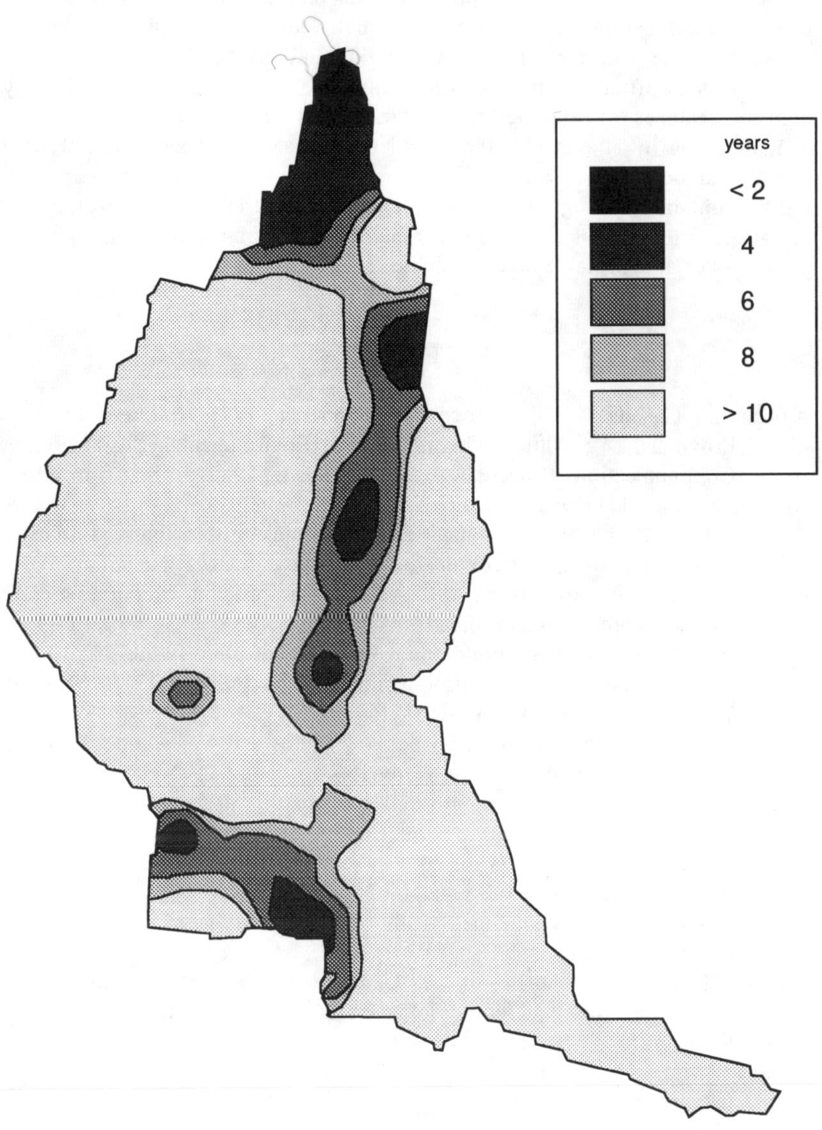

Fig. 61–9. Predicted time (yrs) for atrazine to reach 3 ppb at the 90 cm depth in the Grand River watershed.

CONCLUSIONS

Although the LEACHM - pedotransfer function - geostatistics - GIS methodology is still under development, the preliminary results are very encouraging. The required input data was extractable, or derivable (via pedotransfer functions) from information archived in the NSDB and AWDB databases. The pedotransfer function, kriging and ILWIS manipulations were effective and sufficiently robust to accommodate small map scales and fairly high percentages of missing data. Predictions of potential ground water contamination by atrazine for the Grand River watershed seem plausible and compare favourably with recent ground water survey results. It is consequently felt that this methodology will ultimately prove very useful in the development of agricultural practices and guidelines that maintain agrochemical inputs to the groundwater at acceptable and sustainble levels.

REFERENCES

Agriculture Canada. 1990. Report to Ministers of Agriculture. Federal-Provincial Agriculture Committee on Environmental Sustainability, Communications Branch, Agriculture Canada, Ottawa, Ont., Canada.

Agriculture Canada. 1992. Ontario farm groundwater quality survey, Winter 1991/92. Federal-Provincial Environmental Sustainability Initiative, Research Branch, Agriculture Canada, Ottawa, Ont., Canada.

Baier, W., and G.W. Robertson. 1965. Estimation of latent evaporation from simple weather observations. Can. J. Plant Sci. 45: 276–284.

Beven, K.J. 1991. Modeling preferential flow: an uncertain future? Pp. 1–11. *In* T.J. Gish and A. Shirmohammadi (ed.) Preferential flow: Proc. National Symposium. ASAE, St. Joseph, MI.

Bresler, E. 1973. Simultaneous transport of solutes and water under transient unsaturated flow conditions. Water Resour. Res. 9: 975–986.

Brooks, C.E.P. 1943. Interpolation tables for daily values of meteorological elements. Qual. J. Res. Meteorol. Soc. 69: 160–162.

Brown, D.M. 1976. Heat units for corn in southern Ontario. Ontario Ministry of Agriculture and Food, Guelph, Ont., Canada.

Canadian Water Quality Guidelines. 1989. Water Quality Branch, Environment Canada, Ottawa, Ont., Canada.

Dierckx, J., J.R. Gilley, J. Feyen, and C. Belmans. 1988. Simulation of soil water dynamics and corn yields under deficit irrigation. Irrig. Sci. 9:105–125.

Feddes, R.A., P.J. Kowalik, and H. Zaradny. 1978. Simulation of field water use and crop yield. Centre for Agric. Publ. and Docum., Wageningen, Netherlands.

Gupta, S.C., and W.E. Larson. 1979. A model for predicting packing density of soils using particle size distribution. Soil Sci. Soc. Am. J. 43: 758–764.

Hendrickx, J.M.H., and L.W. Dekker. 1991. Experimental evidence of unstable wetting fronts in homogeneous non-layered soils. Pp. 22–31. *In* T.J. Gish and A. Shirmohammadi (ed.) Preferential flow: Proc. National Symposium. ASAE, St. Joseph, MI.

Hutson, J.L., and R.J. Wagenet. 1989. LEACHM, Leaching Estimation And Chemistry Model. Version 2. Center for Environmental Research, Cornell Univ., Ithaca, NY.

Jabro, J.D. 1992. Estimation of saturated hydraulic conductivity of soils from particle size distribution and bulk density data. ASAE 35: 557–560.

Jury, W.A., W.A. Spencer, and J. Farmer. 1984. Behaviour assessment model for trace organics in soil: III. Application of screening model. J. Environ. Qual. 13:573–579.

McBride, R.A., and E.E. Macintosh. 1984. Soil survey interpretations from water retention data: I. Development and validation of a water retention model. Soil Sci. Soc. Am. J. 48: 1338–1343.

Millette, J.A., and M. Torreiter. 1992. Nonpoint source contamination of groundwater in the Great Lakes Basin: A review. Centre for Land and Biological Resources Research, Research Branch, Agriculture Canada, Ottawa, Ont., Canada.

Nimah, M.N., and R.J. Hanks. 1973. Model for estimating soil water, plant and atmospheric interrelations. I. Description and sensitivity. Soil Sci. Soc. Am. Proc. 37:522–527.

Ontario Ministry of Agriculture and Food. 1993. Guide to weed control. Publ. No. 75. Queen's Printer for Ontario, Toronto, Ont.

Shields, J.A., C. Tarnocai, K.W.G. Valentine, and K.B. MacDonald. 1991. Soil landscapes of Canada: Procedures manual and user's handbook. Land Resource Research Centre, Research Branch, Agriculture Canada, Ottawa, Ont., Canada.

Thomas, G.W., and R.E. Phillips. 1979. Consequences of water movement in macropores. J. Environ. Qual. 8: 149–152.

Tillotson, W.R., R.J. Robinson, R.J. Wagenet, and R.J. Hanks. 1980. Soil water, solute and plant growth simulation. Utah Agric. Exp. Stn. Bull. No. 502. Utah State Univ., Logan.

Topp, G.C., W.D. Zebchuk, and J. Dumanski. 1980. The variation of in situ measured soil water properties within soil map units. Can. J. Soil Sci. 60: 497–509.

USDA-Soil Conservation Service. 1972. National Engineering Handbook, Section 4, Hydrology, Washington, DC.

United States Environmental Protection Agency. 1987. Atrazine health advisory. USEPA, Office of Drinking Water. Washington, DC.

Van Diepen, C.A., C. Rappoldt, J. Wolf, and H. Van Keulen. 1988. CWFS Crop growth simulation model WOFOST. Documentation version 4.1. Centre for World Food Studies, Wageningen, Netherlands.

Van Genuchten, M.Th. 1980. A closed form equation for predicting the hydraulic conductivity of unsaturated soils. Soil Sci. Soc. Am. J. 44: 892–898.

Veenhof, D.W. 1993. The compression characteristics and susceptibility to compaction of soils from the Regional Municipality of Haldimand-Norfolk. Unpubl. M.Sc. thesis, Univ. of Guelph, Guelph, Ont., Canada.

Walker, A. 1978. Simulation of the persistence of eight soil-applied herbicides. Weed Res. 18: 305–315.

62 Vadose Zone Monitoring of Carbofuran Under Surge and Continuous Furrow Irrigated Conditions

Gary L. Bahr
Daniel C. Whitney

Idaho Department of Agriculture
Division of Agricultural Technology
Boise, Idaho

Within Idaho there are numerous important aquifer systems that are categorized as vulnerable to highly vulnerable (Rupert et al., 1991). Many of these vulnerable aquifers underlie intensive agricultural systems. Within these agricultural areas crop protection chemical utilization tends to be high. Surface water irrigation systems are also widely utilized. The combination of these agricultural practices creates a potential for impact to vulnerable ground water resources.

Proper management of pesticides and irrigation water are essential for ensuring that pesticide movement beyond the crop root zone does not occur. Development and implementation of Best Management Practices (BMPs) related to pesticide use and irrigation water management are essential. BMP monitoring and evaluation tools are key components to determining BMP appropriateness and effectiveness. The development of economical analytical techniques is also important. Ground water pollution potential resulting from pesticide usage and conventional furrow irrigation practices is a concern in areas such as the Snake-Payette Hydrologic Unit Area (HUA) in Western Idaho (Steed et al., 1993). Within this HUA, the Idaho Department of Agriculture (IDA) is part of an interagency effort to implement and evaluate the effectiveness of BMPs for ground water protection.

Snake-Payette Hydrologic Unit Area

The United States Department of Agriculture (USDA) is presently administering a water quality project in southwestern Idaho titled the Idaho Snake-Payette Water Quality Hydrologic Unit Area (HUA). The USDA Soil Conservation Service (SCS) and the University of Idaho Cooperative Extension System (CES) are managing the implementation aspects of this project. More than 500 000 farm acres are contained within the HUA boundaries (Fig. 62–1) in Canyon, Gem, Payette and Washington counties (USDA, 1991). Over 40 different crops are grown within the HUA. Small grains, alfalfa, and sugar beets account for ≈70% of the average annual crop acreage (Table 62–1) (USDA, 1991).

A major percentage of the HUA cropland is furrow irrigated (USDA, 1991). Furrow irrigation consists of applying water to cropland furrows under a continuous flow for a given irrigation set time (Yonts et al., 1991). Irrigation efficiency within the HUA averages about 35% with this method and measurements as low as 20% have been recorded at some farms (USDA, 1991). Extended durations of irrigation sets are common with furrow irrigation (usually 12 or 24 h) resulting in excessive percolation below the crop root zone, particularly at the upper end of the field (Goldhamer et al., 1987; Yonts et al., 1991). Such percolation may cause agrichemicals to move beyond the root zone and into ground water.

A majority of the HUA has been ranked as very vulnerable to ground water contamination (Rupert et al., 1991). Previous ground water quality studies conducted near the towns of Weiser, New Plymouth, and Fruitland indicate the occurrence of ground water contamination by nitrates and Dacthal (Baldwin, 1992, personal comm.). Detections of such pesticides support the vulnerability ranking (USDA, 1991).

An alternate approach to continuous furrow irrigation entails the use of surge irrigation (Bishop et al., 1981). Surge systems consist of a two-way tee valve installed in gated pipe which delivers water to the furrows in a cyclic, pulse-like manner. The surge technique has been used for a number of yrs in other states as a means for improving water advance rates (Walker et al., 1981; Goldhamer et al., 1987). Surge has been found to conserve water, reduce infiltration rates, and prevent deep leaching under certain field conditions (Malano, 1983; Izuno et al., 1985; Kemper et al., 1988).

Surge differs from continuous furrow in that water is applied intermittently from one set of furrows to another during the irrigation set (Miller et al., 1991; Bartholomay & Champion, 1991). The use of surge can result in an increase of irrigation efficiency (Miller et al., 1991; Goldhamer et al., 1987; Kemper et al., 1988) and a more uniform wetting along the furrow (Miller et al., 1987). Miller et al. (1991) showed a decrease in water usage and infiltration of >50% with surge as compared to continuous furrow.

The SCS and CES would like to promote surge irrigation as both a water quantity and water quality BMP to HUA growers (Stack, 1992, personal communication). With respect to pesticide leaching, the benefits of surge irrigation when compared to continuous furrow have not been studied within the HUA. Future studies in Idaho relating the benefits of surge and pesticide leaching potential are essential. Growers need assistance in meeting present and future goals to protect ground water. IDA is interested in supporting this effort through monitoring and BMP evaluation.

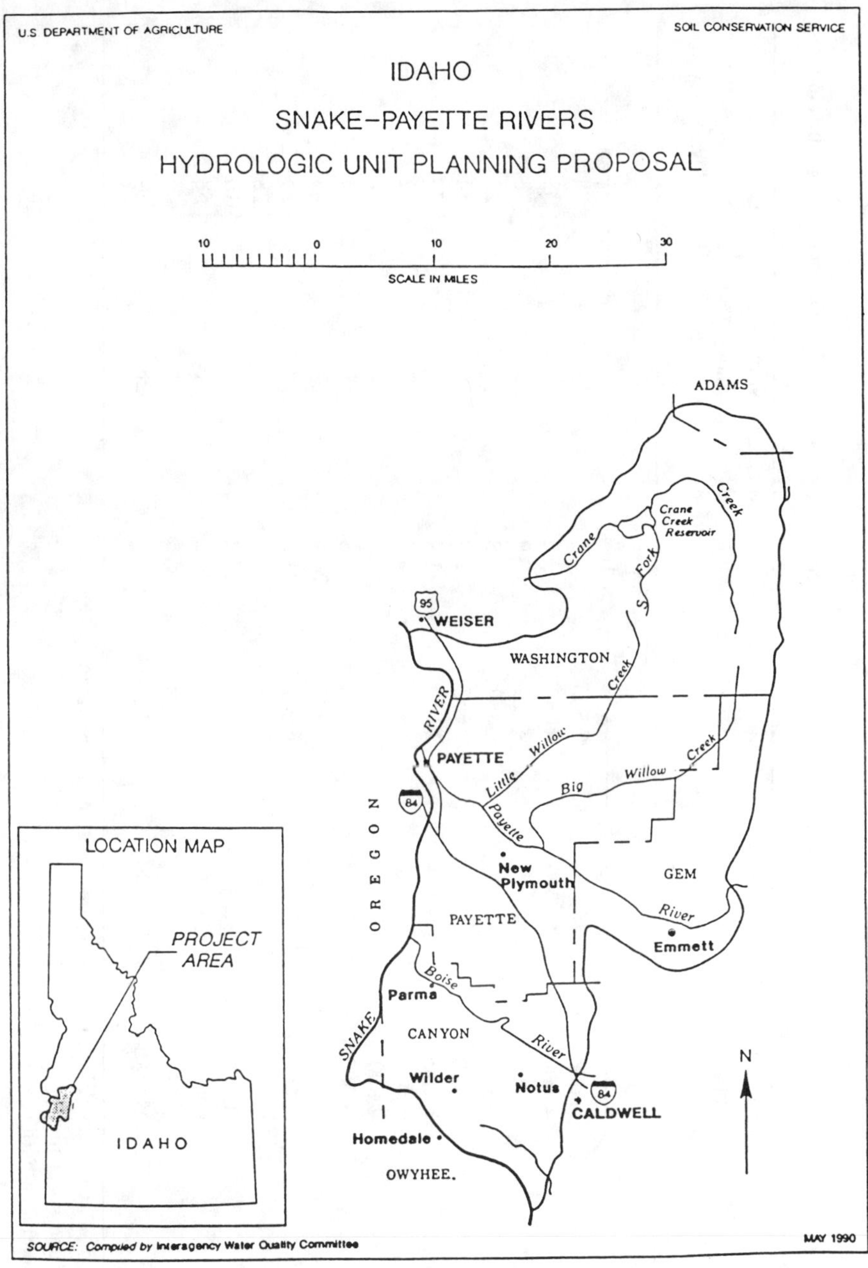

Fig. 62–1. Snake-Payette hydrologic Unit Area (HUA) Boundaries.

Table 62–1. Snake-Payette Rivers Hydrologic Unit Project 1992 cropping practices survey conducted in four southwest Idaho counties.

Crop	Four County Acreage†	Project Acreage	Survey Acreage	Gross Economic Dollar Value	Number of Management Units Surveyed	Estimated Number of Fields Surveyed
	---------- acres ----------			\$ per yr x 10^6	---------- # ----------	
Alfalfa hay	78,600	74,000	2,500	6.7	36	58
Alfalfa seed	24,000	14,000	652	13.5	17	32
Bean-dry edible	21,800	12,100	722	10.9	20	38
Corn-grain	11,900	11,900	458	2.8	12	22
-silage	21,100	14,000	652	2.9	18	34
-sweetcorn	3,600	3,000	1,065	1.9	40	75
-seed	10,500	4,000	951	14.0	17	32
Hops	2,800	1,800	397	8.4	9	9
Mint‡	18,000	13,000	318	19.8	12	23
Onion	7,900	7,700	700	24.9	23	43
Orchard	12,000	7,800	647	21.2	19	35
Potato	9,000	5,000	608	17.7	15	28
Smallgrain§	91,500	82,000	2,429	17.2	57	107
Sugarbeet	41,000	39,000	875	49.2	31	58
Total	354,000	290,000	12,974		328	594

† Canyon, Gem, Payette and Washington counties estimated 1991 acreages

‡ Peppermint and spearmint

§ Smallgrain includes barley, oats and wheat

(from Stieber et al., 1992)

Role of a State Pesticide Management Plan

The United States Environmental Protection Agency (EPA) is currently conducting a re-registration process for pesticides that were registered prior to current scientific and regulatory standards (USEPA, 1991a). The EPA is requiring registrants to submit pesticide specific environmental fate data. The EPA supports the assessment of pesticide leaching potentials and assessments of health and environmental risks from pesticides. Under the EPA's Pesticide and Ground-Water Strategy, use of pesticides which are determined to "generally pose unreasonable effects to the environment" due to ground water leaching will be restricted to those states which develop a State Pesticide Management Plan (SMP) (USEPA, 1991b).

SMPs are to include prevention and response measures, such as BMPs, that address the risks to ground water posed by pesticides (USEPA, 1991b). Evaluation monitoring will be required to assess the effectiveness of BMPs on protecting ground water when such practices are chosen for a SMP labeled pesticide.

Pesticide Monitoring For BMP Evaluations

Monitoring for pesticides in the vadose zone and ground water are key elements of BMP evaluation processes. Soil water monitoring instruments, such as lysimeters, moisture probes, tensiometers, piezometers, and soil probes have been utilized (Biggar & Nielson, 1976; Chow, 1977; Everts, 1989; Everts and Kanwar, 1988; Jordon, 1968; Quin & Forsythe, 1976; Zimmermann et al., 1978).

Lysimeter collection quantities can be low in volume. Gas chromatography and mass spectrometry (GC/MS) analytical methods commonly require hundreds of milliliters of sample for pesticide analyses and are expensive. Immunochemical pesticide analytical techniques have been developed for detecting and measuring a compound. The technique utilizes an antibody which binds only to a given chemical compound and its enzyme conjugate. Approximately one half milliliter of sample is required to perform some analyses. Immunoassays are usually more sensitive, take less time per analysis, and are much less expensive (≈10 times less) than conventional methods (Bushway et al., 1988a).

Immunoassay test kits have been developed for triazines (i.e., atrazine), aldicarb, 2,4-D, carbofuran, cyclodienes (i.e., chlordane), benomyl, and alachlor. Studies conducted by Baumann et. al. (1991) and Bushway et al. (1988b) using the atrazine and chlordane immunoassay test kits, respectively, indicate that immunoassays are a viable method for vadose zone monitoring. Vadose zone monitoring studies utilizing a carbofuran test kit are lacking.

Purpose and Objectives

Vadose zone monitoring is an essential component for evaluating the effect of surge irrigation on ground water quality. The IDA, in coordination with SCS and CES, conducted a vadose zone monitoring study within a grower's sugar beet field near Payette, Idaho which utilized the insecticide carbofuran. Carbofuran is registered in Idaho for use on alfalfa, small grains, and sugar beets which comprises a majority of the annual acreage within the HUA. Surge and continuous furrow irrigation were utilized and managed by the grower.

The purpose of this study was to identify and quantify soil pore water quality changes related to carbofuran by depth and field location within the upper, middle, and lower portion of field under both surge and continuous furrow irrigated conditions. Specific objectives were to: design and evaluate a vadose zone monitoring network utilizing tension lysimeters and soil cores; evaluate an immunoassay analytical technique for soil and soil pore water; evaluate the vadose water quality differences between surge irrigation and continuous furrow irrigation; describe the hydrogeology and soil conditions within the study area; support the development of irrigation BMPS; work cooperatively with the grower; and cooperate with growers and agencies to protect ground water within the HUA.

GENERAL HYDROGEOLOGY AND SOIL CONDITIONS

Hydrogeology

The Snake-Payette Hydrologic Unit Area is located within the Columbia Intermontane Plateau Province (Fig. 62–2). The geology of the HUA generally consists of three stratigraphic sequences: an uppermost unit, a sand and gravel unit, and the Glenns Ferry Formation (USDA, 1991). The uppermost unit consists of a soil mantle of sand and silt varying in thickness that overlies most of the area. Underlying the soil mantle is a sand and gravel deposit. This unit was deposited by the ancestral Snake, Boise, and Payette Rivers during periods of glacial runoff. The sand and gravel was deposited in valleys eroded into the underlying Glenns Ferry Formation. As the ancestral rivers cut into the underlying deposits, sand and gravel material was deposited at successively lower elevations, forming terrace deposits. The unit is up to 200 ft in thickness in some areas. Underlying the sand and gravel unit is a regional deposit of lake sediments referred to as the Glenns Ferry Formation. This unit is at least 5 000 ft thick and consists of sand, silt, and clay, with some interbedded gravel.

According to Steed et al. (1993), there are two different geologic regimes within the HUA: lacustrine and fluvial, and volcanic. The lacustrine and fluvial deposits are extensively utilized for agriculture, whereas the volcanic deposits are utilized as open range. The following is a description of these regimes as presented in Steed et al. (1993). "The lacustrine deposits are geologically located within the Western Snake River Plain which is an elongate feature that stretches from the town of King Hill on the east to the Idaho-Oregon border on the west. The Western Snake River Plain is a fault-bounded depression with normal northwest-trending fault systems forming major segments of both edges of the plain (Malde, 1965). The depression of this area is believed to be aided by the weight of dense Miocene basalts. Above the basalts are sediments from Pliocene stream and lake deposits of the Glenns Ferry Formation and younger alluvium and outwash. The Glenns Ferry Formation is the oldest and deepest

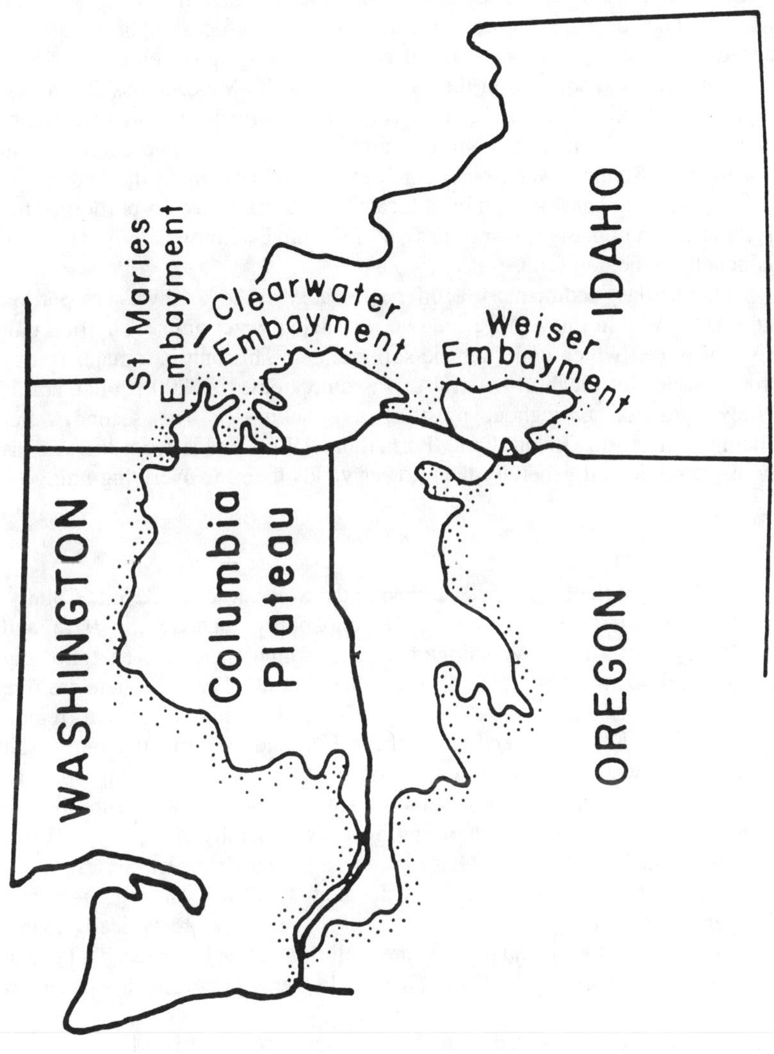

Fig. 62–2. Location map of the Columbia Pateau (taken from Camp et al. (1982).

unit in the lacustrine regime. The top of this unit had been eroded prior to the deposition of the overlying sediments. Above the unconformity produced by this erosion are Quaternary alluvium and outwash from the erosion of the adjacent uplands. Overlying the alluvium and outwash is the soil mantle which is described below.

The volcanic deposits are geologically located on the foot wall of the Western Snake River Plain. They consist of Miocene basalt flows and Miocene stream and lake deposits associated with volcanic episodes. The landforms produced by these deposits are rounded with low topographic relief.

The principle aquifers in the basin occur in the Miocene basaltic rocks, the overlying Tertiary sediments (Glenns Ferry Formation), and Quaternary sediments. Ground water occurs under artesian and water table conditions in these aquifers. Shallow water table conditions exist throughout the hydrologic unit. The Miocene basaltic aquifer is located in the northeastern portion of the hydrologic unit and is the least productive of the three. Consequently, very few wells penetrate the aquifer.

The Tertiary sedimentary aquifer is located in the southwestern portion of the HUA. Within this aquifer are two producing water units. The first unit consists of gravel which underlies the soil mantle. This unit is a major source of ground water for much of the HUA. In some areas, it is 200 ft thick and is generally saturated throughout most of its thickness. The second water producing unit is the Glenns Ferry Formation. Well yields from the Glenns Ferry are variable and generally have lower yields than the overlying unit."

Soils

The soils of the HUA are described in the Soil Survey of Payette County, Idaho (SCS, 1976), the Soil Survey of Canyon County, Idaho (SCS, 1972), and the Soil Survey of Adams–Washington County (unpublished). The following discussion is based on these references and Steed et al. (1993). "There are five general soil regions within the HUA (Fig. 62–3). The first soil region (region 1) is in the northernmost section of the HUA and is derived from basalt residuum. It is well drained and occurs on gently to steeply sloping uplands. A prominent characteristic of these soils is a well developed clayey subsoil with high shrink-swell potential. Soils in region 1 are generally deep. Typical soil classifications include the Gem, Glasgow, Newell, and Brownlee Series.

The next region (region 2) is located southwest of region 1 and interfingers with soil regions 3 and 4. Soils in region 2 are typically loams derived from alluvial sediments. They are well drained and very deep. Typical soil classifications include the Haw, Payette, Power, Purdam, and Van Dusen Series.

Soil region 3 is located further to the southwest and is adjacent to the Payette and Snake rivers in the HUA. The soils in this region are derived from lacustrine sediments and mixed alluvium and are typically silt loams and silty clay loams. The drainage class for these soils ranges from poorly to well drained. Soil classifications of region 3 include the Moulton, Baldock, Chilcoot, and Greenleaf Series.

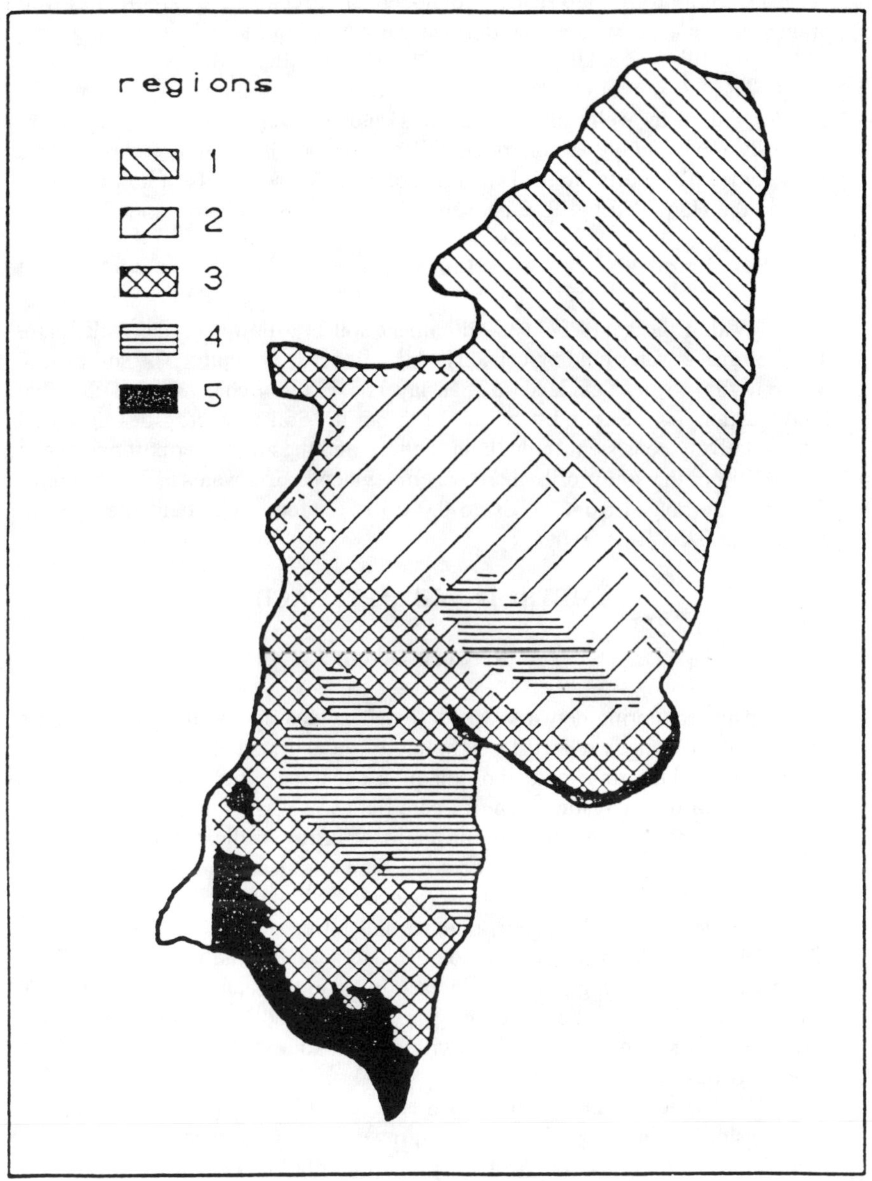

Fig. 62–3. Soil regions within the Snake-Payette HUA [taken from Steed et al. (1993)]

Soil region 4 is located on either side of the upper terraces of the Payette River. This soil region is derived from alluvium and loess. The soils in this region differ from the other soil regions by the dominance of fragmented duripans. Soil profiles are shallow to moderately deep and are classified as well drained in the areas without the duripan horizon. Typical soils in soil region 4 include the Elijah, Lanktree, Vickery, Chilcoot, and Letha Series.

The last region is located in the southern part of the HUA. These soils are derived from well mixed sediments and are very deep. Textures are generally more sandy than the other soil regions and the drainage class ranges from somewhat poorly drained to well drained. Typical soils in soil region 5 include the Harpt, Turbyfill, Cashmere, Cencove, and Feltham Series."

Site Soils

Within the study field the dominant soil is a Power-Purdam silt loam. This soil has a water holding capacity of 2.4 in per ft of depth. The soil profile is deep with no field evidence of pan layers. Textural changes toward a fine sand fraction can be found between 30 to 36 in. Soil characteristics appeared to be relatively consistent throughout the top, middle, and lower portions of the field. Within the soil profile there was no evidence of a water table. Slope of the field site ranged from 0.0010 to 0.0055 ft per foot. The field site average slope is 0.00225 ft per foot.

METHODS AND APPROACH

Network Design

The monitoring network was designed to identify and quantify soil pore water quality changes related to carbofuran by depth and field location within the upper, middle, and lower portion of a sugar beet field under surge and continuous furrow irrigation. The 14 acre field was divided in half. Surge was dedicated to the eastern half and conventional furrow dedicated to the western half of the field. Each irrigation type ran the entire length of the field, 820 ft. Three sets of paired sampling plots were located within each irrigation type and evenly located within the upper, middle, and lower sections of the field (Fig. 62–4). Pairing allows for comparison evaluation across the width of the field. There were six sampling plots within each irrigation type and twelve plots total (Fig. 62–4). The three field sections (i.e., upper, middle, and lower) represent three variable sections of the field with respect to vadose zone wetting front characteristics.

Soil-water quality monitoring was accomplished via soil core sampling and tension lysimetry. The owner–grower stated he had never applied carbofuran on the test field. Soil sampling for carbofuran occurred prior to the 1993 planting season to validate background levels. Fall soil sampling was conducted to compare carbofuran concentrations between soil samples and soil pore water taken from the last lysimeter sampling event.

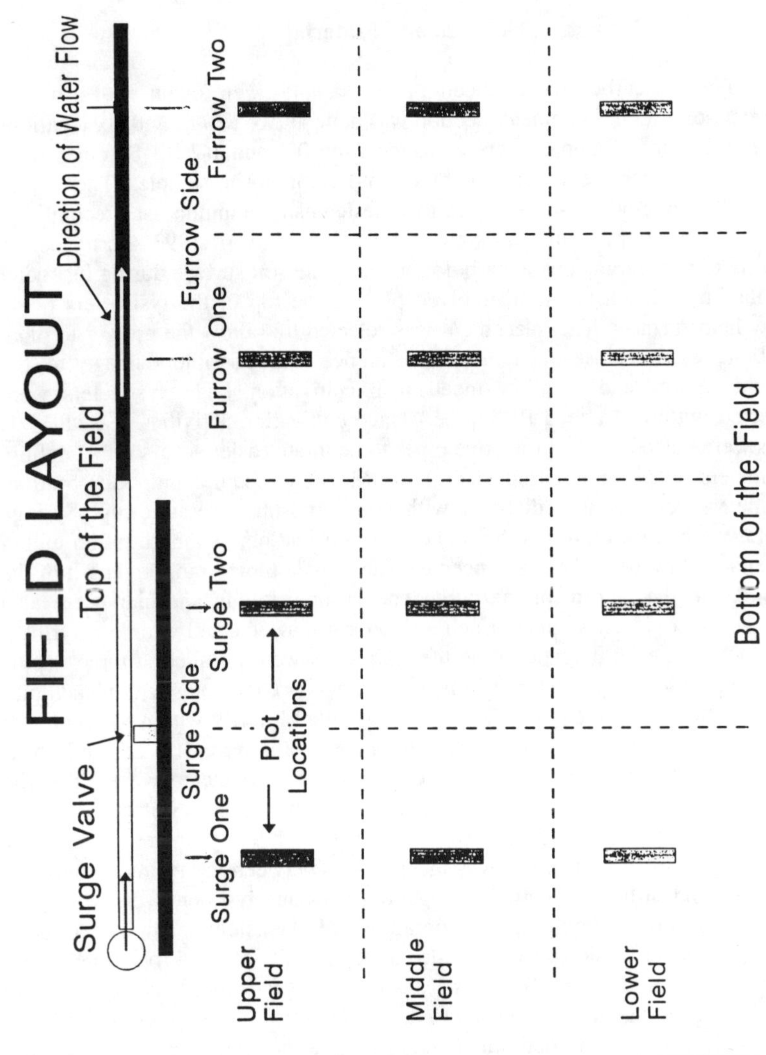

Fig. 62–4. Field layout design for the study.

Sampling with the vacuum lysimeters occurred five times during the 1993 growing season. The first sampling occurred after the first irrigation event and the subsequent samplings occurred after every other event. All soil and soil-pore water samples were collected and handled according to standard EPA QA/QC sampling protocol.

Methods and Materials

Prior to carbofuran application, soil samples were taken with a one in diameter soil probe. Sample collection was done at two depths and six locations within each plot. Samples were collected from 0-1 and 1-2 ft. Samples from each one foot increment were combined into a composite sample. There were a total of six composite soil samples to be tested using immunoassay techniques.

The vacuum lysimeters were installed on 21 April 1993 shortly before planting to avoid transferring carbofuran below the soil surface during lysimeter installation. The installation involved placing the top of the lysimeters ≈6 in below land surface. This placement was selected to: allow the grower to plant his field without instrumentation interference and possible damage to the instrumentation, and to allow mechanical cultivation. The lysimeters were protected within a capped PVC pipe. Once cultivation activities ceased, PVC risers were placed on the protective pipes to facilitate easier access for sampling. Lysimeters were purchased from SoilMoisture Equipment Corporation. Lysimeters were two in in diameter with a one atmosphere ceramic cup. Syringe needles were placed through the rubber stoppers and high strength teflon tubing was attached to the end of each needle. The sample tubing ran the length of the lysimeter to the bottom of the cup. The vacuum tubing and clap were also attached to the rubber stopper. Each lysimeter had a dedicated sampling syringe.

Within each irrigation type the lysimeters were installed at three depths within each plot of the eastern lysimeter row, and at two depths within each plot of the western lysimeter row. They were installed linearly within the center of each plot and spaced three ft apart. The three depths were 1.5, 2.5, and 5.0 ft below the soil surface (Fig. 62–5). The 1.5 and 2.5 ft lysimeters were within the crop root zone for sugar beets is ≈four ft. There were three lysimeters at each depth. The total monitoring network consisted of 90 lysimeters. A buffer consisting of 15 to 16 furrows was incorporated between the lysimeter network rows to avoid influences between irrigation types and lysimeter rows.

SoilMoisture Equipment Corporation Jet Fill tensiometers which measure in-situ soil-water tensions were installed at depths of two ft and four ft near each lysimeter network location (Fig. 62–5). A total of twenty-four tensiometers were installed. These devices were used to detect the wetting front after an irrigation event and determine the optimum time to sample the lysimeters. The granular carbofuran (Furadan® 15G) was applied on 22 April 1993 at planting, and at a label rate of 2 lbs per acre.

The grower managed the irrigation systems. The surge valve was placed at the head of the field and on the eastern half of the field. There were nine conventional furrow irrigation and 10 surge irrigation events through the growing season. All irrigation events were 24 h sets. The first surge irrigation began on

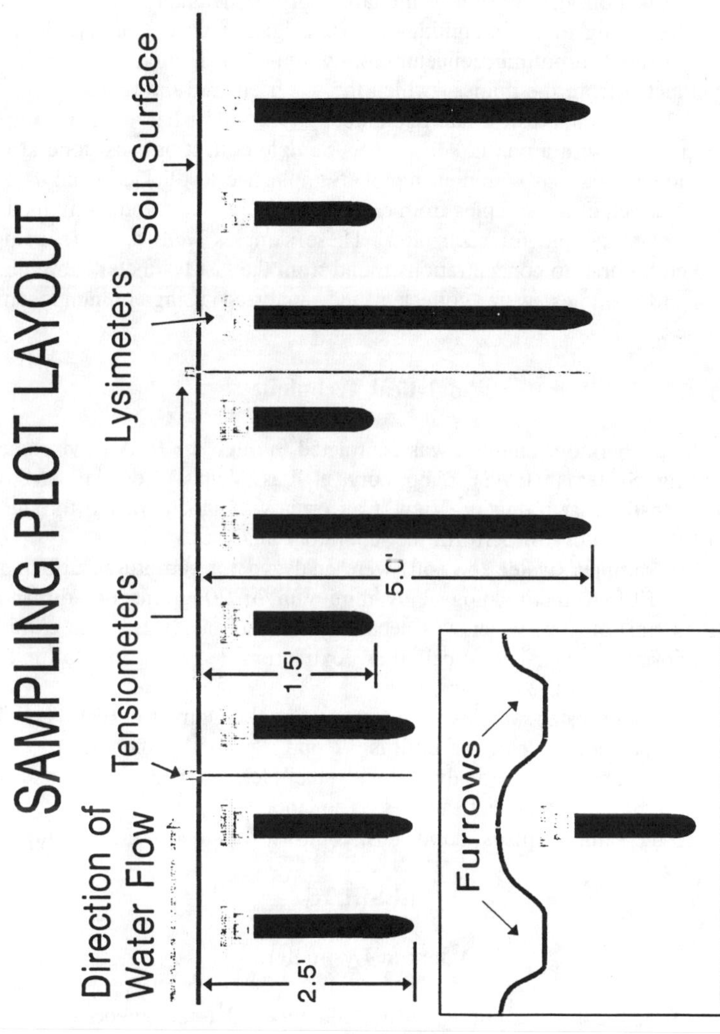

Fig. 62–5. Soil profile view of the lysimeter and tensiometer placement for each sample plot.

24 June and the first furrow irrigation on 25 June. Every other row, which were wheel rows, were irrigated.

Within 24 h after the end of each irrigation event, a 70 psi vacuum was applied to each lysimeter and held for 24 h. Each lysimeter was sampled using a dedicated syringe and samples were placed in 40 ml amber sampling bottles. Lysimeter samples along with sample duplicates, trip, spike, and transfer blanks were preserved on ice and sent to the laboratory immediately.

Monitoring of the incoming surface irrigation water was conducted to identify potential carbofuran concentrations within the irrigation water. Sampling was conducted from the point at which the water entered the field.

After crop harvest in late September 1993, a post harvest soil sampling event was done with a one in soil probe. Sample collection was done at three depths and six locations within each plot. Samples were collected from 0-1, 1-2, and 2-3 ft. Each of the samples from each of the increments were combined into one composite sample for each plot. These samples were taken to compare residual carbofuran to concentrations found from the last lysimeter sampling. A total of 48 samples were collected and analyzed using immunochemical techniques.

Analytical Technique

The laboratory analysis was contracted through the EPA Environmental Monitoring Systems (EMS) Laboratory at Las Vegas, Nevada. Midwest Research Institute at Mountain View, California was subcontracted through the EPA EMS laboratory to perform the laboratory analyses.

All samples (water and soil) were analyzed for carbofuran utilizing the Ohmicron ELISA methodology. A minimum of 20 grams of soil and 10 milliliters (ml) of pore water was necessary for each analysis. The extraction was performed by supercritical fluid extraction (SFE) prior to analysis. Detection limit was 0.2 ng/g.

Soil pore water samples were tested using the Ohmicron ELISA method. The immunochemical detection limit is 0.1 ppb. A small percentage of the soil pore water samples was analyzed with the electron capture method as a verification process for select samples. The laboratory performed QA/QC work related to the sample spikes, duplicates, controls, trips and transfer samples.

RESULTS

Vacuum Lysimeters

From the five sampling events and the 450 total potential lysimeters samples, there were 217 samples available for analysis. Carbofuran was detected in 52 of the 217 pore water samples. Continuous furrow samples accounted for 32 detections while surge samples accounted for 20 detections (Fig. 62–6 and 62–7). The number of samples extracted from the lysimeters located in the surge plots was somewhat equal across the length of the field (Fig. 62–6). The greatest number of lysimeter samples extracted were from the middle field

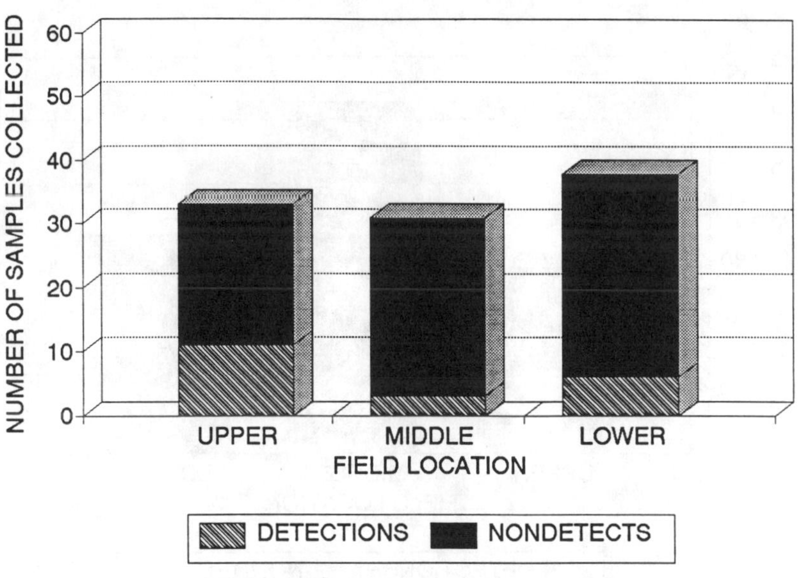

Fig. 62–6. Lysimeter samples collected by field location from the surge irrigation.

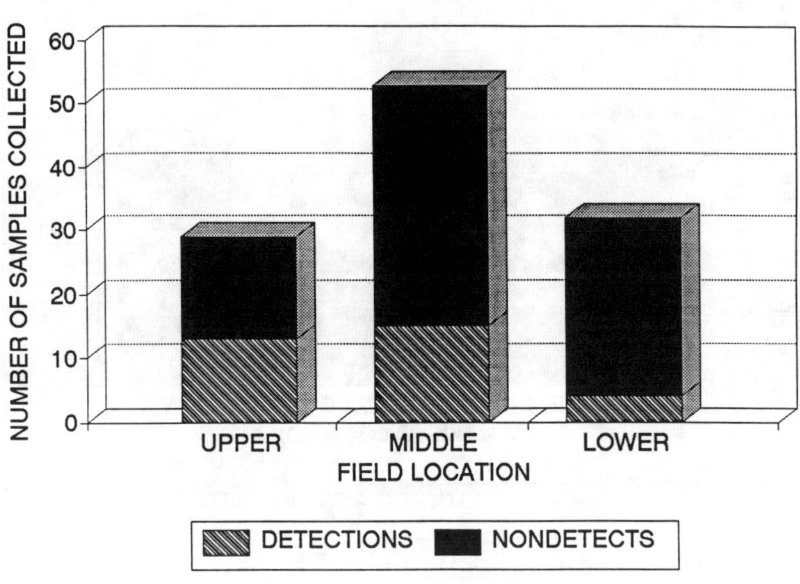

Fig. 62–7. Lysimeter samples collected by field location from the conventional furrow irrigation plots.

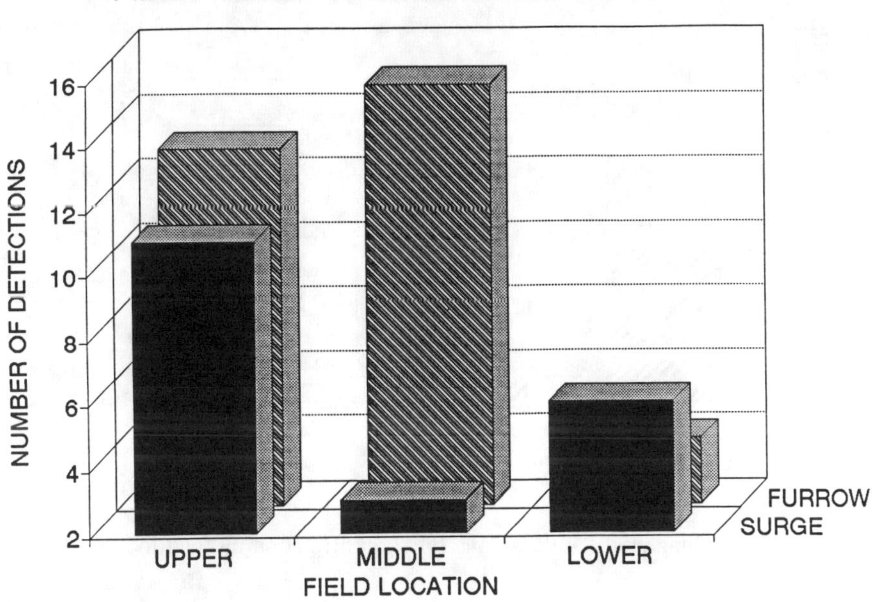

Fig. 62–8. Carbofuran detections from the various field locations from both irrigation types.

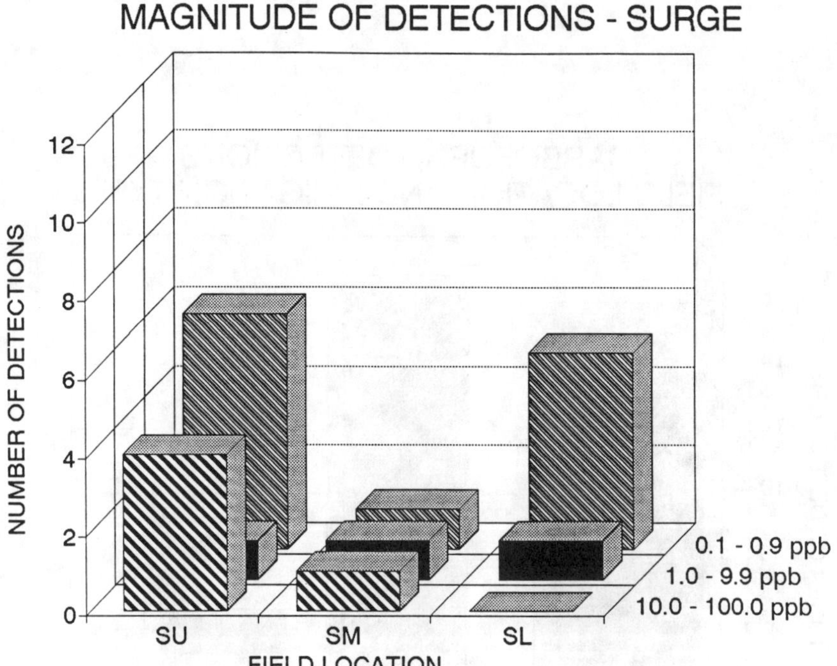

Fig. 62–9. Magnitude of detections for surge irrigation by field location.

locations within the conventional furrow irrigation system (Fig. 62–8). The number of surge detections were lower, as compared to the furrow, from the middle field lysimeter locations (Fig. 62–8).

The concentration levels of the carbofuran detections were relatively low in magnitude. The majority of the detections were in the 0.1 to 0.9 ppb range (Fig. 62–9 and 62–10). The highest detection was 58.9 ppb. The number of the furrow detections between 0.1 to 9.9 ppb were greater in number than the surge detections within this same range (Fig. 62–9 and 62–10).

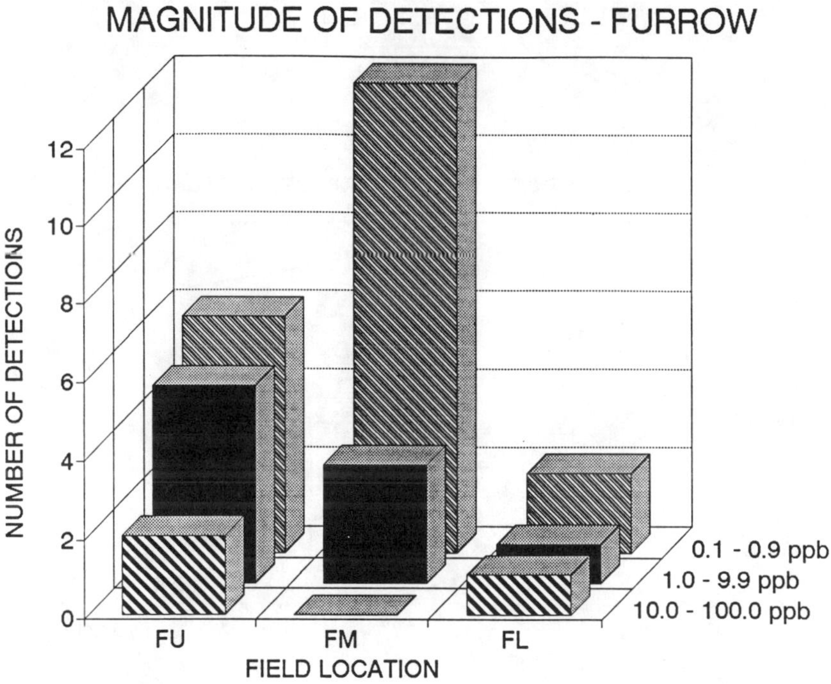

Fig. 62–10. Magnitude of detections for conventional furrow irrigation by field location.

By ranking all pore water detections by lysimeter depth and sampling date, the largest number of detections originated from the 2.5 ft lysimeters during the second and third sampling events (Fig. 62–11). The detections for the 2.5 ft lysimeters were consistently greater in number than the 1.5 ft lysimeters, for four out of the five sampling events (Fig. 62–11). The pattern in Fig. 62–11 indicates that transport of carbofuran was occurring during the irrigation season within the top 2.5 ft of the soil profile. The only detection of carbofuran within a five ft lysimeter occurred during the first sampling event. This detection was at a concentration of 0.01 ppb.

A Wilcoxon rank order test was performed for all carbofuran detections originating from the 1.5 and 2.5 foot lysimeters. To a level of confidence at 95%, the 2.5 foot lysimeter detections were consistently higher in concentration than the 1.5 foot lysimeters detections when placed in ranked pairs (Fig. 62–12). Of the samples that had detections, the 2.5 foot lysimeters consistently had higher numbers and concentrations of detections than the 1.5 foot lysimeters.

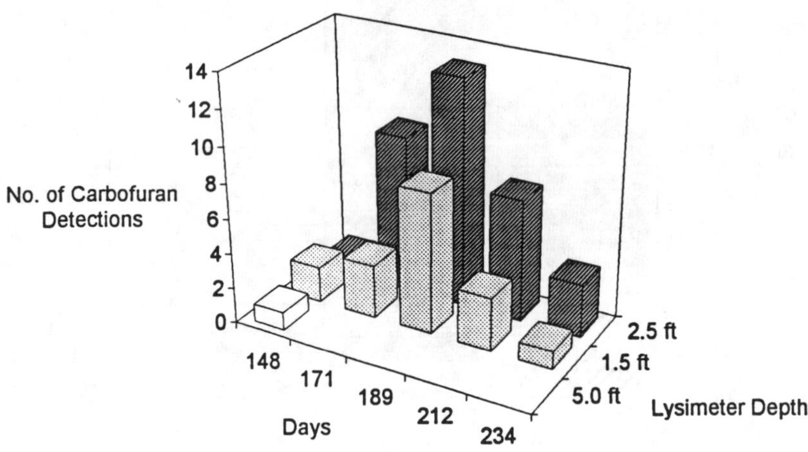

Fig. 62–11. Number of detections by lysimeter depth and time of sampling.

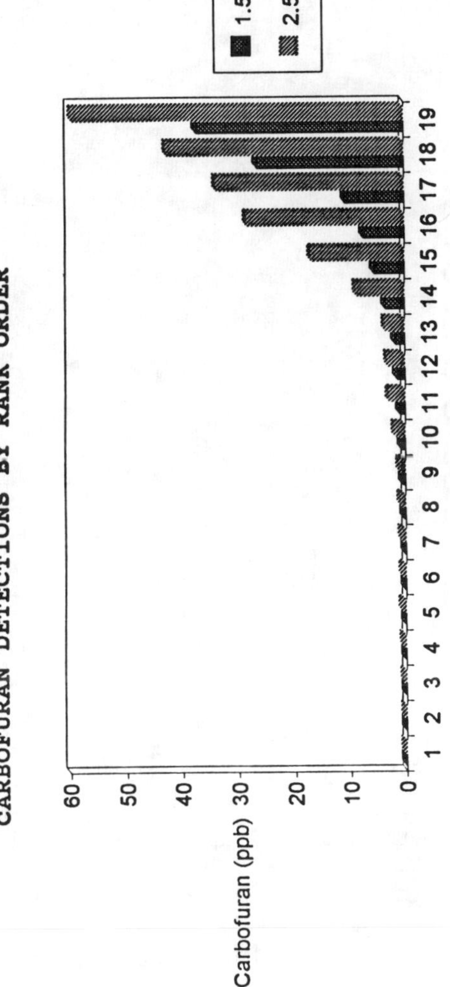

Fig. 62–12. Carbofuran detections by Wilcoxon rank order test for all 1.5 and 2.5 samples detected.

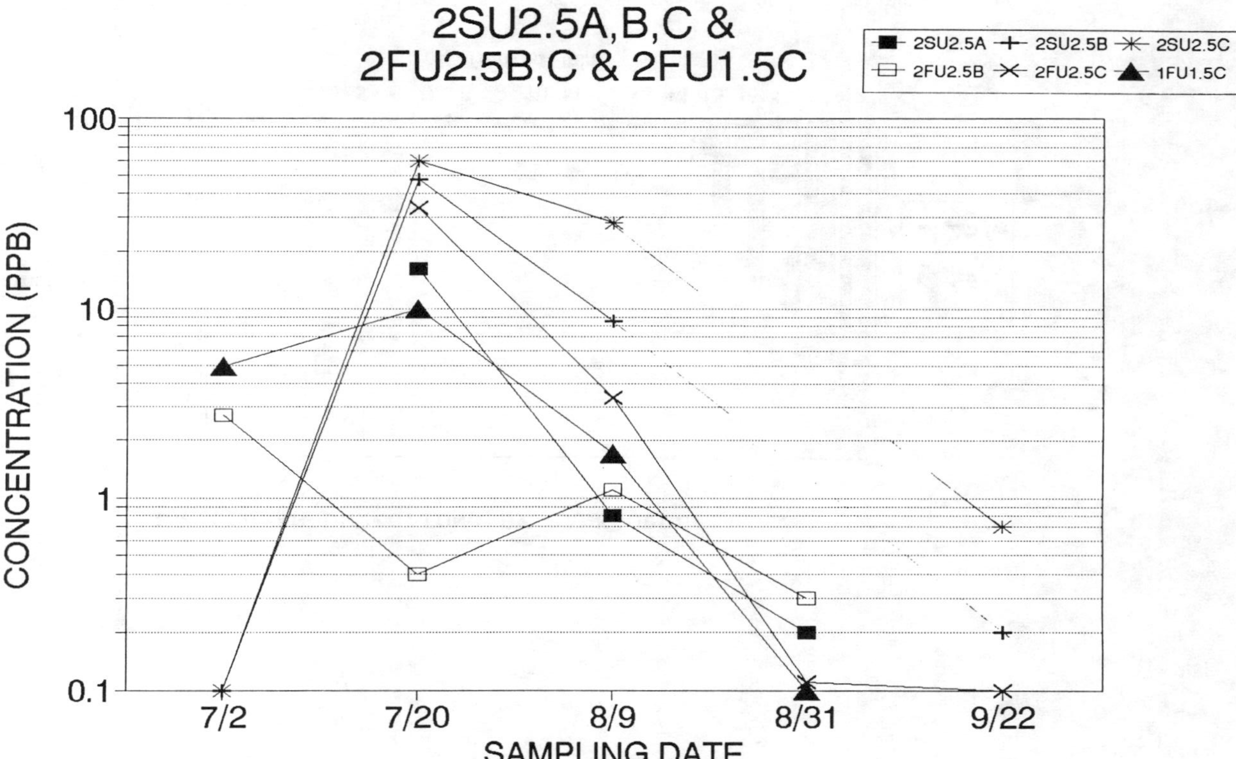

Fig. 62–13. Carbofuran concentrations by sampling event for six select lysimeters located with the upper field plots of both surge and conventional furrow irrigation.

A few of the lysimeters produced pore water samples for all five sampling events. For these lysimeters that did produce each time, there may be an indication that the concentrations were decreasing after the second irrigation event. As an example of this, Fig. 62–13 is a display of concentrations for five sampling events from six 2.5 ft lysimeters. All the lysimeters are from the upper portion of the field. Three are from the surge side and three from the furrow side of the field. In general, the concentrations increase up to the second sampling event and then decrease to trace concentrations by the last sampling event (Fig. 62–13).

Soil Samples

Using immunoassay techniques, the pre-1993 growing season soil sampling results reveal trace concentrations of carbofuran in five of the six samples. The concentrations ranged from 0.2 to 2.0 ng/g. The highest concentration occurred on the surge side in the middle of the field. According to the grower, carbofuran had not been applied to this field in over 10 yrs. Carbofuran may have been present in irrigation water applied to this field in prior yrs.

The post-1993 growing season soil sampling results indicate that carbofuran was present after harvest. Concentrations ranged from nondetect to 44.4 ng/g. The highest concentrations were within the upper one foot of the soil profile. Correlation of this data with the data from the fifth lysimeter sampling is difficult. Carbofuran was detected from the soil samples to three ft indicating a pattern of migration.

Immunoassay Analytical Technique

The carbofuran immunoassay analysis technique was very successful. The MRI laboratory conducted the analysis with consistency and precision throughout the project. The QA/QC processes were very successful. The laboratory's work related to the sample spikes, duplicates, controls, trips and transfer samples were all acceptable. The GC/MS confirmation testing was acceptable as well.

Water Quantity Assessment

A simple water quantity assessment was conducted for the field during the sampling and growing season. Tensiometer readings were taken during the irrigation and lysimeter sampling periods. The tensiometers placed within the two foot zone underwent a pattern of tension change between 24 and 48 h after each irrigation event ended. The readings ranged from 20 to 70 centibars depending on location and soil characteristics. By the 48 h period the tensions often reached 70 centibars, especially when the crop stage was more mature. The four foot tensiometers maintained some consistency at 5 to 15 centibars throughout the growing season.

IRRIGATION WATER BUDGET

Input or Output	Surge (acre-inches)	Furrow (acre-inches)
Applied	31.3	36.7
Estimated Runoff	6.3	11.0
Infiltrated Water	25.0	25.7
+ Effective Rain	2.5	2.5
= Total Infiltrated Water	27.5	28.2
- Crop Use (Agrimet)	26.1	26.1
Soil Storage or Leached Water	1.4	2.1
- Initial Soil Water	0.8	0.8
Adjusted Percolation Estimate	0.4	1.3

Fig. 62–14. An estimated irrigation water budget for the growing season.

There were 9 conventional furrow and 10 surge irrigations during the growing season. A estimated irrigation water budget was developed for the season (Fig. 62–14). Under surge irrigation there was estimated to be 31.3 acre-in of water applied. Under conventional furrow there was an estimated 36.7 acre-in of water applied. Including estimates for runoff (20%), data from a local Agrimet station, and crop data, the adjusted percolation estimates are 0.4 and 1.3 acre-in for surge and furrow irrigation, respectively (Fig. 62–14).

CONCLUSIONS AND RECOMMENDATIONS

Extraction of soil pore water samples from the lysimeters was moderately successful. The timeliness of the vacuum application and extraction was key to the success. Carborfuran was detected in ≈one quarter of the lysimeters that produced sample solution. The majority of the concentrations were below 0.9 ppb. The 2.5 ft lysimeters consistently produced the highest number of detections and concentrations of carbofuran. Movement of carbofuran to 2.5 ft occurred throughout the growing season, but potentially peaked around the third irrigation event. Only one detection of carbofuran (0.01 ppb) was found from a five ft lysimeter. Even with partial success, the lysimeters were valuable tools for assessing the soil pore water concentrations of carbofuran.

The information from the water budget indicates there may be potential for carbofuran leaching under conventional furrow and possibly surge irrigation due to the estimated percolation quantities. Further work to develop water quantity assessments at each plot would be necessary to accurately prove this by correlating lysimeter carbofuran data with percolation data.

By assessing the data available, there is an indication that there may have been less deep leaching of carbofuran at lower concentrations under surge irrigation as compared to continuous furrow. There is little statistical evidence to support this conclusion for this particular study, but the results available indicate a potential for surge to be an irrigation water BMP.

Results indicate that soil and pore water samples can be successfully and cheaply analyzed using immunochemical techniques. Confirmations of ELISA methods were successful using GC/MS. Applications of immunoassay techniques will be beneficial for future pesticide and irrigation water management BMP evaluations.

Working with the grower was a success. The grower managed each irrigation event as unbiased as possible. As the irrigation season progressed, there were some changing of practices that did occur potentially effecting the use of irrigation systems. Regardless, the biggest success may have come with the grower's possible long term interest in utilizing surge irrigation. As an innovator, the grower was a center of focus for summer farm tours. With the cooperation of the HUA agency representatives, the grower was able to display and discuss with others the possible benefits of his surge operation. The project succeeded in supporting the development of irrigation BMPs that minimize the leaching potential of pesticides such as carbofuran. For IDA it was valuable to cooperate with growers and agencies to find solutions for protecting ground water within the HUA.

ACKNOWLEDGEMENTS

The authors wish to thank the U.S. EPA Region 10 Office of Pesticide Programs (OPP) for providing funding for this study through the Regional Applied Research Effort (RARE) grant and the Pesticide Cooperative Agreement to the Idaho Department of Agriculture. Special thanks goes to Garret Wright of EPA Region 10 for coordinating the logistics of this funding, and Gary McCrae of the Boise EPA Field Office for support. Much appreciation goes to Tim Stieber (U. of I., CES) and Tim Stack (SCS) for assistance and coordination within the HUA. Thanks to Dr. Brad King (U. of I., CES) for his input and field assistance. Special appreciation goes to Rick Michaels for his patience and assistance.

REFERENCES

Baldwin, J. 1992. Personal communication. Hydrogeologist, Water Quality Bureau, Div. Env. Qual., Idaho Dep. Health and Welfare, Boise.

Baumann, T., B. Merkel, T. Ruppert, and R. Neissner. 1991. Problems related to pore water sampling for pesticide determination. 7th International Congress of Pesticide Chemistry. Hamburg 5–10.8.9.

Biggar, J.W., and D.R. Nielsen. 1976. Spatial variability of the leaching characteristics of a field soil. Water Res. Res. 12:1:78–84.

Bishop, A.A., W.R. Walker, N.L. Allen, and G.J. Poole. 1981. Furrow advance rates under surged flow systems. J. Irrig. & Drain. Div., ASCE 107

(IR3):257–264.

Bushway, R.J., B. Perkins, S.A. Savage, S.J. Lekousi, and B.S. Ferguson. 1988a. Determination of atrazine residues in water and soil by enzyme immunoassay. Bull. Environ. Contam. Toxicol. 40:647–654.

Bushway, R.J., W.M. Pask, J. King, B. Perkins, and B.S. Ferguson. 1988b. Determination of chlordane in soil by enzyme immunoassay. p. 433–437. *In* Symposium Proc., Oct. 11–13, 1988. Field screening methods for hazardous waste site investigations.

Bartholomay, R.C., and D.F. Champion. 1991. Grand Valley surge demonstration project: A report to the U. S. Dep. of the Interior, Bureau of Reclamation. Colorado State Univ. Coop. Ext. 14 pp.

Camp, V.E., P.R. Hooper, D.A. Swanson, and T.L. Wright. 1982. Columbia River Basalt in Idaho: Physical and chemical characteristics, flow distribution, and tectonic implications. *In* W. Bonnichsen, and R.M. 1Breckenridge (ed.) Cenozoic Geology of Idaho: Idaho Bureau of Mines and Geology Bull. 26, p. 55–75.

Chow, T.L. 1977. Fritted glass bead materials as tensiometers and tension plates. Soil Sci. Soc. Am. J. 41:19–22.

Everts, C.J. 1989. Role of preferential flow on water and chemical transport in a glacial till soil. Unpubl. Ph.D. thesis. Iowa State Univ., Ames, IA.

Everts, C.J., and R.S. Kanwar. 1988. Quantifying preferential flow to a tile line with tracers. Am. Soc. Agric. Eng. meeting paper 88–2635.

Goldhamer, D.A., M.H. Alemi, and R.C. Phene. 1987. Surge vs. continuous-flow irrigation. California Agriculture. Sept.-Oct. 1987. p. 29–32.

Izuno, F.T., T.H. Podmore, and H.R. Duke. 1985. Infiltration under surge irrigation. Trans. ASAE 28(2):517–521.

Jordon, C.F. 1968. A simple, tension-free lysimeter. Soil Science 105:2:81–86.

Kemper, W.D., T.J. Trout, A.S. Humpherys, and M.S. Bullock. 1988. Mechanisms by which surge irrigation reduces furrow infiltration rates in a silty loam soil. Trans. ASAE 31(3):821–829.

Malano, H.M. 1983. Comparison of infiltration process under continuous and surge flow. M.S. thesis. Utah State Univ., Logan.

Malde, H.E. 1965. Snake River Plain. *In* H.E. Wright, Jr. and D.G. Frey. (eds.) The Quaternary of the United States. Princeton University Press. pp. 255–263.

Miller, J.G., C.C. Shock, T.S. Stieber, and L.D. Saunders. 1991. Surge irrigation of Bliss spring wheat, 1991. Am. Soc. Agron. 83rd Annual Meeting, Denver, CO. Oct. 27–Nov. 1. p. 120–123.

Quin, B.F., and L.J. Forsythe. 1976. All-plastic suction lysimeters for the rapid sampling of percolating soil water. New Zealand J. Sci. 19:145–148.

Rupert, M., T. Dace, M. Maupin, and B. Wicherski. 1991. Ground water vulnerability assessment Snake River Plain, Southern Idaho. Idaho Dep. Health and Welfare, Div. Env. Qual., Boise. 25 pp.

Soil Conservation Service. 1972. Soil Survey of Canyon Area, Idaho. USDA SCS in cooperation with University of Idaho, College of Agriculture, Idaho Agric. Exp. Stn. 126 pp.

Soil Conservation Service. 1976. Soil Survey of Payette County, Idaho. USDA-SCS in Cooperation with Univ. of Idaho, College of Agriculture, Idaho Agric. Exp. Stn. 97 pp.

Stack, T. 1992. Personal communication. Project Leader, Snake-Payette Rivers Hydrologic Unit Project, Soil Conservation Service, United States Department of Agriculture, Payette, Idaho.

Steed, R., G. Winter, and J. Cardwell. 1993. Idaho Snake-Payette River hydrologic unit ground water quality assessment, West Central Idaho. Ground Water Quality Tech. Rep. No. 3. Idaho Dep. Health and Welfare, Div. Env. Qual. Boise, Idaho.

Stieber, T.D., T.J. Stack, and N. Hutchison. 1992. Idaho Snake-Payette Rivers Hydrologic Unit Area Project, Cropping Practices Survey. 9 pp.

United States Environmental Protection Agency (EPA). 1991a. EPA's Pesticide Programs. 21T-1005. May, 1991. Office of Pesticide Programs, Washington, DC. 25 pp.

United States Environmental Protection Agency (EPA). 1991b. Pesticides and Ground-Water Strategy. 21T-1022. Oct. 1991. Office of Pesticide Programs, Washington, DC. 78 pp.

United States Department of Agriculture (USDA). 1991. Idaho Snake-Payette Rivers Hydrologic Unit Plan of Work. March, 1991. 47 pp.

Walker, W.R., J.C. Henggeler, and A.A. Bishop. 1981. Effect of surge flow in level basins. ASAE Paper No. 81-2555, ASAE, St. Joseph, MI.

Yonts, C.D., D.E. Eisenhauer, and J.E. Cahoon. 1991. Fundamentals of surge irrigation. NebGuide G91–1028. Cooperative Extension, Inst. Agric. and Natural Res., Univ. of Nebraska-Lincoln.

Zimmermann, C., M. Price, and J. Montgomery. 1978. A comparison of ceramic and teflon in situ samplers for nutrient pore water determinations. Estuarine Coastal Mar. Sci. 7:93–97.

63 Soil Variability and Carbon Dioxide Loss After Moldboard Plowing

D. C. Reicosky

USDA-Agricultural Research Center
North Central Soil Conservation Research Laboratory
Morris, Minnesota

Soil variability in the field is partly due to natural differences in soil properties and can be impacted by past management. Soil spatial variability within fields has widely been demonstrated by soil testing results and crop yield differences. Because landscapes and soils vary greatly, land use and soil management must be tailored to fit the specific soil properties to enhance their utilization and maintain their long-term sustainability. Numerous reasons for variation in soil characteristics include soil forming factors, farming practices, and wind and water erosion. Because irregular soilscape units do not match well with symmetrically-shaped legal boundaries, several landscape units are usually managed in the same way. Larson and Robert (1991) discuss "farming by soil" practices that develop a management perspective for the application of fertilizer and herbicides based on soil-specific information. The "farming by soil" technology is one of the major breakthroughs in soil management in a decade or more. The importance of soil C in nutrient supply through mobilization–immobilization and the global C budget requires information on soil C levels and maintenance for the long-term.

Man's activities in agriculture production associated with tillage can have a significant influence on the atmospheric composition. Research on global warming and the greenhouse effect has highlighted the linkage between gas composition, atmosphere and the world weather patterns (Wood 1990; Post et al., 1990). Many natural factors as well as human activities influence atmospheric composition by merely contributing or removing gases and involving chemical reactions. Minimizing agriculture's impact on global increase in atmospheric CO_2 requires that we sequester C and maintain high levels of soil organic matter. Maintaining or even increasing soil organic matter levels for specific soils to maintain economic crop production may require both reduced tillage and crop rotations that maximize residue produced and returned to the soil surface.

The effect of landscape position on soil respiration has been reported by DeJong (1981) who observed a two-fold increase in soil respiration at the foot

---

of the hill as compared to the top of the hill. The highest CO_2 evolution rate was for grassland, intermediate for cereals, and lowest under fallow conditions. Even on uniform-level soils, spatial variability in soil properties can be large. Dugas (1993) made soil chamber CO_2 measurements sequentially at nine positions in a 5 ha field on a Houston Black clay (fine, montmorillonitic, thermic Udic Pellusterts). The coefficient of variation of chamber CO_2 fluxes across the nine positions averaged 40% throughout the d, indicating the need for a large number of chamber measurements to obtain a representative CO_2 flux measurement. Rochette et al. (1991) found spatial variability in soil respiration, described by the coefficient of variation for each series of measurement, was highest in May at 69% and decreased to 25% toward the end of the season. The number of measurements required to estimate soil respiration within 10% of the mean at a 0.05 probability level was 190 before crop emergence and 30 after 70 days.

Reliable averages of soil and vegetation CO_2 evolution are important measurements related to C cycling. The spatial variability associated with flux of CO_2 from soil needs to be addressed to obtain reliable averages, but has only received limited attention. Speir et al. (1984) found that spatial variability of soil biochemical properties including soil respiration was significantly greater than that of chemical properties. Robertson et al. (1988) determined that spatial variability of soil respiration was random with no spatial correlation at scales of 1 to 80 m. The results may appear random due to temporal variations in soil temperature and water content (Singh & Gupta 1977).

Aiken et al. (1991) reported variation in the soil CO_2 efflux was attributed to positional trends, spatial correlation and random effects in a wheat crop under wet and dry soil conditions. Spatial homogeneity for CO_2 efflux was not found in four of the five data sets evaluated. Positional trends accounted for spatial structure in these cases. Residual variability after removing the positional trends was isotropic and randomly distributed. No spatial correlation was observed after removal of the positional trends. They noted, however, that spatial correlation may have been apparent had the positional trends not been removed. The spatial structure was affected by soil water content under wheat. They further concluded that defining spatial structure for soil respiration required the determination of many environmental factors that are dynamically changing as a function of time as well as with position in the landscape.

Recently, Reicosky and Lindstrom (1993, 1995) showed that tillage, particularly moldboard plowing, resulted in a "flush" of CO_2 when gas exchange measurements were made within 5 min. after tillage. Significant amounts of gaseous C were lost as CO_2; comparable to C in the crop residue plowed under. Reicosky and Lindstrom (1994) assumed 4120 kg ha^{-1} of wheat residue contained 45% C. The cumulative CO_2 released from the soil surface during 19 d after moldboard plowing was the same as 134% of the C in the previous crop residue. This analogy does not imply the CO_2 came from the residue, but does provide perspective on loss of CO_2. The relative rate of gaseous C loss as measured relative to C in the current crop suggests that moldboard plowing can result in enough short-term C loss equivalent to the C in the current crop within 14 d.

There is ample evidence that cropping and associated tillage decreases soil organic matter in the long term (Peck 1989, Wagner 1989). The mechanisms for these losses are not clear, however. The magnitude of gaseous C loss after plowing explains much of the sustained C loss in many of agricultural systems (Reicosky & Lindstrom 1994). Moldboard plowing is one of the most disruptive types of tillage, and now appears to have two major effects: (i) to invert and loosen the soil sufficient to allow rapid CO_2 loss and O_2 entry and (ii) to incorporate–mix the crop residues for enhanced microbial attack. Tillage perturbs the soil system and causes a shift in the gaseous equilibrium by enhancing soil-atmospheric exchange of CO_2 and O_2 that enhances oxidation of soil C and organic matter loss. Conservation tillage limits soil disruption and leaves crop residue on the surface thereby limiting contact with the soil and associated microorganisms reducing residue decomposition rates.

The significant flush of CO_2 immediately after tillage reported by Reicosky and Lindstrom (1993, 1994) confirms the role of tillage affecting C flow within agricultural production systems. There is a need for more information on the variation and magnitude of this CO_2 flush based on the interaction of tillage on different soil types within the landscape. The specific objective of this work was to quantify the variation in CO_2 loss immediately after fall plowing on several soil types previously cropped to wheat. A field was selected with four different soil types and transects established across the soils to follow the CO_2 loss immediately after moldboard plowing.

METHODS AND MATERIALS

The experiment was conducted in the fall of 1993 at the USDA-Agriculture Research Service Swan Lake Research Farm located in west central Minnesota, (45°41'14" N and 95°47'57" W). The soils selected in this field (Table 63–1) range from moderately-well to poorly drained; were formed on glacial till under tall prairie grass vegetation. The surface horizon is generally very dark with relatively high organic matter. Many of the soils are developed over subsoils with high calcium carbonate. The cropping history for 80 years was corn, soybean, and spring wheat with conventional tillage.

The study area was selected by establishing two transects, each 195-m long, across an area where there was significant variation in the soil types as indicated by the soils map, hereafter referred to as north and south transects. Two parallel transects 30 m apart were established in an east-west straight line in anticipation of moldboard plowing along the transects. The measurement locations or sites were selected to be in the center of the soil-map unit on the respective transect. The soil series identification was confirmed by an SCS soil scientist[1]. The soil names and relative location are shown schematically in Fig. 63–1.The name identification includes a sequential number because the transects crossed the same soil map unit more than once.

---

[1]The assistance of soil scientist, Gerry Gorton from the regional SCS office at Fergus Falls, MN in describing the soil profiles is gratefully acknowledged.

Table 63–1[2]. Summary of soil taxonomy and properties in the spatial variation study.

SOIL SERIES (Taxonomy)	DRAINAGE	DEPTH of Al	BULK DENSITY	SOIL ORGANIC MATTER	CLAY CONTENT	PH
		(m)	(kg/m^3)	(%OM)	(%)	(-)
BARNES loam (Udic Haploborolls, fine loamy, mixed)	WELL	0.18	1.40 - 1.50	2 - 5	18 - 27	6.1 - 7.8
HAMERLY loam (Aeric Calciaquolls, fine-loamy, frigid)	MOD. WELL	0.20	1.20-1.60	4 - 7	18 - 27	6.6 - 8.4
PARNELL silty clay loam (Typic Argiaquolls, fine, montmorillonitic, frigid)	VERY POOR	0.56	1.20 - 1.30	6 - 10	27 - 40	6.1 - 7.8
VALLERS silty clay loam (Typic Calciaquolls, fine; loamy, frigid)	POOR	0.30	1.20 - 1.35	5 - 8	28 - 35	7.4 - 8.4

[2]SCS DATA from soil interpretation records. Soils formed on glacial till under tall grass prairie.

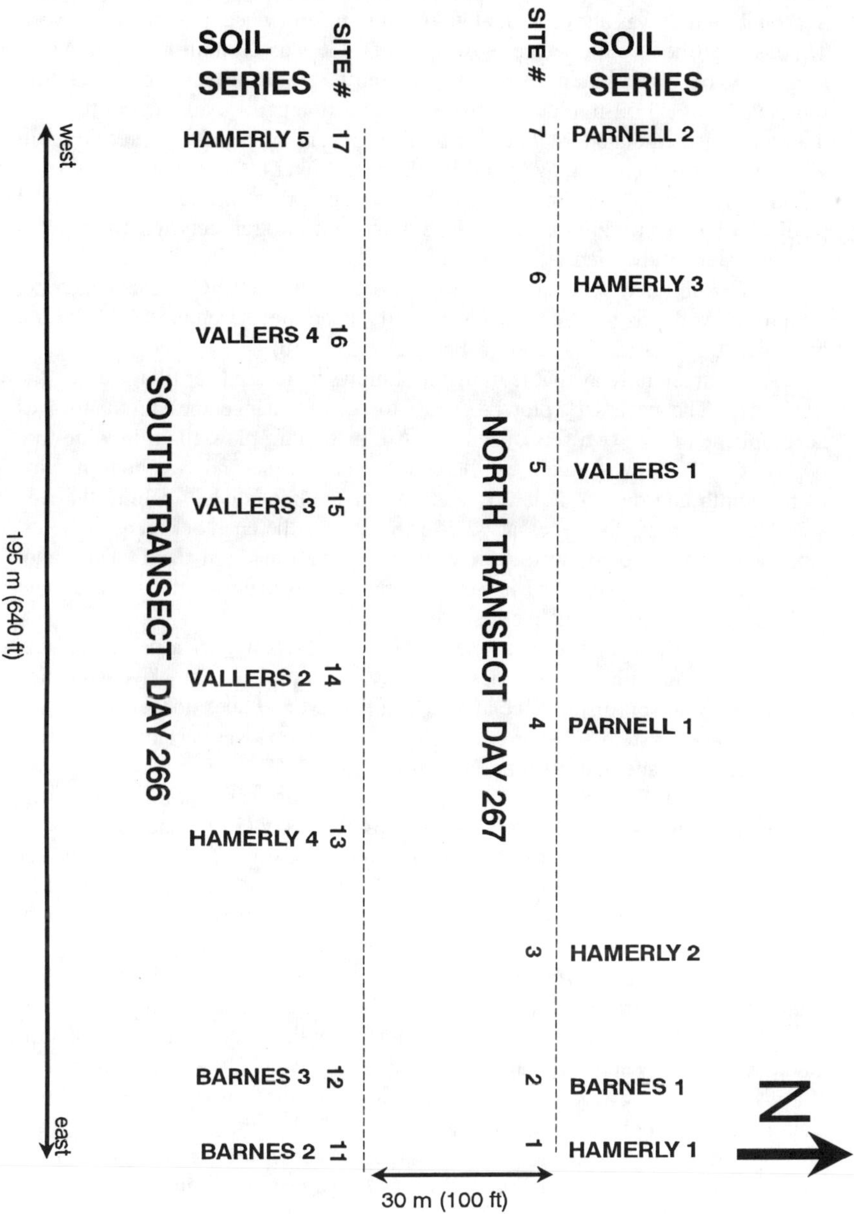

Fig. 63–1. Schematic representation of soil site location within the north and south transects. Soil series and site numbers indicate position in the landscape.

The study area was planted to spring wheat (*Triticum aestivum* L. cv. Marshall) on 21 April 1993 (D 111) and harvested on 17 Aug. 1993 (D 229). Seasonal rainfall was above normal with 524 mm from wheat planting to harvest. The last significant rain before plowing on D 266 was 20 mm on D 262 and 6 mm on D 263. Variation in yield along the transects was measured by harvesting four swaths of a 1.52-m plot combine with 6.09-m lengths over each of the pre-marked measurement sites. The four replicates were averaged to represent grain yield at that site. The average yield for the wheat over the larger field was 1480 kg ha^{-1} while grain yield measured at each of the pre-marked soil sites ranged from 253 kg ha^{-1} to 2655 kg ha^{-1}. Low yields on the transects were related to excess water in low areas.

To minimize weed and volunteer wheat effects on the CO_2 exchange rate, the entire field was sprayed with Ranger[3] (glyphosate) herbicide at 0.8 kg a.i. ha^{-1} on 27 August (D 239) and 13 September (D 256).

Commercially available moldboard plows were used for tillage along the transects. The necessary plowed width for gas exchange measurements was accomplished using two sets of plows. A four bottom plow (0.46 m wide to a depth of 0.22 m) was pulled by the first tractor. A second two bottom plow (same width and depth) pulled by another tractor immediately behind the first was required to get the necessary 2.74-m width for the chamber measurements. Both sets of moldboard plows were each pulled by a medium-sized farm tractor ($\approx$ 70 kw). Plowing resulted in nearly complete inversion of the surface layer and 100% incorporation of the residue.

A no-till (NT) site was selected for gas exchange measurements over undisturbed soil with crop residues as left by the combine. The NT designation only implies no soil disturbance after wheat harvest and does not refer to a long term "tillage system." The previous wheat crop was established using conventional tillage and planting equipment.

The CO_2 flux from plowed and no-till soil surfaces was measured using a large, portable closed chamber described by Reicosky (1990) and Reicosky, et al. (1990) in the same manner as Reicosky and Lindstrom (1993). Briefly, the chamber (volume = 3.25 m^3 covering land area = 2.67 m^2) with mixing fans running, was moved over the plowed surface until the chamber reference points aligned with site reference stakes, lowered and data collected at 1 s intervals for a period of 80 s to determine the rate of CO_2 and water vapor increase. The chamber was then raised, calculations completed and results stored on computer diskette. Data included time, plot identification, solar radiation, photosynthetically active radiation, air temperature, wet bulb temperature and the output of a LI-COR Model 6262 infrared-gas analyzer measuring CO_2 and water vapor concentrations in the same air stream. After the appropriate lag and mixing times, data from a 30 s calculation-window was selected to convert volume concentration of water vapor and CO_2 to a mass basis and then regressed

[3]Mention of a trade name, proprietary product, or specific equipment does not constitute a guarantee or warranty by the U.S. Department of Agriculture and does not imply its approval to the exclusion of other products that may be suitable.

as a quadratic function of time to estimate gas fluxes (Wagner et al., 1994). These fluxes represent the rate of CO_2 and water vapor increase within the chamber and are expressed on unit horizontal land area basis as differentiated from an exposed soil area surface basis resulting from surface roughness. The rough-plowed surface caused occasional large holes around the perimeter of the chamber that did not allow the base of the chamber to form a good seal. These were quickly filled with soil by hand to minimize leakage. No effects of significant leakage were observed once the holes were plugged. The convention of positive fluxes from the soil surface was selected for both CO_2 and water vapor.

The total time for a single measurement including both data collection and computation was about 2 min. Three sequential measurements were made at each site for replication as part of the routine measurement cycle before moving to the next site. Within a single d, four cycles along each transect were used to provide data on the temporal dynamics of CO_2 flux after plowing.

Due to the anticipated rapid decline in the CO_2 flux as a function of time after plowing (Reicosky & Lindstrom 1993), measurements were made with the portable chamber within 30 to 40 s after the pre-marked area was plowed. The sequence of events was as follows: the tractors pulling both sets of plows tilled through the designated experimental site to a pre-determined reference point, then stopped and waited while the chamber was quickly moved over the site and three successive measurements taken. Upon completion of the three measurements, the plows tilled through the next designated experimental site to the next pre-marked location and the chamber moved over the pre-marked area to repeat another series of three measurements. The sequence was repeated across the different soils on the transect until all seven pre-marked sites had been measured. The measurement cycle was then repeated starting on the first site measured that day. The south transect was plowed from west to east and CO_2 flux evaluated on 23 September (D 266). The north transect was plowed on 24 September 1993 (D 267) from east to west. Chamber measurements were made facing into the wind. Within any single d, four measurement cycles were completed on each transect. On D 267, about 24 h after tillage on the south transect only, an additional cycle of measurements was completed. On both days, at the start of each measurement cycle, triplicate measurements were made on the no-till site that had surface residue as left by the field combine. Based on the soils map, the soil was a Parnell, located about 20 m north of plot 7 on the north transect.

The very high flux within 30 s after moldboard plowing caused the CO_2 channel of the infrared gas analyzer to over-range on D 266 that required modification on the calculation method. The majority of the measurements used the standard 30 s calculation window (30 data points) as described by Reicosky and Lindstrom (1993) with the quadratic regression (Wagner et al. 1994). For those data sets where the CO_2 concentration exceeded 500 μ mol mol^{-1}, however, the data were screened to determine the valid range over which the CO_2 concentration was increasing. The calculation procedures were the same as described previously; only the number of data points (the width of the calculation window) was varied by determining initial and final valid data points.

With the extremely high fluxes, the minimum number of data points in any one of the 18 data sets where the CO_2 concentration exceeded 500 μ mol mol^{-1} was 13. In view of this problem, the maximum CO_2 range on the infrared gas analyzer was set to 1,000 μ mol mol^{-1} for the measurements on D 267. The accuracy of using fewer data points may restrict the interpretation of the results on D 266. The trends were reasonable, however, and the results should be as valid as when 30 data points were used.

The cumulative amount of CO_2 evolved after plowing was calculated using numerical integration (trapezoid rule). This method assumes linear interpolation between the measured fluxes over the time interval. The areas for successive time intervals were summed to give a total amount of CO_2 evolved. The cumulative CO_2 flux following moldboard plowing was calculated for ≈3.5 h after tillage on both transects and for about 24 h after tillage on the south transect. The values for 24 h may be subject to error due to the long time between the last two measurements, however they represent a first approximation.

Microclimate data was collected from a standard weather station located ≈200 m from the edge of the experimental area. Measurements included air temperature at 2 m, solar radiation, wind speed, wind direction, relative humidity, and photosynthetically active radiation collected at one-min. intervals and averaged hourly. The change in air temperature and solar radiation may have resulted in differences in the initial flux measured immediately after plowing as the soil temperature may have tended to increase slightly with increase in air temperature. The impact of radiation and air temperature on soil respiration immediately after plowing, however, was not determined by direct measurement.

RESULTS AND DISCUSSION

Microclimate data for two days of this study showed seasonal solar radiation indicative of clear skies during both days. The air temperature went from a minimum of 0.6°C to a maximum to 17°C on D 266 and from a minimum of 2°C to a maximum of 20°C on Day 267 with typical diurnal fluctuations. The relative humidity reached a maximum of 100% during the night and decreased to a low of 40% on D 266 and 42% on D 267. Wind speeds were nominal ranging as high as 4.3 m s^{-1} with variable wind direction from the south, southwest and northwest on both days. Some of the variation in initial CO_2 flux on each soil may be related to the increase in air temperature as a function of time during the day and after plowing. The increase in air temperature from 0900 to 1000 h was 3.0 and 3.1 °C for D 266 and 267, respectively. In this analysis, air temperature was assumed to have little effect on the initial flush of CO_2 .

The grain yield from the pre-marked sites is summarized in Fig. 63–2. The average yields are low due to the cooler and wetter than normal climate experienced in the 1993-growing season. The field-average yield across the remainder of the 24-ha field was 1480 kg ha^{-1}. The yield from each of the sites ranged from 253 to 2345 kg ha^{-1} on the north transect and from 1822 to 2655

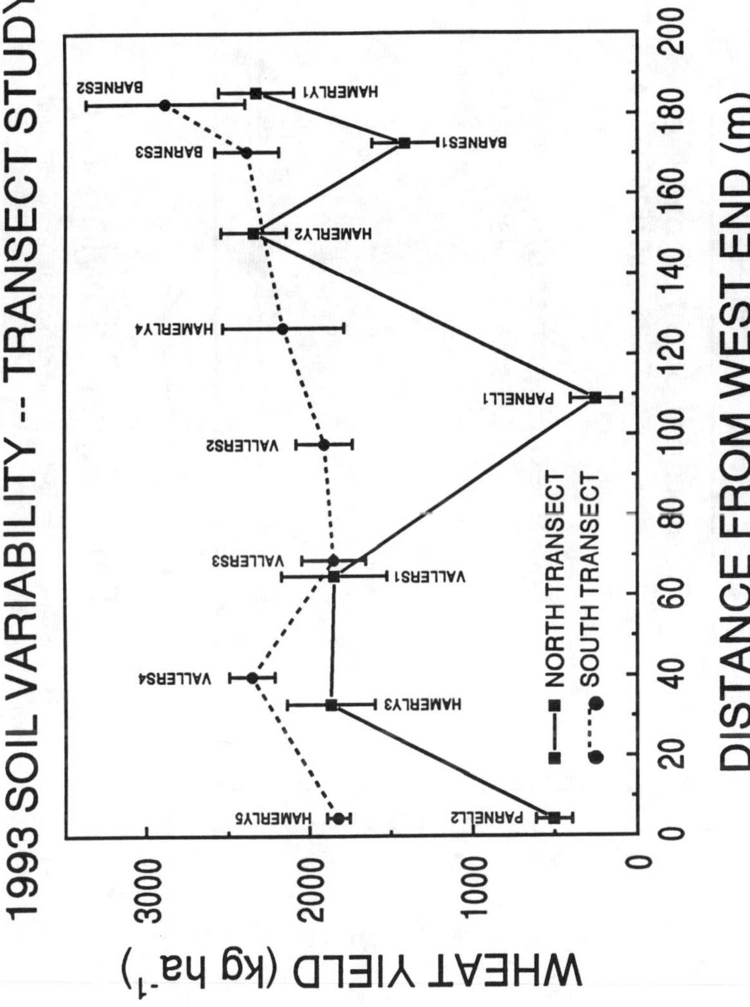

Fig. 63–2. Spatial variation in wheat yield from measurement sites in north and south transects. Error bars represent ± 1 standard error of the mean of 4 replicates.

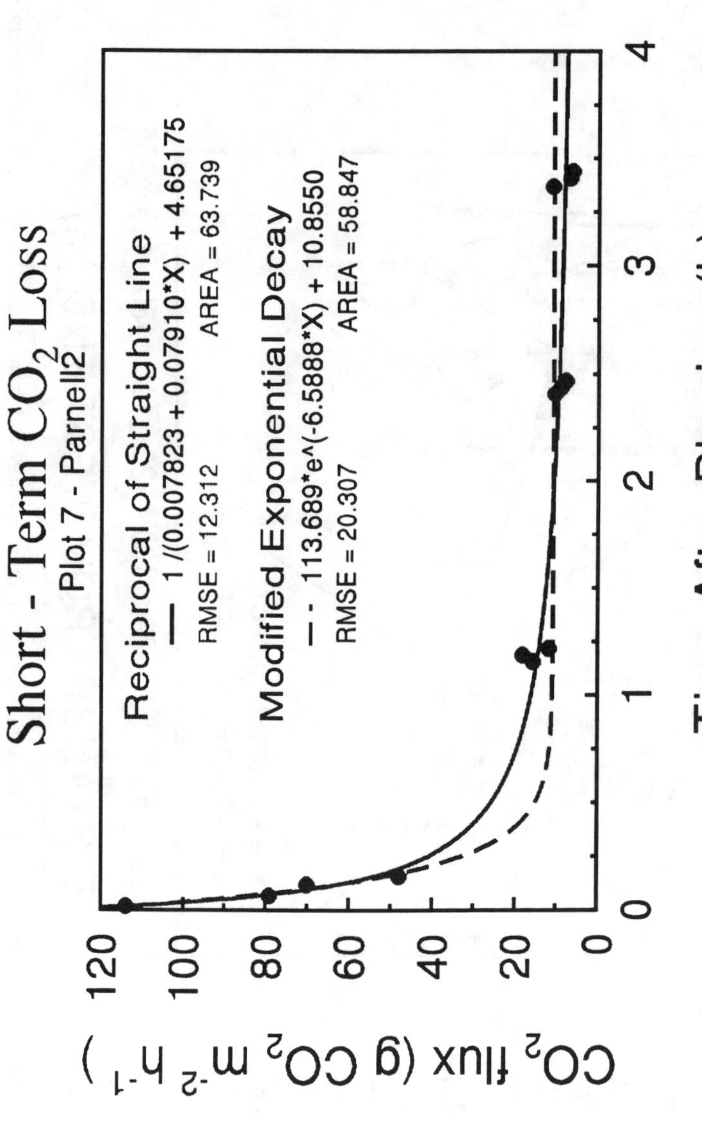

Fig. 63–3. The CO_2 flux as a function of time after plowing for Parnell2 soil on the north transect. Coefficients are for fitted reciprocal linear and modified exponential curves.

kg ha^{-1} on the south transect. Most of the low yields were from low lying poorly drained areas that retained runoff. The Parnell soils had the lowest yields related to their drainage. Yields within the same soil series were variable and ranged from a low of 1822 kg ha^{-1} for Hamerly5 to a high of 2345 kg ha^{-1} for Hamerly2. Similar differences were observed for Vallers and Barnes soils. The results show substantial yield variation related to soil series and position within the landscape that need to be considered in soil management plans.

The change in CO_2 flux as a function of time after plowing is illustrated in Fig. 63–3 for the Parnell2 soil. This represents the change in the CO_2 flux from 1 min. to ≈3.5 h after plowing. The time of the first measurement after plowing was consistent within 30 to 40 s on all sites and was rounded to the nearest minute after plowing for simplicity. The ≈3.5 h time after plowing in the last series of measurements varied slightly due to chamber travel time between sites and is considered approximate. The actual times ranged from 3.40 to 3.62 h. The CO_2 flux in Fig. 63–3 decreased from a high of 114 to 48 g CO_2 m^{-2} h^{-1} within 8 min to about 8 g CO_2 m^{-2} h^{-1} after 3.5 h. All of these fluxes were larger than the no-till site average of 0.38 g CO_2 m^{-2} h^{-1}. The decrease in the CO_2 flux as a function of time after plowing was fitted to a reciprocal linear function and a modified exponential function programmed in SAS (1988). The coefficients for the equations are summarized in Fig. 63–3. Analysis of the data sets showed the reciprocal linear function had the lowest-residual-mean square in 12 of 14 data sets. This curve type fit all the data sets reasonably well and allows a Y intercept and a gradual linear decline in the flux as a function of time after tillage. The data is limited, however, to the ≈3.5 h time interval.

There was a significant difference in the CO_2 fluxes between the plowed areas and the no-till after 24 h on the south transect that indicated significant CO_2 was lost overnight. The change in CO_2 flux from 3.5 h to 24 h after plowing ranged from 1.1 g CO_2 m^{-2} h^{-1} on the Hamerly5 soil to 4.4 g m^{-2} h^{-1} on the Vallers3 soil. The fluxes after 24 h were still 6 to 10 times larger than the 0.18 g CO_2 m^{-2} h^{-1} from the no-till site on the morning of D 267. The decreasing trends in the plowed sites should continue until the next perturbation, i.e. another tillage event or rainstorm that could cause surface sealing or cold temperatures or drier soils that could decrease the CO_2 flux.

The CO_2 fluxes on three different soils in the north transect as a function of time after plowing are illustrated in Fig. 63–4, representing the lowest, middle and highest initial fluxes on D 267. The time trends were similar for all three soils. The data were fitted using the inverse linear function as described previously. The reasonably good fit suggests this function may be used to approximate the initial flush of CO_2, at least for 3.5 h after plowing. The coefficients were extremely variable across all soils and within the same soil series and have no physical meaning. These data and other appropriate functions require further analysis.

The spatial relationship in the CO_2 flux immediately after plowing was of interest. The flux at one min. and ≈3.5 h after tillage for both north and south transect is summarized in Fig. 63–5 and 63–6 as a function of distance from the west end of the transect. The CO_2 fluxes are different for each of the soils and show maximum values at 1 min. after tillage that ranged from a high of 114 g

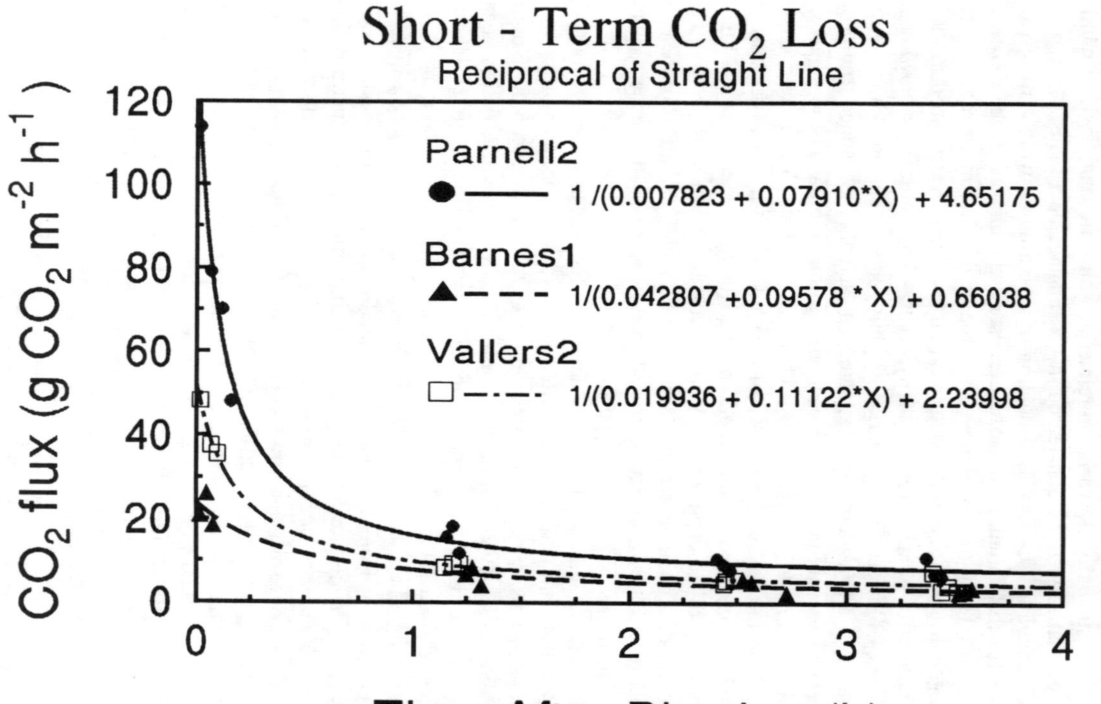

Fig. 63–4. The CO_2 flux as a function of time after plowing for three different soil series on the north transect that represent low, median, and high fluxes measured on D 267.

CO_2 m^{-2} h^{-1} to a low of 50 g CO_2 m^{-2} h^{-1} on the north transect, and a high of 80 g CO_2 m^{-2} h^{-1} to a low of 30 g CO_2 m^{-2} h^{-1} on the south transect. The values at ≈3.5 h after plowing were substantially lower and ranged from 4 to 7 g CO_2 m^{-2} h^{-1} on both transects. The magnitude of initial fluxes and rapid decline in CO_2 flux immediately following plowing suggest significant gaseous C loss, depending on soil type and location within the landscape.

The cumulative CO_2 loss from 1 min. to ≈3.5 h after plowing is summarized in Fig. 63–7 and 63–8 for the north and the south transects, respectively. The cumulative CO_2 loss along the north transect ranged from 67 g CO_2 m^{-2} for Parnell2 to 26 g CO_2 m^{-2} for Barnes1. The cumulative flux for the 3.5-h period shows differences with respect to soil series and with position along the transect for the same series. These values were 30 to 80 times larger than the no-till value of 0.78 g CO_2 m^{-2} for the same period illustrating significant loss of CO_2 in the initial flush immediately after plowing.

The cumulative flux for the first 3.5 h after plowing on the south transect ranged from 60 g CO_2 m^{-2} for Vallers3 to a low of 20 g CO_2 m^{-2} for Barnes2 (Fig. 63–8). The initial measurements on Barnes2, however, were not made until 8 min. after plowing due to operator error and may not be a representative. More realistic is Barnes3 that had 35 g CO_2 m^{-2} evolve during the 3.5-h period after plowing. These cumulative fluxes illustrate substantial variation due to soils and their location within the landscape. All cumulative fluxes were larger than no-till at 0.58 g CO_2 m^{-2} for the same period.

The same method was used to calculate the cumulative CO_2 loss for sites on the south transect where a measurement cycle was made on the morning of Day 2 before starting on the north transect. These data were taken ≈24 h after plowing on D 266 and can be used to approximate the cumulative CO_2 loss for 1 d. The cumulative CO_2 fluxes values ranged from a high of 143 g CO_2 m^{-2} to a low of 77 g CO_2 m^{-2}. The cumulative CO_2 fluxes for 24 h along the south transect were 108, 101, 143, 106, 125, 90, and 77 g CO_2 m^{-2}, respectively for the Hamerly5, Vallers4, Vallers3, Vallers2, Hamerly4, Barnes3 and Barnes2 soils. These values can be compared with the no-till site which lost 6.66 g CO_2 m^{-2} during the same period. The cumulative CO_2 losses during the 24-h-period after plowing were larger than from no-till with substantial soil differences. These results show lingering effects of plow perturbation, however, need to be interpreted with caution due to the long interval with no data points during the night. This is in agreement with observations of Reicosky and Lindstrom (1993), where moldboard plowing showed higher fluxes as long as 3 days after tillage before a 49-mm-rainfall event.

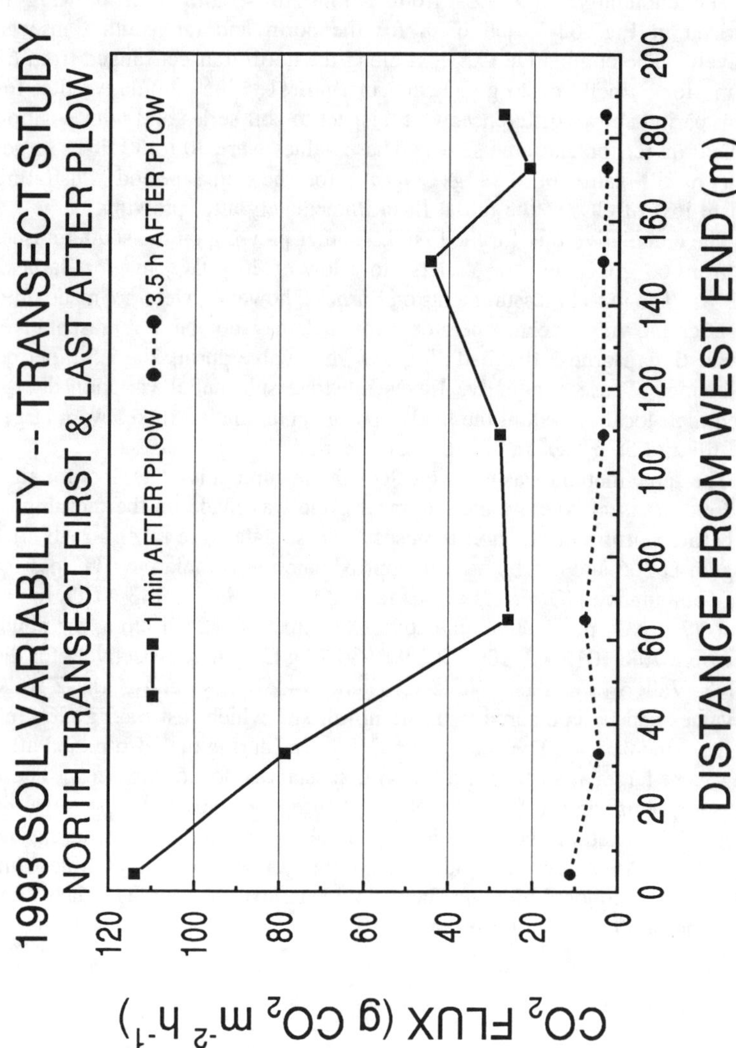

Fig. 63–5. The CO_2 flux at one minute and 3.5 hours after plowing on the north transect as a function of landscape position (distance from the west end).

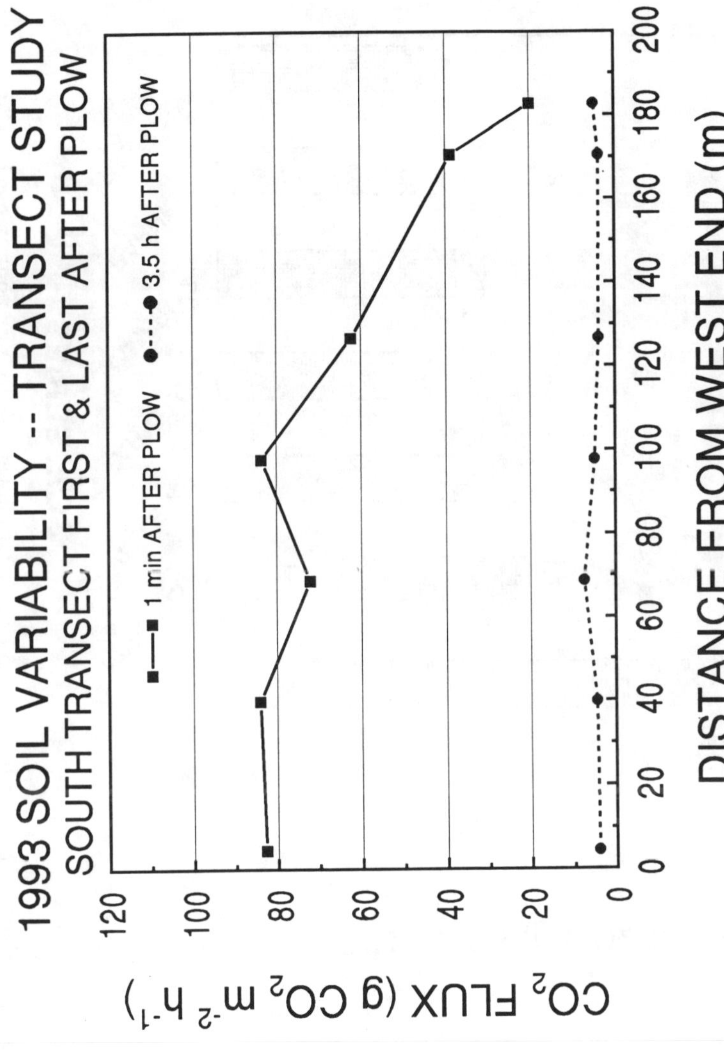

Fig. 63–6. The CO_2 flux at one minute and 3.5 hours after plowing on the south transect as a function of landscape position (distance from the west end).

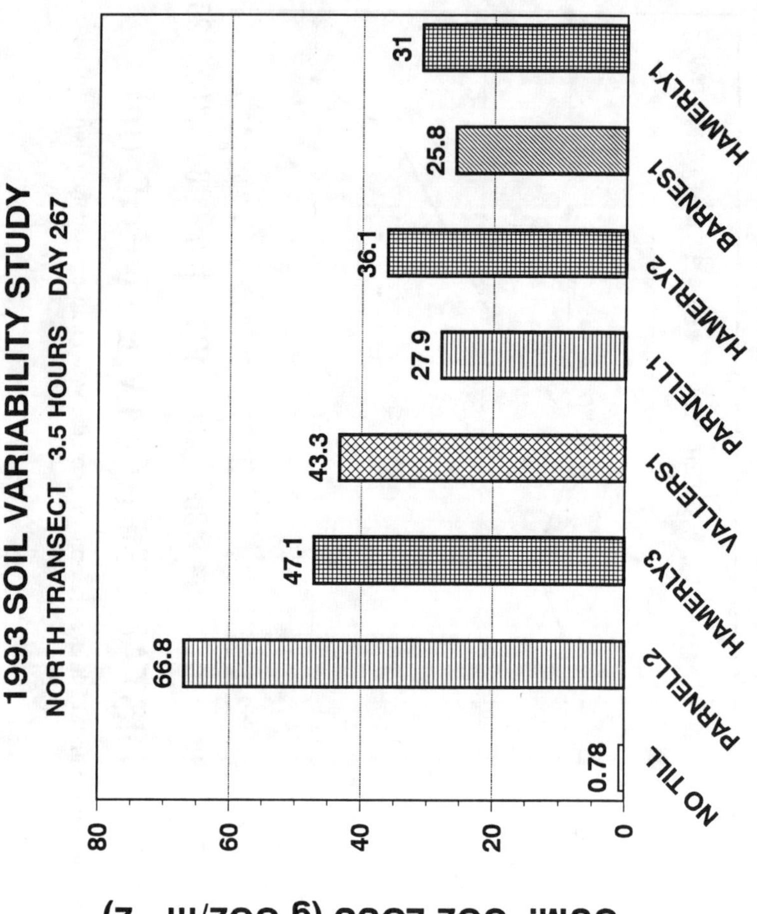

Fig. 63–7. The cumulative CO_2 loss for the first 3.5 hours after plowing for soils on north transect (D 267) using the trapezoid rule for numerical integration.

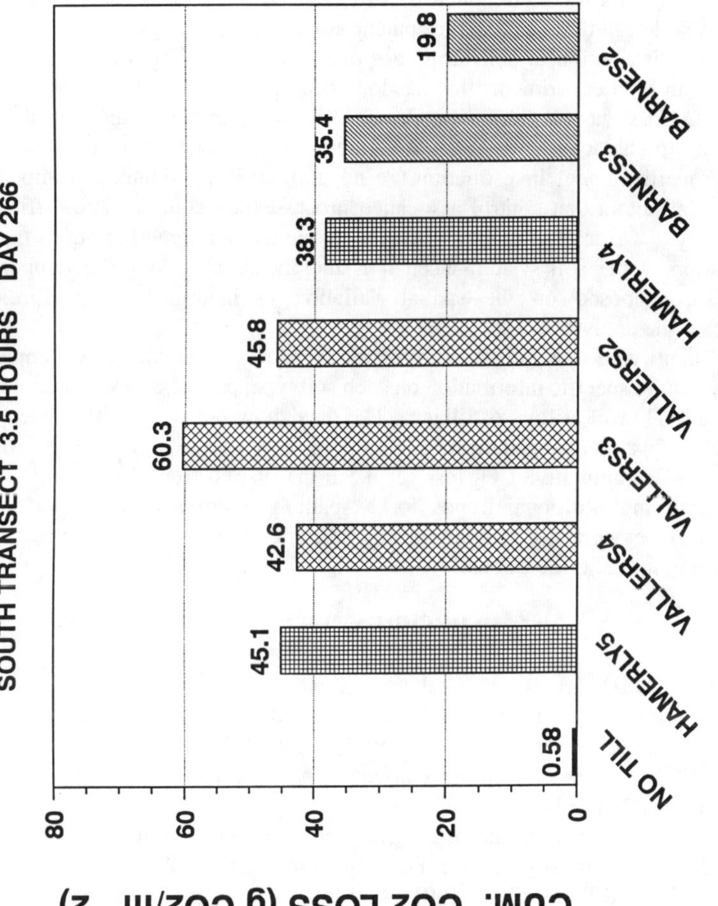

Fig. 63–8. The cumulative CO_2 loss for the first 3.5 hours after plowing for soils on south transect (D 266) using the trapezoid rule for numerical integration.

In summary, these results show a large loss of CO_2 immediately following moldboard plowing that agrees with earlier results of Reicosky and Lindstrom (1993). Measurements within 1 min. after plowing as opposed to 5 min. after plowing increased the maximum initial CO_2 flux nearly three-fold. Enhanced loss of CO_2 and the subsequent entry of oxygen into the soil could shift the gaseous equilibrium and result in enhanced organic matter decomposition. These results help explain the long-term decrease in soil organic matter as a result of plowing on agricultural soils. There was significant spatial variation along the transects that was partly related to soil type, and probably related to soil organic C and water content at the time of tillage.

The distribution of soil properties over space is an important source of variability in field experiments that has long been recognized. In this work, the rate of CO_2 loss was partially dependent on soil position within the landscape. The ability to characterize this spatial variability can be critical in field studies where properties under investigation are not uniformly distributed. Temporal variation in factors that control associated processes are complex and difficult to quantify. Critically important is the dynamic temporal variation from perturbation of the soil system when it is moldboard plowed. The temporal trends superimposed on the spatial variation in landscapes only further complicate the analysis.

Quantifying CO_2 emission from soil as a function of tillage is complex and will require specific information on each soil type, particularly water content and organic C level at time of tillage. The data show positional differences in initial CO_2 flux that ranged from 114 to 35 g CO_2 m^{-2} h^{-1} immediately after plowing and a cumulative CO_2 loss for 3.5 h that varied from 26 to 67 g CO_2 m^{-2}. Quantifying positional trends for a spatially distributed parameter will require further work to provide policy makers with quantitative data for environmental quality decisions.

REFERENCES

Aiken, R.M., M.D. Jawson, K. Grahammer, and A.D. Polymenopoulos. 1991. Positional, spatial, correlated and random components of variability and carbon dioxide efflux. J. Environ. Qual. 20:301–308.

DeJong, E. 1981. Soil aeration as affected by slope position and vegetative cover. Soil Sci. 131:34–43.

Dugas, W.A. 1993. Micrometeorological and chamber measurements of CO_2 flux from bare soil. Agric. For. Meterolo. 67:115–128.

Larson, W.E., and P.C. Robert. 1991. Farming by soil. p. 103–112. *In* R. Lal and F.J. Pierce (ed.) Soil management for sustainability. Soil and Water Conservation Society, Ankeny, IA.

Peck, T.R. 1989. Morrow plots-long term University of Illinois field research plots 1876-present. p. 49–52. *In* J.R. Brown (ed.) Proc, of the Sanborn Field Centennial: A celebration of 100 years of agricultural research. Publ. No. SR 415. Presented June 27, 1989. Columbia, MO.

Post, W.M., T.H. Peng, W.R. Enamuel, A.W. King, V.H. Dale, and D.L. De Angelis. 1990. The global carbon cycle. Am. Scientist. 78:310–326.

Reicosky, D.C., and M.J. Lindstrom. 1993. The effect of fall tillage method on short-term Carbon Dioxide flux from soil. Agron. J. 85(6):1237–1243.

Reicosky, D.C., and M.J. Lindstrom. 1995. Impact of fall tillage on short-term Carbon Dioxide flux. Adv. in Soil Sci. (accepted for publication).

Robertson, G.P., M.A. Huston, F.C. Evans, and J.M. Tiedje. 1988. Spatial variability in a successional plant community: Patterns of nitrogen availability. Ecology 69(5):1517–1524.

Rochette, P., R.L. Desjardins, and E. Pattey. 1991. Spatial and temporal variability of soil respiration in agricultural fields. Can. J. Soil Sci. 71:189–196.

SAS Institute Inc. SAS/STAT™ User's Guide, Release 6.03 Edition. Cary, NC: SAS Institute Inc., 1988.

Singh, J.S., and S.R. Gupta. 1977. Plant decomposition and soil respiration in terrestrial ecosystems. Bot. Rev. 43:449–528.

Speir, T.W., D.J. Ross, and B.A. Orchard. 1984. Spatial variability of biochemical properties in a taxonomically uniform soil under grazed pasture. Soil Biol. and Biochem. 16:153–160.

Wagner, G.H. 1989. Lessons in soil organic matter from Sanborn field. p. 64–70. *In* J.R. Brown (ed.) Proceedings of the Sanborn Field Centennial: A celebration of 100 years of agricultural research. Publ. No. SR-415. Presented June 27, 1989. Columbia, MO.

Wagner, S.W., D.C. Reicosky, and R.S. Alessi. 1995. Model evaluation for calculating photosynthesis and evapotranspiration from closed chamber gas exchange data. Agron. J. 87. (Accepted for Publication)

Wood, F.P. 1990. Monitoring global climate changes: The case of greenhouse warming. Am. Meteorol. Soc. 71(1):42–52.

64 Environmental Concerns Driving Site-Specific Management in Agriculture

Richard Castelnuovo

Environmental Resources Center
University of Wisconsin
Madison, Wisconsin

In the last twenty yrs, government regulation of industrial pollution has improved water quality, but this nation cannot make further environmental progress without tackling agricultural pollution. More people are realizing that farmers rely on environmentally-threatening technologies, but do not bear the same share of responsibility as other polluters. Nowhere is this mounting concern over agricultural pollution better reflected than in the Congressional debate over reauthorization of the Clean Water Act (CWA). Concern over agricultural pollution is also growing outside government. Farmers, producer groups and those doing business with farmers are troubled by expanding standards of legal accountability for agricultural pollution.

In addressing these concerns, everyone from lawmakers to landowners is facing challenges not encountered in regulating industrial pollution. Government officials are realizing that pollution from the nation's two million farms will not be solved by a regulatory or permit program cut from a traditional mold. On the private side, farmers and other groups are searching for management tools to contend with new risks and responsibilities. Site-specific management will form the cornerstone of public programs, as well as private initiatives, designed to respond to the special demands of agricultural pollution. With the expanding use of site-specific management, however, landowners in particular will find themselves in a position once faced by industries using environmental audits. Gathering information to make management decisions, they run the risk of discovering potentially incriminating information and facing difficult questions about access to and use of assessment information.

AGRICULTURAL POLLUTION

The Scope of the Problem

Pollution from agricultural activity presents a complex picture. Initial concern focused on surface water pollution from agricultural runoff containing sediment, pathogens, nutrients and pesticides. Unlike factory discharges to water, this form of pollution does not enter water from distinct points. During the 1980s, reports to Congress condemned agriculture for much of the nation's nonpoint source pollution (EPA, 1986; EPA, 1984).

Despite the indictment against agriculture, federal and state programs have yet to dedicate the necessary resources to address nonpoint source pollution. More easily identified and managed, industrial point source pollution has received greater attention. The pollution from industry comes from limited sources and responds to end-of-the-pipe treatment. In contrast, nonpoint source pollution from farms is diffuse and varies in its impact from one farm to the next. While momentum gathers to tackle agricultural nonpoint source pollution, people are realizing that the keys to success will be different. A successful effort will depend on broad-based cooperation among the agricultural community, an innovative mix of programs and initiatives, and most significantly concepts of site-specific management.

Agriculture also accounts for its share of pollution of the point source variety. Around farm buildings--the farmstead--farmers locate structures and carry out certain activities that may result in point source discharges. In fact, large feedlot operations pose such great risks that they must obtain the same permits as waste-discharging factories. There are other point source culprits: discharges from smaller livestock yards, overflows or leaks from manure storage facilities, leaks from petroleum storage tanks and spills from improperly handled chemicals (Jackson et al., 1993).

Responding to these agricultural point sources of pollution also is more complicated than regulation of point source pollution from industry. They are not confined to limited categories of industry with defined patterns of production. Instead, agricultural point sources are generated by different size operations with dissimilar structures and production activities, and these operations are spread over innumerable sites. In many ways, these are diffuse forms of pollution that could be treated as nonpoint sources. For these and other reasons, regulators and landowners have not directly confronted many questions about point source pollution on the farm. Based on considerations other than the technical nature of the discharge, spills and leaks from manure storage facilities are dealt with under nonpoint source programs, as is the case under as Wis. Stat. Sec. 144.25. In the CWA, Congress has defied logic to classify irrigation return flows as nonpoint source pollution (42 U.S.C. s 1362(14). EPA continues to exclude farm and residential petroleum storage tanks from regulation, even though they are as likely to release their contents as regulated tanks. Among other reasons for this position, EPA has cited the difficulties in enforcing these regulatory measures (BNA, 1990).

Growing public concern over groundwater pollution in 1980s intensified the spotlight on agriculture. Researchers found that agricultural activities played a major role in fouling groundwater. In over 36 states, including Pennsylvania, Florida, Wisconsin, California, and Iowa, the impact of pesticide and fertilizer contamination has been documented. Both nonpoint and point sources are implicated in the problem (Halstead et al., 1990; Lee and Nielsen, 1987).

Since agricultural activities threaten both surface water and groundwater, this raises special problems for managing agricultural pollution. Contamination on the farm may be channelled to either ground or surface sources or both. Variables related to the land, climate and farming practices determine the extent of contamination of these water resources. Under some conditions, ground and surface sources exchange water, so protecting one source without protecting the other may leave both exposed to contamination. In certain cases, practices that protect one source harm another. For example, those farms that employ measures to decrease runoff, such as conservation tillage, may increase infiltration of pesticides and fertilizers into groundwater. Protection of these resources hinges on how farmers balance the pollution risks on their individual farms.

In the end, agricultural pollution issues cannot be intelligently managed within a framework that simply treats environmental protection as separate matters of point and nonpoint source pollution or surface water and groundwater protection. To make inroads into the agricultural pollution problem, it is critical to look at individual farm operations through the lens of site-specific management.

The Consequences of the Problem

The failure to prevent agricultural pollution has far-reaching and profound consequences. Surface water pollution not only harms endangered species, destroys natural habitats and disrupts recreational opportunities, it also directly threatens the quality of drinking water reservoirs for urban centers. Agriculture took the blame for the nation's worst drinking water supply crisis--the 1993 crytosporidium epidemic in Milwaukee, WI. Contamination of groundwater threatens the drinking water supply for over 97% of rural American households. Nearly half of all livestock water and irrigation water comes from groundwater. According to 1980 statistics, groundwater is an indispensable source for 40 percent of the population using public water supplies (Lee and Nielsen, 1987).

The health effects of agricultural chemicals are not well-investigated; however, researchers are gathering evidence linking pesticides to cancer and birth defects and central nervous system disorders. It is well established that nitrate-contaminated water causes "blue baby syndrome"--methemoglobinemia (Halstead et al., 1990). Petroleum fuels contain a number of potentially toxic compounds, including common solvents, such as the known carcinogen benzene, toluene and xylene, and additives such as ethylene dibromide (EDB) and organic lead compounds (Jackson et al., 1993).

Health hazards are only part of the picture; environmental contamination imposes economic hardship. Where feasible, remedial action is expensive, particularly when groundwater is contaminated. Moreover, evidence suggests groundwater pollution may not be satisfactorily remediated at any price (GAO, 1991; Travis and Doty, 1990). As part of legislation to expand Wisconsin's petroleum tank cleanup fund to include unregulated farm tanks, Wisconsin officials estimated cleanup costs ranged from $30,000 to 90,000 for most sites to $250,000 for the worst ten% (DILHR, 1993). Without containment, chemical spills can cost over $200,000 for investigation and remediation (Reichenberger, 1992). Even in the case of agricultural property, contamination can require cleanups that cost more than the value of farm property.

Alternatives to cleaning up contaminated groundwater are not necessarily less expensive, and are not always effective. Excluding regular maintenance costs, installation costs for household filtering systems may exceed $800, while costs for municipal systems may exceed $25 million. Digging new wells in search of uncontaminated drinking water can cost several thousand dollars per household and several hundred thousand dollars per municipality (Flemming, 1987). Costing a fraction of remedial actions, prevention is a sound investment; it is the only reasonable, cost-effective approach to ensuring future supplies of drinking water (Silvas, 1988).

STRATEGIES TO COMBAT AGRICULTURAL POLLUTION

The Role of Pollution Prevention

Pollution prevention has emerged as the guiding principle in a new generation of measures to protect the environment. On the latest anniversary of EarthDay, EPA Administrator, Carole Browner, announced that prevention will be the central ethic of her agency (EPA, 1993). Pollution prevention programs started with industry. Using site-specific audits, companies assessed pollution risks and then instituted actions to reduce these risks.

The agricultural sector is an ideal candidate for pollution prevention. Agricultural pollution does not lend itself to technological solutions developed to treat and control discharges from factories. Nor can effluent limits be monitored to ensure that discharges are within limits. As applied to agriculture, pollution prevention will require farmers to reduce the use of hazardous substances, find alternatives to environmentally-threatening production processes, and modify practices or structures to prevent releases of threatening substances.

EPA has recognized pollution prevention as a useful strategy for agriculture (54 Fed.Reg. sec. 3845-46) and has joined with the United States Department of Agriculture (USDA) to implement an agricultural pollution prevention plan. The Agency Review Draft of the EPA/USDA Agricultural Pollution Prevention Plan (8/19/93) proposes actions to minimize environmental degradation from agriculture through improved management of animal waste, nutrients, pests and ecologically important land. To achieve maximum results, a prevention plan must be tailored to conditions of a particular farm, each with its own pollution risks demanding specialized management.

The Role of Site-Specific Management

In agriculture, pollution prevention has been customarily carried out through design and operating principles known as "best management practices" (BMPs). BMPs cover practices, procedures or facilities designed to prevent leaks, spills or other releases that threaten the environment. Broad ranging, they include adoption of conservation tillage, routine pumping of manure storage facilities, or construction of containment around above-ground petroleum and agricultural chemical storage tanks.

Though the scale of implementation may be larger than an individual farm, BMPs are most effective if implemented on a site-by-site basis. The success of preventive efforts depends on a multitude of diverse and site-specific factors including soil conditions, climate, and age of storage facilities. Production and management techniques, such as fertilizer/pesticide application, can have significantly different impacts on operations carried out in the same area. Insight into site-specific conditions is gained from an assessment. In the case of agricultural pollution, the assessment identifies potential sources of the pollution on a farm and analyzes the pollution risks from these sources. The level of investigation can range from generalizations drawn about a region to detailed evaluations conducted on a particular farm. The key is building a functional base of information pertinent to an individual farm. The more accurate the knowledge about individual conditions, the more effective the plan for management. Of particular significance is accurate information to design approaches that balance the impacts on both surface water and groundwater.

An ongoing program is necessary to fine-tune management strategies to meet the demands of a site. Though it is not an easy task, operators must monitor management changes to evaluate environmental, economic and technical impacts. They may rely on technical data from sophisticated sensors or their own observation. In any case, they must be alert to changes in conditions that call for refinement of management practices and must be prepared to respond with new actions.

Pollution prevention on the farm requires the use of concepts and technologies developed to manage and plan on a site level. Simple and inexpensive changes in practices, such as refined application of fertilizer and pesticides to minimize the risk of overapplication, could cut pollution risks and save farmers money over the long haul. Many aspects of site-specific management, however, presents technical and financial challenges that must be overcome to ensure effective implementation (Silvas, 1988). With advances in this field of management such as improved methods for evaluating and monitoring environmental conditions, costs should decrease, and more farm operators should gain access to these techniques.

APPLYING CONCEPTS REQUIRING SITE-SPECIFIC MANAGEMENT

Federal Programs

From the halls of government to the desks of lenders, many sectors in this society are reconsidering their views about the impact of agriculture on the environment and are rethinking the proper response to agriculture pollution. Traditionally, farmers were persuaded to change their practices with incentives and education, but these methods are not adequate to meet public demands for environmentally-responsible behavior. While voluntary programs will not disappear from the landscape, they are giving way to other strategies that force farmers to take action. Federal agricultural programs, however, secure farmer participation, one trend is clear. BMPs appear to be the preferred method to attack agricultural pollution. As a key to their implementation, site-specific management will grow in importance. A review of the following programs provides insight into the future.

Voluntary programs

The Section 319 program is typical of the traditional approach to agriculture. Added by the 1987 amendments of the CWA, Water Quality Act of 1987, Pub. L. No. 100-4, this new program furnished incentives for states to develop plans for controlling significant sources of nonpoint pollution and improving water quality. BMPs were the centerpiece of section 319, building on the significant administrative role they assumed prior to section 319. From its focus on surface water, the amendments offered the possibility of protecting groundwater. States were supposed to identify BMPs and measures to reduce pollution loading, 'taking into account the impact of the practice on groundwater quality' [42 U.S.C. Sec. 1329(b)(2)(A)].

The Agricultural Water Quality Incentive Program, authorized by the 1990 Farm Bill, Food, Agriculture, Conservation, and Trade Act of 1990, Publ. L. No. 101-624 (codified at U.S.C.A. 3838-3838(f), provides technical and financial assistance to farmers who sign 3–5 yr contracts to implement management plans to protect water quality.

The Farmstead Assessment System, an interagency program funded by the USDA-Extension and Soil Conservation Services and EPA, follows in the voluntary tradition, tackling a pollution problem largely overlooked by other programs--pollution from structures and activities on the farmstead. A single mishap on the farmstead--a chemical spill or petroleum leak--can negate yrs worth of careful field management practices. Building on partnerships with the private sector, the program encourages rural residents to assess pollution risks on their farmsteads and take voluntary actions similar to best management practices to prevent pollution of their groundwater (Jackson et al., 1993).

Conditional benefits

The 1985 Farm Bill, Food Security Act of 1985, Pub. L. No. 99-198 (codified at 16 U.S.C. 3801-3845), introduced new methods to increase use of conservation measures--"sodbusting," "swampbusting," and conservation compliance--designed to protect certain environmentally-sensitive land. Though the law did not technically mandate compliance, farmers who depended on price support and other agricultural benefits were not in any position to refuse to participate (Hamilton, 1993). To continue to receive federal benefits, they had to adopt management plans to protect the environment. Early programs required farmers to develop soil conservation plans based on site-specific potential for erosion.

Mandatory participation

In the Coastal Zone Act Reauthorization Amendments of 1990, Pub. L. 101-508 (codified at 16 U.S.C. Secs. 1451-1464), Congress expanded the power of participating states to protect coastal zones from agricultural pollution. In covered watersheds, the act specifies "Mandatory Management Measures" covering different agricultural activities. EPA Guidance stresses the implementation of management measures using practices that account for conditions on specific sites. It also recognizes that groundwater must be protected in the process of protecting coastal waters (Un.S. Environmental Protection Agency, Office of Water, 1993).

Future of federal programs

The CWA reauthorization should serve as a yardstick for direction of future regulation of agriculture. From all indications, Congress will adopt site-specific management techniques to combat nonpoint source pollution. One version of the reauthorization proposes voluntary measures to prevent agricultural runoff, and provides for development of site-specific management plans on farms. Those who favor stronger mandatory measures are demanding that landowners in impaired or threatened watersheds adopt on-site water quality improvements. One measure calls for landowners in high priority watersheds to implement "site-level programs", using appropriate management measures. The pressure for more mandatory measures is a sign that society intends to re-define the relationship between the farmer and the environment.

Changes in the philosophy underlying environmental protection will accelerate the use of BMPs. The most effective programs will apply site-specific management to implement them. Watershed protection is a leading initiative in the new wave of environmental protection measures. More complex than regulating the influx of pollutants, watershed management covers finer issues of drinking water quality, habitat preservation, and biodiversity. A comprehensive approach, watershed management depends on actions by a variety of stakeholders. To do their part in protecting a watershed, farmers need to know how their operations are threatening an area and what they can do to manage the problems.

With shifting attitudes toward agriculture, government programs affecting agriculture will be re-examined with new standards in mind. When Congress faces reauthorization of the Safe Water Drinking Act, it will consider ways to strengthen the wellhead protection program. Incorporating site-specific management, lawmakers could help communities by forcing farmers and other landowners to identify threats to public wells and take action to eliminate activities that may lead to contamination. From the preliminary discussions about the 1995 Farm Bill, the new law may signal an important shift in agricultural policy, phasing out government subsidies that promote environmentally-unsound practices and substituting payments designed to encourage good stewardship practices. Site-specific management will unquestionably assume greater importance in this scheme. Expanding on this idea, one legal commentator has proposed "incorporating the BMPs currently used by the SCS into those decisions granting bankruptcy relief, providing farm benefits such as subsidized credit and price supports, and evaluating the collateral for private banks" (Barnes, 1993). Farmers undoubtedly will bear more obligations relating to management of their operations. Anticipating these mounting burdens, federal lawmakers proposed legislation (Site-Specific Agricultural Resource Management Act of 1993, 1993 H.R. 1440) calling for one agency to incorporate government conservation and environmental requirements into an integrated resource management plan for individual farms.

State Protection

On the state level, government officials are tackling groundwater pollution from agriculture by devising a variety of programs that employ site-specific management. Historically, groundwater protection has been left to the states. Federal government intervention in the area has been incomplete and inadequate (Silvas, 1988). Facing agricultural threats of contamination from chemicals to petroleum products, states have recently stepped up their efforts to protect groundwater.

In Connecticut, those farming land located within aquifer protection areas must adopt an approved farm resources management plan under section22a-354m of the General Statutes. A technical team helps farmers select components for their management plans and set up schedules for plan implementation. State regulations cover planning issues related to manure management, pesticides, fertilizers, and underground and aboveground storage tanks.

To protect groundwater against chemical contamination, Minnesota has adopted Chapter 103H Groundwater Protection to promote use of best management practices for agricultural chemicals and practices, using traditional tools of education and demonstration projects. The practices span the range of protective actions including "practices to prevent site releases, spillage, or leaks." Sec. 103H.101. Those farmers in areas sensitive to groundwater contamination may protect themselves against liability if they follow soil and water conservation district plans to protect the groundwater from degradation.

Signed into Michigan law in November 1993, Act 247 represents one of the most ambitious efforts to protect groundwater and surface water through a stewardship program. Focused on protection against pesticide and fertilizer contamination, the voluntary program uses regional teams to offer educational opportunities, technical assistance, and private well water sampling. Receiving technical support from these teams, participants volunteer to prepare stewardship plans elaborating site-specific practices to protect groundwater from contamination by pesticides and fertilizers. With a plan in hand, farmers are then eligible for financial assistance to carry out site-specific management. The availability of financial assistance is a vital step in removing a major hurdle in implementing management practices.

Private Initiatives

Farmers and Producer Groups

In terms of liability risks, agricultural property is coming out from under the shadow of commercial property. Though contaminated agricultural property will never rival "Superfund" sites, these sites are presenting complex environmental problems of concern particularly for landowners. Site-specific management affords landowners opportunities to protect themselves.

Landowners are required by law to clean up pollution from their property by laws such as the Wisconsin Spill Law. Wis. Statutes Sec. 144.76. Costsharing is sometimes available (e.g. Minn. Statutes, Sec. 18E.04-Reimbursement for cleanup of chemical spills), but these programs only defray some of the costs. For cleanup of leaks from farm and residential fuel tanks, an area of significant need, few states offer assistance. As noted earlier cleanup can be prohibitively expensive. Unremedied, environmental problems may cut farmers off from needed financing and may prevent them from selling their property (Stein & Siegfried, 1992). Site-specific measures to prevent environmental problems can help landowners avoid these financial consequences (Anon., 1990), as well as yield other rewards such as protection of drinking water quality for their families and livestock, more cost-effective farming operations, and a better reputation as stewards of the land.

Without a nationwide commitment to provide education and technical assistance, landowners face a daunting task in trying to manage to protect the environment. Government standards and new technologies are becoming more complicated everyday and less accessible to the farmers who must devise management plans. Government-sponsored programs such as the Farmstead Assessment System and the Michigan stewardship program are examples of efforts to bridge these gaps by providing farmers the knowledge and tools to manage their property to prevent environmental problems. Based on initial results from the Farmstead Assessment System program, these voluntary programs are proving to be sound investments in environmental protection (Jackson et al., 1993). In addition, producer groups are offering their members assistance hoping to stave off more aggressive government intervention. For example, the Minnesota Turkey Growers recently developed BMPs to help members locate and maintain turkey farms, handle turkey manure, and dispose of dead turkeys. The National Pork Producers Council is urging members to support standardized BMPs for the pork industry (Hamilton, 1993).

Agricultural Business Community

Benefiting from agriculture, the agricultural business community is a logical group to share the burdens of agricultural pollution. With their financial resources, these businesses are being held accountable for agriculture's environmental problems and are facing new risks relating to environmental cleanups. Agricultural lenders, those involved in real estate sales, petroleum marketers, and others are changing the way they do business with farmers to avoid liability. Hoping to nip problems in the bud, they are educating farm customers and supporting improved management. In certain cases, they are refusing to do business with farmers whose property presents environmental risks. Through these means, they are forcing landowners who want to continue a business relationship to rethink management of their land to avoid environmental complications.

Agricultural lenders are at the forefront of these changes. Once a problem confined to commercial lenders, environmentally-contaminated property is a growing concern for agricultural lenders (Duncan, 1992; Knox, 1991). Each yr agricultural lenders report increasing losses from contaminated property (Cox, 1993). Instances of contamination from "routine" farm activities can spell as much trouble for lenders as those from a catastrophic episode:

> [One] lender foreclosed on a farmhouse and storage barn used as collateral for an agricultural loan. Pre-sale inspection revealed underground diesel fuel storage tank, waste pesticide containers, and soil staining in and around the storage shed. Subsequent environmental investigation and cleanup . . . cost approximately $1,000,000 as compared to [the collateral's] appraised value of . . . $27,000 (ABA, 1990).

For many loans, agricultural lenders face greater environmental risks than credit risks (Boehlje, 1992).

Embracing measures such as environmental questionnaires and assessments, lenders have fought back. Using these tools, lenders are avoiding contaminated collateral before they lend or foreclose. They are still working on ways, however, to protect their interests during the life of a loan. They must work around legal limitations that preclude their participation in a lender's affairs. Lenders successfully pressured EPA, however, to create safe harbors to permit them to effectively police their security interests. Under EPA standards, lenders can require borrowers to adopt management procedures to ensure compliance with the law and environmental cleanups (Gillon et al., 1992). To better monitor borrower's management, lenders are seeking more latitude in overseeing their customers' operations. In California, lenders secured a law permitting them to inspect borrower's property if they reasonably believe it is not managed to protect against contamination (Cal. Civ. Code Sec. 2929.5).

With lenders tightening their scrutiny over rural property, landowners cannot afford poor oversight of their property. To avoid contamination, they will turn to comprehensive management supported by site-specific techniques.

PROTECTING INFORMATION FROM USE AGAINST LANDOWNERS

While site-specific management is an effective choice to prevent pollution, it is not without risks. Collecting information is essential to site-specific management, but landowners must consider how they will respond to the discovery of information that could be used against them. Those interested in the information include prosecutors from government agencies and private citizens injured by a farmer's activities. To eliminate the barriers to the use of site-specific management, landowners must receive assurances concerning the use of assessment information. Such assurances can take the form of information confidentiality or limitations on the use of the information.

In some cases, landowners cannot circumvent statutory duties to report the release or presence of hazardous substances, e.g. in Michigan a petroleum tank owner must report leaks to DNR (Sec. 299.836, Mich. Stats.). Michigan also requires property owners to disclose information about environmental violations at the time of sale or transfer of property (Michigan Stats., Sec. 299.610c). Not every problem represents a reportable incident.

Currently, there is no comprehensive protection to shield information collected during an assessment with confidentiality. For those who can afford an attorney, such as an agri-business firm, threats to confidentiality can be reduced by conducting property assessments through an attorney whose legal advice can be incorporated into the process. Assessments cloaked with attorney-client privilege offer some protection against disclosure (Hogan & Bromberg, 1990).

Courts and enforcement agencies may extend some level of confidentiality to landowners wishing to prevent the use of information as evidence. In the past, courts have honored the assertion of a privilege to prevent disclosure of assessment records containing analysis and opinion of a self-critical evaluation. See, e.g., Bredice v. Doctor's Hosp., 50 F.R.D. 249 (D.D.C. 1970) aff'd, 479 F.2d 920 (D.C. Cir. 1973).

A landowner is also protected when government agencies agree to limit their use of assessment information. On the federal and state levels, administrative agencies have agreed not to rely on assessment information for routine enforcement use. The EPA Environmental Auditing Policy Statement, 51 Fed. Reg. 25004 (July 7, 1986), announced several measures to encourage parties to adopt sound environmental management practices, including audits. As a matter of policy, the EPA stated, it "will not routinely request environmental review reports," because such routine requests "could inhibit auditing in the long run, decreasing both the quantity and quality of audits conducted" (51 Fed. Reg. 25007). The EPA explained that its investigative efforts would normally rely on other sources of information, such as monitoring or reporting (51 Fed. Reg 25007).

Convinced about the value of this approach, the EPA encourages state and local regulating agencies "to adopt these or similar policies, in order to advance the use of effective environmental auditing in a consistent manner" (51 Fed. Reg. 25004). To encourage site-specific management, states do not need to protect the assessment information from use, they can protect landowners who are following farm management plans from liability. A policy of New Jersey Environmental Prosecutor's Office states that those who perform audits may avoid criminal prosecutions (Anon., 1992). Under Michigan's new environmental stewardship program, a participant cannot be sued for negligence in contaminating groundwater.

CONCLUSION

Through agricultural and nonpoint source pollution programs, government officials took the first steps to meet the challenge of controlling sources of water pollution from the country's two million farms. To stem the special pollution problems of soil erosion and fertilizer and pesticide runoff, these programs seized on BMPs as part of a system of site-specific management. Both federal and state governments will expand the use of these concepts to further environmental goals such as pollution prevention, watershed protection, and wellhead protection. Those in the private sector--farmers, and their organizations, suppliers and service providers--will increasingly use site-specific management to avoid environmental problems. Issues surrounding the protection of assessment information, however, may inhibit the growth of site-specific management unless steps are taken to remove these deterrents.

REFERENCES

Anon. Feb. 1990. Environmental audits. Successful Farming 91(2): 16.

American Bankers Association (ABA). 1990. Agricultural Lenders Guide To Environmental Liability. *In* K. Duncan, 1992. The back forty: Can lenders prudently lend against farm real estate? (CERCLA liability for foreclosures and receiverships). California Bankruptcy Journal 20:23–44.

Barnes, R. 1993. The U.C.C.'s insidious preference for agronomy over ecology in farm lending decisions. Univ. of Colorado Law Review 64:457–512.

Boehlje, M. 1992. Environmental Issues and Agriculture Lending. Paper on file with Farm*A*Syst, Univ. of Wisconsin-Madison, Madison, WI.

Bureau of National Affairs (BNA). 1990. Underground storage tanks: heating oil, motor fuel tank rules not needed, but new tanks should have protection, EPA says. Environ. Reporter, Current Developments 21:478–81.

Duncan, K. 1992. The back forty: Can lenders prudently lend against farm real estate? (CERCLA liability for foreclosures and receiverships). California Bankruptcy Journal 20: 23–44.

Flemming, M. 1987. Agricultural chemicals in ground water: Preventing contamination by removing barriers against low-input farming. Am. J. Altern. Agric. 2(3): 124–131.

Gillon, P., T. Kleissler, and J. Barnette. 1992. EPA's CERCLA rule on security interest exemption offers benefits to lenders. Inside Litigation 6(6): 9–12.

Government Accounting Office (GAO). Dec., 1991. Water pollution more emphasis needed on prevention to protect groundwater. GAO/RCED–92–47, Washington, DC.

Halstead, J., S. Padgitt, and S. Batie. 1990. Ground water contamination from agricultural sources: Implications for voluntary policy adherence from Iowa and Virginia farmers' attitudes. Am. J. Altern. Agric. 5(3): 126–133.

Hamilton, N. 1993. Feeding our future: Six philosophical issues shaping agricultural law. Nebraska Law Review 72: 210–257.

Hogan, E., and M. Bromberg. 1990. The hidden hazards of the environmental audit. Practical Lawyer 36: 15–26.

Jackson, G., R. Castelnuovo, E. Nevers, and D. Knox. 1993. Preventing water pollution through farmstead management of pesticides and toxic substances. Presented at the American Chemical Society Annual Meeting and Conference. August 22–27, 1993, Chicago, Il.

Knox, A. Jan. 1991. Should your farm have toxic clearance? Farm Journal 115(1): 20D–20E.

Lee, L., and E. Nielsen. 1987. The extent and cost of groundwater contamination by agriculture. J. Soil and Water Conserv. 42(4):243–248.

New Jersey Environmental Prosecutor's Office. May 15, 1992. Voluntary Environmental Audit/Compliance Guidelines, New Jersey (referred to in text as "Audit").

Note. 1978. Agricultural non-point source water pollution control under Sections 208 and 303 of the Clean Water Act: Has forty years of experience taught us anything? North Dakota Law Review 54: 589–618.

Piper, S., C. Young, and R. Magleby. 1990. Benefit and cost insights from the Rural Clean Water Program. J. Soil and Water Conserv. 45(5):203–209.

Reichenberger, L. 1992. Guard against spills. Farm Journal 116(6): 14–15.

Stein, S., and D. Siegfried. 1992. Underground tanks can sink the firmest of deals. Real Estate Today 25:57–59.

Silvas, D. 1988. Groundwater pollution from agricultural activities: Policies for protection. Stanford Environmental Law Journal 7:117–186.

Travis, C., and C. Doty. 1990. Can contaminated aquifers at Superfund sites be remediated? Environ. Sci. Technol. 21: 1464–1466.

United States Environmental Protection Agency, Office of Water. Jan., 1993. Guidance Specifying Management Measures For Nonpoint Pollution in Coastal Waters, EPA 840–B–92–002. Washington, DC. (referred to in text as "Guidance").

United States Environmental Protection Agency. Spring, 1993. EPA Pollution Prevention News, EPA 742–N–93–001. Washington, DC.

United States Environmental Protection Agency, Office of Water. 1986. National Water Quality Inventory, 1986 Report to Congress. Washington, DC.

United States Environmental Protection Agency, Office of Water Program Operations. 1984. Report to Congress: Nonpoint Source Pollution in the U.S. Washington, DC.

Wisconsin Department of Industry, Labor and Human Relations (DILHR). 1993. Fiscal Estimate: 1993 Senate Bill 15.

SESSION VI

TECHNOLOGY TRANSFER

65 Virtual Agriculture: Developing and Transferring Agricultural Technology in the 21st Century

D. A. Holt
S. T. Sonka

Agricultural Experiment Station
University of Illinois
Urbana, Illinois

"Our ability to get better at the innovation process--to drive new products from idea to market faster and with fewer mistakes--is the key to winning this (product) war." Robert G. Cooper

In this paper, we draw an analogy between the agriculture of the future and the memory of computers. With appropriate software, computers on a network can operate so that the memory (data storage and retrieval capacity) of each computer becomes part of the collective memory of the network. To the individual user, this appears as one memory, one source of, and repository for, information. The collective capability is referred to as "virtual" memory.

In the future, developing and successfully commercializing new agricultural products and services, including those associated with site-specific agriculture, will require organizing increasingly complex partnerships. These partnerships will bring together specialists with many different perspectives and kinds of expertise. The success of agricultural initiatives will depend on the capability of the specialists and the degree to which their activities are coordinated, integrated, and focused.

Research and development (R&D) efforts enabling and supporting complex commercial initiatives in agriculture also will be organizationally complex. They will require unusually close relationships in which public institutions and agencies are full partners in commercialization efforts. Like other suppliers/partners in these efforts, public institutions and agencies will provide unique capabilities and facilities. Differences in organizational cultures may make it difficult for some organizations to work well together.

In this environment of "time-based" competition and "just-in-time" R&D, it will be necessary to conceive, design, develop, and implement the technology transfer process as an integral part of the R&D process. Fully integrating technology transfer with other research and development functions will enable

R&D teams to anticipate problems of implementing new technology, so that those problems can be addressed early in the R&D process.

Like the collective memory of networked computers, partners in the complex value-added partnerships of the future, including the partners with responsibility for research and technology transfer, will have to function as one; hence the term "virtual" agriculture. Virtual agriculture is not only analogous to networked computers, it will depend heavily on networked computers. The information infrastructure must play a new role, enabling frequent, easy, powerful, and inexpensive communication and coordination within and among agricultural partnerships. The information infrastructure will also increase the importance of information search and acquisition among R&D activities.

INCREASING SPECIALIZATION IN AGRICULTURE

Prior to the dramatic changes in U. S. agriculture of the past century, the farmer was the producer, processor, distributor, retailer, and consumer of agricultural products and provided most of the inputs and support services. Because each farm was self-sufficient, agriculture could proceed with relatively little communication and coordination outside the individual farm family.

Now, agriculture is driven by the logistical demands of serving large urban populations. It is made up of many highly specialized activities, involving many different individuals and groups (Fig. 65–1). These diverse specialists conceive, design, develop, implement, and manage the activities that once were accomplished almost entirely by farmers. Farmers themselves have become specialists, functioning with other specialists in the industry of agriculture.

Increased awareness and concern about environment, natural resource, food safety, safety in the workplace, social impact, and other issues make agriculture even more complex and result in new specialties and more specialists. To serve the diverse needs and desires of individual consumers and consumer groups and compete in highly competitive agricultural markets, specialized participants in the industry of agriculture, including consumers, must communicate with each other and coordinate their activities much more than their predecessors. This trend will continue in the future.

ADVANTAGES AND DISADVANTAGES OF SPECIALIZATION

Specialization greatly increased agricultural productivity and efficiency over the past century. Now, only a little more than 20% of the U. S. workforce is required to provide agricultural products and services, both domestically and internationally. This is in stark contrast to agriculture in undeveloped nations, in which a very large proportion of a nation's human and physical resources may be required to meet agricultural needs.

In general, a specialist in a task or process performs it better than a non-specialist. Also, specialists may gain some economies of scale in their specialty, thus realizing cost savings, some of which can be passed on to customers. Companies willing to commit to a specialist as a sole-source supplier enable that specialist to gain economies of scale. The company can then reap some of the

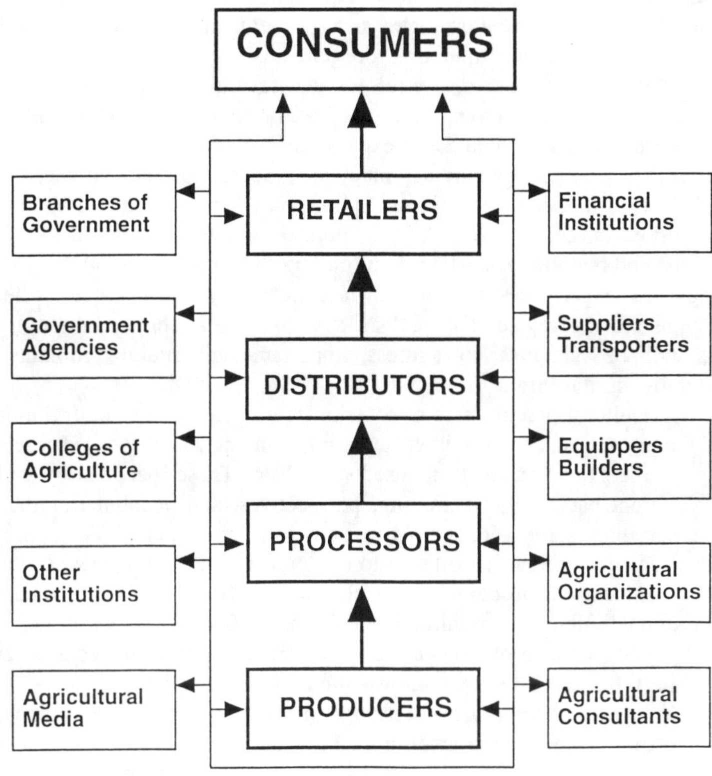

Fig. 65–1. Material and information relationships in the industry of agriculture.

benefits of their supplier's increased efficiency. By serving several customers, perhaps in diverse industries, specialists may be able to spread their market risks.

Specialization has disadvantages, as well. Various environmental, natural resource, and food safety problems may have arisen from a lack of communication, coordination, and cooperation among diverse specialists involved in agriculture. Real and imaginary conflicts of interest may develop among diverse agricultural specialists. An agricultural institution, agency, or private firm made up of narrowly-focused specialists may lose a broader institutional focus, and lack a shared vision of the organization's role in industry and responsibility to society. It may lack a collective sense of identity and become confused on mission and objectives. It may develop some unhealthy internal competition.

DO WE NEED SPECIALISTS OR GENERALISTS?

For practical purposes, there is no such thing as a generalist in modern agriculture. Realistically, no individual can learn enough about the enormously complex industry of agriculture or even major segments of it to be considered a general expert. People trained to be generalists are likely to acquire only superficial knowledge. Persons with only general training and experience will find it particularly difficult to be successful in 21st century agriculture, until they develop specialized knowledge and experience.

Self-proclaimed generalists may be a source of bad advice. I am reminded of my father's admonition, "a little bit of knowledge is a bad thing." It is important for people working in the industry of agriculture to recognize the limitations and boundaries of their own training and experience and draw on the training and experience of others when necessary. Only the collective knowledge and experience of specialists, assembled and integrated through team effort, will be adequate to address important agricultural problems and opportunities in the future.

Agricultural practitioners often criticize universities for not training and employing generalists. Universities increasingly hire narrowly-focused specialists who train others to be narrowly-focused specialists. These specialists often have much different backgrounds, training, perspectives, and vocabularies than the practitioners they serve and, for that matter, from each other. The specialists may do research of great importance to agriculture, but the linkages to practice may be complex and obscure. The technical details of a specialist's research may be unintelligible to agricultural practitioners. Likewise, even though their research may have profound effects on practical systems (e.g., genetic engineering), the specialist may know little about practical systems.

Colleges of Agriculture increasingly hire specialists who specialize at the molecular or macroeconomic level in the hierarchy of systems. But most of the practitioners function at the field, enterprise, farm, or agribusiness level. In my opinion, agricultural colleges should consider hiring more people who specialize in research and education focused at the levels of systems managed by agricultural practitioners. These people could bring such powerful analytic tools as industry analysis, network process theory, game theory, logistics, systems analysis, systems design, systems engineering, business systems, and operations research to bear on the management of practical systems. Having one or more such people in every subject matter area would help the other specialists communicate with practitioners and integrate the work in their specialty more closely with practice.

INCREASED NEED FOR COORDINATION IN MANUFACTURING

Some of the opportunities and problems associated with developing and transferring information and technology for site-specific management may be anticipated by studying trends in manufacturing industries. In many ways, agriculture is a manufacturing industry. Many of its component institutions are manufacturing firms. Much of the technology of site-specific management of agricultural systems must be manufactured. Manufacturers have responded to growing competitive pressure and global markets by adopting new organizational approaches. These include "out-sourcing," "just-in-time manufacturing," and "computer-integrated manufacturing." The need for extraordinary coordination is a consistent theme in these powerful approaches to manufacturing.

Increasingly, manufacturing firms purchase components and services from specialized suppliers (out-sourcing) and merely assemble the final product. For the manufacturing process to work, each component must fit in the final product and perform its unique role effectively. The components must arrive in time, arrive in appropriate quantities, and meet exact specifications required for the manufacturing process, so that it is unnecessary for the manufacturer to maintain large inventories of parts or products (just-in-time manufacturing).

In the out-sourcing situation, each of the suppliers has marketing, design, manufacturing, accounting, and business administration functions, among others, that must be closely coordinated with those of its customer, the manufacturer. In this situation, the high degree of interdependency among manufacturers and suppliers mandates an extraordinarily high level of communication, cooperation, and coordination.

The entire manufacturing process may be coordinated through an elaborate computer hardware and software system linking marketing, design, manufacturing, accounting, business administration, and other functions (computer-integrated manufacturing). Communication, cooperation, and coordination will be even more important in the vast, complex, geographically dispersed, and decentralized world of virtual agriculture.

VALUE CHAINS IN AGRICULTURE

Producing a consumer product in modern agriculture may involve producing a long sequence of intermediate products. Each is used as an ingredient or raw material to make the next product in the sequence. The term 'value-added' is used to describe such processes, because value is added at each step. In the language of economics, each step in the value-added sequence is an "economic stage," a link in a long "value chain." For each product there is a market. In the case of a common food product, the hot dog, one sequence of products includes inbred corn seed, hybrid corn seed, corn grain, feed, swine breeding stock, feeder pigs, fattened pigs, pork, wieners, and, the final product, hot dogs.

Instead of all products in the corn-to-hot dog chain being produced by one farm family, many different farm and agribusiness firms are involved. These include seed firms, corn growers, swine breeders, feeder pig producers, commercial swine producers, meat packers, meat processors, meat distributors, and meat retailers, including restaurants. Supporting each of these groups are financial institutions, suppliers, equippers, builders, transporters, consultants, farm and trade organizations, branches of government, government agencies, land-grant universities and their colleges of agriculture, other institutions, and the media.

This brief description is actually considerably oversimplified. We did not describe the economic stages involved in providing some of the inputs, such as fertilizer for corn production, protein supplement for livestock production, non-meat ingredients of wieners, or buns and condiments for hot dogs. Nor did we list inputs required to produce, market, and utilize co-products or by-products associated with any of the intermediate products. To put it another way, we did not describe the converging and diverging branches of the value chain linking corn and hot dogs.

The value chains leading to commercially successful site-specific management technology converge on the corn-to-hot-dogs value chain and on similar chains leading to other agricultural end products and services. Many different component technologies, including sensors, positioning technology, expert systems and other control software, microprocessors, and automatic control mechanisms, must be integrated to provide the tools for site-specific management. Each item of information and technology must "fit" in the site-specific management system. R&D on these technologies must be closely coordinated with R&D within associated value chains.

Some liken successive economic stages to a river and refer to "upstream" and "downstream" processes. In our example, growing corn is upstream and processing meat is downstream from feeding pigs. Others view producing the sequence of products as a vertical process, starting at the bottom and moving upward through successive products of increasing value and culminating in the value realized by the consumer of the final product. When two or more economic stages are accomplished by one agribusiness firm, some "vertical integration" has occurred. Or, two or more firms may cooperate and coordinate closely to accomplish two or more economic stages. To distinguish this from vertical integration, the term "vertical cooperation" is used. The cooperating firms may be referred to collectively as a "value-added partnership."

INCREASING NEED FOR FUNCTIONAL INTEGRATION

Forces at play in today's economic and social environment create a need for greater communication and coordination among specialists in value-added partnerships. Those forces include increasing competition in global and domestic markets; more demanding and particular consumers; and major advances in our ability to tailor agricultural products, through biotechnology and other approaches, to specific uses and individual needs and desires. In general, farmers and agribusiness firms are trying to escape the dog-eat-dog, cost-based competition of commodity markets. They want to produce differentiated products and services that appeal to consumers because of unique quality characteristics. Like manufacturers, they are seeking to achieve differentiation and sustainable competitive advantage through alliances.

The hot dog provides a good example. Many people love the taste of pork in the form of hot dogs. But some fear that hot dogs, traditionally very high in saturated fat, are not healthy food. Suppose it is possible to produce corn or soybean varieties with exceptionally low levels of saturated fatty acids. When this is fed to pigs, it should lower the saturated fatty acid content of the fat in pork. If the pigs are genetically engineered to have less fat, the effect should be enhanced. This could result in healthier wieners and hot dogs.

Research suggests that the healthier hot dog scenario is possible, but several things would have to happen. Of course, at least some consumers would have to prefer the healthier hot dogs enough to pay more for them. Also, the special, low-saturated-fatty-acid seed, grain, pigs, pork, wieners, and hot dogs will have to be kept separate or at least identified separately from normal types throughout the whole process. Identity-preserved is an important buzzword among those who hope to commercialize grains that confer special characteristics on products manufactured from them.

The crop varieties with lower levels of saturated fatty acids may be genetically different to the extent that they require modified fertilization, establishment, pest control, harvest, and storage techniques. If special crop varieties must be managed differently from other special varieties and from conventional varieties, a new level of complexity is added to site-specific management. Those with knowledge about such differences will gain competitive advantage in the highly competitive agricultural environment of the future.

MECHANISMS OF MARKET COORDINATION

The whole process of producing healthier hot dogs will have to be closely coordinated. Imagine how complex the markets will become if there are many more different types of corn and soybeans, strains of livestock, and versions of each traditional food product. Much more site- and situation-specific information will have to flow within the partnerships involved in producing and delivering each of these products. Some new and improved mechanisms of market coordination will be needed.

At present, most agricultural markets are coordinated through the mechanism we refer to as the "open market." Producers at each level in a value chain submit their products to a market. Quantity and price are informally negotiated among buyers and sellers, based on their perceptions of supply and demand. In the future, more markets will be coordinated by contracts and agreements. Some firms or organizations will seek to coordinate entire value chains through contracts. They will contract with: seed companies to produce special inbreds and varieties; farmers to produce hybrid seed, special grain, and unique feedstuffs; and animal breeders to produce special breeds of animals.

To complete the process, they will need to contract with commercial livestock producers and feeders to produce special types of livestock, packers to produce special meats, and processors to process those special meats into unique food products. They may even contract with retailers to sell the special products. In this way, everybody involved in producing a differentiated product, like those healthy hot dogs, may be organized into a vertical, value-added partnership. Different products will require different degrees of organization and coordination, some less and some more than the example given here.

Why would an organization be motivated to organize such a complex contractual relationship? What would they have to gain for their efforts? Suppose a firm developed or licensed inbred lines of corn that confer special characteristics, like unusually low levels of saturated fatty acids, on products made from corn or animal products produced by animals that are fed corn. To capture the most value from their investment in that development, the firm would need to receive a royalty or other fee for the use of its intellectual property at all stages of processes leading from corn to final products, not just at the seed production stage.

If its inbred lines of corn were not protected by patent but were sufficiently secret to qualify as trade secrets, the firm might not wish to transfer ownership of the material to anyone, only to allow and control its use. Through the contract approach, the firm can extend its intellectual property rights so as to maintain control not only of the special inbred lines of corn but also of intermediate and final products derived from them. Intellectual property rights will often be important issues in value-added partnerships, allowing the partners to maintain a differentiated position in vertically coordinated markets, hopefully long enough to recover initial investments and realize profits.

In some cases, protecting intellectual property rights will be important in successfully commercializing site-specific management technology. If customers see more value in differentiated products associated with site-specific management, they will buy more of them and pay more for them. This additional economic benefit will be divided as fairly as possible among the partners, including the ones who developed the technology and organized the production process. It won't be long, however, before some competitor will produce an even better version of the product and capture some of the market. This brings us to the need for more highly organized and coordinated research and development (R&D) efforts.

VIRTUAL R&D

We made the case that the process that produces and delivers agricultural products and services is more complex than it was in the agriculture of the past. The R&D process by which those products and the means to produce them are conceived, designed, and developed is also more complex. A research effort leading to a differentiated product may have to deal with all the economic stages involved in producing the final product. It should include all the research and development functions, including basic, developmental, and adaptive research and technology transfer (see functional definitions at the end of this paper).

The technology transfer function should be conceived, designed, developed, and implemented as an integral part of and fully integrated with the overall R&D process. The importance of addressing technology transfer problems and opportunities early in a R&D process is revealed by the experience of private firms that survive by successfully commercializing information and technology. Their experience suggests that over 90% of commercial initiatives (not just ideas, but initiatives that have been launched) fail. Also, they find that over 90% of commercial initiatives that fail are perceived as conceptually sound and technically feasible by knowledgeable people, but fail in implementation (personal communication; Dr. Steven Sonka, University of Illinois Agricultural Economics Department). Integrating technology transfer with other R&D

functions enables R&D teams to anticipate problems of implementation and address them with specific R&D activities.

There will often be not only scientific and technical but also economic, social, environmental, legal, and perhaps even political constraints to commercializing a new product. Site-specific management will be required to address some of these issues. Also, the commercial success of a product may depend to some extent on success in commercializing co-products and/or by-products. All of these issues will have to be addressed in the R&D effort. Many different kinds of scientific and technical expertise will be needed just to provide information and technology for one value chain.

Just as manufacturers gain advantage by "just-in-time" manufacturing, vertical partnerships will gain advantage by just-in-time R&D. The many pieces of information and technology required to provide a new or improved product for consumers will have to be provided on schedule. They will have to fit in the overall process and perform their unique functions. Partners will have to work closely to speed the process of innovation, so as to produce a superior product sooner than competitors ("time-based competition") or achieve important environmental and social goals before serious damage is done.

It will be necessary in most situations to coordinate the innovation process more closely. It will probably be preferable to conduct the research and development functions simultaneously instead of sequentially, in a parallel, functionally integrated manner (Fig. 65–2) (Holt, 1991). This is referred to in manufacturing circles as concurrent engineering. Even more organizationally complex versions of this approach include the stage-gate process (Fig. 65–3) described by Cooper (1990), in which stages of parallel R&D are delineated by "gates." The gates represent points at which decision makers measure and evaluate progress and make "go" or "no-go" decisions.

Drucker (1992) discusses further refinements of the parallel approach. He reports that some Japanese firms launch parallel efforts to achieve practical goals that will be implemented sequentially. For example, a Japanese R&D team may be working simultaneously on a new product, a 25% improvement on that product, and an innovative replacement for the product.

In order to achieve maximum productivity and efficiency in integrated R&D efforts, it is important to: clearly define the overall goal of the effort; keep each individual component focused on that goal; make sure the effectiveness of each component is measured in terms of contribution to achieving the overall goal; identify bottlenecks; minimize constraints imposed by bottlenecks; and strive for continuous incremental improvements in performance. This goal-oriented approach is similar to that shown to be advantageous in manufacturing (Goldratt & Cox, 1986).

The key components of virtual agriculture and virtual R&D are the partnerships, that is, the teams or task forces that conduct the R&D and implement the resulting information and technology. The study of teams, how to organize and manage them, and how best to exploit their collective capabilities, is emerging as another specialty, a subdiscipline of management. A lengthy study of teams (Katzenbach & Smith, 1993) revealed that teams form most effectively and often spontaneously around the most demanding performance

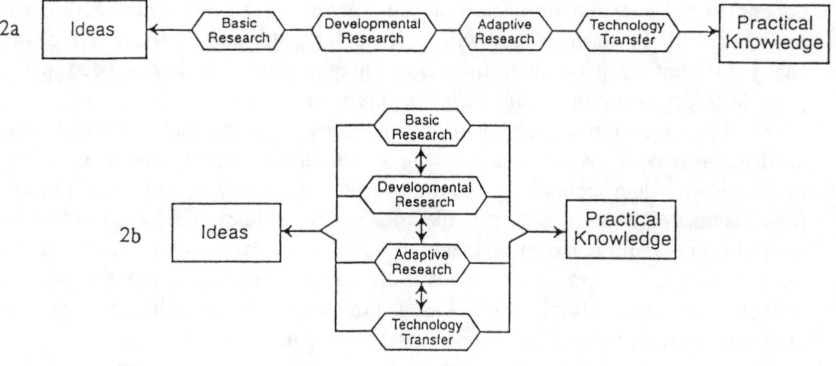

Fig. 65–2. Organizational models of agricultural research and development: 2a--the linear model; 2b--the parallel model.

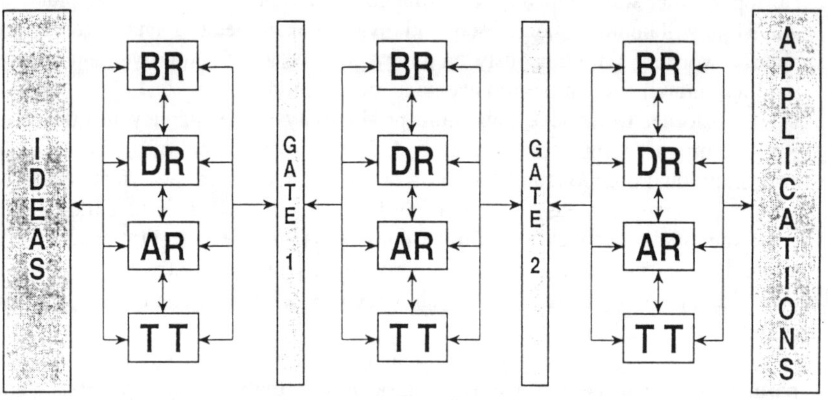

Fig. 65–3. The stage-gate system of R&D organization. BR = basic research; DR = developmental research; AR = adaptive research; and TT = technology transfer. Adapted from Cooper, 1990.

challenges. As a corollary, organizations with the highest organizational performance standards tend to spawn the most effective teams. Note that I distinguished between organizational and individual performance standards.

Keeping complex vertical partnerships coordinated requires many individual transactions among participants. There is a cost in time and money associated with each transaction. Managers need to consider transaction costs, because such costs can exceed the additional value achieved by creating a more elaborate organization. Participants in a vertical partnership will often be far apart, working in organizations that may be scattered over the nation and the world. Because transaction costs would be too high, they will not be able to coordinate their work adequately by using only face-to-face communication mechanisms, like meetings. Information technology can be used to minimize transaction costs.

Rapid and efficient flow of valuable information within and between market and R&D partnerships will be essential to their success and their ability to achieve sustainable competitive advantage. In fact, the rate and efficiency with which information can be generated and transferred within value-added partnerships will be the basis of competition in virtual agriculture. It simply will not be possible to achieve the necessary levels of communication among R&D partners with current management approaches and information technology. Partners, including the research and education partners, in each value-added initiative will need to be in constant communication. It will be necessary to achieve the same levels of communication that are possible among people working in the same room.

Access to a high-performance information infrastructure will be essential for communication within R&D teams. In addition, the infrastructure will so enhance information search and acquisition capabilities that these activities will become a larger and more essential component of R&D. While this is a positive development, Chaudhari (1993) fears that it will result in a decrease in emphasis on basic or "exploratory research." Private firms, he explains, will ask, why spend money on basic research or peripheral exploratory research if the information is available elsewhere? He sees universities taking an even greater share of responsibility for basic research. In the university setting, this research will be less effectively linked to market needs, unless extraordinary efforts are made to facilitate industry/university linkages.

VERTICAL COOPERATION AND ORGANIZATIONAL CULTURE

It is increasingly clear that organizations develop unique organizational cultures. Some are managed from the top down; others are quite decentralized. Some private firms are cost-focused, trying hard to turn out products that appeal to the cost-conscious, sometimes compromising on quality to achieve the low-cost position. Other firms may focus on quality, hoping the perceived value of the product will overcome any customer concerns about higher price. Cultural change, which is usually quite difficult, might be required for firms with contrasting cultures to work together effectively in a value-added partnership.

Because value-added opportunities are created at the primary production level, farmers will be important partners in virtual agriculture. Obviously, farmers will be important users of the products of site-specific management. Culturally, most U. S. farmers and some agribusiness people have been cost-focused, as cost is the basis of competition in commodity markets. Valued cultural attributes of U. S. farmers include rugged individuality, self-reliance, and respect for individual achievement, e.g., the "self-made" man (Atherton, 1961). People with these cultural tendencies may have difficulty working well in situations with high levels of interdependency. Also, some farmers consistently consider themselves vulnerable to and victimized by their suppliers. Large organizations that purchase farm products, such as packing houses and grocery chains, are regarded by some farmers as predators.

On the other hand, U. S. farmers' cultural tradition of "live and let live" is consistent with a modern industrial tendency toward decentralization and cultural diversity. Also, farmers traditionally have been very cooperative with and helpful toward neighbors. In the early days of agricultural mechanization, farmers developed relatively elaborate organizations for sharing expensive equipment and labor (e.g., the threshing ring), thereby achieving important economies of scale. To participate successfully in these informal organizations required a great deal of personal responsibility and integrity, qualities in which farmers take pride. In the future, farmers may gain economies of scale by sharing expensive items of site-specific management technology.

Some of the aforementioned cultural traits, plus farmers' necessary capacity for hard work and timeliness, should enable progressive farmers to function well in virtual agriculture. Fully integrating public sector scientists into these partnerships may be somewhat more challenging (Armstrong, 1992). The university tradition of unrestricted exchange of information may not be completely compatible with preserving intellectual property rights, which will be essential to the success of some value-added partnerships.

Also, in some university circles, more prestige accrues to those who conduct basic research and theoretical studies than to those engaged in developmental research, adaptive research, technology transfer, or agricultural practice. People who are uncomfortable in situations without those status distinctions may find it difficult to work in the highly interdependent and functionally integrated environment of virtual agriculture. Armstrong identifies this as a major challenge facing those who wish to foster greater university involvement in national technology competitiveness initiatives.

VERTICAL COOPERATION AS A FIELD OF STUDY

Achieving necessary levels of market and R&D coordination in agriculture is increasingly complex and difficult because of the number and diversity of interests and concerns involved in each value chain. The study of this topic is leading to new concepts, new organizational paradigms, and new understanding of the economics of organizations (Barry et al., 1992). It is known that creating these highly coordinated partnerships provides more advantages in some cases than others. For example, the need for and success of such partnerships is greater when at least some partners need highly specialized equipment and expertise that cannot readily be shifted to other uses. To illustrate, farmers who invest in effective, highly specialized, site-specific management equipment and expertise may compete more effectively than others for contract production opportunities.

Specialists are motivated to participate in value-added partnerships to minimize transaction costs, among other important advantages. In the open market approach, each individual sale/purchase of intermediate and final products is negotiated separately, as a separate transaction. Using contracts, vertical partners determine the conditions of sale/purchase for many future transactions in one negotiation.

Many transactions are required to organize and conduct R&D programs that are fully integrated over R&D functions, disciplines, and economic stages within value chains. Public sector R&D in the U. S. tends toward single investigator grants supporting narrowly focused research projects. This avoids the cost of transactions required to organize and conduct functionally integrated research, but fails to establish the linkages and achieve the integration required to move concepts rapidly from theory to practice.

If publicly supported scientists are to contribute maximally to developing new agricultural products and services, those scientists will have to be more fully integrated into vertical partnerships. This will inevitably increase transaction costs. Information technology can be used to help minimize R&D transaction costs and maximize benefits relative to costs of creating such elaborate organizations.

CREATING THE VIRTUAL AGRICULTURE INFRASTRUCTURE

Great benefits can accrue from a hardware and software infrastructure that facilitates frequent, easy, and inexpensive communication within and between value-added partnerships. The information "super-highway," that is, the backbone network, will be a key component of this infrastructure. The challenge for telecommunications and information specialists is to develop that infrastructure and tailor it to the unique needs associated with market and R&D coordination. Advanced information technology, including the rapidly emerging computer networks, will enable virtual R&D and virtual agriculture.

FUNCTIONAL DEFINITIONS OF IMPORTANT R&D FUNCTIONS

Basic research produces basic knowledge of the structure and function of physical, chemical, biological, economic, and social systems and materials.

Developmental research produces specific products and practices with potential practical and/or commercial utility.

Adaptive research involves testing, comparing, and refining the products and practices generated in developmental research, evaluating them for their utility in specific situations, measuring interactions among them, and thereby generating information with which users can integrate them into effective, practical, operational systems.

Technology transfer is defined broadly as the processes by which the products of research are transferred to and implemented by consumers and users of the technology and information. Transfer formats include formal and informal instruction, continuing education, in-service training, and decision support.

REFERENCES

Armstrong, J.S. 1992. University research: new goals, new practices. Issues in Science and Technology, Winter, 1992–93, pp. 50–53.

Atherton, L. 1961. The Cattle Kings. University of Nebraska Press, Lincoln, Nebraska.

Barry, P.J., S.T. Sonka, and Kaouthar Lajili. 1992. Vertical coordination, financial structure, and the changing theory of the firm. Amer. J. Agr. Econ. 74:1219–1225.

Chaudhari, P. 1993. Corporate R&D in the United States. Physics Today, December, 1993, pp. 39–40.

Cooper, R.G. 1990. Stage-gate systems: a new tool for managing new products. Business Horizons, May-June, 1990.

Drucker, Peter F. 1992. Japan: new strategies for a new reality. The Wall Street Journal. October 2, 1991. p. A12.

Goldratt, E.M., and J. Cox. 1986. The goal: A process of ongoing improvement. North River Press, Inc. Croton-on-Hudson, New York.

Holt, D.A. 1991. Organizational paradigms of agricultural research and development. Proc. Fortieth Annual Meeting of the Agricultural Research Institute. Published by the Agric. Res. Inst., Bethesda, MD.

Katzenbach, J.R., and D.K. Smith. 1993. The wisdom of teams: Creating the high-performance organization. Harvard Business School Press. Boston, Massachusetts.

66 Boundary Line Determination Technique (BOLIDES)

Ewald Schnug
Jürgen Heym
Donal P. Murphy

Federal Agricultural Research Center
Institute of Plant Nutrition & Soil Science
Braunschweig-Voelkenrode
Germany

The complete implementation of the Local Resource Management (LRM) concept (Schnug et al., 1993) relies on the exploitation of spatially distributed crop yield, plant, and soil data processed in a Local Resource Information System (LORIS). The collection of the required yield data is now routine on some farms in western Europe (Murphy et al., this volume), and reduced cost procedures for collecting appropriate soil data have been proposed (Murphy et al., 1994; Schnug et al., 1994). For crop fertilization, the full exploitation of the soil analytical techniques is hindered by the lack of accepted calibration to categorize soils according to the degree to which crop growth is limited by their nutrient status as determined by these lower cost laboratory analyses. Conventionally, nutrient analyses are made on bulked soil and plant samples taken over a large area with each nutrient assessed by a separate analytical procedure. The analyses used are calibrated within the geographic region concerned using extensive field trials. These long and costly calibration procedures are based mostly on the mathematical or graphical analysis of data relating nutrient level to relative crop response in conventional univariate field trials (Dahnke & Olson, 1990). With the LRM concept, the availability of crop yield and soil data for a very large number of distinct locations within a field enables the use of the boundary line to calibrate new analytical techniques. The analyses of soil or plant samples taken to map nutrient status over a field are calibrated using the field's own crop yield data. Thus, the analytical procedure is 'locally calibrated', and the problems of imprecise conventional calibration (Evanylo & Sumner, 1987) carried out at much lower yield levels, and interference from other limiting factors, are circumvented.

The principle of the boundary line approach has been described by Webb (1972), and it was used to determine critical nutrient levels (Moller-Nielsen & Frijs-Nielsen, 1976; Evanylo & Sumner, 1987). The points relating nutrient status to crop yield (or relative crop yield) are plotted in a two dimensional space.

The line describing the highest yields observed over the range of nutrient values measured is known as the boundary line since it lies on the upper edge of the body of the data. This line describes the response to variation in the test parameter where all other factors are, within the constraints of the field concerned, as close as possible to non-limiting in terms of crop yield. Data points below this line relate to samples where some other factor limits crop response to the nutrient. As an example, a boundary line drawn on data relating crop yield and soil pH is given in Fig. 66–1. A clear optimum pH for the site is identified. The slope of the line at sub-optimal levels is related to the response to added liming. Similar boundary lines can be drawn for any data set where crop yield and nutrient status are measured for a given situation, site or point in a field. To date, due to the lack of statistical procedures, boundary lines are drawn 'by hand', as in the case of Fig. 66–1. This is not a reproducible method, limiting the acceptance of the boundary line approach. BOLIDES is a software program which enables a mathematically correct and reproducible fitting of boundary lines to individual classes of XY scattered data. It calculates optimum values or ranges for plant nutrients.

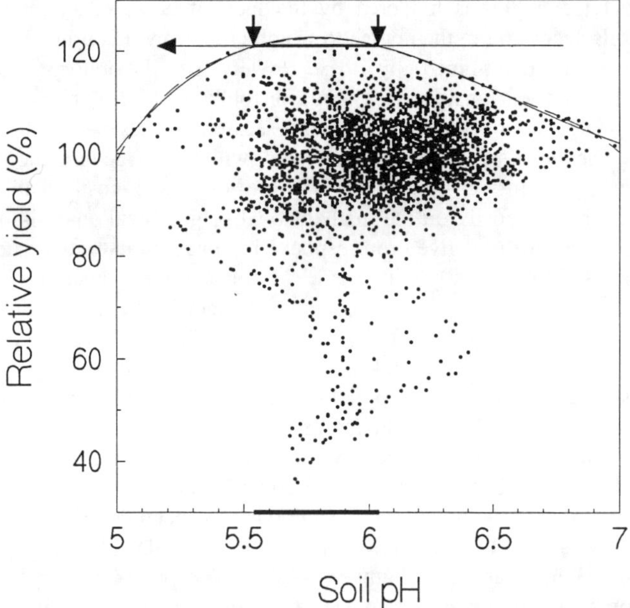

Fig. 66–1. An example of the evaluation of optimum soil pH levels from identifying outliers.

IDENTIFYING OUTLIERS

The main problem in determining the path of the boundary line is the identification and elimination of outliers in the underlying data set (S_o) containing (N_o) data points. These are the small number of points (N_e) that lie isolated from the main body of the data. Their inclusion in the boundary line drawing process completely invalidates the results in the production of fertilizer recommendations. A process which identifies values that are separated from the remaining data by a distance determined in terms of both nutrient status and yield is required. BOLIDES imposes cells on the scattered data with each data point in the underlying set (S_o) centered within its own cell. The size of each cell may be set according to statistical criteria relating to each parameter. In the case of combine harvested crop yield, typical values for the coefficient of variation (c.v.) of combine harvested grain yield may be used to determine cell size on the vertical axis. For the nutrient parameter, the variation in results of the soil test procedure in question, repeated ten times on a given sample, may be used. Thus, for research, the program allows the user to set statistical criteria determining the cell size which then may be reported with the boundary line. When these cells are imposed on the data, a data point that lies in a cell with fewer than a set number of neighbors is identified as an outlier. The defined exclusion procedure normally results in a slightly reduced data set S_1 containing N_1 (N_o-N_e) data points. If no outliers are identified, the two data sets S_o and S_1 are identical. The selection of cell size and the minimum number of data in each cell may greatly affect the number of outliers identified and their location. Figure 66–2 shows the data (as for Fig. 66–1) analyzed with varying cell size (Δ pH, Δ yield) and cell data point content criteria (ncell). Outlying data are marked with boxes and excluded in the boundary line determination steps. Cell size varied along the pH axis (0.10 and 0.05 pH units) and two ncell levels were used (3 and 4). When ΔpH is 0.10, changing the ncell from 3 to 4 has no effect on the upper boundary outliers, and therefore, no effect on the boundary line. Reducing ΔpH from 0.10 to 0.05 increases the number of outliers identified, especially where ncell is 4.

LOCATING UPPER BOUNDARY DATA POINTS

A schematic of the steps used to locate the upper boundary data points is given in Fig. 66–3. The data set S_1 (N_1 data points) that excludes the outliers is further reduced to a data set S_2 (N_2 data points). S_2 consists of the upper boundary data points of S_1. To identify the data points of S_2, BOLIDES first determines the minimum nutrient point (mnp), the maximum nutrient point (MNP), and the maximum yield point (MYP) of the data set S_1 (the subset of S_o). In step 2, BOLIDES searches the upper boundary data points between the mnp and the MNP. Starting with the mnp, the next upper data point is the first data point in S_1 further along the nutrient axis with a higher yield value. Further upper data points along the nutrient axis and on the left of the MYP are identified similarly. In step 3, the boundary line between the MYP and the MNP is identified. Starting at the MNP, BOLIDES locates the upper boundary data

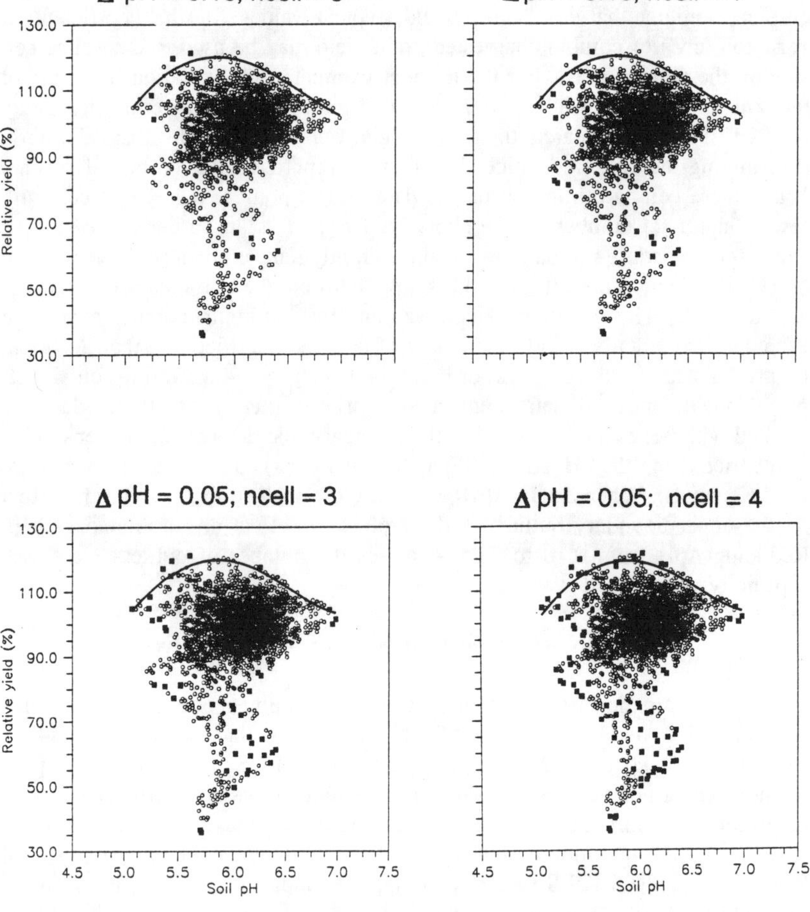

Fig. 66–2. Outlier identification and boundary lines as affected by cell size and number of data per cell.

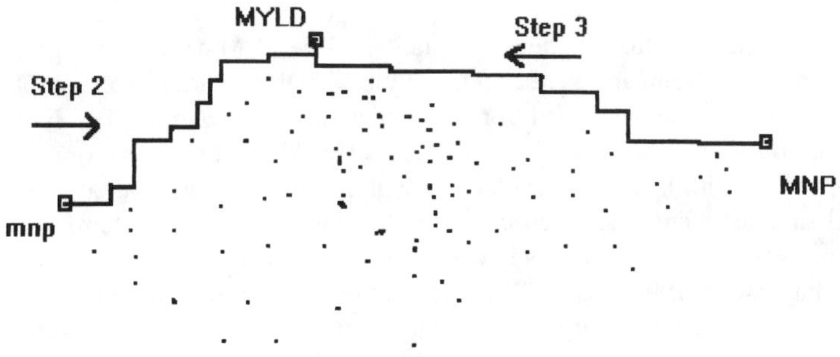

Fig. 66–3. A schematic of the identification of upper data points.

points by the same procedure as in step 2, but going in the opposite direction.

FITTING THE BOUNDARY LINE

BOLIDES uses a 4th order polynomial to fit a curve to the data set S_2 according to the least squares method. The higher order polynomial is used because of the importance of identifying the slope of the curve at each side of the peak. A curve steeply sloping up to the peak is indicative of situations where large responses to the addition of nutrients to deficient areas are expected. The first derivative of the fitted polynomial gives predicted yield response to fertilization in relation to nutrient test level.

DISCRIMINATING VARIABLES

BOLIDES presents the data points relating yield to nutrient status together with the boundary line and its equation. If another variable is having a significant effect on the response to the nutrient, its presence is indicated by two or more distinct concentrations of points, each with potentially its own boundary line response to the nutrient. In practical situations, this third variable is likely to be a permanent feature such as soil organic matter or clay content. Where the data set relates the relative yield of several crops to nutrient status, crop species may also be a discriminating variable. BOLIDES allows data on a third discriminating variable to be used to categorize the data points and produce an individual boundary line for each category. Where the third variable is quantitative (e.g. soil clay content), the data may be categorized in relation to their distribution.

BOUNDARY LINE ANALYSIS OF CROP YIELD AND SOIL DATA (VINDUM, DENMARK)

Methods

Data from the long term evaluation of the LRM concept being conducted at Vindum in Denmark were analyzed by BOLIDES. Surface soil samples taken every 30 m were analyzed for plant available phosphorus (0.2 N H2SO4), potassium (ammonium acetate), pH (2.5:1 CaCl2 solution–soil), organic matter (loss on ignition), and clay content through rubidium (Murphy et al., 1994). The soil samples were positioned using the Global Positioning System. These data were synchronized with winter wheat yield data taken every two seconds from the harvest of 1992 and 1993 using a Dronningborg combine harvester fitted with 'Flowcontrol' (Murphy et al., this volume). The yield data from the two years were synchronized with each other, and with the soil data, to a 10 x 10 m grid using the inverse distance technique. The two years' yield data were combined according to the equifertile concept (Schnug et al., 1994) to give the spatial variation in the average. The resultant 1159 data points were analyzed using BOLIDES for the determination of the optimum pH and available potassium and phosphorus. The analysis of the data for potassium are reported here. In determining the boundary line, soil potassium values were plotted against the yield expressed as a percentage of the mean yield. Five percent (5%), 0.2 pH units and 2 mg/100g were used to determine cell size in relation to the relative yield, pH and soil test value, respectively. A data point was identified as an outlier if its cell contained less than four data points.

Results

Figure 66–4 shows the relationships between relative yield and the soil fertility parameters together with the absolute frequency of the data for clay. Following scrutiny of the distribution of clay data and using knowledge of the site concerned, the data were categorized according to the following clay classes: class 1: <6%; class 2: 6-9%; class 3: >9%. For each clay class, the number of data points in S_0 and S_1 for pH, available phosphorus and potassium are given in Table 66–1. The S_1 data sets for potassium, and their boundary lines are given in Fig. 66–5. Table 66–2 gives the maximum yield value and corresponding nutrient values with the coefficients of the boundary lines given in Fig. 66–5.

CONCLUSIONS

The exploitation of the boundary line approach using BOLIDES satisfies calibration requirements. Optimum nutrient levels are identified (Table 66–2). Coefficients of the polynomials of boundary lines in Fig. 66–5 with the maximum yield values and their corresponding nutrient levels are determined.

Table 66–1. The number of data points in the S_o and S_1 data sets for phosphorus, potassium and pH.

	S_o	pH	Phosphorus	Potassium
Clay Class				
< 6% clay	70	63	47	60
6-9% clay	729	722	715	718
> 9% clay	360	360	349	354

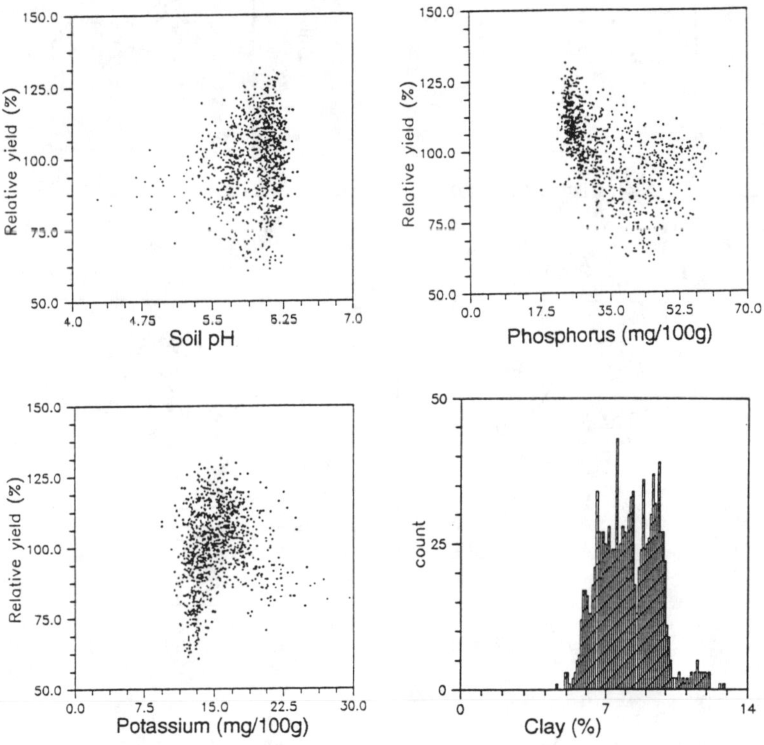

Fig. 66–4. Relative crop yield and soil analysis data for pH, phosphorus and potassium, and the absolute frequency of soil clay contents.

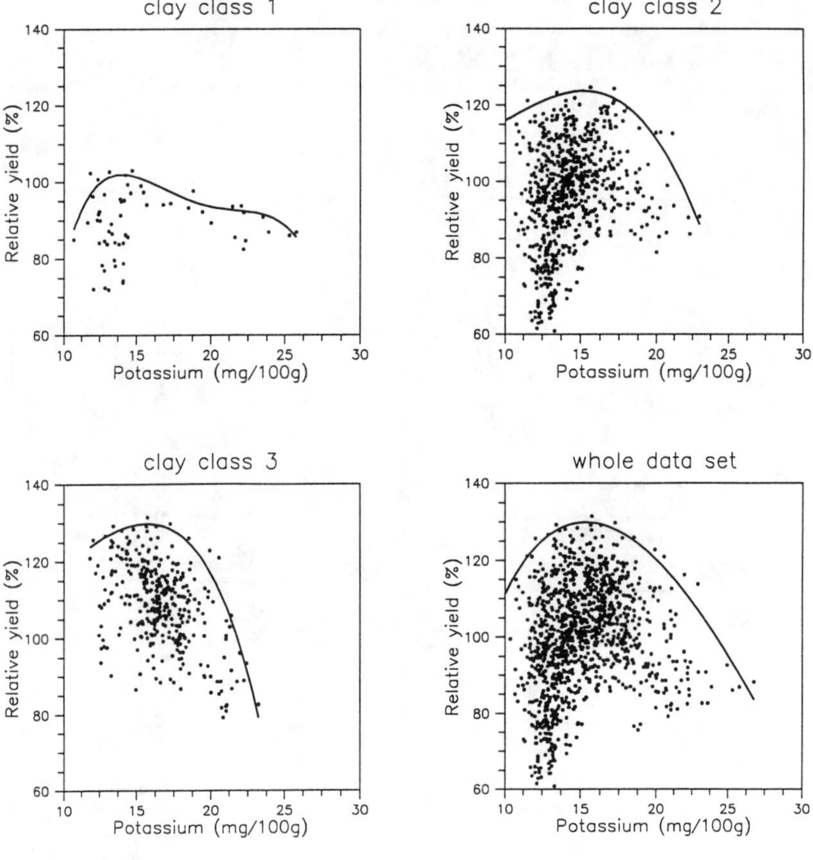

Fig. 66–5. The $S_{.1}$ data sets and their upper boundary lines for potassium values measured on the three clay classes, and the combined data set for all clay classes with its upper boundary line.

Table 66–2. Coefficients of the polynomials of boundary lines in Fig. 26-5 with the maximum yield values and their corresponding nutrient levels.

Clay class 1.
Relative yield =-645.8 + 165.8x + -13.5x2 + 0.473x3 + -0.0061x4
Maximum yield = 101.96; Available potassium at maximum yield = 14.0

Clay class 2.
Relative yield =-172.1 + -20.4x + - 2.3x2 + -0.096x3 + -0.0012x4
Maximum yield = 123.63; Available potassium at maximum yield= 15.2

Clay class 3.
Relative yield = 17.8 + 21.2x + - 1.8x2 + -0.092x3 + -0.0020x4
Maximum yield = 129.80; Available potassium at maximum yield= 15.7

All data.
Relative yield =-100.6 + 38.8x + - 2.2x2 + -0.053x3 + -0.0005x4
Maximum yield = 129.93; Available potassium at maximum yield= 15.3

The first derivative of the curve quantifies the response to the changes in the nutrient parameter. BOLIDES allow conventional statistical criteria to be used to identify outlying data. The presence of a third discriminating variable is indicated and, where the data are available, is considered by BOLIDES to produce several boundary lines for different soil, site or crop categories.

In examining the boundary line for the potassium data, a clear optimum is identified. Over the whole data set, yields within 5% of highest observed were confined to soil potassium levels 13-17 mg/100g. The analysis indicates that this target potassium level is not affected by soil clay content; the curve peaked at 14.0 for clay class 1, and 15.7 for clay class 3. The main advantage of the boundary line approach for LRM is that it enables the local calibration of a wide range of analytical techniques. The optimum nutrient status so determined is related directly to the circumstances in which the results are to be used. Thus, the cost of establishing a data base of agronomic resources is reduced, and the relevance of the data to the specific circumstances is increased.

In employing the boundary line concept, we agree with Webb (1972); the boundary line is based on biological rather than mathematical principles. The line does not confirm the reasons for the lower yields observed, only that they occur. An intuitive understanding of the soils and crops from which the data come (based on conventional studies of crop responses to nutrient availability and soil conditions) is essential if uncertainty associated with the use of the line is to be minimized.

REFERENCES

Dahnke, W.C., and R.A. Olson. 1990. Soil test correlation, calibration and recommendation. p 45–71. *In* Westerman, R.L. (ed.) Soil testing and plant analysis. SSSA, Madison, WI.

Evanylo, G.K., and M.E. Sumner. 1987. Utilization of the boundary line approach in the development of soil nutrient norms for soybean production. Comm. Soil Sci. Plant Anal. 18: 1355–1377.

Moller-Nielsen, J., and B. Frijs-Nielsen. 1976. Evaluation and control of the nutritional status of cereals. II. Pure effect of a nutrient. Plant and Soil 45: 339–351.

Murphy, D.P., E. Schnug, and S. Haneklaus. (this volume). Yield mapping - a guide to improved techniques and strategies.

Murphy, D.P., E. Schnug, and S. Haneklaus. 1995. Innovative soil sampling and analysis procedures for the local resource management of agricultural soils. Proc. of the 15th Congress of the International Society of Soil Science (in press).

Schnug, E., S. Haneklaus, and D. Murphy. 1994. Equifertiles - an innovative concept for the efficient sampling in the local resource management of agricultural soils. Aspects of Applied Biology 37. (in press).

Schnug, E., D.P. Murphy, S.H. Haneklaus, and E.J. Evans. 1993. Local resource management in computer aided farming: A new approach for sustainable agriculture. p 657–663. *In* Fragoso, M.A.C. and van Beusichem, M.L. (ed). Optimization of Plant Nutrition. Kluwer Academic Publishers, Netherlands.

Webb, R.A. 1972. Use of the boundary line in the analysis of biological data. J. Hort. Sci. 47: 309–319.

67 The Status of Information-Based Agrichemical Management Services in Wisconsin's Agrichemical Supply Industry

Steven Wolf
Peter Nowak

Department of Rural Sociology
University of Wisconsin-Madison
Madison, Wisconsin

As suppliers of crop inputs such as fertilizer and pesticides, agrichemical dealers are important components of contemporary crop production systems. While often not acknowledged, dealers also provide a significant amount of input related information to their customers. The private agribusiness sector, and local dealers in particular, play a critical role with respect to the information transfer function in farming systems (Ford & Babb, 1985). Current developments in this industry indicate that dealers will play an increasingly important role as cropping systems become more site specific through more intense use of information.

Site specific agriculture is defined here as a process of enhancing the correlation between variance in agroecological parameters with variance in management as pursued in a defined socioeconomic setting. The benefits of site specific agriculture will accrue to farmers through more efficient resource allocation, to society through pollution prevention, and to agribusinesses through marketing of analytic procedures, technological hardware, and management tools associated with site specific agriculture.

Intense informational requirements are one of the defining features of site specific agriculture. The analytic procedures, technological hardware, and management tools often associated with site specific agriculture are dependent on increasingly large and complex information streams that must be managed in a systematic fashion. In order to practice site specific agriculture, farmers will have to develop their own informational capabilities or contract with someone else to provide this knowledge and service. The cumulative outcome of this process represents an important topic of further research with respect to the structure of agriculture, the nature of farmer–agribusiness relations, and role of the farmer in agriculture. Nowak et al. in this volume address farmers' ability to fill their own informational requirements with respect to site specific agriculture. In this paper the focus is on the participation of agrichemical supply firms in supporting farmers' practice of site specific agriculture.

The growing importance of application of information to cropping systems is exemplified by the emergence of the crop consultant, and in particular, the independent crop consultants (i.e., those firms or individuals not affiliated with any product sales organization). It is estimated that this profession now markets consulting services on 53% of cotton acres, 53% of vegetable acres, 21% of corn acres and 13% of soybean acres. Overall, independent crop consultants provide knowledge-based analytical services on 16% of U.S. cropland (Nowlin, 1993).

Agrichemical dealerships originated to a large extent to fill the role of middle-man between agrichemical manufacturers and farmers. Based on a historic emphasis on product sales, dealerships currently rely heavily on profits from product sales. Informational services have not evolved into important and discrete profit generating functions (Hoffman, 1993). Services that are offered by dealers often generate no tangible revenues as service fees are commonly embedded in product margins (Schuster, 1993). The degree to which the product supply function is de-coupled from the information supply function has proven to be a contentious issue to date (Simmonds & Berston, 1991) and may significantly shape the manner in which site specific agriculture is practiced on many farms. The role of information as applied to agricultural production systems and the information delivery infrastructure appear to have entered a transition stage characterized by rapid development. At issue in this paper is the extent to which agrichemical dealers are participating in this transition by offering the type of knowledge-based analytical services which undergird site specific agriculture.

THE ROLE OF DEALERS IN SITE SPECIFIC AGRICULTURE

Site specific agriculture can be viewed as a heterogenous package or bundle of techniques, tools, and management requisites (Nowak, 1993). Generic statements about this bundle can be misleading. Site specific agriculture can vary between reliance on information-intense analytical procedures up to where these analytical practices are coupled with state-of-the art, capital-intense hardware. In this study we focus on the low capital facet of site specific agriculture. This can be viewed as the type of activities associated with the more information intense practices found in integrated crop and pest management. This "softer path" of site specific agriculture has been selected because farming system changes associated with integrated crop management type activities are more compatible, in relative terms, with the nature of Wisconsin agriculture. For our purposes, Wisconsin agriculture can be characterized by medium-sized, diversified farms on physiographically rolling terrain with highly variable soils.

It is important to note, however, that some portions of Wisconsin and other farming regions represent settings where other, more capital intensive and modernistic site specific agricultural technologies may make rapid inroads. In other words, grid soil sampling results patched to a variable rate applicator through a satellite link where the results are mapped by a combine with on-the-go yield sensing equipment may be appropriate for some farming systems, but for the bulk of Wisconsin's farms such innovations appear to be incompatible. The short-term impact of site specific agriculture on Wisconsin agriculture will be through more efficient agrichemical management through integrated crop management practices.

The critical elements for successful application of low-capital, site specific agrichemical management are: i) human capital and ii) infrastructural capacity. We will argue that while many Wisconsin farmers are in poor position with respect to these two elements, many dealers are quite well situated.

Patterns of human capital allocation (Huffman, 1988) on farms makes practice of site specific agriculture problematic without support from off-farm sources. Farmers are believed to allocate human capital resources to promote return on investment and assume desirable risk postures according to patterns of personal preference. A farmer's investment, for example, in becoming trained to scout effectively for weeds, and then maintain a crop scouting regimen, must be weighed against the opportunity costs of such an expenditure (Huffman, 1988). Existing knowledge gaps and labor constraints would indicate that resource allocation tradeoffs required to independently (i.e., without off-farm support) practice site specific agriculture are substantial and are likely to be viewed unfavorably by the many farmers.

The weaknesses of farms with respect to infrastructural capacity are clear. Activities such as manure analysis, soil testing, and certain pest identifications or nutrient deficiencies require laboratory facilities. Activities such as spot spraying and pesticide banding may require specialized equipment. Based on the current status of record keeping on farms, development of detailed record keeping and record retrieval systems may pose a challenge to many farmers.

The imprecision exhibited in contemporary agrichemical management in Wisconsin testifies to the degree to which farmers' management is not now site specific (Shepard, 1993). A variety of economic technological (Pocock, 1993), and socio-political (Buttel, 1993) factors are associated with and perpetuate sloppy agrichemical management (NRC, 1993). Despite vast public expenditures directed at regulatory, enforcement, cost sharing, technical assistance, and educational programs, the fact that farmers are not able to use agrichemicals without generating substantial off-site costs (NRC, 1993) indicates that constraints to efficient agrichemical management exist unaddressed within farm management structures. In sum, the current inadequate capacity of farmers to supply their own information-intense analytic services, coupled with the poor record of agrichemical management as evidenced by environmental problems indicates that farmers are not likely to adopt site specific agriculture without significant external support.

The preceding paragraphs have argued that farmers, as a group, are poorly positioned to respond to the challenges of implementing site specific agriculture. This is not the case with the local agrichemical dealer. They represent local businesses with proven access to farmers and their operations (Ford and Babb, 1985; Funk and Downey, 1981). Many of these agrichemical dealers are expected to build on their current relationships by moving beyond their traditional roles of product sales, delivery, re-packaging and custom application. As information markets become more formalized dealers are expected to reallocate their financial and human capital to develop the infrastructural capacity required to provide informational services to farmers (Hoffman, 1993).

Development of informational inputs for cropping systems, like products such as seed, fertilizer and pesticides, would appear to be a logical development for agrichemical dealers. Yet providing information in the context of site specific agriculture has its challenges. Dealers must: i) compete with a growing independent crop consulting industry; ii) overcome some farmers' and governmental perceptions (Simmonds & Berston, 1991) of conflict of interest associated with agrichemical consulting while simultaneously selling the very same products; and iii) integrate site specific services with preexisting facilities and equipment inventories, personnel, and business plans. A diffuse infrastructure similar to that now required to provide agrichemical products and product-related services (e.g., delivery and custom application) will be required. The infrastructural emphasis, however, will be on trained personnel rather than spray booms and pesticide loading and mixing facilities. Dealers' previous experience in managing people, information, and technology should support their engagement in an information-based service market.

Strengthening of dealer-farmer relations has been strongly advocated by professional associations representing agrichemical dealers (Hoffman, 1993). Dealer associations are promoting an active role for dealers with respect to information provision and orientation toward service to i) protect their industry's position against incursion by the independent crop consulting industry, ii) diversify their members' earning opportunities in an era of greatly constricted agrichemical product profit margins, and iii) assume a proactive, public relations stance by projecting a positive image for their industry with respect to the environmental performance of cropping systems. Whether attention devoted to service and information delivery by dealer association publications will serve to trigger or accelerate changes in the agrichemical supply industry is unknown, but the volume of coverage is an indication that these issues may have substantial implications for the industry.

A significant factor which foreshadows growth in dealers' information-based service offerings is the need to identify new profit centers at a time when many local agribusinesses are failing. The maturation of the industry has reduced fertilizer and pesticide product margins to critical levels. It is estimated that 20 to 30% of the estimated 12 000 U.S. agrichemical supply businesses will go out of business over the next 10 yrs (Hoffman, 1993). As a result of these pressures, dealers are expected to move to develop and capture lucrative spin-off markets associated with the agrichemical product supply function, namely the service market.

Service sales represent new potential profits as well as a means by which firms can retain and attract new customers for their products. As agrichemical supply firms struggle to maintain and expand their market share at a time when low product margins have eroded price differentials between firms, services are expected to become the differentiating feature between competing firms.

The factors emphasized here which indicate dealers may be substantially involved in site specific agriculture are: i) the emergence of service as a survival mechanism for dealers, ii) dealers' traditional access to farmers and farm decision making, and iii) the capabilities of dealers with respect to personnel, information and technology management.

The Environmental Working Group has developed a series of case studies focussing on dealers' service orientations and pollution prevention activities (Hoffman, 1993). The Hoffman study represents a micro-level analysis of the behavior of 6 firms. The study described in this paper is complementary to Hoffman's research in that the research questions are similar, yet the scope of the study is quite different. Here we evaluate at the state level the degree to which Wisconsin agrichemical dealers currently provide site specific information to farmers. These data will provide a baseline against which future development of information-based services can be evaluated. It will also serve to characterize the variation in the population of Wisconsin dealers. Included in this characterization is an assessment of constraints and opportunities for expanded or accelerated development of dealers' site specific agriculture technology transfer capabilities.

METHODS

The purpose of the study was to evaluate the degree to which dealers engage in site specific agriculture or support their customers' practice of site specific agriculture. To this end a mail survey was conducted of all businesses in the state which held, and made use of, both a 1993 commercial pesticide sales license (Wis. Stats. 94.685) and a 1993 restricted use pesticide commercial applicator license (Wis. Stats. 94.703). The Wisconsin Department of Agriculture, Trade and Consumer Protection pesticide bureau's database was used to identify respondents. Each independently licensed location in the state which met these criteria was considered a discrete entity. Those businesses which operated multiple outlets (i.e., branch or satellite facilities) were asked to complete separate questionnaires for each outlet. As defined by these criteria there were 228 eligible respondents.

The individual most knowledgeable of overall business activities at each dealership was asked to complete the mail questionnaire. Respondents were instructed to consult staff agronomists for information when appropriate. Information was collected on dealership characteristics, income sources, patron characteristics, business priorities, and past, current, and future service offerings. Respondents were also asked to identify operative constraints to expansion of information-based agrichemical management service offerings. Technical terms were clearly defined in the survey to reduce errors associated with inconsistent interpretation of terminology. The final response rate was 77%.

A non-response bias test was conducted through telephone interviews with non-respondents. Of the 49 non-respondents, 30 (61%) agreed to participate in the bias test. There was no statistically significant variation for items examined between the distribution of non-respondents' responses as compared to respondents. The finding of no difference combined with the high response rate supports the assumption that the results are generalizable to the state's agrichemical industry.

All valid responses were entered into SPSS (Statistical Package for the Social Sciences). All data was entered into the system twice to minimize data entry errors.

RESULTS

Dealer Characteristics

The ownership structures of the dealerships were represented by three major types; 21% were individually owned, 7.4% were public corporations, and 71.6% were member owned cooperatives. Ninety-two percent (92%) of the cooperatives were affiliated with a regional cooperative. The remainder of the cooperatives were independent cooperatives. Dealerships had been in business between 2 and 100 yrs with an average of 40.4 yrs in the business. The number of full-time employees at these dealerships varied between 0 and 75 persons with an average of 11.4 full-time employees. In addition, these dealerships employed an average of 2.5 part-time, yr-around employees and 6.4 seasonal or peak-time employees. Each dealership has an average of 5.0 employees who hold a commercial pesticide applicators license.

Dealerships were asked to check one of twelve categories that best represented their 1992 gross income. These income categories varied between less than $500,000 to more than $10,000,000. Approximately half (52.0%) of the dealerships grossed $2,000,000 or less in 1992. On the other end of the spectrum, there were 9.9% who grossed more than $8,000,000 in 1992. The average 1992 gross income was in the $2,000,000 to $3,000,000 dollar category. Gross income data is illustrated in Fig. 67–1.

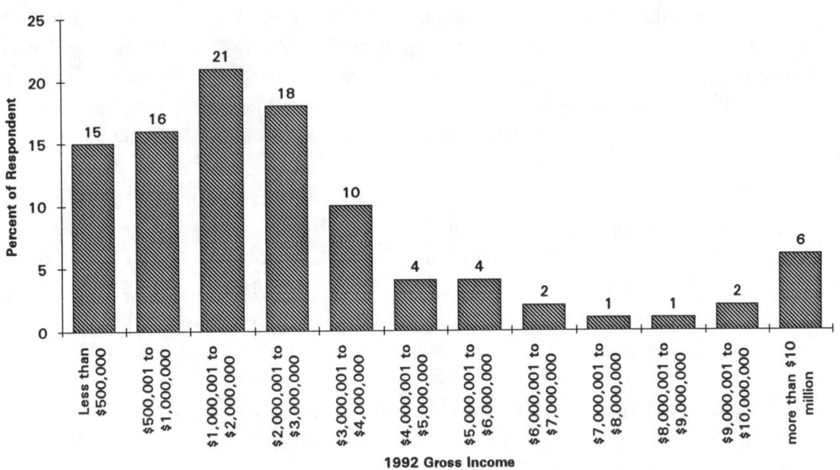

Fig. 67–1. 1992 Gross income of Wisconsin agrichemical supply firms ($).

Label	Mean (%)	Std Dev
Agricultural fertilizer product sales	35.2	17.9
Agricultural pesticide product sales	23.0	14.8
Other product sales	23.6	25.2
Custom application services	7.2	6.4
Grain transport, storage & drying service	3.5	9.1
Crop/soil fertility testing services	.9	1.3
Crop/pest scouting services	.5	1.2
Other services	.8	3.9
Miscellaneous	5.4	14.0

Fig. 67–2. Distribution among sources of income for Wisconsin agrichemical supply firms (% of 1992 gross income).

Dealers' Income Distribution

On average, 35.2% of the 1992 gross dealership income resulted from agricultural fertilizer sales. This compares to an average of 23.0% of the 1992 gross dealership income from agricultural pesticide sales. Other product sales (e.g., seeds, feed, petroleum) accounted for an average 23.6% of gross income. Consequently, sales of products accounted for an average of 81.8% of 1992 gross dealership income. Custom application services accounted for an additional 7.2% of gross income. Grain related services such as drying and storage represented, on average, 3.6% of dealers' gross income. A total of 1.5% of dealers' income was derived from soil fertility testing services and crop–pest scouting services. The fact that these site specific analytic services represent a very small fraction of dealers' income indicates that at this time such services are tangential to dealers' primary activities. The distribution among these sources of dealers' 1992 gross income is illustrated in Fig. 67–2. Standard deviations are included to provide an indication of variance in the population.

Dealers were asked about pest management and fertility management services they offered in 1990, 1993 and would offer in 1995. Of the 20 services measured in the survey, 14 were selected as having potentially positive economic and environmental implications. That is, use of these 14 services offers the opportunity to increase production efficiency while decreasing the probability of nutrients or pesticides "leaking" into the larger environment. Each of the 14 services represent site specific agrichemical management practices or serve as prerequisites to site specific management. The past, present and future offerings of these 14 pest and fertility management services are presented in Fig. 67–3. These data indicate a trend toward increased service offerings.

Service	Percent of Firms in 1990	in 1993	in 1995
Soil Testing for whole field average	90%	91%	88%
Pesticide record keeping for farmers	71%	82%	85%
Custom herbicide spot spraying	71%	79%	77%
Plant tissue testing	62%	76%	74%
Soil testing for within field variation	56%	62%	63%
Manure analysis	50%	59%	59%
Nitrate-nitrogen soil testing	36%	53%	55%
Weed mapping	35%	47%	57%
Formal weed scouting	35%	45%	56%
Estimate yield reductions due to weeds	26%	40%	46%
On-farm sprayer calibration	28%	38%	40%
Establish "no treatment" check plots	25%	33%	41%
Use pest population models	19%	27%	36%
On-Farm manure spreader calibration	3%	7%	13%
Number of Services Offered - Mean	6.0	7.4	7.9

Fig. 67–3. Wisconsin agrichemical supply firms' information-intense service offerings (1990-1993-1995).

To some degree, the pattern of service offerings indicates traditional agricultural priorities. Services which are production oriented (i.e., yield maximizing) are widely available and have a lower dependence on information and trained personnel. Information-intense services which potentially enhance production efficiency and improve the environmental performance of cropping systems are less likely to be available to farmers.

The services most frequently added by dealers between 1990 and 1993 were nitrate-nitrogen soil testing, plant tissue testing, and estimation of yield reductions due to weeds. Looking forward to 1995, the service offerings reportedly being added most frequently by dealers were weed mapping, formal weed scouting, and use of pest population models. This growth in services data reflect some firms adding widely available services in order to remain competitive, while some firms appear to be extending their repertoire of service offerings to include services which up to this time have been largely unavailable.

Constraints to Service Development

In order to assess dealers' recognition of the potential for development of information-intensive services a number of questions were posed to dealers with specific reference to weed scouting services. Eighty percent (80%) of dealers agreed that custom weed scouting could be profitable for farmers, while an additional 18% were not sure (2% disagreed with the proposition). Only 8% agreed that farmers were currently effectively scouting weeds (11% were not sure). Only 4% saw farmers as having the necessary resources (i.e., time and knowledge) to effectively scout their own fields (14% were not sure). Importantly, when asked whether they agreed that weed scouting was profitable for dealers, 43% agreed, 25% disagreed, and 32% were unsure. These statistics indicate that dealers overwhelmingly perceive farmers as both potential beneficiaries of weed scouting and unprepared to be self-sufficient in this regard. At the same time, there is wide disagreement regarding the potential profits associated with dealer-based scouting services. Dealers inability or unwillingness to capitalize on what appears to be well recognized demand suggests that constraints, of some kind, prohibit dealers from realizing profitable weed management services.

In order to better understand impediments to supplying informational services to farmers, dealers were asked about constraints to development and expansion of site specific crop management services. The leading factors in descending frequency were dealers' concerns that potential profits based on services did not justify expansion or investment, fixed costs associated with product sales prevented investment in service delivery, and lack of customer demand for such services. These responses indicate that factors both internal and external to dealerships may impact dealers with respect to service provision (Wolf and Nowak, 1994). Factors such as debt load associated with product inventory, storage and delivery systems which make additional investment problematic is an example of an internal constraint. An example of an external factor is represented by the need to increase farmers' recognition of the value of such service, and based on this recognition, farmers' willingness to compensate dealers' sufficiently to ensure profitability.

The transition from service for free to service for a fee is an important consideration in understanding dealers' position relative to service delivery. Traditionally, dealers' costs for services such as product delivery and weed scouting prior to custom herbicide application have been embedded in product costs (Schuster, 1993). As profit margins on products have been reduced through extensive competition between firms, there has been a trend toward separating the cost of services from the product cost in order to offer customers a "no-frills", low cost option. This trend has alienated some customers and represents and impediment to development of dealer-based service for a fee programs.

Thirty nine percent (39%) of Wisconsin dealers offering intensive soil sampling capable of measuring within field soil variation were found to be generating no revenue from such service. For formal weed scouting, defined as three or more visits per field per yr with at least one visit prior to determining weed management strategy, 47% were not generating revenue from such service. It is assumed that the costs for these services are buried in product margins. Until dealers can recoup the costs for site specific services, the availability and quality of these services will be limited.

Dealers' Relations with Independent Crop Consultants

The crop consulting industry, especially independent crop consultants (ICC), represent an alternative source of information-based analytic services for farmers. As such, independent crop consultants will compete with dealers for farmers' service dollars. The development of the crop consulting industry and the relationships between firms associated with agrichemical sales organizations and those "independent" (NAICC, 1994) of product sales firms will influence the manner in which site specific agriculture is practiced on many farms. If regulatory developments (Simmonds and Berston, 1991) or farmers' concerns regarding conflicting dealer interests serve to decouple information delivery from product supply, a very different, perhaps non-competitive, relationship between dealers and consultants is anticipated.

A series of questions were developed to assess the current relationships between Wisconsin dealers and ICCs. Eighty one percent (81%) of Wisconsin dealers reported that they were aware of ICCs operating in their trade areas. Of the firms aware of ICCs in the area, 61.4% reported no impacts on revenues, 16% reported revenues had declined, and 23.6% experienced an increase in revenue due to the presence of the ICCs. When asked about the consistency of the ICC's crop management recommendations with respect to the recommendations coming from their dealerships, 58.7% reported mixed results (agree on some, disagree on others), 37.1% agreed most of the time, and 4.2% found themselves in disagreement most of the time.

CONCLUSION

This paper has argued that the agrichemical dealer will play an important role in the development of site specific agriculture. This role represents an expansion beyond the typical perception of the dealership investing in capital-intense equipment and making it available to local farmers on a custom basis. Instead, it has been argued that dealers possess both the human capital and infrastructural capacity to offer the information-intense, analytical services that serves as the foundation of site specific agriculture. Consequently, dealers have the opportunity to offer differing levels of site specific agriculture based on the characteristics of local farming systems.

While the potential for dealer involvement in site specific agriculture is there, results from the analysis indicates that little of this potential is currently being realized. Barriers to achieving this potential appear to fall in three categories. First are the internal capabilities of the dealership. Firms that have traditionally relied on the generation of profits from the sales of products are having difficulty integrating the concept of profits based on the sale of services. Long-term investment strategies, current debt structures and training of personnel have all emphasized product storage, handling and sales. Both the motivation and capabilities required for developing a business plan designed to integrate more "service-for-a-fee" profit lines appears to be lacking in some dealerships.

The second category of reasons for dealers not achieving their potential in site specific agriculture is external to the firms. This concerns both the marketing experience for developing a demand among current and future customers, and the competition from other providers of information-intense services. Dealers appear to rely heavily on the manufacturers of the products in their inventory for marketing, sales and retailing expertise. The manufacturers develop and implement marketing campaigns giving the dealership little opportunity to develop these skills. Coupled with this is the competition from independent crop consultants. It is expected that there would be a small proportion of farmers in the trade areas of the dealerships who would understand the importance and need for information-intense services. The independent crop consultants appear to be positioned to capture the existing demand in the market as evidenced by their widespread presence in the state (81% of the dealers reported ICCs operating in their trade areas).

The third and final category of reasons for dealers not achieving their potential in site specific agriculture is also external to the firms. This concerns the role of environmental regulation. Most environmental regulation to date has focused on the agrichemical dealer as a potential point source of pollution. Dealers have had to invest significant resources in protecting surface and groundwater from point-source type activities (e.g., mixing, handling, loading and storage of agrichemicals). Little environmental regulation, however, has focused on dealers' relations with farmers and nonpoint source pollution potential. The few programs in this area (e.g., the SP-53) has purposively excluded the dealer from involvement. Yet environmental regulations focused on the farmer generate a demand for enhanced service and support from dealerships. An analysis of the implementation of the atrazine rule in Wisconsin, the most stringent in the nation, found agrichemical dealers to be beneficiaries of this regulation due to increased demand for record-keeping, custom application and sales of more expensive corn herbicides (Nowak & Wolf, 1994). An indirect benefit which accrued to dealers as a result of Wisconsin's atrazine regulations was the intensification of farmer-dealer relations. As farmers make the transition away from atrazine, which is recognized as a very straight-forward, easy to use herbicide, to more complex, management intensive herbicides, dealers are being called upon to assist farmers as they learn to effectively use unfamiliar products.

In sum, there exists a significant potential to more fully involve the agrichemical dealership in site specific agriculture. This involvement, however, is based on understanding dealer's current positions, and the very real constraints dealerships face in trying to expand services that compliment a site specific agriculture.

REFERENCES

Buttel, F. 1993. Socioeconomic impacts and social implications of reducing pesticide and agricultural chemical use in the Unites States. pp 153–181. *In* D. Pimentel and H. Lehman (ed.) The pesticide question: environment, economics and ethics. Chapman and Hall, New York.

Ford, S.A., and E.M. Babb. 1985. Farmers sources of information. Agribusiness. 5 (5) pp. 465–476.

Funk, T., and W. Downey. 1981. What influences the farmers buying decisions? Fertilizer Progress. Nov.-Dec. 1981. p 24–27.

Hoffman, W. 1993. Stemming the flow: Agrichemical dealers and pollution prevention, case studies from the Great Lakes Basin. Environmental Working Group, Washington, DC.

Huffman, W.E. 1988. Human capital for agriculture. Pp. 499–517. *In* R. J. Hildreth, et al. (ed.) Agriculture and rural areas approaching the twenty-first century. Iowa State University Press, Ames, IA.

NAICC. 1994. National Alliance of Independent Crop Consultants: Professionalism in Agriculture. Conference Proceedings. January 27–29, 1994. Memphis, TN.

National Research Council. 1993. Soil and water quality: An agenda for agriculture. Committee on long range soil and water conservation, Board on Agriculture. Washington, DC.

Nowak, P. 1993. Social issues related to soil specific crop management. Pp. 269–285. *In* P. C. Robert, et al. (ed.) ASA, CSSA, SSSA, Madison, WI.

Nowak, P., and S. Wolf. 1994. Results of a dealer survey including impacts of the Atrazine Rule. Proceedings 1994 Fertilizer Aglime and Pest Management Conference. University of Wisconsin, Madison, WI.

Nowlin, B. 1993. NAICC report. Ag Consultant. Fall, 1993. p 13. Pimentel, D., H. Acquay, M. Biltonen, P. Rice, M. Sila, J. Nelson, V. Lipner, S. Giordano, A. Horowitz, and M. D'Amore. 1993. Assessment of environmental and economic impacts of pesticide use. pp 47–84. *In* D. Pimentel and H. Lehman (ed.) The pesticide question: environment, economics and ethics. Chapman and Hall, New York.

Pocock, J. 1993. "SmartBox" proves savvy on insecticides. Wallaces Farmer. December, 1993. p 56.

Shepard, R. 1993. Beyond superficial targeting: designing educational strategies for water quality programs. Ph. D. Dissert. Inst. Environ. Studies, Land Resources Program, University of Wisconsin, Madison, WI.

Shuster, D.J. 1993. A study of technology transfer assistance by agricultural cooperatives. M.S. Thesis, Dep. Agric. Econ., University of Wisconsin, Madison.

Simmonds, B., and D. Berston. 1991. ASCS mulls consulting dilemma: Will new government programs leave dealer-based consultants out in the cold? Agrichemical Age. March, 1991. pp 8–10.

68 How Can You Transfer Site-Specific Crop Management Technology to Farmers?

Grant Mangold

ag/INNOVATOR
Linn Grove, IA

I've been covering aspects of VRT, GPS, GIS and SSM technologies and attempting to be part of the transfer of these concepts and practices as they have developed, for the past eight years. I do know that we are ALL involved in the communications aspect of transfer of technology--as consumers, communicators, and compilers of information. And in this case, the farmer (as end-user) does the most compiling, as we'll discuss a bit later. As part of the communication approach of farm magazines and newsletters, such as my Bytes & Beans column in Soybean DIGEST and my ag/INNOVATOR newsletter, one goal is to clarify complex concepts and simplify them for practical application on the farm. With that in mind, I'd like to consider briefly three aspects: i) the course or process of communication, ii) the content of communication, and iii) the challenges of communicating site-specific crop management technology.

Processes of Communication

Communication involves a two-way interactive process. As a rule of thumb, we can say the purpose is to impart information for a desired change in activity. An example could be the cartoon, which says "Before computers, he would get up at five to work the north forty. Now he gets up to par the front nine." Ag computing developers and others have the goal of transferring ag computing technology to make the farmer more efficient.

There's another rule of thumb that, for this presentation, I'll call the "first rule of thirds." It follows research indicating that, in general, one third of the population responds primarily to what they hear (verbal), another third to what they see (visual), and another third to what they touch (physical). The best learning takes place when we can involve all three of these gates of perception.

If your purpose is to impart information to effect a desired change of activity, you need a plan. First, identify the needs of the audience. Then develop a strategy for your product or service to fill those needs. Finally, evaluate and modify both the strategy and the product or service to ensure that it meets the needs of your audience. That's part of the two-way, interactive continuum of communication. A 12-year-old friend in my community received an archery set for Christmas. I was amazed when I saw arrows smack in the middle of all his targets. "Scott, how did you become such an accomplished archer so soon?" I asked. "Well," he responded, looking at the ground and scuffing his feet, "first I shoot the arrow, then I draw the target." Without a plan, whatever we aim at can become our target. Without a plan--and a means to evaluate and modify that plan, how can we be sure that our communication strategy (not to mention our product or service) is reaching our audience and having the desired effect?

Another rule of thumb for communication, which we can call the "third rule of thirds:" An effective communication strategy will communicate three benefits, three ways, three times. Generally, people find it difficult to process more than three promised benefits or "whys" when we consider new courses of action. We may respond primarily to what we see, hear, or touch. And we often need to hear, see, or touch the message or product three times before we make the desired response. That's the whys, hows, and whens of communication--three things, three ways, three times. In terms of the transfer of agricultural technology, numerous studies have shown that ag producers learn primarily from other producers. They want to see it in action on a farm near them to help them deal with the "it won't work on my farm" reaction. They want to interact with other producers who make it work. That's part of the importance--and benefit--of onfarm research, field days, and demonstrations. I find it interesting that many players in site-specific crop management began with strategic partnering with interested growers.

Content of Communication

The second major point of this paper is the content of communication for site-specific farming. Producers want to know at least three things: is it economical? is it ecological? is it practical? It must be cost-effective--though they may be willing to wait for an economic return. They're also interested in the potential of the technology to help them be even better stewards of the environment. In time, this factor may over-ride the elusiveness of economic considerations, as regulations and other pressures may push agriculture into using site-specific management--and figure out exactly how to make it profitable later.

There are other factors in the course and the content of communication to consider. The human brain is created as the ultimate computer. And as John Barrett and researchers at Purdue explain it, we are constantly collecting, accepting, and rejecting information about a wide range of ideas--for immediate or later use. Consequently, after making a decision, we may be searching for additional information before we change our activity or put that decision into action.

Peter Nowak and researchers at the University of Wisconsin have identified several factors to account for that time lag (as well as explain why some farmers never adopt new technology). Basically, we are either unable or unwilling to make the change or turn our choice into action. That's why the "continuum of communication" (perhaps using the three rules of thumb discussed above) is so important. You never know when the missing piece of information will be supplied that causes the target audience to respond the way you want them to. You need to keep repeating your message, using evaluation and modification, to gain your desired effect.

Basically, producers who are in that time lag between decision and action are asking themselves: do I have the tools and talents (or access to them) to make this new technology work in my operation? He needs to be equipped and supported as he adopts new technology. With that in mind, you can see how important it is to evaluate your strategy of communication--as well as the design of your product or service--to make sure you are meeting the needs of the target audience. How well does your site-specific or GPS/GIS product or service meet the real needs of farmers and dealers? How easy is it to use? What's your plan to evaluate and modify it?

As we consider transferring computer-intensive technology to farmers, we need to be aware that most estimates put computer ownership on the farm at from 30% to 40% of the population. But that's ownership--there's no good guess as to how many computers are actually used by farmers. Along with that, many farmers may not be keeping adequate records (either crop or financial) to make the most beneficial use of site-specific management. That is changing--and those most likely to adopt SSM technologies are often more advanced along these lines. Also, those farmers who are now adopting new computer technologies don't fit the pattern of the first computer users of 10 years ago.

Perhaps today's progressive farmers are increasingly recognizing the truth in this quote by Neil Havermale of Farmers Software Association: "If you don't USE information age technology, you'll be ABUSED by it." This leads to the third major point of this paper: the challenges of communicating site-specific crop management--challenges due to its computer- and technology-intensive nature. Let's look at the POWER of information, of presumption, and of innovation.

Power of Information

Something that happened at the airport illustrates the power of information. The man in front of me had been very rude to the ticket agent, and when it was my turn I complimented her for exhibiting such patience. "Oh it was just the power of information," she replied. "He's going to New York...and his bags are going to Seattle."

One of the biggest challenges in the transfer of SSM technology will be dealing with all the data--not only the sheer volume of data but also the format of it. You can fill up your hard drive pretty fast with the raw numbers dealing with variability in soil types, yield, varietal response, weather, equipment performance, and numerous other factors. The farmer must deal with ALL of these variables each year as he seeks to run a profitable, environmentally responsible farming operation. And how varieties interact with the weather that year makes a big difference in his bottom line--as well as affecting how he understands variability in that field when he changes varieties in a couple of years.

As we mentioned earlier, the farmer is the ultimate compiler of all that information, whether or not it's been filtered by consultants or computers. That's a lot of data or facts to compile--and computers will be necessary to help the farmer deal with it all. But the data is the important thing--and must become information that can be managed by the farmer. That's the critical paradigm shift we're most interested in for the information age--converting data into information, then adding knowledge and wisdom. Ultimately, the producer is the key person in the process who can add the right knowledge and wisdom for the farm he operates. He also is the risk-taker, whether or not he owns the land. But the land-owner faces the environmental responsibility or liability for that farm.

For all these reasons, the question of who owns the data, who has title to the data, and who has access to the data will be crucial considerations in the communication and transfer of site-specific crop management technology. As the risk-taker or liable party, the owner/operator must have control over the data, not only to make the appropriate decisions about site-specific management of the farm, but also to collect and analyze the data. Beyond that, he must be able to transfer the data to new computer tools and technologies as they become available. The historical dataset for a given farm may well become it's most important "improvement" or value-adding asset. That's why the "open architecture" of site-specific technologies is so important--and will become more important as we go through the information age and beyond.

Another challenge is the "exaggerated expectations" of site-specific technologies. Scientists always seem to want more data and funding, but developers and entrepreneurs want more customers.

Power of Presumption

It's all to easy to get excited about the technology--and build over-stated expectations of capabilities. Another airport story illustrates this "power of presumption." I had some time before my flight so I stopped at a vending machine and bought a little package of chocolate covered cookies. Then I sat down at one of those little tables to read the Conservative Chronicle. Then I noticed a rustling sound, and as I glanced over the top of the newspaper, I noticed a man opening a pack of cookies! To let him know they were mine, I put down the paper, reached over and took a cookie, and looking him in the eye, chomped it firmly. Then I went back to my paper. Soon I could hardly believe that I was hearing the rustling of the cellophane. I glared over the paper as the man gingerly pushed the package--with the last cookie--toward me. Then he got up and left. Of all the nerve, I fumed. Just then they called my flight. As I reached in my coat pocket to get my ticket, I pulled out....my package of cookies!

The power of presumption works both ways--for both the communicator and consumer of a product or idea. Again, our goal is to communicate tools and technolcgies that will be practical as they help farmers operate more cost-effective, more ecologically-responsible farms.

Another challenge along these lines is to prevent producers from feeling "trapped by technology" as they go through the coming changes in agriculture. Some say only those who understand or adopt new technology will be winners after the year 2000.

Farmers need access to technology--and particularly the data on their own fields--as they seek some maneuvering room as they get squeezed further in the farm production value chain from genetics to consumer end-product. There are those who would prefer farmers remained technologically dependent...data deficit...and beholden to others for the information and tools of decision-making. I believe farmers must be empowered to harness the power of information to make better decisions.

In closing, let me summarize our discussion of the course or process of communication, the content of communication, and the challenges of communicating site-specific technology.

Power of Innovation

Basically there are two types of innovators and pioneers: the dreamers who imagine what things could be like...and the do-ers who stay awake and make it come true. But there's another quality of innovation about the pioneers of new technology, and I'll close with this true story from the Oregon Trail about a pioneer I'll call Peg-leg Pierre. Peg-leg was a loner on the wagon train, and deathly afraid of Indians. The kids used to sneak up on him at night around the campfire and scare him out of his wits with their whoops and hollers... One day it seems old Peg-leg's wagon got stuck in the mud flats while crossing a wide river. The rest of the wagon train hadn't noticed, and had traveled a good distance away. While some of the men were riding back to help pull him out of the mud, just what they feared happened. A band of renegade Indians circled Peg-leg's wagon, and the other pioneers feared the worst. But they could hardly believe their eyes as they galloped closer--for there was old Peg-leg, standing up in his wagon with the canvas thrown back, shooting at the Indians! Needless to say, this demonstration of courage made him a hero, and never again was he teased about being afraid of Indians.

Well, long after he settled in the Willamette Valley, Peg-leg confided on his deathbed what really happened that day in the mud flats that made him a hero. His wooden peg-leg got stuck in a knot-hole in the bottom of the wagon--and he had no choice but to stand there and shoot at the renegades!

You can draw your own analogy from that story, but it shows me the power of innovation. Agriculture is faced with numerous challenges today--both economically and ecologically. It appears that site-specific crop management technologies will help us meet those challenges--but it will require a good bit of innovation to make it practical, so farmers not only survive but prosper.

69 A Decision Aid For Determining Planting And Replanting Management Of Grain Sorghum

R. L. Vanderlip
R. W. Heiniger
S. W. Welch
D. L. Fjell

Agronomy Department
Kansas State University
Manhattan, Kansas

Recommendations of management practices for planting or replanting grain sorghum [*Sorghum bicolor* (L.) Moench] must include some consideration of yield variability caused by weather conditions and of the unique management and risk situation of the individual sorghum producer. Recent improvements in modelling sorghum growth have resulted in a tool suitable for quantifying yield variability and determining production risk. The challenge is to extend the outcome of such a simulation analysis to producers in a manner that is simple to use and relevant to their individual situation. A sorghum plant growth model was used to simulate yields over a wide range of planting dates, plant populations, maturity classes, and soil types using long-term climatic data. These yield data were processed to produce probability distributions, which form the database for the decision aid. Users enter their location, soil type, and planting management for (i) an original planting and (ii) an alternative planting on the same or (for replanting) a later date. The decision aid selects the yield data for these situations, and combines them with an economic analysis based on the producer's costs, and displays the range of yields and gross returns expected. Using this decision aid, a sorghum producer can: (i) compare yields and gross returns across a range of climatic conditions; (ii) compare yields and returns for a number of different management practices; (iii) determine whether or not to replant a damaged stand; and (iv) based on his risk preference, determine the "best" management practices to use when planting or replanting on a given soil type.

INTRODUCTION

Grain sorghum [*Sorghum bicolor* (L.) Moench] is a popular crop in the midwestern US where rainfall is limited and unpredictable. Unfortunately, the sorghum plant's ability to adapt to a wide range of environmental conditions and the variable climate of this region make it difficult to determine what management practices should be used to obtain the best economic return. This is particularly true when considering whether or not to replant damaged stands. Selection of management practices should include some consideration of yield variability caused by weather conditions and of the unique management and risk situation of the individual sorghum producer. Plant growth models have the potential to provide quantitative information about the impact of climatic risk on the selection of production strategies (Arkin & Dugas, 1984). The challenge is to extend the outcome of such a simulation analysis to producers in a manner that is simple to use and relevant to their individual management and financial situation.

Two key factors must be considered in choosing management practices for planting or in deciding whether or not to replant an existing stand of grain sorghum. First, the producer needs to know the yield expectations of the sorghum crop. Before planting or replanting decisions can be made, a thorough knowledge of optimum planting dates, plant populations, and maturity classes for a particular location and soil type is essential (Benson, 1990). Normally, the choice of management strategies is based on treatment averages obtained from short-term field studies or on trial-and-error experiences (Arkin & Dugas, 1984). Although these sources can be valuable, they do not allow producers to determine yield expectations over the range of possible management practices or climatic conditions at a given location. Plant growth models offer an alternative approach, with the potential of providing useful quantitative yield information for many different management practices over the range of historical climatic conditions experienced at that site.

Second, the producer must be able to examine the risk associated with different planting or replanting strategies. Anderson et al. (1977) stated that the extension of technological advice will be more effective if due recognition is given to the impact of risk. Rather than focusing on estimating treatment means, whole probability distributions must be explored, if risk-averse producers are to be served. Ideally, the "best" management choice would be the one with the highest average return and the lowest variability. In agricultural systems, however, these are not generally synonymous; higher means often are associated with increased variability. In the Great Plains, the major source of variability is the weather. Based on the management practices used, the weather plays a role in determining the variability in yield potential of the original and replanted stands. Therefore, management decisions must be based on yield data spanning the range of environments and conditions that could impact on the decision process. The yield probabilities obtained from a simulation model could be used along with an economic function to determine the net returns to different planting and replanting strategies over the range of climatic conditions at the location (Arkin et al., 1978). Based on their ability to accept risk, sorghum producers could evaluate the distribution of net returns from each strategy and then select the options that meet their financial and management goals.

Recent improvements in modelling sorghum growth with SORKAM (Rosenthal et al., 1989) have resulted in a tool suitable for quantifying yield and its variability over a wide range of management practices (Heiniger, 1994). Simulating yields for different management practices using long-term climatic data allows yield expectations to be determined over the range of historic weather conditions. Heiniger (1994) demonstrated that, when combined with an economic analysis, these yield data could be used to assess climatic risk. The key to using this information for making specific recommendations for a producer would be to combine planting costs with yield data for a specific location, soil type, and management preference. Then the producer could analyze options, given the possible gains or losses in yields and net returns. Our objective was to develop a computer-based decision aid that would allow a sorghum producer to examine yields and economic returns over a range of historical weather patterns and management situations and then make adjustments to management practices and risk strategies to take advantage of, or to guard against, the most likely scenario.

MATERIALS AND METHODS

The decision aid was developed by combining an extensive database of yield information with an economic module that allows the sorghum producer to input costs and grain prices. Grain yields for the database were simulated using an improved version of the SORKAM sorghum growth model (Heiniger, 1994). SORKAM is a physiologically based source-sink model that simulates plant growth and development (Rosenthal et al., 1989). The SORKAM model accounts for the effects of temperature and moisture on plant growth, but not for the effects of nutrients, insects, or disease. SORKAM has been validated for a number of locations in Kansas. An extensive validation analysis has found that this model can accurately simulate yields from different planting dates, plant populations, and sorghum hybrids based on their maturity classification (Heiniger, 1994). In this analysis, the SORKAM model was able to capture from 76 to 92% of the variability in observed sorghum yields. Climatic data inputs for the yield simulations were obtained from the Kansas State University Weather Data Library (Brown, 1978). Soil, cultivar, and location-specific inputs for the SORKAM model for different locations and soil types have been described by Heiniger (1994).

With 33 years of climatic data (1959–1991), yield probabilities for several Kansas locations were determined based on the soil type selected. At each location, yields were simulated for seven planting dates and, within each planting date, for all combinations of six plant populations and three maturity classes (Table 69–1). This produced a total of 126 yield distributions, each with 33 data points. Yield databases using different soil moisture levels (full, three-quarters full, one-half full, and one-quarter full) at each planting date also were developed. Therefore, for each soil type at each location, yield data consist of five databases containing 126 yield distributions each with 33 data points. Yield distributions for planting dates and plant populations that differ from those in the database are interpolated by using the two closest distributions.

The decision aid interface featuring context sensitive help was adapted from a decision support system developed for analyzing crop selection in Australia (Jamieson et al., 1992). When the decision aid is executed, the user is asked to select his location and soil type from the list currently in the decision aid (Fig. 69–1). The user then inputs information identifying management practices for his original planting scenario. This information includes the planting date, the plant population, and the maturity class of the hybrid used. The decision aid uses this information to select or calculate the yield distribution for this planting. Additional input required for describing alternate management practices depends on whether the user is analyzing a replanting option or a different management combination for the original planting. The user can also input the current soil moisture status, which will aid in identifying possible risk or opportunity. Based on this information, the decision aid selects the yield distribution for comparison with the original scenario.

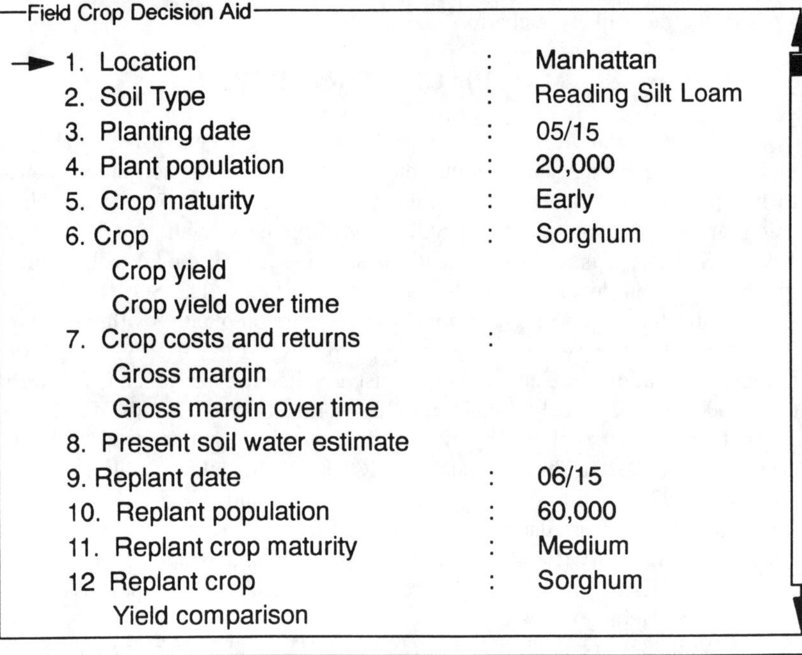

Fig. 69–1. Management input screen showing the information that the user inputs to describe the location, soil type, and original and alternative management practices. Items are changed either by selecting choices from a menu or by entering them from the keyboard..

Table 69–1. Parameter values used in simulating sorghum planting dates, plant populations, and maturity classes for each soil type and soil moisture level in the decision aid model.

Parameter	Units	Simulated Values
Plant date	day of year	15 April
		1 May
		15 May
		1 June
		15 June
		1 July
		15 July
Plant population	plants ha^{-1}	20,000
		50,000
		100,000
		150,000
		200,000
		300,000
Hybrid maturity	early (15 leaves)	early
	medium (17 leaves)	medium
	late (19 leaves)	late

The economic module is based upon the Kansas State University Farm Management Guides for making cost-return projections for grain sorghum (McReynolds & Schlender, 1986). For the original planting, the module calls for keyboard input of variable costs per acre and grain prices per bushel (Fig. 69–2). Depending on whether the user asks for a replanting analysis or analysis of an original planting opportunity, the module calls for additional input. In the case of replanting, the module adds the variable costs associated with replanting to those for the original planting. In analyzing different management practices, the producer can provide a new set of costs and grain prices for the new management practice being considered. Using the proper yield distribution, the decision aid calculates costs and returns for each of the 33 years for both the original and alternate yield distributions.

The decision aid uses the yield distributions and the economic information to provide output in the form of grain yields and gross margins. Initially, the user could view a graph showing the whole range of yield expectations for the original planting management. A table showing quantitative yields at certain points in the distribution also could be selected. Another graph provides information about average yield expectations for the selected population and hybrid maturity when planted at different dates during the growing seasons. Similar displays for gross margins expected from this original planting are available.

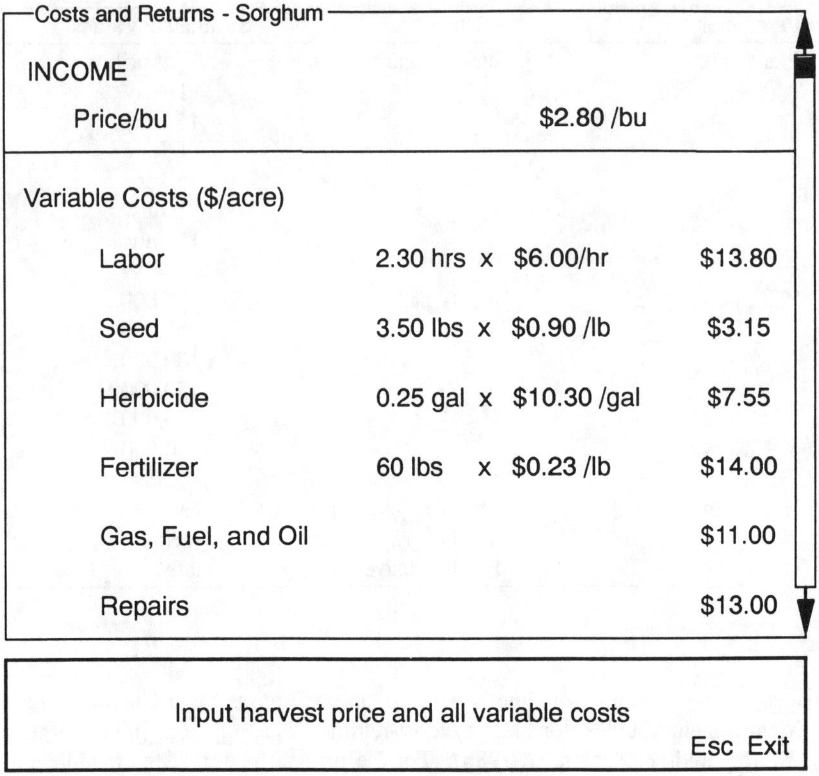

Costs and Returns - Sorghum

INCOME				
Price/bu			$2.80 /bu	
Variable Costs ($/acre)				
Labor	2.30 hrs	x	$6.00/hr	$13.80
Seed	3.50 lbs	x	$0.90 /lb	$3.15
Herbicide	0.25 gal	x	$10.30 /gal	$7.55
Fertilizer	60 lbs	x	$0.23 /lb	$14.00
Gas, Fuel, and Oil				$11.00
Repairs				$13.00

Input harvest price and all variable costs

Esc Exit

Fig. 69–2. Cost and return screen showing the input form for entering costs and the harvest price for the original planting. Scroll bars allow the user to view different sections of the cost analysis form.

The value of such a decision aid becomes apparent once the alternate crop management practices have been selected and the economic information has been entered. The user then can select a graph or table showing the comparison between the original and alternate yield distributions or gross margins. This gives the user access to the whole spectrum of yields or gross margins. From this output, the user can evaluate and select the optimal management strategy based on risk preference. If desired, the user can select other practices for evaluation, or revise the economic information to assess the effects of changing costs or grain prices. The flexibility built into the system gives the user almost unlimited potential for making comparisons and identifying management strategies that suit the user's particular needs.

RESULTS

A Case Study

The utility of the decision aid can be demonstrated best by considering a hypothetical situation for a sorghum producer located at Manhattan, Kansas. In this case, the producer has experienced stand reductions and is interested in determining whether or not to replant this stand and what management to use when doing so. Upon executing the decision aid, the producer selects his location (Manhattan) and soil type (Reading silt loam [fine montmorillic typic ustoll (0–1% slope)]). Next, the producer inputs the original planting date (15 May), remaining plant population (20 000 plants acre^{-1}), and hybrid maturity (early). The costs incurred in the original planting operation (\$85 acre^{-1}) and the expected harvest price (\$2.00 bu^{-1}) are input.

A quick review for this situation reveals that yields typically range from 56.0 to 82.1 bu acre^{-1} (Fig. 69–3) and expected gross margins from \$27.04 to \$79.00 acre^{-1} (Fig. 69–4). Given good weather, the sorghum producer decides it is possible to replant by 15 June. The anticipated population for the replanting opportunity (60 000 plants acre^{-1}) and the maturity class of the hybrid (medium)

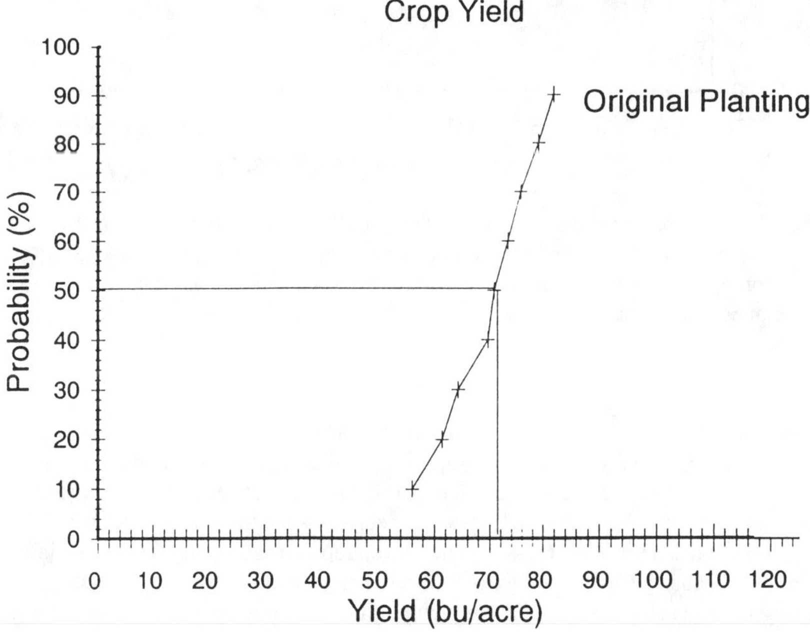

Fig. 69–3. Grain yields for grain sorghum planted on 15 May with 20 000 plants acre^{-1} remaining from an early hybrid at Manhattan, KS. Users can move the cursor along the frequency distribution to determine the yield level for the selected probability.

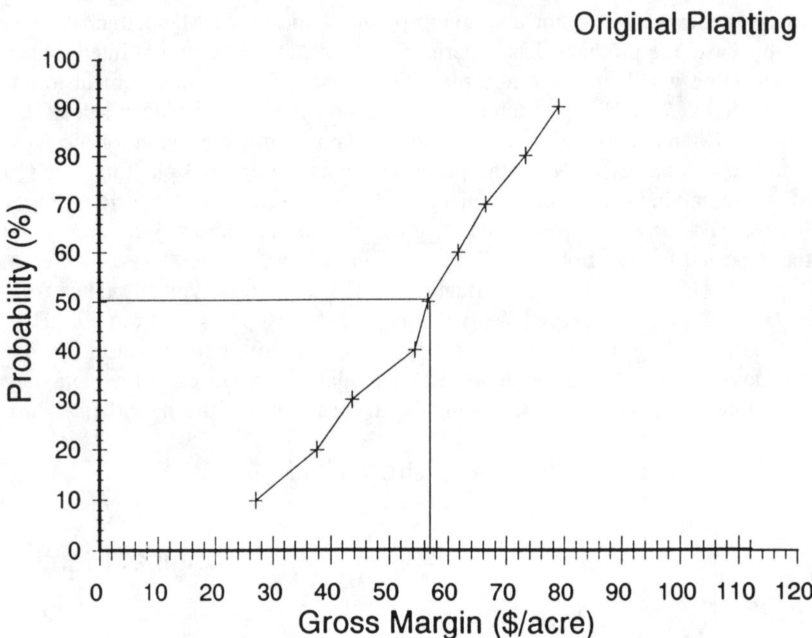

Fig. 69–4. Gross margins calculated from yields for sorghum planted on 15 May with 20,000 plants acre^{-1} remaining from an early hybrid and inputs from the economic module which described the costs and returns associated with the management practices used.

are entered along with the costs for additional tillage and planting operations ($20.20 acre^{-1}). By selecting the graph for yield comparisons, the producer now can observe the difference in yield expectations between the original and replanting scenarios. Replanting yield expectations for this case range from 57.5 to 106.0 bu acre^{-1} (Fig. 69–5). Therefore, replanting appears to be a viable option because little chance exits of decreasing yields. However, when gross margins are considered, a different picture emerges. With gross margins ranging from $9.80 to $106.82 acre^{-1}, the economic comparisons indicate a 20% chance of losing money by replanting and an 80% chance of increasing income (Fig. 69–6). In this case, the producer must consider risk in deciding whether or not replanting is a viable option. Other replanting management strategies or cost assessments could be tested to determine if risk could be reduced or profits increased.

DISCUSSION

The computer-based decision aid described has several advantages over traditional methods of recommending management practices. First of all, it has the capability of quantifying the level of risk present in the system. This gives the sorghum producer the ability to analyze options in light of risk preferences. Producers who are interested in sustainability or are risk averse will identify different options than will producers who are risk neutral or risk seeking. Second, the producer can interact dynamically with the model by imputing different cost projections or selecting different management practices. This allows the producer to explore alternatives based on unique needs and problem-solving skills. For the first time, management recommendations can be tailored to the individual farming situation. Finally, because the yield databases for each soil type and location are unique, a sorghum producer can micro-manage cropping practices by soil type for efficient production. With knowledge of soil types, a producer would have the ability to select management strategies that suit the soil properties. This makes the decision aid an indispensable tool for recommending farming practices based on information from Geographic Information System databases.

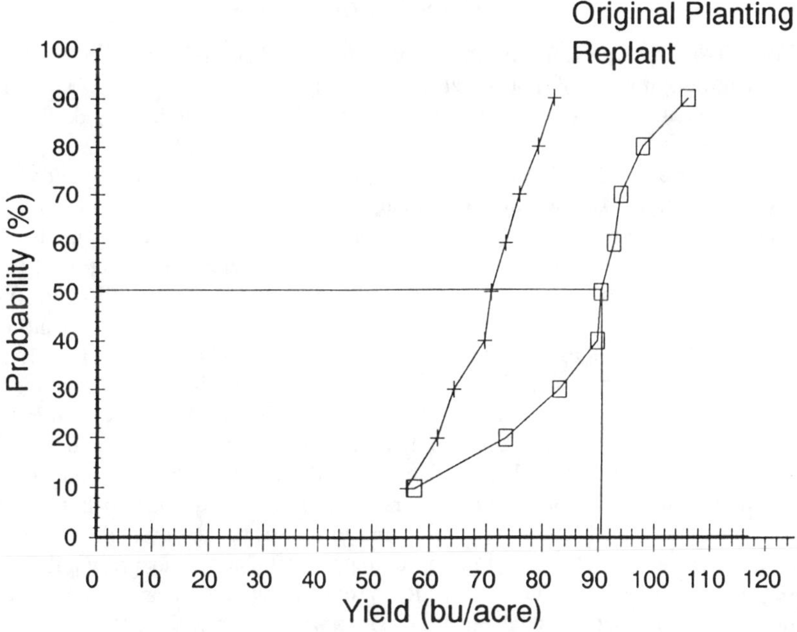

Fig. 69–5. Comparisons between yield distributions for a 15 May planting with 20 000 plants $acre^{-1}$ remaining from an early hybrid and a 15 June planting seeded at 60 000 plants $acre^{-1}$ using a medium hybrid.

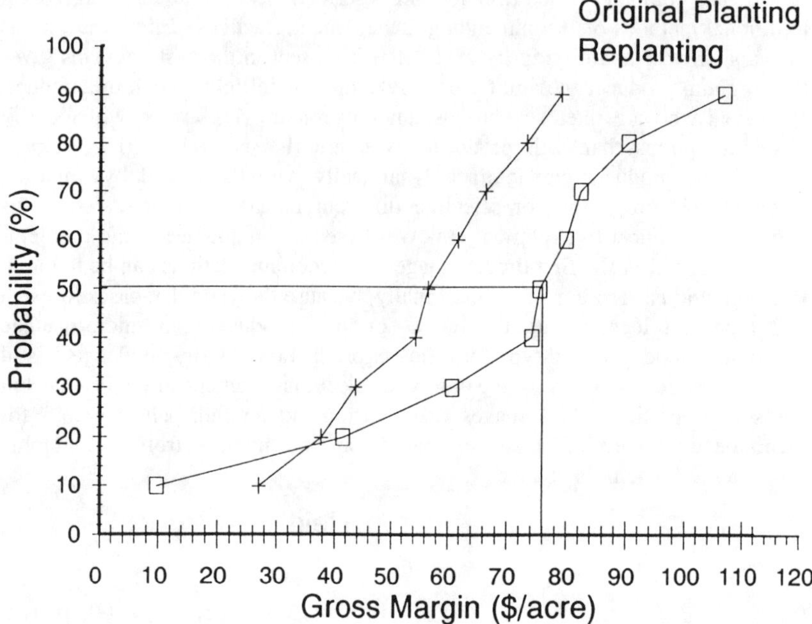

Fig. 69–6. Comparisons between gross margin distributions for a 15 May planting with 20 000 plants acre^{-1} remaining from an early hybrid and a 15 June planting seeded at 60 000 plants acre^{-1} using a medium hybrid.

Using this decision aid, a sorghum producer can: (i) compare yields and gross margins across the range of climatic conditions for a given location, soil type, hybrid maturity, planting date, and plant population; (ii) compare yields and gross margins for different combinations of management practices; (iii) determine whether or not to replant a damaged sorghum stand; and (iv) determine the "best" management practices to use when planting or replanting. Furthermore, all of these decisions can be made with full knowledge of the risks involved with each strategy. The producer can utilize this risk information to take advantage of opportunities or to guard against potential losses either by adjusting farm or financial management or by purchasing crop insurance.

This study describes a new approach to making management recommendations for sorghum producers. The abilities to quantify risk, to tailor recommendations to producers' needs, and to micro-manage by soil type make this system a powerful tool for improving the efficiency and profitability of sorghum enterprises. The information provided by this decision aid allows farmers to make better management decisions when faced with difficult choices. In the future, the addition of different crop choices and the integration of yield databases sensitive to weed, pest, and fertility management practices would further enhance the utility of this tool for decision making.

REFERENCES

Anderson, J.A., J.L. Dillion, and B.J. Hardaker. 1977. Agricultural decision analysis. Iowa State University Press. Ames, IA.

Arkin, G.F., C.W. Richardson, and S.J. Maas. 1978. Forecasting grain sorghum yields using probability functions. Trans. ASAE 21:874–877.

Arkin, G.F., and W.H. Dugas, Jr. 1984. Evaluating sorghum production strategies using a crop model. p. 289–295. *In:* V. Kumble (ed.) Agrometeorology of sorghum and millet in the semi-arid tropics. Proc. of the Intern. Symposium. 15–20 Nov. 1982. ICRISAT Center. Patancheru, A.P. 502324, India.

Benson, G.O. 1990. Corn replant decisions: a review. J. Prod. Agric. 3:180–184.

Brown, M.J. 1978. Sources of weather and climatic information for Kansas. Bulletin 614. AES. Kansas State Univ., Manhattan, KS.

Heiniger, R.W. 1994. Validation and use of a plant growth model to determine replant guidelines for grain sorghum [*Sorghum bicolor* (L.) Moench]. Ph.D. Dissertation. Kansas State Univ., Manhattan, KS.

Jamieson, A., G.L. Hammer, and H. Meinke. 1992. A computer based decision aid for sunflower, sorghum and wheat. 13th International Sunflower Conference. 13–18 Sept. Pisa, Italy.

McReynolds, K.L., and J.R. Schlender. 1986. Dryland grain sorghum in central Kansas. KSU farm management guide MF-575. Coop. Ext. Service. Manhattan, KS.

Rosenthal., W.D., R.L. Vanderlip, B.S. Jackson, and G.F. Arkin. 1989. SORKAM: a grain sorghum growth model. TAES Computer Software Documentation Series. MP-1669. TAES. College Station, TX.

70 Commercial Collection and Processing Of Geo-Referenced Soil Samples

Joe W. Tevis

Consulting Engineer and
Spatial Data Systems
Sioux City, Iowa

The fundamental objective of agricultural spatial management is to conserve and improve the efficiency of resources and inputs while controlling their potential impact on the environment. Systems using a maps based approach to satisfy these objectives require three separate but interacting components. They are i) collection of agricultural spatial data, ii) processing and interpretation of spatial data, and iii) spatially sensitive distribution of the agricultural inputs.

Agricultural Spatial Data

Agricultural spatial data is data describing the within field distribution of agronomic properties which are known to influence plant response or environmental impact potential. Sources include punctual field collection, remote sensing, government data bases, and implement mounted sensors. A list of common examples of each of these sources is given in Table 70–1.

Data Processing and Decision Support

After the spatial data is collected it is then processed in order to assign each point in the field an attribute value which is a function of the data. The result is one attribute layer for each spatial data set. For low resolution punctual data such as soil samples, it is necessary to interpolate the data prior to or as part of the creation of attribute layers from the spatial data. Methods which have been used to interpolate soil properties include Kriging and its many variations, inverse distance or "pseudo-kriging", and others reviewed by Laslett and McBratney (1990).

Table 70–1. Sources of Agricultural Spatial Data.

Punctual Field Collection	Remote Sensing	Government Data Bases	On-board Sensors
soil samples	satellite imagery	SSURGO (SCS) Soil Survey Geographic Data Base	soil pH
crop scouting	aerial photography	ISPAID (Extension) Iowa Soil Properties and Interpretations DataBase	organic matter
plant tissue	aerial videography	DEM (USGS) Digital Elevation Model	crop yield
soil penetrometer	ground penetrating radar	DLG (USGS) Digital Line Graphics	

Additional data layers called derived layers can be generated from one or more attribute layers. The most common derived layer is the application rate map but they are also used to generate intermediate layers similar to a database query.

One simple but useful method for generating either a numeric or character derived layer is to apply a threshold function to a specified attribute layer (Tevis & Searcy, 1991[b]). An example of a character derived layer using a threshold function would be to classify a soil pH layer into "low", "medium", or "high" areas. Boolean sentences can also be applied to single or multiple layers to create new derived layers (Motz et al., 1993). Fertilizer application maps have been generated by applying an Expert System to several agronomic attribute layers (He et. al., 1992). Other methods for generating derived layers currently being investigated include fuzzy logic and neural networks.

Spatial data from multi-spectral remote sensing, particularly satellite imagery, produces one spatial data set or layer for each reflectance band recorded. Raw satellite data must be processed to produce information which can be utilized in agricultural management. These layers can be combined in specific ways to create vegetative indices which have been correlated to crop yield. It is also possible to process the bands to generate a classification layer to facilitate soil mapping (Su et al., 1989 and van Deventer et al., 1991).

Spatially Sensitive Distribution of Agricultural Inputs

The final step in a maps based spatial management system is to use application rate maps to direct the distribution of each input. The methods used to generate each field's rate maps and the maps themselves should be compatible with the level of management being applied to the field. There are three ways with the level of management being applied to the field. There are three ways in which spatial data and accompanying decision support can be utilized each of which represents a different level of spatial management.

The simplest approach is to use the spatial data to make improved uniform recommendations based on the area or integrated average of the interpolated response surface. The second level of spatial management is to make variable rate recommendations according to predetermined zones in the field. This can be accomplished with conventional equipment and an increased management effort. The highest level of spatial management requires making continuously variable application across the field. This methods requires Variable Rate Technology (VRT) equipment capable of dynamic control of application rates as a function of a previously generated and digitally stored rate map.

Each of these approaches has the potential for improved efficacy of agricultural inputs and reduced environmental impact over conventional management methods. The most appropriate method and the potential improvement for each field is a function of: i) the degree of variability, ii) management skills, iii) available machinery, and iv) the environmental impact potential.

Development of Spatial Data Systems Consulting

Products and Services

The goal of Spatial Data Systems is to develop products and services to aid in the spatial management of agricultural resources and inputs. The emphasis is on the first two components of a maps based system, specifically the collection and interpretation of agricultural spatial data. It was decided that these products and services should be provided on a consulting basis and that clients should not have to make additional machinery investments to utilize the service. This constraint served to expand the potential market but was also consistent with the position that VRT is not always necessary or the most appropriate level of spatial management. The guidelines and philosophies used to shape Spatial Data Systems services and products are summarized in Table 70–2. The products and services which are currently being offered are listed in Table 70–3.

Table 70–2. Development guidelines of spatial data system products and services.

1.	Emphasis on spatial data collection and interpretation.
2.	Machinery investments by clients not required.
3.	Adaptable to varying farm size, management objectives and management skills.
4.	Usable by clients having access to VRT equipment
5.	Off-the-shelf hardware used in data acquisition

Table 70–3. Current Products and Services.

1. GPS Field Survey
    - field boundaries
    - non-production areas
    - landmarks

2. Geo-referenced soil sampling
    - sampling pattern design
    - sample collection
    - handling and shipping

3. Spatial Data Processing and Interpretation
    - spread sheet report of lab results
    - maps of soil property spatial distributions

4. Field scale GIS Software, LandSteward 1.0
    - display of spatial distribution layers
    - map printing
    - crop scouting component
    - herbicide database

Equipment

Global Positioning System All positioning data reported in this paper were collected using the Garmin GPS SRVY II which is a commercially available "off the shelf" receiver. It is a multi-plexing single channel receiver capable of tracking up to eight satellites with a one second update. The SRVY II tracks the L1 and C/A code signals and the internal memory can store up to 18 hours of pseudo-range data or 200,000 points of 3D position data.

Positions were recorded in the static differential mode using two of the Garmin GPS SRVY II receivers. The base unit was equipped with the standard cylindrical antenna mounted on the receiver. The mobile or rover unit was equipped with a ground plane antenna with magnetic mount and six foot cable. The pseudo-range data was corrected using the post-processing software provided by Garmin.

Spatial Data Processing Software Spatial data processing and map generation was performed with the Spatial Data Systems software package LandSteward Pro 1.0. The software given to each client, LandSteward 1.0, contains the same source code but several of the more technical utilities have been disabled. A description of the more important features of the LandSteward Pro software is given in a following section.

GPS FIELD SURVEY

The field survey is the first step in preparing a field for spatial management. Survey position data is geo-referenced using Global Positioning System receivers. The primary objective of the survey is to define the boundaries of the field. This data is converted to plane or rectangular coordinates which can then be used to make area and distance calculations required for crop management planning.

The GPS field survey also records position data defining non-productive areas and landmarks in the field. Non-productive areas include terraces, waterways, building sites, creeks, and tree groves. Including these areas in the field database will result in a better estimate of the tillable area and also serve as convenient reference points when using the maps to direct field applications. Landmarks such as trees and gates can also serve as reference points but they may also have other uses. For example the location of tile inlets can help design manure and herbicide application plans with reduced environmental impact. An example of the mapping generated from a GPS field survey is shown in Fig. 70–1.

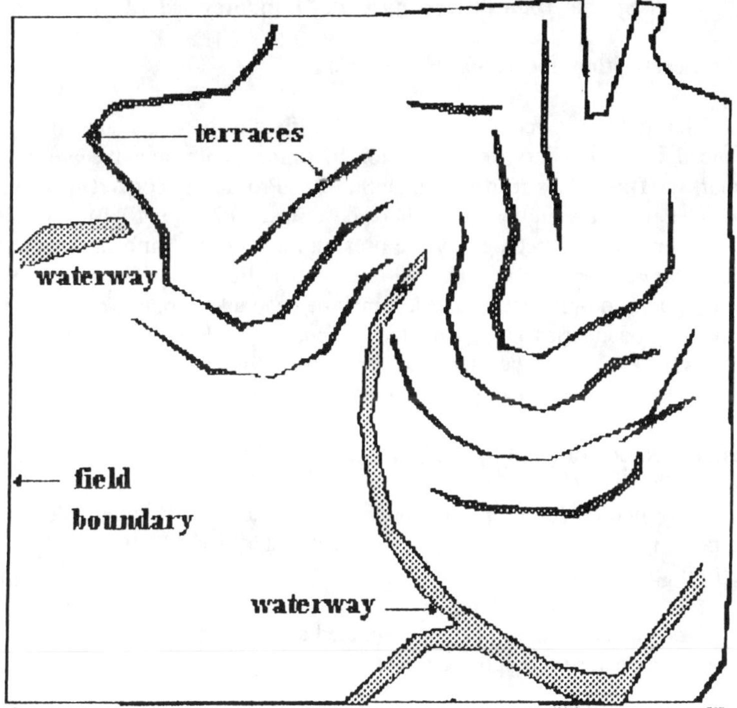

Fig. 70–1. GPS field survey map for Northwest Iowa field.

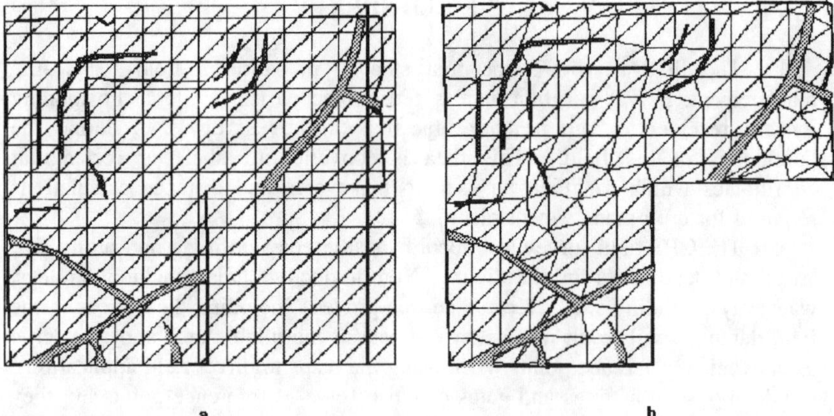

Fig. 70–2. Field Discretization - a) computer generated, b) optimized interactively.

Spatial Data Processing Using LandSteward 1.0

Field Discretization and Grid Optimization

The field file generated in the GPS field survey contains the geo-referenced field boundary vector, vectors defining non-productive areas, and landmarks. This file is read into LandSteward Pro and is converted to plane coordinates using the arc-length methods described by Torge (1980). The field is then discretized as specified by the user into a raster of uniform rectangular or triangular cells as shown in Fig. 70–2a. In some fields, especially those with many irregular non-productive areas, it may be desirable to optimize the grid so that the cell boundaries align with the boundaries of features in these areas. LandSteward Pro allows the user to perform this optimization interactively. The result is a finite element like grid as illustrated in Fig. 70–2b.

Interpolation of Punctual Spatial Data

Most interpolation schemes for estimating the value of a variable $Z(t)$ at some point t_0 as a function of a spatially distributed variable, $Z(t_i)$, follow the same four steps. They are:

1. Assume that $Z(t_0)$ can be approximated as the following finite sum where λ is a vector of weighting terms

$$Z(t_0) = \sum_{i=1}^{n} \lambda_i \, Z(t_i). \tag{1}$$

2. Define an appropriate merit function, **M**, which is a function of the weighting vector λ.
3. Determine the λ which minimizes (or maximizes) the merit function.
4. Estimate $Z(t_0)$ by substituting λ into (1).

The most fundamental difference in interpolation schemes is found in step 2, the definition of the merit function. In LandSteward Pro the user may select from two spatial data interpolation methods each with several options. For research projects with a high sampling pattern Kriging interpolation is available including semivariance estimation and semivariogram estimation. For commercial applications the merit function based on the inverse distance is more appropriate. This is called pseudo-kriging in LandSteward 1.0 Options include specifying either a kriging radius or a minimum number of kriging points. Pseudo-kriging also has the option of using a classification layer to modify the weighting terms as a function of local distribution statistic. This method is discussed in a later section.

Generation of Attribute Layers

Before agricultural spatial data collected from a field can be utilized it must be converted to an attribute layer and stored in a field scale GIS. This process essentially involves the assignment of each management cell an attribute value as a function of the spatial data. LandSteward has several methods of performing this cell assignment and are outlined in Table 70–4.

Using Classification Layers in Spatial Data Processing

Unlike most fields used for research many of the fields in Northwest Iowa have a number of distinctively different areas or sub-fields. These sub-fields may be defined by soil type, slope, drainage characteristics, crops or combinations of any of these field features and conditions. Classification layers describing these sub-fields or classes can be helpful in many aspects of processing and interpretation of agricultural spatial data collected from these fields.

Technologies being developed such as satellite imagery and aerial videography have successfully been used in a research setting to identify soil type differences, land use variation, and crop response or biomass. In the absence of these high tech methods, however, it is still possible to create useful classification layers using commonly available information. Spatial Data Systems utilizes observational data collected during the field survey, SCS soil maps, and client notes on past management practices and crop performance. Classification layers can be created interactively using LandSteward Pro. The specific rules used to define the classes are dependent on the intended application of the layer.

Table 70–4. LandSteward cell assignment methods.

Method	Description
cell averaging	The spatial data is first interpolated such that each cell has 1-3 interpolated points. Then the average of these values is assigned to the cell.
block kriging	Each cell is assigned an attribute value which is essentially the integrated average (numerical approximation) of several kriged points in the cell. Requires previously generated semi-variogram.
pseudo-block	Similar to block kriging except that pseudo-kriging is used instead of kriging therefore does not require a semi-variogram.
trends	First the spatial data is detrended using median-polish detrending. Then the cell is assigned a value equal to the local low frequency trend. (Tevis, et. al, 1991[a])

Soil Sample Pattern Design

Spatial management of agricultural inputs requires that the soil sampling produces data which defines not only the magnitude of important soil properties but also their spatial distribution. When available, a priori knowledge of the spatial distribution can be used to design efficient sampling patterns by better insuring that each important spatial trend is detected and by reducing the number of redundant samples. For example, recent research indicates that soil fertility is often closely related to past management practices. Therefore, in the absence of more specific information, it would be reasonable to use a past management classification layer as a template or pattern for collecting P and K soil samples. Whereas a soil map would be a better choice when the objective is to determine the magnitude and spatial distribution of soil pH.

Example results of using a past management classification layer to design a soil sampling pattern is summarized in Table 70–5. The samples were taken from a field in Northwest Iowa which is typical of many fields in the area. The field has ten terraces, several waterways, and other non-productive areas all of which served to create seven easily identifiable classes. A sampling pattern was designed which assured that at least two samples were taken from each class. Sampling densities within the classes varied from 1 sample/1.2 hectares to 1 sample/3.1 hectares.

Table 70–5. Fertility soil sample results sorted by past management class.

Class	N	Area (Hectares)	P1, mg kg^{-1} (weak Bray)	P2, mg kg^{-1} (strong Bray)	K, mg kg^{-1} (potassium)
1	6	18.7	22, 4^a 5^b	27, 4^a 6^a	173, 4^a 5^b
2	2	2.3	31	42	244
3	4	9.2	23	28	275
4	4	7.8	45, 1^a 6^a	54, 1^a 6^a	226, 1^a
5	3	7.0	58, 1^b 6^b	70, 1^b 6^b	381, 1^b 6^b
6	6	14.3	30, 4^a 5^b	36, 4^a 5^b	167, 5^b
7	3	4.4	24	29	146

N^a < 10% probability that the data from class N is drawn from the same parent distribution.

N^b < 15% probability that the data from class N is drawn from the same parent distribution.

The validity of the classification was tested by performing the Kolmogorov-Smirnov Test for the data from each class pair. The Kolmogorov-Smirnov Test is a non-parametric procedure for testing the null hypothesis that two data sets have been drawn from the same parent population (Press, et al., 1988). The result is reported as the probability that the two data sets have the same parent population. Class pairs having significance levels of < 10% and 15% are noted in Table 70–5. The remaining data pairs had significance levels of between 30–90%.

Interpolation of Punctual Spatial Data

The general form of the equation used to interpolate agricultural spatial data such as soil samples is given by (1). For Kriging interpolation the weighting terms are determined which minimize the estimation variance as defined by the semivariogram for the field. When Kriging is not practical, as in most commercial applications, it is common practice to make the weighting terms a function of the inverse of the lag where the lag is the distance from the point being estimated to the known data point.

As conventionally applied, inverse-distance interpolation selects weighting terms based on lag distance alone, without consideration of the local distribution statistics. This approach has the potential of over-weighting data which are close to the point being estimated but are taken from an area having a statistically different parent population.

Spatial Data Systems has developed an interpolation method which uses appropriate classification layers to modify the weighting terms as a function of the local distribution statistics. The method is based on the Kolmogorov-Smirnov Test described in the previous section and is a current interpolation option in the LandSteward Pro data processing software.

The first step in the method is to sort the punctual data by class using a selected classification layer. Next the Kolmogorov-Smirnov Test is applied to each class pair and the resulting matrix of probabilities is stored. Interpolation then proceeds using the inverse-distance method with one modification. Before weighting terms are calculated, the classes of each estimated point-punctual data point pair are compared. If the two points are from different classes, and thus potentially different parent distributions, the lag distance is modified according to (2) where ks[i][j] is the probability that the data from the ith and jth classes were drawn form the same parent population. *mzw* is an additional influence coefficient having a value of 0–1 to be specified by the user at runtime.

Variable Application by Management Zone

Classification layers can also be used in spatial management systems which do not have access to application equipment with VRT technology. For this use of classification layers, the criterion or rules for defining the classes may

$$lag = lag \, / \, ks[i][j] \, / \, mzw \tag{2}$$

be different than for the previous two applications. Here the objective is to classify the field into management zones which may conveniently be or currently are being managed as sub-field units. The boundaries of the classes are defined by natural or man-made features which already exist in the field. These include terraces, waterways, set-aside boundaries, and crop rotation boundaries.

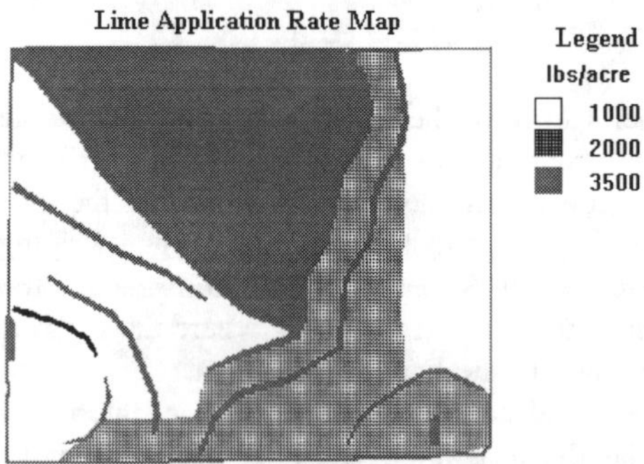

Fig. 70–3. Lime application rates for management zone classes.

An example of using classification layers to generate application rate maps is shown in Fig. 70–3. This map defines the application of lime for pH control for three distinctly different areas in the field. In this field conventional management would likely have resulted in the highest rate being applied to the entire field.

Discretizing the management of agricultural inputs in this way requires that the spatial data processing and interpretation be sensitive to the classification layer. The fundamental objective is to use the pertinent spatial data to define a representative homogenous model for each management zone class. A simple approach would be to assign each class an attribute value which is the arithmatic mean of the data within the class. This may be sufficient for some data but such linear combinations often allow outliers (very high or low data) to disproportionately influence the outcome. An alternative is to first assign each management cell an attribute value then assign each class attribute using area averaging as defined by (3).

$$\textit{class attribute} = \frac{\sum_{1}^{N} A_i a_i}{\sum_{1}^{N} A_i} \qquad (3)$$

N = *number of cells in class*
A_i = *area of the ith cell*
a_i = *attribute of the ith cell*

Table 70–6. Data means and area averages for management zone classes in Northwest Iowa field.

		P^{-1} mg kg^{-1} (weak Bray)		P^2 mg kg^{-1} (strong Bray)		K mg kg^{-1} (potassium)	
Class	N	data mean	area avg.	data mean	area avg.	data mean	area avg.
1	6	22	23	27	28	173	177
2	2	31	27	42	33	244	196
3	4	23	25	28	30	175	181
4	4	45	42	54	51	226	225
5	3	58	49	70	59	381	312
6	6	30	30	36	37	167	173
7	3	24	29	29	34	146	153

The effect of area or integrated averaging is to reduce the impact of outliers without completely loosing the information they may contain. The results of the method are illustrated in Table 70–6 which lists both the arithmatic mean of the data and the area average of each management zone class for fertility data from a Northwest Iowa field.

Closing Comments

At a recent farm show, after patiently describing my service to an "average" farmer his parting comment was "When farming comes to this, that's when I quit". Granted this was an extreme reaction but it did point out a failure of those of us championing this new way of managing our resources. Specifically, we are not doing a good job explaining the benefits of this new technology and probably because we don't know ourselves. More time and effort should be spent evaluating site-specific systems both from a economic and environmental perspective. In other words, technology has far surpassed our understanding of how it should be used.

Along these lines I have a few suggestions and observations which I think would serve to advance the development and acceptance of site-specific or spatial management.

1. It is important that we recognize that not all fields are good candidates for site-specific management. More should be done to classify fields according to their potential benefits from this higher level management.
2. More effort should be made to make new technology available to a larger percentage of farmers, and not to a select few of the larger operators.
3. Current amendment recommendations need to be reviewed. It seems that they have a built in safety factor which may not be appropriate or necessary for spatial or site-specific managment systems.

REFERENCES

He, B., C.L. Peterson, and R.L. Mahler. 1992. An expert system linked with a GIS database. ASAE 1992 International Winter Meeting. Paper No. 92–3556.

Laslett, G.M., and A.B. McBratney. 1990. Further comparison of spatial methods for predicting soil pH. Soil Sci. Soc. Am. J., 54:1553–1558.

Motz, D.S., S.W. Searcy, and P.E. Neuhaus. 1993. PC-MAPS a tool for site specific crop management. ASAE 1993 International Winter Meeting. Paper No. 93–3556.

Press, W.H., B.P. Flannery, S.A. Teukolsky, and W.T. Veterling. 1988. Numerical recipes in C. Cambridge Univ. Press, Cambridge.

Su, H., M.D. Ransom, and E.T. Kanemasu. 1989. Detecting soil information on a native prairie using Landsat TM and SPOT satellite data. Soil Sci. Soc. J. 53:1479–1483.

Tevis, J.W., D.A. Whittaker, and D.J. McCaulley. 1991a. Efficient use of data in the kriging of soil PH. ASAE 1991 International Summer Meeting.

Paper No. 91–7047.

Tevis, J.W., and S.W. Searcy. 1991b. Generation and digitization of management zone maps. ASAE 1991 International Summer Meeting, Paper No. 91–7048.

Torge, W. 1980. Geodesy. Walter de Gruyter, New York.

van Deventer, P.A., R.A. Filho, W.J. Elliot, A.D. Ward, and P.S. Thenkabail. 1991. Soil mapping using Landsat Thematic Mapper Data. ASAE 1991 International Summer Meeting, Paper No. 91–7046.

71 Field Experiences Supporting Precision Farming Technology

Ted S. Macy

Applications Mapping, Inc.
Frankfort, Illinois

Applications Mapping, Inc., has been heavily involved in the technology transfer associated with precision agricultural practices for the past two years. During that time, the author has worked with farmers that have fully embraced automated variable rate application of all materials with their own equipment and others that prefer to acquire the services through the local farm input suppliers. This paper relates the author's experiences with providing software, hardware, training, and support to both of these market segments.

Applications Mapping also works with fertilizer dealers that have purchased fully automated variable rate application equipment, as well as those that have begun providing manually controlled variable rate application as the first step in the growth path to precision agriculture for themselves and their clients. This paper discusses some of the issues involved in supporting site-specific practices at the commercial fertilizer dealer level.

And finally, Applications Mapping installed and supported over twenty yield monitors during the fall of 1993. This paper relates experiences supporting these installations and provides examples of the type of information generated, and questions raised by this activity.

72 Correlation Between Physical Variability and Management Decisions From a Representative Sample of Wisconsin Corn Producers

P. Nowak
R. McCallister
S. Wolf

University of Wisconsin
Madison, Wisconsin

The purpose of this study is to examine the extent a representative sample of Wisconsin corn producers change management decisions in response to variation in soils and landscape. The study is based on the 1987 USDA National Resources Inventory (NRI). All operators of Wisconsin NRI points where corn was produced between 1985 and 1987 were identified (N - ≈1200) through logical selection on the NRI data file. The names and addresses of these farmers were determined by examining NRI records in 62 county and state USDA offices. These names and addresses were obtained with a confidentiality agreement with USDA. These operators will be contacted post-harvest 1994 with a mail survey instrument including an aerial photo of the field where the NRI sample point is located. Respondents will be asked a series of questions about soil features and management decisions about the NRI point highlighted on the aerial photo, the encompassing fields, and other fields in corn production. This data will then be linked with case-specific SOILS 5 information available in the NRI data base, and extrapolated data (e.g., the number of soil mapping units within the outlined field) taken from county soil surveys. The analysis of this multi-layered data set will address questions such as the extent farmers vary fertilizer and pesticide inputs in response to variation in soil characteristics, if management varies by the soil's capability to attenuate agrichemicals, what features of the soil are most important in "triggering" a management response to physical variation, and how farmers characterize soil variation. Understanding the how and why farmers currently respond to variation in physical settings needs to serve as the foundation for future site-specific management activities.

APPENDIX I

WORKSHOP II - SITE-SPECIFIC CROP MANAGEMENT SYSTEMS

Dean Fairchild, Moderator

EQUIPMENT PRESENTATIONS

A. SOIL SAMPLING/MAPPING

1. Concord, Inc., Fargo, N.D. - Virgil Mahlum
   Soil Sampling, Detectspray, Airseeders

2. Applications Mapping, Inc. - Ted Macy, Frankfort, IL.
   Soil Sampling/G.P.S., Mapping

3. Micro-Track Systems, Inc., Mankato, MN -
   Scott Veldman and staff
   G.P.S., Soil Sampling, Yield Monitors

B. NAVIGATION/POSITIONING

1. Mobile Data Communications, Cedar Rapids, IA. -
   Rick Engelby and staff
   G.P.S. and Parallel Tracking

2. Accutrak, Regina, Saskatchewan, Canada
   Navigation, Guidance Systems, Parallel Tracking

C. SITE-SPECIFIC APPLICATION EQUIPMENT

1. Dep. of Soil Science, Univ. of Minnesota, St. Paul, MN -
   Dr. Gary Malzer
   Variable Ammonia and N-SERVE

2. SOILTEQ, INC., Minnetonka, MN -
   Roger Knutson and staff
   Variable Fertilization Applications and Soil Mapping

3. Tyler, Inc., Benson, MN. - Don McGrath and Staff
   Variable Fertilizer Application, Organic Matter Sensor

4. Deere & Co., E. Moline, IL - Wayne Smith

D. OTHER

1. Larson Systems, Inc., Ames, IA. -
   Don Larson

APPENDIX II

PANEL OF PRODUCERS WHO HAVE ADOPTED SITE-SPECIFIC MANAGEMENT IN THEIR CURRENT OPERATIONS

Moderator:	**Ted Macy**	**Frankfort, IL**
Participants:	**Don Bot**	**Cottonwood, MN**
	Lynn Child	**Quincy, WA**
	Ken Dalenberg	**Mansfield, IL**
	Gary Wagner	**Climax, MN**

The following are summaries of the presentations by the participants:

Don Bot - I farm, with my brothers, a 1300 acre corn/soybean operation. It is all planted no-till; corn in 22.5 inch rows and drilled soybeans.

We have grid sampled on 2.5 acre cells (330 x 330 feet), analyzed for pH, P, K, and Zn, and put results in map format.

We have computerized air-flow spreaders, modified our planting equipment (John Deere 750 No-Till drills) with (International) seed modules. We achieve variability by replacing ground drives with hydraulic motors. We have the capability of varying seeding rates, urea rates, and the P-K mixture.

We did some initial work with a yield monitor. We need that input back to develop the (planting rate, etc.) maps with which we are working.

We use a cold flow unit to apply anhydrous ammonia and vary the rate (fall) applied.

We would like to use variable rate technology in soybean planting so that we could plant (at the same time) two varieties. One variety would be tolerant to iron chlorosis - a problem on some of our calcareous soils.

Lynn Child - We farm in the Columbia basin with irrigation. We use rill irrigation so there is much land leveling. Potatoes are our principal crop.

We grid sample in the fall on 200 x 200 feet (cells), principally for P and K analysis. We develop kriged maps on a 5-level (of P and K) basis which are input to a PROM-chip for the SOILECTION application machine. Our P levels are quite variable but with 'zoned' application, we have been able to 'level' P and K.

Sampling cost for lab analysis and map generation is about $10 per acre. Compared to about $2 per acre for general blanket application. Fertilizer application cost is about $7.50 per acre.

Using variable rates on our potato field, about 25 percent of the field received no P or K. However, the total fertilizer cost was about the same if one compared variable rate application to a blanket (uniform) rate (of 70 pounds R and 200 pounds K per acre). But with variable rate we feel we are putting the fertilizers where it's needed most. By using full application of fertilizer we can be more timely in spring planting.

I see much improved uniformity in potato growth and market quality and we have seen some production increase. Other factors such as improved pest control have likely contributed to quality and quantity improvement.

Ken Dalenberg: We farm about 1700 acres of corn and soybeans in east-central Illinois.

We began using some aspects of site-specific management in the mid-80s; more specific soil sampling; use of airflow seeding. In 1989 we started grid sampling (one sample per 3.3 acres). In 1992 we began using the Soil Teq program with a Global Positioning System for both soil testing and fertilizer application. Since then we have also used yield monitoring.

Our soils are variable ranging from Drummer (clay loam) in the low areas to Dana (silt loam) on the ridges. This variation is evident in the pH range - from 5.9 to 6.6. As a result we have used VRT to apply lime only where needed.

Our P1 (phosphorus) tests ranged from 39 to 117 pounds per acre and K tests from 312 to 489. We used VRT to bring P1 levels to 50-55 and K levels to about 350. We believe this kind of application reduces fertility costs per acre.

Our (93) yields ranged from 83 to 177 bushels per acre. This yield range was related to the kinds of soil but not in the way one would normally expect. This was a very wet year. Highest yields were on the upper slope soils. Low-lying soils had yields greatly reduced from excess water. The yield map also showed us varietal differences in relation to the environmental conditions.

We believe VRT provides a way to re-allocate costs and inputs to those areas which will give us greatest net returns. As one analyzes application costs, other input costs, the economics of scale may indicate those operations that the farmer can do and those that a (service) supplier might perform.

When we consider the rapid changes in VRT technology and the relatively high costs of certain equipment, we believe the role of the supply company is critical.

As a farmer I would want to retain the option to manipulate yield information, varietal selection and other management options.

Gary Wagner: My two brothers and I manage about 3200 acres in sugar beets, small grains, and sunflowers in the Red River valley (of Minnesota and North Dakota). It is all non-irrigated acreage. Sugar beets is our main cash crop.

We are new to the technology. This past year we used the Ag Leader 2000 yield monitor to produce a yield map on about 200 acres of our spring wheat. The wide variation we found was not explained by fertility differences but was - in our judgement - related to compaction. Due to different soil conditions in the field, we had tilled some areas that were too wet. The resulting compaction reduced yields in areas which were clearly evident on the yield map.

We are intensely interested in nitrogen management in our sugar beets so grid sampling is very beneficial. Nitrogen affects sugar content of the beets. Too much will reduce percent sugar. Raising percent sugar by one-half percent may increase value of crop by $35-$45 per acre. We need to know the true N variability in our fields.

With the global positioning system we are able to locate patches of wild oats in the spring grain. This can (and will) result in significant savings in herbicide appreciation costs - probably as much as $10-$12 per acre.

The following are questions from the audience and answers given by the panelists.

Q. What kind of data management assistance and support systems are you looking for?

A. Dalenberg: We are dependent on people with computer expertise. Supply companies and crop consultants are doing much of our data manipulation. But I would like to have input in deciding the driving factors in yield change.

Wagner: Because of the work level in our operations we do depend on crop consultants.

Child: We depend on a variety of consultants. I would like some insecticide and entomological recommendations. Perhaps there are liability considerations.

Bot: We have not used consultants but we do need user-friendly computer programs.

Q. What is your soil sampling program?

A. Child: There is a 20-foot radius around each sample point with 8-10 samples (composited) in that area. A 200 x 200 foot grid seems to be statistically satisfactory for us.

Bot: We use dead reckoning to find the mid-point of the sample area. We take 5 samples (per area) to a 6-inch depth.

Wagner: We use roughly 5-acre grids. We sample to a depth of 2 feet, sometimes 4 feet.

Dalenberg: Our sampling is on 3.3 acre grids using GPS to locate cells. Five cores are mixed at each point.

Q. Why would farmers adopt this technology? What would be a major fear?

A. Bot: Economics. A fear might be the extra time needed to learn the technology.

Child: It gives me a better picture of field variability which I use for variable fertilization. A fear is that we will be inundated with information that we can't handle.

Wagner: Economics. Profit. A fear is that government will use the information to control inputs, field operations.

Q. Have you quantified the economic return?

A. Bot: We could not make a good comparison last year because weather destroyed 1/4 of our crop. We need to see a profitable return in 3 years or less.

Child: Not quantified but do have detailed records of costs. Believe I can see a slight upward trend on potatoes. Don't know on our other crops.

Dalenberg: I would like to have 2 or 3 years in a database using the yield monitor. After 4 years of using VRT for fertilizer application I have a tremendous return but expect that to level off as our buildup program continues.

Wagner: Believe there are benefits but need a year or two (more) to put in actual dollars.

Q. Do you manage differently on rented or owned land?

A. Dalenberg: We have everything from owned to cash rent to crop share to bushel rent and we use it on all.

Wagner: No. We want to use the best methods that will assure us that we can keep farming all land that we have. We feel it is necessary to build up a good database on all land we operate.

Child: We farm rented land comparably, if not better, than our own.

Bot: We are absorbing the additional cost of grid sampling (on rented land) but we treat it pretty much the same.

Q. How do you decide on a seeding rate?

A. Bot: This year we made it on soil type but I think we will make it on a compromise between yield goal and soil type.

Q. How do you decide on nitrogen rates?

A. Dalenberg: We have not used variable rates of N on corn - as yet.

Bot: We correlate yield goal with the amount of N to apply.

Q. Have you become better observers?

A. Dalenberg: Yes, it does make us better observers. But with all the technology there are on-site management decisions to be made.

Bot: Yes. Because of the competition for land, I have to manage more efficiently and profitably.

Q. Do you see a potential for weed control in site-specific management?

A. Dalenberg: We are using VRT for weed chemicals....trying to be environmentally responsible and using only those rates required.

Wagner: We can record the chemicals applied. A crop consultant can locate the weed areas and that can be programmed. There is a benefit to weed control in site-specific farming.

Q. Does this technology work for small farm as well as large?

A. Dalenberg: It really makes no difference if you are farming 160 or 3000 acres.

Child: I think farm size is a relative item. What is 'average' today may be 'small' tomorrow.

Bot: I think the advantage is to the educated farmer, not the size of operation.

APPENDIX III

EXHIBITORS

Application Mapping, Inc., Frankfort, IL

Concord, Inc., Fargo, ND

Farmers Software Association, Ft. Collins, CO

Larson Systems, Ames, IA

MicroTrak Systems, Inc., Mankato, MN

Midwest Technologies, Springfield, IL

Mobile Data Communications, Cedar Rapids, IA

Monsanto, Infielder, St. Louis, MO

Potash & Phosphate Institute, Monticello, IL

RDI Technologies, Spicer, MN

SoilTeq, Inc., Minnetonka, MN

Spectrum Technologies, Inc., Plainfield, IL

Willmar Manufacturing, Willmar, MN

APPENDIX IV

PARTICIPANT LIST

Rob Abel	TRW/ESL, 970 Stewart Drive, Sunnyvale, CA 94086
Bill R. Aiken	950 N Meridian, Indianapolis, IN 46204
Ken Akerman	Farmers Software Association, Box 660, Fort Collins, CO 80522
C. Eugene Allen	Vice President, Univ. of Minnesota, St. Paul, MN 55108
Raymond R. Allmaras	Dep. of Soil Science, Univ. of Minnesota, St. Paul, MN 55108
C. R. Amerman	USDA-ARS, Room 233, Bldg 005, BARC-W, Beltsville, MD 20705
David Anderson	Unisys, Box 64525, MS U2T20, St. Paul, MN 55164-0525
Linda Anderson	S332 Turner, 1102 S. Goodwin, Urban, IL 61801
Noel Anderson	Concord Inc., 2800 7th Ave N, Fargo, ND 58102
Terry Anvik	Holly Sugar Corp., 5320 Mark Dabling Road, Colorado Springs, CO 80918
Cindy Arnevik	Faribault County, 412 N Nicollet, Blue Earth, MN 56013
Pat Arthur	Farm Industry News, 7900 International Drive, #300, Minneapolis, MN 55423
Robert E. Ascheman	Ascheman Associates, Inc., 2921 Beverly Drive, Des Moines, IA 50322

Jeff Ashland	Crop Watch Ltd., Box 211, Spirit Lake, IA 51360
Leonard L. Bashford	Dep. Biol. Systems Eng., Univ. of Nebraska, Lincoln, NE 68583
Walter C. Bausch	USDA-ARS Water Mgmt. Research, AERC CSU Foothills Campus, Fort Collins, CO 80523
Robert H. Beck	Cenex/Land O'Lakes, P O Box 64089, St. Paul, MN 55164
Hugh Beckie	Agriculture CANADA, Box 1240, Melfort, Saskatchewan S0E 1A0 CANADA
Jeff Beckman	Cannon Valley Coop, 806 2nd Street, Kenyon, MN 55018
Michael D. Beeson	Great Plains Agronomics, 7977 Co Road 15, Wahpeton, ND 58075
Doug Beever	Sherritt, Inc., 3500 Manulife Place, 10180 101st St., Edmonton, Alberta T6H 5T6 CANADA
Denny Bell	Agri-Logic, Route 17, Box 198B, Brazil, IN 47834
Jay C. Bell	Dep. of Soil Science, Univ. of Minnesota, St. Paul, MN 55108
Brian Bennett	MVTL Labs-Soils, 35 W. Lincoln Way, Nevada, IA 50201
Andy Bensend	A. B. Seed, 440 14th, Dallas, WI 54733
Robert K. Berg	SDSU Research Farm, Route 3, Box 93, Beresford, SD 57003
Julie Berling	Miller Meester Advertising, 17 N Washington Ave., Minneapolis, MN 55401

Stuart Birrell	Dep. of Agric. Engineering, Univ. of Missouri, Columbia, MO 65211
Alfred M. Blackmer	Dep. of Agronomy, Iowa State University, Ames, IA 50011
Tracy Blackmer	Dep. of Agronomy, Univ. of Nebraska, Lincoln, NE 68583
Simon Blackmore	Cranfield University, Silsoe College, Silsoe, Bedford, ENGLAND
Vladimir Bobrov	Tyler, E Hwy 12, Box 249, Benson, MN 56201
Robert Boehle	Boehle Consulting, Inc., 1403 E Lafayette, Bloomington, IL 61701-6940
German Bollero	Dep. of Agronomy, Univ. of Illinois, 1102 S. Goodwin, Turner Hall, Urbana, IL 61801
Steve Borgelt	Dep. of Agric. Eng., Univ. of Missouri, 232 Ag Eng - MU, Columbia, MO 65211
Andrew Bosworth	Dep. of Soil Science, Univ. of Wisconsin, 1525 Observatory Drive, Madison, WI 53706
Don Bot	Route 2, Box 43, Cottonwood, MN 56229
Johan Bouma	Agricultural Univ. Soil Science, Box 37, 6700A Wageningen, THE NETHERLANDS
Randall A. Brady	Ag Consulting & Remote Sensing, 621 N. Parkway, Santa Cruz, CA 95065
David D. Breitbach	USDA-Soil Conservation Service, 375 Jackson St., Suite 600, St. Paul, MN 55101

Charles Bremer Pioneer Hi-Bred International, Box 1536, O'Fallon, IL 62269

Kalyn Brix-Davis Dep. of Plant Sciences, South Dakota State Univ., Ag Hall 206, Brookings, SD 57007

Boydell Broughton University of Sydney, Romaka, Pallamallawa, New South Wales, AUSTRALIA 2394

Thomas Bruulsema Dep. of Soil Science, Univ. of Minnesota, St. Paul, MN 55108

Don Bullock Dep. of Agronomy, Univ. of Illinois, Urbana, IL 61801

Rob Bullock Software Solutions of Illinois, Box 380, Shelbyville, IL 62565

Lowell Busman Southern Experiment Station, Univ, of Minnesota, Waseca, MN 56093

Michael Cahn Dep. of Agric. Eng., Univ. of Illinois, 376 AESB, 1304 W. Pennsylvania, Urbana, IL 61801

Cynthia A Cambardella USDA-ARS, 2150 Pammel Drive, Ames, IA 50010

Gaylon S. Campbell Dep. of Agronomy, Washington State Univ., Pullman, WA 99163–6420

Ronald H. Campbell HarvestMaster Inc., 1740 N. Research Park Way, Logan, UT 84321

Gregg R. Carlson Montana State University, N. Ag Res. Ctr., Star Route 36, Box 43, Havre, MT 59501

Paul Carter Pioneer Hi-Bred International, 4445 Corporate Drive, W. Des Moines, IA 50266

Richard Castelnuovo	Univ. of Wisconsin, B142 Steenbock Library, Madison, WI 53706
Allan Cattanach	Dep. of Agric. Eng., North Dakota State Univ., 227 Walter Hall, Fargo, ND 58105
Jonathan Chaplin	Dep. of Agric. Eng., Univ. of Minnesota, St. Paul, MN 55108
H. H. Cheng	Dep. of Soil Science, Univ. of Minnesota, St. Paul, MN 55108
Gary Chesney	Loral, 1300 S Litchfield Road, Goodyear, AZ 85338
Lynn A. Child	LC Farms, Inc., 6243 Road P.W.W., Quincy, WA 98848
Neal B. Christensen	Farmland Industries, 7441 O Street, Suite 300, Lincoln, NE 68510
L. G. Clark	Infielder Crop Records, 800 N Lindburgh Blvd., St. Louis, MO 63104
Tom Colvin	USDA-ARS, 2150 Pammel Drive, Ames, IA 50011
Steve Copeland	USDA-ARS, Univ. of Minnesota, Dep. of Soil Science, St. Paul, MN 55108
Bruce Cowgur	Midwest Technologies, Inc., 2733 E Ash Street, Springfield, IL 62703
Luke A. Crawford	Agris Corporation, 300 Grimes Bridge Road, Roswell, GA 30075
Tom D'Avello	Soil Conservation Service, 1902 Fox Drive, Champaign, IL 61820
Eric Dahlen	Soil Teq, Inc., 5720 Smetana Dr, Ste 100, Minnetonka, MN 55343–9688

Dave Dahlgren	Cargill, Inc., CGD/NWG, 101 3rd Ave SW, Clarion, IA 50525
Ken Dalenberg	Scattered Acres Farm, Route 1, Box 173, Mansfield, IL 61854
John A. Daniel	USDA-ARS, P O Box 1430, Durant, OK 74702
J. Glenn Davis	Dep. of Soil Science, Univ. of Minnesota, St. Paul, MN 55108
Mark DeBower	Cargill, Inc., Box 5645, Minneapolis, MN 55440
Reinder DeJong	CLBRR-Agri CANADA, Central Exp. Farm, Bldg 74, Ottawa, Ontario K1A 0C6 CANADA
Joel Dekkers	RDI Technologies, Green Lake Mall, Spicer, MN 56288
Ken Denholm	Agriculture CANADA, 70 Fountain St., Guelph, Ontario N1H 3H6 CANADA
Dan Desutter	Double D Enterprises, Route 1, Box 88, Attica, IN 47918
Earl Dick	Cargill, Inc., Box 9300, Dept #19, Minneapolis, MN 55440
Stephanie Dickson	Top Soil Testing Service Co., 2312 Boysenberry Lane #3, Springfield, IL 62707
Craig L. Dobbins	Dep. of Agric. Econ., Purdue Univ. 1145 Krannert Building, Rm 640, W. Lafayette, IN 47907-1145
J. D. Dolezal	Larson Systems, 113 Welch Ave., Ames, IA 50014-7228
Robert H. Dowdy	Dep. of Soil Science,. Univ. of Minnesota, St. Paul, MN 55108

Jean-Yves Drolet	Consultants BPR, 4655 Blvd. W. Hamel, Quebec, Ontario G1P 257 CANADA
Scott Drummond	Dep. of Agric. Eng., Univ. of Missouri, 259 Agric. Eng. Building, Columbia, MO 65211
Harold R. Duke	USDA-ARS, AERC-CSU, Fort Collins, CO 80523
Wayne P. Dulaney	USDA/ARS/NRI/RSPL, BARC-West, Bldg 007, Route 115, Baltimore, MD 20705
Myron Dunlavy	Willmar Manufacturing, Box 957, Willmar, MN 56201
Georg Duristein	Satcon, Bundesstv 7, Obertheves, GERMANY 97537
Damon Engelby	Mobile Data Communications, 425 2nd St SE, Skywalk #3, Cedar Rapids, IA 52401
Rick Engelby	Mobile Data Communications, 425 2nd St SE, Skywalk #3, Cedar Rapids, IA 52401
Steve Ernst	Unisys-Advanced Programs-Agric., Box 64525, St. Paul, MN 55164
Dean Fairchild	Soil Teq Inc., 5720 Smetena Drive, Suite 100, Minnetonka, MN 55343
Michael Farmer	Unisys-Advanced Programs-Agric., Box 64525, St. Paul, MN 55164
Dick Feller	Holly Sugar Corp., Drawer 1778, Hereford, TX 79045
Richard Ferguson	Dep. of Agronomy, Univ. of Nebraska, Box 66-SCREC, Clay Center, NE 68933
Rod Fischer	Dep. of Mech. Eng., South Dakota State Univ., Brookings, SD 57007

Paul Fixen	Potash and Phosphate Institute, Box 682, Brookings, SD 57006
Norm A. Flore	Western Coop Fertlizers Ltd., Box 2500, Calgary, Alberta T2P 2NI CANADA
Dennis D. Francis	USDA-ARS, 241 Kelm Hall, Lincoln, NE 68583
Robert Frank	University of Illinois, Marion Extension Service, 108 Airway Drive, Marion, IL 62966
David W. Franzen	Shilds Soil Service Inc., Paxton, IL 60957
Mike Freeberg	Centrol of Twin Valley, 102 E Main, Box 367, Twin Valley, MN 56584
Thomas Fry	Pioneer Hi-Bred Inc., Route 1, Box 158, Maynar, MN 56260
Roger Frye	Software Solutions of Illinois, Box 380, Shelbyville, IL 62565
Dennis J. Fuchs	Southwest Experiment Station, Univ. of Minnesota, Box 428, Lamberton, MN 56152
Earl Fuller	Dep. of Agric. & Appl. Econ., Univ. of Minnesota, St. Paul, MN 55108
Dale E. Gandrud	Gandy Company, Box 528, Owatonna, MN 55060
Pierre Yves Gasser	Ag Knowledge, 1-117 Charlotte St., Ottawa, Ontario K1N 8K4 CANADA
Larry Gaultney	E. I. DuPont, Box 30, Elkton Road, Newark, DE 19714
Steve Gerrish	Agritechnics Inc., Box 256, Elkhorn, WI 53121

Coby Gilbreath	Holly Sugar Corp., Drawer 1778, Hereford, TX 79045
George L. Gille	Dep. of Agric., Missouri State Univ., Maryville, MO 64468
Daniel Gimenez	Dep. of Soil Science, Univ. of Minnesota, St. Paul, MN 55108
Tom Goddard	Alberta Ag, Food & Rural Development, 7000-113 St., #206, Edmonton, Alberta T6H 5T6 CANADA
Joseph E. Goebel	1903 Cox Street, Eagle Pass, TX 78852
Shannon Gomes	Cedar Basin Crop Consulting, 102 Rainbow Dr., Waverly, IA 50677
Garry Good	DMI Inc., Box 65, Goodfield, IL 61761
Phil Goodrich	Dep. of Agric. Eng., Univ. of Minnesota, St. Paul, MN 55108
Roger Graham	Agris Corporation, 300 Grimes Bridge Road, Roswell, GA 30075
Duane A. Griffith	Dep. of Agric. Econ., Montana State Univ., Bozeman, MT 59717
Larry Gunderson	Brown Nicollet Comm. Health, 3301 S Washington, St. Peter, MN 56082
Rex Guthland	Willmar Manufacturing, Box 957, Willmar, MN 56201
David Hagert	Agri Data Inc., Route 1, Box 56A, Emerado, ND 58228
Tom Hall	Pioneer Hi-Bred International, 11152 Aurora Ave., Des Moines, IA 50322

Mike Hansen	Cottonwood County Courthouse, Windom, MN 56101
Greg Hanson	MVTL Labs-Soils, 1126 N Front St., New Ulm, MN 56073
Lowell Hanson	Field Control Systems, Box 10851, White Bear Lake, IL 55110
Charles L. Harms	VigoroAgribusiness, 1220 Potter Dr., Room 108C, W. Lafayette, IN 47906-1383
John Hatlestad	Unisys, Box 64525, MS U2T20, St. Paul, MN 55164-0525
Elwood Hatley	Penn State Univ., 116 ASI Bldg., University Park, PA 16802
Neil Hauermale	Farmers Software Association, Box 660, Fort Collins, CO 80522
Joe Hauwiller	Soil Teq, 5720 Smetana Drive, Minnetonka, MN 55343
John Hebblethwaite	Monsanto, 800 N Lindbergh Blvd., St. Louis, MO 63167
Ronnie W. Heiniger	Dep. of Agronomy, Kansas State Univ., 15600 Cedar Meadows Road, Wamego, KS 66547
Randy Hemb	MVTL Labs-Soils, 1126 N Front St, New Ulm, MN 56073
Tom Henry	Box 1, Westhope, ND 58793
Gary W. Hergert	University of Nebraska-WCRED, Rt 4, Box 46A, North Platte, NE 69101
Tracy Hershey	Pioneer Hi-Bred, 7200 NW 62nd Ave., Johnston, IA 50313

Chad Hertz	Univ. of Illinois, 305 Mumford, 1301 W Gregory Dr., Urbana, IL 61801
Paul Hewitt	Cargill, Inc., #19, Box 9300, Minneapolis, MN 55440
Dave Hilde	American Crystal Sugar Co., 101 N 3rd St., Moorhead, MN 56560
Andy Hill	Dekalb Agra Inc., 4743 Co. Rd. 28, Waterloo, IN 46793
Steve Hoffman	Hoffman Crop Consulting Inc., 8426 Borgwardt Lane, Manitowac, WI 54220
Bill Holmes	RR2, Box 187, Oran, MO 63771
Don Holt	Univ. of Illinois, 211 Mumford, 1301 W Gregory Dr, Urbana, IL 61801
C. Steven Holzhey	USDA-SCS, Federal Bldg, #152, 100 Centennial Mall N, Lincoln, NE 68508-3866
Gary Hopp	Amana Farms Inc., Box 189, Amana, IA 52203
Robert H. Hornbaker	Univ. of Illinois, 305 Mumford, 1301 W Gregory Dr, Urbana, IL 61801
Donald D. Howard	Dep. of Plant & Soil Science, Univ. of Tennessee, 605 Airways Blvd., Jackson, TN 38301
Jerry Huffman	DowElanco, 2009 Fox Drive, Suite D, Champaign, IL 61820
Dave R. Huggins	Southwest Experiment Station, Univ. of Minnesota, Box 428, Lamberton, MN 56152

David F. Hughes	Dep. of Soil & Atmos. Science, Univ. of Missouri, 266 Agric. Eng. Bldg., Columbia, MO 65211
Dan Humberg	Dep. of Agric. Eng. Dep., South Dakota State Univ. Box 2120, Brookings, SD 57007
John W. Hummel	USDA-Crop Protection Res. Unit, 1304 W Penn Ave., Urbana, IL 61801
Michael D. Hutter	Northern Ag Management, Box 474, Westhope, ND 58793
Darrol M. Ike	Agro-Info, 732 Northwood Drive, Delano, MN 55328
Jan Jarman	Anoka Sand Plain Project, Box 250, 12433 Pine Street, Becker, MN 55308
Dan Jaynes	NSTL-USDA-ARS, 2150 Pammel Drive, Ames, IA 50011
Jay D. Johnson	Prairie Crop Pro-Tech, 220 Hillside Ave., Waterloo, IA 50701
Kelly Johnson	Better Buy Agro, Box 1595, Assiniboia, Saskatchewan, CANADA
Paul W. Johnson	USDA-SCS, Box 2890, Washington, DC 20013
Steven Johnson	Unisys, Box 64525, MS U2F22, St. Paul, MN 55164-0525
Carol Jones	C.E. Jones & Associates Ltd., 204 26th Bastion Square, Victoria, B.C. V8W 149, CANADA
Joseph Kaplan	2910 E Franklin Ave, #1219, Minneapolis, MN 55405

John C. Keng	AgCANADA, Research Station, Box 3000 Main, Lethbridge, AB T1J 4B1 CANADA
Jeff Kenkel	Infield Crop Record Systems, 800 N Lindburgh Blvd., St. Louis, MO 63104
Bhairav Raj Khakural	Dep. of Soil Science, Univ. of Minnesota, St. Paul, MN 55108
Sam Kincheloe	IMC Fertilizer Inc., One Nelson C. White Parkway, Mundelein, IL 60060
Bradley King	University of Idaho, Aberdeen Research & Ext., 1693 S 2700 W, Box AA, Aberdeen, ID 83821
Newell R. Kitchen	Dep. of Atmos. Science, Univ. of Missouri, 250 Ag Engineering Building, Columbia, MO 65211
Roger Knapp	Route 2, Box 139, Webster, SD 57274
Roger Knutson	Soil Teq Inc., 5720 Smetana Drive, Suite 100, Minnetonka, MN 55343-9688
Cory Koesterman	Ag Data Inc., Route 1, Box 56A, Emerado, ND 58228-9733
John J. Kubik	John Deere, 1800-158th St., East Moline, IL 61244
Lyle Kuhlmann	Agriland Elevators Inc., 302 Byron Ave., Byron, MN 55920-0127
Greg LaBarge	Ohio State Univ. 135 Courthouse Plaza, Wauseon, OH 43567
John A. Lamb	Dep. of Soil Science, Univ. of Minnesota, St. Paul, MN 55108

Michael N. Langman	Dep. of Agric., Univ. of Nova Scotia, Box 550, Truro, Nova Scotia B2N 5B3 CANADA
Greg LaPlante	Centrol of Twin Valley, 102 E Main, Box 367, Twin Valley, MN 56584
Bob Larson	Cargill, Inc., Box 908, Cheyenne Wells, CO 80810
David W. Larson	RDI Technologies, 7102 Fairview Dr., Dekalb, IL 60115
Don Larson	Larson Systems, 113 Welch Ave., Ames, IA 50014-7228
William E. Larson	Dep. of Soil Science, Univ. of Minnesota, St. Paul, MN 55108
Bruce Lear	Cargill, Inc., Public Affairs, 1101 15th Street NW, Washington, DC 20005
Dale Leikam	Farmland Ind. Agronomy Dep., 3315 N Oak, Kansas City, MO 64116
Jerome N. Lensing	Pioneer Hi-Bred, 1919 W 57th St, Sioux Falls, SD 57106
Phalai Li	Dep. Agric. Eng., Univ. of Minnesota, 226 Ag Engineering Building, St. Paul, MN 55108
Grant Lien	Tyler Limited Partnership, 6644 Knox Ave S, Richfield, MN 55423
Michael J. Lindstrom	North Central Soil Conservation Research Lab, North Iowa Ave., Morris, MN 56267
William A. Lindstrom	Northrup King Co., 1120 Maple St. S., Northfield, MN 55057

Tom Lombard	Dep. of Agric. Agronomy Services, South Dakota State Univ., 523 E Capitol Foss Building, Pierre, SD 57501
Dan Long	Northern Ag Research Center, Montana State Univ., Star Rte 36, Box 43, Havre, MT 59501
Allan Lorence	Rawson Control Systems, 24 6th St NW, Oelwein, IA 50662
William Lueschen	Southwest Experiment Station, Univ. of Minnesota, Box 428, Lamberton, MN 56152
Russell Lunde	Simplot of CANADA Limited, Box 1095, Tuxfordia, Saskatchewan CANADA
Rick Lundry	Cargill, Inc., CGD/NWG, Bingham Lake, MN 56118
Steve Lupkes	Pioneer Hi-Bred, Box 98, Constantine, MI 49042
Tim T. Macha	ConsulTech, Route 1, Box 148, Slaton, TX 79364
Ted Macy	Applications Mapping Inc., 21 Ash St, Suite 5, Frankfort, IL 60423
Virg Mahlum	Concord Inc., 2800 7th Ave N, Fargo, ND 58102
Mike Mailander	Louisiana State University, 167 Doran, Baton Rouge, LA 70803–4505
Gary L. Malzer	Dep. of Soil Science, Univ. of Minnesota, St. Paul, MN 55108
Grant Mangold	Ag/Innovation, 7014 W Hwy C-13, Box 1, Lin Grove, IA 51033

John Mann	Soil Teq Inc., 5720 Smetana Drive, Ste 100, Minnetonka, MN 55343–9688
Janice Mattson	George Mattson Farms Inc., Box 382, Chester, MT 59522
Dewayne M. Mays	SCS, Federal Bldg., Room 152, 100 Centennial Mall N, Lincoln, NE 68508
Alexander McBratney	Dep. of Agric. Chem & Soil Sci., University of Sydney, Bldg A05, Sydney, New South Wales, Australia 2006
Don McGrath	Tyler Limited Partnership, E Hwy 12, Box 249, Benson, MN 56201
Tom McGraw	Minnesota Crop Monitors, Rte 2, Box 150, Buffalo Lake, MN 55314
Colin McKenzie	Alberta Special Crops Research, SS #4, Brooks, Alberta T1R 0N1 CANADA
Bob McLeese	USDA-SCS, 1902 Fox Drive, Champaign, IL 61820
Brennan Melauaphy	Taralan, 648 N St., #101, Geneva, IL 60134
Russell Memory	Flexi-Coil Ltd., 1000 71st St E, Saskatoon, Sasketchewan S7N 0W0 CANADA
Gerald Miller	University of Minnesota, Extension Service, St. Paul, MN 55108
Mark Miller	Crop Production Consulting, R & S Box 167, Fergus Falls, MN 56537
Barry Moellman	Dicky-john Corporation, Box 10, Auburn, IL 62615

Bruce Montgomery Minnesota Dep. of Agriculture, 90 W Plato, St. Paul, MN 55107

Robert J. Monson Ag-Chem Equipment Co., 5720 Smetana Drive, Minnetonka, MN 55343

Thomas Morris Dep. of Plant Science, Univ. of Connecticut, 1376 Storres Rd, Box U-67, Storrs, CT 06269

John Morrison USDA/ARS, 808 E Blackland Road, Temple, TX 76502

Dave A. Mortensen Dep. of Agronomy, Univ. of Nebraska, 362C Plant Science, Lincoln, NE 68583-0915

Donal Murphy FAL, Institute for Soil Science, Bundesallee 50 38116

Kaye Murphy Infielder Crop Records, 800 N Lindburgh Blvd., St. Louis, MO 63104

Allen Myers Ag Leader Technology, Box 234B, Ames, IA 50010

George Nelson Unisys, Box 64525, MS U2T2O, St. Paul, MN 55164-0525

Keith Ness Minnesota Pollution Control Agency, 520 Lafayette Road N, St. Paul, MN 55155-4194

Donald R. Nielsen University of California, Land, Air & Water Resources, Davis, CA 95616

Gerald A. Nielsen Dep. of Plant & Soil Science, Montana State University, 826 Leon Johnston Hall, Bozeman, MT 59717-0312

Sheilah Nolan Alberta Ag, Food & Rural Dev., 7000-113th St, #206, Edmonton, Alberta T6H 5T6 CANADA

Mark Nordwald	Cargill Inc., CGD/NWG, Minneapolis, MN 55440
Mike Nordwald	Cargill Inc., 4800 Main St, Suite 530, Kansas City, MO 64112
Peter Nowak	University of Wisconsin, 350 Ag Hall, 1450 Linden Drive, Madison, WI 53706
Clayton Ogg	USEPA, 401 M Street SW, PM 2124, Washington, DC 20460
Robert Olieslagers	Katholieke Universiteit Leuven, Kardinaal Mercierlaan 92, 3000 Heverlee, BELGIUM
Don Olson	University of Minnesota, Extension Service, St. Paul, MN 55108
Ron Olson	Applications Mapping, 21 Ash St, Suite 5, Frankfort, IL 60423
John Oolman	Dep. of Plant Science, South Dakota State Univ., Ag Hall, Brookings, SD 57007
Emilio Oyarzabal	Monsanto, 4123 77th St. Circle, Urbandale, IA 50322
Doug Page	Profit-Trac Sampling Inc., Box 519, Dodge Center, MN 55927
Dana Palmer	6102 Kenosha Drive, Lubbock, TX 79413
Ron J. Palmer	University of Regina, 3737 Wascana Pkwy-Regina, Saskatchewan CANADA S4S-0A2
Glen Pask	Flexi-Coil Ltd., 1000 71st St E., Saskatoon, Saskatchewan S7K 3S5 CANADA
Gavin B. Patterson	4103A 108th St SE, Everett, WA 98208

Dennis W. Paulinski	Unisys, Box 64525, St. Paul, MN 55164-0525
Bernie Paulson	McPherson Crop Management, Box 228, St. Clair, MN 56080
Thayer B. Pendleton	InterAg Ltd., 300 Grimes Bridge Road, Roswell, GA 30075
James L. Petersen	University of Nebraska, WCREC Agronomy/Soils, Route 4, Box 46A, North Platte, NE 69101
Arthur E. Peterson	Dep. of Soil Science, Univ. of Wisconsin, 1525 Observatory Drive, Madison, WI 53706-1299
Todd A. Peterson	University of Nebraska, Agronomy/SE R.E.C., 113 Mussehl Hall, Lincoln, NE 68583-3858
Terry Pickett	Deere & Co., 1800-158th Street, East Moline, IL 61244
David Pieczarka	Agway Inc., Box 4741, Syracuse, NY 13221
Francis J. Pierce	Dep. of Crop & Soil Sciences, Michigan State Univ. 564 PSSB, East Lansing, MI 48824
Jeff Polenske	Centrol of Eastern Wisconsin, 1229 W 8th, Appleton, WI 54914
Kim Polizotto	Potash Corp. Saskatchewan, 3030 Salt Creek Lane, Arlington Hgts., IL 60005
Dennis Printz	Holly Sugar Corp., Drawer 178, Hereford, TX 79045
Norman D. Prochnow	USDA/SCS, 208 2nd Ave. SW, Jamestown, ND 58402-2096
Bob Racek	Cargill Inc., CGD/NWG - Box 190, Blue Earth, MN 56013

Gyles W. Randall Southwest Experiment Station, University of Minnesota, 35838 120th St., Waseca, MN 56093–4521

Derek W. Rapp Monsanto, The Agricultural Group, 800 N Lindbergh Blvd., St. Louis, MO 63167

Stephen L. Rawlins USDA/ARS, Route 2, Box 2953A, Prosser, WA 99350

Sisir K. Ray North Dakota State University, Box 5626, Fargo, ND 58105-5626

Yella Reddy Crop Technology Co., 1507 Harre Lane, York, NE 68467

Harold F. Reetz, Jr. Potash & Phosphate Institute, Route 2, Box 13, Monticello, IL 61856

George Rehm Dep. of Soil Science, Univ. of Minnesota, St. Paul, MN 55108

Larry Reichenberger Farm Journal, Box 117, Andale, KS 67001

Donald C. Reicosky USDA/ARS/MWA, North Iowa Avenue, Morris, MN 56267

Keith Reid Ontario Ministry of Ag & Food, Rte 3, Walkerton, ONT. N06 2VD CANADA

Ron Riffey Applications Mapping Inc., 21 Ash St., Ste 5, Frankfort, IL 60423

Pierre Robert Dep. of Soil Science, Univ. of Minnesota, St. Paul, MN 55108

Leon D. Rohrbaugh Micro-Trak Systems, Inc., PO Box 3699, Mankato, MN 56002

Ron Rosenbaum WOHIC Soil & Forage Lab, 106 N Cecil Street, Bonduel, WI 54107

Kendall Rostenberg	InterDok, 173 Halstead Ave, Box 326, Harrison, NY 10528
Eugene Roytburg	Dep. of Agric. Eng., Univ. of Minnesota, St. Paul, MN 55108
Jim Ruhland	Centrol Inc., Box 198, Cottonwood, MN 56229
C. Ford Runge	Dep. Agric. & Applied Econ., University of Minnesota, St. Paul, MN 55108
Ed Runge	Dep. Soil & Crop Science, Texas A&M Univ., College Station, TX 77843-2474
Dick Rust	Dep. of Soil Science, Univ. of Minnesota, St. Paul, MN 55108
E. John Sadler	USDA/ARS, Box 3059, Florence, SC 29502
John Sawyer	Growmark Inc., 1701 Towanda Ave, Box 2500, Bloomington, IL 61702-2500
Jim Schepers	USDA/ARS Agronomy, 113 Kein Hall, Lincoln, NE 68583-0915
Berlie L. Schmidt	USDA-CSREES, 329 Aerospace Bldg, Washington, DC 20250
Walter Schmidt	Ohio State University, 952 Lima Ave, Box C, Findlay, OH 45840
Stefan Schneider	Geo-Konzept, 15642 Garnet Way, Apple Valley, MN 55124
Todd Schnobrich	Cargill Inc., Box 9300, Minneapolis, MN 55440
William Schoenwecker	Omni Advertising, 1358 Courthouse Blvd., Inver Grove Heights, MN 55077

Kevin R. Scholl	Mowers Soil Testing PLUS, 117 E Main, Toulon, IL 61434
Bert Schou	ACRES, Box 249, Cedar Falls, IA 50613
Harold Schramm	Spectrum Technologies Inc., 12010 S Aero Drive, Plainfield, IL 60544
Mark Schrock	Dep. of Agric. Eng., Kansas State Univ., 147 Seaton Hall, Manhattan, KS 66506
Joseph Schumacher	Dep. of Plant Sciences, South Dakota State Univ., Mechanical Eng/Crothers Hall, Brookings, SD 57007
Ronald Schutz	Robins, Kaplan, Miller & Ciresi, 2800 LaSalle Plaza, 800 LaSalle Ave., Minneapolis, MN 55402–2015
Richard A. Schwartzbeck	National Crop Insurance Service, 7301 College Blvd, Suite 170, Overland Park, KS 66210
Stephen W. Searcy	Texas A&M University, West Campus Route 4A, College Station, TX 77843
Mark W. Seeley	Dep. of Soil Science, Univ. of Minnesota, St. Paul, MN 55108
Michael D. Shoup	Amana Farms, Inc., Box 189, Amana, IA 52203
Sid Siefken	Mobil Data, 425 2nd St SE, Skywalk #3, Cedar Rapids, IA 52401
John Simons	Cargill Limited, 300-240 Graham, PO Box 5900, Winnipeg, Manitoba R3C 4B5 CANADA
B. B. Singh	Scientific Crop Advisory, Box 844, Hastings, NE 68902-0844

Paul Skinner	Vineyard Investigations, 26 Laurel Lane, Napa, CA 94558
Andrey Skotnikov	Tyler Inc., E Hwy 12, Box 249, Benson, MN 56215
Wayne F. Smith	Deere & Co., Precision Farming Grp, 1800 158th St., E Moline, IL 61244
John Soderholm	RDI Technologies, Green Lake Mall, Suite 1, Spicer, MN 56288
Michael Sololub	Soil Science Dep., University of Saskatchewan, Saskatoon, Sask. S7N 0W0 CANADA
Ronald W. Steffen	Dickey-john Corp., Box 10, Engineering, Auburn, IL 62615
Darrell Steiner	Ag Chem Equipment, 13625 Geyser Path, Apple Valley, MN 55124
Ward Stienstra	Dep. of Plant Pathology, Univ. of Minnesota, St. Paul, MN 55108
Steve A. Stiles	Brookside Labs Inc., Info Sys. Mgr, 308 S Main St, Box 456, New Knoxville, OH 45871
Todd Stockdale	All American Cooperative, Eyota Branch, Box 148, Eyota, MN 55934
Kenneth A. Sudduth	USDA-ARS, Ag Engineering Bldg., University of Missouri, Columbia, MO 65211
C. L. Swart	Belmond Labs Inc., Route 2, Box 203, Belmond, IA 50421
Larry Swenson	North Dakota State University, Waldron Hall, Room 103, Fargo, ND 58105

Alan B. Telck	Holly Sugar Corp., Box 764, Sheridan, WY 82801
Joe W. Tevis	Spatial Data Systems, 3144 Norman Drive, Sioux City, IA 51104
Dan Theobald	Micro-Trak Systems, Inc., PO Box 3699, Mankato, MN 56002
Wayne Thompson	Dep. of Soil Science, Univ. of Minnesota, St. Paul, MN 55108
Mike Thurow	Spectrum Technologies Inc., 12010 S Aero Drive, Plainfield, IL 60544
Mark Tomer	Dep. of Soil Science, Univ. of Minnesota, St. Paul, MN 55108
Greg Townsend	Pioneer Hi-Bred International Inc., 400 SE Delaware, Ankeny, IA 50021
Shrini K. Upadhyaya	Dep. of Bio & Agric. Eng., Univ. of California, Davis, CA 95616
Bill Urbanowicz	Mid-Ohio Chemical, 394 Catalpa Lane, Mount Gilead, OH 43338
Brad Van Koeten	101 N Miles, Fremont, IA 52561
Richard L. Vanderlip	Dep. of Agronomy, Kansas State Univ., Throckmorton Hall, Manhattan, KS 66506
Chris VanKessel	Dep. of Soil Science, Univ. of Saskatchewan, Saskatoon, Saskatchewan S7N 0W0 CANADA
Gary E. Varvel	USDA-ARS, 123 Keim, East Campus, Lincoln, NE 68583
Ed Vasey	Crop Watcher, 2802 Maple St NE, Fargo, ND 58102
Byron Vaughan	Harris Labs, 624 Peach Street, Lincoln, NE 68502

J. Verhagen — Agricultural University Soil Science, 6700A Wageningen, THE NETHERLANDS

Jeffrey Vetsch — University of Minnesota, 35835 120th St, Waseca, MN 56093

Thomas G. Wadzinski — Cargill Hybrid Seeds, Route 3, Box 751, Aurora, IL 60506

G. Wagner — Route 1, Box 123, Climax, MN 56523

Darel Walker — Cargill Inc., Box 430, Koots, IN 46347

David Wall — Minnesota Pollution Control, Water Quality, 520 Lafayette Road, St. Paul, MN 55155

John Walter — Successful Farming, 1716 Locust Street, Des Moines, IA 50309-3026

Chang Wang — Agriculture CANADA, Centre for Land & Biol. Resources, C.E.F. Ottawa K1A 0C6 CANADA

Raymond C. Ward — Ward Laboratories Inc., Box 788, Kearney, NE 68848-0788

Darryl D. Warncke — Dep. of Crop & Soil Sciences, Michigan State Univ., E. Lansing, MI 48824-1325

Gloria Watro — City of Crookston, 124 N Broadway, Crookston, MN 56716

Paul Wayland — Cargill, Inc., P. O. 69, Jeffers, MN 56145

Bruce Weller — ICI Seed, 501 N Oakridge Road, Brandon, SD 57005

Monte Weller — Dekalb Agra Inc., 4743 Co. Rd. 28, Waterloo, IN 46793

Dean E. Wesley Key Agricultural Services, 114 Shady Lane, Macomb, IL 61455

Kirk A. Wesley Key Agricultural Services, 114 Shady Lane, Macomb, IL 61455

Tad L. Wesley Key Agricultural Services, 114 Shady Lane, Macomb, IL 61455

Brett Whelan University of Sydney, Agric., Chem. & Soil Sci. Dep., Building A05, Sydney, New South Wales, Australia 2006

Heri Widiyatmanto Dep. of Agronomy, Ohio State Univ., 676 Harley Drive #6, Columbus, OH 43202

Dave Wiebold Amana Farms Inc., Box 189, Amana, IA 52203

Steve Wiedman Mowers Soil Testing Plus Inc., Box 518, Toulon, IL 61742

Lori Wiles CSU Water Management Unit, AERC, USDA-ARS, Fort Collins, CO 80523

Ray Williford USDA/ARS, Box 36, Stoneville, MS 38776

David Willis Agassiz Crop Management Inc., Route 5, Box 126A, Thief River Falls, MN 56701

Bob Wilson InterMountain Canola, 2300 N Yellowstone, Idaho Falls, ID 83401

John Wilson Montana State University, Earth Sciences/GIAC, 200 Traphagen Hall, Bozeman, MT 59717

Steven Wolf University of Wisconsin, 350 Ag Hall, Madison, WI 53715

Dick Wolkawski Dep. of Soil Science, Univ. of Wisconsin, Madison, WI 53706

Nyle Wollenhaupt Dep. of Soil Science, Univ. of Wisconsin, Madison, WI 53706

Herb Wright Geophyta, 2720 C.R. 254, Vickery, OH 43464

Nathan Wright Geophyta, 2685 C.R. 254, Vickery, OH 43464

Tong Wu Robins, Kaplan, Miller & Ciresi, 2800 LaSalle Plaza, 800 LaSalle Ave., Minneapolis, MN 55402-2015